3rd European Symposium on Fire Safety Science 2018

Journal of Physics: Conference Series
Volume 1107

Nancy, France
12 - 14 September 2018

Editors:

Pascal Boulet
Bart Merci
Patrick Van Hees

ISBN: 978-1-5108-7518-0
ISSN: 1742-6588

Printed from e-media with permission by:

Curran Associates, Inc.
57 Morehouse Lane
Red Hook, NY 12571

Some format issues inherent in the e-media version may also appear in this print version.

Printed by Curran Associates, Inc. (2019)

For permission requests, please contact the Institute of Physics at the address below.

Institute of Physics
Dirac House, Temple Back
Bristol BS1 6BE UK

Phone: 44 1 17 929 7481
Fax: 44 1 17 920 0979

techtracking@iop.org

Additional copies of this publication are available from:

Curran Associates, Inc.
57 Morehouse Lane
Red Hook, NY 12571 USA
Phone: 845-758-0400
Fax: 845-758-2633
Email: curran@proceedings.com
Web: www.proceedings.com

Table of contents

Volume 1107

3rd European Symposium on Fire Safety Science

12–14 September 2018, Nancy, France

Accepted papers received: 9 October 2018
Published online: 14 November 2018

Preface

Keynote Lecture

Material behavior in fire

Fire suppression

Evacuation

IOP Conf. Series: Journal of Physics: Conf. Series **1107** (2018) 011001 doi:10.1088/1742-6596/1107/1/011001

Proceedings of the 3rd European Symposium on Fire Safety Science

Pascal BOULET[1], Bart MERCI[2] and Patrick VAN HEES[3]

[1]*Université de Lorraine, CNRS, LEMTA, France*
[2]*Ghent University, Department of Fluid Mechanics, Heat and Combustion, Ghent, Belgium*
[3]*Lund University, Division of Fire Safety Engineering, Lund, Sweden*

pascal.boulet@univ-lorraine.fr

EDITORIAL

This special issue is based on the papers selected for presentation during the *3rd ESFSS (European Symposium on Fire Safety Science)*, held in Nancy (France) from 12 to 14 September 2018. Following the conference held in Cyprus (2015), the 3rd ESFSS was the third edition of a series of symposia organized in Europe, with the participation and support of the International Association for Fire Safety Science (IAFSS). The aim was to gather researchers from within Europe and beyond to have exchanges and discussions about fire safety science. In this frame, a workshop *Agenda 2030 for a Fire Safe World* was also organized by IAFSS on the afternoon of Tuesday, September 11, prior to the symposium, with the aim to discuss the definition of a Fire Safety Mission and the identification of specific topics for fire science research.

Around 120 contributions have been received during the conference preparation, involving both submissions for oral and poster presentation. A selection, based on two peer-reviews of full papers, finally conducted to the acceptance of 35 contributions for oral presentation and 55 for poster presentation, plus 11 posters in a special session dedicated to work-in-progress.

The conference was attended by 140 scientists, coming not only from Europe but from around the world, namely from Australia, Belgium, Canada, China, Czech Republic, Finland, France, Germany, India, Indonesia, Ireland, Japan, Portugal, Russia, Slovenia, Spain, Sweden, Turkey, United Kingdom, USA.

The present volume contains all the papers of the contributions presented during the seminar. They are organized into five main topics, in accordance with the conference program:

- Material behavior in fire
- Fire dynamics, structure in fires
- Wildland fires
- Fire suppression
- Evacuation and Human behavior

The conference organizers gratefully acknowledge the members of the scientific committee and the experts who carried out the reviews of the papers. They are also grateful to PROGEPI and the local organizing committee for the preparation of the conference. The support by the IAFSS, LEMTA laboratory, the French research group GDR Feux, the Université de Lorraine, the CNRS and the Région Grand Est is also acknowledged.

Symposium chair
Prof. P Boulet, U Lorraine, France
Prof. B Merci, Ghent U, Belgium
Prof. P Van Hees, Lund U, Sweden

Scientific committee
Prof. D Alvear, U Cantabria, Spain
Prof. L Bisby, U Edinburg, UK
Prof. G Boustras, European U Cyprus, Cyprus
Prof. P Boulet, U Lorraine, France
Prof. W K Chow, Hong Kong Poly Tech U, China

IOP Conf. Series: Journal of Physics: Conf. Series **1107** (2018) 011002 doi:10.1088/1742-6596/1107/1/011002

Peer review statement

All papers published in this volume of *Journal of Physics: Conference Series* have been peer reviewed through processes administered by the proceedings Editors. Reviews were conducted by expert referees to the professional and scientific standards expected of a proceedings journal published by IOP Publishing.

IOP Conf. Series: Journal of Physics: Conf. Series **1107** (2018) 022001 doi:10.1088/1742-6596/1107/2/022001

Thermal decomposition of solid fuels. Objectives, challenges and modelling

Thomas ROGAUME

Institut Pprime, UPR 3346 CNRS, Université de Poitiers, ISAE-ENSMA, France

thomas.rogaume@univ-poitiers.fr

ABSTRACT

The progress of computing means, the complexity of the situation concerning fires, the need to describe with more accuracy all the different phases of the fire behaviour (such as the ignition, its growing, its extinction…) has led to a great effort in the past few years of the international research community to improve the performance of the numerical simulation codes. Those codes consist of different sub-models in order to simulate the different physical, thermal and chemical aspects of the fire, such as the thermal decomposition.

This one plays a significant role on the fire evolution since it represents the fuel quantity that supplies the combustion process. Then, the requirement of a more detailed characterization and description is one actual and important challenge. This is the challenge of this keynote lecture which is dealing with thermal decomposition of solid fuels in the conditions of fire, addressing both the experimental investigations and the modelling development aspects.

After a presentation of the objectives and challenges of thermal decomposition, this lecture details the different steps required for its description, based on an experimental and numerical scaling-up approach. Examples of application are done. Special focus is highlighted on different scientific bolts and issues for thermal decomposition modelling.

KEYWORDS

Thermal decomposition, solid fuels, model of pyrolysis, scaling-up, heat transfer, mass transfer, Arrhenius, thermo-physical properties.

IOP Conf. Series: Journal of Physics: Conf. Series **1107** (2018) 022001 doi:10.1088/1742-6596/1107/2/022001

INTRODUCTION

The buildings, the furnishings, the material used in different applications (for example transport) have evolved considerably in the past few years, generating new challenges and requirements concerning fire safety science (complexity). Then, all around the word, fire safety engineering met a strong development. This one uses models of fire simulation, for example Fire Dynamics Simulator [1] or FireFoam [2]. Finally, the requirement of a better description of all the steps of a fire, particularly the non-established events (such as ignition, fire growing, extinction…) is principal and conductible to several works to improve those models.

The Computational Fluid Dynamics (CFD) models are composed by several sub-models, each one has specific functions: heat transfer, radiation, flows and turbulence, combustion… and pyrolysis. Specifically, the sub-model of pyrolysis permits describing the chemical evolution (reactions) of a small volume element (dV) as a function of temperature and the atmosphere (%O_2). It expresses the relation between temperature, atmosphere and the kinetic of thermal decomposition of a volume element dV of one material, which is considered at each time step homogeneous in terms of temperature and composition. Thus, the sub-model of pyrolysis permits describing the kinetic of mass loss of one solid material and then the quantity of fuel brought to the combustion. Currently nicknamed the « source term », it has a large influence on the ignition process (limits of flammability), on fire growing, on flame propagation and characteristics, on the gaseous emissions, etc.
However, the pyrolysis model is just a part of the description of the thermal decomposition process of a solid fuel, and it must be coupled to other ones in order to describe heat transfers (temperature), mass transfers (%O_2, devolatilization transfer, reactive mixture, etc.) and the boundary conditions of the volume element considered for the pyrolysis modelling.

In this context, the objective of the present contribution is to describe how to model the thermal decomposition of a solid fuel. The development of the sub-model of pyrolysis and its characteristics will be presented, as well as its coupling with other sub-models, permitting taking into account the strong coupling between the condensed and the gaseous phases and to describe the thermal decomposition of a solid. This is realized based on the scaling-up approach presenting by Torero [3] and addressing both experimental and numerical aspects. Finally, the presentation of the different steps of development of a model of thermal decomposition permits explaining the actual scientific challenges and issues.

In the first part, the challenges and the phenomenology associated with the thermal decomposition of a solid are presented. Next, the steps of the development of the model of pyrolysis are detailed, treating of the reaction mechanism, the kinetic law and the function of conversion. An attention is brought on the Arrhenius law and on the determination of the kinetic parameters associated to this one. The modelling of the heat and mass transfers requires the knowledge of the values of different thermal, physical and chemical parameters. Their investigations and limits are presented. At the end, the scaling-up validation of the model is addressed.

The different aspects are illustrated by examples of applications on several materials, at different scales.

I. DEFINITION AND PHENOMENOLOGY OF THE THERMAL DECOMPOSITION

The thermal decomposition of a solid represents its mass loss and then the quantity of combustible gases emitted and that permits the combustion to take place (or not). Thus, the coupling between the thermal decomposition of the condensed phase and the combustion into the gaseous phase is very important: thermal decomposition controls the fuel gases but is dependant of the oxygen diffusion and the heat transfers permitting the heating of the material, for its decomposition. This strong coupling is presented on figure 1 and is summarized on figure 2, inspired from [4]:

IOP Conf. Series: Journal of Physics: Conf. Series **1107** (2018) 022001 doi:10.1088/1742-6596/1107/2/022001

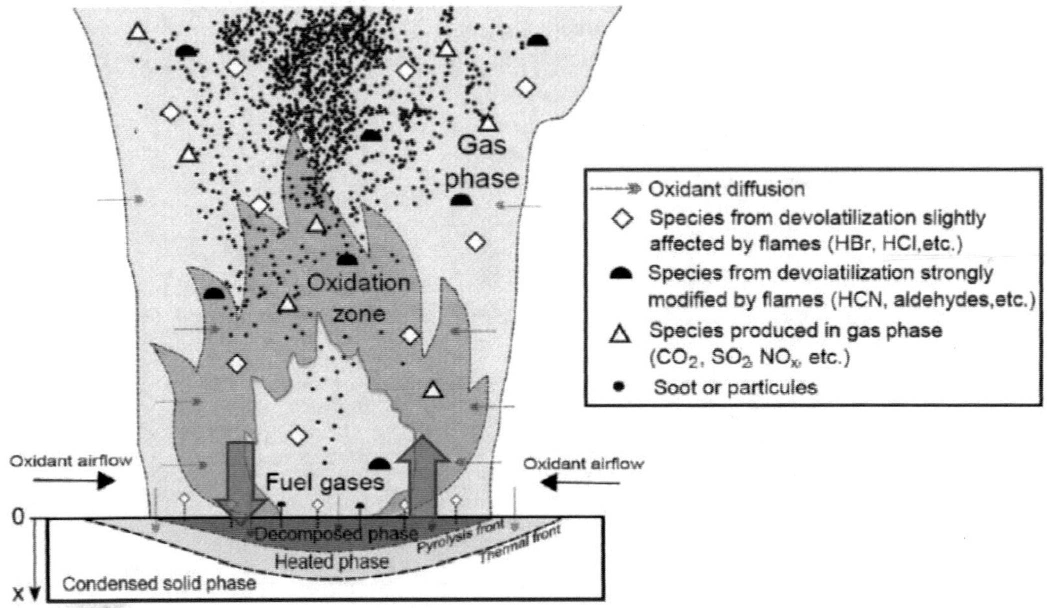

Figure 1: illustration of the coupling between the thermal decomposition and the combustion into the gaseous phase.

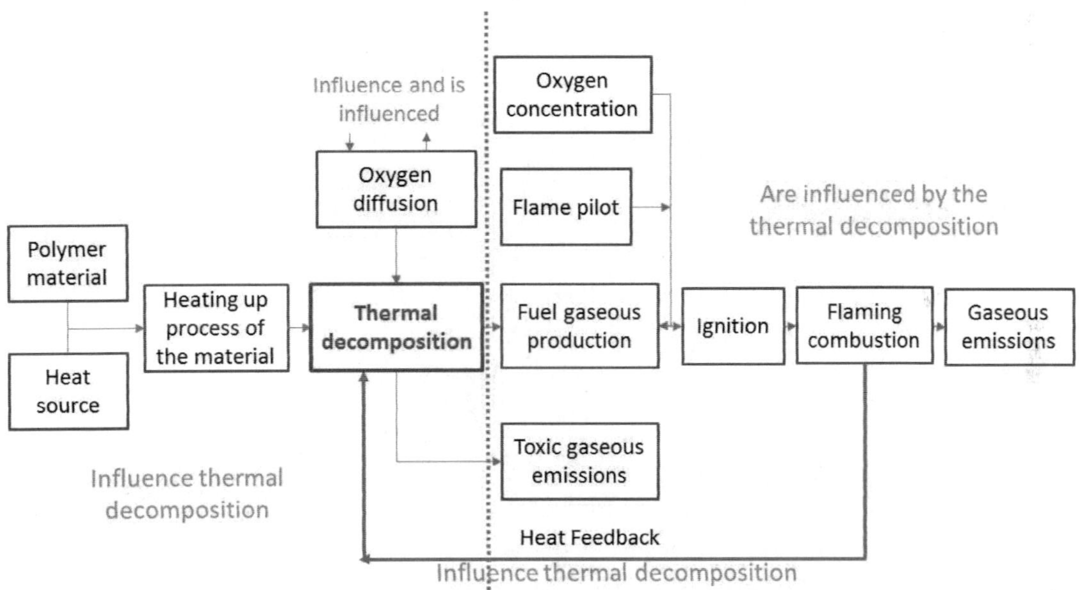

Figure 2: presentation of the coupling between the condensed and the gaseous phases during the combustion of a solid fuel. Inspired from [4]

It is shown that the thermal decomposition is dependant of the local oxygen concentration, the local temperature as well as the structure and the properties of the material (solid fuel) studied. Its description reclaims 3 parts: a sub-model of pyrolysis; the determination of physical, chemical and thermal properties; and the coupling with other sub-models to describe heat and mass transfers. Those 3 aspects will be addressed.

However, due to the multitude and the complexity of the phenomenon to characterize and to model, it is classical to study independently the gas phase and the condensed one. Concerning the condensed phase, the interaction with the gas one is just considered using an external heating source (classically an external heat flux) and a local diffusion of oxygen. Torero [5] proposes a monodimensional representation of the phenomenon occurring into the solid, figure 3. When a solid, initially at ambient temperature is exposed to an external heat flux ($\dot{q}_i$ ou $\dot{q}_e^{"}$)

part of this heat flux permits the heating of the material by conduction ($\dot{q}_c$) and a part is lost by reradiation of the surface ($\dot{q}_r$ *or* $\dot{q}''_{SR}$). Its temperature starts to increase in depth and as a function of time. If the temperature is sufficient to break the link between the condensed molecules (energy of linking), the solid starts to decompose at a flow rate ($\dot{m}_p$). The initial mass fraction of solid fuel ($Y_{F,s}$) is consumed to produce gaseous components ($Y_{F,g}$), while some oxygen can diffuse into the solid matrix ($Y_{O_2,g}$). Heat and mass transfers into the condensed phase are dependant of its permeability χ.

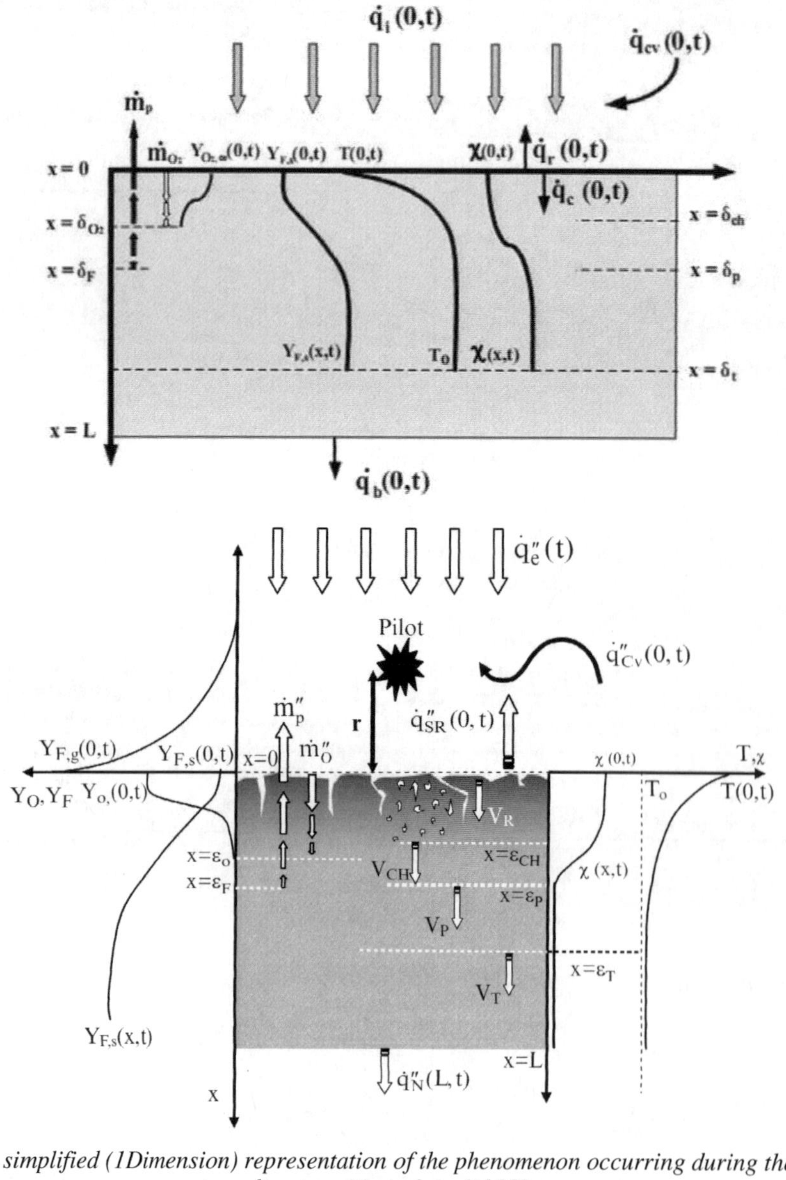

Figure 3: simplified (1Dimension) representation of the phenomenon occurring during the thermal decomposition of a solid [5]

The energy balance of the controlled volume is defined by the equation:

$$\frac{\partial[\rho_S C_S T]}{\partial t} = \frac{\partial}{\partial x}\left[k_S \frac{\partial T}{\partial x}\right] + \frac{\partial[\dot{m}_P'' C_{P,P} T_P]}{\partial x} - \frac{\partial[\dot{m}_O'' C_{P,O} T_O]}{\partial x} + \frac{\partial[\rho_S V_R C_S T]}{\partial x} + \dot{q}_{RAD}'''$$
$$+ \sum_{i=1}^{i=N} \Delta H_{P,i} \rho_S \left[A_i Y_O^{m_i} Y_S^{n_i} e^{-E_i/RT}\right]$$

With the last term describing the energy due to the n reactions of thermal decomposition.

From figure 3, it is shown that the thermal degradation depends of as a function of time [5] - 1D approach:
- Temperature T(x,t).
- Local mass fraction of combustible, Y_S(x,t).
- Local mass fraction of oxygen, Y_{O_2}(x,t).
- Mass fraction of residual solid fuel, $Y_{F,S}$(x,t).
- Permeability, χ(x,t).
- Thickness of oxygen diffusion into the solid, δ_{O_2}(x,t).
- Thickness of the reactive zone, δ_F(t).
- Kinetic parameters values of each reaction, A_i the pre-exponential factor, n_i the order of the reaction i, m_i the mass of i, E_i the energy of activation.

The mass loss rate due to the thermal decomposition is expressed locally by [5].

$$\dot{m}_p'''(x,t) = Y_{F,s}(\mathbf{x},t) \sum_{i=1}^{N} \left[A_i Y_{0_2}^{m_i}(\mathbf{x},t) Y_s^{n_i}(\mathbf{x},t) e^{-E_i/RT(\mathbf{x},t)} \right]$$

And for each volume element dV [5]:

$$\dot{m}_p''(\mathbf{x},t) = \int_0^L \chi(\mathbf{x},t) \left[Y_{F,s}(\mathbf{x},t) \sum_{i=1}^{N} \left(A_i Y_{0_2}^{m_i}(\mathbf{x},t) Y_s^{n_i}(\mathbf{x},t) e^{-E_i/RT(\mathbf{x},t)} \right) \right] dx$$

The objective is thus to represent the thermal decomposition, which reclaims a model of pyrolysis, physical, chemical and thermal properties, and the description of coupling, heat and mass transfers, as presented figure 4 (inspired from [6]).

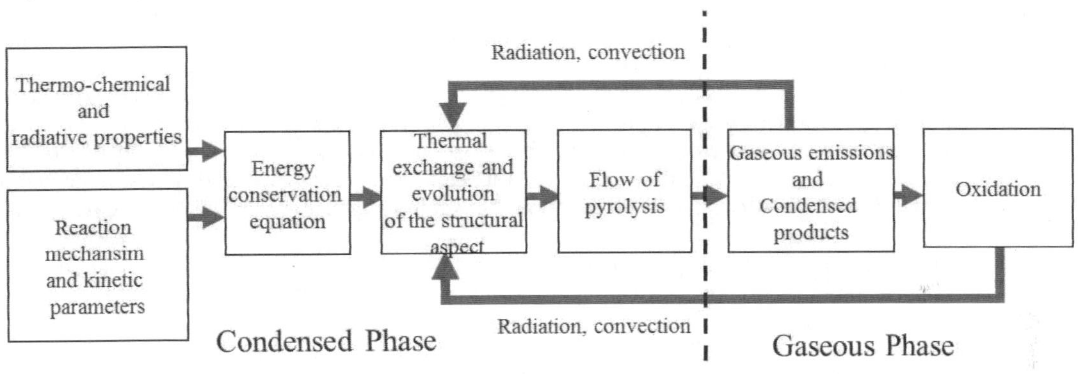

Figure 4: thermal decomposition of a solid fuel. Aspects to be represented (inspired from [6]).

In a first part, we are interested in the model of pyrolysis.

II. THE MODEL OF PYROLYSIS

The sub-model (or model) of pyrolysis expresses the chemical evolution (reactions) with the relation between temperature, atmosphere and the kinetic of thermal decomposition of a volume element dV of the material studied which is considered at each time step homogeneous in term of temperature and of composition.

Three main approaches are classically used in order to determine a model of pyrolysis [7]:
- Modelistic approach – Model fitting method: it uses a define reaction mechanism with an Arrhenius formulation. Then it requires the definition of A, E_a and n for each reaction and those terms are defined as properties of the reaction.
- Isoconvertionnal approach – Free model method: it permits d determining the evolution of the activation energy as a function of the degree of conversion of the reaction. E_a is dependant of α (the progress of the reaction) and T. It does not use a reaction mechanism (just one reaction) but is based on an Arrhenius form. The evolution of E_a permits presenting the mass loss rate. This approach is available in the case of 1 reaction of thermal decomposition, or when the steps are clearly separated (no interaction) and chronologic (not parallel).

- The hybrid approach is a combination of the modelistic and the isoconversionnal one. Each peak of mass loss rate is treated with a Kissinger method.

Recently (in the past 15 years), great efforts have been made by the international scientific community in order to develop more detailed and performant models of pyrolysis, for example by Rein & al. [8], Lautenberger & al. [9-11], Stoliarov & al. [12-14], Matala & al. [15], Rogaume and Fateh & al. [16, 18-21], Bustamante Valencia [17]. Those different groups consider the model fitting method, which is used the most in the different software. Our interest will focus on the rest of this contribution through this approach. With the modelistic approach, an imposed and more or less complex model is used to represent the detail kinetic of thermal decomposition and model of pyrolysis required:

- A reaction mechanism:

$$m_1 \xrightarrow{\dot{\omega}, \nu} m_2$$

- A kinetic law and a function of conversion. The rate of reaction of the solid phase is the multiplication of k(T), the law of variation of the rate and f(α), the function of conversion. The variation of the rate is classically represented by a modified Arrhenius function. Thus:

$$\frac{d\alpha}{dt} = k(T) f(\alpha) = A \exp\left(-\frac{E}{RT}\right) f(\alpha)$$

with T the temperature, α the progress of the reaction, E the energy of activation, A the pre-exponential factor and R the constant of the perfect gases.

2.1. Determination of the reactional mechanism

Four main methods are used to determine the reactional mechanism of thermal decomposition of a solid:

- The lumped parameter approach (LPA) considers that the material is one, homogeneous. The thermal decomposition is the one of this material.
- The constituent approach considers that a material is composed of several constituents and that its thermal decomposition is the sum of the one of each component – Example: wood is composed of cellulose, hemicellulose and lignin, so its mechanism of thermal decomposition is the one of the cellulose, the one of the lignin and the one of the hemicellulose. This approach is quite simple, but it neglects the facts that can cause some interactions between the constituents. This is the case for the wood…
- The functional approach. The thermal decomposition is described as functional groups, due to the well-known fragmentation of the polymers. This last one is very precise but time consuming and it requires a well knowledge of the polymer decomposition steps [7].
- The last one consists of describing all the elementary reactions of thermal decomposition, as it is classically done for the reactions into the gas phase. One example is presented into [22]. However due to the complexity of the solid structures and of the thermal decomposition it is inapplicable for safety science engineering and multicomponent materials for example.

Then, it currently appears that the LPA approach is the more acceptable and it has been used by several authors [4, 8-21]. It relies on the use of results of thermal decomposition of the solid studied from thermogravimetric analysis, which can be coupled to detailed gas analysis, like a Fourier transformed infra-red spectrometer, gas chromatography or mass spectrometer. During those experimental investigations, very small samples of solid (few mm and mg) are heating up under a controlled atmosphere (inert, oxygen %) from the ambient temperature to several hundred degrees (1000°C) at a controlled heating rate, classically from 5 to 20°C/min. The mass loss is measured as a function of the time and (knowing the heating rate) the temperature, as well as the gaseous emissions (if possible). The mass loss rate is secondly deducted.

The evolutions of the mass loss and the mass loss rate obtained for different experimental conditions permit the determination of the reactional mechanism.

For example, the thermal decomposition of a polyurethane foam is studied [16]. Figure 5 presents respectively the evolution of the mass loss and of the mass loss rate, for two atmospheres and four heating rates [16].

IOP Conf. Series: Journal of Physics: Conf. Series **1107** (2018) 022001 doi:10.1088/1742-6596/1107/2/022001

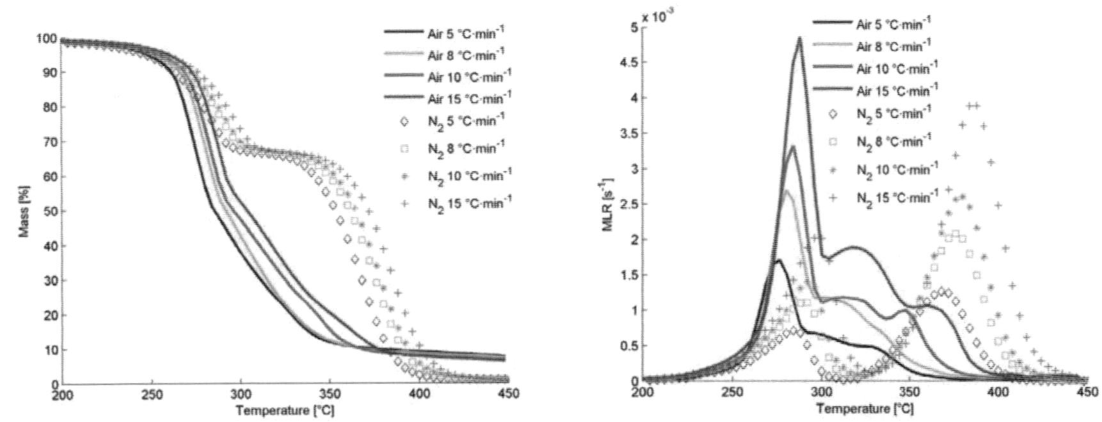

Figure 5: evolution of the mass loss and the mass loss rate during the thermal decomposition of a PU foam at TGA scale [16].

From those results, it is shown that the thermal decomposition of the PU foam is taking place by 2 global reactions of pyrolysis (under inert atmosphere) and 3 more reactions under air (oxidation). Those reactions are presented into the table 1 [16].

No	Type of reaction	Temp. [°C]	Reactives		Products solid or liquid		Products gas
1	Pyrolysis	200 – 340	PPUF	→	υ_1·Polyol	+	τ_1·[Isocyanate]
2	Pyrolysis	340 – 450	Polyol	→	υ_2·Residue	+	τ_2·[Polyol + H_2CO + H_2O + CH_4]
3	Oxidation	200 – 275	PPUF + O_2	→	υ_3·Polyol	+	τ_3·[Polyol + CO_2 + H_2O]
4	Oxidation	220 – 300	Polyol + O_2	→	υ_4·Char	+	τ_4·[Polyol + H_2CO + CH_4+ CO + CO_2 + H_2O]
5	Oxidation	300 – 450	Char + O_2	→	υ_5·Residue	+	τ_5·[Polyol + H_2CO + CH_4+ CO + CO_2 + H_2O]

Table 1: reactions of thermal decomposition of a PU foam [16]

The main challenge consists of determining a more realistic mechanism, and classically several can be proposed [17]. The evolutions of the gaseous emissions are thus of best importance. The accuracy of modelistic approaches does not depend on the resolution technique, but rather on the description of the apparent chemical reaction pathway. It is then interesting to focus on the complexity level of the reaction mechanism, which has been done by Marquis & al. [23]. Several mechanisms complexity with various levels of sophistication (number of linked reactions) are studied using a new statistical technique. The analysis shows the possibility to simplify the initial reaction mechanism's complexity, without influencing the accuracy of forecasts.

From the reactions analysis, the reactional mechanism of the PU foam proposed by [16] is (Figure 6):

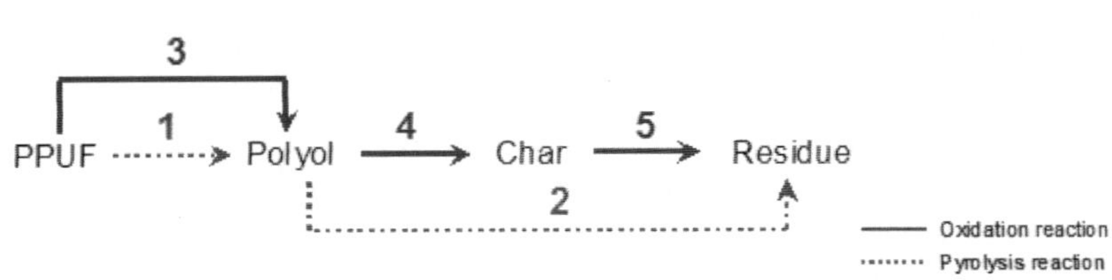

Figure 6: mechanism of thermal decomposition of a PU foam [16]

2.2. Determination of the kinetic law

The rate of reaction of the solid phase is the multiplication of k(T), the law of variation of the rate and f(α), the function of conversion. The variation of the rate is classically represented by a modified Arrhenius function.

$$\frac{d\alpha}{dt} = k(T)f(\alpha) = A\exp\left(-\frac{E}{RT}\right)f(\alpha)$$

with T the temperature, α the progress of the reaction, E the energy of activation, A the pre-exponential factor and R the constant of the perfect gases.

The model-fitting method is expressed in terms of the degree of conversion: equal to 0 at the beginning of the test and to 1 when all the mass has been decomposed. The degree of conversion is defined as:

$$\alpha = \frac{m_0 - m_i(t)}{m_0 - m_\infty}$$

With m_0 the mass at the beginning, $m_i(t)$ the mass at an arbitrary time, m_∞ the mass at the end of the process.
Different functions of conversion can be used [7], some are presented into the table 2:

Model		f(α)
Reactional Order	ordre 0	1
	1[er] ordre	$1-\alpha$
	2nd ordre	$(1-\alpha)^2$
	ordre 3	$(1-\alpha)^3$
	ordre *n*	$(1-\alpha)^n$
Exponential law		α
Power function	P = 3/2	$\frac{3}{2}\alpha^{1/3}$
	P = 2	$2\alpha^{1/2}$
	P = 3	$3\alpha^{2/3}$
	P = 4	$4\alpha^{3/4}$
	P = *n*	$n\alpha^{(1-1/n)}$

Tableau 2: Different functions of conversion into the model of pyrolysis, from [7]

Classically the function $f(\alpha) = (1-\alpha)^n$ is used, since it is adequate to represent only the kinetic aspect, with n which may be linked to a degree of complexity of the solid studied, for example the degree of polymerization.

> *Special attention: sometime, physical phenomena are indistinguishable from the kinetic, and it is necessary to use a different conversion functions. It can be the case considering the oxygen diffusion into the solid, which impacts the kinetic of decomposition (cf Jander):*
>
> $$f(\alpha) = \left[1-(1-\alpha)^{\frac{1}{3}}\right]^2$$

Concerning the rate of the reactions, it is described by a modified (for solid application) law of Arrhenius. Then, the rate of each reaction is: $\dot{\omega} = k(\dot{T}).f(\alpha)$:

$$\dot{\omega} = Ae^{-\frac{E_a}{RT}}m^n$$

with A the pre-exponential factor, E_a the energy of activation, R the constant of perfect gases, T the temperature, m the mass, n the order of the reaction.

IOP Conf. Series: Journal of Physics: Conf. Series **1107** (2018) 022001 doi:10.1088/1742-6596/1107/2/022001

Special attention: the law of Arrhenius has been developed for gas application and is based on the theory of the collisions. Then, its application for the solid thermal kinetic representation can be questionable. The works of Batiot [21] have then permitted establishing firmly its application and its form, using a detailed upscaling approach, from the atoms to the solid at matter scale.
Moreover, Batiot & al. [24] have conducted a sensitivity analysis on the different parameters of this law (A, E_a, n) on the mass loss on the mass loss rate response. They show the compensation effect between A and Ea but also the existence and the uniqueness of the solutions obtained for those parameters, in accordance with the works of Marquis & al. [25]. Because those parameters are micro-scale properties, there cannot be determined without using inverse optimisation methods, described below.

Each reaction of the reactional mechanism has a rate $\dot{\omega}_i$ defined by an Arrhenius law of the form:

$$\dot{\omega}_i = A_i e^{-\frac{E_i}{RT}} \left(\frac{m_i}{m_o}\right)^{n_i} Y_{O_2}^{\delta}$$

From those expressions, the total mass loss rate of the solid (MLR_i or $\frac{dm_t}{dt}$) is expressed by the sum of the mass loss rate due to the i reactions of the reactional mechanism:

$$MLR_i = \frac{dm_i}{dt} = \sum_{j\epsilon J} v_j \omega_j - \sum_{k\epsilon K} \omega_\epsilon$$

With m_i the mass of the intermediate compounds, and i the number of intermediate species, J the number of reactions of thermal decomposition of the mass I and K the number of reactions permitting the gain of mass i. If m_t the total mass, so

$$\frac{dm_t}{dt} = \sum_{i=1}^{l} \frac{dm_i}{dt}$$

The rate of the reaction revealed four unknown parameters for each reaction: A the pre-exponential factor, E_a the energy of activation, n the order of the reaction and υ the stoichiometric coefficient. The determination of those parameters cannot be done experimentally [21], and is realised using inverse methods of optimization. Different methods are proposed, there are compared into [26-27]. Let us quote by example: the genetic algorithms (GA) [28], the Genetic Algorithms Simulated Annealing, the Stochastic Hill Climber, the Shuffled Complex Evolution [29] or the particle swarm [30]. There are applied to the mass loss and (not systematically) to the mass loss rate evolutions. There use an evaluation function permitting to estimate the convergence between the numerical solutions obtained with the optimised parameters and the experimental ones. Different evaluation functions ϕ are proposed as a function of the authors [8, 17, 31] and it is shown that this one have an influence on the solutions obtained [17, 21].

Considering the example of application presented figures 5 and 6 and the table 1, the total mass loss rate of the polyurethane foam [16] is:

$$\frac{dm}{dt} = \sum_{i=1}^{4} MLR_i = \frac{dm_{PPUF}}{dt} + \frac{dm_{Polyol}}{dt} + \frac{dm_{Char}}{dt} + \frac{dm_{residue}}{dt}$$

So:

$$\frac{dm}{dt} = (v_1 - 1)\omega_1 + (v_2 - 1)\omega_2 + (v_3 - 1)\omega_3 + (v_4 - 1)\omega_4 + (v_5 - 1)\omega_5$$

With each reaction having a rate:

$$\dot{\omega}_i = A_i \, exp\left(-\frac{E_i}{RT}\right) m_i^{n_i} y_{O_2}^{\delta}$$

During this study, the 4i so 20 unknown parameters have been determined by the Genetic Algorithms method [16]. The values obtained are presented into the table 3.

Reaction	Parameter	Value	Range		Units
			High	Low	
PPUF pyrolysis	E_1	169.9	190	150	kJ·mol^{-1}
	A_1	6.09 x 10^{13}	1 x 10^{22}	1 x 10^{7}	s^{-1}
	n_1	0.91	1	0.1	_
	υ_1	0.69	0.9	0.1	kg·kg^{-1}
	τ_1	–	9 x 10^{9}	1.5 x 10^{9}	_
Polyol pyrolysis	E_2	243.9	260	100	kJ·mol^{-1}
	A_2	4.42 x 10^{17}	1 x 10^{19}	1 x 10^{7}	s^{-1}
	n_2	1.26	1.5	0.1	_
	υ_2	0.10	0.81	0.1	kg·kg^{-1}
	τ_2	4.9 x 10^{9}	9 x 10^{9}	1.5 x 10^{9}	_
PPUF oxidization	E_3	214.1	240	161	kJ·mol^{-1}
	A_3	3.07 x 10^{18}	1 x 10^{20}	1 x 10^{7}	s^{-1}
	n_3	0.48	3	0.2	_
	υ_3	0.44	0.7	0.1	kg·kg^{-1}
	τ_3	8.9 x 10^{4}	1.5 x 10^{5}	3 x 10^{4}	_
Polyol oxidization	E_4	213.6	240	161	kJ·mol^{-1}
	A_4	1.26 x 10^{18}	1 x 10^{22}	1 x 10^{7}	s^{-1}
	n_4	0.95	3	0.3	_
	υ_4	0.56	0.7	0.1	kg·kg^{-1}
	τ_4	8 x 10^{5}	2.2 x 10^{6}	2 x 10^{4}	_
Char oxidization	E_5	160.8	240	160	kJ·mol^{-1}
	A_5	4.30 x 10^{12}	3 x 10^{15}	1 x 10^{11}	s^{-1}
	n_5	1.64	3	0.5	_
	υ_5	0.25	0.8	0.1	kg·kg^{-1}
	τ_5	3.4 x 10^{6}	9 x 10^{6}	1.7 x 10^{5}	_

Table 3: kinetic parameter values of the different reaction of thermal decomposition of a PU foam [16]

The figure 7 presents the comparison between the experimental and the numerical mass loss rate evolutions [16].

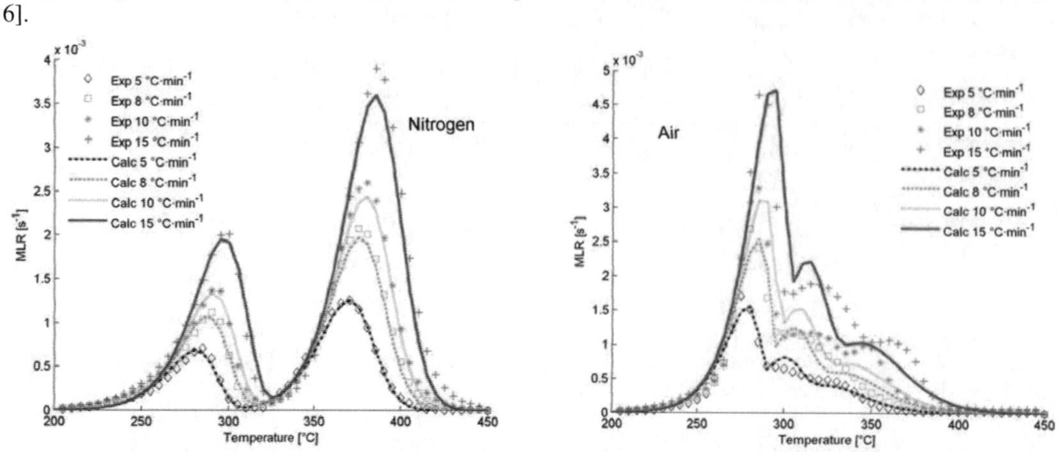

Figure 7: comparison of the experimental and the numerical mass loss rate evolutions [16]

The results show a very good agreement, the pyrolysis model (and then the reactional mechanism and the kinetic parameters) are validated at this particular scale, the matter one. However, just the kinetic aspect of the thermal decomposition is represented and it is now necessary to be able to represent the thermal behaviour of a real fuel, considering the thermal and mass transfers. A scaling-up approach is used, it requires notably the determination of several thermo-physical properties of the fuel studied.

III. DETERMINATION OF THE THERMO-PHYSICAL PROPERTIES

In order to be able to simulate the heat and mass transfers during the thermal decomposition of a solid, it is necessary to determine different thermo-physical properties:

- The thermal conductivity k. This can be done by the flash method, or using guarded hot plate.
- The density ρ
- The specific heat C_p, using a differential scanning calorimetry
- The enthalpies of the reactions ΔH, using a differential scanning calorimetry
- The radiative properties (emissivity, absorptivity), using and infra-red and a thermal camera approach, as done in [32-33].
- The heat of combustion, using a bomb calorimeter.

However, it is very important to note that those properties must be determined for all the intermediate compounds formed throughout the thermal decomposition. Taking the example of the PU foam, which means that k, ρ, Cp, ΔH, ε… must be determined for the initial PU foam (PPUF), the "Polyol", the Char and the Residue. The determination of the certainty of those properties is one of the main actual problems faced by the international community, due to several reasons:

- The experimental limitations: indeed, for several parameters, the experimental investigations are not adequate and show major limitations. For example:
  - It is very difficult to measure the conductivity k on porous matrix
  - It is not possible to measure k and C_p after a certain temperature, the one of thermal decomposition. Indeed, the gaseous emissions generated by the thermal decomposition will affect irreversibly the set-up used.
    Some authors used a TGA/DSC apparatus for the determination of C_p and ΔH, however it is shown that the precision of the measure is quite low and the C_p value has to be corrected by the mass loss.
- The possibility to isolate each intermediate compounds in order to be able to realize the measurements of those properties. In fact, it is very difficult to isolate and to extract the condensed materials formed during the thermal decomposition process, with notions of superposition of the reactions and so of the compound formed for same temperature range. A solution should be to decompose the solid until the temperature of formation of each intermediate compound, if possible, and to stop the test. This may permit isolating each compound. Nevertheless, when cooling, the structure of the material will be modified, and that will make an important impact on the measurements.

New experimental approaches are actually developed by the team of S. Stoliarov in the University of Maryland, using for example the micro-calorimeter set-up to determine the heat of combustion or of thermal decomposition [34-35], or the Controlled Atmosphere Pyrolysis Apparatus (CAPA) to deduce the conductivity [36-37]. However, if the first method gives encouraging results, it does not permit isolating and determining the properties values for each intermediate compounds and it can, then, just be applicable in case of one mechanism of thermal decomposition with only one reaction. Concerning the second one (CAPA), it is limited to specific materials and conditions (gasification).

Then it appears to be very difficult to experimentally determine certain of those parameters, as k, ρ, C_p, ΔH. Two alternative solutions are actually used:

- The parameters are determined by inverse optimization methods, for example using Gpyro [10]. Nevertheless, this solution cannot be considered satisfactory, because this is like the addition of some degrees of liberty, some mathematical fitting methods with more variables, the thermochemical properties. Moreover a compensation phenomenon can be observed due to the large unknown parameters (the kinetics and the thermochemical ones). Finally, the values obtained can have no physical sense.
- Classically « weighting (average) law » are used to determine the value of the thermos-physical properties, between the properties of the initial material and the ones of the final one:
  - Or, equivalent properties are taken for all the intermediate compounds, as for just one equivalent material.

- Or, a linear evolution of the properties between the initial and the final materials is considered. An example is presented into [38].

Special attention: the determination of the thermo-physical properties for the total range of temperature of the thermal decomposition, and for all the intermediate compounds into the kinetic mechanism is essential for the modelling of the thermal decomposition. However, it is one of the main bottlenecks actually, requiring huge efforts of the scientific community.

IV. MODELLING OF THE THERMAL DECOMPOSITION – SCALING-UP

As presented before, notably on figures 1 and 2, the challenge remains the modelling of the thermal decomposition at real scale, then considering the interactions between both the condensed and the gaseous phases. This requires the description of the heat and mass transfers between these two. The scaling-up approach presented by [5] and adopted by [4, 17, 39] is used: it is based on experimental investigations to determine the initial and boundary conditions for the simulation, and to furnish results permitting a comparison with the numerical ones, for the validation (or not) of the model.

Several bench scales are classically used with the principle consisting of the addition of complexity in increasing scale:

- Firstly, at small scale, using the cone calorimeter [40], the Fire Propagation Apparatus [41] or the Controlled Atmosphere Pyrolysis Apparatus [36-37]. This permits to work in one-dimension, with heat and mass transfers but without a flame propagation.
- At product scale, using for example the radiant panel, the IMO-LIFT [42], the Medium Burning Item [43], the Single Burning Item [44], permitting some investigation in two and three dimensions, with a flame propagation.
- At real scale, like realised for example by [45-46].

The modelling approaches use dedicated codes, like Gpyro [10-11] or Thermakin [12-13], or consists of the implementation of the model of pyrolysis and the thermo-physical properties previously presented in a generalized CFD code, like FDS [1] or Firefoam [2] and to simulate the experimental configurations studied.

Several studies consider those different approaches and just some examples are given. Considering the PU Foam previously studied at matter scale, the works of [47] have permitted its characterization using the cone calorimeter coupled with several gas analysers. Some results are presented on figure 8 for the mass loss rate (MLR) and the Heat Release Rate (HRR) [47].

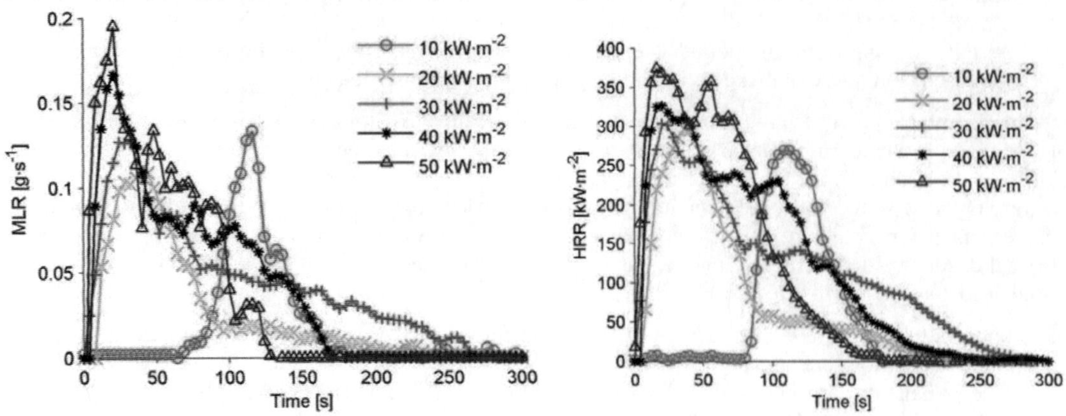

Figure 8: MLR and HRR evolution during the thermal decomposition of a PU Foam using the cone calorimeter [47]

Dealing with the PU foam in their study of the behaviour of a railway seat, Guillaume & al. [39] have simulated, at the specific condition of $50kW/m^2$, the cone calorimeter experiments. In this study, the model of pyrolysis and the thermo-physical properties have been introduced into the FDS code. The results obtained concerning the Heat Release Rate are presented on figure 9 [39]. The results show a good agreement permitting to validate the model at this scale.

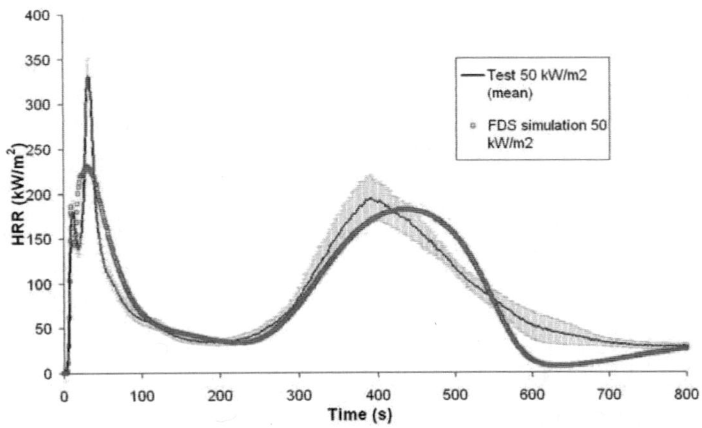

Figure 9: numerical and experimental evolutions of the HRR evolution during the thermal decomposition of a railway seat using the cone calorimeter [39]

If classically, the mass loss rate, the mass loss rate and the heat release rate are considered for the validation of the model, the work of Fateh & al. [48] shows that it is necessary to consider other parameters, like the temperature evolutions in the depth of the material and the thickness evolution of the intermediate compounds.

At a higher scale, Guillaume & al. [39], in their scaling-up validation of the model - to represent the thermal decomposition of a railway seat, have modelled some tests using the single burning item. The figure 10 [39] is presenting the comparison of the numerical and the experimental HRR evolutions at the Single Burning Item scale. A good agreement is obtained.

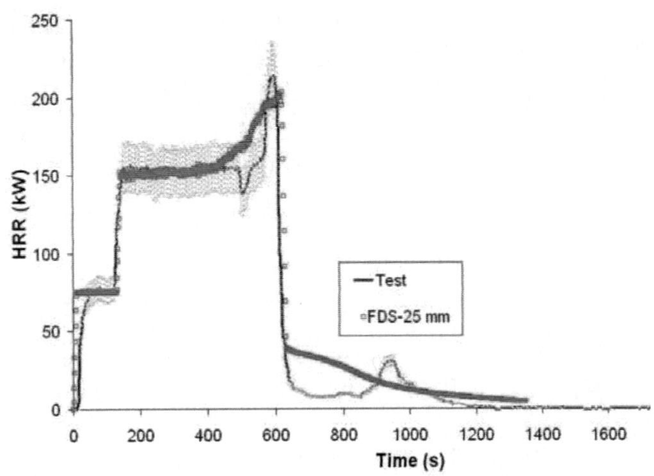

Figure 10: Numerical and experimental evolutions of the HRR evolution during the thermal decomposition of a railway seat using the single burning item [39]

CONCLUSION

The description of the thermal decomposition of the solid fuels is of the best interest and importance. It has a large influence on the ignition process, the kinetic of the fire development, its extinction and on the characteristics of the flame and its propagation.

The thermal decomposition is controlled by the phenomena taking place into the condensed phase but is dependent of the heat and the mass transfers associated to the gaseous phase. Thus, its description requires the development of a model of pyrolysis which simulate the kinetic of decomposition of the solid phase, the determination of the thermo-physical properties of the condensed compounds, and, finally, the description of the heat and mass transfers which determine the conditions of the degradation. Then, different steps are required to model the thermal decomposition of a solid.

Different methods are used to develop a model of pyrolysis, however, the modelistic one is used the most. It requires a mechanism of thermal decomposition and a kinetic law, which - is composed by a function of conversion and by a law of variation of the rate, adopting a modified Arrhenius form. The development and the validation of the model of pyrolysis is realized at matter scale, using thermogravimetric analysis.

The determination of the values of the thermo-physical properties is necessary for the modelling of the heat and mass transfers into the condensed phase but it remains an important challenge.

The modelling of the thermal decomposition is realised by a scaling up approach, permitting furnishing experimentally the initial and boundary conditions as well as results for the model application and development.

ACKNOWLEDGMENTS

The author wishes to thank: the “Institut des Risques Industriels Assurantiels et Financiers” for their technical and human support into the HESTIA platform ; his colleagues, particularly Benjamin Batiot, Jocelyn Luche, Marc Poisson and Franck Richard ; his past and present PhD students.

REFERENCES

[1] FDS – Fire Dynamics Simulator and Smokeview. NIST, 100 Bureau Drive, Gaithesburg, MD, USA http://www.fire.nist.gov/FDS

[2] FireFOAM, http://www.openfoam.org/. FM Global, http://code.google.com/p/firefoam-dev/.

[3] Torero J., Scaling up fire. Proceedings of the Combustion Institute, 34:99-124, 2013.

[4] Marquis D., Caractérisation et modélisation multi-échelle du comportement au feu d’un composite pour son utilisation en construction navale. Thèse de doctorat, Université de Nantes, Ecole des Mines de Nantes, 2010.

[5] Torero J.L., Flaming ignition of solids fuels, SFPE Handbook of fire protection Engineering 4th edition, National fire protection association, 2.260–2.276, 2008.

[6] Marquis D. & al. Impact de la connaissance des propriétés matériaux dans les simulations incendies – GDR Feux, LEMTA Nancy 2011.

[7] Witkowski A, Stec A. Hull R., Thermal decomposition of polymeric materials. SFPE Handbook of fire protection engineering, 5th edition, 167-254, 2016.

[8] Rein G., Lautenberger C., Fernandez Pello A.C., Torero J.L., Urban D.L., Application of genetic algorithms and thermogravimetry to determine the kinetics of polyurethane foam in smoldering combustion. Combustion and Flame, 146, p. 95-108, 2006.

[9] Lautenberger C., Rein G., Fernandez-Pello A.C. – The application of a genetic algorithm to estimate material properties for fire modeling from Bench-Scale Fire Test Data. Fire Safety Journal, vol. 41, p. 204-214, 2006.

[10] Lautenberger C. – Gpyro – A generalized pyrolysis model for combustible solids. Users’ Guide (2009) http://code.google.com/p/gpyro/.

[11] Lautenberger C., Fernandez-Pello A.C. Generalized Pyrolysis Model for Combustible Solids. Fire Safety Journal, vol. 44, pp. 819–839, 2009.

[12] Stoliarov S., Crowley S., Walters R.N., Lyon R.E. – Prediction of the burning rates of charring polymers. Combustion and Flame, vol. 157-11, p. 2024-2034, 2010.

[13] Stoliarov S., Leventon I.T, Lyon R.E., Two-dimensional model of burning for pyrolyzable solids. Fire and Materials, vol. 38, pp. 391-408, 2014.

[14] Stoliarov S. Parameterization and Validation of Pyrolysis Models for Polymeric Materials. Fire Technology, vol. 52-1, pp 79–91, 2016.

[15] Matala A. Estimation of solid phase reaction parameters for fire simulation (Master’s thesis). Helsinki University of Technology, Finland, 2008.

[16] Rogaume T., Bustamante Valencia L., Guillaume E., Richard F., Luche J., Rein G., Development of the thermal decomposition mechanism of polyether polyurethane foam using both condensed and gas phase release data. Combustion Science and Technology, vol. 183, pp. 627-644, 2011.

[17] Bustamante Valencia L. Experimental and numerical investigation of the thermal decomposition of materials at three scales: application to polyether polyurethane foam used in upholstered furniture. PhD thesis, Université de Poitiers, Ecole Nationale Supérieure de Mécanique et d'Aérotechnique, 2009.

[18] Fateh T. Rogaume T., Luche J., F. Richard, Jabouille F. Kinetic and mechanism of the thermal degradation of a plywood by using thermogravimetry and Fourier-transformed infrared spectroscopy analysis in nitrogen and air atmosphere. Fire Safety Journal, vol. 58, pp. 25-37, 2013.

[19] Fateh T. Rogaume T., Luche J., F. Richard, Jabouille F. Modeling of the thermal decomposition of a treated plywood from thermo-gravimetry and Fourier-Transformed Infrared Spectroscopy experimental analysis. Journal of Analytical and Applied Pyrolysis, vol. 101, pp. 35-44, 2013.

[20] Fateh T., Rogaume T., Richard F., Joseph P. Experimental and modelling studies on the kinetics and mechanisms of thermal degradation of poly (methyl methacrylate) in nitrogen and air. Journal of Analytical and Applied Pyrolysis, vol.120, pp. 423-433, 2016.

[21] Batiot B. Etude et modélisation de la cinétique de décomposition thermique des matériaux solides. Application à la dégradation du bois en cas d'incendie. Thèse de doctorat, Université de Poitiers, Ecole Nationale Supérieure de Mécanique et d'Aérotechnique, 2014.

[22] Gascoin N., Navarro-Rodriguez A., Gillard P., Mangeot A., Kinetic modelling of high polyethylene pyrolysis : Part1. Comparison of existing methods. Polymer degradation and stability, vol. 97, pp. 1466-1474, 2012

[23] Marquis D., Batiot B., Guillaume E., Rogaume T., Influence of reaction mechanism accuracy on the chemical reactivity prediction of complex charring material in fire condition. Journal of analytical and applied pyrolysis, vol. 118, pp. 231-248, 2016.

[24] Batiot B., Collin A., Richard F., Rogaume T. Sensitivity and uncertainty analysis of Arrhenius parameters in order to describe the kinetic of solid thermal degradation during fire phenomena. Fire Safety journal, vol. 82, pp. 76-90, 2016.

[25] Marquis D., Guillaume E., Camillo A., Rogaume T., Richard F. Existence and uniqueness of solutions of a differential equation system modelling the thermal decomposition of polymer materials. Combustion and Flame vol. 160, Issue 4, pp. 818-829, 2013.

[26] Lautenberger C., Fernandez-Pello C., Optimization Algorithms for Material Pyrolysis Property Estimation, Proceedings of 11th IAFSS, 2011

[27] Siarry P., Dréo J., Pétrowski A., Taillard E. Métaheuristiques pour l'optimisation difficile. Ed Eyrolles, ISBN13: 978-2-212-11368-6, 2003.

[28] Houck C., Joines J., A genetic algorithm for function optimization: a matlab implementation. North Carolina State University Edward P. Fitts Department of Industrial and Systems Engineering Report NCSU IE TR 95-09, 1995.

[29] Duan Q.Y., Gupta V.K., Sorooshian S., Shuffled Complex Evolution Approach for Effective and Efficient Global Minimization. Journal of Optimization Theory and Applications, Vol. 76:501-521, 1993.

[30] Eberhart R., Kennedy J. A new optimizer using Particle Swarm Theory. Sixth international Symposium on Micro Machine and Human Science, 1995.

[31] Esperanza M., Garcia A., Font R., Conesa J. Pyrolysis of varnish based on a polyurethane. Journal of Analytical and Applied Pyrolysis 52: 151-166, 1997.

[32] Boulet P., Parent G., Acem Z., Rogaume T., Fateh T., Zaida J., Richard F., Characterization of the radiative exchanges when using a cone calorimeter for the study of the plywood pyrolysis. Fire safety journal vol. 51, pp. 53–60, 2012.

[33] Acem Z., Brissinger D., Collin A., Parent G. , Boulet P., Quach T.H.Y., Batiot B., Richard F., T. Rogaume T., Surface temperature on a carbon composite sample during thermal degradation. International Journal of Thermal Sciences, vol. 112, pp.427-438, 2017.

[34] Stoliarov S., Raffan-Montoya F., Walters R., Lyon R.E., Measurement of the global kinetics of combustion for gaseous pyrolyzates of polymeric solids containing flame retardants. Combustion and flame, vol.173 Issue 1, 2016.

[35] Ding Y., McKinnon M.B., Stoliarov S.I., Fontaine G., Bourbigot S. Determination of kinetics and thermodynamics of thermal decomposition for polymers containing reactive flame retardants: Application to poly(lactic acid) blended with melamine and ammonium polyphosphate. Polymer Degradation and Stability, Volume 129, pp. 347-362, 2016.

IOP Conf. Series: Journal of Physics: Conf. Series **1107** (2018) 022001 doi:10.1088/1742-6596/1107/2/022001

[36] Semmes M, Liu X, McKinnon MB, Stoliarov SI, Witkowski A. A model for oxidative pyrolysis for corrugated cardboard. In: Proceedings of the Eleventh International Symposium on fire Safety Science; 2014.

[37] Swann J.D., Ding Y., McKinnon M.B., Stoliarov S.I., Controlled atmosphere pyrolysis apparatus II (CAPA II): A new tool for analysis of pyrolysis of charring and intumescent polymers. Fire Safety Journal, Vol. 91, pp. 130-139, 2017.

[38] Linteris G.T., Lyon R.E., Stoliarov S.I., Prediction of the gasification rate of thermoplastic polymers in fire-like environments. Fire Safety Journal, Volume 60, pp. 14-24, 2013.

[39] Guillaume E., Camillo A., Rogaume T., Application and Limitations of a Method Based on Pyrolysis Models to Simulate Railway Rolling Stock Fire Scenarios. Fire Technology, 50, pp. 317-348, 2014.

[40] ISO 5 660-1:2002, Reaction to fire tests – Heat release, smoke production and mass loss rate – Part 1 Heat release rate (cone calorimeter method). Second edition 2002.

[41] ISO 12136:2011, Reaction to fire tests -- Measurement of material properties using a fire propagation apparatus.

[42] ISO 5658-2:2006, Reaction to fire tests - Spread of flame - Part 2: Lateral spread on building and transport products in vertical configuration.

[43] ISO 21367:2007, Plastics - Reaction to fire - Test method for flame spread and combustion product release from vertically oriented specimens.

[44] EN 13823:2014, Reaction to fire tests for building products – Building products excluding flooring exposed to the thermal attack by a single burning item.

[45] The Dalmarnock Fire Tests: experiments and modelling, (Eds.) G. Rein, C. Abecassis Empis, R. Carvel, University of Edinburgh, November, 2007, ISBN 978-0-9557497-0-4.

[46] Guillaume E., Didieux F., Thiry A., Belivier A., Real-scale Fire Tests of One Bedroom Apartments with Regard to Tenability Assessment, Fire safety journal, vol. 70, pp. 81-97, 2014.

[47] Bustamante Valencia L., Rogaume T., Guillaume E., Rein G., Torero J.L., Analysis of principal gas products during combustion of polyether polyurethane foam at different irradiance levels. Fire Safety Journal, vol. 44, issue 7, pp. 933-940, 2009.

[48] Fateh T., Rogaume T., Richard F., Modeling of the pyrolysis of plywood exposed to heat fluxes under cone calorimeter. 11th International Symposium on Fire Safety Science, 2014.

Maze running into intumescence: mechanistic aspects in polypropylene

Serge Bourbigot[1], Johan Sarazin[1], Fabienne Samyn[1] and Maude Jimenez[1]

[1]*R_2Fire group/UMET – UMR CNRS 8207, ENSCL, University of Lille, France*

serge.bourbigot@ensc-lille.fr

ABSTRACT

The concept of intumescence was applied to make flame retarded polypropylene (PP). This paper examines two types of intumescence in PP) based on expandable graphite (EG, physical expansion) and on modified ammonium polyphosphate (AP760, chemical expansion). Reaction to fire of PP containing EG and AP760 was first evaluated by cone calorimetry. The incorporation of intumescent additives at relatively low loading (10 wt%) in PP permits the reduction by 70% of pHRR. The mode of action occurs via the formation of an expanded carbonaceous layer in all cases. The protective coating acts mainly as heat barrier in the case of the formulations containing AP760 or as heat dissipater with EG. The incorporation of small amount of EG in PP-AP760 modifies heat transfer in the coating creating a strong anisotropy. Upon expansion graphite worms align normal to the surface increasing the transverse heat conductivity (lower efficiency of the heat barrier) and hence, decreasing the fire performance (decrease by only 30% of pHRR). Kinetic analysis was then performed to quantify the thermal stability of the intumescent systems. It reveals that the intumescent additives do not modify the reactional scheme of the PP thermal decomposition but they increase slightly the thermal stability of the intumescent systems.

KEYWORDS:

Intumescence; polypropylene; fire chemistry; kinetic analysis; heat transfer

NOMENCLATURE LISTING

A	frequency factor (1/s)
E	activation energy (kJ/mol)
t	time (s)
k	kinetic constant (1/s)
$f(\alpha)$	reaction model

Greek symbols

α	degree of conversion (kg/kg)

IOP Conf. Series: Journal of Physics: Conf. Series **1107** (2018) 032001 doi:10.1088/1742-6596/1107/3/032001

INTRODUCTION

The word "intumescence" comes from Latin 'intumescere', which means 'to swell up'. From the dictionary, intumescence is defined as the act or process of swelling or enlarging. Some definition also mentions the action of heat to expand the body. French Writers like Jules Verne or François-René de Chateaubriand used the word 'intumescence' with different meanings: a hilly area on volcano (J. Verne in The Mysterious Island, 1874) or the waves' swell (F.R. de Chateaubriand in Mémoires d'Outre-Tombe, 1848). Taken 'as is', those definitions are not directly linked to the flame retardancy of polymeric materials or to the fire protection. Nevertheless, the above definitions describe well the behavior of an intumescent material used for fire retardancy. When heating beyond a critical temperature, an intumescent material begins to swell and then to expand. The result of this process is a foamed cellular charred layer on the surface, which protects the underlying material from the action of the heat flux or the flame. Visually, the swelling and the expansion looks like 'black waves' swollen at the surface of the material and the final char exhibits hemispheric shape with a roughed or smooth surface. The paper focuses on intumescence and its application in polymeric materials.

The concept of intumescence enables to make flame retarded polymeric materials exhibiting high performance [1] but there is a lack of data (qualitative and quantitative) able to fully describe its mechanism. It is one of the goal of this paper to bring new insights on the mechanism of intumescent systems. This paper examines two types of intumescence in polypropylene (PP) based on expandable graphite (EG, physical expansion) and on modified ammonium polyphosphate (AP760, chemical expansion). Some examples of using ammonium polyphosphate and expandable can be found elsewhere [2-4]. Reaction to fire of PP containing EG and AP760 is first evaluated by cone calorimetry and the efficiency of the intumescent barrier is measured at the same time. Kinetic analysis is then performed to quantify the thermal stability of the intumescent systems.

EXPERIMENTAL

Materials and processing

Commercial grade of isotactic polypropylene (PP) was used in this study: PP (PPH9060) was supplied by Total petrochemicals (Feluy, Belgium). PPH9060 has a melt flow rate (MFR) for a load of 2.16 g at 230°C of 25 g/10 min. Expandable graphite (EG) is the commercial grade ES350F5 from Graphitwerk Kropfmühl (Germany) with an average grain size of 300 μm. EG is a synthesized intercalation compound of graphite that expands or exfoliates when heated. A wide variety of chemical species can be used to intercalate graphite materials (e.g. sulfate, nitrate, various organic acids …) [5]. Sulfate was used in this grade to make graphite bisulfate. Modified ammonium polyphosphate (AP760) is the commercial grade of Clariant (Germany) with the brand name Exolit AP760. It is an intrinsic intumescent system containing 20 wt% phosphorus and 14 wt% nitrogen acting in synergy. It is mainly based on ammonium polyphosphate as acid source and tris(hydroxyethyl) isocyanurate (THEIC) as char former.

The strategy was to blend PP with flame-retardants (FRs) in a twin-screw extruder. The total loading of FRs in PP was 10 wt% varying the ratio AP760/EG (wt/wt) at 10:0; 5:5; 9:1 and 0:10. This loading was selected because it provides an acceptable performance according to cone calorimetry. Compounding was performed using HAAKE Rheomix OS PTW 16 twin-screw extruder. The extruder is a co-rotating intermeshing twin screw with a barrel length of 400 mm and screw diameter of 16 mm (L/D = 25) with 10 zones. PP and FRs were incorporated using two gravimetric side feeders into the extruder. Polymer flow rate is fixed to extrude about 500 g/h with a screw speed of 300 rpm. The temperature profile of the extruder from feeder to die was set at 200/200/200/200/200/170/185/180/200/200°C.

Cone calorimetry

FTT (Fire Testing Technology) Mass Loss Calorimeter (MLC) was used to carry out measurements on samples following the procedure defined in ASTM E 906. The equipment is identical to that used in oxygen consumption cone calorimetry (ASTM E-1354-90), except that a thermopile in the chimney is used to obtain heat release rate (HRR) rather than employing the oxygen consumption principle. Our procedure involved exposing specimens measuring 100 mm x 100 mm x 3 mm in horizontal orientation. External heat flux of 35 kW/m² was used for running the experiments. This flux corresponds to common heat flux in mild fire scenario. MLC was used to determine heat release rate (HRR). When measured at 35 kW/m², HRR is reproducible to within ±10%. The cone data reported in this paper are the average of three replicated experiments. In addition to those measurements, thermocouples were embedded on the backside of the materials in horizontal position. They measured temperature as a function of time during a regular cone experiment.

Kinetic analysis

ThermoGravimetric Analysis (TGA) was carried out using a Netzsch TG 209 F1 Libra at five heating rates (1, 2, 5, 10, 20 and 50°C/min) from 30°C to 800 °C in nitrogen flow (50 cm^3/min). Samples of exactly 10 mg (±0.3 mg) were put in open alumina pans. Typically, two replicates were run for each sample, and the average was reported. Both the onset (5% mass fraction loss) and peak mass loss rate have an uncertainty of 0.9°C (2σ). We corrected for each heating rate the buoyancy force (calibration with empty pan). Kinetic analysis and modeling of the degradation of the samples was made using an advanced thermokinetic software package developed by Netzsch Company [6]. For kinetic analysis, it is assumed that the material decomposes according to the Eq. (1):

$$A_{solid} \rightarrow B_{solid\ or\ liquid} + C_{gas} \quad (1)$$

The rate expression dα/dt, where α is the degree of conversion is assumed to be defined by Eq. (2):

$$d\alpha/dt = k(T).f(\alpha) \quad (2)$$

where k is the kinetic constant, k = A.exp(-E/RT) according to the Arrhenius law, A is the frequency factor, E is the activation energy and f(α) is the so-called "reaction model". All reactions are assumed to be irreversible. In the case of decomposition and since the evolved gases were continuously removed by the fluid flow in the TGA chamber, this is a reasonable assumption. It is also assumed that the overall reaction (Eq. (1)) is the sum of individual reaction steps (formal or true step) with constant activation energy, as generally accepted in chemistry. The model can then include competitive, independent and successive reactions. The equations were solved with multivariate kinetic analysis (determination of the parameter via an hybrid regularized Gauss-Newton method or Marquardt method).

RESULTS AND DISCUSSION

Reaction to fire

EG and AP760 and some of their combinations (AP760/EG at 5:5 (wt/wt) and 9:1 (wt/wt)) were incorporated in PP at 10 wt% total loading. The four formulations were evaluated by cone calorimetry at an external heat flux of 35 kW/m² and compared to neat PP (Fig. 1-a).

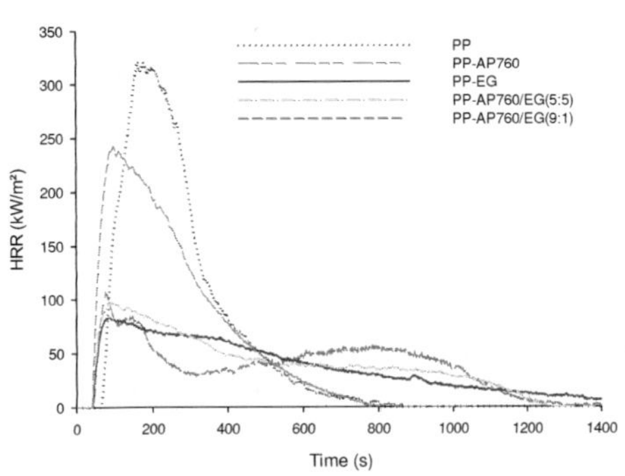

(a)

(b)

Fig. 1. (a) HRR curves as a function of time of intumescent PP (external heat flux = 35 kW/m²); (b) Intumescent coatings formed from PP-EG (up) and PP-AP760 (bottom).

Except for the formulation containing 1 wt% EG, the intumescent PPs exhibit peaks of heat release rate (pHRR), which are decreased by about 70%. The formation of an expanded charred layer is observed at the surface of the materials evidencing an intumescent phenomenon. The formulation containing 1 wt% EG exhibits a pHRR only decreased by 30%. An intumescent phenomenon is also observed but its efficiency is not as high as the other intumescent PPs. It is noteworthy the formulations containing EG behave differently compared to PP-AP760 (Fig. 1-b). The intercalation compounds decompose rapidly into gaseous products, which blast off the graphite flakes. Those flakes make then worms forming an entangled network at the surface of the material. On the contrary, the intumescent char formed from PP-AP760 exhibits high expansion (higher than those with EG)

IOP Conf. Series: Journal of Physics: Conf. Series **1107** (2018) 032001 doi:10.1088/1742-6596/1107/3/032001

and a smooth surface. It evidences therefore two distinct mechanisms: (i) the physical expansion of graphite flakes forming a protective entangled network and (ii) the formation of an intumescent char by a series of chemical reactions. Those two types of protective coating provide the same level of protection according to the cone scenario.

During the cone calorimetry experiments, thermocouples were placed on the backside of the materials to measure the temperature as a function of time and to quantify the efficiency of the intumescent barrier (thermocouple embedded in the material in horizontal position). Fig. 2-a shows temperature changes as a function of time. Up to 50s, the slopes of the curves are similar. At higher times (t>50s for PP-EG and t>80s for PP-AP760 and PP-AP760/EG(5:5)), the slopes of the intumescent PPs (except PP-AP760/EG(9:1)) are smaller than that of neat PP (curve of PP was stopped when the thermocouple is no longer in the matter because of the complete decomposition of PP). It corresponds to the formation of an intumescent layer reducing heat transfer from the flame to the backside of the substrate and so, it limits the temperature rise on the backside. Note the network formed from PP-EG is the most efficient compared to the char layers formed by PP-AP760 and PP-AP760/EG(5:5). The final temperatures are close (525°C vs. 490°C at 1100s) but the temperature rise of PP-EG is much smaller than the others at the beginning of the experiment (50s<t<225s). PP-AP760/EG(9:1) exhibits an unexpected behavior because its backside temperature increases extremely rapidly to 650°C. Its backside temperature reaches a pseudo steady state at 670°C for 350s (50s<t<400s) and then decreases to 525°C. This temperature drop occurs when the combustion stops: it removes the additional heat feedback of the flame and hence it decreases the total heat flux affecting the material.

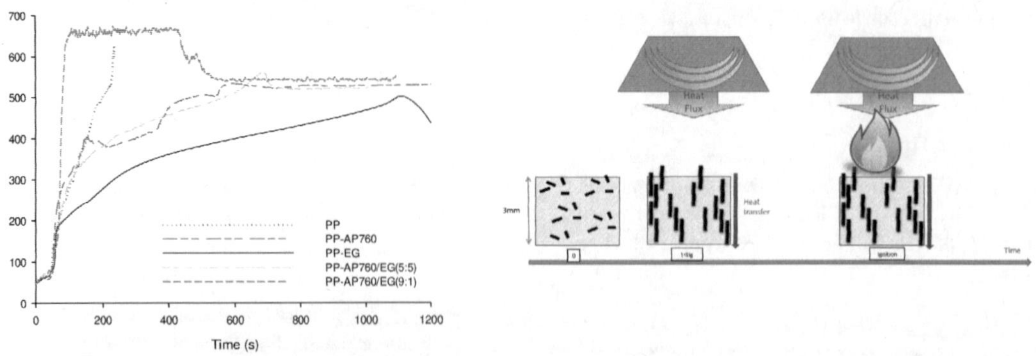

(a) **(b)**

Fig. 2. (a) Backside temperature as a function of time of intumescent PPs during cone calorimetry experiment (external heat flux = 35 kW/m²); (b) Schematic description of the effects of graphite worms in PP-AP760/EG(9:1) on heat transfer through the charred network.

The results suggest the involvement of different mechanisms. PP-A760 forms an expanded char at the surface, which reduces the heat transfer from the flame to the material. The incorporation of small amount of EG in a conventional intumescent system (PP-AP760) strongly affects its performance. Visual observation shows graphite worms grows up at the surface but there is no entanglement of the worms. The worms exhibit an alignment normal to the surface (Fig. 2-b) and hence, they might enhance heat conductivity (transverse heat conductivity) through the material as shown by temperature measurement. In the other cases, EG can form an entangled network, which flattens because of the expansion. Hence, in-plane heat conductivity is enhanced (parallel to the surface) and the worms play the role of heat dissipater (this effect is limited when EG is combined with AP760). Even if the cone heater irradiates the whole surface, it is reasonable to assume heat dissipation occurs on side surfaces of the samples. Those two types of behaviour leads to huge difference in terms of performance and it evidences the anisotropic character of EG when used as FR.

Kinetic analysis

The previous section has shown the fire performance of PP was strongly enhanced incorporating AP760 and EG. The goal of this section is to examine whether the thermal stability of the intumescent PPs could be modified. The approach was to perform kinetic analysis on the thermal decomposition of the neat PP and of the 4 intumescent formulations. Before starting any fitting procedure, it is necessary to define a model (combination of reactions) and to preset starting values for the kinetic parameters. A convenient approach is to use model-free analysis as a preliminary step of the kinetic analysis. A model-free analysis provides the plot of the activation energy versus the fractional weight loss. For the 5 materials, it reveals that the plot of activation energy exhibits a plateau at conversion degree higher than 0.2 of about 250 k/mol and a more complicated shape at low conversion degree (plots not shown). This indicates that the decomposition does not take place as a one-step reaction but as multi-step reactions.

TG curves of neat PP (Fig. 3-a) suggest a single step of decomposition but the simulation of the curves including only one single step does not give satisfactory results. The combination of two successive reactions considering an Avrami-Erofeev function followed by an auto-catalyzed order function gives an excellent fit (Fig. 3-a). The number of steps and the resulting calculated kinetic parameters (Table 1) are consistent with the conclusions of the Friedman analysis. The thermal decomposition of PP is a typical radical chain mechanism where initiation, H-abstraction, β-scission and radical recombination reactions are the relevant classes [7]. The first reaction described by an Avrami-Erofeev function (small contribution on the whole decomposition) is attributed to chain initiation reactions, which produce radicals and require lower activation energy. The second reaction is attributed to β-scission reactions, which are known to be the main decomposition reactions of PP. The autocatalysis is justified by the formation of allyl radicals and allyl H atoms during the decomposition of PP [8], which can enhance the decomposition of PP. Our reactional scheme makes then sense with the decomposition pathway of PP.

Intumescence occurs mainly in the condensed phase and it involves numerous chemical reactions. So, kinetic analysis of the reaction in the physical-chemical sense is not possible. That is why we have considered that the overall process of the intumescence could be described by multi-step processes. For the 4 intumescent PPs, TG curves were simulated using the same reactional scheme as the neat PP suggesting the decomposition of the intumescent PPs is mainly governed by the thermal decomposition of the neat PP. All the fits are excellent (Fig. 3-b shows the simulated curves of PP-AP760/EG(9:1) as typical example; the other simulations are not shown) and the calculated kinetic parameters are shown in Table 1. The activation energies of the intumescent PPs are slightly higher than that of the neat PP indicating some enhancement of the thermal stability of the intumescent PPs.

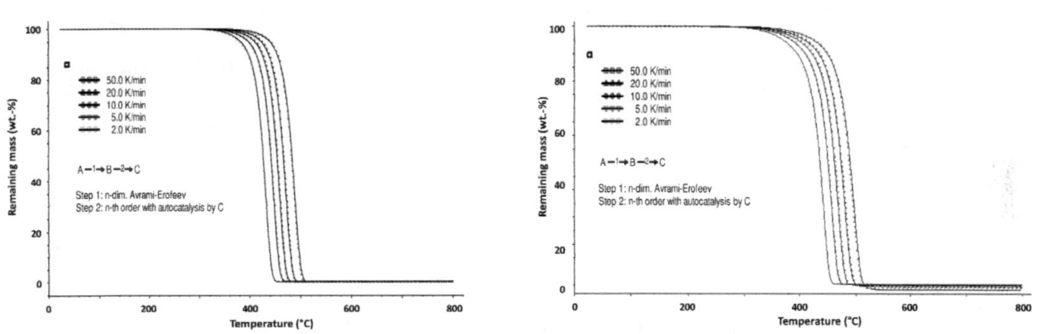

(a) **(b)**

Fig. 3. Experimental (dots) and simulated (line) TG curves at different heating rates in N_2 flow of neat PP (a) and PP-AP760/EG(9:1) (b).

Table 1. Computed kinetic parameters of the decomposition of neat PP and of the 4 intumescent PPs

Reaction N°	log(A) (A: s^{-1})	E (kJ/mol)	Reaction order or Dimension
		PP	
1	11.9 ± 0.3	182 ± 5	0.87 ± 0.07
2	15.3 ± 0.3	245 ± 3	0.91 ± 0.11
		PP-AP760	
1	13.5 ± 0.3	204 ± 4	0.53 ± 0.02
2	16.9 ± 0.1	270 ± 2	0.90 ± 0.05
		PP-EG	
1	20.4 ± 1.9	250 ± 15	0.20 ± 0.06

Reaction N°	log(A) (A: s^{-1})	E (kJ/mol)	Reaction order or Dimension
2	17.2 ± 0.1	276 ± 2	1.10 ± 0.05
		PP-AP760/EG(5:5)	
1	14.9 ± 0.6	219 ± 9	0.58 ± 0.02
2	16.1 ± 0.1	259 ± 2	0.90 ± 0.03
		PP-AP760/EG(9:1)	
1	13.3 ± 0.4	205 ± 5	0.53 ± 0.02
2	16.5 ± 0.1	264 ± 2	0.77 ± 0.05

CONCLUSION

The incorporation of intumescent additives at relatively low loading (10 wt%) in PP permits the reduction by 70% of pHRR. The mode of action occurs via the formation of an expanded carbonaceous layer in all cases. The protective coating acts mainly as heat barrier in the case of the formulations containing AP760 or as heat dissipater with EG. The incorporation of small amount of EG in PP-AP760 modifies heat transfer in the coating creating a strong anisotropy. Upon expansion graphite worms align normal to the surface increasing the transverse heat conductivity (lower efficiency of the heat barrier) and hence, decreasing the fire performance (decrease by only 30% of pHRR). Kinetic analysis on the intumescent PPs reveals that the intumescent additives do not modify the reactional scheme of the PP thermal decomposition but they increase slightly the thermal stability of the intumescent systems.

ACKNOWLEDGEMENT

This work has received funding from the European Research Council (ERC) under the European Union's H2020 – the Framework programme for Research and Innovation (2014-2020) / ERC Grant Agreement n°670747 – ERC 2014 AdG/FireBar-Concept.

REFERENCES

[1] Alongi J., Han Z. and Bourbigot S. (2015) Intumescence: Tradition versus novelty. A comprehensive review. *Progress in Polymer Science* 51:28-73, DOI 10.1016/j.progpolymsci.2015.04.010

[2] Duquesne S., Delobel R., Le Bras M. and Camino G. (2002) A comparative study of the mechanism of action of ammonium polyphosphate and expandable graphite in polyurethane. *Polymer Degradation and Stability* 77:333-344, DOI

[3] Lu Y., Zhang Y. and Xu W. (2011) Flame retardancy and mechanical properties of ethylene-vinyl acetate rubber with expandable graphite/ammonium polyphosphate/dipentaerythritol system. *Journal of Macromolecular Science, Part B: Physics* 50:1864-1872, DOI 10.1080/00222348.2011.553168

[4] Guo C., Zhou L. and Lv J. (2013) Effects of expandable graphite and modified ammonium polyphosphate on the flame-retardant and mechanical properties of wood flour-polypropylene composites. *Polymers and Polymer Composites* 21:449-456, DOI

[5] Focke W.W., Badenhorst H., Mhike W., Kruger H.J. and Lombaard D. (2014) Characterization of commercial expandable graphite fire retardants. *Thermochimica Acta* 584:8-16, DOI 10.1016/j.tca.2014.03.021

[6] Opfermann J. (2000) Kinetic analysis using multivariate non-linear regression. I. Basic concepts. *Journal of Thermal Analysis and Calorimetry* 60:641-658, DOI 10.1023/A:1010167626551

[7] Lattimer R.P. (1993) Direct analysis of polypropylene compounds by thermal desorption and pyrolysis-mass spectrometry. *Journal of Analytical and Applied Pyrolysis* 26:65-92, DOI 10.1016/0165-2370(93)85019-U

[8] Ranzi E., Dente M., Faravelli T., Bozzano G., Fabini S., Nava R., Cozzani V. and Tognotti L. (1997) Kinetic modeling of polyethylene and polypropylene thermal degradation. *Journal of Analytical and Applied Pyrolysis* 40-41:305-319, DOI 10.1016/S0165-2370(97)00032-6

IOP Conf. Series: Journal of Physics: Conf. Series **1107** (2018) 032002 doi:10.1088/1742-6596/1107/3/032002

Modelling of fire performance of Cross Laminated Timber (CLT) panels

Trong Tuan TRAN[1], Mourad KHELIFA[1], Ali NADJAI[2], Marc OUDJENE[1] and Yann ROGAUME[1]

[1]*University of Lorraine, LERMAB, EA 4370, 27 rue Philippe SÉGUIN - 88051 Épinal - FRANCE*
[2]*Fire Safety Engineering and Technology (FireSERT), Ulster University, United Kingdom*

E-mail addresses: trong-tuan.tran@univ-lorraine.fr (Presenting author),
mourad.khelifa@univ-lorraine.fr (Corresponding author)

ABSTRACT

A numerical method has been developed to incorporate the effects of heat transfer in a CLT panel exposed to fire. The procedure has been added into the software package Abaqus [1] as a user-defined subroutine (Umatht), and has been verified using both time-and spatially dependent heat fluxes in two- and three-dimensional problems. The aim is to contribute to the development of simulation tools needed to assist structural engineers and fire testing laboratories in technical assessment exercises. The accuracy of the used thermal properties and the finite element models was validated by comparing the predicted results with an available fire test in literature. It was found that the model calibrated to results from standard fire conditions provided reasonable predictions of temperatures within CLT panel exposed to fire.

KEYWORDS

Fire performance; Thermal properties; Fire test; Abaqus; CLT; Finite element models.

1. INTRODUCTION

A major hurdle to the development of the use of timber panels in construction is their perceived combustibility, which raises questions about their fire safety. Indeed, the flammability of cladding panels, particularly in high rise buildings, poses a serious safety issue. Fig. 1 shows a facade panel of the Grenfell tower in London exposed to a real fire on the 14th June 2017, which, unfortunately, left more than 80 dead. In this case, the use of CLT panels in construction could constitute a solution because the formation of a char layer acts as an insulator and protects the inner core of the virgin timber, at the same time, it does reduce the effectiveness of the cross-area to carry loads as shown in Fig. 2 and reported in refs [2-5]. However, their use in construction will remain marginal until the fire safety issue is resolved. In addition, there is also a lack of experimental evidence on their fire performance. Moreover, there is very little data available the fire performance of CLT panels and most of it is empirical. Although they are used as structural elements, there are virtually no design methods available, which limits their use commercial use. The aim of the present work therefore work is to study the fire performance of these panels.

Fig 1. Facade of the Grenfell tower in the aftermath of the fire.

(a) (b)

Fig. 2. Charcoal protective layer of (a) CLT panel [2] and (b) solid timber beam [5].

Few researchers have simulated the fire behavior of CLT panels using numerical procedures based on finite element models. Frangi et al. [6] presented and discussed the results of an extensive testing program on the fire performance of CLT panels under the standard fire as provided by ISO 834-1 [7]. The fire tests were performed in the laboratory on small specimens exposed on one side to fire. In [6], a particular attention was given to the effect of the adhesive on the fire behavior of CLT panels compared to solid timber panels.

The thermal models, available in the literature, can be roughly developed using procedures based on purely experimental observations in the absence of an understanding of the mechanism. This situation stems, first, from the complex fire behavior of the CLT panels used in construction and, second, from its lack of information on the characterization procedure of different thermal proprieties of timber material. In our opinion, there are no advanced finite element models considering the variation of thermal proprieties of timber material during the heating process.

The objective of the study is to develop a three-dimensional thermal model for predicting the thermal behavior of CLT panels exposed to fire. The developed model here can predict temperature profiles and char formation during the fire exposure. The developed 3D finite element model is incorporated into the Abaqus software [1] using a user subroutine. Thus, the fire behavior of CLT panels is investigated by the Umatht procedure. The finite element model is calibrated and validated by comparing its results with experimentally measured data.

2. NONLINEAR HEAT TRANSFER FORMULATION

In this section, the governing 3D heat conduction equations for timber material are derived. The material properties in the heat transfer equations, such as thermal conductivity λ, density ρ and specific heat c_p are assumed to be dependent on the temperature T.

2.1. Heat constitutive equations

The internal volumetric heat energy rate $\dot{\bar{Q}}_i$ produced inside the solid V can be expressed as:

$$\dot{\bar{Q}}_i = \int_V \frac{\partial}{\partial t}(\rho Q_i)dV = \int_V \rho(T)\frac{\partial Q_i}{\partial t}dV \quad (1)$$

$\dot{Q}_i$ is the internal heat energy density per unit mass, and ρ is the density. The rate of the internal energy density can be rewritten as:

$$\frac{dQ_i}{dt} = \frac{dQ_i}{dT(x_i,t)}\frac{dT(x_i,t)}{dt} = c_p(T)\frac{dT(x_i,t)}{dt} \quad (2)$$

$T(x_i,t)$ is the current spatial temperature, t is the time and x_i denotes the location coordinates in 3D space.

Substituting Eq. (2) into Eq. (1), the internal energy density yields becomes:

$$\dot{\bar{Q}}_i = \int_V \frac{\partial}{\partial t}(\rho Q_i)dV = \int_V \rho(T)c_p(T)\frac{dT(x_i,t)}{dt}dV \quad (3)$$

Fourier's law assumed that the heat flux rate transferred through surface A is proportional to the heat conductivity times the spatial gradient of the temperature. Thus, the total external energy rate $\dot{Q}_e$ can be obtained by:

$$\dot{\bar{Q}}_e = \int_A q.n_i dA = \int_A \left[\lambda(T)\frac{dT(x_i,t)}{dx_i}\right].n_i dA = \int_V \frac{d}{dx_i}\left[\lambda(T)\frac{dT(x_i,t)}{dx_i}\right]dV \quad (4)$$

where q is the surface heat flux vector, λ is the thermal conductivity and n_i is the unit vector outward normal to the surface.

Combining Eq (3) and (4), the Partial Differential Equation (PDE) can be rewritten as:

$$\rho(T)c_p(T)\frac{dT(x_i,t)}{dt} = \frac{d}{dx_i}\left[\lambda(T)\frac{dT(x_i,t)}{dx_i}\right] \quad (5)$$

In this study, the thermo-physical properties (λ, c_p and ρ) are functions of temperature. Thus the PDE becomes a nonlinear, and a more complex numerical technique is required for general solution, e.g. timber under high temperatures [8-12]. We note that the different thermal properties (λ, c_p and ρ) values of the timber material, used in this study, are those given and defined by Thi et al. (2016) in [12].

2.2. Finite element implementation

In solving the governing equations, Eq. 5, we must determine the thermal strain and the temperature. There is one user subroutine provided by Abaqus for more flexible and wider engineering application. The subroutine is Umatht, which is used to define the thermal constitutive equation (heat transfer equation, Eq. 5).

Variables necessary to be defined in Umatht for solving the heat transfer equation are:

- internal thermal energy per unit mass U;
- variation of U with respect to temperature $\frac{\partial U}{\partial T}$;
- variation of U with respect to the spatial gradients of temperature $\frac{\partial U}{\partial(\partial T/\partial x_i)}$ (i = 1,2,3);
- heat flux vectorf is defined so that: $q = -f.\underline{n}$, where q is the heat flux per unit area flowing into the solid;
- variation of f with respect to temperature $\frac{\partial f}{\partial T}$; and the variation of f with respect to the spatial gradients of temperature $\frac{\partial f}{\partial(\partial T/\partial x_i)}$ (i = 1,2,3). These terms need to be calculated at the end of each time increment.

IOP Conf. Series: Journal of Physics: Conf. Series **1107** (2018) 032002 doi:10.1088/1742-6596/1107/3/032002

3. RESULTS AND DISCUSSION

A CLT beam tested by Fragiacomo et al. [13] is simulated with the developed finite element model. The geometrical details, the loading and the boundary conditions are shown in Fig. 3. The fire test was carried out according to ISO 834-1[7]. The average initial density ρ_0, the moisture content ω, the coefficient of convection h_c, and the emissivity ε, are respectively equal to 450 kg/m^3, 12%, 25 W/(m^2.K) and 0.8. The beam is meshed with 8400 C3D8T solid elements.

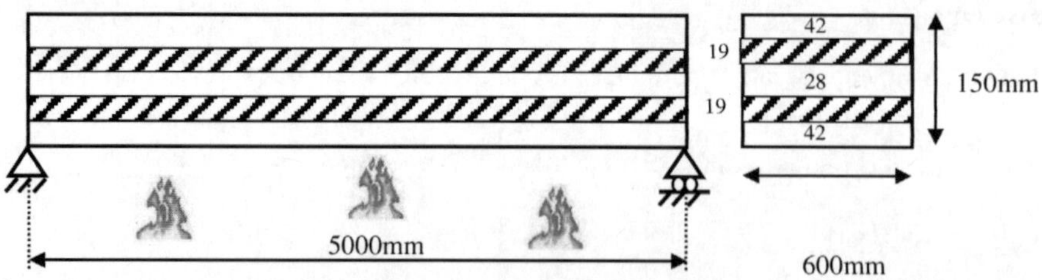

Fig. 3. CLT beam in flexure in a fire environment [2,13].

The temperature distribution across the cross section at different times of exposure is shown on Figs. 4. The formation of the charred layer starts to at temperatures above 300°C. The temperature profile shows clearly that the charred layer is uniform across the section. The residual cross section reduces with time of exposure. The beam fails in a catastrophic manner when the thickness of the residual section is not sufficient to resist the applied moment. It can be seen that temperature gradients are higher at the exposed surface (fire side). At the beginning, the formation of the char is faster and then due to the reaction of the char gradually decreases with time.

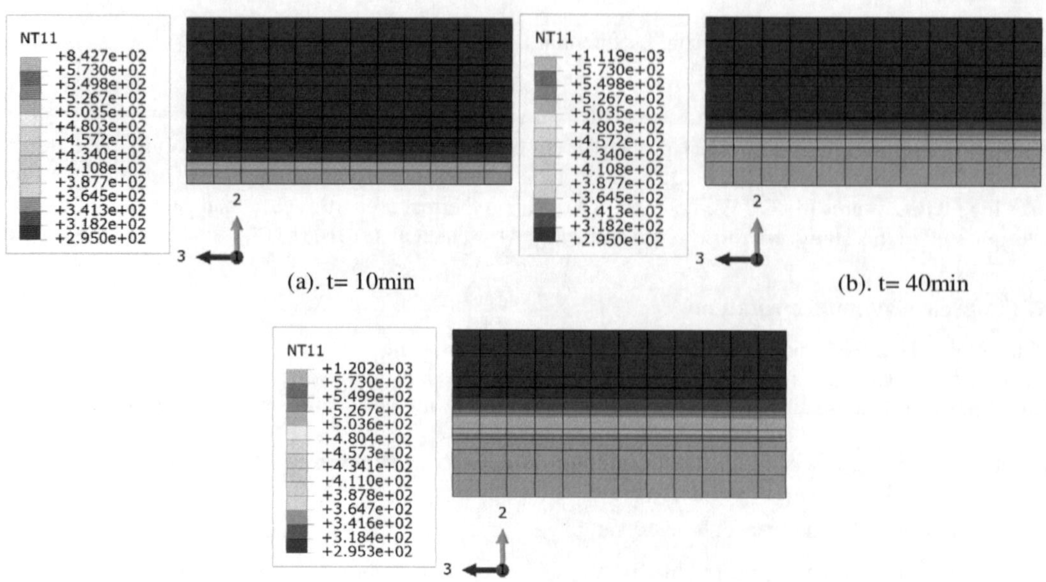

Fig. 4. Temperature profiles T[K] across the cross section at different times of exposure.

Fig 5 shows the predicted temperatures at two different positions of the thermocouples. Fig.5a shows that the temperature remains stable, about 20°C, for a very short time of less than five minutes. Beyond this time, the temperatures start to increase progressively. The same observation is also made at the position (at 22 mm below the exposed surface) of the second thermocouple as illustrated in Fig. 5b. It can be seen that the predicted temperatures agree quite well with the experimental ones

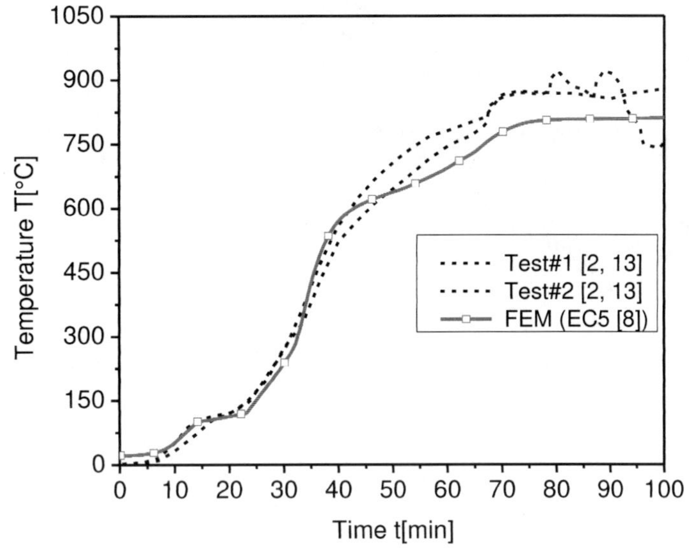

(a). At 21 mm below the exposed surface.

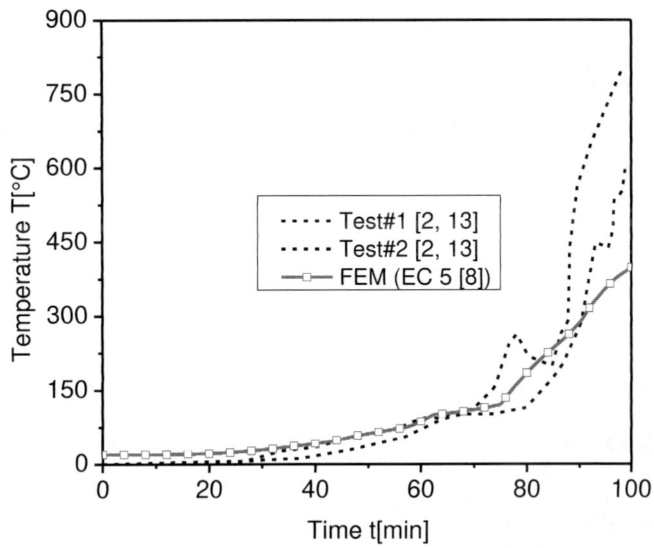

(b). At 22 mm below the exposed surface.

Fig. 5. Predicted and measured temperatures at two different thermocouple locations.

Fig 6 shows a comparison between the experimental formation of char and that predicted for CLT beam at fire exposure time t=100 min. In the absence of experimental data for different charring depths, in the presented work in refs [2, 13], only the formation zones of char are compared (see Fig. 6). It can be seen that the predicted char zones are similar to those obtained experimentally by Fragiacomo et al.[13] and Menis A.[2].

(a) (b)

Fig. 6. Formation of char after the fire test: (a) numerical and (b) experimental [2,13].

4. CONCLUSION

A 3D finite element model is developed to solve the partial differential equations of heat conduction numerically. Accordingly, the user subroutine Umatht has been implemented in Abaqus code to define their thermal constitutive relations in the finite element calculation. According to the predicted results, the validity and effectiveness of Umatht are confirmed. This study lays a foundation for the modelling of fire performance of laminated timber members used in engineering structures.

Further investigations are now in progress to extend these finite element models with the hope of simulating the thermomechanical behavior of timber structures using various geometrical and loading conditions and to provide more information for engineering applications used in timber construction.

REFERENCES

[1] Abaqus. *Finite Element Modelling Programme and Standard User's Manual.* Version 6.14. SIMULIA, 2016.

[2] Menis A. *Fire resistance of Laminated Veneer Lumber (LVL) and Cross-Laminated Timber (XLAM) elements*. PhD Thesis. Università degli studi di Cagliari, Italy, 2012.

[3] Lane W.P. *Ignition, charring and structural performance of laminated veneer lumber*. Fire Engineering Research Report 05/3. University of Canterbury, Christchurch, New Zealand, 2005.

[4] Khelifa M., Khennane A., El Ganaoui M., Rogaume Y. *Analysis of the behavior of multiple dowel timber connections in fire*. Fire Safety J, 2014;68:119–128.

[5] Samaké A., Taazount M., Audebert P., Palmili P. *Thermo-hydric transfer within timber connections under fire exposure: Experimental and numerical investigations*. Appl Therm Eng, 2014; 63:254-265

[6] Frangi A., Fontana M., Hugi E., Jöbstl R. *Experimental analysis of cross-laminated timber panels in fire*. Fire Safety J, 2009;44:1078–1087.

[7] ISO 834-1. *Fire-resistance tests. Elements of building construction*. Part 1: General requirements. International Organization for Standardization, Geneva, 1999.

[8] Eurocode 5 - *Design of timber structures. Part 1-2: General - Structural fire design*. CEN 2004 (European Committee for Standardization), EN 1995-1-2, Brussels, Belgium.

[9] Knudson R.M, Schniewind A.P. *Performance of structural wood members exposed to fire*. Forest Prod J, 1975; 25(2):23–32

[10] Fredlund B. *Modelling of heat and mass transfer in wood structures during fire*. Fire Safety J, 1993;20:39–69.

[11] Janssens ML. *Modeling of the thermal degradation of structural wood members exposed to fire*. Fire Mater, 2004;28:199–207.

[12] Thi V.D., Khelifa M., El Ganaoui M., Rogaume Y. *Finite element modelling of the pyrolysis of wet wood subjected to fire*. Fire Safety J, 2016; 81:85–96

[13] Fragiacomo M., Menis A., Clemente I., Bochicchio G., Ceccotti A. *Fire resistance of Cross-Laminated Timber panels loaded out of plane*. J Struct Eng, 2013; 139 (12).

IOP Conf. Series: Journal of Physics: Conf. Series **1107** (2018) 032003 doi:10.1088/1742-6596/1107/3/032003

Study of the influence of heat transfer of a CLT beam through FEM

Juan Enrique Martínez-Martínez[1], Mar Alonso-Martínez[2], Felipe Pedro Álvarez Rabanal[3] and Juan José del Coz Díaz[4].

[1,2,3,4] *Department of Construction and Manufacturing Engineering, GICONSIME, University of Oviedo, Spain*

quique@constru.uniovi.es

ABSTRACT

The heat transfer through wood in a fire situation is controlled by several processes, such as species, rate of heating and reaction kinetics. The article examines the challenges related to the thermal analysis of Cross-Laminated Timber (CLT) beams under non-standard fire conditions using ANSYS software. In this work, three numerical simulations are developed using Finite Elements (FE). The boundary conditions applied to simulate non-standard thermal conditions are: heat flux, radiation and convection, and temperature. And these boundary conditions are useful to compare heat transfer through Cross-Laminated Timber (CLT). The thermal properties of the wood used are based on the Eurocode 5. To validate numerical models, existing experimental results are used. The results show that the heat transfer through the CLT beam could be adjusted only by conduction. Improving the adjustment with respect to radiation and convection condition.

KEYWORDS: *Cross-Laminated Timber; CLT; numerical simulation; heat transfer.*

IOP Conf. Series: Journal of Physics: Conf. Series **1107** (2018) 032003 doi:10.1088/1742-6596/1107/3/032003

INTRODUCTION

Cross-laminated timber (CLT) is a product made of several layers (lamellae) of softwood timber, stuck together. Each lamella is glued crosswise using a polymer adhesive. This way of construction provides certain structural advantages, such as an increase in strength and stiffness in two directions [1].

The extended use of CLT in the modern construction industry makes it necessary to understand its structural behaviour under fire conditions. The combustibility of timber is one of the main reasons that standards restrict the use of wood. Most of the research carried out in fire field focuses on the charring rates [2].

In fire conditions, when the wood reaches temperatures above 220 ºC, pyrolysis takes place. In this process, wood is transformed into char wood. This char layer has low conductivity values, which causes a reduction of the carbonization range and protects the internal layers from the increase in temperature. However, the char layer is not able to support any load. The Eurocode 5 [3] takes into account a reduction in the cross-section caused by charring.

Both the European and Spanish standards provide methodologies for the design of wooden elements with appropriate behaviour in fire. Many different studies have been carried out using the European standard to determine the temperature distribution in a wood element subjected to an increase in temperature.

Numerical simulation is a powerful tool to investigate the performance of wood structures subjected to fire. Validated numerical simulation is covered in Eurocode 5 [3] part 1.2 in "advanced calculation methods". The general procedure to determine the mechanical resistance of structural timber includes several steps, such as the determination of temperatures in the timber, of the resistance of cross-sections using the temperature field, and of the resistance of the structure.

However, one of the problems encountered when performing a thermal analysis on wood elements is that, due to simplifications of the model, the thermal properties of timber must be calibrated. Pyrolysis is a key aspect to predict the temperature distribution in any section of the wood element. The thermal conditions applied, and the influence of heat on the thermal properties of the timber are also taken into account in order to fit the experimental and numerical results. Most analytical models use simplified approaches such as the Eurocode 5 [3] part 1.2, which explains the changes in the thermal properties of timber as temperature increases.

Most of the fire modelling research based on finite element method (FEM) [4-7] uses the standard ISO fire curve [8]. This research has achieved improvements in obtaining the charring rate and accurate time and temperature-dependent thermal properties. However, much needs to be done to provide more accurate data under more realistic fire conditions. In the work done by Lineham et al [1], a CLT beam is exposed to the action of fire. This boundary condition differs from the ISO 834 standard curve [8]. In this work, to analyse the heat transfer, three thermal analyses were performed and compared with the thermal results obtained by Lineham et al [1]. The objectives of this paper are to compare heat transfer of an element subjected to a non-standard condition using three boundary conditions: heat flux, convection and radiation, and temperature, and to define a strategy to set up a thermal analysis.

NUMERICAL MODELING

The finite element method (FEM) is successfully used as an advanced calculation method following to Eurocode 5 [3]. In this paper, FEM is used to determine the temperature distribution of timber elements under fire conditions.

The numerical modelling presented in this work uses FEM to simulate the thermal test of two CLT beams carried out by Lineham et al [1]. Following Lineham's work, two different lay-ups configuration were studied, one with three layers and the other with five Fig. 1.

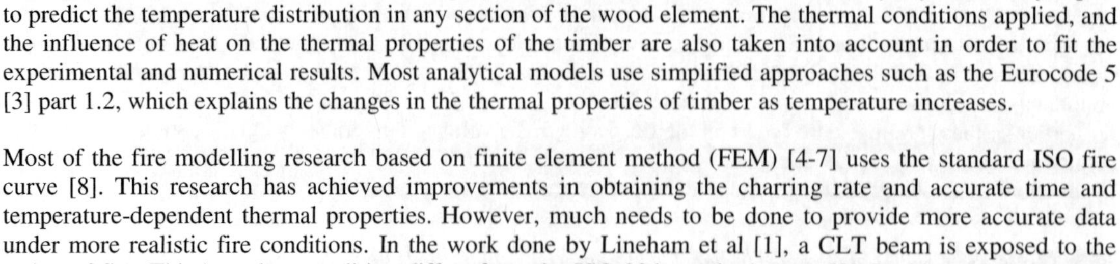

Fig. 1. Geometrical configuration.

To study the temperature distribution, the geometry used was the same as the exposed part of the tests carried out by Lineham et al [1]. The geometry is shown in Fig. 2. To compare the effects of three different heat conditions a 2D transient thermal analysis of CLT beams was developed using the software ANSYS. These three heat conditions were applied to the model for two configurations: three-layer geometry and five-layer geometry. The thermal boundary conditions were applied in one-dimensional (1D) heat transfer to the CLT beam.

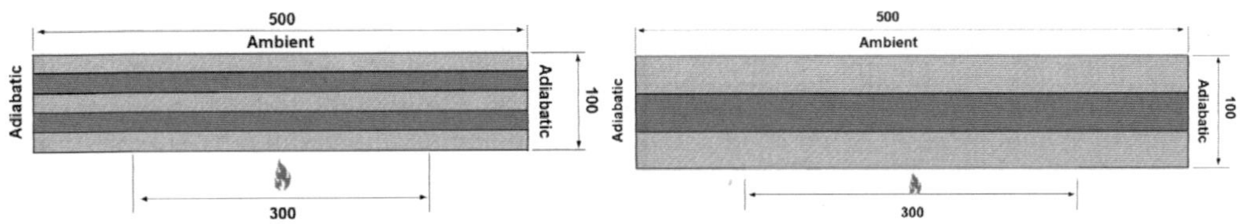

Fig. 2. Geometrical configuration

The three heat developed models are explained below:

In the first model, a constant heat flux of 27 KW/m^2 was applied to the exposed surface. This heat flux matches the real boundary condition applied in the experimental tests. The incident heat flux in the experimental tests was measured using a Gardon gauge.

In the second model, convection and radiation were applied. According to Eurocode 1 [4] emissivity (ε) was equal 0.8 and convection coefficient (h) was equal to 25 W/m^2 K. The ambient temperature applied was variable.

The third model applied a variable temperature to the exposed surface. Therefore, the numerical model took into account the heat transfer only by conduction through its layers. If the heat flux exchange is known, the temperature distribution at the exposed surface of the wood can be expressed with [5]:

$$\dot{q} = q_{Gardon} \tag{1}$$

$$\dot{q} = k_{wood} \cdot (^{T_S - TC3}/_{e}) \tag{2}$$

where $\dot{q}$ represents the heat flux measured using a Gardon gauge; k_{wood} is the thermal conductivity of timber; T_S is the exposed surface temperature and *TC3* is the value at the thermocouple inside the CLT beam (Fig 3(a) and Fig 3(b)); and *e* is the distance between the exposed surface and the thermocouple *TC3.*

For all the models, a convection coefficient of 9 W/m^2 K and an initial ambient temperature of 12ºC were applied to the non-exposed surface. The value of the convection coefficient is proposed in [6]. In addition, two adiabatic surfaces at the cross section are defined where no thermal condition is applied. (Fig. 2).

To compare the temperature distribution between the simulation and the real tests, ten coordinate systems were created. The coordinate systems and the thermocouples were placed in the same position (Fig. 3(a), Fig. 3(b)).

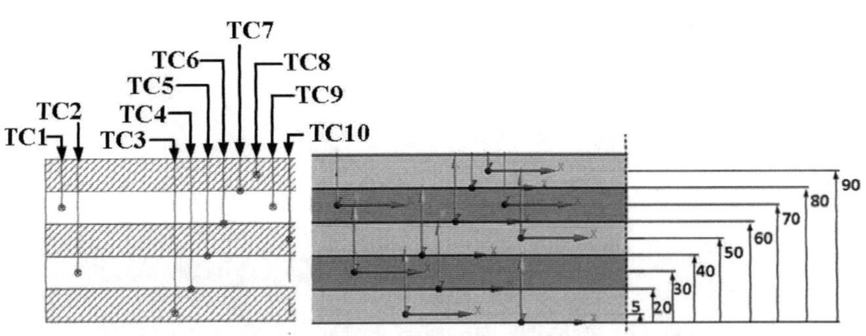

Fig 3(a). Placing the coordinate systems for 5 layers. (Distance of thermocouples from the exposed surface in mm).

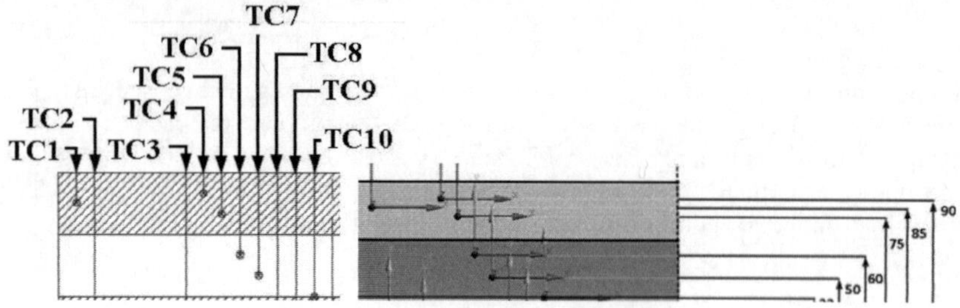

Fig 3(b). Placing coordinate systems for 3 layers. (Distance of thermocouples from the exposed surface in mm).

Different authors have studied the relationship between the thermal properties of timber and the increase of temperature [9-11]. However, the thermal properties (conductivity, specific heat and density) used in this work were associated with Annex B of Eurocode 5 [3]. The use of these properties simplifies the analysis of heat transfer in timber. Furthermore, the specific heat takes into account the latent heat of vaporization of water in the timber [6].

Finally, the following assumptions were considered in the numerical simulations:

- Thermal properties: density, specific heat and conductivity were defined as isotropic properties.
- Effects of moisture migration and char layer formation were taken into account in the values proposed in Eurocode 5 [3].

RESULTS

In order to evaluate the computational efficiency of the FE model the time needed to solve the problems is presented in Table 1.

Table 1. MAPDL elapsed time (minutes)

	Heat Flux	Radiation+Convection	Temperature
3-Layer configuration	39	5	7
5-Layer configuration	70	8	14

The comparison between the experimental results and the thermal simulations are shown below. Two of the three boundary conditions studied, convection and radiation, and temperature bring the numerical temperature distribution close to the experimental. However, in the first model, the use of the real heat flux as a thermal condition overestimated the temperatures of all the thermocouples.

The following figures (Fig. 4. and Fig. 5.) show the adjustment of the temperature distribution between the experimental and numerical results at different depths. In order to simplify the graphs, the values of only three of the ten thermocouples are plotted for convection and radiation (R+C Sim.), and temperature (Temperature Sim.).

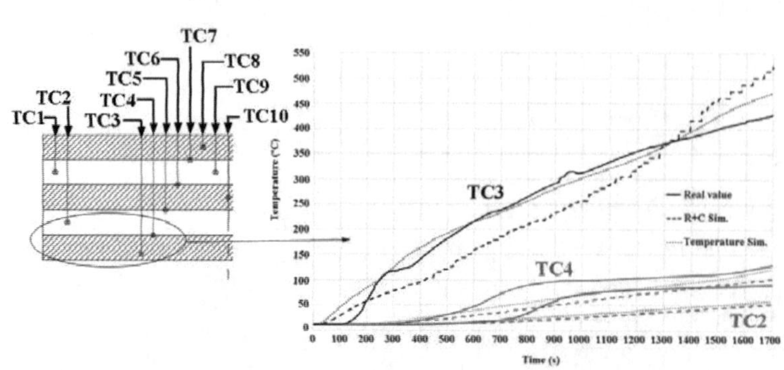

Fig 4. Graphic comparison between experimental and numerical results for TC2, TC3 and TC4 in 5-layer geometry.

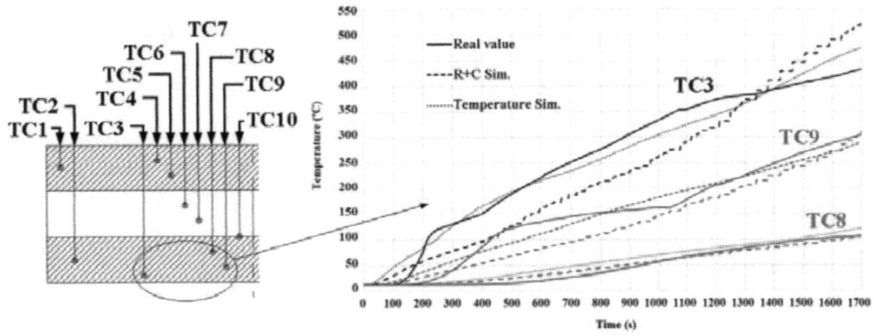

Fig 5. Graphic comparison between experimental and numerical results for TC3, TC8 and TC9 in 3-layer geometry.

CONCLUSIONS

In this research, simulations were performed under different thermal conditions in CLT beams. These simulations consisted of three different boundary conditions applied to the exposed surface. The results obtained show that the thermal boundary conditions have a great influence on the temperature distribution.

Heat flux is not able to simulate the thermal distribution through the CLT beam when it is measured using Gardon gauge. This heat flux sensor is designed to measure the total radiation flux density. However, heat flux defined in Ansys is referred to the rate of heat flow across a unit area.

In a fire situation, radiation and convection are the main conditions to be taken into account. Therefore, these conditions optimize the temperature distribution.

The temperature distribution in a CLT beam can be adjusted by a variable temperature applied on the exposed surface when the boundary conditions are known.

The results obtained in this research are an original contribution to analyse the thermal properties of CLT beams. The simulated temperatures do not stabilize at 100 ºC (as shown in Fig. 4.), because localized moisture is not taken into account. The further the thermocouple is from the exposed surface, the worse the simulation results. Thus, as a future line, it is necessary to investigate the influence of parameters, such as the moisture evaporation, mass transfer, pyrolysis, and different thermal properties of contacts between layers, to develop optimized thermal simulations.

The computational time is an important factor when a numerical simulation is developed. Differences between each calculation time show the importance of developing an efficient numerical simulation, as thermal analyses are usually coupled with structural analyses.

ACKNOWLEDGEMENTS

The authors of this article are grateful to BRE Centre for Fire Safety Engineering at the University of Edinburgh for the experimental data and the technical support. Finally, thanks to Swanson Analysis Inc. for the use of the university research version of the ANSYS v 18.1 program in this paper. This work was partially financed by the Asturias Government and FICYT and also co-financed with FEDER funds under Research Project FC-15-GRUPIN14- 004.

REFERENCES

[1] Lineham S. A., Thomson D., Bartlett, A., Bisby, L., Hadden, R. Structural response of fire-exposed cross-laminated timber beams under sustained loads. Fire Safety Journal. 85, 23-24, 2016

[2] Bartlett, A., Hadden, R., Bisby, L. & Law, A. Analysis of cross-laminated timber charring rates upon exposure to non-standard heating conditions. Fire and Materials. San Francisco, CA: Interscience Communications Ltd. 2015.

[3] European Committee for Standardisation (CEN). Eurocode 5. Design of timber structures. Part 1-2: General – Structural fire design; EN1995-1-2. Brussels, Belgium, 2004.

[4] European Committee for Standardisation (CEN). Eurocode 1. Actions on structures. Part 1-2: General actions – Actions on structures exposed to fire; EN 1991-1-2. Brussels, Belgium, 2003.

[5] Thi, V.D., Khelifa, M., Oudjene, M., Ganaoui, M. El, Rogaume, Y. Finite element analysis of heat transfer through timber elements exposed to fire. Engineering Structures. 143, 11–21, 2017

[6] N. Werther, J.W.O. Neill, P.M. Spellman, A.K. Abu, P.J. Moss, A.H. Buchanan, et al., Parametric study of modelling structural timber in fire with different software packages. 7th Int. Conf. Struct. Fire, Zurich, Switzerland, 2012.

[7] König J., Walleij L., One-dimensional charring of timber exposed to standard and parametric fires in initially unprotected and post-protected situations. Report I 9908029, Swedish Institute for Wood Technology Research, Stockholm, Sweden, 1999.

[8] International Organization for Standardization (ISO). ISO 834-1. Fire-resistance tests - Elements of building construction - Part 1: General requirements; 1999-09.

[9] Buchanan A.H., Structural Design for Fire Safety, John Wiley & Sons Ltd, Chichester, 2001.

[10] König J., "Effective thermal actions and thermal properties of timber members in natural fires", Fire and Materials 30(1), 51-63, 2006.

[11] Effect of increased charring on the narrow side of rectangular timber cross-sections exposed to fire on three or four sides Fire Mater., 35 (2011), pp. 593-605

Assessment of fire behaviour of polyisocyanurate (PIR) insulation foam enhanced with lamellar inorganic smart fillers

Eleni ASIMAKOPOULOU[1], Jianping ZHANG[1], Maurice McKEE[1], Kinga WIECZOREK[2], Anna KRAWCZYK[2], Michele ANDOLFO[2], Marco SCATTO[3], Michele SISANI[3] and Maria BASTIANINI[3]

[1] *FireSERT, Belfast School of Architecture and the Built Environment, Ulster University, United Kingdom*

[2] *SELENA Lab, Poland*

[3] *Prolabin and Tefarm SRL, Italy*

e.asimakopoulou@ulster.ac.uk

ABSTRACT

Modern day energy codes are driving the design and multi-layered configuration of exterior wall systems with significant emphasis on achieving high performance insulation towards improving energy performance of building envelopes. Use of highly insulating polyisocyanurate (PIR) based insulation materials enhanced with eco-friendly lamellar inorganic fillers contributes to meeting energy performance requirements, environmental challenges and cost reduction without undue compromise of the overall building fire safety. Towards that end, the aim of the current work is to assess the fire behaviour of PIR foams enhanced with lamellar inorganic smart fillers, namely Layered Double Hydroxides (LDHs) and ZrP. Experimental results indicate that fire reaction properties and thermal stability of foam samples enhanced with three types of lamellar inorganic smart fillers are evaluated using cone calorimeter (CC) and thermogravimetric (TGA) analysis. The initial degradation temperature of PIR-layered filler samples increases compared with neat PIR foam, indicating that incorporation of flame retardants decelerates the degradation of PIR foam and as result increases the thermal stability of PIR foam. Thermal decomposition of the PIR samples occurs in two distinct stages associated with the degradation of the urethane-urea linkages of the hard segment, releasing low calorific capacity products and the degradation of polyol derived products with higher calorific capacity than those derived from isocyanate. Increasing the filler content results in increased char formation and decreased peak Heat Release rates (HRR).

KEYWORDS:

polyisocyanurate insulation; layered doubled hydroxides; zirconium phosphate; fire performance; thermogravimetric analysis; cone calorimeter analysis

INTRODUCTION

Modern day energy codes are driving the design and multi-layered configuration of exterior wall systems with significant emphasis on achieving high performance insulation towards improving energy performance of building envelopes. Insulation in walls may comprise of either noncombustible material such as fiberglass or mineral wool, or frequently encountered a wide range of highly insulating combustible foam plastic materials. Most commonly used insulation materials include polymers such as extruded polystyrene, expanded polystyrene, polyurethane foam (PUF) and polyisocyanurate (PIR) with or without flame retardants [1]. Appropriate use of the above materials requires that they meet both energy performance requirements, environmental challenges and cost reduction without undue compromise of the overall building fire safety. Numerous studies [1-3] have established that thermal decomposition of polymeric foams, in both inert and oxygen atmospheres, is a complex process consisting of numerous decomposition pathways that strongly depend on the reactivity of organic compounds employed in its synthesis. PIR foams are part of the polyurethane (PUR) rigid foam family and their main characteristic is that they contain a high percentage of cyclic isocyanurate chemical linkages and use of polyester polyol instead of the standard polyether polyol used in PUR's. PIR are based on the reaction of polycyclotrimerization of diisocyanates or isocyanate terminated prepolymers to form triazine-trione ring structured isocyanurate rings [4] that, from the thermodynamic point of view, are more thermally stable than urethane bonds found in PUR foams as it dissociates at approximately 200°C as opposed to 350°C for polyisocyanurates [5].

Regarding the flammability of polymer insulation-related flame retardants, the substitution of commonly used halogen-based flame retardants for eco-friendly "greener" ones such as mineral fillers like Layered Double Hydroxides (LDHs) and zirconium phosphate nanocomposites [2, 3], is currently of great interest for avoiding the release of corrosive and toxic volatile compounds from combustion. In contrast to halogenated flame-retardants, layered compounds attract significant attention and are explored as second-generation fire-retardant materials to alternatively be used as efficient and more environmental friendly additives to various polymers with the ability to improve their flame retardancy and thermal stability and smoke suppression [6]. LDHs consist of synthetics layered compounds containing positively charged metal hydroxide layers with change balancing anions located in between. More specifically, LDHs are described by the generic formula $[M^{II}_{1-x}M^{III}_{x}(OH)_2\ A^{n-}_{x/n}]$ x yH_2O, where M^{II} is the divalent cation, M^{III} is the trivalent cation, A is the interlayer anion with n- charge and x and y are fraction constants [7]. LDHs may be formed with all dipositive metal ions from Mg^{2+} to Zn^{2+}, and almost all transition metal trivalent ions and are classified as members of the hydrotalcite ($MgAlCO_3$) supergroup [8]. Their increased fire retardancy is observed in both gas and solid phases as they develop non-flammable gases diluting flammable gases and promoting surface charring. Their mechanism during thermal degradation mainly includes the release of water and carbon dioxide, diluting combustion gases, and reducing endothermic decomposition of metal hydroxides. Specifically, this latter inorganic-reinforced carbonaceous residue thermally protects the underlying polymer as the formation of this residue and slows down the combustion process of PUF [2, 6]. At elevated temperatures, LDHs have been shown to release water and go through endothermic decomposition. Different authors have recently studied several types and contents of LDH in polymeric-based insulation materials [2, 3]. Despite their effectiveness, LDHs have until now limited commercial success as fire retardants because of their difficulty to disperse and distribute in polymers, which limits their effectiveness [2] and most available studies concern their incorporation in PUF. A recent study [9] firstly investigated the potential synergistic effect between organically modified nanoclay LDH and flame retardants on PIR nanocomposites on improving the fire retardancy and fire behaviour.

Alpha zirconium phosphate (α-ZrP) structure, with chemical formula $Zr(HPO_4)_2$ x H_2O, is composed by stacking layers made of planes of zirconium atoms bonded, on both sides, to monohydrogen phosphate groups. Each phosphate group is bonded to three Zr atoms of the plane, while each zirconium is octahedrally coordinated by six oxygens of six different monohydrogen phosphate groups. The water molecules are located in the zeolitic cavities of the interlayer region [10]. Zirconium phosphate can be easily functionalized through intercalation reactions and has been widely studied in literature as nanostructurant filler for polymer composites. Thanks to the highly tunable properties and unique cation exchange properties, ZrP exhibits several advantages over conventional natural clays (e.g. montmorillonites), including higher purity and ease of intercalation and/or exfoliation [11-13]. One of the most interesting aspects of this layered material is its capacity to reduce the rate of combustion of the polymers [14] and its flame retardancy properties are proven to reduce the total heat evolved and the cone calorimeter measured heat release rate [15,16]. Even reduced amounts of nanoparticle α-ZrP addition in virgin materials greatly affect their thermal stability and burning behavior, as inhibition of thermal decomposition and higher char yields formation can be observed [16].

The focus of the current work is to study potential synergistic effects between PIR and layered fillers with "smart" flame retardants on improving the flame retardancy and fire behavior of rigid insulation foam samples. Towards this end, fire reaction properties and thermal stability of foam samples enhanced with three types of lamellar inorganic smart fillers has been investigated using thermogravimetric analysis (TGA) and cone calorimeter (CC). The effects of LDH concentration on the thermal degradation and flammability of PIR are also examined.

EXPERIMENTAL

Materials

PIR samples with a constant isocyanate index (NCO/OH) equal to 3.0 were obtained from lab-scale prototyping experiments conducted at SELENA Labs. Laboratory samples were produced by high pressure impingement mixing type of foam machinery, operating at constant processing parameters. Reacting mixtures were poured centrally into horizontal 35cm x 35 cm x 5 cm mold, heated at 50°C, with steel facers attached to the bottom and the lid. Three new types of PIR foams samples with layered fillers, namely PIR-$MgAlCO_3$ filler (PIR-LDH1), PIR-MgAlStearate filler (PIR-LDH2) and PIR-ZrP octadecylamine filler (PIR-ZrP3) are investigated and assessed versus a pure PIR sample. LDH2 and ZrP3 are organically modified layered fillers, in which organic modification acts also as compatibilizer between filler and polymer matrix.

Test methods

Cellular structure of the samples was evaluated using scanning electron microscopy (SEM) and crystal structures of samples were analyzed by X-ray diffraction (XRD). The thermal stability was evaluated in terms thermogravimetric analysis (TGA) under N_2 (inert gas) environment in a Mettler Toledo TGA apparatus. About 10 mg foam sample was placed in an alumina pan with no lid. The heating rate was 20°C/min with a N_2 flow of 150 ml/min. The following parameters were determined: initial degradation temperature, $T_{5\%}$ (temperature at 5% weight loss), corresponding maximum temperature, T_{max}, and weight loss, W, for each degradation step and char residue at 1000°C. Cone calorimeter (CC) tests were performed with a Dark Star Research Ltd (UK) apparatus according to the ISO 5660-1 [16] at 50 kW/m^2. The dimensions of the samples are 100mm x 100mm x 24mm and the density of the samples is approximately 50 kg/m^3. Experimental results includes time to ignition (TTI), Combustion Time (CT), total HRR (THR), peak HRR (p-HHR), average HRR (Av-HRR), average mass loss rate (Av-MLR), smoke production rate (SPR), smoke and CO yield. Additionally, the unexposed surface temperature was measured using a type K thermocouple.

RESULTS AND DISCUSSION

SEM and XRD

To verify that the LDHs and ZRP are well dispersed in PIR. SEM and XRD analysis were firstly carried out. Figure 1 shows the SEM morphological evaluation of neat PIR (left) and PIR-LDH1 (center) and XRD spectra of neat PIR and PIR-ZrP3 (right). The SEM results depicts the regular and cellular structure of plain PIR and PIR-LDH1 verifying that addition of LDH fillers does not modify the cellular structure. The XRD spectra of PIR-ZrP3, compared to those of neat PIR indicate that ZrP layer has been exfoliated in the matrix as a great amount of polymers has entered into the gallery space separating ZrP layer during in situ intercalation polymerization process.

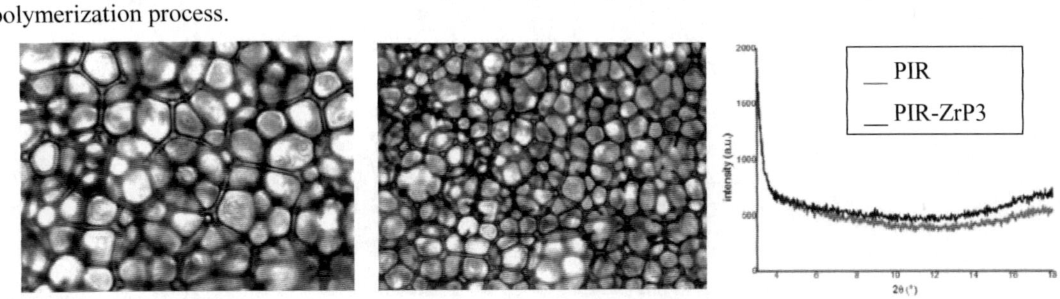

Fig. 1.SEM morphological evaluation of neat PIR (left) and PIR-LDH1 (center) and XRD spectra of neat PIR and PIR-ZrP3 (right)

Thermogravimetric analysis (TGA)

Figure 2 presents the weight loss and weight loss rate of all formulations. It can be observed that the degradation of all PIR foams in inert atmosphere occurs in two steps. The results are summarized in Table 1. The initial

IOP Conf. Series: Journal of Physics: Conf. Series **1107** (2018) 032004 doi:10.1088/1742-6596/1107/3/032004

degradation temperature, $T_{5\%}$, corresponds to 255°C for pure PIR. It can be seen form Table 1, that $T_{5\%}$ increases compared with net PIR foam, indicating that incorporation of flame retardants decelerates the degradation of PIR foam and as a result increases the thermal stability of PIR foam higher temperature range. The first pyrolysis step takes place from 200 to 400°C and the maximum rate degradation temperature, $T_{max,1}$, is substantially decreased with the layered filler addition and only in PIR-6%ZrP3 remains approximately at the same levels. This first step is related to the degradation of the urethane-urea linkages of the hard segment, releasing low calorific capacity products; residue weight values of this first reaction are denoted as W_1. The second stage of decomposition corresponds to the degradation of polyol derived products with higher calorific capacity than those derived from isocyanate [3] and lower residue weight, W_2. The second degradation step of the pure PIR foam takes place between 400 and 600°C, and the maximum rate degradation temperature, $T_{max,2}$, is 477°C with the final char residue 19.5% of the initial mass. With LDH/ZrP fillers, the temperatures are generally decreased due to degradation of the fillers at lower temperature. Increasing filler content results in increased char formation.

Table 1. TGA data of pure PIR, PIR-LDH1, PIR-LDH2 and PIR-ZrP3.

Sample	Density		Stage 1		Stage 2		
		$T_{5\%}$	$T_{max,1}$	W_1	$T_{max,2}$	W_2	Char res.
	kg/m3	°C	°C	wt.%	°C	wt.%	wt.%
PIR	50.4	255	373	64.5	477	36.2	19.5
PIR-2%LDH1	50.0	258	329	81.1	422	58.1	29.9
PIR-2%LDH2	50.5	273	363	74.2	426	56.1	30.5
PIR-2%ZrP3	54.0	271	355	77.5	426	59.0	30.7
PIR-4%LDH1	48.0	266	329	82.2	411	60.0	33.3
PIR-4%LDH2	49.5	272	339	75.5	416	59.9	33.1
PIR-4%ZrP3	45.0	266	366	74.7	409	61.6	30.3
PIR-6%LDH1	54.0	257	333	80.8	407	60.0	33.1
PIR-6%LDH2	48.0	274	337	81.3	424	57.4	32.6
PIR-6%ZrP3	55.0	278	376	76.9	482	50.8	31.9

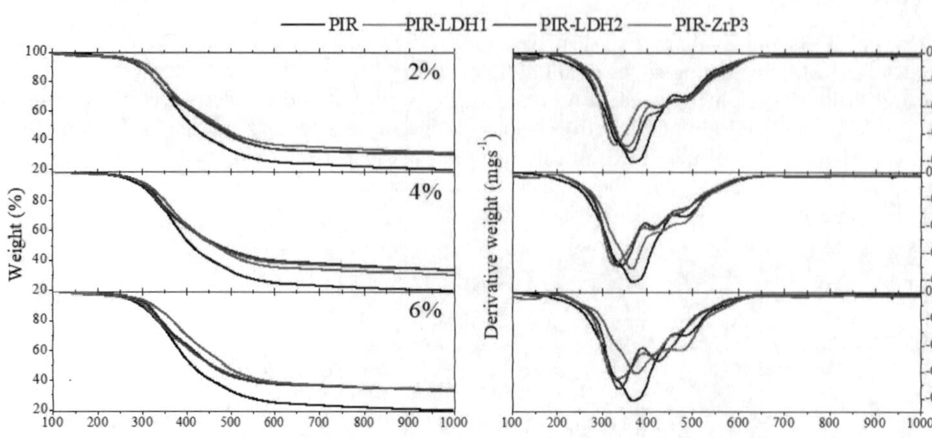

Fig. 2. TG (left) and DTG (right) of pure PIR, PIR-LDH1, PIR-LDH2 and PIR-ZrP3 in N_2 atmosphere.

Cone Calorimeter analysis

Almost all samples instantly ignited as shown in Table 2 summarising all the cone calorimeter data. Figure 3 shows the heat release rate and smoke production rate histories of all formulations. The presence of layered filler contributes to a decrease of p-HRR and HRR data correlate well with addition of fire retardants, demonstrating the beneficial effect of layered filler addition for fire performance. The decrease in the HRR of this second stage, is more pronounced, achieving a reduction of up to 40% for PIR-2%LDH3. PIR-4%ZrP3 results in higher char residue, as the surface char layer reduced both the oxygen and heat fluxes towards the polymer surface thus limiting the volatile compounds and therefore the Av-MLR [9]. The ultimate fire retardancy mechanism is turning polymer into protection barrier chars that absorb energy and shield the rest of the PIR from radiation. The carbonation layer protects the polymer by not giving access to oxygen to sustain combustion. The smoke yield and CO yield of all fillers-containing formulations are similar to and in some cases less than those of neat PIR indicating that they don't promote production of toxic gases.

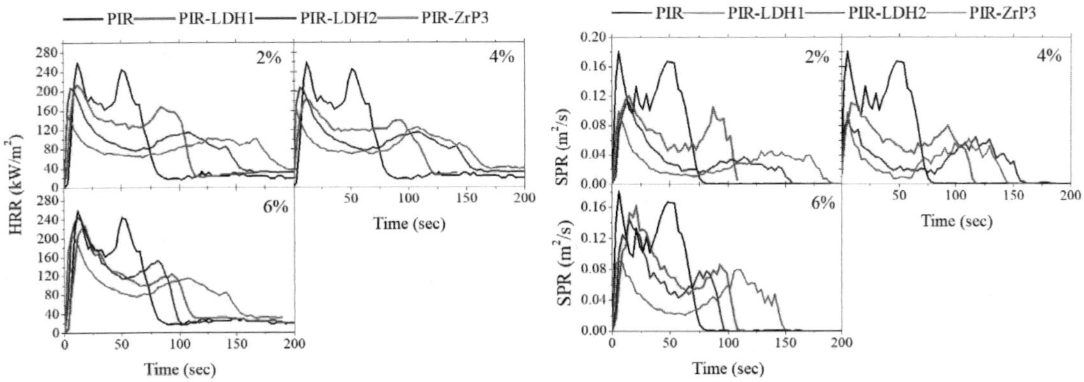

Fig. 3. HRR (left) and smoke production (right) rates of pure PIR, PIR-LDH1, PIR-LDH2 and PU-ZrP3.

Table 2. Flammability and smoke emission behavior of pure PIR and PIR-layered filler samples.

Sample	HF	TTI	CT	THR	p-HRR	Av-HRR	Av-MLR	Smoke yield	CO yield
	kW/m²	s	s	MJ/m²	kW/m²	kW/m²	g/s	-	-
PIR	50	1	102	17.85	259.1	135.2	0.1247	0.098	0.0329
PIR-2%LDH1	50	1	100	17.15	213.4	164.9	0.1039	0.069	0.0273
PIR-2%LDH2	50	1	150	17.68	206.7	119.8	0.0697	0.054	0.0382
PIR-2%ZrP3	50	5	192	16.67	152.0	86.8	0.0588	0.049	0.0334
PIR-4%LDH1	50	1	116	15.20	182.4	131.1	0.0722	0.086	0.0186
PIR-4%LDH2	50	1	150	17.96	206.7	149.5	0.1208	0.056	0.0349
PIR-4% ZrP3	50	1	202	18.50	176.0	92.0	0.0437	0.062	0.0646
PIR-6%LDH1	50	7	104	16.98	226.6	163.3	0.1037	0.079	0.0317
PIR-6%LDH2	50	1	95	14.73	244.4	155.0	0.0869	0.081	0.0230
PIR-6% ZrP3	50	1	150	18.22	196.7	132.4	0.0753	0.059	0.0301

CONCLUSIONS

Fire reaction properties and thermal stability of PIR form with MgAlCO3, MgAlStearate and ZrP octadecylamine fillers have been evaluated using TGA and cone calorimeter (CC). It was found that layered fillers promote the formation of reinforced char layer, providing an effective barrier against heat and oxygen, release noncombustible gases, and simultaneously effectively suppress smoke and gases during the combustion process. PIR-4%ZrP3 resulted in higher char residue and decreased p-HRR values. The use of different layered fillers does not lead to increase smoke generation or CO production. Increasing filler content resulted in increased char formation for all samples and decreased peak Heat Release rates (HRR) for LDH1 and LDH2. Further analysis at other heating rates under both oxygen and inert atmospheres will be further analyzed to derive a kinetic model to simulate the pyrolysis degradation mechanism PIR-layered filler foams. Additional combination of different layered fillers will be investigated in the future to further tune the fire resistance properties of PIR nanocomposites.

ACKNOWLEDGEMENTS

This work is conducted within the EENSULATE project is financially supported by EU H2020 project (EEB-01-2016, Grant No. 723868).

REFERENCES

[1] O'Connor, D., "The Building Envelope: Fire Spread, Construction Features and Loss Examples", *SFPE Handbook of Fire Protection Engineering* (5th *ed*) Hurley M.J. (ed.), National Fire Protection Association, Quincy, MA 02269, 2016, p **3242**/3512.

[2] Li, Y.C., Yang, Y.H., Shields, J.R., Davis, R.D. (2015) Layered double hydroxide-based fire resistant coatings for flexible polyurethane foam, *Polymer* 56:284-292, http://dx.doi.org/10.1016 /j.polymer. 2014.11.0 23

[3] Gomez-Fernandez, S., Ugarte, L., Pena-Rodriguez, C., Zubitur, M., Corceuera, M.A., Eceiza, A. (2016) Flexible polyurethane foam nanocomposites with modified layered hydroxides, *Applied Clay Science* 123:109-120, https://doi.org/10.1016/j.clay.2016.01.015

[4] Chen, K., Tian, C., Liang, S., Zhao, X., Wang, X. (2018) Effect of stoichiometry on the thermal stability and flame retardation of polyisocyanurate foams modified with epoxy resin, *Polymer degradation and Stability* 150:105-113, https://doi.org/10.1016/j.polymdegradstab.2018.02.015

[5] Kuranska M., Cabulis U., Auguscik M., Prociak A., Ryszkowaska J., Kirpluks M. (2016) Bio-based polyurethane-polyisocyanurate composites with an intumescent flame retardant, *Poymer Degradation and Stability* 127:11-19, http://dx.doi.org/10.1016/j.polymdegradstab.2016.02

[6] Wang, D.Y., *Novel fire-retardant polymers and composite materials* (1st *ed*) Woodhead Publishing Series in Composite Science and Engineering, Woodhead Publishing, 2016.

[7] Nalawade, P., Aware, B., Kadam, V.J., Hirlekar, R.S. (2009) Layer double hydroxides: A review, *Journal of Scientific and Industrial Research* 68:267-272.

[8] Mills, S.J., Christy, A.G., Genin, J.M.R., Kameda, T., Colombo, F. (2012) Nomenclature of the hydrotalcite supergroup: natural layered hydroxides, *Mineralogical Magazine* 76:1289-1336, https://doi.org/10.1180/ minmag.2012.076.5.10

[9] Gao, L., Zheng, G., Zhou, Y., Hu, L., Feng, G., Xie, Y. (2013) Synergistic effect of expandable graphite, melamine polyphosphate and layered double hydroxide on improving the fire behavior of rosin-based rigid polyurethane foam, *Industrial Crops and Products* 50:638-647, https://doi.org/10.1016/j.indcrop.2013.07.050

[10] Clearfield, A., David Smith, G. (1969) Crystallography and structure of alpha-zirconium bis (monohydrogen orthophosphate) monohydrate, *Inorganic Chemistry* 8:431-436, https://doi.org/10.1021/ic50073a005

[11] Sue, H.J., Gam, K.T., Bestaoui, N., Spurr, N., Clearfield, A. (2004) Epoxy nanocomposites based on the synthetic a-zirconium phosphate layer structure, *Chemistry of Materials* 16:242-249, https://doi.org/10.1021/cm030441s

[12] Sun, L., Boo, W.J., Browning, R.L., Sue, H.J., Clearfield, A. (2005) Effect of crystallinity on the intercalation of monoamine in a-zirconium phosphate layer structure, *Chemistry of Materials* 17:5606-5609, https://doi.org/ 10.1021/cm051160i

[13] Sun, L.Y., Boo, W.J., Sun, D.Z., Clearfield, A., Sue, H.J. (2007) Preparation of exfoliated epoxy/ a-zirconium phosphate nanocomposites containing high aspect ratio nanoplatelets, *Chemistry of Materials* 19:1749-1754, https://doi.org/10.1021/cm062993r

[14] Liu, X.Q., Wang, D.Y., Wang, X.L., Chen, L., Wang, Y.Z. (2011) Synthesis of organo-modified α-zirconium phosphate and its effect on the flame retardancy of IFR poly(lactic acid) systems, *Polymer Degradation and Stability* 96:771-777. https://doi.org/10.1016/j.polymdegradstab.2011.02.022

[15] Alongi, J., Franche, A. (2010) Flame retardancy properties of a-zirconium phosphate-based composites, *Polymer Degradation and Stability* 95:1928-1933, https://doi.org/10.1016/j.polymdegradstab.2010.04.007

[16] ISO 5660-1, *Reaction to fire tests – Heat release, smoke production and mass loss rate – Part 1: Heat release rate (cone calorimeter method)*, International Organization for Standardization, Geneva, Switzerland, 2002.

IOP Conf. Series: Journal of Physics: Conf. Series **1107** (2018) 032005 doi:10.1088/1742-6596/1107/3/032005

An engineering model for ignition and extinction of wood flames using bench-scale data

Paul LARDET[1], Véronique GEORGES[1], Lucas TERREI[2] and Mahjouba NICHANE[1]

[1]*Université Paris-Est, Centre Scientifique et Technique du Bâtiment (CSTB), France*
[2]*LEMTA, UMR CNRS 7563, Université de Lorraine, France*

paul.lardet@cstb.fr

ABSTRACT

Predicting the fire spread in wooden structure buildings widely uses engineering tools such as computational fluid dynamics (CFD) simulation softwares. These softwares mix several physical models to address a wide range of phenomena, such as heat and mass transfer, pyrolysis, combustion or fluid dynamics. These models must be kept as simple as possible to limit the modelling costs.
The behaviour of wooden materials under fire conditions has been extensively studied for several years. The existing studies mainly focuses on the characterization of ignition of these materials. Several physical parameters have been examined, notably ignition time and temperature or critical ignition flux. In particular, ignition temperature has been shown a very elusive quantity (Babrauskas [1]). Some more recent work (Shen [2]) stated on the complexity to use theoretical models to predict ignition phenomena at an engineering level.
Wood extinction has been much less studied. This issue raises as a major question in recent work for building fire safety studies, as it is a key parameter to demonstrate robustness against fire. It is commonly agreed that self-extinction of wood occurs in most cases when the external thermal aggression stops (Emberley [3]). The physical phenomena leading to extinction are still under study.
This work proposes a criterion-based model of wood products ignition, pyrolysis and extinction with a very simple approach. The proposed criterion is built on test data measured on cross laminated timber and plywood panels, and identifies a burning region in the domain (Surface temperature) vs. (Net heat flux at the surface). This model has been incorporated into a CFD simulation software and used to model a large-scale fire of a plywood facade. It produced good outcomes for plywood fire spread predictions with very few modelling costs.

KEYWORDS:

Ignition; extinction; cone calorimeter; modelling; heat transfer; CFD.

INTRODUCTION

The prediction of wood based material ignition and extinction when exposed to fire conditions is of prime interest for fire engineering studies. Indeed, the fire hazards in wooden structures are directly related to the wood contribution to fire. Moreover, the fire resistance studies of wooden structures rely on the capacity of wood fire to extinct when the external thermal flux decreases.
The present study focuses on piloted ignition and flame extinction. Indeed, this work aims at giving engineering criteria to predict ignition and extinction: As piloted ignition is easier to achieve than spontaneous ignition, a criterion based on the former gives worst case estimations maximizing the safety level. Smouldering extinction could be of interest for a fire safety study, as smouldering also transforms wood into char and destroys its resistance. It is not studied here because this phenomenon is very different from flaming and cannot be included in the same disappearance criterion. Its potential influence is discussed in Crielaard [4].
This work intends to give a simpler criterion than pyrolysis model but for both wood ignition and extinction phenomena, whereas prescribed heat release rate models cannot predict extinction. This criterion can then be used in a Computational Fluid Dynamics (CFD) software in order to more accurately predict flame ignition and extinction in fire resistance studies.
Although moisture content is known to have major influence on ignition (Moghtaderi [5]), it is not studied here. It could be included in a future work.

METHODS AND MATERIALS

Specimens preparation

Two types of wood specimens were selected from spruce cross laminated timber (CLT) and from okoume and birch plywood. Okoume and birch samples were taken from the raw material of the test setup used by Duny [6]. The samples were 100 x 100 mm wide. CLT samples were 50 mm thick and plywood samples were 18 mm thick.

All the samples were conditioned in controlled environment at 50 % relative humidity and 23°C.

They were wrapped with aluminum foil except for the top side exposed to the radiative flux. The specimens were framed by a steel edge frame as specified in ISO 5660-1.

Thermocouples type K with wire thickness of 1 mm and type FC08K with pliable metal for surface junction were used at the exposure surface of wood. The sensors were directly stapled into wood as showed on the picture below. This kind of measurements can affect the thermocouple reading. Some further work with measurements based on infrared devices is still in progress.

Fig. 1. Sample and sensor preparation.

Apparatus

The cone calorimeter was used according to ISO 5660-1. During the experiment, the fan flow rate was kept about 24 l/s and the quantity of heat release rate (HRR) as a function of time was determined directly by the oxygen depletion with a gas analyser. The cone temperature was correlated for different preliminary calibration tests, using a fluxmeter at the position of the sample.

Procedure

The specimen was exposed horizontally 25 mm below the cone heater. The incident heat flux $\dot{q}_{inc}$ was set to values between 20 and 75 kW/m².
The surface temperature T_s and heat rate release were recorded during the test. A sample was considered ignited in this study when a visible flame was first observed. Spark plug was used as a pilot source until the appearance of flame. From the ignition time, the test was run for 2 or 10 min and then stopped by decreasing the external heat flux to 10 kW/m². The extinction time was considered when the visible flame disappeared. The phenomenon of smouldering was noticed but not studied. Table 1 sums up the procedure at each step.

IOP Conf. Series: Journal of Physics: Conf. Series **1107** (2018) 032005 doi:10.1088/1742-6596/1107/3/032005

Table 1. Measurements during each test step.

Step	***Description***	**Data measurements**
Ignition	Sample placed under the cone heater. $\dot{q}_{inc}$ constant	Ignition temperature (°C)
Combustion	Several minutes exposure. $\dot{q}_{inc}$ constant	HHR (kW/m²) Burning duration (s)
Extinction	$\dot{q}_{inc}$ decreased to 10 kW/m² after 2 or 10 minutes of exposure.	Extinction temperature (°C) Cone heater temperature (°C)

EXPERIMENTAL RESULTS

Ignition step

The surface temperatures at ignition T_{ig} are presented in Table 2. Each line corresponds to a single test.

Table 2. Ignition temperatures.

CLT	
$\dot{q}_{inc}$ (kW/m²)	**T_{ig} (°C)**
20	422
25	416
25	341
25	328
50	279
50	225
75	268
75	149

Birch plywood	
$\dot{q}_{inc}$ (kW/m²)	**T_{ig} (°C)**
15	352
25	434
50	288
75	266

Okoume plywood	
$\dot{q}_{inc}$ (kW/m²)	**T_{ig} (°C)**
15	348
25	311
50	295
75	277

T_{ig} is significantly reduced when $\dot{q}_{inc}$ increases. This trend is slightly stronger for CLT.

Combustion step

The heat release rate was recorded during combustion for the three wooden materials. For plywood specimens, a complete burning was performed. Figure 3 depicts HRR variation during combustion with $\dot{q}_{inc}$=50 kW/m².

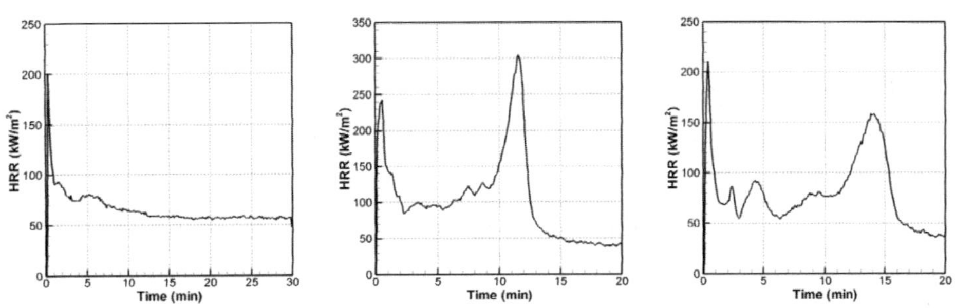

Fig. 2. HRR measurements – from left to right: CLT, birch plywood and okoume plywood.

HRR first reaches a peak at ignition, then drops and keeps a steady level during the combustion step. Based on these results, heat release rate can be considered as not depending on incident heat flux during most of the combustion duration. For plywood series, the total burning time can be deduced from the measurements. A second HRR peak occurs just before extinction, when residual wood and char is almost completely cracked leading to an increase of material surface participating to the pyrolysis (Delichatsios [7]).

Extinction step

In this step, extinction is studied by decreasing $\dot{q}_{inc}$. The heat flux decrease rate is not adjustable but imposed by the inertia of the cone heater. $\dot{q}_{inc}$ value is deduced from the cone temperatures measurement. Table 3 presents the surface temperature at extinction T_{ext}. Each line corresponds to a single test.
Extinction temperatures are higher than ignition temperatures. No clear trend is visible compared to incident heat flux.

Table 3. Extinction temperatures.

CLT		
Exposure time (min)	$\dot{q}_{inc}$ **(kW/m²)**	T_{ext} **(°C)**
2	25	554
	26	547
	21	522
10	31	541
	32	629

Birch plywood		
Exposure time (min)	$\dot{q}_{inc}$ **(kW/m²)**	T_{ext} **(°C)**
2	16	448
	19	486
	27	554

Okoume plywood		
Exposure time (min)	$\dot{q}_{inc}$ **(kW/m²)**	T_{ext} **(°C)**
2	21	450
	30	585
	38	601

ANALYSIS

Net heat flux formulation

The present work aims at identifying surface heat transfer phenomenological criterion at ignition and extinction. As both these phenomena are related to wood heating, a net heat transfer balance must be considered to identify the main phenomena.

For each test series presented above, the net heat flux $\dot{q}_{net}$ at ignition or extinction is estimated with Eq. 1. Flame irradiance $\dot{q}_f$ has been estimated following Eq. 2 (Law [8]).

$$\dot{q}_{net} = h(T_g - T_s) + \varepsilon(\dot{q}_{inc} + \dot{q}_f - \sigma T_s^4) \tag{1}$$

$$\dot{q}_f = (1 - e^{-0.3t_f})\,\sigma T_g^4 \tag{2}$$

σ being the Stefan-Boltzman constant. The other parameters are defined in Table 4 with the hypothesis taken for each case.

Table 4. Hypothesis for $\dot{q}_{net}$ calculations

Parameter	**Symbol**	**Ignition**	**Extinction**
Gas temperature (°C)	T_g	50.0	800.0
Heat transfer coefficient (W/m²/K)	h	20.0	20.0
Emissivity (-)	ε	0.9	0.9
Flame thickness (m)	t_f	-	0.1
Flame irradiance (kW/m²)	$\dot{q}_f$	-	2.2

The resulting values are presented in Fig. 3.

For CLT, a clear trend can be identified with a strong ignition temperature decrease when the net heat flux increases. This trend is less clear for plywoods. Extinction temperatures are higher than ignition temperatures. It comes from the insulating effect of the char layer on the pyrolysis.

The net heat flux at ignition for low values of incident heat flux is close to zero. For a low irradiance, the time to ignition is high, and surface heat transfer is close to equilibrium. It is therefore consistent to measure quasi-null values.

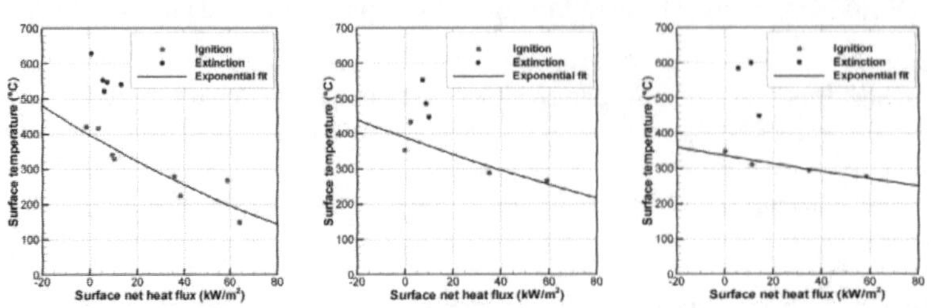

Fig. 3. T_{ig} vs. $\dot{q}_{net}$ correlation – from left to right: CLT, birch plywood and okoume plywood.

It has been chosen to correlate the absolute temperature at ignition to the net heat flux with an exponential law as presented in Eq. 3. T_0 and $\dot{q}_0$ are the fitting parameters. Resulting curves are described in Fig. 3.

$$T_{ig} = T_0 \cdot e^{-\dot{q}_{net}/\dot{q}_0} \tag{3}$$

For engineering purpose, extinction points have conservatively not been considered for the correlation. It leads to a worst-case estimation of the burning time, because it underestimates the temperature to extinction.

Phase space interpretation

The T_s vs. $\dot{q}_{net}$ plane can be interpreted as a phase space of the local surface heat balance. This space is divided by the criterion line fitted above. The sample surface follows a trajectory in this space during fire, and burning occurs for the parts of the trajectory above the criterion line. The trajectories are governed by Eq. 1. As the surface temperature increases, the net flux decreases.

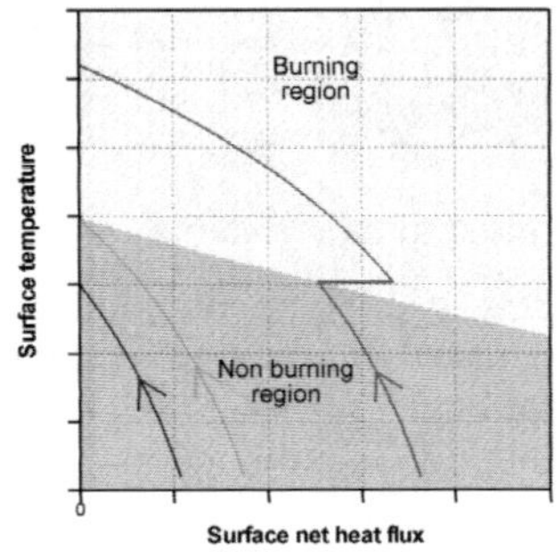

Fig. 4: Trajectory examples in heat transfer space phase

Fig. 4 presents three theoretical trajectories with constant incident heat flux, and other thermal parameters identical to those of Table 4. The first trajectory in blue occurs for a low incident heat flux value. As this value is inferior to the critical heat flux, the trajectory never crosses the criterion line, and no ignition happens.

The second trajectory in green occurs at critical incident flux. The trajectory reaches the criteria line at $\dot{q}_{net} = 0$. This point is never physically reached, as it requires an infinite heating time. Therefore, the T_0 parameter of the correlation is the ignition temperature under critical heat flux conditions, as described in [5].

The third trajectory in red occurs for a high incident heat flux value. This trajectory deviates when crossing the criterion line, because of the heat flux generated by the combustion.

EXPERIMENTAL DATA CORRELATION

Numerical model

A comparison to real scale tests has been used to validate the previous criteria line. The examined experimental data come from a series of façade fire tests in Duny's research [6]. Three tests were performed: one with an inert façade, and two with plywood façade, respectively okoume plywood and birch plywood identical to the one tested in our work. The test setup comprised a small fire room opening on a 5 m tall façade. The source fire came from 1.5MW wood stake for 20 minutes.

The Duny test setup has been modelled with Fire Dynamics Simulator (FDS) v6.5.2, following the rules given in his work: The wood stake HRR was specified from the inert façade test measurements, and a 10 cm mesh was used.

FDS was modified to take into account the criterion line correlated above. The criterion calculation from net heat flux is introduced during the predictor step, and the comparison to surface temperature is done during the corrector step. A constant heat release rate per unit area has been imposed, based on the mean value measured during the burning step of the cone tests.

Results comparison

The simulation results have been processed to identify the burning duration at each point of the façade, defined as the duration for which an area has experienced a surface temperature above the criterion line. Fig. 5 compares the calculated burning time for both plywood materials with the visual aspects of the façade after burning.

The unburnt area (light grey in the simulation) corresponds to the places where the wood has not ignited. The charred area (dark grey in the simulation) corresponds to the places where the wood surface has ignited and for a burning time lower than 9'30'' for birch and 11'30'' for okoume. These times were deduced from the burning durations measured with the cone tests. The completely burnt area (black in the simulation) corresponds to the zone where the wood surface presents a burning time higher than 9'30'' for birch and 11'30'' for okoume.

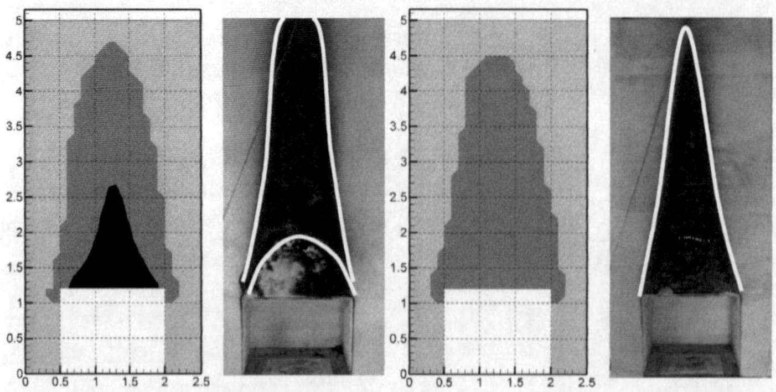

Fig. 5: Façades after burning – From left to right: Birch simulation and test, okoume simulation and test

Both facades burning are relatively well represented. The partially burnt area is slightly underestimated, whereas the completely burnt area is slightly overestimated. No complete burning is visible on okoume plywood façade during the tests (Koutaiba [9]) nor in simulations.

CONCLUSIONS

The three test series performed on wooden material gave a unique phenomenological criterion for engineering prediction of ignition and extinction. This criterion has been incorporated in FDS and proved to predict burning areas and thickness on plywood facades, with very few additional modelling costs.

It must be consolidated with two major points. First, a deeper physical analysis would improve the understanding of the cone-scale test results and the fitting of the criterion. Second, it would be interesting to correctly define the ignition and extinction criteria with precise measurement techniques. The vertical orientation with the CLT could be studied to see if a difference exists in the ignition temperature or in the correlation between surface temperature and net flux. Surface temperature could be measured over a larger area rather than with a punctual measurement with a thermocouple. The sample exposure to the external heat flux could also vary to observe the impact on the extinction time and depth of the char layer.

REFERENCES

[1] Babrauskas, V (2001). Ignition of wood. A review of the state of the art. Interflam 2001, 71-88

[2] Shen, D.K, Fang, M.X, Chow, W.K (2006). Ignition of wood based materials by thermal radiation, International Journal on Engineering Performance-Based Fire Codes, 8(2), 69-89

[3] Emberley, R, Inghelbrecht, A, Yu, Z, Torero, J (2016). Self-extinction of timber. Proceedings of the Combustion Institute. 36. 10.1016

[4] Crielaard, R (2015). Self-extinguishment of Cross-Laminated Timber. Master's thesis Report, Faculty of Civil Engineering and Geoscience, Delft University of Technology

[5] Moghtaderi, B, Novozhilov, V, Fletcher, D.F, Kent, J.H (1997). A new correlation for bench-scale piloted ignition data of wood. Fire Safety Journal, 29(1), 41-59

[6] Duny, M (2016). Etudes numérique et expérimentale des phénomènes de propagation d'un incendie le long d'une façade. PhD thesis, Ecole Nationale Supérieure de Mécanique et d'Aérotechnique

[7] Delichatsios, M, Paroz, B, Bhargava, A (2003). Flammability properties for charring materials. Fire Safety Journal, 38(3), 219-228

[8] Law, M (1978). Fire safety of external building elements. The design approach. Engineering Journal, 15, 59-74.

[9] Koutaiba, E.M, Dhima, D, Duny, M, Garo, J.P, Wang, H.Y, Quentin, J (2017). Experimental Study on Vertical Wooden Façade Combustion. Procedia Engineering, 210, 520-527

IOP Conf. Series: Journal of Physics: Conf. Series **1107** (2018) 032006 doi:10.1088/1742-6596/1107/3/032006

Velocity and temperature measurements in a vertical wall flame

Andres VALENCIA[1] , Bertrand LECORDIER[2], Alexis COPPALLE[2]

[1]*Department of Fire Protection Engineering, University of Maryland, USA*
[2]*UMR 6614 CORIA, INSA Rouen, France*

Alexis.coppalle@coria.fr

ABSTRACT

To study the combustion of vertical solids, a laboratory setup has been built in order to produce stationary and easily repeatable experiments. It consists of a rectangular porous burner fed with a mixture of methane and ethylene, which produces a sooting and luminous flame. That system has been already used to determine the soot concentration in a vertical flame [1,2].
Velocity and temperature measurements were performed inside the reactive boundary layer. The average velocities (horizontal and vertical), as well as their fluctuations, were measured thanks to the 'Particle Image Velocimetry' laser diagnostic. Temperature profiles were obtained using fine thermocouples placed at different heights.
The visual appearance of the flame shows that it is fully turbulent from a height of about 20cm. In this turbulent zone of the flame, the profiles of the vertical velocity and the temperature evolve similarly according to the height. Near the wall, the velocity and the temperature increase rapidly and reach a maximum located a few millimeters to the wall. For the velocity, this maximum increases with the height in the flame because of the entrainment of the hot gases by the effects of gravity. For the temperature, such evolution of the maximum is not observed. Far away the wall, the velocity and temperature profiles widen toward the outside of the boundary layer, because of the dilution with the air. On the other hand, the horizontal velocity fluctuations are high compared to the mean values and at each height the profiles have a maximum which is more shifted from the wall.
A heat flux sensor was placed just above the burner to determine the total heat flux received. The mean value is 23.6 kW / m2. The latter value will be analyzed and compared, as well to the velocity and temperature measurements, to other values obtained for the same wall flame configuration [3-5].

KEYWORDS: *heat transfer; fluid dynamics; wall fire; PIV, temperature*

[1] Hebert, D., Talbaut, M., Coppalle, A.. 2D soot concentration and burning rate of a vertical pmma slab using laser-induced incandescence. Proceedings of the combustion institute, 34:2575–2582, 2013.

[2] Valencia, A.,Talbaut, M., Yon, J.,Godard, G.,Gobin, C.,Coppalle, A. Soot and velocity mapping and 2D soot sheet dimensions in a buoyant wall-fire. Proceedings of the Combustion Institute 36 (2017) 3219–3226

[3] de Ris, J.L., Markstein, G.H. , Orloff, L.,Beaulieu, P.A. Similarity of turbulent wall fires. Fire Safety Science –Seventh International Symposium, International Association for Fire Safety Science, 2002, pp. 259–270.

[4] Ren, N.,Wang, Y., Trouvé, A., Large Eddy Simulation of vertical turbulent wall firesProcedia Engineering 62 (2013) 443 – 452

[5] Ren, N.,Wang, Y., Vilfayeau, S., Trouvé, A., Large eddy simulation of turbulent vertical wall fires supplied with gaseous fuel through porous burners, Combustion and Flame 169 (2016) 194–208

INTRODUCTION

An important case of fire is the propagation of a flame on a vertical combustible wall. For such configuration, the fluid is mainly driven by gravity effects, and the velocity is low. The emission of combustible vapors is due to the thermal degradation of the solid and the heat release is provided by their combustion in the flame zone. These phenomena are coupled by the heat fluxes exchanged at the wall by convection and radiation. Many works have been devoted to the study of vertical heated walls and wall-flames [3-9]. However for these last ones, few of them have allowed to study experimentally the fluid flow and its temperature near the wall and in the reactive boundary layer. In order to deal with this, a rectangular porous burner fed with a mixture of methane and ethylene was built in order to produce a sooting and luminous vertical flame. In the first part of the paper, the burner device, the velocity (PIV), temperature and total flux measurements are described. In the second part, the results are presented, discussed and compared to some values found in other works.

EXPERIMENTAL SETUP

Burner

The fig.1 shows the vertical burner used in this work. The burner has been used in previous research, and is described in detail in references [1,2]. In resume, it consists of a porous burner installed vertically with dimensions of 20 cm of width, 40 cm of height and 2 cm of thickness. This burner is fed with a mixture of Methane (0.243 g/s) and Ethylene (0.512 g/s). A steel wall cooled with water at ambient temperature have been placed around the burner, and an ignition line supplied Methane (0.017 g/s) on the bottom of the burner, in order to stabilize the flame.

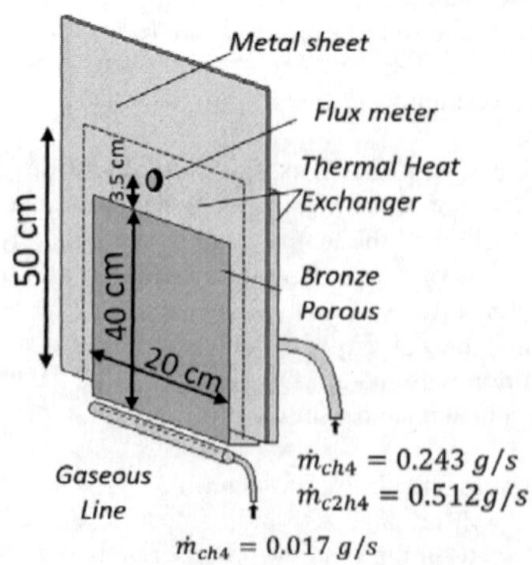

Figure 1: burner layout.

PIV measurements

The PIV setup is shown in fig. 2. A dual-head 532 nm Nd:Yg laser (Quantel Twins BSL) was used with a repetition rate of 10 Hz and a pulse duration of about 7 ns. The optical arrangement and the dichroic mirror shown on fig.2 have been set up in order to generate two overlapped and Gaussians laser sheets separated by 200 μs, and propagated vertically through the flame. These sheets had a dimension of ~800 μm x 45 mm (FWHM) in the x and y axis respectively. The laser sheet were located at 8 cm from one side, and were in contact with the wall in order to measure the near-wall region, and in the *y* direction only the most homogeneous portion (3.3 cm) of the laser sheets was used. The PIV signal was recorded with a 2048 ×2048 pixel CCD camera (JAI RM/TM- 4200CL) equipped with a 50 mm Nikon lens. The dimensions of the observed field have been set with a resolution of 71 μm/pixel. Because of the low velocity in the flame (about 2 m/s), the time between pulses was set to Δt = 200 μs. The images were recorded with the DynamicStudio (Dantec Dynamics) software application, and the velocities vectors have been calculated with a home-made algorithm. The instantaneous minimum velocity measurable with the optical device and the PIV algorithm used is ~6 cm/s. Low-velocity seeded air was injected in the flame through three injectors; two of the injectors, located at y = 1 cm and at the bottom of the burner, ensured a sufficient particle density throughout the flame. The third seeding injector, placed in the ignition line located below the burner, improved seeding near the wall at the bottom of the burner. In this way, the flame dragged particles from the bottom to the top.

Six different acquisitions were performed, with about 430 pairs of images for each case. So the mean and RMS values of the velocity components were calculated from this dataset. However, as seen in fig. 2, the particle seeding in the boundary layer is not homogeneous. Inside each pair of image used to determine the particle displacements, the percentage of valid points is greater than 23% for y> 5mm and, in some cases, less for y<5mm.

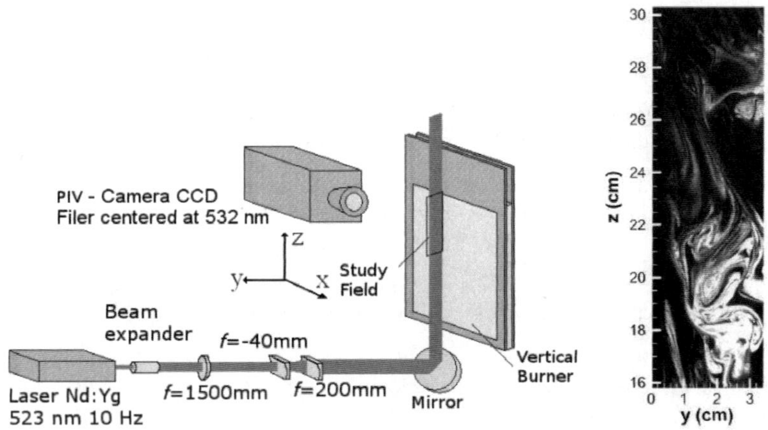

Figure 2: schematic view of the PIV system (right) and example of a raw seeded image detected by the camera.

Temperatures in the flame

Four thermocouples (type K) have been used to measure at the same time the temperature in the flame at heights h= 17-20-25.5-28 cm, as shown in fig. 3-a. Each thermocouple is tied to a steel rod, and there is 4 cm bare-wire length, corresponding to the distance between the junction and the end of the rod. The thermocouple tree is fixed and the burner is moved in order to measure the temperature at different y locations in the reactive boundary layer. The displacement accuracy is about 0.5mm. The size of thermocouple junction is 0.5 mm, so the time-response is too long to be able to measure the temperature fluctuations. So only the mean temperatures are given. When the junction is inside a combustion zone, the gas temperature is high and the junction emits radiation. Based on the reference [10], the mean maximum radiation correction is estimated to be 80k. For each experiment, the temperature at one location is averaged over 30s, and the mean profiles given below correspond to the average of four experiments. The reproducibility between these dataset is 100 °C (= standard deviation).

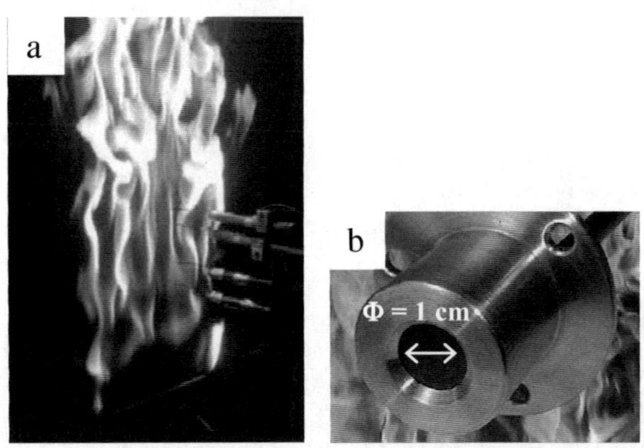

Figures 3: (a) view of the 4 thermocouples (height =17-20-25.5 and 28cm) on their support in order to measure the temperatures in the flame, (b) view of the heat flux sensor.

Heat flux measurement

An heat flux sensor (CAPTEC manufactor) has been installed just above the burner as shown in fig 1. The sensible surface diameter is 1 cm and covered with black paint (fig. 3-b). It is based on a differential measure and not sensitive to the sensor temperature. It gives the total flux (convective and radiative). A mean value has been calculated by averaging the data obtained during the four experiences of temperature measurement. The value is 23.6 +/- 3.3 kW/m2.

RESULTS AND DISCUSSIONS

Velocities

Fig. 4-a shows the profiles of the average vertical velocity $\bar{V}$ at 4 different heights. As expected, the velocity is very low close to the wall of the burner and it increases with the distance to the wall. However inside the boundary layer, the velocity profiles show the influence of the buoyancy forces. Compared to the forced convection case, the maximum is located inside the boundary layer and close to the wall, as seen for the case of an heated wall [9], or in wall-flame [2,3,8]. Three main zones can be identified:

1. Near-wall zone (y < ~5 mm): In these first millimeters, there is a sharp increase of the velocity, their fluctuations are low, as shown below in fig. 5-b. This zone can be qualified as laminar, the transfers of energy, species and momentum take place mainly by diffusion, and they are influenced by the viscosity.

2- Overlapping zone (~5 mm <y < ~10 mm): This region is identified as the combustion zone, the heat release accelerates the flow thanks to the buoyancy forces. Thus the maximum of velocity is located in this zone.

3-Far from the wall (y > ~ 10 mm): In this zone, the fluxes (energy, quantity of movement and species) are mainly turbulent. There is a slow decrease of the velocity when y increases, and this decrease spreads over several centimeters away from the wall.

The combustion is sustained by fuel injection throughout the burner height. This causes a continuous acceleration of the flow. This is suggested by the maximum of V which increases with height, as can be seen in fig. 4-a. An increase in the thickness of the dynamic boundary layer with the height is also observed. This is the consequence of the turbulence which generates velocity fluctuations, with the formation of large vortices, allowing more and more air to be mixed in the boundary layer. The order of magnitude of vertical velocities reported here are in agreement with previous results obtained in wall flames [3,8].

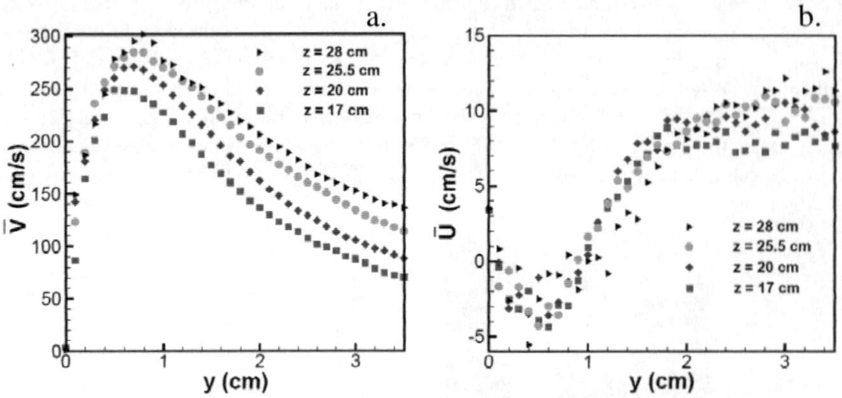

Figures 4 : Average velocity profiles for 4 different heights in the flame (17, 20, 25.5 et 28 cm).

a.: Vertical, *b.*: Horizontal.

Fig. 4-b shows the average horizontal velocity profiles $\bar{U}$ at 4 heights in the flame. There is more than one order of magnitude between the values of two velocity components. Close to the wall (0 <y <1 cm), the means of the U velocity are low, close to zero. Knowing that the instantaneous minimum velocity measurable with the optical device and the PIV algorithm used is 6 cm / s, some of the instantaneous U values calculated in this area have to be interpreted with caution. However for the average, the low values shown in fig. 4-a are mostly the result of a balance between large positive and negative fluctuations, as will be shown in the next figure. For y> 1 cm, the average horizontal velocity is positive, which corresponds the hot gas expansion towards the outside of the boundary layer.

Fig. 5-a-b show the profiles of the RMS values of the vertical and horizontal velocities at 4 different heights in the flame. The RMS values obtained for y <0.5 cm, close to the wall, are too sensitive to the small number of validated velocity vectors, as previously explained. Therefore, profiles of RMS values are only given for y > 0.5 cm. The variations of the RMS values have strong similarities with those observed for the average values. Indeed for $RMS\ V\ (y, z)$, we note that the velocity fluctuations increase with the height. At 28 cm height, the maximum value is 80 cm/s, corresponding to a turbulence intensity $\frac{RMS\ V}{\bar{V}}$ equal to 40%. With increasing height, the $RMS\ V$ profiles spread progressively in the horizontal direction. For $RMS\ U$, strong fluctuations are observed, the maximum RMS value is equal to 40 cm / s at 28 cm height, which is 4 times the average value, obtained in the same zone of the boundary layer (y> 2 cm). Similar profiles have already been observed, numerically for a reactive case [5], but also experimentally for a case without combustion [9] (only $RMS\ V$). For these two cases, $RMS\ V$ values have a sharp increase with increasing distance from wall, and the maximum is observed near the wall.

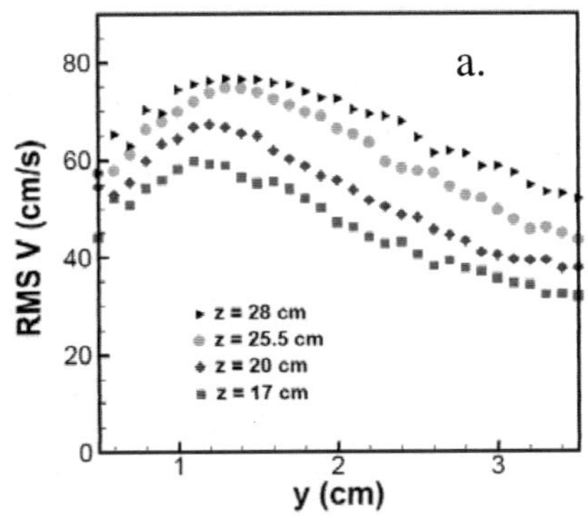

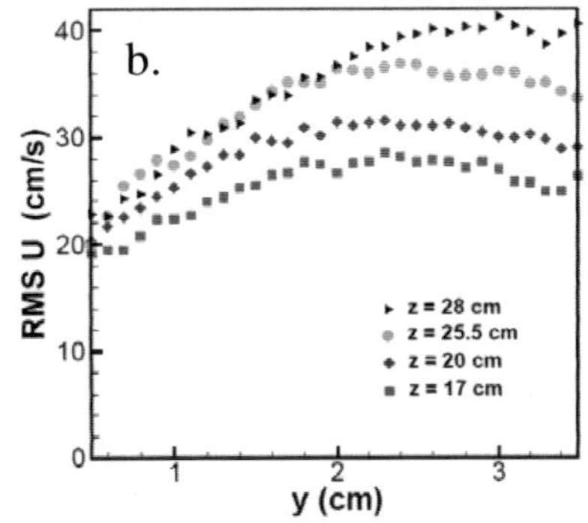

Figure 5 : Velocity RMS profiles at 4 different height in the flame (*17, 20, 25,5 et 28 cm*).

a.: Vertical, *b.*: Horizontal.

Temperatures

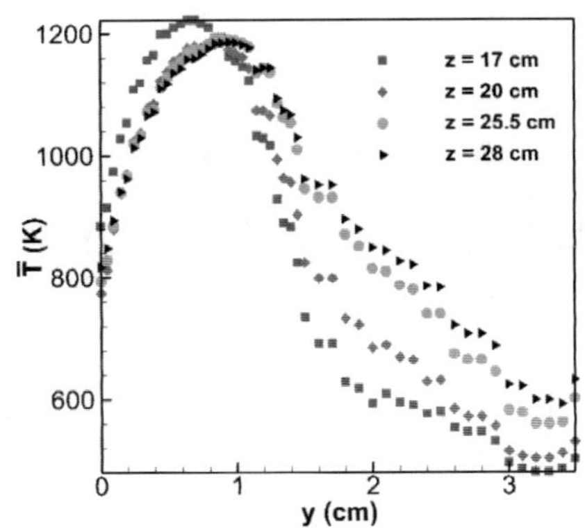

Figure 6 : Horizontal profiles of the mean temperatures at 4 heights in the flame:
17, 20, 25.5 and 28 *cm*.

On fig.6 is reported the mean temperature as a function of the distance from the wall, and at 4 heights in the flame. These profiles show a lot of similarities with those of the average vertical velocity as shown in fig 4. Three main zones can be also defined. Near the wall, there is a sharp increase of the temperature with no large difference between the different heights. Further from the wall (overlapping zone ~5 mm <y < ~15 mm), the temperature variations with the distance to the wall are small, and the maximum are observed here. This region is identified as the combustion zone, with a high heat release. Contrary to the vertical velocity profiles reported in fig. 4, the maximum temperature does not increase with the height. However, like for the velocity , the width of the temperature profile in the y direction is increasing with heigh.. Finally, far from the wall (y > ~ 15 mm), the temperature increases with height, as already observed for the vertical velocity. In this zone, the flow is highly turbulent and intermittent.

Comparison with other studies

As said previously, this wall-flame has been already used to study soot concentrations in the flame [1-2]. However results provided here are focused on velocity and temperature, and it is interesting to compare them to values obtained in other works. However, the comparison that will be presented here has to be view as qualitative. There are only a few studies for which experimental or numerical values of velocities and temperatures in the boundary layer and near the wall are available. The comparison uses the experimental results obtained by de Ris [3], with a higher-size burner and propylene as combustible, the numerical results of the simulation of this experiment [4,5], and the experimental results provided by Ahmad [8], with a smaller-size burner with ethanol as combustible. Firstly the comparison is about the maximum of velocity in the reactive boundary layer. A velocity scale may be defined as $(2gz)^{1/2}$. So the ratio $V_{max}^* = \frac{V_{max}}{(2gz)^{1/2}}$ is a mean to compare results obtained at different heights. At each top of the burners, the ratio is $V_{max}^* = 1.3$ in the present study at z=0.3 m, $V_{max}^* = 0.9$ in the de Ris's work at z=0.77 cm and $V_{max}^* = 1.8$ in Ahmad's work at z=0.1 m. Obviously, this scaling does not succeed for grouping these data. Further studies are necessary to take also in account the fuel flow and heat release rates, the former being very different (equal to 9.4 g/(m^2.s) in the present study, 17 g/(m^2.s) in the reference [3], and 2.9 g/(m^2.s) in reference [8]).

At each top of the burners, the maximum temperatures are 1220K, 1550K and 1550K for the present study, the Ris's and Ahmad's works respectively. This difference may be explained by the flow field which is more or less laminar in the Ahmad experiment, leading to a low mixing rate with air and higher temperatures, and by the higher fuel flow rate in the de Ris's experiment, providing a higher heat release rate and greater temperatures.

Finally, the total heat flux measured just above the top of the porous burner (z=0.435m) is 23.6 kw/m2 in this experiment, 22.3 kw/m2 at z=0.4m in the de Ris's experiment, and 26.9 kw/m2 at z=0.112m in the Ahmad's one. These values are close, indicating that the dependency of the total heat flux on the mass loss rate (or the heat release rate) is small at the lower part of the wall-flame.

CONCLUSION

A vertical flame has been produced by a 40cm high porous burner, fed with a mixture of methane and ethylene. The measurements of the velocity inside the flame zone show that the maximum is close the wall at about 1 cm, that the flow is highly turbulent in the horizontal direction (rmsU/U values greater 3 are possible). The temperature and velocity profiles at each height show similarities. Near the wall (y < ~5 mm), there is a sharp increase. The maximum are observed further from the wall (y ~5 mm <y < ~15 mm), in a zone being identified as the combustion zone, with a high heat release rate. Contrary to the vertical velocity of the flow, which is accelerated by the buoyancy forces, the maximum temperature does not increase with the height. The data presented here are suitable for the development of analytical and CFD models of turbulent wall fires.

REFERENCES

[1] Hebert, D., Talbaut, M., Coppalle, A.. 2D soot concentration and burning rate of a vertical pmma slab using laser-induced incandescence. Proceedings of the combustion institute, 34:2575–2582, 2013.

[2] Valencia, A.,Talbaut, M., Yon, J.,Godard, G.,Gobin, C.,Coppalle, A. Soot and velocity mapping and 2D soot sheet dimensions in a buoyant wall-fire. Proceedings of the Combustion Institute 36 (2017) 3219–3226

[3] de Ris, J.L., Markstein, G.H. , Orloff, L.,Beaulieu, P.A. Similarity of turbulent wall fires. Fire Safety Science – Seventh International Symposium, International Association for Fire Safety Science, 2002, pp. 259–270.

[4] Ren, N.,Wang, Y., Trouvé, A., Large Eddy Simulation of vertical turbulent wall firesProcedia Engineering 62 (2013) 443 – 452

[5] Ren, N.,Wang, Y., Vilfayeau, S., Trouvé, A., Large eddy simulation of turbulent vertical wall fires supplied with gaseous fuel through porous burners, Combustion and Flame 169 (2016) 194–208

[6] Pierre Joulain. Convective and radiative transport in pool and wall fires : 20 years of research in poitiers. Fire Safety Journal, 26(2) :99 – 149, 1996.

[7] Marcelo Annarumma, Laurent Audouin, Jean-Michel Most, and Pierre Joulain. Wall fire close to an horizontal surface: Numerical modelling and experimental validation. Fire and Materials, 16(1) :1–13, 1992.

[8] T. Ahmad and G.M. Faeth. Turbulent wall fires. Symposium (International) on Combustion, 17(1) :1149 – 1160, 1979.

[9] T. Tsuji, Y. Nagano: Velocity and Temperature Measurements in a Natural Convection Boundary Layer Along a Vertical Flat Plate, Experimental Thermal and Fluid Science 1989; 2:208-213

[10] NACA-TN-3766, Radiation and Recovery Corrections and Time Constas of several Chromel-Alumel Thermocouple Probes in High-Temperature, High-Velocity Gas Stream, 1956

IOP Conf. Series: Journal of Physics: Conf. Series **1107** (2018) 032007 doi:10.1088/1742-6596/1107/3/032007

Intermediate-Scale Tests Of Ventilated Facades With Aluminium-Composite Claddings

Eric GUILLAUME[a]*, Talal FATEH[b], Renaud SCHILLINGER[a], **Roman CHIVA[a]**, Sebastian UKLEJA[b], Roy *WEGHORST*[c]

[1] *Efectis France, France*

[2] *Efectis UK-Ireland, United Kingdom*

[3] *Kingspan B.V., Netherlands*

Roman.chiva@efectis.com

ABSTRACT

The fire behaviour of a building façade is dependent on the overall system's performance, rather than the performance of the individual components. A façade system includes the cladding and the insulant's characteristics, but also the cavities, cavity barriers, mounting and fixings, substrate and any singularities, such as window frames. Aluminium-Composite-Materials (ACM's) are increasingly commonly used as cladding when designing building facades nowadays, but different types of cladding can achieve very different levels of performance.

Assessment of a particular façade system's fire performance can be undertaken using large-scale testing such BS8414-1, but there is currently no commonly accepted method of extending the scope of a large scale tested system, to account for variances from the configuration that has been tested; it is therefore proposed that intermediate testing could allow a route for assessing the significance of different system configurations and components to extend the scope of large scale tested systems.

The test protocol proposed for such intermediate level testing, is a façade fire propagation test according to the ISO 13785-1 standard, with additional heat release rate and gases analysis using FTIR. Tests results have been compared to real-scale data available to estimate measurement relevance of such intermediate-scale tests.

In order to appreciate the variability of different system configurations, as well as the effect of different interactions between panels and insulants, a test campaign of 9 combinations of cladding and insulants was performed using intermediate level testing, this involved 3 different Aluminium Composite Material (ACM)-based cladding systems in combination with 3 different insulants within each system.

KEYWORDS:

Flame spread; fire growth; façade fire; ignition.

INTRODUCTION

The global market for façade insulation and especially ventilated facades, is growing quickly and likely to double in size by 2024 [1]. In this growing market, the proportion of Aluminium-Composite-Materials (ACM) is currently estimated as 25% of the market share for US and the same level for Europe [2]. Such global and increasing use of ACM's requires proper risk evaluation. Façade systems have been implemented on a great many commercial and residential buildings, providing a good level of energy efficiency, weather resistance and improved aesthetics. ACM cladding is generally followed by a cavity, then the insulant, over the face of a building structure (often either a steel frame or masonry or concrete framed substrate). The construction products used as external façade assemblies may include combustible insulants and combustible claddings. As stated in Table 1, there have been several significant fires over the last decade involving rainscreen facades; a large proportion of these have involved to the use of a combustible ACM-based cladding. Such fires are described in the references [3], [4] and [5] and have involved massive fatalities and property losses. According to the table, propagation through the cladding and penetration from the outside to the inside are the most important parameters driving these fires and their consequences.

Table 1. Recent façade fire events involving ACM claddings.

Date	Place	Circumstances and consequences
1/10/2010	Wooshin Golden Suites, Busan, South Korea	No fatalities, 5 injured. Fire from apartment, propagated by the façade.
14/5/2012	Roubaix, France	Dwelling building fire, 1 fatality, 20 apartments (over 94) destroyed. Initiated from apartment fire, propagated through decorative ACM panels on balconies
17/7/2012	Polat Tower, Istanbul, Turkey	Fire caused by a faulty air conditioning unit, no fatalities.
18/11/2012	Tamweel (Al Seef) Tower, Dubai, UAE	No fatalities. The building was made uninhabitable by the fire, and is expected to be reconstructed
3/4/2013	Hotel and Business Center, Grozny, Chechnya	Fire completely destroyed the plastic trimming used on the building's exterior, but the interior remained untouched.
25/11/2014	Lacrosse Tower, Melbourne, Australia	No fatalities or serious injuries. Levels six to twenty-one were affected by fire and many more were affected by water damage.
21/2/2015	Marina Torch Tower, Dubai, UAE	7 injuries. The fire started in the middle of the building and spread rapidly due to falling flaming debris and high winds. External cladding was charred from the 50th floor (over 82) to the top of the tower
19/5/2015	Dwelling building, Baku, Azerbaijan	16 fatalities, 63 injuries. Fire propagated on façade after a renovation.
1/10/2015	Nasser Tower, Sharjah, UAE	19 injuries. Fire started on the third storey and moved up through façade.
31/12/2015	Hotel The Address, Dubai, UAE	No fatalities. Fire started outside the 20th floor of the hotel but didn't spread inside.
14/06/2017	Grenfell tower, London	71 fatalities. Initiated from an apartment fire, rapid propagation to the façade and penetration from the outside to the other storeys
4/8/2017	Marina Torch Tower, Dubai, UAE	Second fire on the same tower. No fatalities. Debris from the fire falling to the ground and starting a second fire in the streets below.
22/12/2017	Dwelling building, Jecheon, South Korea	29 fatalities. Initiated from a car fire in underground car park, then propagated to the cladding

After the Grenfell tragedy in June 2017 [5], the UK government commissioned seven large-scale BS8414-1 [7] tests in order to determine which types of insulation could safely be used with different cladding types. These tests were performed by the BRE, according to criteria from BR135 document [8]. The test campaign involved 3 types of cladding: ACM with a polyethylene dominated core from here on referred to "ACM PE" (this is the type that was used on Grenfell tower), Fire retardant ACM cladding, with a better fire performance, hereafter referred to as "ACM FR" and ACM cladding with a mineral core filling of limited combustibility, hereafter referred to as "ACM A2". These cladding types were initially tested in combination with a PIR insulant as used at Grenfell and a Mineral wool insulant. Test results[9][10][11][12][13][14][15]are summarized in Table 2. None of the ACM PE based compositions passed the test. Unfortunately, BS8414 tests gives very little quantitative information for further interpretation of fire behaviour of these systems. Moreover, the tests were extinguished as soon as BR135 criterion failed, making a complete fire scenario (growth, possible plateau and decay of the fire) impossible to assess. An additional test was undertaken with ACM-FR and phenolic insulation [16]. The rationale for this was that not all plastic foams are alike. It was conceivable that a phenolic insulant could pass the test with FR-grade cladding, even if PIR did not. The phenolic did perform a little better than the PIR - but it still failed. The phenolic was deemed to have failed the test after 28 minutes. The equivalent PIR test lasted just 25 minutes, both met all temperature criteria set out in BR135 however they both failed due to "flames on top of the rig". Looking at the reports test pictures, it is clear that, although, it passed, the mineral wool insulation with the ACM FR barely only passed the criteria "flames on top of the rig".

Table 2. Results of BRE tests after Grenfell disaster.

BS8414-1 tests	With combustible insulant (PIR)	With mineral wool insulant
Aluminium with mineral core (ACM A2)	Pass	Pass
Aluminium with fire-retardant core (ACM FR)	Fail at 25 min	Pass
Aluminium with polyethylene core (ACM PE)	Fail at 8 min	Fail at 7 min

FM Global recently undertook a series of tests on ACM-based panels, according to a 16 feet parallel panel test stated in ANSI/FM4880 [17]. Results were then compared with NFPA 285 [18]. A large number of façade systems were evaluated. The results were sufficient for FM Global to take a view that the ANSI/FM 4880 parallel panel test method was discriminant enough to identify hazardous assemblies. Compared to NFPA 285, several compositions that complied with this standard didn't pass the parallel panel test, mainly because of lower heat exposure in the NFPA 285 test compared to realistic façade fire scenarios. Large-scale and real-scale façade experiments such BS8414-1 are however expensive and take a long to prepare. As a consequence, there is a need for an intermediate test method able to be correlated with a large scale reference test to account for variations from what is tested at a larger scale. An example of application is to extend reference tests with small changes in geometry or thicknesses of a component, and to validate this extended application using a scientifically based assessment. The test campaign described hereafter in this publication intends to cover this need, consider sensitivity and identify whether an intermediate-scale test method can be discriminant enough.

DESCRIPTION OF TEST SET-UP

A series of t 9 tests were undertaken based on the ISO 13785-1 standard [19]. This test involves assessment of a medium-scale mock-up of facades. In this test façade samples are fixed over a calcium-silicate board maintained on a steel frame. All the equipment is placed with wind screens on 3 faces made of Fire-rated plasterboards. The test arrangement is similar to a one third scale of BS 8414-1 test, with a 100 kW gas burner placed at the bottom of the back wing of the sample. The burner is a sand-diffusion propane burner of 100 kW, with a length of 1200 mm and a width of 100 mm and a height of 150 mm. Its upper surface is placed 250 mm below the lower edge of the sample. The complete system is then placed under a large hood to collect effluents. For this testing, several deviations from ISO 13785-1 standardized test protocol were made:

- Test duration was systematically 30 minutes with burner on;
- Tests were performed under a large calorimetric hood. The heat release rate and smoke effluents rate was measured continuously according to ISO 24473[20];
- Smoke was collected for FTIR analysis of the effluents, according to ISO 16405 [21] and ISO 19702 [22].

Depending on the energy release, two different sizes of calorimetric hood were used. The medium one was a 3 m x 3 m hood used as per ISO 9705 standard [23]. This allows good measurement conditions for heat release rates from 100 kW to approximately 3.5 MW. The larger one was a 9 m x 9 m hood, which allows good measurement conditions for heat release rates from 500 kW to 20 MW.

DESCRIPTION OF SAMPLES

Cladding

Three different Aluminium Composite Materials (ACM) panels used as cladding were tested. References of the products are as follows: a) Alpolic A2 limited-combustibility cladding; Hereafter designed as "ACM-A2", b) Alpolic/fr-RF Reduced-combustibility cladding; Hereafter designed as "ACM-FR", c) Reynobond PE standard cladding with polyethylene core ; Hereafter designed as "ACM-PE"

Insulants

Three different insulants were used in combination with each of the cladding types for the test combinations, as follows: a) Kingspan K15 phenolic foam (50mm thick), hereafter designed as "K15", b) Celotex RS5000 PIR(50mm thick), hereafter designed as "PIR", c) Mineral wool Rockwool Duoslab (100mm thick) hereafter designed as "MW". Information regarding these insulants, such as density or thermal conductivity is available on product datasheets from their respective manufacturers. The different thicknesses of combustible insulants or mineral wool were chosen to achieve similar levels of thermal performance. As the total thickness is

IOP Conf. Series: Journal of Physics: Conf. Series **1107** (2018) 032007 doi:10.1088/1742-6596/1107/3/032007

different between organic foams and mineral wool compositions, burner position was adjusted in order to produce similar thermal attack to the cavity.

Mounting and Fixing

The insulant and the cladding were assembled on Calcium silicate boards (860 kg/m^3) and as follows:

- Cladding made of panels of 779 x 508 mm (3 x 2 panels for back wall and 3 panels for side wall of test wing). Gaps between cladding panels are 20 mm wide;
- Cavity of 50 mm with intumescent cavity fire barrier above second rank of panels. At the position of the cavity barrier, the thickness of the cavity is reduced to 24 mm in this zone;
- Vertical frame made of aluminium profiles. Lower edge of the test frame is covered by a 2 mm thick aluminium L profile, with a 20 mm airgap at the bottom panel and the angle.

Test Sequence

The 9 different combinations were tested at Efectis UK-Ireland, Belfast, in indoor facilities (no wind, constant temperature). The tests using ACM-FR and ACM-A2 compositions were performed under Efectis UK-Ireland intermediate scale hood (3 m x 3 m), similar to the hood described in ISO 9705-1 standard [23]. The tests using ACM-PE claddings were performed under Efectis UK-Ireland large hood (9 m x 9 m), with a limit of 20 MW. All tests were only performed once.

TEST RESULTS

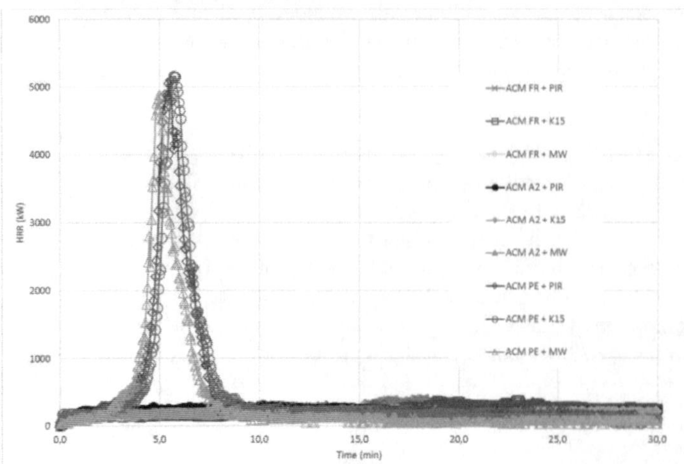

Fig 1. Heat release rate.

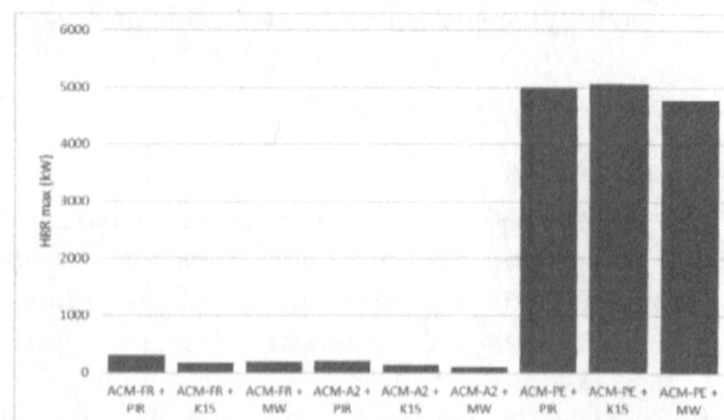

Fig 2. Maximum heat release rate (contribution of burner removed).

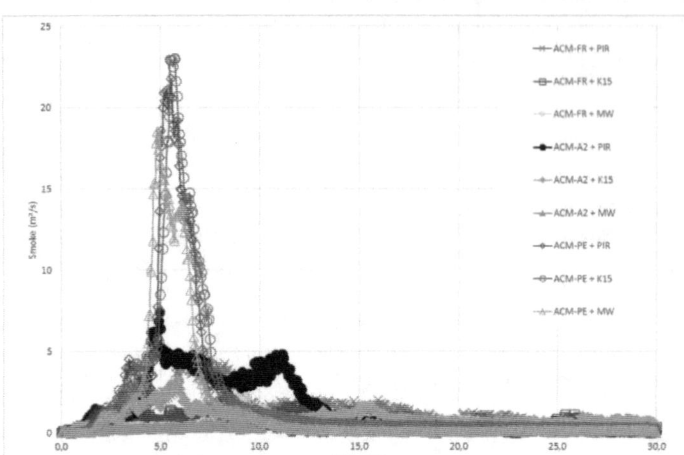

Fig 3. Rate of smoke production.

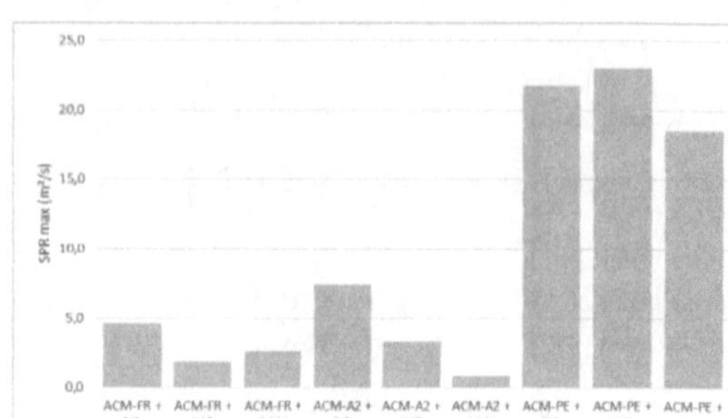

Fig 4. Maximum rate of smoke production.

Roughly, the tests involving ACM-FR and ACM-A2 present close patterns vs. time, with limited degradation sometimes visible inside the cavity but limited by the fire barrier. Maximum heat release rate from the tests with ACM-PE grew to almost 5 MW. This was similar for each of the ACM-PE tests and was more than 16 times higher than for all the other tests. This peak appears early, after only 4 minutes. The Peak was of a short duration, only a few minutes. For the test involving the ACM-PE+MW, the heat release rate curve has a similar trend observed for the tests involving ACM-PE + K15 and ACM-PE + PIR, but the decay started earlier, meaning a probable small contribution from combustible insulants in the decay phase. This probably explains the difference seen between these three tests for total heat released. The peak of heat released is very intense but so short that in a test averaged for 30 minutes, the difference between the tests using ACM-FR or ACM-A2 and the ones using ACM-PE is visible but less important. The results from the tests using ACM-FR or ACM-A2 show globally the same trend, with a low heat release during the whole duration of the test, but not exceeding 300 kW. Composition ACM-A2 + MW gives the best results as expected, but the compositions with ACM-A2 + K15 and ACM-FR + MW gave close results. On total heat release, higher values for tests using PIR compared to tests using K15 (about 5% more) probably indicate that PIR contributes more than K15 to energy released. They both contribute significantly more than MW. In conclusion for heat release, the largest contributor to the peak was the type of cladding when ACM-PE was used, with a little influence from the nature of the insulant during the decay phase. For the ACM-A2 and ACM-FR compositions, the insulants behaved broadly in a similar manner, especially compared against the performance of compositions integrating ACM-PE claddings. Results on rate of smoke production reveal again the peak observed for tests with ACM-PE claddings, but the difference is less visible than for the heat release rate. It means that the combustion phase corresponding to this phase, from 4 minutes to 8 minutes when ACM-PE is used, produces less smoke in proportion than the rest of the test. This is probably linked to a very intense combustion that re-burns all the smoke. PE combustion produces less smoke in proportion. This is also visible in the results for the total amount of smoke produced, where the difference between the tests was of less importance. In these results, tests that produce the smallest quantity of smoke were on ACM-FR + K15, ACM-FR + MW and ACM-A2 + MW compositions. Tests using PIR, seem to release more smoke than those with K15 and MW, but the difference was small. Gases evolved was quantified using non-dispersive infrared analyser for carbon dioxide (CO_2) and electrochemical cell for carbon monoxide (CO). Other species have been measured using FTIR analyser and semi-quantitative analysis is presented. FTIR analysis performed during the tests has proven the presence of unburnt hydrocarbons in the smoke. Evidence of methane, ethylene, acetylene and propane was highlighted during intense combustion in tests using ACM-PE claddings, and visible for methane and ethylene. No traces of hydrogen cyanide were found in any test, or any other nitrogen-containing species. This means that contribution of insulants, and especially PIR, is not visible in terms of species production, even in the largest fires involving ACM-PE claddings.

CONCLUSIONS

Tests performed are discriminant between solutions. They highlight that, for tested compositions, the cladding is the most important parameter driving global fire behaviour. ACM-PE based cladding systems gave very different results from the other solutions tested. This was especially visible in heat release rates, where fire intensity was very high, whatever the insulant used in the system. The contribution of the insulant was only remarkable in these tests during the decay phase. The cavity barrier was largely ineffectual in the 3 tests with ACM-PE cladding, as the integrity of the cavity was not ensured. Additional gas analyses highlighted a very well-ventilated condition of combustion with the ACM-PE based cladding compositions, probably enforced by the test setup and entrainment inside the cavity. Carbon monoxide release was low in proportion. Hydrocarbon release was remarkable, but no other species was detected within the limits of detection achievable in this setup. For the solutions with ACM-FR and ACM-A2 claddings, the results for all the insulants trended in a similar manner in these test conditions, especially when compared to the performance of ACM PE based compositions. In these tests and tested compositions, the cladding was the governing part of the constructive system and the contribution of insulants remained low in terms of energy or gases evolved. This confirms BRE tests according to BS 8414-1 and supplements the results with gases evolved and heat released from the beginning to 30 minutes without external action such extinction. These tests highlight also, that the BRE tests might have developed much more higher with the ACM-PE cladding, if they had not been extinguished. For such constructive systems, use of intermediate-scale tests is a very powerful tool to complete any reference real-scale test, for example in the case of extended applications. Unfortunately, intermediate scale tests as well as the reference BS8414-1 test don't cover all the details of a façade constructive system and single points, such as window frames, could have an important effect. Their effects on fire behaviour of the facade requires further evaluation.

REFERENCES

[1] Facades Market Analysis By Product (Ventilated, Non-Ventilated), By End Use (Commercial, Residential, Industrial), By Region (North America, Europe, Asia Pacific), And Segment Forecasts, 2014 – 2025. Published Nov 2017, Report ID: GVR-1-68038-221-1.

[2] Cladding Market Analysis By Product (Steel, Aluminum, Composite Panels, Fiber Cement, Terracotta, Ceramic), By Application, Competitive Landscape, And Segment Forecasts, 2014 – 2025. Published Jun, 2017, Report ID: GVR-1-68038-477-2.

[3] Valiulis, J., "*Building Exterior Wall Assembly Flammability: Have we forgotten the past 40 Years?"* In Fire Engineering Magazine, November 2015.

[4] White, N., Delichatsios, M., "*Fire Hazards of Exterior Wall Assemblies Containing Combustible Components. FPRF final report"*, project FE2568, 2004 Quincy, MA, USA.

[5] White, N., Delichatsios, M., Ahrens, M., Kimball, A., "Fire hazard of exterior wall assemblies containing combustible components". *Proceedings of 1st International Seminar for Fire Safety of Facade, 2013* Paris, pp. 77-88.

[6] UK Department for Communities and Local Government, "*Interim report into the Review of Building Regulations and Fire Safety*", issued 18 December 2017 https://www.gov.uk/government/news/interim-report-into-the-review-of-building-regulations-and-fire-safety

[7] BS 8414-1:2015+A1:2017. "*Fire performance of external cladding systems. Test method for non-loadbearing external cladding systems applied to the masonry face of a building*".

[8] S. Colwell, T. Baker, "BR135*": Fire performance of external Thermal Insulation for Walls of Multistorey Buildings (*3rd ed)", IHS/BRE.

[9] UK Department for Communities and Local Government, "*Collection: Grenfell Tower*", last update 22 November 2017 https://www.gov.uk/government/collections/grenfell-tower

[10] BRE. "*Fire test report*", DCLG BS 8414 test no.1. 7 August 2017.

[11] BRE. "*Fire test report*", DCLG BS 8414 test no.2. 3 August 2017.

[12] BRE. "*Fire test report*", DCLG BS 8414 test no.3. 8 August 2017.

[13] BRE. "*Fire test report*", DCLG BS 8414 test no.4. 11 August 2017.

[14] BRE. "*Fire test report*", DCLG BS 8414 test no.5. 14 August 2017.

[15] BRE. "*Fire test report*", DCLG BS 8414 test no.6. 25 August 2017.

[16] BRE. "*Fire test report*", DCLG BS 8414 test no.7. 21 August 2017.

[17] Agarwal, G., "Research Technical Report", *Evaluation of the Fire Performance of Aluminum Composite Material (ACM) assemblies using ANSI/FM 4880.* December 2017, FM Global, Norwood, MA02062, USA.

[18] NFPA 285. *Standard Fire test Method for Evaluation of Fire Propagation Characteristics of Exterior Non-Load-Bearing Wall Assemblies Containing Combustible Components*, 2012.

[19] ISO 13785-1:2002. *Reaction-to-fire tests for façades - Part 1: Intermediate-scale test.*

[20] ISO 24473:2008. *Fire tests - Open calorimetry - Measurement of the rate of production of heat and combustion products for fires of up to 40 MW.*

[21] ISO 16405:2015. *Room corner and open calorimeter - Guidance on sampling and measurement of effluent gas production using FTIR technique*.

[22] ISO 19702:2015. *Guidance for sampling and analysis of toxic gases and vapours in fire effluents using Fourier Transform Infrared (FTIR) spectroscopy.*

[23] ISO 9705-1:2016. *Reaction to fire tests - Room corner test for wall and ceiling lining products - Part 1: Test method for a small room configuration.*

Simulation of the pyrolysis of charring polymers: influence of the porous media properties

Jianwei SHI[1, 2], Germain BOYER[1] and Jean-François THOVERT[2]

[1]*PSN-RES, SA2I, Institut de Radioprotection et de Sûreté Nucléaire (IRSN), France*
[2]*Institut Pprime, CNRS, France*
jianwei.shi@irsn.fr

ABSTRACT

Recent large-scale experiments of cable tray fires have shown that the composition of the cable-sheath materials have a strong influence on the fire growth and spread. Thus, the correct prediction of the fire scenario requires the characterization of the sheath materials thermal and thermokinetic parameters. Yet, this task may be tricky for the materials whose pyrolysis generally leads to the formation of a porous residual. Whereas the physical constants related to the material degradation kinetics and to the virgin material properties can usually be determined using thermogravimetric analysis and differential scanning calorimetry, the porous residue properties are difficult to obtain and rarely available for pyrolysis models validation.

The present paper deals with the simulation of reference cone calorimeter experiments carried out on PVC samples. An extension to porous media of the pyrolysis model of the CALIF^{3}S-ISIS fire simulation software, developed at IRSN and validated on non-charring polymers, is used for the simulations. Special attention is paid to the modelling of the thermal diffusion effects in the porous residuals. On the one hand, based on the intrinsic conductivity of the different media (virgin material, intumescent intermediate state, char and gases), several effective conductivity models are tested to evaluate the relevant range for the global material conductivity. On the other hand, it is observed that the porous residuals of PVC generally contain large pores due to the material intumescence. That is why a radiative conductivity is included in the model to account for the radiative transfer in the pores. Without precise knowledge of the PVC residuals morphology, a Penetrable Sphere Model (PSM) is chosen to represent the porous medium geometry. This model is parametrized by a constant porosity deduced from the experimental data, and a variable pore size. It allows deducing a porous medium radiative conductivity which is added to the usual effective conductivity.

It is shown that, in the particular case of PVC pyrolysis, the uncertainty on the purely conductive effects induces a very limited dispersion on the simulation of the Heat Release Rate (HRR), since all the considered models yield very similar results. Conversely, the sensitivity of the radiative conductivity and therefore of the pyrolysis kinetics to the PSM pore size is much larger. The optimal pore size with respect to the HRR temporal evolution is of the same order as the pore size observed in some PVC residuals. Finally, pyrolysis simulations are performed with this optimized parameter for different cone heater induced heat fluxes and sample sizes. A good overall agreement is found between the experimental and the numerical HRR evolution, except for the thinnest sample for which the numerical HRR is overestimated. More generally, this work shows that further effort on the morphological description of the porous residuals and on the modelling of the radiative transfer is required if one aims at performing predictive pyrolysis simulations.

KEYWORDS

Fire chemistry; pyrolysis; porous media; heat transfer; Penetrable Spheres Model; Rosseland approximation

INTRODUCTION

In recent years, fire spreading along cable trays in Nuclear Power Plants (NPP) has been widely studied in large scale facilities [1,2]. The influence of large scale parameters such as the cable trays loading, spacing, width, the cables arrangement and the presence of a wall has been demonstrated. But recent large-scale experiments carried out in the framework of the OECD PRISME program led by French "Institut de Radioprotection et de Sûreté Nucléaire" (IRSN) have shown that the cable-sheath materials may also have a strong influence on the fire growth and spread [2,3]. In particular, the comparison between 5-horizontal trays experiments containing halogenated (i.e. PVC) or non-halogenated (mainly PE/EVA, compounds with mineral fire retardants) materials has shown that the initial HRR growth and the peak HRR are larger for PVC cable-sheaths than for non-halogenated cable-sheaths, but that the effective heat of combustion is larger for the latter than for the former. Yet, two approaches aiming at predicting cable tray fire scenarios have been developed in the last ten years. On the one hand, large-scale models such as the FLASH-CAT model [1] consist in predicting the fire Heat Release Rate from the large scale geometrical characteristics of the cable tray configuration. This requires a correct prediction of the horizontal and vertical spreading rates by accounting for the influence of the cable-sheath materials thermal properties. On the other hand, comprehensive pyrolysis models have been used together with computational fire dynamic simulations to study cable fire scenarios [4]. They require the characterization of the cable-sheath materials thermal and thermokinetic parameters. Unfortunately, although they can be predictive for non-charring polymers such as PMMA [5,6], the prediction of the degradation rates of polymers such as PVC or PE/EVA compounds with fire retardants stumbles over the correct characterization of the thermal properties of the porous residuals generated during the pyrolysis process. Therefore, these unknown parameters are generally set using an optimization procedure [7]. In particular, all the diffusive effects occurring in the heat transfer balance may be tied together in a single elliptic term which includes the pure conduction and a Rosseland-like radiative conductivity.

The purpose of the present paper is to evaluate the sensitivity of pyrolysis models to the uncertainty on the pure conductivity and radiative conductivity of porous pyrolysis residuals. An extension of the pyrolysis model of the CALIF^{3}S-ISIS [8] fire simulation software developed at IRSN and validated on non-charring polymers [6] is considered here. It will be tested against the PVC characterized and tested by Stoliarov *et. al.* [7]. Therefore, after a description of the extended pyrolysis model for porous materials and of the hypotheses on unknown parameters [6,8], the sensitivity of the pyrolysis model to parameters involved in usual pure conductivity and radiative conductivity submodels will be analysed. Then, a comparison between the numerical results and reference cone calorimeter experiments will be presented.

MODEL DESCRIPTION

General features

The pyrolysis model developed in the CALIF^{3}S-ISIS fire simulation software [8] allows simulating the thermal degradation of solid materials, involving N_R parallel or sequential reactions and a set of N_w solid materials and N_g gaseous species. Each solid material M_j is composed of a constitutive solid species S_j of intrinsic density ρ_j, a micro-porosity φ_j and a macro-porosity Φ_j. Let $\rho_j^0 = \rho_j(1-\varphi_j)$ denote the microporous solid density (mass of solid species S_j over the volume occupied by S_j and the micro-pores), and $\tilde{\rho}_j$ denote its partial density (mass of solid species S_j over an arbitrary volume element). Thus, if $X_j = \tilde{\rho}_j/\rho_j^0$ denotes the microporous solid volume fractions ($j = 1$ to N_w), the total macro-porosity reads $\Phi = 1 - \sum_j X_{w,j}$. Let also $\tilde{X}_j = X_j(1 - \Phi_j)$ be the volume fraction of material M_j, with $\sum_j \tilde{X}_{w,j} = 1$, which is used to ensure volume conservation. Then, the gas transport in the porous material is modelled by the Darcy law, the continuity equation and the transport equation related to each gaseous species G_j ($j = 1$ to N_g). The material permeability and the species diffusion can be kept constant since the model shows a negligible sensitivity to these parameters on a wide range of values. The total enthalpy conservation equation considered for non-porous materials is extended by including the terms related to the gas transport, namely accumulation, convection and pressure work.

Thermal degradation scheme and material properties

A two-step degradation scheme [7] is used to describe the pyrolysis of PVC, but a formulation of the pyrolysis gases is proposed according to several studies of gas analysis [9]. Indeed, the first step of the thermal degradation of PVC ($(C_2H_3Cl)_n$) consists in the ejection of Chlorine acid (HCl) and heavy hydrocarbons, mainly Benzene (C_6H_6). In the second step, a wide range of hydrocarbons is produced; among them are acetylene (C_2H_4) and hydrogen (H_2). Therefore, the retained degradation scheme is

$$\begin{cases} S_1 \rightarrow \mu_1 S_2 + \nu_1 G_1 + \nu_2 G_2 \\ S_2 \rightarrow \mu_2 S_3 + \nu_3 G_1 + \nu_4 G_3 + \nu_5 G_4 \end{cases} \quad \text{with} \quad \begin{cases} S_1 \equiv (C_2H_3Cl)_n,\ S_2 \equiv (CH)_{1.73n}(HCl)_{0.14n},\ S_3 \equiv C_n \\ G_1 \equiv HCl, G_2 \equiv C_6H_6, G_3 \equiv C_2H_4, G_4 \equiv H_2 \end{cases} \qquad (1)$$

Here S_1 denotes the virgin PVC, S_2 is an intermediate pseudo-species and S_3 is the char residual. μ_j and ν_j are the yields of solid and gaseous products j evaluated according to the species conservation and the characterized products yields [7]: μ_1=0.44, ν_1=0.504 and ν_2=0.056 for the first reaction, and μ_2=0.47, ν_3=0.184, ν_4=0.33 and ν_5=0.016 for the second one. The considered Arrhenius constants, heats of pyrolysis associated with each step, the virgin material specific heat, density and conductivity are those reported in Ref. [7], as well as each material's surface emissivity. Note that the present pyrolysis model allows recovering the correct mass loss evolution measured by thermogravimetric analysis depicted in Ref. [7]. The specific heat and intrinsic conductivity of S_2 are taken equal to those of the virgin material, and one choses the properties of coal for S_3.

Micro- and macro-porosity

Without further knowledge of the morphology of the solid materials resulting from the PVC degradation, one assumes arbitrarily that the micro-porosity is caused by mass loss and that macro-porosity is due to intumescence, the latter being probably caused by a strong gas ejection across a partially melt material. Thus, intumescence affects the intermediate material and the char residual in the same way. If β_j denotes the intumescence factor of material M_j, one sets $\beta_1 = 1$ and $\beta_2 = \beta_3 = 7$ [7]. In these conditions, since the PVC density is $\rho_1 = 1430$ kg/m³, the microporous solid densities and macro-porosities are: $\rho_1^0 = 1430$ kg/m³, $\rho_2^0 = \mu_1\rho_1^0 = 629.2$ kg/m³, $\rho_3^0 = \mu_1\mu_2\rho_1^0 = 295.7$ kg/m³, $\Phi_1 = 0$ and $\Phi_2 = \Phi_3 = 1 - 1/\beta_2 = 0.857$.

Conductive heat transfer models

Various models can be proposed to deduce an effective conductivity $\lambda_{eff,c}$ from the microporous solids conductivities λ_j and volume fractions X_j and from the gas conductivity λ_g and volume fraction X_g (i.e. macro-porosity Φ). The arithmetic and harmonic volumetric averages $\lambda_\parallel$ and $\lambda_\perp$ correspond to stratified structures oriented parallel or perpendicular to the heat flux, and provide the largest and the lowest possible values for $\lambda_{c,eff}$ for given volume fractions, regardless of the morphology [10]. The tighter upper and lower Hashin & Strickman bounds [10,11] can also be applied if isotropy is assumed. As a rule, the arithmetic average $\lambda_\parallel$ is used in the following. Numerical tests of the sensitivity to the various formulations are presented in Fig. 2. In this study, the conductivity λ_g of gaseous species is about 0.1 Wm^{-1}K^{-1} [12]. The conductivity reported in [7] for the PVC is applied for the virgin material, $\lambda_1 = 0.17$ Wm^{-1}K^{-1}. For the intermediate microporous solid, one assumes that the solid species is PVC and that the micropores are filled with gases, which yields $\lambda_2 = 0.13$ Wm^{-1}K^{-1}. For the final microporous residual, a similar argument, this time with coal as a solid species, yields λ_3=0.22 Wm^{-1}K^{-1}.

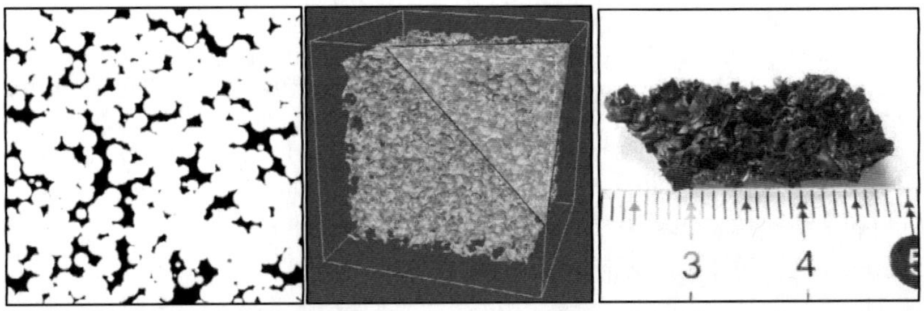

Fig. 1. PSM reconstruction example (left: 2D; center: 3D) with porosity 0.857 and sample of PVC carbon residue observed in CFS PRISME tests [2] (right).

Morphology and radiative transfer models

The geometrical shape of the PVC carbon porous residuals is very complex and the conductive and radiative heat transfer may depend on their geometrical features, such as porosity, specific surface area, characteristic length and isotropy. Since measuring those properties experimentally is extremely difficult, they can be modelled by the means of a conceptual representation. For lack of direct information on the morphology of the PVC carbon residual, the well-known Penetrable Sphere Model (PSM) [13] is chosen to represent it, since the spherical void inclusions are reminiscent of the gas bubbles observed in the experimental samples. Pictures of a PSM sample with a porosity of 0.857 are shown in Fig. 1, along with a photo of a carbon residue cut off from a PVC sample after its pyrolysis.

Considering the size of the pores in the PVC residuals, the radiative transfer in the porous medium may increase the heat diffusion and must be accounted for. One assumes that the solids are opaque and gray materials, and that the radiative transfer in the intumescent intermediate species and char can be modelled by a Rosseland approximation, as proposed by Roudani *et al.*[14] for alveolar media. The local average temperature $\bar{T}$, specific surface area S and surface emissivity ε of the solid interfaces are taken into consideration in Roudani's model, as well as the conductivity coefficient $\tilde{\lambda}_c$ and a form factor ω_s. The Rosseland radiative conductivity $\lambda_{eff,r}$ reads

$$\lambda_{eff,r} = \frac{8\left(1-\tilde{\lambda}_c\right)^2 \varepsilon\sigma\bar{T}^3}{\omega_s S} \tag{2}$$

For a PSM with in such a high porosity as considered in this study, the conductivity coefficient $\tilde{\lambda}_c$ is nearly 0, and ω_s is set equal to 0.5. The surface area reads $S = 3(1-\phi)\log\left(\frac{1}{1-\phi}\right)/R$, where R is the sphere radius. Finally, the total effective conductivity is the sum of the conductive part and the radiative part: $\lambda_{eff} = \lambda_{eff,c} + \lambda_{eff,r}$.

NUMERICAL RESULTS

For the simulations, the independence with respect to the time step δt and cell size δx is ensured with $\delta t = 0.1$ s and $\delta x = 0.012$ mm. The heat release rate per unit area is related to the sum of the heat release rates per unit area of all three gaseous fuels by $\dot{q}" = \sum \dot{q}_j" = \sum \Delta h_j \dot{m}_j"$ where Δh_j is the heat of combustion value reported in [15], namely 21.6, 41.5 and 141.8 MJ/kg for C_6H_6, C_2H_4 and H_2 respectively. The experimental heat release rate per unit area is deduced from the measured oxygen consumption during the cone calorimeter experiment. This measure is usually preferred from mass loss-derived measures which are usually noisier. After ignition, a time dependent correction is added to the initial value of external heat flux (EHF) to account for the intumescence, which leads to a surface movement towards the heater of the PVC sample. An additional 15kWm^{-2} of incident radiative heat flux is also applied in the total EHF to simulate the presence of the flame.

Sensitivity analysis to conductivity parameters

The sensitivity to the radiative conductivity is considered first. In this case, the only unknown parameter in Eq. 4 is the spherical pore radius. All other things being equal, the surface area decreases and the radiative conductivity increases with R. The sensitivity of the HRR to the PSM radius and the comparison to the experimental results obtained for 6 mm PVC samples under 75 kW/m^2 are shown in Fig. 2. The optimal sphere radius for the intermediate and char species is found to be 2.8 mm, which is larger than the 1 mm pores observed on the experimental residue in Fig. 1. An example of test without radiative heat transfer is also presented for comparison, and shows that the radiative transfer must be accounted for to simulate the pyrolysis of PVC. Conversely, comparisons of different effective conductivity models, namely the arithmetic and harmonic averages as well as the upper and lower Hashin & Strickman bounds are also depicted in Fig. 2. The limited influence of the effective conductivity model on PVC pyrolysis is due to the very large porosity of the intermediate and carbon residue, and therefore to the small conductive transfers compared to the radiative contribution: $\lambda_{eff,r}$ ranges from 0.07 to 2.99 $\text{Wm}^{-1}\text{K}^{-1}$, compared to the maximum species conductivity, $\lambda_3 = 0.21$ $\text{Wm}^{-1}\text{K}^{-1}$ for the final microporous residual.

Comparison to other cone calorimeter experiments

Without further information about the experimental mass loss rate of PVC pyrolysis, the experimental HRR under various conditions and the corresponding simulation results are compared in Fig. 3. This HRR is also decomposed into the parts associated with the first and the second reaction. The optimal pore size previously determined is used to evaluate the radiative conductivity. As observed earlier, the HRR curves exhibit a first peak just after ignition (10 s) associated with the first pyrolysis reaction, which mostly consists in the ejection of chloride acid and a small fraction of heavy fuels. The two subsequent are associated with the second pyrolysis reaction by which most of the flammable volatiles are released. The numerical time evolution of the heat release rate is in a good qualitative agreement with the experimental one for all cases. In particular, for tests performed on 6 and 9 mm thick samples the initial HRR peak and the average HRR are well reproduced; for the sample tested under 50 kW/m^2, the experimental HRR curve is delayed due to the initial cycles of extinction. However, the peak HRR is clearly overestimated for the 3 mm thick sample test. In all cases, the partial heat release rates associated with the first and the second pyrolysis steps exhibit a sharp decay. The final decay, sharper numerically than experimentally, is usually explained by a smoldering process which possibly occurs in the late stage of the material degradation. But more generally, the abrupt ending of both pyrolytic reactions may be due to an oversimplification in the two-step degradation model; possible missing intermediate reactions, which are difficult to be observed in thermogravimetric analysis, could yield smoother transitions.

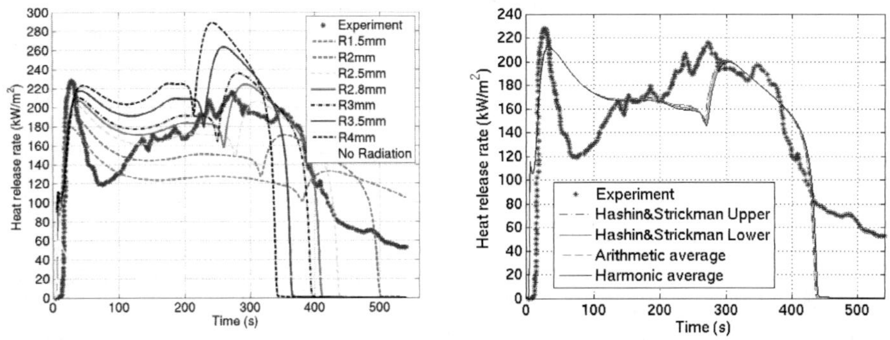

Fig. 2. Comparison of heat release rate per unit area between cone calorimetry experiment and numerical simulations. Left: influence of the PSM pore radius in the Rosseland radiative conductivity approximation. Right: influence of the pure conductivity model. PVC sample thickness: 6 mm. Imposed heat flux: 75 kW/m^2.

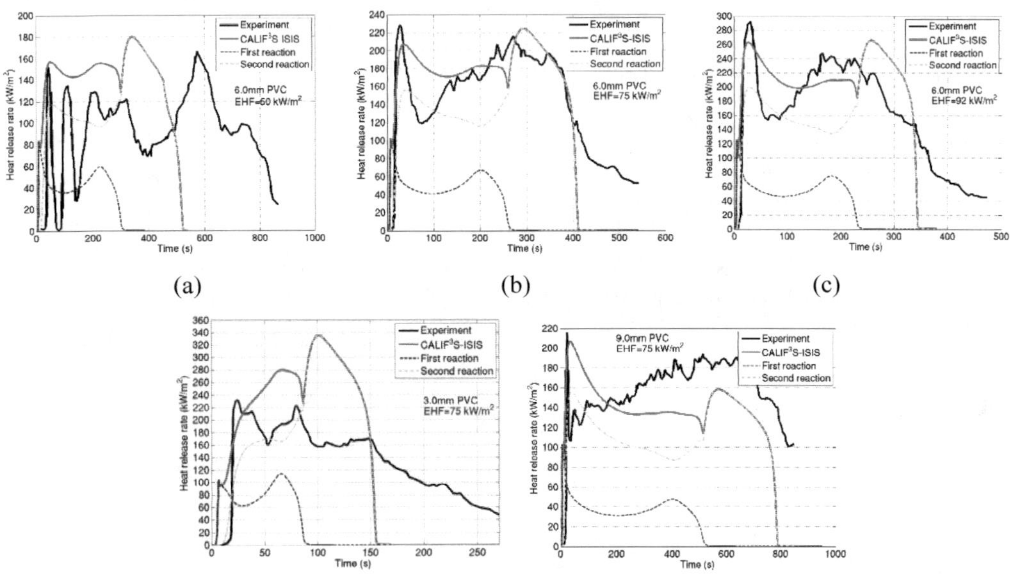

(a) (b) (c)

(d) (e)

Fig. 3. Comparison of heat release rate per unit area between cone calorimetry experiments and numerical simulation results for 3 mm (d), 6 mm (a, b, c) and 9 mm(e) sample thickness and 50 kW/m^2 (a), 75 kW/m^2 (b, d, e) and 92 kW/m^2 (c) imposed heat flux.

CONCLUSION

The present study is an attempt to assess the sensitivity of pyrolysis models to the pure thermal conductivity and radiative conductivity variations resulting from the uncertainties on the PVC porous residuals properties. In this respect, the HRR results show little sensitivity to the pure effective conductivity model, but a strong sensitivity to the pore radius which parametrises the Penetrable Sphere Model underlying the radiative conductivity estimation. Without further knowledge of the residuals geometry, the optimized pore size value allows a good agreement between the experimental and simulated HRR time evolution.

As a consequence, the residue morphology of such porous residuals should be better described if one aims at predicting the pyrolysis rate of such materials. In highly intumescent materials such as PVC, the radiative transfer seems to be dominant due to the large pore size and the relatively low gas conductivity. Note that, considering the large porosity and pore size and the possible semi-transparency of the solid species, the Rosseland approximation used to model the radiative transfer is not necessarily valid, and more relevant models should be used in a predictive approach. Despite the fact that the obtained radiative conductivity is model-dependent, the importance of radiative heat transfer during such process is evidenced. Further attention should be paid in this topic if a more accurate prediction is expected. Conversely, non-intumescent compounds filled with mineral fire retardants are expected to induce highly conductive mineral residuals with limited radiative effects if the pores are small. In this case, more attention should be paid on the modelling of the pure effective conductivity.

REFERENCES

[1] McGrattan, K.B., Lock, A., Marsh, N., Nyden, M. R., Price, M., Morgan, A.B., Galaska, M. and Schenck, K. (2010) "Cable heat release, ignition, and spread in tray installations during fire (CHRISTIFIRE) - Volume 1: Horizontal trays", *NUREG Report CR-7010*, Office of Nuclear Regulatory Research.

[2] Zavaleta, P., Charbaut, S., Basso, G. and Audouin, L. Multiple horizontal cable tray fire in open atmosphere, *Proceedings of the Fire and Materials 2013 Conference*, volume 13, 2013, pp. 57-68,

[3] Zavaleta, P. and Audouin, L. (2016) Fire spreading from a real open-doors electrical cabinet to overhead multiple cable trays into a confined and mechanically-ventilated facility, *International Conference on Fire Research and Engineering (Interflam)*, volume 14, pp. 1563-1574

[4] Matala, A. and Hostikka, S. (2011) Pyrolysis modelling of PVC cable materials, *International Symposium of Fire Safety Science*, International Association for Fire Safety Science, pp. 917-930, https://doi.org/10.3801/iaffs.fss.10-917

[5] Stoliarov, S.I., Crowley, S., Lyon, R.E. and Linteris, G.T. (2009) Prediction of the burning rates of non-charring polymers, *Combust. and Flame*, 156:1068–1083. https://doi.org/10.1016/j.combustflame.2008.11.010

[6] Boyer, G. (2017) Fully coupled CFD simulation of the pyrolysis of non-charring polymers: A predictive approach, *International Symposium of Fire Safety Science*, pp. 208–217, https://doi.org/10.1016/j.firesaf.2017.03.070

[7] Stoliarov, S.I., Crowley, S., Walters, R. N. and Lyon, R.E. (2010) Prediction of the burning rates of charring polymers, *Combust. Flam*, 157:2024–2034, https://doi.org/10.1016/j.combustflame.2010.03.011

[8] ISIS development team, ISIS 5.0.0: Physical modelling, *Technical report*, IRSN/PSN-RES/SA2I/LIE, 2017.

[9] ZHOU, H. (2017) Combustible Solid Waste Thermochemical Conversion, *Springer thesis*, Tsinghua University, https://doi.org/10.1007/978-981-10-3827-3

[10] Torquato, S., *Random Heterogeneous Materials*, Springer, 2002

[11] Hashin, Z. and Shtrikman, S., (1963) A variational approach to the theory of the elastic behavior of multiphase materials, *J. Mech. Phys. Solid*, volume 11, pp. 127-140, https://doi.org/10.1016/0022-5096(63)90060-7

[12] McBride, B.J., Gordon. S. and Reno, M.A. (1993) Coefficients for Calculating Thermodynamic and Transport Properties of Individual Species, *NASA Technical Memorandum 4513.*

[13] Thovert, J.F. and Adler, P.M. (2011) Grain reconstruction of porous media: Application to a Bentheim sandstone, *Phys. Rev*. E 83, https://doi.org/10.1103/physreve.83.056116

[14] ROUDANI, C. Modélisation à l'échelle microscopique des transferts thermiques radiatifs en milieu poreux, *PhD thesis*, ENSMA, 2008.

[15] Tewarson, A., "Generation of Heat and Chemical Compounds in Fires", DiNenno P.J. (ed.), *SFPE Handbook of Fire Protection Engineering (3nd ed)*, 2002.

Response of Asymmetric Slim Floor Beams in Parametric-Fires

Naveed Alam[1], **Ali Nadjai[1]**, Chrysanthos Maraveas[2], Konstantinos Daniel Tsavdaridis[3], Faris Ali[1]

[1]Fire Safety Engineering and Technology (FireSERT), Ulster University, United Kingdom
[2]Department of Architecture, Geology, Environment & Constructions, University of Liège, Belgium
[3]School of Civil Engineering, University of Leeds, United Kingdom

a.nadjai@ulster.ac.uk

ABSTRACT

State-of-the-art slim floor systems are a newest addition to the composite construction industry and several types are currently being used for building and construction purposes. Asymmetric slim floor beams are a type of slim floor systems which consist of a rolled section with a larger bottom flange. The larger bottom flange induces asymmetry and offers an efficient use of the material strength as a composite beam. It also offers a larger area to support the steel decking and pre-cast slab units during the construction of floor. Experimental and analytical investigations on response of asymmetric slim floor beams have shown that these beams offer a higher fire resistance in comparison to the conventional composite systems with down-stand steel beams. Previous investigations on these beams have been conducted in standard fire exposure conditions, hence, their response to natural fire scenarios still deems further examination. This study addresses response of asymmetric slim floor beams in natural fire exposure conditions. For this purpose, finite element models developed and verified by the authors are employed to study the thermal and structural response of slim floor beams in fast and slow parametric-fire exposures. Results obtained show that the asymmetric slim floor beams behave differently in parametric-fires in comparison to that in standard fire exposure conditions. Asymmetric slim floor beams continued to support the loads for the whole duration of parametric fires without undergoing excessive deflections and offering a better fire resistance. Unlike in case of the standard fire where the temperatures keep on increasing throughout the duration, temperatures on the slim floor beams decrease after reaching a maximum point in parametric-fires. It was found that for fast parametric-fires, the thermal gradient across the section is more severe as compared to that for the slow parametric-fires at earlier stages of fire exposure. In case of the fast parametric-fires, the rise and fall of temperatures on the slim floor beams are rapid while in case of the slow parametric-fire, these variations in temperatures are subtle. It was observed that the structural response of slim floor beams in standard and parametric fires depends on the average temperature across the steel section. Deflections predicted for the beams were found to be directly related to these average temperatures. Outcomes of this study will benefit in understanding the response of asymmetric slim floor beams in natural fire conditions and will aid to develop simple fire design methods for future use.

KEYWORDS:

Asymmetric slim floor beams, fire resistance, natural fires, finite element modelling, structural response, thermal response.

IOP Conf. Series: Journal of Physics: Conf. Series **1107** (2018) 032009 doi:10.1088/1742-6596/1107/3/032009

INTRODUCTION

Response of slim floor beams at elevated temperatures deems for detailed investigations because of their abundant use in high rise buildings and car parks [1]. Slim floor beams offer shallower depths as compared to the traditional composite beams, as a result, they help in reducing the depth of floors and height of the structures. These beams consume lesser construction materials due to their reduced height which reduces the cost of construction. Combination of slim floor beams with composite steel decking offers a faster method of construction as well as an easy passage for electric and mechanical networks [2]. Most part of the steel section in such beams is encased within the floor concrete except for the bottom flange, hence, these beams offer a higher fire resistance being protected from flames and direct exposure to heat [3]. Response of slim floor beams has previously been investigated through several experimental [4] and analytical investigations [5] [6]; they are yet limited to standard fire exposures.

This study addresses the response of asymmetric slim floor beams (ASBs) in parametric fires; a representation of the natural fires, through analytical investigations. During previous investigations by the authors, behaviour of the slim floor beams at elevated temperatures has been replicated and verified against the test data and the verified finite element (FE) models are further used to perform sensitivity studies [5][6]. Here, these verified FE models are used to simulate the response of slim floor beams against natural fires in terms of parametric-fires given in the Eurocodes [7]. As the steel beam is partially protected by the concrete and only the bottom flange is exposed to fire, significant thermal gradients are developed across the section and thermal bowing is evident. The parametric fires selected during this study are a representation of a wide range of natural fires expected in an office compartment.

AIMS AND OBJECTIVES

This research is instigated with the purpose to explore the response of asymmetric slim floor beams in parametric-fires, a representation of the natural fires. To achieve this, thermal and structural response of asymmetric slim floor beams is investigated in two natural fire scenarios, a fast fire and a slow fire. Response of asymmetric slim floor beams is then analysed in comparison with that in standard fire exposure conditions and differences in their behaviour are highlighted. During slow fire exposures, it is expected that more uniform temperatures will develop, while in fast heating exposures, extreme thermal gradients are expected. As the temperature profile of the steel beam is important and affects their fire resistance, behaviour of the slim floors exposed to natural fires must be investigated.

RESPONSE OF ASBs IN NATURAL FIRES

The Natural Fires

Most of the investigations on the response of slim floor beams are performed against the standard fire exposures. In this study, response of the slim floor beams is assessed against the natural fires in terms of parametric-fires as recommended by the Eurocodes [7]. The parametric-fire curves representing the natural fires are in terms of a fast and a slow parametric-fire as shown in Fig. 1.

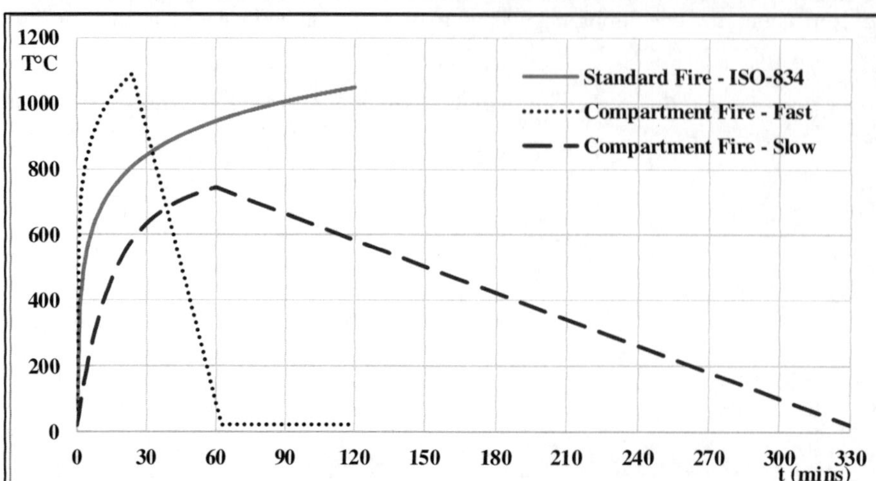

Fig. 1. Parametric and standard fire curves

For both fire scenarios, the design value of fire load density ($q_{t,d}$) is taken as 200 MJ/m^2, considering the compartment to be a representative of an office building. The 'b' factor, which is the representation of the density, specific heat and thermal conductivity of compartment boundaries, is taken as 1120 $J/M^2s^{1/2}K$. Values of the design fire load density and that of factor 'b' are kept similar for both fast and slow parametric-fires. Value of the opening factor for the slow fire is taken as $0.02m^{1/2}$; minimum value recommended in the Eurocodes [9] while that for the fast fire is taken as $0.1m^{1/2}$. These parametric fires cover a wide range of the natural fires thus enabling a general applicability of the findings from this research. Parametric fire curves and the standard fire curves used in this research are produced using the equations proposed in the Eurocodes, [7].

Finite Element Modelling

Response of asymmetric slim floor beams to natural fires is investigated through finite element modelling (FEM) using ABAQUS [8]. Their response at elevated temperatures is replicated and verified by the authors previously using the FEM and various parametric studies have also been performed using these verified models [5][6]. In this study, the same FE models are used to simulate the response of a slim floor beam assembly 5000 mm long having a span of 4500 mm between the supports. The beam assembly consists of an ASB-280 rolled steel section and a composite slab constructed using normal weight concrete and deep steel decking. The beam assembly is 308 mm deep and has a width of 950 mm as shown in Fig. 2. This beam assembly is similar to the one used previously during an experimental investigation [4]. During the FEM, half of the beam assembly is modelled while the support and boundary conditions are kept similar to those reported for the test except for the fire exposure conditions [4]. Non-linear thermal and mechanical properties of concrete and steel are adopted from the Eurocodes [9][10]. Further details of the two-step FEM method used in this study are available in previous research publications by the authors [5][6].

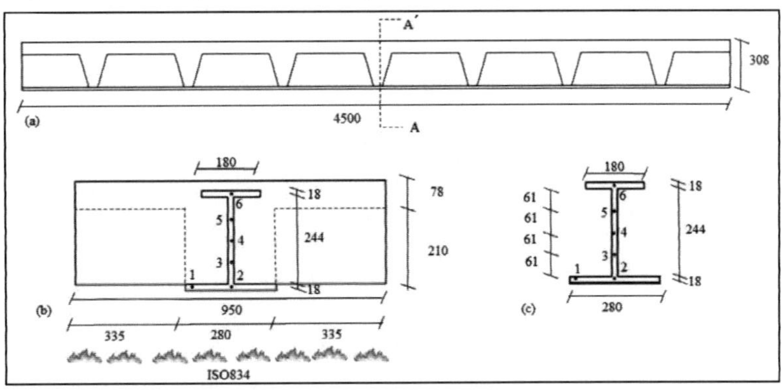

Fig. 2. Details of the test assembly, (a). Side elevation, (b). The section at AA´, (c). Cross-section of steel beam

Three investigations, one each against the standard fire, the slow parametric-fire and the fast parametric-fire are performed. Applied loads during the structural analyses are kept equivalent to a degree of utilization of 0.55 of the steel beam section. This degree of utilization gives a representation of the maximum expected external loads in any fire case scenario [11].

Results and Discussion

Thermal predictions obtained from the FEM analysis are depicted in Fig. 3. Thermal predictions for the flanges are shown in Fig. 3(a) where location 6 represents mid-point of the top flange while location 1 represents the one on edge of the bottom flange. Thermal predictions for the top flange are similar for all fire exposures and remain within 200°C for 120 mins. The top flange being far from exposed surfaces is least influenced by the fire exposure type. On the other hand, parts of the beam closer to the exposed surface are most influenced by the exposure type. In case of the standard fire exposure, temperatures on thermocouple 1, on the bottom flange, keep on increasing and reach 550°C, 800°C, 950°C and 1000°C after 30, 60, 90 and 120 mins of heating, respectively. For the fast parametric-fire, temperatures at this location are 850°C and then decrease to 450°C, 230°C, and 150°C for the same duration of exposure. In case of the slow parametric-fire, temperatures on the bottom flange gradually increases to 600°C after 76 mins and slowly decrease afterwards to 587°C after 120 mins. For thermocouple position on the middle of the steel web, position 4, the predicted temperatures continue to rise for standard fire exposure as shown in Fig. 4(b). For fast-parametric fire, these temperatures increase for the first 65 mins and then gradually decrease while for the slow and natural fire exposure conditions, temperatures at this location continue to rise for 120 mins as shown in Fig. 4(b). These temperature differences highlight the effect of exposure conditions on the thermal behaviour of slim floor beams and suggest their behaviour is highly fire-exposure dependent.

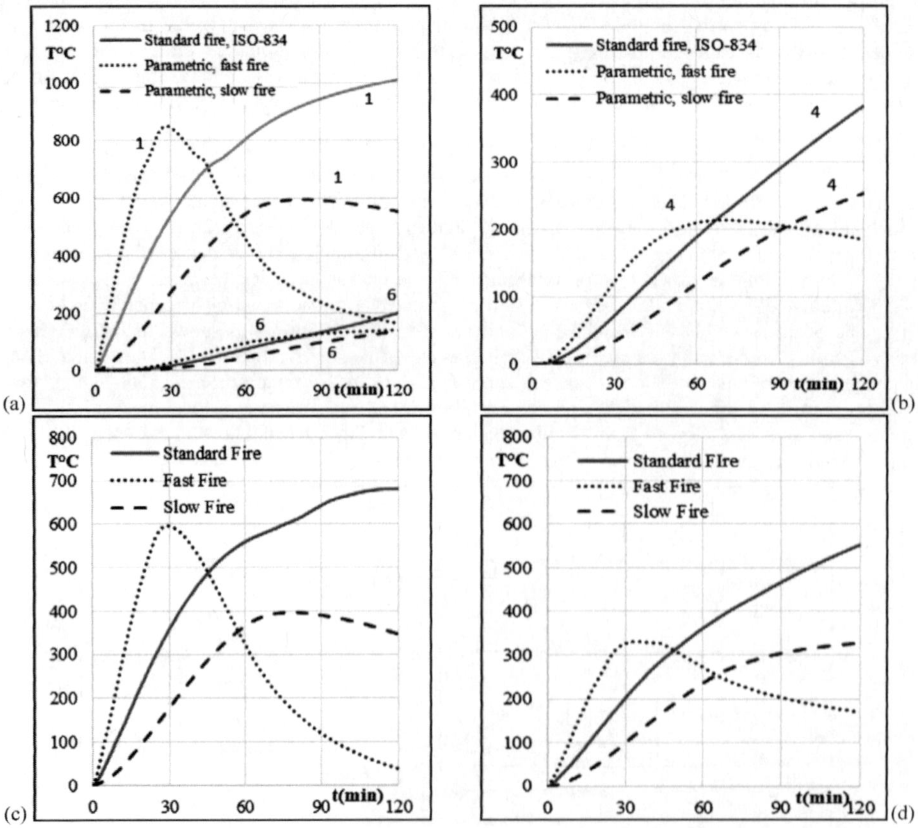

Fig. 3. Comparison for thermal results for standard and parametric fires: (a) thermal predictions for flanges, (b) thermal comparisons for the web, (c) thermal gradient and (d) average temperatures across the steel-section

Thermal gradient across the section of the slim floor beam was momentous as it has been reported previously for various slim floor beam types [4][5][6]. Based on the fire scenarios, distinct thermal gradients were observed across the beam during this study. The thermal gradients in terms of the temperature difference between thermocouple positions 2 and 6, middle of the bottom and the top flange, are presented in Fig. 3(c). For the first 45 mins of fire exposure, the thermal gradient for fast fire is the greatest with a maximum temperature difference of 598°C after 30 mins. This thermal gradient gradually decreases and reduces to 50°C after 120 mins of fire exposure. In case of the standard fire, thermal gradient across the section increases throughout the duration of heating. For the slow parametric-fire, this thermal gradient is the least of all three fire exposure types for the initial 56 mins. Beyond this point, the thermal gradient for slow fire becomes higher than that predicted for the fast parametric-fire as shown in Fig 3(c).

Like the thermal gradient, average temperatures on the steel section were also found to be variable for different fire exposures. The average temperatures shown in Fig. 3(d) are taken from the predicted temperatures for six thermocouples shown earlier in Fig. 1(c). In case of standard fire, the average value of temperatures keeps on rising for the whole duration of fire exposure while for the parametric fires, these temperatures decrease after reaching a maximum value. For the first 48 mins of heating, the average temperatures are higher for the fast parametric-fire and then gradually reduce onwards. Like the thermal gradient, average temperatures are the lowest for the first 65 mins of heating for the slow parametric-fire. Average temperatures for the slow fire keep on increasing and overtake those predicted for the fast parametric-fire beyond 65 mins. The thermal gradients and average temperatures on the section can also be seen in Fig. 4 after 120 mins of fire exposure. For standard fire, the thermal gradient as well as the predicted temperatures across the section, are the highest (Fig. 4(a)) while for fast parametric-fire, these are the lowest (Fig. 4(b)). As heating part of the curve for the slow parametric-fire is longer in comparison to that of the fast-parametric fire, at 120 mins, the temperature gradient and the predicted temperatures are higher for slow fire in comparison to the fast parametric-fire as shown in Fig 4.

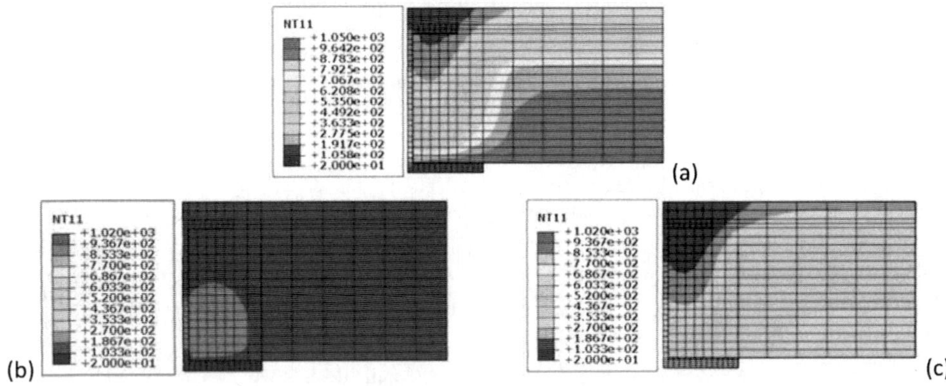

Fig. 4. Temperatures contours after 120 mins of fire exposure for: (a) standard fire, (b) fast fire and (c) slow fire

Structural response of the slim floor beam assembly is analysed in terms of the mid-span deflection predicted for the standard and the parametric fire exposures. The failure criteria adopted here are deflection-based and in terms of the maximum deflection and maximum rate of deflection given by the British Standards [12]. It is seen in Fig. 5 that the slim floor beam assembly in standard fire conditions exceeds the deflection rate limits after 74 mins of heating. In case of the fast parametric-fire, the maximum mid-span deflection was predicted to be 94 mm after 38 mins of heating. Beyond this point, the deflection reduced and became 62.5 mm after 60 mins and then further reduced to 25.5 mm after 120 mins. For the slow parametric-fire exposure, the predicted mid-span deflection was 28 mm after 30 mins while the maximum predicted deflection was 59 mm after 83 mins. This deflection reduced to 54 mm after 120 mins of heating as shown in Fig. 5. For both parametric-fire exposures, neither limits of the deflection were reached, and the slim floors outlasted the whole duration of fire without failure. These results show that the structural behaviour of slim floor beams evidently has a direct dependence on the average temperatures. For instance, the average temperatures across the section keep on increasing for standard fire and the predicted mid-span deflection follows the same. Likewise, the average temperatures as well as the deflections are higher for the first 33 mins for the fast parametric-fire and reduce afterwards. Similar relationships are seen for the slow parametric-fire as well. This suggests that the structural response of slim floor beams is a function of the average temperatures across the steel beam section.

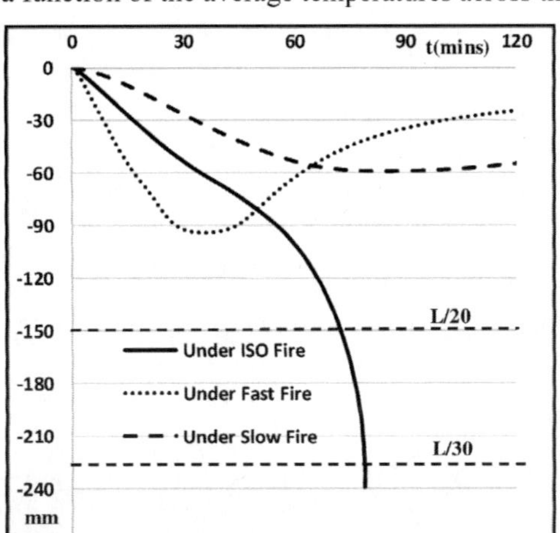

Fig. 5. Mid-span deflections for different fire exposure conditions

IOP Conf. Series: Journal of Physics: Conf. Series **1107** (2018) 032009 doi:10.1088/1742-6596/1107/3/032009

CONCLUSIONS

This paper presents the findings of a computational study conducted to analyse the response of slim floor beams in natural fire exposure conditions in terms of the parametric-fires. The fire scenarios used during this study represent a wide range of possible fires, hence findings from this study have a general applicability. It is found that the exposure conditions have a significant influence on response of the slim floor beams in fire. In standard fires, temperatures of the assembly continue to rise throughout its duration while in parametric fires, these temperatures reduce after reaching a maximum value. Although the thermal gradient was high across the beam section for all fire exposures, the magnitude of this gradient was variable from one exposure type to another. The slim floor beam assembly outlasted the whole duration of the parametric-fires which suggests that these beams offer an improved fire resistance in real fire conditions. Variance in the thermal and structural response of the investigated slim floor beam against different fire exposures suggests that their behaviour in fire is highly exposure-type dependent. It is interesting to see a direct relation between the average temperatures on the steel beam and mid-span deflection. This affiliation suggests that the structural response of slim floor beams is a function of the average temperatures on the beam.

Authors of this research are currently working on devising simple fire design methods for slim floor beams. These fire design methods are expected to provide a relationship between the fire rating, the average temperatures, and degree of utilization of slim floor beams.

REFERENCES

[1] Lu, X. & Mäkeläinen, P., (1996), Slim Floor Developments in Sweden and Finland. *Structural Engineering International*, 6(2), pp.127–129. http://dx.doi.org/10.2749/101686696780495789.

[2] D. L. Mullet, (1998), *Composite Floor Systems*, The Steel Construction Institute, 1st edition London.

[3] Newman, G.M., (1995), Fire resistance of slim floor beams. *Journal of Constructional Steel Research*, 33(1-2), pp.87–100. http://dx.doi.org/10.1016/0143-974x(94)00016-b.

[4] D. E. Wainman, (1996), Preliminary Assessment of the Data arising from a Standard Fire Resistance test Performed on a Slimflor Beam at the Warrington Fire Research Centre on 14th February, 1996, *Technical Note*, SL/HED/TN/S2440/4/96/D.

[5] Maraveas, C., Swailes, T. & Wang, Y., (2012), A detailed methodology for the finite element analysis of asymmetric slim floor beams in fire. *Steel Construction*, 5(3), pp.191–198. http://dx.doi.org/10.1002/stco.201210024.

[6] Alam, N., Nadjai, A., Ali, F., Nadjai, W., (2018). Structural response of unprotected and protected slim floors in fire. *Journal of Constructional Steel Research*, 142, pp.44–54. http://dx.doi.org/10.1016/j.jcsr.2017.12.009.

[7] European Committee for Standardization (2009), *EN1991-1-2, Eurocode 1 – Actions on structures – Part 1–2: General Rules –Structural Fire Design*, European Committee for Standardization.

[8] ABAQUS, (2016), Finite Element Modelling Programme and Standard User's Manual, Version 6.14. SIMULIA.

[9] European Committee for Standardization, (2008), *BS EN 1992-1-2 Eurocode 2: Design of concrete structures part 1-2 General rules – Structural fire design*, European Committee for Standardization.

[10] European Committee for Standardization (2009), *BS EN 1993-1-2 Eurocode 3: Design of steel structures, Part 1-2, General rules – Structural fire design*, European Committee for Standardization.

[11] Bailey, C.G., (1999). The behaviour of asymmetric slim floor steel beams in fire. *Journal of Constructional Steel Research*, 50(3), pp.235–257. http://dx.doi.org/10.1016/s0143-974x(98)00247-8.

[12] British Standards Institution. (1987). *BS 476-20, Fire tests on building materials and structures. Method for determination of the fire resistance of elements of construction (general principles).* British Standards Institution

Numerical Study of the Decomposition Mechanism of Intumescent Coating

Talal FATEH[1*] Salah BENKORICHI[1], Lucas ROBINET[1,2], and Jean-Louis CONSALVI[2]

[1] *FireSERT, School of the Built Environment and the Built Environment Research Institute, Ulster University, Newtownabbey BT37 0QB, UK*

[2] *Aix-Marseille Université, IUSTI/UMR CNRS 7343, 5 rue E. Fermi, 13453 Marseille Cedex 13, France*

*t.fateh@ulster.ac.uk

ABSTRACT

Intumescent coating represents an efficient way to increase the fire resistance of construction materials. During its degradation, intumescent materials expand and form a char layer that acts as an efficient thermal shield between the heat source and the materials.

Encouraging predictions of thermal degradation and expansion of intumescent coatings in experimental conditions representative of fires were reported in the last decade by considering models involving conservation equations for mass and energy for the intumescent material coupled to approximate sub-models for describing the swelling mechanism. These degradation models used generally simplified decomposition mechanisms involving one, two or three reactions. In addition, the dehydration process that occurs in the early stage is usually ignored.

The objective of this article is to assess the relevance of several multi-step reaction mechanisms to describe the mass loss and the mass loss rate of ICWB, an intumescent coating, in TGA at different heating rates in inert atmosphere. The kinetic parameters of the different reactions were optimized using the Shuffled complex evolution (SCE) technique.

Results demonstrate that the drying process, which occurs up to 130°C for ICWB, cannot be neglected since the associated mass loss represents about 5%. In addition, the drying process can be described by a single reaction. The use of a simple mechanism, based on one or two reactions to model the degradation of dry intumescent coating, is insufficient and leads to significant discrepancies on both mass loss and mass loss rate. Considering three reactions improve considerably the predictions but discrepancies are still observed for temperature higher than about 420°C. Model results show that a four-step is required to capture accurately all the details of the degradation of dry intumescent coating under inert atmosphere. Simulations were run under air, showing that an additional reaction for oxidation is required to extend the four-step reaction mechanism for oxidative atmosphere, leading to a five-step reaction mechanism.

KEYWORDS:

Structures in Fires, Intumescent Coatings, Inert Atmosphere, Decomposition, Thermogravimetric Analysis, Multi-Step Reaction Mechanism

IOP Conf. Series: Journal of Physics: Conf. Series **1107** (2018) 032010 doi:10.1088/1742-6596/1107/3/032010

INTRODUCTION

Thin films of fire-retardant intumescent coatings have become the main choice as passive fire protection for structural steel and other construction materials in buildings and other facilities. These coatings can expand up to one hundred times their initial thickness when degrade under thermal exposure, forming an insulating char layer over the underlying substrate. This limits the rate of temperature rise of the construction materials and prolongs their loading bearing capability when exposed to fire.
The expansion of intumescent coating when exposed to a heat source is a complex phenomenon that involves a complex mixture of gas, solid and liquid phases. Detailed degradation models based on approximate solutions of mass, momentum, and energy equations at the scale of an individual bubble were developed and found in qualitative agreement with experimental observations [1]. However no quantitative validation of such detailed modelling was reported to date. On the other hand, less sophisticated degradation models, based on the conservation equations of mass and energy for the intumescent material coupled to approximate sub-models for describing the swelling mechanism, have produced encouraging predictions in configurations relevant for fire safety applications [2-6]. These degradation models use generally a simplified decomposition mechanism for the intumescent material. Staggs et al. [5] used a two-step reaction mechanism with one reaction decomposing coating into gas and another reaction decomposing coating into char. Lautenberger and Fernandez-Pello [6] considered also a two-step reaction mechanism where the first reaction converts coating into char and gas and causes intumescence and the second converts char into ash and gases. A natural way to construct the decomposition mechanism is to start from the accepted assumption that intumescent materials are composed of three active components, namely an acid source, a blowing agent, and a charring agent, bound in a binder polymer [7] and to relate the phenomena observed experimentally to the degradation of these three species. The degradation of the acid source, of the blowing agent and of the charring agent are then associated to the melting stage with the formation of a viscoelastic phase, to the swelling stage where a large amount of gaseous products is formed and trapped in the coating to cause the molten matrix to swell and to the charring stage with the formation of the final charred solid residue, respectively. This picture naturally led to the development of a three-step reaction mechanism to model the degradation of the intumescent materials [2-4]:

$$S_1 \rightarrow G_1 \quad (1)$$

$$S_2 \rightarrow G_2 \quad (2)$$

$$S_3 \rightarrow \nu_C S_C + \nu_v G_3 \quad (3)$$

where S_1, S_2 S_3, and S_C refer to the acid source, the blowing agent, the charring material and the final charred residue, respectively. G_1, G_2, and G_3 are the volatile species released from the reactions.
The aim of this study is to assess the capability of the simple reaction mechanisms to model the decomposition of intumescent coating and, if necessary, to develop a more complex one. The corresponding kinetic parameters will be determined from TGA data obtained by considering several heating rates under inert and oxidative atmospheres [8]. This kind of approach was already applied to determine kinetic parameters of charring materials undergoing melting and thermal degradation [9, 10].

METHODS

Experiments

TGA experiments were conducted on ICWB, an intumescent coating material devoted to steel protection [8]. More details about ICWB can be found in Ref. [8]. The main components of ICWB are: i) a synthetic binder, Pentaerythritol ($C(CH_2\,OH)_4$), with a fusion temperature between 250°C and 300°C under both nitrogen and air. It also plays the role of carbon supplier, ii) A fire retardant additive, Ammonium Polyphosphate [NH_4 $PO_3]_n$, which also plays the role of acid source. Its decomposition starts around 200°C under both nitrogen and air, and iii) A blowing agent, Melamine ($C_3H_6N_6$) which also plays the role of nitrogen derivative. Its melting point is around 400°C under both nitrogen and air.

TGA experiments were carried out at different heating rates of 5, 10, 15, 20 and 25°C/min, over a temperature range from 20°C to 600°C [8]. The influence of the atmosphere was investigated by considering either air or nitrogen (at a flow rate of 50ml/min). Samples were ground to a fine powder using a mechanical grinder prior to the run to make 10mg homogeneous samples in each case. All temperatures are measured by a type K thermocouple.

Numerical model

Zero-dimension simulation were performed with Gpyro [11] version 0.8186 to examine thoroughly the thermal degradation of ICWB under multi-step reaction mechanisms. Gpyro solves transient equations of the mass and species evolution of a "lumped" particle having negligible gradients of temperature species as occurs in idealized thermal analysis experiments. For a given reaction, the reaction rate is given by:

$$\dot{\omega}''_i = Aexp(-E_a/RT)m_i^n \tag{4}$$

where A is the pre-exponential factor, E_a the activation energy, n the reaction order, and m_i the mass of the reactant. The Shuffled Complex Evolution (SCE) optimization algorithm [12], is used to optimize these kinetic parameters.

RESULTS AND DISCUSSIONS

Figure 1 represents the cumulative mass loss (CML) and the mass loss rate (MLR) evolutions as a function of temperature under inert atmosphere (N_2) for the different heating rates. The experimental evolutions are consistent with previous studies [2-4]. The thermal decomposition is mainly observed between 200 and 500°C and occurs with the different stages enumerated in the introduction: i) Stage 1 (up to 130 °C): drying process; ii) Stage 2 (from 130 °C to 180 °C): melting with no significant expansion and mass loss. Therefore, this stage cannot be considered as a reaction step; iii) Stage 3 (from 180°C to 300 °C): decomposition of the blowing agent with a rapid expansion of coating. The mass loss associated with this stage ranges is about 16%; iv) Stage 4 (from 300°C to 600°C): char formation where maximum rate of mass loss occurs. These results show that all the processes leading to the formation of the char layer occur below 600°C which is selected as upper temperature to calibrate the reaction mechanism. The proposed reaction step should be related to the decomposition mechanism of Ammonium Polyphosphate/ Pentaerythritol mixtures as described in Ref. [13]. This study showed that a relatively large amount of non-flammable volatile products (H_2O, NH_3) is released, the quantity depending on the initial decomposition.

The results reported in Ref. [8] showed that mass loss during the drying process (Stage 1) is about 5 %. As a consequence, this process, occurring up 130°C for ICWB, cannot be ignored as usually done in modeling studies. In the present study, it is introduced via a single reaction:

$$R1: Wet\ Char \rightarrow \nu_{H2O}H2O + (1-\nu_{H2O})\ Dry\ Char \tag{5}$$

The optimized kinetic parameters are provided in Table 1.

Table 1. Estimated kinetic parameters for dehydration.

Reagent	Product	Kinetic parameters values			
		$\text{Log}_\text{A}(\text{s}^{-1})$	$E_a(\text{kJ. mol}^{-1})$	n(-)	υ(-)
Wet-Char	Dry–Char	6.43	66.2	5	0.945

In the following, the focus will be only on reactions that occur after dehydration under different atmosphere conditions (N_2 and air), and R1 and the kinetic parameters reported in Table 1 will be used to model the drying process.

A two-step reaction mechanism was first considered for the degradation process (stage 3 to 4) of dry ICWB (not shown). This mechanism was found to overestimate significantly the mass loss during the swelling process. In addition, although it was able to capture the highest peak of MLR, it was found to miss the first

IOP Conf. Series: Journal of Physics: Conf. Series **1107** (2018) 032010 doi:10.1088/1742-6596/1107/3/032010

and third peak. Consequently, a two-step mechanism is not sufficient do describe the degradation process of dry ICWB and a three-step mechanism should be considered (stage 3 to 4):

$$R2: Dry\ Char \rightarrow \upsilon_1 \alpha - Char + (1 - \upsilon_1) Gas \tag{6}$$

$$R3: \alpha - Char \rightarrow \upsilon_2 \beta - Char + (1 - \upsilon_2) Gas \tag{7}$$

$$R4: \beta - Char \rightarrow \upsilon_3 \gamma - Char + (1 - \upsilon_3) Gas \tag{8}$$

The predicted evolutions of CML and MLR based on the three-step reaction mechanism (Eq. 6-8) are also plotted in Fig. 1 as a function of temperature. The related fitted parameters are given in Table 2. Only the CML for a heating rate of 20°C/min was reported to make the figure readable, the level of agreement between numerical and experimental CML being consistent for the other heating rates. Results in Figs 1a show that CML is well reproduced by the numerical model despite an underestimation for temperature higher than 420°C. Figure 1b shows that that the MLR is well predicted during the blowing (Stage 3) stage. In particular, the first and second peaks are accurately reproduced. However, the model ignores completely the final peak which explains the above-mentioned discrepancies of the CML. As a consequence, the results in Fig. 1 demonstrate that although it provides a quite accurate description of the degradation process, the three-step mechanism is not able to reproduce all the details of the degradation mechanism, especially for temperature higher than about 400°C.

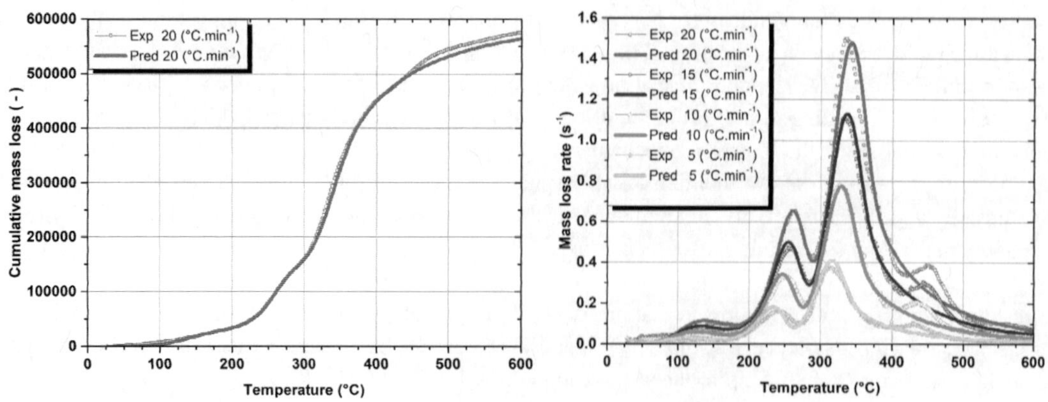

(a) (b)

Fig. 1: Evolution of the cumulative mass loss (a) and mass loss rate (b) of three-step reaction mechanism for the degradation process under nitrogen for 4 different heating rates.

Table 2. Estimated kinetic parameters of three-step reaction mechanism for the degradation process under nitrogen.

Atmosphere	Reagent	Product	Kinetic parameters values			
			$\text{Log}_A(s^{-1})$	$E_a(\text{kJ.mol}^{-1})$	n(-)	υ(-)
N_2	Dry-Char	α–Char	9	110	1	0.877
N_2	α–Char	β–Char	11.5	154	1.18	0.750
N_2	β–Char	γ–Char	10.6	152	5	0.623

In order to capture these details, a 4-step reaction mechanism is considered. This mechanism extends the previous one (Eqs. 6-8) by adding another reaction (R5):

$$R5: \gamma - Char \rightarrow \upsilon_4 \varepsilon - Char + (1 - \upsilon_4) Gas \tag{9}$$

All the reaction parameters are summarized in Table 3. It should be pointed out, that all the parameters for reactions R2 to R4 were recalibrated. Figure 2 compare the experimental and predicted CML and MLR.

Results show clearly that the 4-reaction mechanism capture all the details of the degradation process with both CML and MLR being almost perfectly reproduced. These results suggest that a five-step mechanism should be used to model the thermal degradation of intumescent coating with one reaction for the drying process and four reactions for the degradation of the dry material.

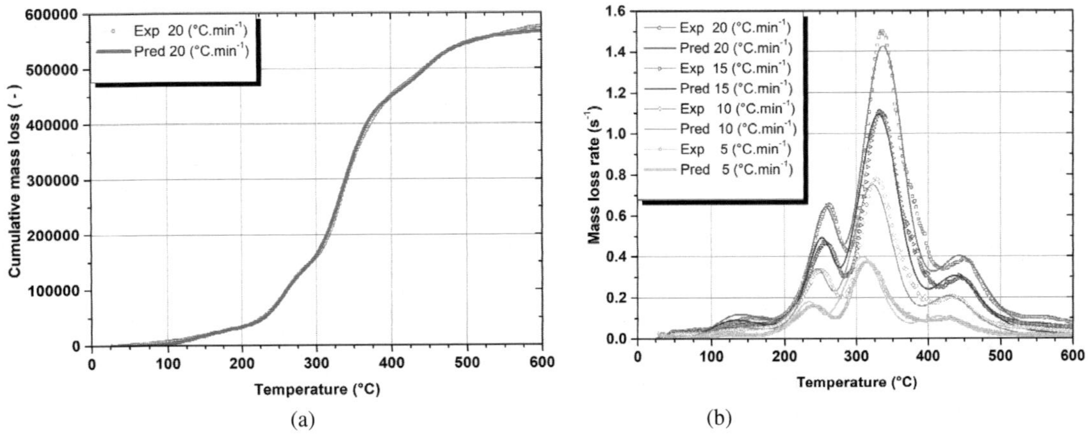

Fig. 2: Evolution of the cumulative mass loss (a) and mass loss rate (b) of four-step mechanism under nitrogen for 4 different heating rates.

Table 3. Estimated kinetic parameters of four-step reaction mechanism under nitrogen.

Atmosphere	Reagent	Product	Kinetic parameters values			
			$Log_A(s^{-1})$	$E_a(kJ.mol^{-1})$	n(-)	υ(-)
N_2	Dry-Char	α–Char	10.7	125	1.1	0.889
N_2	α–Char	β–Char	10.1	139	1.6	0.698
N_2	β–Char	γ–Char	14.1	123	4.96	0.851
N_2	γ–Char	ε–Char	11.1	178	1.86	0.813

CONCLUSIONS

This study investigates the capability of several multistep reaction mechanisms to describe the thermal degradation of ICWB, an intumescent material, exposed to different heating rates under both inert atmosphere (N_2). The following conclusions can be drawn:

1) The drying process, usually neglected in numerical modeling of intumescent coatings, cannot be disregarded since it can be related to mass loss of about 5% and is strongly endothermic. Results reported in this study shows that it can be described by a single reaction.

2) A two-step reaction mechanism is unable to predict accurately the degradation of the dry intumescent coating, leading to significant discrepancies on the mass loss and mass loss rate.

3) A three-step reaction mechanism, usually used in literature, improves significantly the two-step reaction mechanism. It reproduces accurately the mass loss and mass loss rate up to about 400°C. However, it is not able to describe accurately the degradation process for higher temperature.

4) Model results show that a four-step reaction mechanism is required to describe capture accurately all the trends of the degradation process of dry intumescent materials.

Only the results and discussions relative to inert atmosphere have been presented in this article. The effects of oxygen have been discussed in Ref. [8], showing that they become significant at temperature higher than 400°C. Simulations were also run under air showing that one additional reaction for oxidation is required to update the four-step reaction mechanism provided by Eqs. 6-9 for oxidative atmosphere, leading to a five-step reaction mechanism.

In the current study, two concentrations of oxygen have been studied (zero and 21% of oxygen). Regrettably, these conditions are not always available in the real case of fire. It is then necessary to complete the present investigation, with an investigation of the thermal degradation under different intermediate concentrations of oxygen [14].

REFERENCES

[1] Butler, K.M., Baum, H.R. and Kashiwagi, M., "Three-Dimensional Behavior Modeling of Intumescent Behavior in Fires" *Fire Safety Science -- Proceedings of the Fifth International Symposium*, International Association for Fire Safety Science, 1997, pp. 523-534.

[2] Di Blasi, C. and Branca, C. (2001) Mathematical Model for the Nonsteady Decomposition of Intumescent Coatings, *AIChE Journal* 47(10):2359-2370, http://doi:10.1002/aic.690471020.

[3] Di Blasi, C.(2004) Modeling the Effects of High Radiative Heat Fluxes on Intumescent Material Decomposition, *Journal of Analytic Applied Pyrolysis* 71:721–737, https://doi.org/10.1016/j.jaap.2003.10.003.

[4] Zhang, Y., Wang, Y.C., Bailey, C.G. and Taylor, A.P. (2012) Global Modelling of Fire Protection Performance of Intumescent Coating under Different Cone Calorimeter Heating Conditions, *Fire Safety Journal* 50:51–62, https://doi.org/10.1016/j.firesaf.2012.02.004.

[5] Staggs, J.E.J., Crewe, R.J. and Butler R. (2012) A Theoretical and Experimental Investigation of Intumescent Behaviour in Protective Coatings for Structural Steel, *Chemical Engineering Science* 71:239–251, https://doi.org/10.1016/j.ces.2011.12.010.

[6] Lautenberger, C. and Fernandez-Pello, C. (2009) Generalized Pyrolysis Model for Combustible Solids, *Fire Safety Journal* 44:819-839, https://doi.org/10.1016/j.firesaf.2009.03.011

[7] Jimenez, M., Duquesne, S. and Bourbigot, S. (2006) Intumescent fire protective coating: toward a better understanding of their mechanism of action, *Thermochimica Acta* 449:16–26, https://doi.org/10.1016/j.tca.2006.07.008.

[8] Fateh, T., Guillaume, E. and Joseph, P. (2017) An experimental study of the thermal performance of a novel intumescent fire protection coating, *Fire Safety Journal* 92:132-141, https://doi.org/10.1016/j.firesaf.2017.05.021.

[9] Li J. and Stoliarov S.I. (2014) Measurement of kinetics and thermodynamics of the thermal degradation for charring polymers, *Polymer Degradation and Stability* 106:2-15, http://dx.doi.org/10.1016/j.polymdegradstab.2013.09.022

[10] Li J., Gong J. and Stoliarov SI. (2015) Development of pyrolysis models for charring polymers, *Polymer Degradation and Stability* 115: 138-152, https://doi.org/10.1016/j.polymdegradstab.2015.03.003.

[11] Lautenberger, C. (2007) A Generalized Pyrolysis Model for Combustible Solids (PhD. thesis) University of California, Berkeley.

[12] Ding, Y., Wang, C., Chaos, M., Chen, R. and Lu, S. (2016) Estimation of beech pyrolysis kinetic parameters by Shuffled Complex Evolution, *Bioresource Technology* 200:658–665, http://dx.doi.org/10.1016/j.biortech.2015.10.082.

[13] Camino, G., Costa, L., Trossarelli, L., Costanzi, F. and Pagliari, A. (1984) Study of the Mechanism of Intumescence in Fire Retardant Polymers: Part VI--Mechanism of Ester Formation in Ammonium Polyphosphate-Pentaerythritol Mixtures, *Polymer Degradation and Stability* 12: 213-228.

[14] Fateh, T., Richard, F., Rogaume, T. and Joseph, P. (2016) Experimental and modelling studies on the kinetics and mechanisms of thermal degradation of polymethyl methacrylate in nitrogen and air, *Journal of Analytical and Applied Pyrolysis* 120: 423-433. https://doi.org/10.1016/j.jaap.2016.06.014

Thermal oxidative decomposition estimation combining TGA and DSC as optimization targets for PMMA

Alain ALONSO IPIÑA[1], Mariano LÁZARO URRUTIA[1], David LÁZARO URRUTIA[1] and Daniel ALVEAR PORTILLA[1]

[1]*GIDAI Group, University of Cantabria, Spain*

alonsoia@unican.es

ABSTRACT

Thermal analysis techniques play a key role to determine and characterize solid phase thermal decomposition. In this sense, Simultaneous Thermal Analysis (STA, i.e. TGA and DSC tests carried out simultaneously) are widely employed, since it provides information about how mass is lost and energy released while the temperature of the sample increases. Fire computer models combined with methods numerical methods are widely used to represent the results from tests and to achieve the values of the kinetic and thermal parameters.

Previous works looked forward achieving those parameters using, as unique optimization target, the mass loss curve (TGA) or its derivative (DTGA). As the study of heat release rate is a decisive element to characterize the material properly, most recent works were adding additional measures. These extra measurements concern the heat transfer and the energy required or released during temperature programmed heating, such as heat rate release, heat of gasification, or the surface temperatures of the samples. The information about the energy is provided by the Differential Scanning Calorimetry curve (DSC).

Despite of the employment of the information provided by the DSC, this information usually is not used as a target to approach the DSC simulated curve to the experimental one as TGA does. Based on the lack of use of the DSC curve as numerical approaching process to set the kinetic properties, we decide to explore the possibility of adding this as a new target in the process. Therefore, kinetic and thermal properties might be achieved fitting experimental and simulated curves simultaneously, which should allow us to take into account the decomposition process and their energy released.

Results obtained in the present work reveal the major challenge of getting a set of parameters, which can fit DSC curve. The level of accuracy reached when only TGA is utilized as target to approach is higher than the level of accuracy of DSC curve. This fact makes increase the value of the errors when both curves are used as targets to approach. In other words, an approach to both curves simultaneously cannot be directly made. With this consideration in mind, this paper proposes an alternative methodology in order to fit TGA curve considering the optimization of the DSC curve.

The methodology proposed in the present work is applied to the analysis of poly(methyl methacrylate) (PMMA).

KEYWORDS:

Thermal decomposition, solid phase chemistry, numerical methods, optimization methods, TGA, DSC, CFD, PMMA.

INTRODUCTION

For fire computer modelling purposes, most of the thermal and kinetic properties cannot be obtained directly from the experimental tests and in order to find the values of these properties it is necessary to make an approach by applying different methods.

Previous works, performed using numerical optimization methods, have claimed their effectiveness to obtain the kinetic and thermal parameters of different materials. The work of Lautenberger and Fernandez-Pello [1] compiles briefly some contributions to this area classified by type of method, scale and laboratory apparatus used and tested material.

Focus on numerical methods applied to thermal analysis, early works employed the variation of TGA exclusively as a target to approach. We can highlight the works of Rein et al. [2], Capote et al. [3] or Hasalová et al. [4].

In order to improve the characterization of the material assessing the energy released, some works were adding this information. There are several ways to in include this feature e.g. analysing the heat of gasification which defines the ignition resistance and fire response [5], employing the DSC curve to adjust a numerical model [6] [7], determining the heat transfer basis among other curves in DSC [8],defining the heat transfer or setting the limits of the heat of pyrolysis of the different reactions during pyrolysis process [9], obtaining the values of the enthalpy values for oxidation reactions [10], identifying exothermic or endothermic peaks [11] or determining the enthalpy during the oxidation process [12].

These previous works [6-9] employed the information provided by the DSC curve with the aim to achieve a more comprehensive set of properties that model the thermal decomposition of the material. However, we observed that DSC curve is not used as a target to approach to the experimental one. Based on the lack of treatment of the DSC curve in this way, our work explores adding the DSC curve as a new target to approach, using numerical methods. Therefore, kinetic and thermal parameters may well be achieved fitting experimental and simulated TGA and DSC curves simultaneously. In that way, this methodology should allow us to take into account the decomposition process and their energy released. The material analysed in this work is the polymer poly(methyl methacrylate) (PMMA).

THERMAL DECOMPOSITION MODEL

Thermal decomposition and pyrolysis models describe how the material decomposes with the temperature. This work uses the FDS pyrolysis model software [13]. This model is based on the Arrhenius equation Eq. 1 to represent a material that undergoes one or more decomposition reactions. The Arrhenius equation provides the relationship between the reaction rate (r_{ij}) (1/s) and the temperature (T) (K).

$$r_{ij} = \frac{d\propto}{dt} = A_{ij} f(\propto) e^{\left(-\frac{Ea_{ij}}{RT_s}\right)} \quad (1)$$

The term A is the pre-exponential energy (1/s), E_a is the activation energy (J/mol), ($f(\propto)$) is the reaction model, coefficient α is the extent of conversion.

The thermal decomposition process usually takes place through multiple reactions. A scheme of these reactions needs to be defined in the model to represent properly the thermal decomposition. PMMA is made by a macromolecule (ethylene and one methyl group replacing one hydrogen atom), and a second hydrogen atom is replaced by an acetyl group, creating the basic monomer unit [14]. Thermal decomposition of PMMA can be simplified and represented with two main reactions: main chain random scission, followed by and the homolytic scission of the methoxycarbonyl side group [15]. Considering this, a reaction scheme made of two consecutive reactions was used to model the thermal decomposition. Eq. 2 and Eq. 3 show the scheme of reaction employed:

$$A + O_2 + Q \xrightarrow{Reaction\ 1} v_p^1\ B + v_f^1\ Gas \quad (2)$$

$$B + O_2 + Q \xrightarrow{Reaction\ 2} v_f^2\ Residue + v_r^2\ Gas \quad (3)$$

IOP Conf. Series: Journal of Physics: Conf. Series **1107** (2018) 032011 doi:10.1088/1742-6596/1107/3/032011

To sum up the previous reaction scheme, material A reacts (reaction 1) (Eq. 2) and produces a fictitious amount (v_p^1) of material B and releases an amount of fuel gas (v_f^1). Material B will be the reactive material of reaction 2. Material B reacts (reaction 2) (Eq. 3) and produces an amount of residue (v_f^4) and an amount of fuel gas (v_r^4).

OPTIMIZATION ALGORITHM

This work applies a numerical method to achieve the suitable values of the parameters that characterize the decomposition process, known as Shuffled Complex Evolution (SCE) [16].

Once the reaction scheme is established (Eq. 2, 3) it is necessary to determine which parameters are required to model the decomposition process. In accordance with Eq. 1 the triplet kinetic (A, E_a, n) are needed. Other parameters such as density (ρ), specific heat (C_p), conductivity (k), heat of reaction (H_r) and emissivity (ε) were included in order to determine the effect of each one in the approaching process. The reduced mass and size of the samples may lead to discard the influence of parameters such as C_p, k or ε. Nevertheless, the works of Lázaro et al. [17] and Comesaña et al. [18] exposed the effects of the boundary conditions, particularly the variation of the sample mass and the effect of the thermal lag when STA test is performance. For this reason, C_p, k and ε were also considered. Density (ρ) was included as a factor (v_p^i) i.e. the amount of material that does not react and will become in reactive mass for the next reaction. Density has to be taken into account in this way since the volume of the simulated sample does not change.

SCE algorithm provides to FDS the values of the INPUT parameters that FDS will use to simulate the decomposition process and obtain the TGA and DSC curves. SCE will converge the value of the function towards a global minimum. The function to minimize its value is the Evaluation error function Eq. (6), which assess the error produced in the approach of TGA and DSC signals.

$$Evaluation\ error\ function = \sqrt{\left(\alpha \cdot X_{TG_{error}}\right)^2 + \left(\beta \cdot X_{DSC_{error}}\right)^2} \qquad (4)$$

The term (x_i) is the mathematical operator used to evaluate the difference between experimental ($x_{i,exp}$) and simulated curves ($x_{i,sim}$). Eq. 5 show this operator:

$$X_{error} = \frac{\sqrt{\sum_i (x_{i,exp} - x_{i,sim})^2}}{\overline{x}_{exp}} \qquad (5)$$

The coefficients α and β are used to give relative weight to both signals in overall error, i.e. whether we assume the DSC curve is more relevance or not, and figure out its influence on the global error. Coefficients α and β are linked such as α + β = 1. In that way, the more influence of DSC curve, the less influence of TGA curve.

SCE algorithm requires an initial range of values for the parameters to start the process. During the process, the algorithm never goes from those values range. The values are collected from the bibliography, increasing and decreasing them a percentage in order to provide a certain degree of flexibility.

Table 1. Range of values of the parameters to start the optimization process.

Variable	**Activation energy (A) (1/s)**	**Pre-exponential factor (E_a) (kJ/kmol)**	**Reaction order (n) (-)**	**Specific heat (C_p) (kJ/kg K)**	**Conductivity (k) (W/m K)**	**Emissivity (ε) (-)**	**Heat of reaction (H_r) (kJ/kg)**	**(v_p)(-)**
Upper value	1.00e+15	200000	5.00	3.00	1.50	1.00	-1000	0.99
Lower value	1.00e+9	90000	0.50	0.20	0.01	0.50	1200	0.01

RESULTS AND DISCUSSION

To begin with, a STA test according the standard ASTM-E1131 was executed with an oxygen concentration of 21%. The heating rate was 30 K/min from 50 °C to 800 °C. Once TGA and DSC curves were obtained, we use several strategies to fit both curves simultaneously. All laboratory tests were repeated 3 times to ensure the repeatability.

The first approach assessed the TGA curve exclusively, i.e. coefficients α and β from Eq. 4 have a value of 1 and 0, respectively. This approach was made in order to test the validity of the scheme reaction model proposed in Eq. 2 and Eq. 3. Results obtained validate the scheme reaction proposed since an accurate TGA simulated curve was obtained.

To check the influence of DSC over the approaching process, three distinct attempts were carried out: #1, the coefficients had a value of $\alpha = 0.5$ and $\beta = 0.5$; #2, $\alpha = 0.7$ and $\beta = 0.3$, and #3, $\alpha = 0.9$ and $\beta = 0.1$. Attempt #1 gives same relevance in the error equation (Eq. 4).

Attempt #1 gives same relevance in the error equation (Eq. 4). The results revealed that the TGA error increased and the DSC error did not improved as it was supposed. These results showed that employment of the DSC curve in numerical approaching process could be not useful as it was expected. Attempts #2 and #3 were executed to verify that in the approaching process, the more influence of the DSC curve, the higher values of the total error.

Due to the effect of the reaction rate (r_{ij}) over both curves, the only parameter in Eq. 6 that allows modifying the energy released or aborted by the sample (DSC curve) without changing the TGA curve is the heat of reaction ($H_{r,ij}$). Hence, a new attempt (#4) was executed changing exclusively the heat of reaction of each reaction, once the TGA curve was fitted properly, as in in attempt #1 had been done.

$$\dot{q}'''_{s,c}(x) = -\rho_0 \sum_{i=1}^{N_m} \sum_{j=1}^{N_{r,i}} r_{i,j}(x) H_{r,ij} \quad (6)$$

This attempt #4 reached more accurate results for both curves, than the previous attempts. Next Fig. 1 shows the results obtained using this approaching strategy.

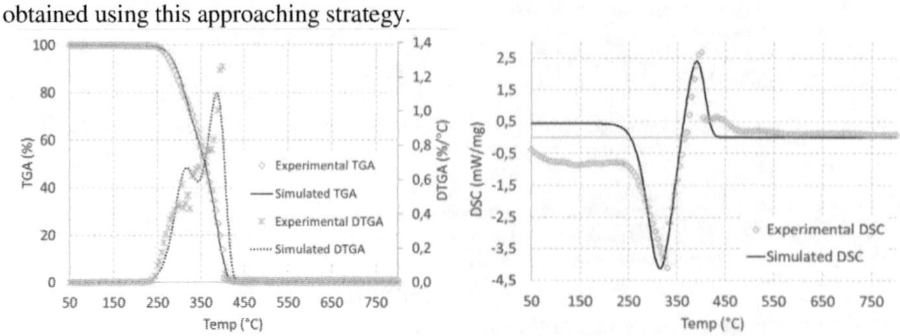

Fig. 1. STA approaching results (attempt #4) (heating rate 30 K/min): left) TGA and DSC curves, right) DSC curves.

As this implies a different approach, Eq. 4 is not expected to assess the errors, because TGA simulated curve is obtained previously and then DSC curve is adjusted. Therefore, Eq. 5 was used to evaluate the errors by replacing in each case the value of mass for the TGA or energy for the DSC. Next table 2 collects the errors obtained.

Table 2. Errors made by the approaching attempt No.4.

	Error in TGA	Error in DTGA	Error in DSC
Attempt #4	0.351	14.933	25.272

Fig 1. and Table 2 reveal the errors obtained for the curves related to the variation of the mass have less values, so the approach of the TGA or DTGA curves is more accurate than the approach of the DSC. The errors of the Table 2 are the best errors (lowest values) obtained for all approaching attempts explained in this section.

To validate the set of parameters obtained in the approaching process, we employed this set to simulate two new decomposition process under different heating rates (10 K/min and 50 K/min). Then the simulated results were compared with their corresponding STA experimental tests. Next Fig 2. shows the comparison for a heating rate of 10 K/min and Fig. 3 for 50 K/min. Table 3 displays the errors for both heating rates.

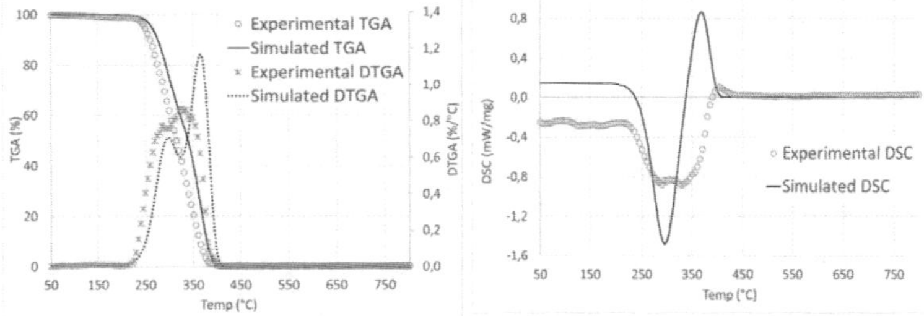

Fig. 2 STA validation results (heating rate 10 K/min): left) TGA and DTGA curves, right) DSC curves.

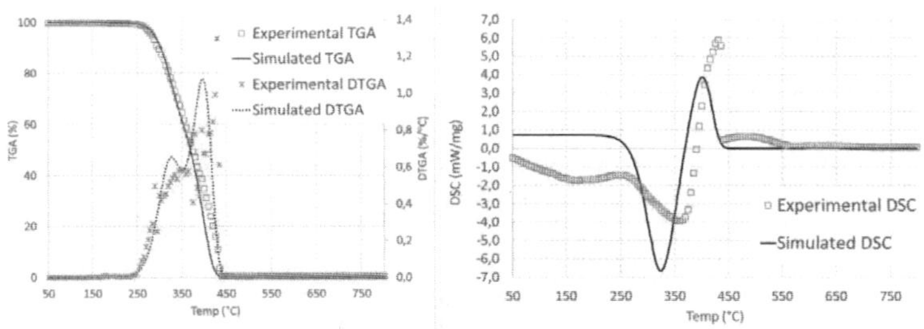

Fig. 3. STA validation results (heating rate 50 K/min): left) TGA and DTGA curves, right) DSC curves.

Table 3. Errors for the validation results.

Heating rate (K/min)	Error in TGA	Error in DTGA	Error in DSC
10	1.962	23.924	25.889
50	0.892	16.540	44.107

In general, it would appear that the set of parameters obtained is valid to simulate other heating rates, especially when TGA curves are compared. However, the variability of the values of the DSC curve makes the approach to these curves has less accuracy. The errors produced are higher when the set is applied at 10 K/min.

Next Table 4 collects the values of the parameters obtained in the best approaching (attempt #4)

Table 4. Parameters obtained by the approaching process

Parameter	Material A	Material B	Residue
Emissivity (ε)	1.00	1.00	1.00
Conductivity (k)	0.94	0.21	0.04
Specific heat (C_p)	0.87	0.98	0.38
Reaction order (n)	1.92	0.87	
Activation energy $Log_{10}(A)$	11.66	11.75	
Pre-exponential factor (E_a)	149459	169647	
Heat of reaction (H_r)	-643	500	
v_p	0.55	0.00	

CONCLUSIONS

As far as including energy in optimization process is concerned, DSC curve was considered as objective to optimize simultaneously with TGA curve. According to the results, DSC simulated curve cannot be fitted as the same quantity as TGA curve since DSC approaching makes unfit the TGA approaching. The more influence of DSC is taken into account in the numerical method and in the Eq. (5) the more error is produced in the approaching of the TGA curves. To obtain the DSC curve without unfit the TGA one, the best method is to achieve first a proper adjustment to the TGA curve, without considering the DSC curve, and then adjust the values of the heat of reaction of each reaction.

This reached such an accurate approximation to the TGA curve from other heating rates are used. These findings cannot be extrapolated to the comparison of DSC curves, and more investigation is need regarding this issue.

ACKNOWLEDGMENTS

The authors would like to express their thanks to the Nuclear Safety Council for the cooperation and co-financing of the project "Simulation of fires in nuclear power plants" and the Spanish Ministry of Economy and Competitiveness for the PYRODESIGN Project grant, Ref.: BIA2012-37890, financed jointly by ERDF.

REFERENCES

[1] Lautenberger, C. & Fernandez-Pello, A., 2011. *Optimization Algorithms for Material Pyrolysis Property Estimation*. Fire Safety Science, 10, pp.751–764. http://dx.doi.org/10.3801/iaffs.fss.10-751.

[2] Rein G, Lautenberger C, Fernandez-Pello A, Torero J, Urban D. *Application of genetic algorithms and thermogravimetry to determine the kinetics of polyurethane foam in smoldering combustion*. Combustion and Flame [Internet]. Elsevier BV; 2006 Jul;146(1-2):95–108. http://dx.doi.org/10.1016/j.combustflame.2006.04.013

[3] Capote J, Alvear D, Abreu O, Lazaro M, Puente E. *Pyrolysis Characterization of a Lineal Low Density Polyethylene*. Fire Safety Science. International Association for Fire Safety Science; 2011;10:877–88. http://dx.doi.org/10.3801/iafss.fss.10-877

[4] Hasalová L, Ira J, Jahoda M. *Practical observations on the use of Shuffled Complex Evolution (SCE) algorithm for kinetic parameters estimation in pyrolysis modeling*. Fire Safety Journal; 2016 ; 80:71–82. http://dx.doi.org/10.1016/j.firesaf.2016.01.007

[5] Stoliarov SI, Walters RN. *Determination of the heats of gasification of polymers using differential scanning calorimetry*. Polymer Degradation and Stability; 2008 ; 93(2):422–7. http://dx.doi.org/10.1016/j.polymdegradstab.2007.11.022

[6] Li J, Stoliarov SI. *Measurement of kinetics and thermodynamics of the thermal degradation for non-charring polymers*. Combustion and Flame, 2013;160(7):1287–97. http://dx.doi.org/10.1016/j.combustflame.2013.02.012

[7] Li J, Stoliarov SI. *Measurement of kinetics and thermodynamics of the thermal degradation for charring polymers*. Polymer Degradation and Stability; 2014 ;106:2–15. http://dx.doi.org/10.1016/j.polymdegradstab.2013.09.022

[8] McKinnon, MB, Stoliarov, SI, Witkowski A. *Development of a pyrolysis model for corrugated cardboard*. Combustion and Flame; 2013: 160(11): 2595-2607. https://doi.org/10.1016/j.combustflame.2013.06.001

[9] Li KY, Pau DSW, Hou YN, Ji J. *Modeling Pyrolysis of Charring Materials: Determining Kinetic Properties and Heat of Pyrolysis of Medium Density Fiberboard*. Industrial & Engineering Chemistry Research; 2013, 13;53(1):141–9. http://dx.doi.org/10.1021/ie402905z

[10] Huang S, Sheng JJ. *An innovative method to build a comprehensive kinetic model for air injection using TGA/DSC experiments*. Fuel; 2017;210:98–106. http://dx.doi.org/10.1016/j.fuel.2017.08.048

[11] Hsieh W, Cheng W, Chen L, Lin S. *Non-isothermal dehydration kinetic study of aspartame hemihydrate using DSC, TGA and DSC-FTIR microspectroscopy*. Asian Journal of Pharmaceutical Sciences; 2017, http://dx.doi.org/10.1016/j.ajps.2017.12.001

[12] Pu W, Pang S, Jia H. *Using DSC/TG/DTA techniques to re-evaluate the effect of clays on crude oil oxidation kinetics*. Journal of Petroleum Science and Engineering; 2015;134:123–30. http://dx.doi.org/10.1016/j.petrol.2015.07.014

[13] McGrattan K, Hostikka S, Floyd J, Weinschenk C, McDermott R, Overholt K. *Fire dynamics simulator. User's guide*. NIST special publication 1019 6th Edition, 2017. https://doi.org/10.6028/nist.sp.1018e6

[14] Cardarelli, F. *Materials Handbook - A concise desktop reference*. Materials & Design. 2001, 22(3), p.237. http://dx.doi.org/10.1016/s0261-3069(00)00075-3.

[15] Costache MC, Wang D, Heidecker MJ, Manias E, Wilkie CA. *The thermal degradation of poly(methyl methacrylate) nanocomposites with montmorillonite, layered double hydroxides and carbon nanotubes*. Polymers for Advanced Technologies. 2006; 17 (4), 272–80. http://dx.doi.org/10.1002/pat.697

[16] Duan Q, Gupta VK, Sorooshian S. *A shuffled complex evolution approach for effective and efficient global optimization*. J OPTIMIZ THEORY APP, 1993. 76:501-521. https://doi.org/10.1007/bf00939380

[17] Lázaro D, Lázaro M, Alonso A, Alvear D. *Effects of boundary conditions variation on chemical reactions during STA measurements*. Proceedings of the International Conference Research and advanced Technology in fire Safety. 2017

[18] Comesaña R, Gómez MA, Álvarez MA, Eguía P. *Thermal lag analysis on a simulated TGA-DSC device*. Thermochimica Acta. 2012,547:13–21. http://dx.doi.org/10.1016/j.tca.2012.08.008

IOP Conf. Series: Journal of Physics: Conf. Series **1107** (2018) 032012 doi:10.1088/1742-6596/1107/3/032012

Modeling heat transfers across a silicone-based intumescent coating

Tatenda NYAZIKA[1], Maude JIMENEZ[1], Fabienne SAMYN[1] and Serge BOURBIGOT[1]

[1] *Univ. Lille, CNRS, ENSCL, UMR 8207, Unité Matériaux et Transformations (UMET), France*

tatenda.nyazika@ed.univ-lille1.fr

ABSTRACT

Silicone-based intumescent coatings (SIBC) containing expandable graphite (EG) have been designed and studied in our laboratory and good physical barrier properties have been reported for both cellulosic and hydrocarbon fire scenarios. The aim of this work is to develop a preliminary 2D numerical model of this coating behavior during a fire, which is, to the best of our knowledge, not reported yet in the literature for such coating.

The fire performance of SIBC containing EG was evaluated in pure radiative heating conduction using a cone calorimeter heating element at 50kW/m^2. The intumescent phenomenon of the SBIC containing EG was considered as a heat transfer problem with a moving boundary. Expansion was assumed to be homogenous and occurring in one dimension and a 2D numerical heat transfer model was developed, taking into account the moving boundary using the Arbitrary Langrangian-Eulerian Method (ALE) implemented in COMSOL Multiphysics V.5.3. A water cooled sample holder with a well-defined temperature condition was used to avoid the assumption of the adiabatic conduction at the backside of the sample. The model was consistent with experimental results and it was able to reasonably predict the temperature profiles. However, it did not capture the temperature increase at times higher than 450s because of cracking in the silicone-based intumescent coating.

KEYWORDS: *modeling, heat transfer, intumescence, silicone-based coating*

NOMENCLATURE LISTING

ρ	density (kg/m^3)
c_p	specific heat capacity (J/kgK)
k	thermal conductivity (W/mk)
T_r	average cone surface temperature (K)
T_a	ambient temperature (K)
h_{ho}	horizontal heat transfer coefficient (W/m^{2}k)
h_{ve}	vertical heat transfer coefficient (W/m^{2}k)
F	view factor (-)
T	Temperature (K)
subscripts	
c	copper plate
cs	calcium silicate insulation block
$intu$	intumescent coating
s	steel plate
$pyro$	pyroceram
max	maximum
Greek	
ε	emissivity (-)
σ	Stefan Boltzmann constant (Wm^{-2}K^{-4})
λ	Thermal diffusivity (m^2/s)
α	Conversion rate

IOP Conf. Series: Journal of Physics: Conf. Series **1107** (2018) 032012 doi:10.1088/1742-6596/1107/3/032012

INTRODUCTION

The strength and stiffness of steel are decreased when this material is exposed to high temperatures, which can lead to a disaster [1]. Intumescent coatings are often used as efficient passive fire protective methods on metallic structures. When exposed to heat, they begin to swell and expand to form an insulative char layer which limits mass and heat transfers between the flame and the substrate [2]. If the efficiency of such a coating can be experimentally measured, the modeling of its behavior is challenging. In the literature, numerical models have been reported to predict the performance of intumescent materials. Complex models [3-5] have been developed for the char layer formation but these models come with limitations. In the approach presented by Di Blasi and Branca [4], the gas-producing step is considered as the most important step in the intumescence phenomenon process, which is only suitable for intumescence resulting from the reaction between an acid source, a carbon source and a blowing agent.

This work aims at modeling the performance of a silicone based intumescent coating (SBIC) containing expandable graphite (EG) developed in our laboratory using a phenomenological approach. SBIC exhibits good physical barrier properties in the case of both cellulosic and hydrocarbon fire scenarios thanks to the fast formation of an expanded structure having low thermal conductivity [6-8]. With EG, the expanded structure is obtained through the decomposition of the intercalation compounds inserted between the carbon layers of graphite which cause graphite exfoliation. Most intumescent paints currently on the market are epoxy or water based paints, the models are consequently exclusively based on these systems [9, 10] . To the best of our knowledge, models for such SBIC are not yet reported in literature and it is noteworthy that the approach considering the gas-producing step is not applicable in our case.

In this study, the fire behavior of SBIC exposed to an external pure radiative heat flux of 50kW/m^2 is investigated. A 2D numerical heat transfer model taking into account the moving boundary using the Arbitrary Lagrangian-Eulerian Method (ALE) [11, 12] implemented in COMSOL Multiphysics V.5.3. is proposed. The thermal conductivity versus temperature, the expansion versus time and the kinetic parameters have been characterized and used as inputs for the model. The results are then discussed and a comparison between the experimental and predicted data is done.

EXPERIMENTAL

Materials

The silicone-based coating (formulation details found in a previous paper [14]) was applied on 100 x 100 x 3 mm^3 steel plates after proper preparation of the surface including degreasing and sand blasting steps. Teflon moulds were used to obtain a repeatable coating thickness of 4mm. The coated plates were left for 24h at room temperature and then cured in an oven at 40°C for 6h as a customized procedure.

Fire Test

A cone calorimeter heating element was used to provide a constant incident radiative heat flux of 50kW/m^2 on the surface of the sample. The coated steel plate was placed on top of a 100 x 100 x 50 mm^3 calcium silicate (calsil) block which was then placed on a 100 x 100 x 10 mm^3 copper plate. This multi-layer system (coated steel plate-calsil-copper plate) was mounted on a water-cooled sample holder as shown in Fig 1. The water-cooled sample holder temperature was maintained at 40°C using a cooling system (chiller). Therefore, a constant temperature boundary condition was imposed on the experimental set-up which is implemented in the numerical heat transfer model as a Dirichlet boundary condition.

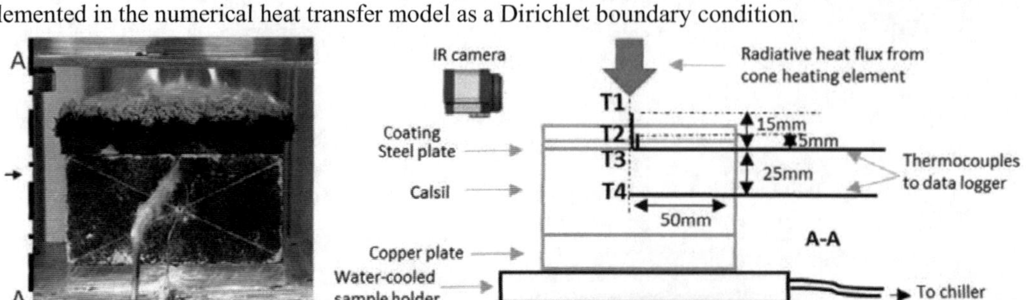

Fig. 1. Experimental set-up.

During the experiment, K-type thermocouples were used to record the temperature as a function of time at different locations as shown in Fig. 1. T2 was embedded in the SBIC and T3 was welded at the backside of the steel plate. This set-up allowed evaluating the temperature gradient across the system. The mathematical model described in the next section is based on this set-up.

2D MODEL - MATHEMATICAL FORMULATION

The 'intumescence' phenomenon was considered as a heat transfer problem with a moving boundary in our mathematical formulation. To incorporate the expansion of the SBIC, the solver assembles the discretized model on a deformed mesh using the ALE description [11, 12]. The complete mesh comprised of 4900 quadrilateral elements with 900 quadrilateral elements in the intumescent layer and a maximum element size of 0.002m and a minimum element size of 7.5E-6m. Expansion was assumed to be homogeneous and occurring in one dimension. As a first approach, the goal was not to model all the phenomena involved in intumescence but to consider the fundamental parameters affecting its thermal insulation performance i.e. thermal conductivity and its expansion [6-8]. The following 2D numerical model was built up: four distinct layers were selected i.e. (i) the copper plate layer (y ∈]0, c[), (ii) the calsil layer (y ∈]c, cs[), (iii) the steel plate layer (y ∈]cs, s[) and (iv) the SIBC layer (y ∈]s, l(t)[). Perfect thermal contact was assumed between the layers. Heat is transferred by pure conduction and the boundaries are fixed throughout the test in the first three layers, Eq. 1.

$$\rho_i c_{pi} \frac{\partial T}{\partial t} = \nabla.(k_i \nabla T) \qquad \text{where } i\text{=c,cs,s} \tag{1}$$

In the fourth layer, i.e. the intumescent layer, the exposed surface moves upwards as the coating expands while the interface steel/coating is fixed (Eq.2), l(t) represents the expansion of the intumescent coating in the y-direction as a function of time.

$$y \in]s, l(t)[, \rho_{intu} c_{pintu} \frac{\partial T}{\partial t} = \nabla.(k_{intu} \nabla T) - \rho_{intu} H_r A e^{-\frac{E_a}{RT}} (1-\alpha)^n \tag{2}$$

In Eq. 1 and Eq. 2, the term on the left hand side (LHS) represents the rate of change of internal energy of the material and the first term on the right hand side (RHS) indicates the heat transfer by conduction. The last term in Eq. 2 accounts for internal heat energy released due to the endothermic decomposition process of the coating which is modelled assuming a nth order reaction. To solve the governing equations, initial and boundary conditions need to be specified.

Initial and boundary conditions

The surface of the intumescent coating is exposed to radiative heat flux from the cone calorimeter heating element. This is a typical case of surface-to-surface radiation and the view factor has to be considered in boundary condition at the exposed surface. The formula for the view factor used in this study was proposed by Wilson et al. [13] who studied the uniformity of radiant heat fluxes in the cone calorimeter. The effect of expansion on surface radiation heat exchanges is also taken into account. The heat feedback from combustion on the surface is assumed to be negligible as compared to the external radiation [14]. The boundary condition on the exposed surface is shown in Eq. 3. The term on the LHS represents the total heat flux received by the coating. The first term on the RHS represents the surface-to-surface radiation of the material and the cone heating element, the second term and last term are the heat loss by radiation and convection on the surface, respectively.

$$-\vec{n}.(-k_{intu} \nabla T) = \varepsilon\sigma\left(F{T_r}^4 + (1-F){T_a}^4\right) - \varepsilon\sigma T^4 + h_{ho}(T_a - T) \tag{3}$$

The boundary condition on the sides of the respective materials is presented in Equ. 4. The first term of the RHS represents heat lost by radiation and the second one is the heat lost by convection.

$$-\vec{n}.(-k\nabla T) = \varepsilon\sigma(T_a^4 - T^4) + h_{ve}(T_a - T) \tag{4}$$

At the backside of the copper plate, a Dirichlet boundary condition is imposed as T=40°C. The heat transfer coefficients are determined using the assumption of natural convection on a hot plate. The initial condition is set at T=40°C i.e. the initial temperature of all the materials. After determining the initial and boundary conditions, the thermophysical properties of SBIC were measured (the properties of the other materials were given by the suppliers) and the governing equations were solved in 2D using the Finite Element Method in COMSOL Multiphysics V.5.3.

IOP Conf. Series: Journal of Physics: Conf. Series **1107** (2018) 032012 doi:10.1088/1742-6596/1107/3/032012

THERMOPHYSICAL PROPERTIES

The thermal diffusivity of the coating was measured from 30°C to 150°C using a Laser Flash Apparatus (LFA) from Netzch (LFA 467 Hyperflash). The apparatus uses a non-contact method by heating one side of the sample by a xenon lamp as a flash source and on the other side, an infrared detector records the temperature profile. The temperature profile is analyzed using a mathematical model (penetration model) included in the analysis software and thermal diffusivity is determined.

The specific heat capacity was obtained by a comparative method using measurements from the LFA. The coating and a reference material (pyroceram) with known properties were measured under the same conditions. By comparing the maximum temperatures obtained, the specific heat capacity of the sample can be determined using Eq. 5.

$$c_{pintu} = \frac{T_{max\_pyro}}{T_{max\_intu}} \cdot \frac{(\rho \cdot l)_{pyro}}{(\rho \cdot l)_{intu}} \cdot c_{p\_pyro} \quad (5)$$

The thermal conductivity was derived from Eq. 6 using the parameters obtained previously.

$$k_{intu} = \lambda_{intu} \cdot c_{pintu} \cdot \rho_{intu} \quad (6)$$

The emissivity of the coating was measured at room temperature using a Vertex 70v vacuum FT-IR spectrometer from Bruker equipped with an integrating sphere.

Kinetics of decomposition: thermogravimetric analysis (TGA) were carried out at different heating rates (1, 5 and 10°C/min) using a SDT Q600 from TA instruments. The Arrhenius parameters were determined using a thermokinetic software package from Netzsch [15].

Expansion measurement

The expansion of the intumescent coating was monitored by an infrared camera (FLIR camera A40) to obtain a good contrast during the experiment. The captured images were analyzed using Image J software (NIH, Bethesda, MD) and the expansion of the coating was determined as a function of time [16, 17]. Using this approach, the expansion was assumed to be homogeneous and occurring in one dimension.

RESULTS AND DISCUSSION

Expansion, thermal conductivity and fire performance

Fig. 2 presents the evolution of expansion (a) and thermal conductivity (b) as a function of time. The temperature versus time curves of the uncoated and coated steel plates are also plotted (Fig 2.c) to demonstrate the coating's barrier properties. The expansion of the coating when exposed to a pure radiative heat flux of $50kW/m^2$ occurs in four steps (Fig. 2(a)). First, between 0 and 300s, the expansion increases quite steadily with a high expansion rate, it then stabilizes at around 450s up to 560%. The expansion increases again to reach its maximum (725%) after 560s, it then remains steady afterwards until the end of the experiment. The average expansion rate over the whole expansion was 0.05mm/s. The obtained expansion profile was used as an input in the model.

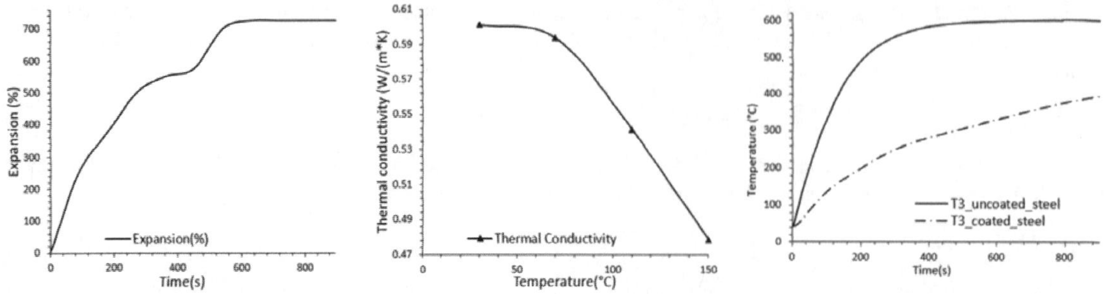

(a) (b) (c)

Fig. 2. (a) Expansion as a function of time of the SBIC; (b) the thermal conductivity of the virgin SBIC and (c) Temperature profiles at the bottom of the steel plate for virgin and coated steel.

Fig. 2(b) shows that the thermal conductivity obtained by using the comparative method is in the same order of magnitude as that found for a typical SBIC [7, 8] decreasing from around 0.6 W/m.K at ambient temperature to 0.48 W/m.K at 150°C. It has been reported that for another SBIC (not the same as that used in this work), the thermal conductivity decreases from 1 W/m.K at ambient temperature to 0.35 W/m.K at 300°C and increases up to 0.45 W/m.K at 700°C [8]. In the current study, we assumed that, for temperatures higher than 150°C, the conductivity was constant, which is acceptable considering the values reported by Bourbigot et al. [8]. The temperature (Fig. 2(c)) of the uncoated steel plate rises up to a steady state at approximately 600°C whereas that of the coated steel rises up to a maximum of around 400°C demonstrating the physical barrier properties of the coating. Thanks to the previous measurements, the barrier properties can be attributed to some extent to the expansion rate and low thermal conductivity.

Comparison of experimental and predicted data

Fig. 3 compares the predicted and experimental temperature profiles of the coating at different locations. On the experimental results, the temperature profile T1 starts at around 110s because when the intumescent coating embeds the thermocouple. From this point, the temperature decreases up to around 450°C (after 450s), owing to the barrier effect of the expanding insulative char layer. For temperature profiles T2 and T3, from 0s to 140s (<220°C), conduction governs the heat transfer phenomena during this period and it is controlled by the radiative heat source, thermal diffusivity and expansion of the coating. The onset of thermal degradation is observed from 220°C corresponding to around 140s (TGA analysis not shown here), corresponding to around 140s on Fig. 3. For T1, T2 and T3, from 450s, the temperatures profiles tend to increase rapidly even though we would expect them to reach a steady state because the char layer is near full expansion. This is explained by the formation of some visible cracks at the surface of the coating. It was also claimed by Gardelle et al. [6] that in a pure radiative heating case, SIBC's cracks were due to the high vibration of the Si-O bond in infrared field. Finally, for the temperature profile T4, heat is transferred by pure conduction inside the calsil block and the temperature tends to increase steadily up to 140°C for the duration of the test which makes sense because of the thermal insulation effect of both the coating and the calsil as well as the cooling effect of the sample holder.

For T4, the numerical results are in excellent agreement with the experimental data for the duration of the test whereas for the other thermocouples, agreement is only observed from 0s to around 450s. The agreement is attributed to the accuracy of the determined thermophysical properties and the appropriate boundary conditions. The assumption of a constant thermal conductivity at temperatures higher than 150°C and a nth order reaction can be validated because for T1, T2 and T3, the temperature profiles are consistent at temperatures higher than 150°C between 0s to 450s. The model fails to capture the temperature increase from 450s because of the phenomenon explained previously which is not included in the model. Work is in progress to include this phenomenon.

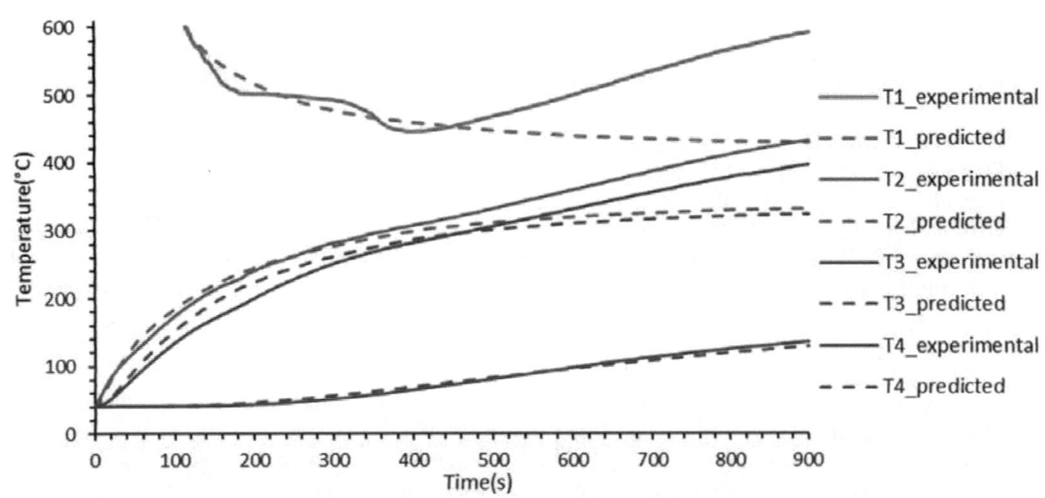

Fig. 3. Predicted and experimental temperature profiles of the SBIC at different locations.

CONCLUSION

This work, using a phenomenological approach, aimed at modeling the heat transfers across a silicone based coating containing expandable graphite exposed to an external radiative heat flux of $50kW/m^2$. The first model developed was consistent with experimental results and it was able to reasonably predict the temperature profiles. However, it did not capture the temperature increase at times higher than 450s because of cracking in the silicone-based intumescent coating. Further work on a new model is in progress to include this phenomenon.

ACKNOWLEDGEMENTS

This work has received funding from the European Research Council (ERC) under the European Union's H2020 – the Framework programme for Research and Innovation (2014-2020) / ERC Grant Agreement n°670747 – ERC 2014 AdG/FireBar-Concept.

REFERENCES

[1] Quintiere, J.G. and F.A. Williams. (2014) Comments on the National Institute of Standards and Technology Investigation of the 2001 World Trade Center Fires, *Journal of Fire Sciences* 32:281-291, http://dx.doi.org/10.1177/0734904114528457.

[2] Bourbigot, S., et al. (2004) Recent Advances for Intumescent Polymers, *Macromolecular Materials and Engineering* 289:499-511, http://dx.doi.org/10.1002/mame.200400007.

[3] Butler, K.M., H.R. Baum, and T. Kashiwagi. "Three dimensional Modeling of Intumescent behavior in Fires," *Fire Safety ScienceProceedings of the fifth International Symposium*. 1997. Melbourne, Australia.

[4] Di Blasi, C. and C. Branca. (2001) Mathematical Model for the Nonsteady Decomposition of Intumescent Coatings, *AIChE J.* 47:2359–2370.

[5] Di Blasi, C. (2004) Modeling the effects of high radiative heat fluxes on intumescent material decomposition, *Journal of Analytical and Applied Pyrolysis* 71:721-737, http://dx.doi.org/10.1016/j.jaap.2003.10.003.

[6] Gardelle, B., et al. (2013) Thermal degradation and fire performance of intumescent silicone-based coatings, *Polymers for Advanced Technologies* 24:62-69, http://dx.doi.org/10.1002/pat.3050.

[7] Gardelle, B., et al. (2013) Characterization of the carbonization process of expandable graphite/silicone formulations in a simulated fire, *Polymer Degradation and Stability* 98:1052-1063, http://dx.doi.org/10.1016/j.polymdegradstab.2013.02.001.

[8] Bourbigot, S., B. Gardelle, and S. Duquesne. "Intumescent silicone-based coatings for the fire protection of carbon fiber reinforced composites," *Fire Safety Science -- Proceedings of Eleventh International Symposium*. 2014. International Association for Fire Safety Science.

[9] Zhang, Y., et al. (2012) Global modelling of fire protection performance of intumescent coating under different cone calorimeter heating conditions, *Fire Safety Journal* 50:51-62, http://dx.doi.org/10.1016/j.firesaf.2012.02.004.

[10] Bhargava, A. and G.J. Griffin. (1999) A Two Dimensional Model of Heat Transfer Across a Fire Retardant Epoxy Coating Subjected to an Impinging Flame, *Journal of Fire Sciences* 17:188-208, http://dx.doi.org/10.1177/073490419901700304.

[11] Donea, J., et al., *Arbitrary Lagrangian–Eulerian Methods*, in *Encyclopedia of computational mechanics, fundamentals*. 2004, John Wiley & Sons, Ltd: UK.

[12] *COMSOL Multiphysics Reference Manual v. 5.3*, 2017.

[13] Wilson, M.T., B.Z. Dlugogorski, and K.E. M. (2003) Uniformity of radiant heat fluxes in cone calorimeter, *Fire Safety Science -- Proceedings of the Seventh International Symposium*:815-827.

[14] Shi, L., et al. (2015) Modeling the Pyrolysis and Combustion Behaviors of Non-Charring and Intumescent-Protected Polymers Using "FiresCone", *Polymers* 7:1979-1997, http://dx.doi.org/10.3390/polym7101495.

[15] Opfermann, J. (2000) Kinetic Analysis Using Multivariate Non-linear Regression. I. Basic concepts, *Journal of Thermal Analysis and Calorimetry* 60 641–658.

[16] Bourbigot, S., B. Gardelle, and S. Duquesne. "Intumescent silicone-based coatings for the fire protection of carbon fiber reinforced composites," *Fire Safety Science - Proceedings of the Eleventh International Symposium* 2014. DOI: 10.3801/iafss.fss.11-781.

[17] Bourbigot, S. and S. Duquesne. "Quantitative approach of intumescence by numerical simulation," *243 rd American Chemical Society National meeting*. 2012. San Diego.

IOP Conf. Series: Journal of Physics: Conf. Series **1107** (2018) 032013 doi:10.1088/1742-6596/1107/3/032013

Comparative study of the thermal decomposition behaviour of different amino acids and peptides

Alexander Schaberg[1], Robert Wroblowski[1] and Roland Goertz[1]

[1]*Chemical safety and fire defence, University of Wuppertal, Germany*

schaberg@uni-wuppertal.de

ABSTRACT

Thirteen nitrogen-containing compounds, including 2 peptides and 11 amino acids, were analysed for their thermal decomposition behaviour using TGA/DSC coupled with MS and FTIR.
Various decomposition courses have been found. Glycine, cysteine, cystine, proline, D- and L-threonine decomposed in two stages, whereas arginine, asparagine, glutamine, histidine and tryptophan decomposed in three stages. The peptides of glycine both decomposed in four decomposition steps, except for the glycylglycine under oxidative conditions, which decomposed in three stages. Proline proved to be exceptional, as it needed only one decomposition step under nitrogen to decompose (99.5% mass loss).
The decomposition products of amino acids and peptides described in literature could be confirmed in this work. These were mainly carbon dioxide, water, ammonia, hydrocyanic acid and nitrogen oxides. In traces, also isocyanates and nitriles could be detected, but no clear identification of these substances could be made due to superposition of the signals. For example, in the absorption spectrum at 3000 cm^{-1}, a superposition of the hydrocarbons and amines occurred and in the two sulphur-containing amino acids cysteine and cystine at 2100 cm^{-1}, a superposition of the resulting sulfur compounds could be observed.
It has also been shown that with increasing chain length of the peptides, in this case of glycine, the molecule becomes thermally more stable. Concerning cystine, something similar was observed for cystine, this is due to the strong sulphur bonding of the monomers. In addition, glycine and its peptides have been shown to have doubled or tripled the volume of the resulting carbon foam formed as length increased, a feature interesting for future research involving intumescent materials.

KEYWORDS:

fire chemistry; evolved gas analysis; thermogravimetric analysis; material characteristics; combustion

IOP Conf. Series: Journal of Physics: Conf. Series **1107** (2018) 032013 doi:10.1088/1742-6596/1107/3/032013

INTRODUCTION

In thermal decomposition by combustion, a vast variety of compounds is formed influenced by both combustion parameters and source material. In order to estimate how proteins or peptides behave thermally depending on their structure, one has to understand how amino acids, which are the building blocks of proteins and peptides, behave thermally. Amino acids are organic compounds of simple design with at least one amino group (-NH_2) and one carboxyl group (-COOH) (fig. 1). This gives them a remarkable physicochemical reaction behaviour, utilised by every life form known to humankind. [1]

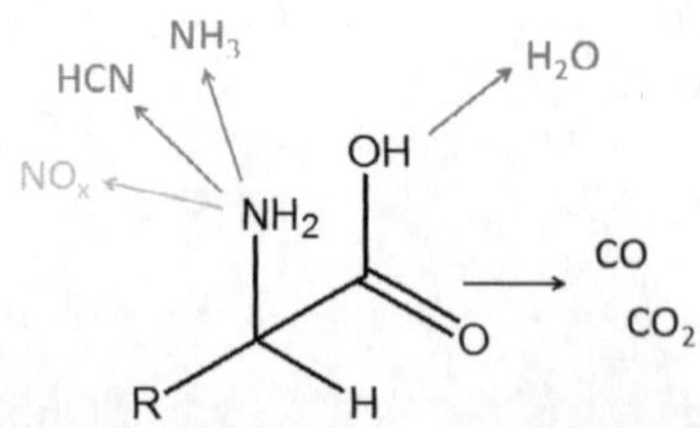

Fig. 1. General formula of amino acids and main anorganic decomposition products.

All compounds, except glycine, have at least one chiral centre, while naturally occurring α-amino acids are predominantly L-configured. Therefore, unless specified otherwise, the amino acids referred to in the following text are of the L-configured form.
Their behaviour under heat has been studied over a wide temperature range, though mostly under pyrolytic conditions. A study of glycine, alanine and serine in the temperature range of 200-300 °C in vacuum found mostly water, carbon dioxide, ammonium and methylamine for glycine and even less for the remaining amino acids. The condensed phase for glycine yielded formamide, acetamide, N-methylacetamide, n-butylamine and propanamide. [2] Another study of valine, leucine and isoleucine showed a wider emission spectrum, as displayed in table 1, under similar conditions, utilizing a gas chromatographic system.

Table 1. Decomposition products of valine, leucine and isoleucine. [3]

valine	**leucine**	**isoleucine**
ammonia carbon dioxide carbon monoxide propane propene isobutane isobutylene acetone isobutyraldehyde isobutylamine N-isobutylidene-isobutylamine diisobutylamine	ammonia carbon dioxide carbon monoxide isobutane isobutylene isopentane 3-methyl-1-butylene acetone isobutyraldehyde isovaleraldehyde isobutylamine isoamylamine N-isobutylidene- isoamylamine N-isoamylidene- isoamylamine diisoamylamine	ammonia carbon dioxide carbon monoxide butane butylene isopentane 2-methyl-1-butylene 2-butanone methylbutyraldehyde 2-methylbutylamine N- / 2-(methylbutylidene) -2-methylbutylamine amine, bis (2-methylbutyl)

Another study [4] of poly-L-leucine and poly-L-proline performed under pyrolytic conditions showed a conversion of the nitrogen inventory to HCN (58%) and NH_3 (31%) at 700 °C, with an even higher ratio of over 2 parts HCN per ammonia at further elevated temperatures in the poly-L-leucine. Poly-L-proline even yielded 10 parts HCN per ammonia, proving the high diversity and dependence on the chemical structure of the amino acid.

In addition, HCNO was identified as an effluent, whereas NO or N_2O did not form or where below the detection limits of the used IR-spectrometer.

Although many interesting findings have been reported as mentioned before, none of those studies were conducted under combustion-like conditions, e.g. under typical atmospheric conditions or (synthetic) air. Therefore, our aim was to study differences under pyrolytic and combustion conditions to further the understanding of the fire effluents and behaviour of natural products. As amino acids (or peptides) make up often-used textile household and clothing products like feathers, leather, wool and silk, their general behaviour and especially the occurrence of toxic gases are of high interest and play a major role in understanding how effluents might be prevented or why high concentrations of ammonia and cyanides may occur.

To measure the combustion behaviour, a thermogravimetric analyser, coupled with differential scanning calorimetry (TGA/DSC) and evolved gas analysis (EGA) was utilised. Using this setup, the change in mass of the compound under influence of an applied heat rate under pre-set atmospheric conditions and flow rates can be studied. In addition, monitoring heat fluxes between the compound containing crucible and a reference crucible, thus giving information of the energetic changes, is possible. The effluents of the analytes are driven towards the EGA and can be identified by their charge-to-mass-ratio using mass spectrometry (MS) and their response to infrared irradiation by a Fourier-transform infrared spectrometer (FTIR). [5-7]

EXPERIMENTAL

The experiments were performed in a TGA/DSC 1 (Mettler Toledo). The starting temperature was set to 30 °C. With a heating rate of 10 K/min, the temperature was ramped to 800 °C under a flow of 40 mL/min synthetic air or nitrogen. The compounds (analytical grade, 99% purity, Sigma Aldrich/Alfa Aesar/VWR) were placed in 70 μL aluminium oxide crucibles. To analyse the effluents, a ThermoStar GSD 320 (Pfeiffer Vacuum) in SEM-mode with 65 eV ionisation energy and heated capillary (150 °C) and a Nicolet™ iS™50 (Thermo Scientific) with TGA-IR-module (heated transfer line, 200 °C and heated gas cuvette, 220 °C) were used.

In order to prove the usability in the TGA/DSC, preliminary experiments over a Bunsen burner were carried out, which also showed interesting clues of the general combustion behaviour of the amino acids and peptides. Therefore, 10 mg of each compound was transferred in similar crucibles as used in the TGA/DSC experiments.

RESULTS AND DISCUSSION

In the following text, a short overview of the different combustion and pyrolysis characteristics of the studied compounds is given, as well as an in-depth review of the data obtained for glycine and the resulting peptides.

Glycine, glycylglycine and glycylglycylglycine

The most basic amino acid (glycine, see fig. 1 with R = H) and its peptide derivatives showed interesting behaviour under oxidative conditions in the preliminary experiments. Once exposed to direct heating, a black carbon foam formed rapidly (fig. 2). The volume of this foam seemed to have a direct relation to the amount of linked glycine monomers – although the total mass used for each test stayed the same. In addition, the foam remained stable when further exposed to heat for another 20-30 seconds. These intumescing features were also present in similar extend with histidine; most other amino acids did form a much less stable foam during combustion which disappeared within seconds.

Fig. 2. Decomposition of glycine (left), glycylglycylglycine (mid) and histidine (right).

IOP Conf. Series: Journal of Physics: Conf. Series **1107** (2018) 032013 doi:10.1088/1742-6596/1107/3/032013

When those compounds are subjected to TGA/DSC, the following data can be obtained (see table 2). A typical TGA-curve is also displayed for glycine (fig.3).

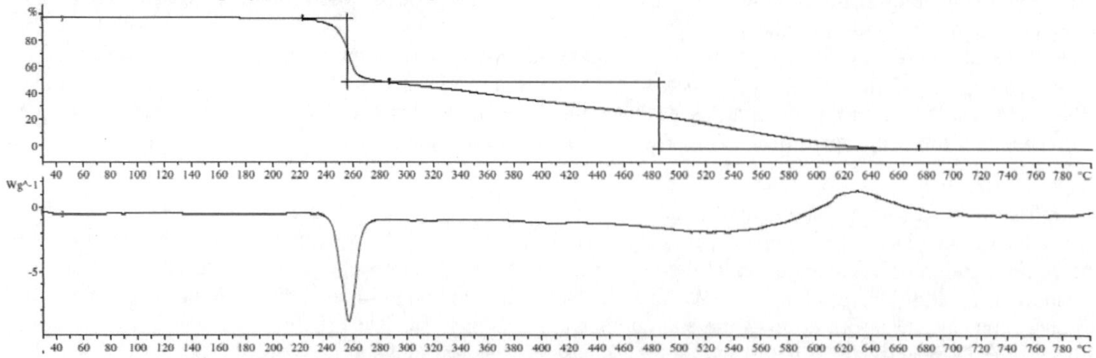

Fig. 3. TGA/DSC curve of glycine.

Table 2. Decomposition of glycine and its derivatives.

Glycine		
Synthetic air	**Decomposition step 1**	**Decomposition step 2**
Mass loss (in % of remaining mass)	50	100
Effluents	NH_3, H_2O, HCN, NO, NO_2, CO_2, isocyanates	NH_3, H_2O, HCN, CO, CO_2, isocyanates, nitriles
Temperature range (in °C)	220-280, $\Delta H > 0$	280-650, $\Delta H < 0$

Glycine		
Nitrogen	**Decomposition step 1**	**Decomposition step 2**
Mass loss (in % of remaining mass)	50	30, 20% remaining
Effluents	NH_3, H_2O, NO_2, CO_2, isocyanates, nitriles	CO_2, NO_2, isocyanates, buten
Temperature range (in °C)	220-280, $\Delta H > 0$	280-800, $\Delta H > 0$

Glycylglycine			
Synthetic air	**Decomposition step 1**	**Decomposition step 2**	**Decomposition step 3**
Mass loss (in % of remaining mass)	19	20	100
Effluents	NH_3, H_2O, HCN, NO, CO_2	NH_3, H_2O, CO_2	NH_3, H_2O, HCN, CO, CO_2, isocyanates, nitriles
Temperature range (in °C)	220-250, $\Delta H > 0$	250-310, $\Delta H < 0$	310-650, $\Delta H > 0$

Glycylglycine				
Nitrogen	**Decomposition step 1**	**Decomposition step 2**	**Decomposition step 3**	**Decomposition step 4**
Mass loss (in % of remaining mass)	19	25	13	22, 21% remaining
Effluents	NH_3, H_2O, NO_2, CO_2	NH_3, H_2O, NO_2, CO_2	NH_3, H_2O, HCN, NO_2, CO_2, isocyanates, nitriles	NH_3, H_2O, HCN, NO_2, CO, CO_2, isocyanates, nitriles
Temperature range (in °C)	220-250, $\Delta H > 0$	250-320, $\Delta H > 0$	320-380, $\Delta H > 0$	380-800, $\Delta H < 0$

Glycylglycylglycine				
Synthetic air	**Decomposition step 1**	**Decomposition step 2**	**Decomposition step 3**	**Decomposition step 4**
Mass loss (in % of remaining mass)	20	13	23	100
Effluents	NH_3, H_2O, HCN, NO, NO_2, CO_2, isocyanates	NH_3, H_2O, HCN, NO, NO_2, CO, CO_2, isocyanates	NH_3, H_2O, CO_2, isocyanates, nitriles	NH_3, H_2O, HCN, NO_2, CO, CO_2, isocyanates
Temperature range (in °C)	240-260, $\Delta H > 0$	260-310, $\Delta H < 0$	310-440, $\Delta H < 0$	440-670, $\Delta H < 0$
Glycylglycylglycine				
Nitrogen	**Decomposition step 1**	**Decomposition step 2**	**Decomposition step 3**	**Decomposition step 4**
Mass loss (in % of remaining mass)	20	16	16	21, 27% remaining
Effluents	NH_3, H_2O, NO_2, CO_2	NH_3, H_2O, NO_2, CO_2	NH_3, H_2O, NO_2, CO_2, isocyanates, nitriles	NH_3, H_2O, HCN, CO_2, isocyanates, nitriles
Temperature range (in °C)	240-260, $\Delta H > 0$	260-310, $\Delta H > 0$	310-400, $\Delta H > 0$	400-800, $\Delta H < 0$

Glycin and the different peptides show an interesting behaviour in many aspects studied by TGA/DSC. For example, a longer chain results in higher thermal stability, whereas the small, easily formed inorganic products remained relatively similar, with the main difference resulting from the surrounding atmosphere. The amount of complex structures in the effluents seems to be a function of the chain length, with a more frequent occurrence and higher concentration of isocyanates, nitriles and HCN in the biggest peptide and at higher temperatures. Under inert atmosphere, neither glycine nor glycylglycine showed the formation of HCN.

Remaining amino acids

The remaining amino acids (arginine, asparagine, cysteine, cystine, glutamine, histidine, proline, D and L-threonine, tryptophane) showed, as could be expected by more complex compounds, a more diverse spectrum of effluents. Those involved complex alkanes and alkenes (methane, butane, butylene, and pentane), aromatic compounds (benzene, toluene), amines and other nitrogen containing compounds (e.g. isobutylamine, urea, HONO, aromatic nitriles). Furthermore, sulphur-containing amino acids yielded SO_2, H_2SO_3, thiocyanates and thioles. Most amino acids required 2-4 decomposition steps in total. Proline decomposed in an explosive manner, losing 99.5% mass between 220-290 °C. All amino acids were fully oxidised under synthetic air without any residue, under nitrogen a variable amount of carbon remained. For the threonines, the remains had a mass of less than 1%, while histidine had a residue of 40% under inert conditions. Many amino acids began decomposing as their respective melting point was reached, but cystine and tryptophan began pyrolysing 20-40 °C before melting.

Cysteine proved to be less stable than its dimer (cystin), having double its mass loss rate at a lower initial temperature. This can be traced back to the strong sulphur bonding between the monomers.

Looking at the D- and L-configured isomers of threonine, no differences in effluents were found, but the decomposition mechanism seemed to differ slightly – both began their mass loss at roughly 225 °C, but the L-threonine showed a second, small step at 240 °C, which may be caused by better intermolecular bonding or interactions.

CONCLUSIONS

The experiments showed the high diversity even within the small group of relatively similar compounds, those are also shown in figure 4. The first and main products of decomposition were carbon dioxide, water and ammonia. Carbon monoxide occurred as a trace in most analyses; nitrogen oxides were even less frequently

IOP Conf. Series: Journal of Physics: Conf. Series **1107** (2018) 032013 doi:10.1088/1742-6596/1107/3/032013

detected. In addition, isocyanates and nitriles were found at elevated temperatures, just like the hydrocyanic acids.

Although EGA is able to monitor analytes of high volatility during a TGA/DSC-analysis, further equipment is needed to show the even more diverse field of lower volatile compounds. Therefore, gas chromatography and a so-called VCI-oven will be employed to analyse the effluents of a simulated smouldering fire including amino acids. Preliminary experiments with tryptophane already showed high amounts of heterocycles, nitriles and mixtures of both, which may be specific for the used amino acid. Furthermore, this type of analysis is capable of separating the different isocyanates and nitriles obtained by TGA/DSC, which cannot be separated and thus quantified by EGA due to superpositions. In general, the EGA of most researched compounds within this paper showed a much smaller amount of effluents under inert conditions. However, those decompositions products were often much more complex than those under synthetic air, which correspondents with findings under smoldering fire conditions, as simulated by the VCI-apparatus.

To also monitor the complete nitrogen inventory of the compounds, further experiments will be carried out under argon atmosphere to measure the generation of pure nitrogen by decomposition.

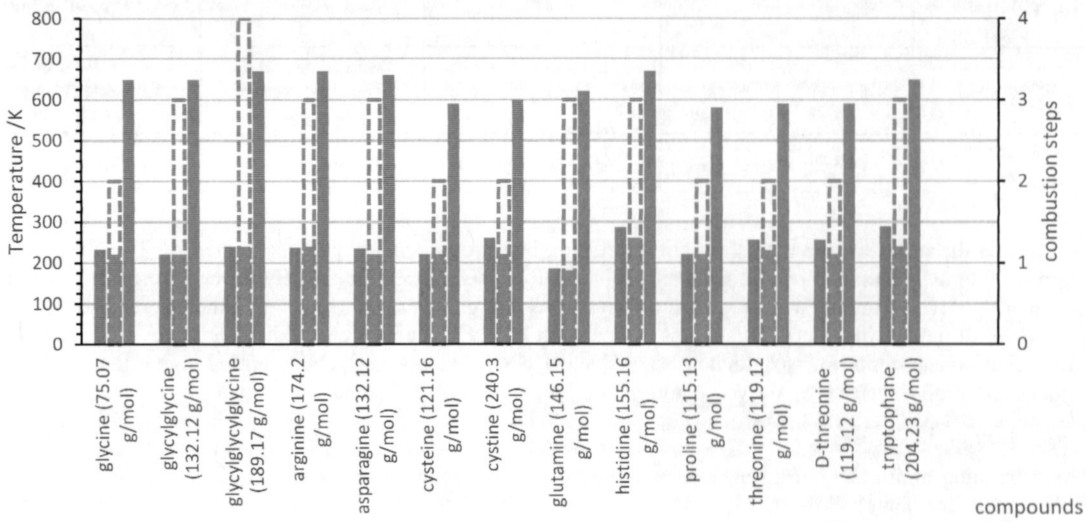

Fig. 4. Decomposition of all featured compounds, showing their melting point (blue), begin (grey) and end (orange) of combustion temperatures and decomposition steps (dotted green) under synthetic air

REFERENCES

[1] Jabuke, H.D., Jeschkeit, H., *Aminosäuren, Peptide, Proteine*, Weinheim Verlag Chemie, Weinheim, 1982

[2] Yablokov, V.Y. et al. (2009) Study of the Rates of Thermal Decomposition of Glycine, Alanine and Serine, *Zhurnal Obschei Khimii* 79:1344-1346

[3] Lien, Y.C., Nawar, W.W. (1974) Thermal decomposition of some Amino Acids. Valine, Leucine and Isoleucine, *Journal of Food Science* 79:911-913

[4] Hansson, K.-M. et al. (2003) Pyrolysis of poly-L-leucine under combustion-like conditions, *Fuel* 82:653-660

[5] Brown, M.E., *Introduction to Thermal Analysis: Techniques and Applications*, Kluwer Academic Publishers, New York, 2001

[6] Gross, J.H., *Massenspektrometrie Ein Lehrbuch (2nd ed)*, Springer-Verlag, Berlin/Heidelberg, 2013

[7] Hesse, M., Meier, H. and Zeeh, B., *Spektroskopische Methoden in der organischen Chemie (7th ed)*, Georg Thieme Verlag, Stuttgart, 2005

IOP Conf. Series: Journal of Physics: Conf. Series **1107** (2018) 032014 doi:10.1088/1742-6596/1107/3/032014

High-Throughput Fire Tests and Weathering-Induced Degradation Behaviour of Intumescent Coatings

Lars-Hendrik Daus[1], Manfred Korzen[1] and Bernhard Schartel[1]

[1]*Bundesanstalt für Materialforschung- und Prüfung (BAM), Germany*

lars-hendrik.daus@bam.de

ABSTRACT

In this work, the weathering-induced degradation of intumescent coatings was investigated by a systematic and comprehensive approach. A mechanism is revealed that is proposed to be responsible for the loss of function of intumescent coatings induced by weathering.

First, the thermal decomposition of artificially weathered intumescent coatings was examined. To get a better understanding of the weathering and ageing phenomena, the degradation behaviour of the single ingredients during the weathering process was investigated, as well as their chemical and physical interactions. For the systematic approach, the materials that are essential for intumescence (ammonium polyphosphate, pentaerythritol, titanium dioxide, melamine and the binder) are treated with moisture, elevated temperature and UV radiation.

Thermogravimetry (TG) and IR spectroscopy were used to compare the initial samples with their different grades of weathering. We demonstrate that ammonium polyphosphate, melamine and the binder are mainly responsible for the ageing process. Further, it was demonstrated that TG and IR spectroscopy are suitable measuring methods to detect the effects of weathering on intumescent coatings.

Finally, a small-scale fire test procedure is introduced. Based on the reduction of the sample size, up to 50 samples can be tested in a single fire test. The results of this fire test have the same quality as the results from standard intermediate fire tests corresponding to DIN 4102-8.

KEYWORDS:

Intumescent coatings; fire resistance; weathering resistance

INTRODUCTION

Among different kinds of fire protection materials for steel constructions, intumescent coatings are the ones preferred by architects today and thus dominate the corresponding market. With the application of intumescent coatings, the time to failure for structural steel is enhanced in the case of fire, with the steel construction remaining visible [1]. A great deal of research has been performed to optimize intumescent coatings, but there are still some shortcomings to be resolved. Solvent-free coatings, especially, are still sensitive to ageing and weathering [2]. The intumescent coatings work through the mechanism of swelling when they are heated up. When the binder gets soft, its ingredients cause chemical reactions so that the coating forms a multicellular foam. Because of the low heat conductivity of the inflating gases, the coating generates an effective insulation layer between the underlying steel and the heat source. Furthermore, the hot surface may reflect or radiate the heat so that in the case of fire the heating rate of the steel construction is reduced [1, 3].

There are three essential components for intumescent coatings: a carbon source, an acid donor and a blowing agent. The binder which bonds the ingredients together can act as a carbon source as well. In some cases, additional inorganic ingredients are added to improve the thermal resistance strength. These inorganic additives undergo chemical reactions at high temperatures, which improve the thermal stability of the char. One of the compositions that is currently common contains pentaerythritol (PER) as a carbon source, ammonium polyphosphate (APP) as an acid donor and melamine (MEL) as a blowing agent [4]. With this composition, polyvinyle acetate (PVAC) is often used as a binder. As it also acts as a white pigment, titanium dioxide (TiO_2) is often used as an inorganic filler. Activated by heat, the carbon source, together with the acid donor, creates a high-viscose carbonaceous network (which is also phosphoric, due to the APP) by alcoholysis and esterification. Due to its thermal decomposition, the blowing agent releases nonflammable gases and the resulting bubbles produce a multicellular foam [4]. At high temperatures, when the organic compounds are already evaporated, TiO_2 reacts chemically with the remaining phosphate in the APP to form a ceramic structure containing titanium pyrophosphate [5].

Some research has been done to investigate the effects of weathering on intumescent coatings [1, 5]. Until now, only the coating has been examined as a whole system. Our approach is to reveal the degradation mechanisms and main impacts of weathering by studying the degradation behaviour and interaction among the single ingredients. For the experiments, the samples are divided in two categories. The weathering-induced degradation and reactions of the ingredients are investigated on the basis of pure and mixed dry chemicals, whereas the ageing-related performance of the intumescent coatings is evaluated by coatings that have been applied to steel panels. The samples are tested at different states of the artificial weathering process by fire tests and TG. Furthermore, our task is to improve the resistance to weathering of intumescent coatings by systematically varying the formulations. The different grades of weathering and the systematic variations of the formulation call for a high number of samples. In order to deal with the high number of samples, the size of the steel panels is reduced to the scale of 75 mm x 75 mm x 5 mm. The small specimen size facilitates the storage of a sufficiently high number of samples in the ageing chamber and enables high-throughput fire tests so that up to 50 samples are evaluated simultaneously in one fire test.

HIGH-THROUGHPUT TESTING

Samples and Sample Preparation

First, a primer is spread over the surface of the steel panel using a paint roller. After 12 hours of curing, the application of the first layer can be started. The coating is applied with in several steps using a squeegee. The maximum thickness of each layer must not exceed 500 μm in order to prevent cracks in the surface of the coating. After curing, the new layer is shrunk to a dry film thickness of about 250 μm. This means that a layer with a dry film thickness of about 1 mm consists of four layers. The tolerance of the resulting dry film thickness is about +/- 20 μm.

The fire tests are performed in an oil-powered furnace with a chamber size of 100 cm x 100 cm x 100 cm. A hanger welded on to the chamber connects the back side of the panels with a metal wire. This metal wire passes through a vermiculite board and is fixed on the backside of the board with screws. A pair of thermocouples (Type K, chromel-alumel, 0.5 mm diameter) is also welded to the back surface of the samples to measure the temperature during the fire test. The furnace has two open windows, in which the vermiculite boards with the mounted samples were placed (Fig. 1 (a)).

Evaluation of the Testing Procedure

The high-throughput fire tests of the small-scale samples yield results equal to those obtained with an intermediate-scale fire test conforming with the specifications of DIN 4102-8 [6]. The dimensions of the intermediate-size steel panel are 500 mm x 500 mm x 5 mm and the chamber of the corresponding furnace is 170 cm long, 63 cm high and 40 cm wide. In both furnaces, the flame proceeds parallel to the surface of the samples. The temperature program is the UTTC in accordance with DIN EN 1363 -1 [7].

The good comparability (Fig. 1 (b)) of the result can be attributed to the fact that the thermal behaviour of the many small panels arranged close to each other is similar to the thermal behaviour of the one large panel that is used in the intermediate fire tests. The reason for the lower temperature of the intermediate-scale sample is that the distance of the small-scale sample to the furnace flame and to the thermal element of the furnace (20 cm distance) is greater than the distance of the intermediate-scale sample to the thermal element (10 cm distance) and the furnace flame. Thus the intermediate-scale sample is more strongly affected by heat radiation than the small-scale sample. In addition, the interiors of the two furnaces have different fluid properties, related to their different internal dimensions. The quality of the results, in combination with the high number of samples that are tested in one fire test, qualifies this testing procedure for future fire tests.

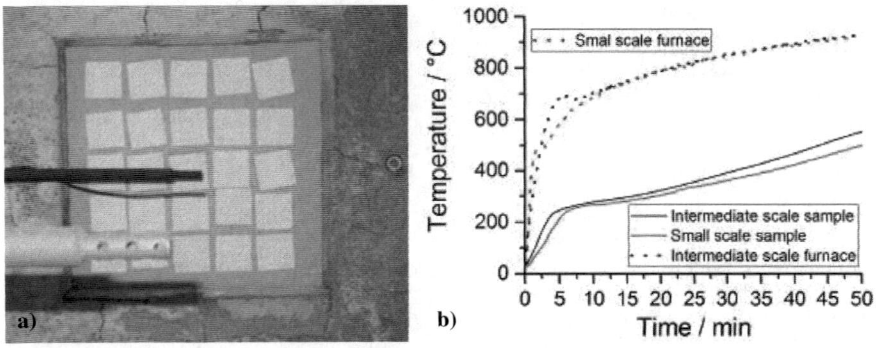

Fig. 1. High-throughput fire testing, arrangement of the small-scale samples in the furnace (a); comparison of the high-throughput small-scale fire test with an intermediate-scale fire test (b).

AGEING BEHAVIOUR

Material and Methods

In this work, two intumescent coatings, based on a common formulation, are treated with artificial weathering. Both formulations contain ammonium polyphosphate, pentaerythritol, melamine, PVAC and titanium dioxide. For the first formulation, common phase-II APP is used. For the second formulation, the common phase-II APP is replaced by an APP coated with melamine formaldehyde resin. For closer investigation of the weathering-induced reactions, the principal ingredients of the intumescent coating and selected combinations of the ingredients were treated with artificial weathering as well.

The single ingredients are: APP, PER, MEL, TiO_2 and PVAC.

The combinations consist of mixtures of the dry chemicals:

APP + MEL, APP + PER and APP + TiO_2;

TiO_2 + MEL and TiO_2+ PER;

TiO_2 + APP + MEL and TiO_2 + APP + PER

The coatings, the ingredients and the combination of the ingredients were treated with cyclic artificial weathering, including a humid and a warm period:

A period of 6 hours at 40 °C and 95 % relative humidity was followed by a period of 6 hours at 70 °C and 20 % relative humidity. When UV radiation is mentioned, it was applied with the power of 35 W / m² in a range from 290 nm to 400 nm. The samples were stored in the weathering chamber as shown in Fig. 2 (a).

The coatings were subjected to the artificial weathering conditions for 15 weeks. UV radiation was applied during the entire period. Every 3 weeks a pair of coated steel panels was removed from the weathering chamber for examination. Hence the coatings were available in the initial state and after 3, 6, 9, 12 and 15 weeks of artificial weathering (Fig. 2(b)). The single ingredients and the combinations were treated with the same conditions for 24 weeks. In order to investigate the influence of UV radiation on the weathering-induced effects of ageing, all ingredients and the combination of ingredients were artificially weathered with and without the exposure to UV radiation.

After the accelerated weathering tests, all dry chemicals were examined using TG and ATR. Furthermore, the weathered single ingredients were used to produce intumescent coatings. These coatings were prepared by replacing one of the usual initial ingredients with the respective weathered ingredient in each case. All coatings were evaluated by TG. Furthermore, the coatings containing the weathered single ingredients were evaluated by fire tests. The TG measurement was taken using ground coatings with a mass of 5 mg. All TG measurements took place under nitrogen atmosphere with a heating rate of 10 K / min.

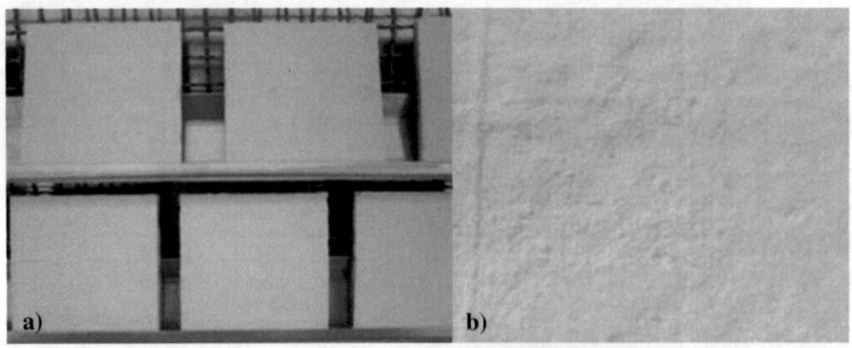

Fig. 2. Samples, first coating in the initial state (a); first coating after 9 weeks of artificial weathering (b) .

Weathering-Induced Effects on the Intumescent Coating Measured by TG

After 9 weeks of artificial weathering, depositions occurred on the surface of both types of coatings (Fig. 2(b)). This observation is consistent with the experiments of Wang [1], as it is APP, MEL and PER that migrate to the surface of the coating. The TG measurements of the weathered coatings (Fig. 3 (a) and (b)) show a mass loss at around 100 °C, which is only observed in the curves of the weathered coatings. As water evaporates at 100 °C, the mass loss can probably be attributed to water that infiltrated the coating during the weathering process. Another difference concerning the mass loss of the samples in different states of weathering is visible between 200 °C and 300 °C. According to the previous work of R. Kunze [8], where a similar coating is analysed by TG, water and ammonia are released in this temperature range and the coating starts to foam up.

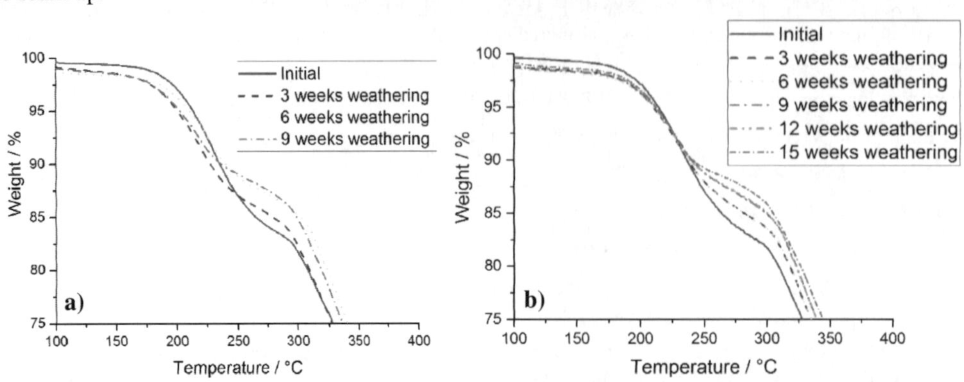

Fig. 3. TG measurements, thermal decomposition of the first coating at different states of accelerated weathering (a); thermal decomposition of the second coating at different states of accelerated weathering (b).

Both substances are released during the thermal decomposition of APP and melamine [4, 9]. The fact that the mass loss is reduced in the temperature range where the intumescence occurs indicates that the performance of the coating has deteriorated. The fact that a smaller quantity of volatile components is released leads to the assumption that the intumescence is affected. For this, two causes are considered. During the weathering process, the ingredients that are necessary for the intumescence migrated to the surface of the coating and are no longer available for the intumescence process [1]. Furthermore, the ATR measurements (Fig. 5 (a)) and the TG measurements (Fig. 5(b)) of the artificially weathered ingredients prove that APP reacts chemically with MEL during the artificial weathering process. Our assumption is that during the artificially weathering process of the intumescent coating, the APP reacts with the MEL inside the coating. We assume that this is another reason for the weathering-induced loss of function of intumescent coatings. It is worth mentioning that there is no further change in the thermal degradation behaviour (measured by TG) of the first coating after 9 weeks of artificial weathering. We assume that the moisture diffusion in the coating and the reaction between APP and MEL is completed after this time. The results of the TG measurements of the second coating (Fig. 3 (b)) show similar behavior, but impact of weathering on the second coatings continues even after 9 weeks of weathering. This leads to the assumption that the second coating that contains the modified APP can resist the damage of weathering for a longer period. This supports the assumption that the chemical reaction of APP and MEL is responsible for the weathering-related loss of function of the intumescent coating, as the surface treatment of the modified APP slows down the chemical reaction of APP and MEL.

Ageing Effects Caused by Weathering of Single Ingredients and Combinations

Neither the APP nor the other ingredients changed their thermal behaviour during the artificial weathering process when they were weathered separately. This was proved by ATR and TG measurements and fire tests. Only the ATR spectra of the binder changed after 6 weeks of artificial weathering (Fig. 4 (a)). The peak in Fig. 4 (a) at 3300 cm^{-1} indicates adsorbed moisture. The thermal degradation behaviour of the artificially weathered APP is shown in Fig. 4 (b) and (c). The coatings for this fire test were prepared by replacing of the ingredients with the artificially weathered version one at a time.

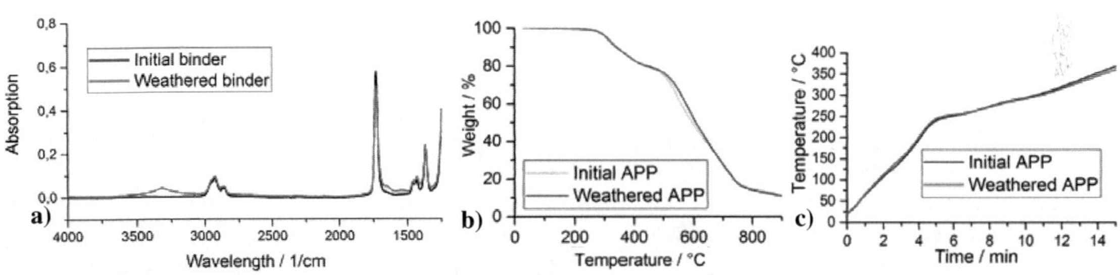

Fig. 4. Examination of the weathered ingredients, comparison of the initial binder with the artificially weathered binder (a); comparison of the initial APP with the artificially weathered APP via TG (b); comparison of the fire tests using an intumescent coating containing initial APP (blue) and artificially weathered APP (red) (c).

As to the combinations of the ingredients, only the combination of APP with MEL showed any difference between the initial and the weathered states. The differences are visable in the ATR spectra (Fig. 5(a).) and in the thermal degradation behaviour that was measured by TG (Fig. 5 (b)).

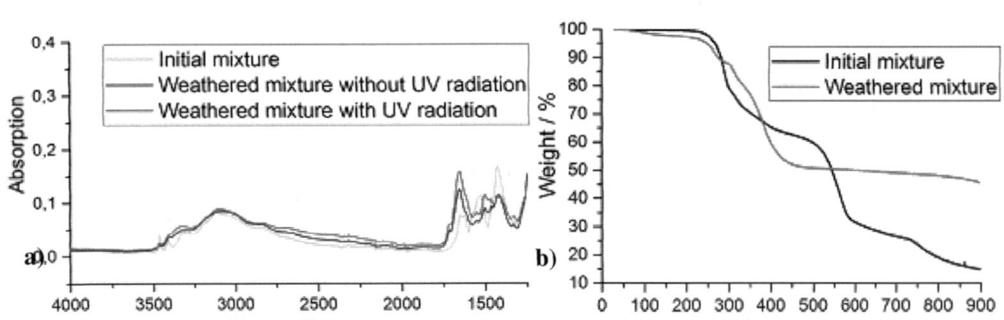

Fig. 5. Comparison of the initial APP-MEL mixture with the artificially weathered APP–MEL mixture, investigation via ATR (a); investigation via TG (b).

Neither the exposition to UV radiation nor the addition of TiO_2 to this mixture made any major difference.

CONCLUSION

Two mechanisms are considered to be responsible for the weathering-induced change of the performance of intumescent coatings. The first is the migration of the ingredients to the surface. The second is the chemical reaction of MEL with APP inside the coating. Both mechanisms are driven by moisture, so that humidity is considered to be the most relevant influence of weathering.

Further studies will be done concerning the weathering-related performance of intumescent coatings to reveal the entire mechanism of the weathering-induced loss of function of intumescent coatings. These studies include fire tests of artificially weathered intumescent coatings and coatings containing the artificially weathered combinations of APP and MEL.

ACKNOWLEDGEMENTS

The financial support for the research presented in this paper provided by the MNPQ program of the German Federal Ministry for Economic Affairs and Energy (grant no. 18/14) is gratefully acknowledged.

REFERENCES

[1] Wang, L.L., Wang, Y.C., Li, G.Q. (2013), Experimental study of hydrothermal aging effects on insulative properties of intumescent coating for steel elements, *Journal of Applied Polymer Science,* 104: 748–753, http://dx.doi.org/10.1016/j.firesaf.2012.10.004

[2] Morys, M., Illerhaus, B., Sturm, H., Schartel, B. (2017), Size is not all that matters: Residue thickness and protection performance of intumescent coatings made from different binders, *Journal of Fire Sciences,* 35: 284–302

[3] Camino, G., Costa, L. (1988), Performance and Mechanisms of Fire Retardants in Polymers – A Review, *Polymer Degradation and Stability*, 20: 271–294

[4] Anderson, A., Lundmark, S, Maurer, F.H.J. (2006), Evaluation and Characterization of Ammoniumpolyphosphate – Pentaerythritol – Based Systems for Intumescent Coatings, *Journal of Applied Polymer Science,* 104: 748–7531

[5] Weil, E. D. (2011), Fire-Protective and Flame-Retardant Coatings – A State-of-the Art Review, *Journal of Fire Science,* 29: 259–296

[6] DIN 4102 -8: 2003, Fire Behavior of Building Materials and Components – Part 8: Small Scale Test Furnace.

[7] DIN EN 1363 - :2012, Fire Resistance Tests – Part 1: General Requirements

[8] Kunze, R., Schartel, B., Bartholmai, M., Neubert D., Schriever R. (2002), TG-MS and TG-FTIR Applied for an Unambiguous Thermal Analysis of Intumescent Coatings, *Journal of Thermal Analyses and Calorimetry,* 70: 897–909

[9] Camino, G., Costa, L. Trossarelli, L. (1985), Study of the Mechanism of Intumescence in Fire Retardant Polymers: Part V – Mechanism of Formation of Gaseous Products in the Thermal Degradation of Ammonia Polyphosphate, *Polymer Degradation and Stability,* 12: 203–211

IOP Conf. Series: Journal of Physics: Conf. Series **1107** (2018) 032015 doi:10.1088/1742-6596/1107/3/032015

The Impact of Bromine- and Phosphorous-Based Flame Retardants on Flame Stability and Heat Feedback from Laminar Wall Flames

Isaac LEVENTON[1], Roland KRAEMER[2] and Stanislav STOLIAROV[3]

[1] *Fire Research Division, National Institute of Standards and Technology, USA*
[2] *Advanced Materials and Systems Research, BASF SE, China*
[3]*Department of Fire Protection Engineering, University of Maryland, USA*

Isaac.Leventon@NIST.gov

ABSTRACT

Flame retardants are widely used to improve the resistance and response of combustible materials to fire. Unfortunately, widespread bioaccumulation of brominated flame retardants has evoked growing concern with regards to the environmental hazards that flame retardants pose, thus industry is moving towards developing safer and equally effective alternatives. To develop new, efficient flame retardants, it is critical to understand the exact mechanisms by which they inhibit flaming combustion. Our research group has developed a systematic methodology to assess the flammability of polymeric materials through microscale combustion calorimetry (MCC) and experimental measurement of flame heat feedback and material burning behavior during upward flame spread. In this work, that experimental framework is used to quantify the mechanisms of action of bromine- and phosphorous-based flame retardants. Here, a series of experiments was conducted in which flame heat transfer and sample mass loss rate were measured as a flame spread over 7 cm tall, 5 cm wide glass-reinforced poly(butylene terephthalate) (PBT) samples manufactured with increasing amounts (12, 16, and 24 wt %) of the brominated flame retardant, poly(pentabromobenzyl acrylate). In an additional series of tests, similar measurements were obtained for PBT samples manufactured with increasing amounts (8, 12, 16, and 20 wt %) of the phosphorous-based flame retardant, aluminum diethyl-phosphinate (trade name: Exolit OP 1230). These tests allow for the study of the impact of flame retardants on key features of the system including: peak flame heat flux, flame stability, and condensed phase decomposition behaviors (i.e. charring). Current measurements identify an optimal loading concentration of Exolit OP 1230 for the PBT samples tested in this work, identify each flame retardant's impacts on char yield and heat of combustion, and indicate that bromine- and phosphorous-based flame retardants (at least at the concentrations tested here) do not affect flame to surface heat transfer in the continuous region of wall flames and do not have a significant effect on flame stability at ignition. These measurements allow for greater understanding of tightly coupled condensed- and gas-phase processes that control flame spread and material burning behavior, thus providing a quantitative connection between material composition and the controlling mechanisms of fire growth.

KEYWORDS:

Fire retardants, wall flames, heat transfer, flame stability, flame spread

IOP Conf. Series: Journal of Physics: Conf. Series **1107** (2018) 032015 doi:10.1088/1742-6596/1107/3/032015

INTRODUCTION

To improve their resistance and response to fire conditions or simply to meet fire safety regulations and standards, flame retardants are often added to combustible materials such as synthetic polymers. Typically, these flame retardants work in the condensed phase by interfering with pyrolysis (e.g. promoting char formation), reducing gas phase combustion efficiency, or both. Market demand for flame retardants is estimated at $2.3 billion with approximately 36% of this market share attributed to brominated flame retardants (BFRs), which are highly effective and relatively low in cost [1] via. [2]. With the rapid expansion in their use, BFRs have been found across the environment in the atmosphere, soil and sediment, bodies of water, and in the tissues of invertebrates, fish, birds, and mammals [3]. This bioaccumulation has evoked growing concern with regards to the environmental hazards that these flame retardants pose [4] and thus industry is moving towards developing safer and equally effective replacements.

To understand how flame retardants reduce material flammability and to develop new, efficient alternatives, it is critical to determine the mechanisms by which they inhibit flaming combustion. One approach is to propose a chemical reaction mechanism through which a flame retardant or a class of flame retardants works. This can allow for the identification of structural chemical features that may be responsible for observed flammability reduction performance; however, this approach does not allow for quantitative prediction of material flammability performance for varied material compositions (e.g. different flame retardant loading) or in response to a range of fire scenarios. Alternatively, one can perform bench scale flammability tests such as ASTM D2863 (Limiting Oxygen Index, LOI) [5] or UL-94 [6]. ASTM D2863 determines the lowest oxygen concentration that will "just support flaming combustion of a material". Although LOI can be quantified in this test, it is not an intrinsic property of the material (it is configuration and scale dependent) thus the material may undergo flaming combustion at significantly lower oxygen concentrations in fully developed fires. UL 94 tests allow for a better qualitative assessment of ignition and flame spread over a material; however, test results are only expressed in terms of a relative ranking scale and it has been shown that no broad, quantitative relationship exists to predict material flammability performance (i.e. burning behavior in the cone calorimeter [7]) using UL-94 ratings [8]. Ultimately, although widely applied, many test standards show limited ability to predict material response outside of particular test conditions [9] and conflicting assessments often arise between different tests [10].

Flame spread is known to be controlled by a positive feedback between coupled processes of solid phase material degradation and gas phase combustion [11]. As gaseous degradation products react with the ambient oxidizer in a diffusion flame, a fraction of the heat produced in this reaction is transferred back to the solid, causing further degradation and production of flammable gases. Quantifying how flame retardants affect these processes is critical to understanding existing and developing new flame retardants and to predicting material flammability performance. Our research group has developed a systematic methodology to assess the flammability of polymeric materials through microscale combustion calorimetry (MCC) [12] and experimental measurement of flame heat feedback and material burning behavior during the early stages of upward flame spread [13, 14]. This experimental framework is used in this work to quantify the mechanisms of action of bromine- and phosphorous-based flame retardants, which are known to act through different mechanisms: primarily in the gas and solid phase, respectively [15]. Here, a series of experiments was conducted on glass-reinforced poly(butylene terephthalate) (PBT) samples manufactured with increasing amounts (8 to 24 wt. %) of either a bromine- or phosphorous-based flame retardant. These tests allow for the quantification of the impact of flame retardants on key features of the system including: peak flame heat flux, flame stability, and mass loss rate during vertical burning, as well as heat of combustion, and char yield.

EXPERIMENTAL SETUP

The PBT used in this study was obtained from BASF in the form of 6 mm thick, 5 cm wide, and 7 cm tall slabs. All samples were reinforced with chopped glass fibers (25 wt. %). Samples were manufactured with either the brominated flame retardant poly(pentabromobenzyl acrylate) (trade name FR 1025) or the phosphorous-based flame retardant, aluminum diethyl-phosphinate (trade name Exolit OP1230). Additional tests examined samples manufactured with two flame retardants (potential synergists): FR 1025 with antimony trioxide or Exolit OP 1230 with Melapur M200. Table 1 identifies the formulation of samples tested in this work.

MCC experiments were conducted in accordance with ASTM D7309 [12] using 3 to 5 mg material samples that were pyrolyzed in nitrogen at a heating rate of 1 K s^{-1}. The heat of complete combustion of gaseous pyrolyzates (ΔH_c) and char yield (μ_{char}) were determined. The chopped glass fibers were assumed not to react in MCC tests and their influence on calculated ΔH_c and μ_{char} was subtracted accordingly. Each material was sampled and tested in the MCC three times to ensure reproducibility.

Table 1. Sample compositions and preheating and ignition times in flame spread experiments.

Flame Retardant Concentration (wt. %)	Sample Preheat Duration (s)	Burner Application Duration, t_{ign} (s)
No Flame Retardant [a]	0	120
8 % Exolit OP 1230	420	20
12 % Exolit OP 1230	420	40
16 % Exolit OP 1230	600	55
20 % Exolit OP 1230	600	70
8 % Exolit OP1230 & 4% Melapur M200	600	No Sustained Ignition
12 % FR 1025	300 or 600[b]	30
16 % FR 1025	600	20
24 % FR 1025	720	10[c]
9% FR 1025 & 3% Antimony Trioxide	720	No Sustained Ignition

[a] Tests on PBT without a flame retardant were conducted in a previous work [13]. Ignition was achieved without external heating, using a non-premixed propane burner that heated the lower 2.5 cm of samples.
[b] Ignition could be achieved with 300 s preheating; however, tests were repeated with a 600 s preheat to allow for direct comparison of mass loss rate after ignition of samples with different FR 1025 loading.
[c] Ignition achieved by a handheld, premixed propane torch applied to the lower 3 cm of samples.

Flame spread experiments were conducted in the apparatus shown in Fig. 1. Samples were insulated at their back surface and all four sides using 6.0 mm thick sheets of Kaowool PM Insulation and secured within a steel holder that exposed only the front surface of the sample slab. Sample preparation is presented in detail elsewhere [13]. Tests began by positioning an electric radiant heater to provide an external heat flux (q''_{ext}) of 10 kW m^{-2} across the sample's front surface. This heat flux, q''_{ext}, remained constant throughout the duration of tests and varied by less than 10 % across the sample's front surface. Samples were then ignited using a premixed methane/oxygen burner, which was kept in place at the bottom edge of the sample (y = 0) just long enough for sustained, uniform ignition across the sample's width. Immediately after sample ignition, the burner was removed and the radiant heater was repositioned to provide q''_{ext} = 20 kW m^{-2}. Samples were allowed to burn until self-extinction occurred. Sample preheating (q''_{ext} = 10 kW m^{-2}) and burner application durations, were selected to best maintain similar ignition conditions between samples; however, as shown in Table 1, higher flame retardant loading concentrations necessitated longer preheating to achieve ignition.

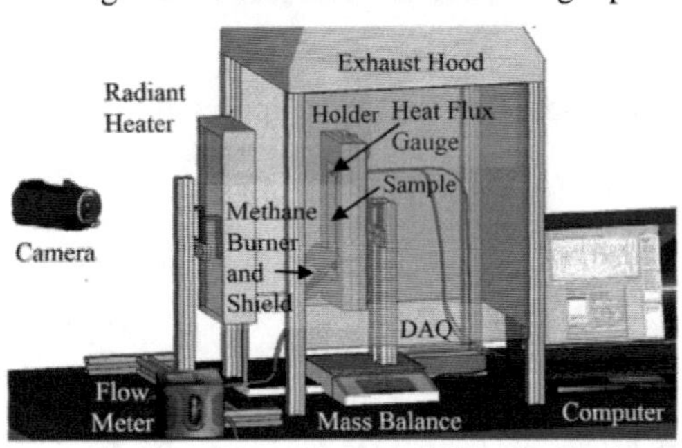

Figure 1. Experimental setup for vertical burning and upward flame spread experiments.

Sample mass loss rate and flame to surface heat flux were measured during flame spread tests. These measurements were obtained in separate experiments to avoid potential interference between the corresponding diagnostic tools. To measure sample mass loss rate, the sample holder was placed on top of a Mettler Toledo XS4002S balance, which recorded sample mass at a frequency of 1 Hz. Mass loss rate, $\frac{dm}{dt}$, was calculated as the numerical derivative of measured sample mass, using a 1 s time step. For flame heat flux measurements, a 0.95 cm diameter, water-cooled (T_{HFg} = 291 K) Schmidt-Boelter heat flux gauge was positioned at the top of the sample (y = 7 cm) such that its face was flush with that of the slab and the surrounding insulation (above). Calibration, mounting, and maintenance of the heat flux gauge is described in detail elsewhere [13,14]. Total heat flux (from the flame and external heater) measured by this water-cooled heat flux gauge, q''_{HFg}, was recorded at 2 Hz using an NI USB-9211A data acquisition module (DAQ) connected to a computer. For each sample and measurement type, tests were successfully repeated 2-3 times (up to six samples were tested for each material, some did not ignite). Measurements from repeated tests were averaged together and this combined dataset was smoothed using a 5 s running average. All tests were videotaped.

RESULTS AND DISCUSSION

Figure 2 shows burning behavior during upward flame spread over PBT with and without flame retardants. Without a flame retardant, PBT flames spread and remain continuous and laminar across the full length of the sample. Soot deposition is observed across the sample's surface, downstream of the pyrolysis front. Extinction occurs only after complete sample burnout, leaving behind a residual structure (char and chopped glass fibers). Samples containing the phosphorous-based flame retardant Exolit OP 1230 developed a thin, flaky char layer at their surface that thickened while burning. This char acts as an insulator and physically separates the flame from the sample. This produces non-uniform burning across the sample's width and local extinction of the flame, with flamelets primarily found attached to cracks in this layer. Removal or destruction of this char layer allows for the formation of a larger flame that burns uniformly across the width of the newly exposed sample. With the bromine-based FR 1025 (not shown in Fig. 2) flames remain continuous and laminar across the sample; however, flame extinction occurs much sooner in experiments beginning with flickering at the base of the flame, followed shortly thereafter by partial liftoff of the base of the flame (oscillations that increase in magnitude), and complete flame extinction. The destabilizing effect of this flame retardant on the base of the diffusion flame (and resulting liftoff) is consistent with previous observations [16,17].

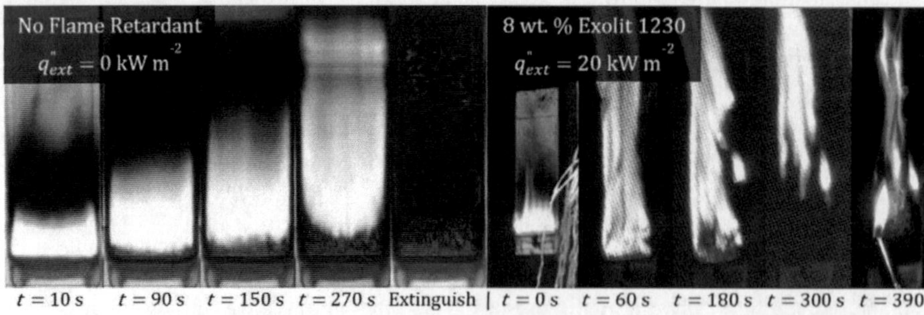

Figure 2. Vertical burning behavior of glass-reinforced PBT samples without (left) and with (right) 8 wt. % of a phosphorous-based flame retardant. Note: timestamps indicate time, *t*, after sample ignition.

Figure 3 plots ΔH_c and μ_{char} of PBT samples studied in this work; error bars for ΔH_c are calculated based on the 5% uncertainty of oxygen consumption calorimetry measurements and for μ_{char} as two standard deviations of the mean. As shown here, increasing the loading of Exolit OP 1230 from 0 to 20 wt. % yields an increase in μ_{char} from 3.1 to 12.2 % but produces a negligible impact on ΔH_c. Increasing the loading of FR 1025 from 0 to 24 wt. % decreases ΔH_c by 38.2% while slightly increasing μ_{char} from 3.1 to 7.2%. The potential synergists Melapur M200 and antimony trioxide appear to have a negligible impact on measured ΔH_c and μ_{char}; however, peak heat release rate measured throughout the duration of MCC experiments decreases by approximately 20 % for samples with 9 wt. % FR 1025 and 3 wt. % antimony trioxide (versus PBT with only 12 wt. % FR 1025).

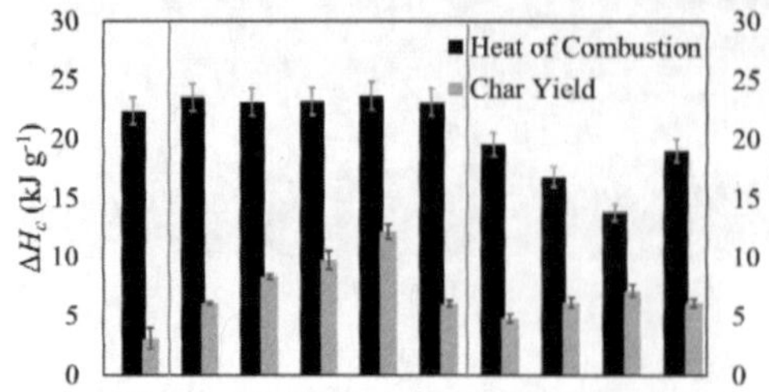

Figure 3. MCC-measured heats of complete combustion of gaseous pyrolyzates (ΔH_c) and char yield (μ_{char}).

Figure 4 shows mass loss rate, $\frac{dm}{dt}$, and total flame heat flux, q''_{HFg}, of PBT samples during upward flame spread tests. Note: for mass loss rate and flame heat flux tests, samples with 12 % FR 1025 were preheated for 600 and 300 s, respectively. With higher loading of FR 1025, peak mass loss rate increases and is observed sooner after ignition; however, with more FR 1025, extinction occurs earlier and total mass loss decreases. Data from PBT with 9 wt. % FR1025 and 3 wt. % antimony trioxide is not shown in Fig. 4 because sustained flaming ignition of this material could not be achieved; however, given similar preheating, these samples ignited briefly and supported approximately the same peak mass loss rate, but 35% lower total mass loss, as PBT with 12 % FR 1025. Increased FR 1025 loading does not significantly alter measured q''_{HFg}. This indicates that flame to surface heat transfer in the continuous region of these laminar wall flames is not impacted by this flame retardant at these concentrations, even if it has been shown to act in the gas phase.

Figures 4c and 4d show measured mass loss rate and q''_{HFg} from tests on PBT with Exolit OP 1230. Note: increased scatter in Figs. 4c, 4d arises due to greater variation in material burning behavior (e.g. local flame extinction and/or attachment across the sample's surface) in repeated tests on PBT with Exolit OP 1230. Compared to results from tests on samples with FR 1025 – mass loss rate in these tests gradually decreased prior to extinction, as a char layer developed. With higher loading of Exolit OP 1230 (from 0 to 16 wt. %) peak mass loss rate and total mass loss both decrease; however, with even greater loading (from 16 to 20 wt. %), both peak mass loss rate and total mass loss *increase*. This indicates that, for this series of PBT tests, 16 to 20 wt. % Exolit OP 1230 may be optimal for reducing material flammability. Depending on the structure and proximity of flamelets to the heat flux gauge (Fig. 2) significant variations in q''_{HFg} are observed. However, when a steady, continuous flame is established over the surface of the heat flux gauge (in Fig. 4d, these measurements are highlighted by a thick, black, dashed outline) q''_{HFg} is similar to that measured for brominated samples and does not demonstrate a distinct dependence on flame retardant loading.

Flame stability can be assessed by calculating the minimum energy release rate needed for sustained flaming of PBT samples as: $Q'_{crit} = \left(\frac{1}{w}\frac{dm}{dt}\Big|_{t=0}\right) \times \Delta H_c$, where w represents flame width (at ignition, $t = 0$, this is equal to sample width, 5 cm). Q'_{crit} equals 5.0, 5.9, and 7.5 kW m^{-1} for PBT with 12, 16, and 24 wt. % FR 1025, respectively. This apparent increase in Q'_{crit} with loading of FR 1025 does not appear to be significant (i.e. variations are within two standard deviations of the mean). For PBT samples with 8, 12, 16, and 20 wt. % Exolit OP 1230, Q'_{crit} equals 5.6, 6.4, 5.7, and 2.4 kW m^{-1}, respectively. Note: for 20 wt. % Exolit OP 1230, Q'_{crit} is determined based on a single experiment. These measurements show that flame stability at ignition does not demonstrate a strong dependence on the loading of either the bromine-based FR 1025 or the phosphorous based Exolit OP 1230. PBT samples with mixtures of two flame retardants (see Table 1) did not ignite; this may indicate that certain flame retardant combinations affect flame stability at ignition.

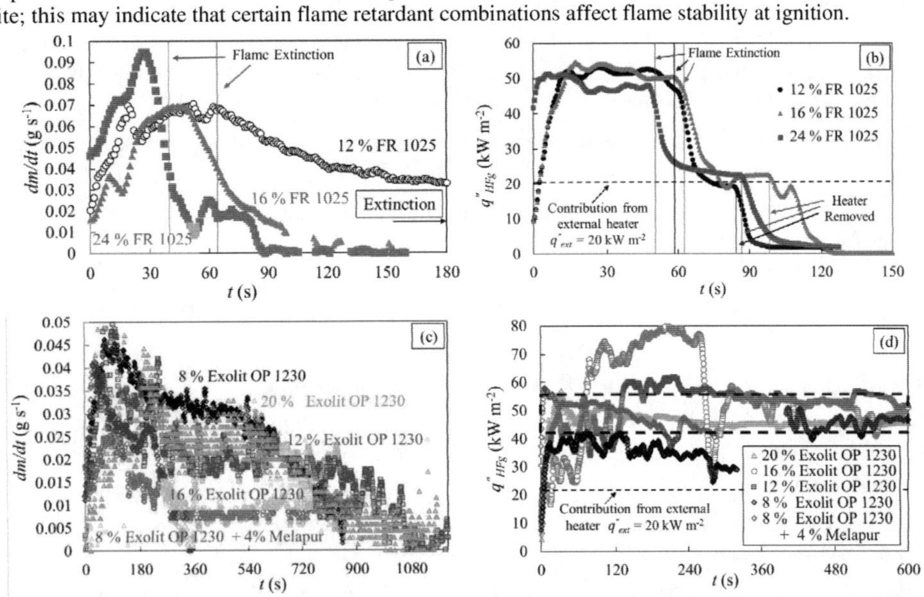

Figure 4. Measured mass loss rate, $\frac{dm}{dt}$, and total heat flux, q''_{HFg}, of PBT samples with bromine- or phosphorous-based flame retardants: FR 1025 (plots a & b) and Exolit OP 1230 (plots c & d), respectively.

CONCLUSIONS

In this work, the mechanisms of action of bromine- and phosphorous-based flame retardants were assessed by measuring key parameters controlling composite material burning behavior including: flame heat flux, sample mass loss rate, heat of combustion, and char yield. Key findings include: flame to surface heat flux in the continuous region of a laminar wall flame is not affected by the flame retardants tested here (even those known to act in the gas phase), quantification of the critical energy release rate, $\dot{Q}'_{crit}$, required for sustained flaming (this measure of flame stability does not demonstrate a clear dependence on flame retardant loading, at least at the concentrations tested here), and the identification of an optimal loading concentration (wt. %) for the phosphorous-based flame retardant, Exolit OP 1230, to reduce peak mass loss rate and total mass loss of PBT samples burning in the vertical configuration.

REFERENCES

[1] Kirk, R.E., Othmer, D.F., *Kirk-Othmer Encyclopedia of Chemical Technology.* New York: Wiley (2007).

[2] Guerra, P., Alaee, M., Eligarrat, E., Barcelo, D., "Introduction to Brominated Flame Retardants: Commercially Products, Applications, and Physicochemical Processes," ed. Eljarrat, E., Barcelo, D., *Brominated Flame Retardants - Handbook of Environmental Chemistry,* Berlin: Springer, Vol:16, p. 1-17 (2010). https://doi.org/10.1007/698_2010_93

[3] De Wiit, C. A., Kierkegaard, A., Ricklund, N., Sellstrom, U., "Emerging Brominated Flame Retardants in the Environment," Eljarrat, E., Barcelo, D. (ed.), *Brominated Flame Retardants - Handbook of Environmental Chemistry,* Berlin: Springer, Vol: 16, p. 241-286 (2011). https://doi.org/10.1007/698_2010_73

[4] Birnbaum, L.S., Staskal, D.F, "Brominated Flame Retardants: Cause for Concern?," Environmental Health Perspectives 112: pp. 9-17 (2004). https://doi.org/10.1289/ehp.6559

[5] ASTM Standard D2863-13, "Standard Test Method for Measuring the Minimum Oxygen Concentration to Support Candle-Like Combustion of Plastics (Oxygen Index)", ASTM International, West Conshohocken, PA, DOI: 10.1520/D3801-10, www.astm.org (2003).

[6] UL Standard 94, "Standard for Safety of Flammability of Plastic Materials for Parts in Devices and Applications testing," Northbrook, IL, Underwriter's Laboratories (2013).

[7] ASTM Standard 1354-15A, "Standard Test Method for Heat and Visible Smoke Release Rates for Materials and Products Using an Oxygen Consumption Calorimeter," ASTM International, West Conshohocken, PA, DOI: 10.1520/E1354-15A, www.astm.org (2003).

[8] Morgan, A. B., Bundy, M. "Cone calorimeter analysis of UL-94 V rated plastics", Fire and Materials 31: pp. 257-283 (2007). https://doi.org/10.1002/fam.937

[9] Quintiere, J., "The Application of Flame Spread Theory to Predict Material Performance," Journal of the Research of the National Bureau of Standards: pp. 61-70 (1988).

[10] Grant, G., Drysdale, D., "Numerical Modelling of Early Flame Spread in Warehouse Fires," Fire Safety Journal 24: 247-278 (1995). https://doi.org/10.1016/0379-7112(95)00022-l

[11] DeRis J.N., "Spread of a Laminar Diffusion Flame," Proceedings of the Combustion Institute 12: pp. 241-252 (1969). https://doi.org/10.1016/s0082-0784(69)80407-8

[12] ASTM Standard D7309, Standard Test Method for Determining Flammability Characteristics of Plastics and Other Solid Materials Using Microscale Combustion, ASTM International, West Conshohocken, PA (2013).

[13] Leventon, I.T., Korver, K.T., Stoliarov, S.I., "A Generalized Model of Flame to Surface Heat Feedback for Laminar Wall Flames," Combustion and Flame 179: pp. 338-353 (2017).

[14] Leventon, I.T., Li, J., Stoliarov, S.I., "A Flame Spread Simulation Based on a Comprehensive Solid Pyrolysis Model Coupled with a Detailed Empirical Flame Structure Representation," Combustion and Flame 162: pp. 3884-3895 (2015). https://doi.org/10.1016/j.combustflame.2015.07.025

[15] Hilado, C. J., *Handbook of Plastics*, Lancaster, PA: Technomic Publishing Co., Fourth Edition, p. 167-195 (1990).

[16] Linteris, G.T., Takahashi, F., Katta, V.R,, "Cup-burner flame extinguishment by CF_3Br and Br_2," Combustion and Flame 149: pp. 91-103 (2007). https://doi.org/10.1016/j.combustflame.2006.12.013

[17] Sohn, C.H., Chung, S.H., Kim, J., "Instability-Induced Extinction of Diffusion Flames Established in the Stagnant Mixing Layer," Combustion and Flame 117: pp. 404-412 (1999).

IOP Conf. Series: Journal of Physics: Conf. Series **1107** (2018) 032016 doi:10.1088/1742-6596/1107/3/032016

PROPTI – A Generalised Inverse Modelling Framework

Lukas ARNOLD, Tristan HEHNEN, Patrick LAUER, Corinna TRETTIN, and Ashish VINAYAK

[1] Civil Safety Research, Forschungszentrum Jülich, Germany
l.arnold@fz-juelich.de
[2] Civil Safety Research, Forschungszentrum Jülich, Germany
tristan.hehnen@cern.ch
[3] Department of Fire and Explosion Safety, University of Wuppertal, Germany
lauer@uni-wuppertal.de
[4] Department of Fire and Explosion Safety, University of Wuppertal, Germany
trettin@uni-wuppertal.de
[5] Civil Safety Research, Forschungszentrum Jülich, Germany
a.vinayak@fz-juelich.de

ABSTRACT

Simulation of pyrolysis involves the knowledge of reaction kinetics and thermo-physical parameters, which are in general not directly measurable. Inverse modelling provides means to deduce the needed parameters from experimental data. This complex process involves the coupling of a simulation model with an optimisation method as well as the handling of a large amount of data. All of these processes are prone to errors and therefore a unified and automated implementation is beneficial for the whole community. In this contribution, a software to carry out this process is proposed. PROPTI is an open source tool written in Python, that is meant to provide a framework for inverse modelling of parameters in computer simulation, with emphasis on pyrolysis modelling in fire simulation. Its generalised formulation allows the usage of any simulation model in combination with various experimental data. The underlying optimisation library allows the utilisation of HPC systems. After presenting the concept of this framework, two examples are shown to illustrate the process and demonstrate the capabilities.

KEYWORDS: *pyrolysis, FDS, inverse modelling, optimisation, HPC, framework*

INTRODUCTION

Pyrolysis describes a thermo-chemical decomposition of solids releasing gaseous species, some of which are combustible. Thus, modelling pyrolysis is essential to simulate burning of solid materials. Often the heat release rate of a fire scenario is used to represent a design fire. This implicitly prescribes the pyrolysis and therefore the fire development. In this approach, environmental conditions, like ventilation and geometry, have no self consistent impact on the fire. To overcome these restrictions, the pyrolysis reaction rate can be computed, based on the material properties, temperature and environmental conditions. One possible estimation of the reaction rate is formulated following the Arrhenius equation, with its material specific set of reaction kinetic parameters.

To compute pyrolysis the material temperature has to be determined. This implies that the reaction kinetic parameters are not sufficient to fully describe the degradation behaviour. Thermo-physical parameters, like density or heat capacity, are necessary, as well as other parameters, like layer thickness. Real materials are complex mixtures of different chemical components. This makes it very difficult to design experiments to directly gain the needed parameters. Therefore, a possible approach to obtain these parameters is an inverse modelling process (IMP). [5]

As a general concept, given experimental data and a model, the target of the IMP is to identify the set of model parameters, which best represent the experimental observation. To achieve this objective, an IMP builds upon an optimisation algorithm. A simulation of the current parameter set is carried out, then the simulation results are compared to the target data. Based on this result the optimisation algorithm proposes a new set of parameters for the next iteration. This loop is repeated until convergence criteria are satisfied. This approach is shown in figure 1.

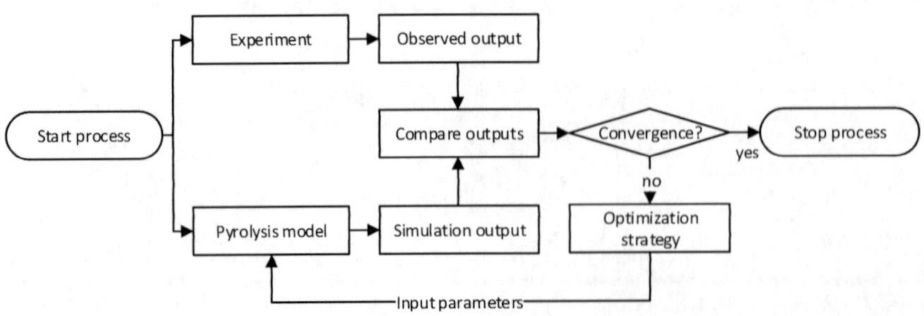

Fig. 1. Inverse modelling process.

A challenge of this approach is to couple the simulation model to an optimisation algorithm, and to process the resulting data. Matala investigated the usage of different optimisation strategies [8], Wang used a genetic algorithm in combination with FDS to investigate road tunnel fires [12], Bal and Rein investigated effects of inverse pyrolsis modelling on the estimated parameters[2], and many more. One of the most notable contributions is Gpyro by Lautenberger [7], which provides a highly detailed (3D) pyrolysis model and boundary conditions for FDS. It also provides means for material property estimation with a collection of optimisation algorithms.

In contrast to the above, the presented framework in this contribution is not aimed to develop a new model for heat transfer or pyrolysis itself. It strives to provide a generic method interface of how to perform inverse modelling as an open source software application. Therefore, users need to spend less time to set up a framework on their own, but can focus on the actual task at hand. Thus, it is highly beneficial for the fire safety community.

A flexible and open source IMP framework called PROPTI [1] is proposed. It focuses on handling the experimental data sets and the communication between simulation software and optimisation algorithms. Different packages of optimisation algorithms can be linked to PROPTI, which allows to chose algorithms in a simple fashion. Up to now, SPOTPY [4] has been linked, but connections to other packages are easily possible. Virtually any simulation software can be used, since PROPTI handles the coupling in a general way. However, due to the authors current research activities, the focus is on Fire Dynamics Simulator (FDS) [10]. An appropriate documentation, examples and tutorials on the usage of PROPTI are provided.

FRAMEWORK

The PROPTI framework is written in Python 3 and consists of a set of object oriented modules and convenience scripts, which ease typical application scenarios. It depends on common Python modules, as well as the optimisation package SPOTPY [4]. The framework is written in a very general way, with no emphasis on any special simulation model or experimental setup. However, there exist formal requirements for the definition of the model's input parameters and the formatting of simulation and experimental data. The optimisation process is able to consider multiple experiments and has no restriction on which parameters shall be optimised. For demonstration purposes, the modelling of thermogravimetric analysis (TGA) experiments with FDS is shown in the following section. Before the IMP is started, a sensitivity analysis can be conducted to quantify the individual impact of each parameter and eventually reduce the number of parameters to be optimised. Optimisation methods can easily be swapped and arbitrary algorithm packages can be linked to PROPTI.

To process experimental data sets and results of the simulation model, they need to be stored as simple text files. This requirement is directly satisfied by FDS if output devices are used, otherwise an additional post-processing has to be triggered. The data sets do not have to be defined at the exactly same evaluation points, as PROPTI will use interpolation methods for data comparison.

Data and Model Definition

As the initial step, a configuration file must be created, that utilises Python syntax. It contains details about where to find the experimental data, which optimisation method to be applied and which parameters are to be examined. Information on the simulation model execution, including the path to the executable, the execution mode (serial or parallel) as well as input file templates must be provided. The experimental data is a set of data files that cover multiple experiments or repetitions. Thus, their relationship to the corresponding simulation setups, i.e. input file templates, has to be defined as well.

Execution of the Optimisation

Once the configuration file is set up, the optimisation procedure can be started. The main phases, which are consequently repeated, are basically:

1. PROPTI calls the optimiser to define a set of current parameter values to be evaluated.
2. PROPTI sets up working directories for each simulation run.
3. PROPTI uses the provided information to replace the place holders in the template file, or multiple files if needed, with the current parameter values.
4. PROPTI executes the models.
5. PROPTI extracts the simulation data and passes it to the optimiser.
6. The optimiser compares the experimental and simulation data and generates a fitness value.
7. The optimiser decides if further evaluations are needed, based on the fitness value.

A further advantage of PROPTI is an additional set of tools, that provides means for automated pre- and post-processing within the IMP. Due to the usage of placeholders and Python syntax, complex logic dependencies between parameters can be represented. Since the optimisation process can be computationally demanding, it benefits from high performance computing (HPC) systems and therefore PROPTI is designed with HPC application in mind. Depending on the algorithm used, PROPTI is able to simultaneously execute simulations with these parameter sets. This concept allows to use shared (via sub-processes), as well as distributed (via MPI) memory systems and is already implemented in PROPTI. It has been successfully used on CERN HPC-clusters and the JURECA supercomputer [6].

Analysis

After the model evaluations are finished, the optimiser provides the parameter set, which best fits the experimental data sets. General means of analysing the simulation results are provided by PROPTI. It is possible to create plots, that show the development of each parameter during the IMP. Furthermore, an overview of the fitness value development is provided. FDS input files with the best parameter set can be generated for each simulation setup.

IOP Conf. Series: Journal of Physics: Conf. Series **1107** (2018) 032016 doi:10.1088/1742-6596/1107/3/032016

EXAMPLES

The examples demonstrate general capabilities of PROPTI on deliberately simple setups. Mock-up material parameters have been generated, utilising a simulation of a thermogravimetric analysis (TGA) experiment from the FDS examples (`tga_analysis.fds`). This fabricated experiment is used to verify the capability of the IMP to reproduce known values.

The selected material consists of three components, which together define a material sample of cellulose. The first component is water, that evaporates around 100 °C, and for simplicity is neglected here. The two remaining components are each described by the same material parameters, like density or heat of combustion, and differ only in reaction kinetics. Here, the FDS parameters `REFERENCE_TEMPERATURE` and `REFERENCE_RATE` have been chosen, as described in the FDS user guide. Obviously, these parameters are easily estimated by fitting a triangle to each peak, as described in literature already [9]. As stated above, these examples are chosen for the sake of demonstrating the procedure.

The first example demonstrates that the combination of PROPTI and the SCE-UA algorithm is able to converge to a known solution. The second example demonstrates how multiple experiments, and simulations thereof, can be used to gain a more robust parameter set. Here, the material is subjected to three different heat fluxes, by using three heating rates in both the experiment and the simulation: 5 K/min, 10 K/min and 15 K/min (`HEATING_RATE`). More sophisticated setups, which include a wider variety of experiments or more repetitions, are possible as well and are part of authors current research.

As an optimisation algorithm the SPOTPY implementation of the Shuffled Complex Evolutionary Algorithm (SCE-UA) [3] was chosen. Four complexes were used, based on the four optimisation parameters and the maximum number of repetitions, i.e. the generated parameter sets, was set to 600. Note, in the second example three different simulation setups are involved, therefore up to 600 times three model executions are needed.

Example 1

After 600 repetitions, which corresponds to 16 generations, the best parameter set is close to the original parameter values. A comparison between the parameters themselves is provided in table 1. Figure 2a provides a comparison between the mock-up experimental data and the model's response using the best parameter set. Figure 2b shows the development of the fitness value during the run of example 1. The fitness value is defined as a negative value and the optimiser tries to maximise it. Each generation is represented as a boxplot. Thus, it is easier to judge the development of the fitness value, specifically for generations that each contain hundreds of simulations.

For assessing the parameter sensitivity, a Fourier Amplitude Sensitivity Test (FAST) [11] is conducted. The output is shown in table 2 with s_1, the individual influence of a single parameter, and s_t, the sensitivity of the model to a single parameter dependent on all other parameters.

As a visualisation of the sensitivity assessment, the development of the parameters with the lowest and the highest values are shown. The parameter that got the lowest significance value during the FAST assessment is the reference rate of component 2. Its development over the IMP is shown in figure 3a. The highest significance value was received by the reference temperature of component 1, shown in figure 3b. These figures represent the expected development, based on the sensitivity analysis.

Table 1. Comparison of known parameters from the prescribed FDS example and the SCE-UA results.

	Known parameter	Example 1	Example 2
Reference temperature component 1	315.0	315.006	315.008
Reference rate component 1	0.0056	0.00559347	0.00559759
Reference temperature component 2	430.0	429.992	430.007
Reference rate component 2	0.0075	0.00750615	0.00749932

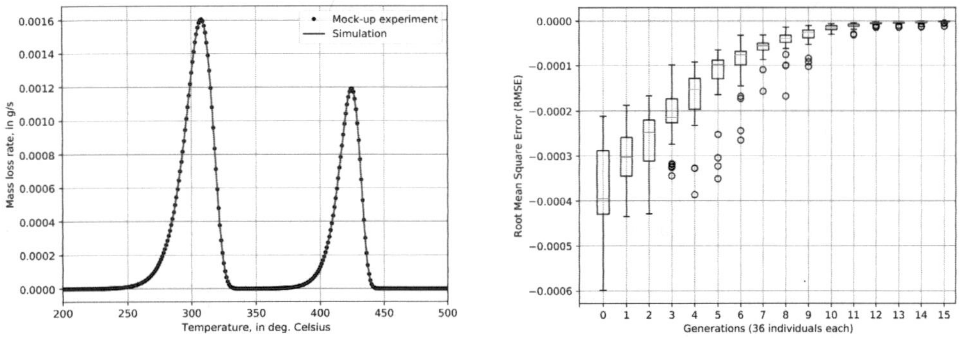

(a) Comparison between the mock-up experimental (original FDS) data and the simulation response of the best parameter set.

(b) Development of the fitness value during the inverse modelling process, where generations are represented as boxplots.

Fig. 2. Results of the first example.

Table 2. Sensitivity of each parameter on the simulation response, computed with FAST.

Parameter	s_1	s_t
Reference temperature component 1	0.352860	0.579744
Reference rate component 1	0.319699	0.402163
Reference temperature component 2	0.028856	0.140844
Reference rate component 2	0.030123	0.068708

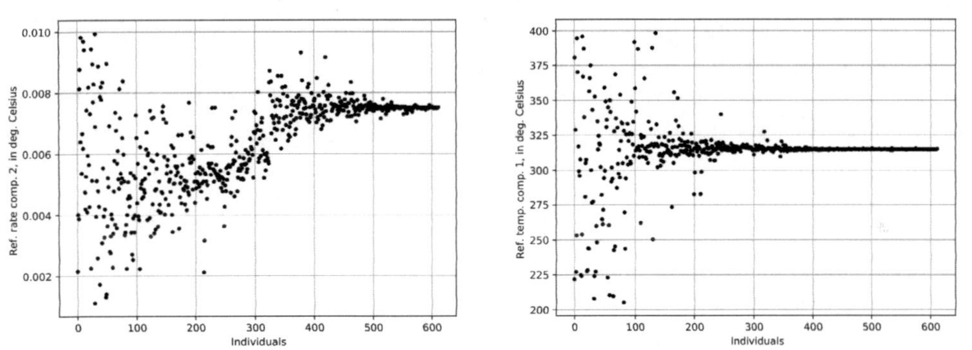

(a) Development of the reference rate of component 2 during the inverse modelling process. Lowest significance value from FAST.

(b) Development of the reference temperature of component 1 during the inverse modelling process. Highest significance value from FAST.

Fig. 3. Parameter development in the first example.

Example 2

The previous example is extended to three different simulation setups for different heating rates. Thus, each parameter set needs to be evaluated under three different conditions and all results are compared to their respective (mock-up) experimental data. In this case, the same parameter set is written into three FDS input files, which represent the different experimental conditions. The simulation setups enable the user to specify the different relations between individual experimental and simulation data series. As a result of the IMP, an optimal parameter set is generated to recreate the material behaviour well over all experimental conditions. The simulation results of the best parameter set are shown in figure 4a, which match the target mass loss rates reasonably well. An overview of the development of the fitness values for example 2 is provided in the boxplots in figure 4b. The fitness value converges again.

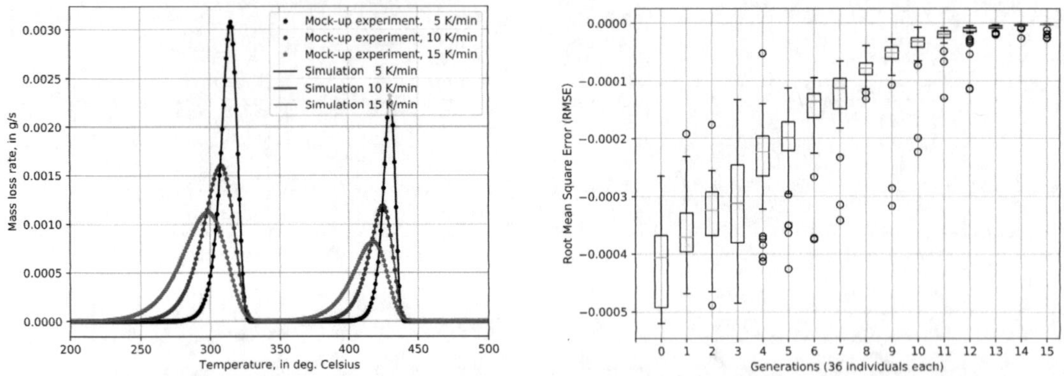

(a) Comparison between the mock-up experimental (original FDS) data and the simulation response of the best parameter set.

(b) Development of the fitness value during the inverse modelling process, where generations are represented as boxplots.

Fig. 4. Results of the second example.

SUMMARY & OUTLOOK

A framework to estimate parameters for pyrolysis modelling, utilising an inverse modelling process is proposed in this work. It provides a guided starting point for scientific and engineering projects to model fire propagation. Its generality allows to use arbitrary simulation software as well as optimisation algorithms. Furthermore, its standardisation supports the knowledge transfer among its users. The basic capabilities and features were demonstrated and verified, based on simple TGA data sets. The authors welcome anyone to make use of this framework and contact the authors in case of questions. As this is an ongoing project, the authors encourage contributions to this project.

ACKNOWLEDGMENTS

The authors gratefully acknowledge the computing time granted (project jjsc27) by the JARA-HPC Vergabegremium and VSR commission on the supercomputer JURECA [6] at Forschungszentrum Jülich.

REFERENCES

[1] L. Arnold, T. Hehnen, P. Lauer, C. Trettin, and A. Vinayak. "FireDynamics/propti," March. Zenodo. Mar. 2018. doi:10.5281/zenodo.1188756.

[2] N. Bal, and G. Rein. "On the Effect of Inverse Modelling and Compensation Effects in Computational Pyrolysis for Fire Scenarios." *Fire Safety Journal* 72: 68–76. 2015. doi:10.1016/j.firesaf.2015.02.012.

[3] Q. Duan, V. K. Gupta, and S. Sorooshian. "Shuffled Complex Evolution Approach for Effective and Efficient Global Minimization." *Journal of Optimization Theory and Applications* 76 (3). Springer: 501–521. 1993.

[4] T. Houska, P. Kraft, A. Chamorro-Chavez, and L. Breuer. "SPOTting Model Parameters Using a Ready-Made Python Package." *PloS One* 10 (12). Public Library of Science: e0145180. 2015.

[5] M. M. Khan, A. Tewarson, and M. Chaos. "Combustion Characteristics of Materials and Generation of Fire Products." In *SFPE Handbook of Fire Protection Engineering*, edited by M. J. Hurley et al., 1143–1232. Springer New York, New York, NY. 2016. doi:10.1007/978-1-4939-2565-0_36.

[6] D. Krause, and P. Thörnig. "JURECA: General-Purpose Supercomputer at Jülich Supercomputing Centre." *Journal of Large-Scale Research Facilities JLSRF* 2: 62. 2016.

[7] C. Lautenberger. "Gpyro3D: A Three Dimensional Generalized Pyrolysis Model." INTERNATIONAL ASSOCIATION FOR FIRE SAFETY SCIENCE. 2014.

[8] A. Matala, C. Lautenberger, and S. Hostikka. "Generalized Direct Method for Pyrolysis Kinetic Parameter Estimation and Comparison to Existing Methods." *Journal of Fire Sciences* 30 (4). Sage Publications Sage UK: London, England: 339–356. 2012.

[9] K. McGrattan, S. Hostikka, R. McDermott, J. Floyd, C. Weinshenk, and K. J. Overholt. *Fire Dynamics Simulator User's Guide*. Apr. 2018.

[10] K. McGrattan, R. McDermott, W. Mell, G. Forney, J. Floyd, S. Hostikka, and A. Matala. "Modeling the Burning of Complicated Objects Using Lagrangian Particles." In *Conference Proceedings of the Twelfth International Interflame Conference*, 743–753. 2010.

[11] A. Saltelli, S. Tarantola, and K.-S. Chan. "A Quantitative Model-Independent Method for Global Sensitivity Analysis of Model Output." *Technometrics* 41 (1). Taylor & Francis Group: 39–56. 1999.

[12] X. Wang. "Fire Dynamics Simulator (FDS) Pyrolysis Model Analysis of Heavy Goods Vehicle Fires in Road Tunnels." Phdthesis, University of Canterbury, Private Bag 4800, Christchurch, New Zealand. 2017.

Experimental study on the onset of swelling for thin intumescent coatings

Andrea LUCHERINI[1], Nemer ABUSAMHA[2], Jupiter SEGALL-BROWN[3] and Cristian MALUK[1]

[1]*School of Civil Engineering, The University of Queensland, Australia*
[2]*Arup, Dubai, UAE*
[3]*Arup, Sydney, Australia*

a.lucherini@uq.edu.au

ABSTRACT

Intumescent coatings (also known as reactive coatings) are widely used to protect load-bearing steel structural elements during fire. Upon heating, these initially thin coatings swell to form a low density, highly insulating foamed char that insulates steel substrate. Within the context of modern fire safety engineering design of steel structural systems, thermal insulation during fire may be key for assuring that steel does not reach 'critical' temperatures that may yield local structural instability and/or progressive failure (i.e. global instability).

Traditionally, the fire performance of intumescent coatings is typically based on compliance (and to some extent their performance) to the standard fire resistance test in furnace, where single coated elements are exposed to the standard temperature-time curve. Numerous research studies have emphasised the influence of heating conditions on the onset of swelling and the overall effectiveness of intumescence. Moreover, some studies have shown that slow growing fires or low heating regimes may cause an incomplete swelling of the intumescent coating, or even melting or delamination. Consequently, the onset of swelling of is key for assuring its effectiveness in providing thermal insulation to the steel substrate during fire. Past researchers have shown that the onset of swelling occurs for temperatures in the range of 200-300°C. There are limited research studies that have look at the effects of varied, non-standard heating regimes on the onset or effectiveness of swelling of thin intumescent coatings.

The study presented herein investigates the onset of swelling for a commercially available thin intumescent coating exposed to a wide range of heating regimes. Experiments are performed by using radiant panels and controlling incident radiant heat flux at the exposed surface of coated steel samples. This allows for the direct and precise control of the thermal boundary conditions at the exposed surface of tested samples. The onset of swelling criteria were defined in two ways: (1) visual observation of swelling during heating or (2) time-history of the protected steel temperature.

The outcomes of this work aim at yielding a sound understanding of the true effectiveness of thin intumescent coatings. Experimental results mainly show that onset of swelling is directly influenced by the heating regime and it occurs for test samples heated at constant incident radiant heat flux above 20 kW/m^2. The onset of swelling for the tested thin intumescent coating occurs for steel temperatures between 150 and 300°C and 'mean coating temperature' between 350 and 450°C. However, the steel temperature at onset of swelling is inversely proportional to the amount of constant incident heat flux at the exposed surface of tested samples, while the 'mean coating temperature' is directly proportional. As a conclusion, the time-history of net heat flux at the exposed surface of tested samples is not the only parameter that governs the onset of swelling for thin intumescent coatings.

KEYWORDS:

Thin intumescent coatings; onset of swelling; incident radiant heat flux; heat transfer modelling; fire testing; H-TRIS.

INTRODUCTION

Intumescent coatings (also known as reactive coatings) are widely used to protect load-bearing steel structural elements during fire. Upon heating, these initially thin coatings swell to form a low density, highly insulating foamed char that insulates steel substrate. Within the context of modern fire safety engineering design of steel structural systems, thermal insulation during fire may be key for assuring that steel does not reach 'critical' temperatures that may yield structural local instability and/or progressive failure (i.e. global instability) [1]. The increasing growth of thin intumescent coatings in the built environment is associated with the low impact on the aesthetics of bare steel elements, the cost-effective on-site or off-site application and the perceived good performance of the product across the industry [2]. Intumescent coatings are usually based on a combination of organic and inorganic components bound together in a polymer matrix [3]. Thin intumescent coatings are typically solvent-based or water-based products and their application is defined in a Dry Film Thickness (DFT), usually within the range of a few millimetres.

Traditionally, the fire performance of intumescent coatings is typically based on compliance (and to some extent their performance) to the standard fire resistance test in furnace, where single coated elements are exposed to the standard temperature-time fire curve [4, 5]. Numerous research studies have emphasised the influence of heating conditions on the onset of swelling and the overall effectiveness of intumescence [1, 3, 6-8]. Moreover, some studies have shown that slow growing fires or low heating regimes may cause an incomplete swelling of the intumescent coating, or even melting or delamination [1, 8]. Consequently, the onset of swelling has a key role for assuring the coating effectiveness in providing thermal insulation to the steel substrate during fire. Past researchers have shown that the onset of swelling occurs for temperatures in the range of 200-300°C [9-11]. Most of the past experimental research has been conducted using small-scale samples (few grams or milligrams), disregarding any heat transfer process within the expanding coating. There are limited research studies that have look at the effects of varied, non-standard heating regimes on the onset or effectiveness of swelling of thin intumescent coatings.

The study presented herein investigates the onset of swelling for a commercially available thin intumescent coating exposed to a wide range of heating regimes. Experiments are performed by using radiant panels and controlling incident radiant heat flux at the exposed surface of coated steel samples. This allows for the direct and precise control of the thermal boundary conditions at the exposed surface of tested samples. The onset of swelling criteria were defined in two ways: (1) visual observation of swelling during heating or (2) time-history of the steel temperature. This enables the definition of a threshold for the onset of swelling, found (based on exploratory work described herein) to be influenced by incident radiant heat flux at the exposed surface, overall net heat at the exposed surface and steel temperature.

EXPERIMENTAL INVESTIGATIONS

Fire test method

Within the scope of this study, a Heat-Transfer Rate Inducing System (H-TRIS) method was used. This fire test method allows for the direct control of incident radiant heat flux at the exposed surface of the test samples, i.e. control of the time-history of incident heat flux [12]. This is possible by controlling the relative position between the target exposed surface of the test sample and an array of radiant heaters coupled with a computer-controlled mechanical linear motion system (see Fig. 1). This H-TRIS is built from a single high-performance gas-fired radiant heater mounted on a supporting frame (see Fig. 1). This experimental setup enables the careful visual inspection of test samples during fire testing (e.g. gauging the rate of swelling of intumescent coatings), technically challenging during a conventional standard furnace test [13].

Experimental campaign

Within the scope of this work, 200x200 mm^2 steel plates with a thickness of 10 mm (section factor A_s/V_s = 100 m^{-1}), were professionally coated with a commercially available solvent-based thin intumescent coating. The initial applied DFT was 1600 μm (± 200 μm). During the experimental campaign, individual uncoated and coated test samples were heated during 30 minutes under constant incident radiant heat flux, varying within 10 and 55 kW/m^2. During testing, individual samples were positioned on custom-built sample holder (Fig. 1) and the unexposed surface (backside) of the steel plate was insulated using a low density insulating material to minimise heat losses at the back of the test sample. Steel temperature was recorded using K-type thermocouples attached at the unexposed surface of each sample, placed at different locations. The transient rate of swelling of the coating was measured by image processing of videos taken using a high-resolution camera and the exposed surface temperature at the coating swelling front was gauged using an Infra-Red camera (Fig. 1).

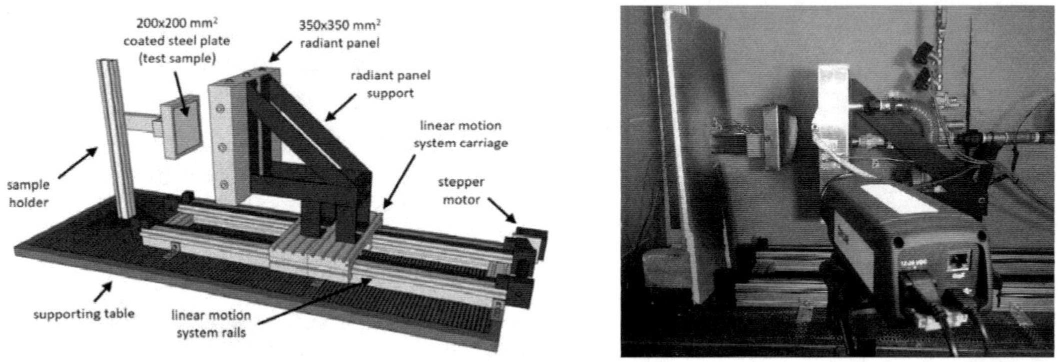

Fig. 1. Schematic illustration (left) and photography (right) of the experimental setup.

Criteria for onset of swelling

In available literature, there is no reference known to authors that has presented a unified, standard definition for the onset of swelling for intumescent coatings during heating. Within the scope of the work described herein, onset of swelling for intumescent coatings was defined in accordance to two criteria:

- *Minimum swelling* - considering the applied DFTs, the minimum swelled coating thickness was defined at 5 mm (Fig. 2).
- *Steel temperature divergence* - when the steel temperature protected by a reactive coating diverges from the steel temperature protected by an inert insulation (coating prior to swelling), mainly because of a rapid change in its thermo-physical properties. Mathematically, this was defined as the minimum slope change in the steel temperature gradient, i.e. the minimum in the second derivative (Fig. 2).

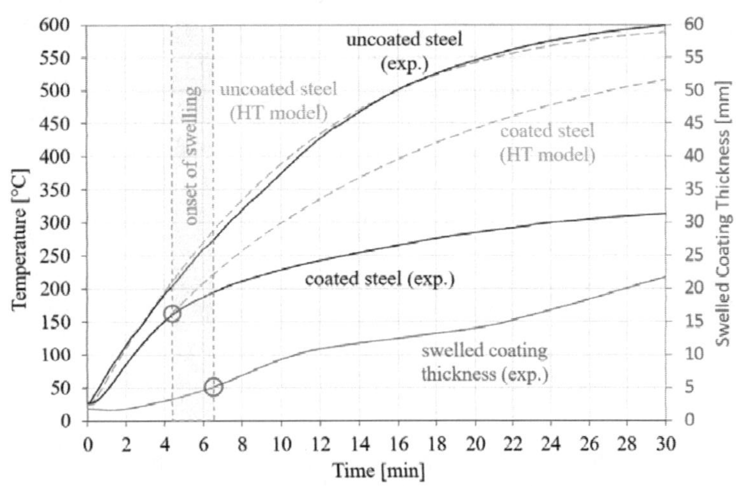

Fig. 2. Steel temperature at the unexposed surface and swelled coating thickness for test samples (uncoated and coated): criteria for onset of swelling applied to experimental data and heat transfer model data for a typical test sample tested at a constant incident radiant heat flux of 40 kW/m^2.

Heat transfer model

An explicit one-dimensional heat transfer model was formulated in order to simulate and understand the thermal boundary conditions and the heat transfer within the tested samples. The model is a finite differences numerical method developed by Emmons and Dusinberre [14, 15] and it solves explicitly a conduction problem by resolving energy-balance equations in the main direction of the heat flow. The model is based on finite differences and discretization into exposed, interior and unexposed elements. The thermal boundary conditions at the exposed surface of the test sample was defined as an incident radiant heat flux $\dot{q}''_{inc}$ with convective $\dot{q}''_{loss,conv}$ and radiative $\dot{q}''_{loss,rad}$ heat losses. Adiabatic conditions were assumed at the unexposed surface. The material properties used were:

Steel

- Thickness: $d_s = 0.01\ m$
- Density: $\rho_s = 7850\ kg/m^3$ [5]
- Specific heat capacity: $c_{p,s}$ [5]
- Thermal conductivity: λ_s [5]
- Emissivity: $\varepsilon_s = 0.75$

Intumescent coating (prior to swelling)

- Thickness: d_c = initial DFT
- Density: $\rho_c = 1550\ kg/m^3$
- Specific heat capacity: $c_{p,c} = 1375\ J/kgK$
- Thermal conductivity: $\lambda_c = 0.10\ W/mK$
- Emissivity: $\varepsilon_c = 0.90$

The adopted explicit heat transfer model was validated using experimental temperature data (Fig. 2) in order to verify the acceptability and rationality of the results. In addition, using the adopted heat transfer model, the time-history of net heat flux at the exposed surface was investigated. The net heat flux $\dot{q}''_{net}$ was obtained by subtracting the estimated heat losses $\dot{q}''_{loss,tot}$ (convective $\dot{q}''_{loss,conv}$ and radiative $\dot{q}''_{loss,rad}$ component) from the incident heat flux $\dot{q}''_{inc}$. Moreover, the *accumulative thermal energy flux* E''_{th} absorbed by the test sample at onset of swelling was calculated as the area under the net heat flux curve (Fig. 3):

$$E''_{th} = \int_0^{t_{act}} \dot{q}''_{net}\ dt \tag{1}$$

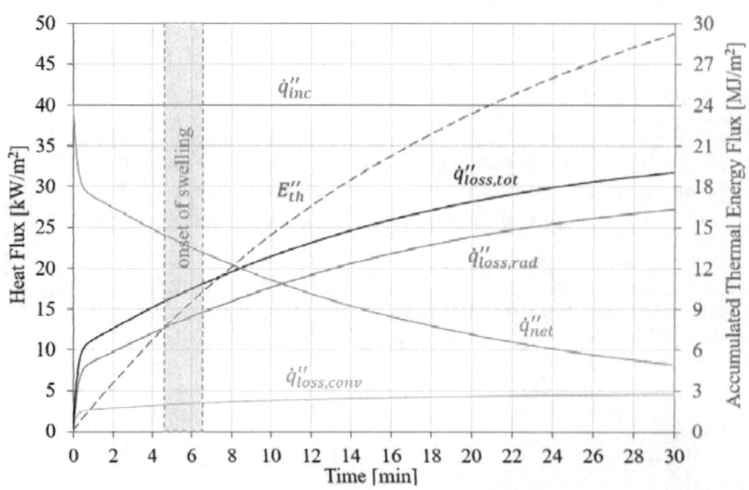

Fig. 3. Incident radiant heat flux, heat losses (convective and radiative component) and accumulative thermal energy flux for a coated test sample tested at a constant incident radiant heat flux of 40 kW/m^2.

RESULTS AND DISCUSSION

Onset of swelling

The time necessary to onset of swelling according to the two previous criteria was evaluated for all the tests conducted under different constant incident radiant heat fluxes. Figure 4 shows how time to onset of swelling decreases with increasing incident radiant heat flux and how the two criteria are comparable. Particularly, the minimum incident heat fluxes to reach onset of swelling was found at 23 kW/m^2 and no swelling was registered for samples tested under 20 kW/m^2. As a conclusion, the critical constant incident radiant heat flux for swelling to occur for this thin intumescent coating was found to be between 20 and 23 kW/m^2 (Fig. 4).

Accumulated thermal energy flux at onset of swelling

The net thermal energy flux accumulated by test samples prior to the onset of swelling was estimated using the adopted heat transfer model (Fig. 5). A clear trend was not visible, but it was found that swelling only occurred above a minimum accumulated net thermal energy flux of 7 MJ/m^2, regardless of the constant incident heat flux imposed on the test sample.

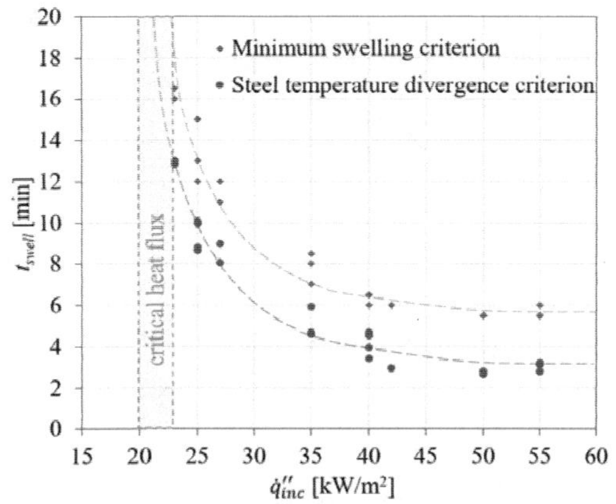

Fig. 4. Time to onset of swelling for different constant incident heat fluxes.

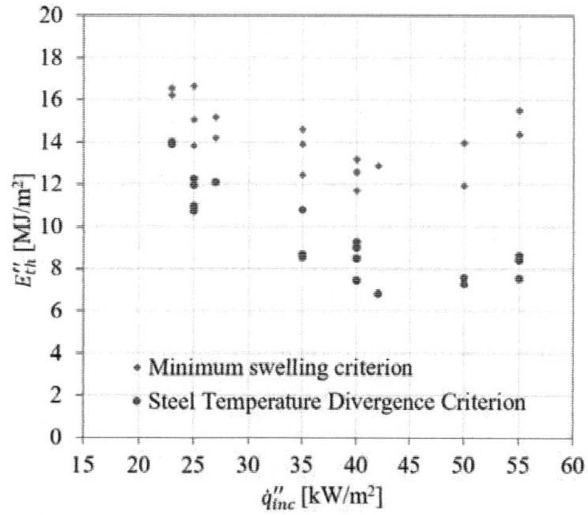

Fig. 5. Accumulated net thermal energy flux at onset of swelling for different constant incident heat fluxes.

Steel and coating temperature at onset of swelling

Figure 6 shows the steel temperature at the time when the onset of swelling was achieved in accordance with the two criteria. Once again, the two criteria show a similar trend. Particularly, thin intumescent coating swelled for steel temperatures between 150 and 300°C and evidenced an increase of steel temperatures at onset of swelling for samples tested at lower heat fluxes.

For samples tested at a constant incident heat fluxes of 25, 40 and 55 kW/m^2, the surface temperature of the coating was measured using an Infra-Red camera: the coating emissivity ε_c was set equal to 0.90. Subsequently, the 'mean coating temperature' was calculated as the average between the surface and steel temperature. Figure 7 shows that the 'mean coating temperature' at onset of swelling varies between 350 and 450°C and evidenced an increase of 'mean coating temperature' for samples tested at higher heat fluxes.

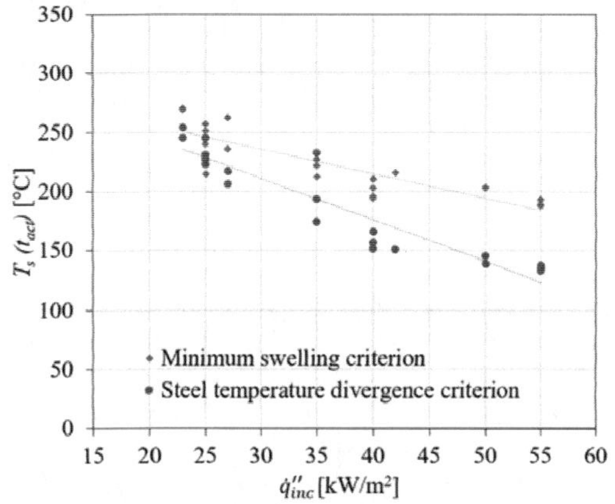

Fig. 6. Steel temperatures at onset of swelling for different constant incident heat fluxes.

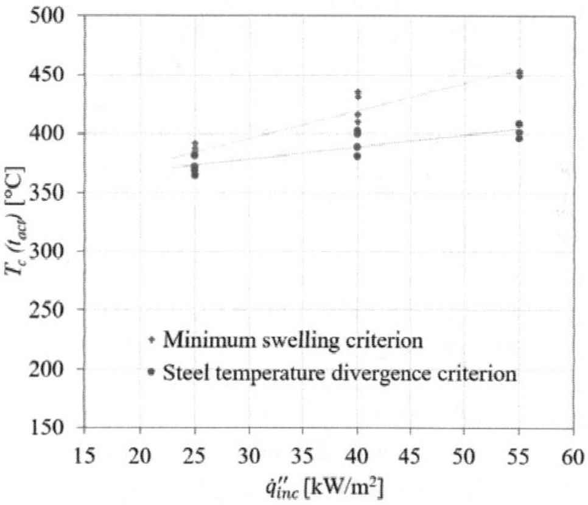

Fig. 7. 'Mean coating temperature' at onset of swelling for different constant incident heat fluxes.

CONCLUSIONS

This paper presented a systematic investigation on the onset of swelling for thin intumescent coatings exposed to various levels of constant incident radiant heat flux. Based on the experimental outcomes of this work, this research study aimed at yielding a sound understanding of the true effectiveness of thin intumescent coatings. From the results analysed herein, the following concluding remarks may be drawn:

- This research study confirms that many aspects of thin intumescent coatings' behaviour, like the onset of swelling, are directly influenced by the heating regime at the exposed surface of tested samples;

- The onset of swelling for thin intumescent coatings can be assessed according to two criteria, one related to the minimum swelled coating thickness and one to the coated steel temperature;
- The onset of swelling occurs for test samples heated at constant incident radiant heat flux above 20 kW/m^2;
- The accumulated net thermal energy flux at the time of onset of swelling remains steady and above 7 MJ/m^2 for the range of heating regimes investigated herein;
- The onset of swelling for the tested thin intumescent coating occurs for steel temperatures between 150 and 300°C and it is inversely proportional to the amount of constant incident heat flux at the exposed surface of tested sample;
- The onset of swelling occurs for the tested thin intumescent coating occurs for 'mean coating temperature' (as defined herein) between 350 and 450°C and it is proportional to the amount of constant incident heat flux at the exposed surface of tested samples;
- The time-history of net heat flux at the exposed surface of tested samples is not the only parameter that governs the onset of swelling for thin intumescent coatings.

Future research needs to deepen the key aspects related to intumescence. In particular, future studies of this work will focus on the influence of DFT and the influence of different time-histories of incident radiant heat flux on the onset of swelling of thin intumescent coatings.

REFERENCES

[1] Elliott A., Temple A., Maluk C. and Bisby L. (2014) *Novel testing to study the performance of intumescent coatings under non-standard heating regimes.* Fire Safety Science – Proceedings of the 11th International Symposium, University of Canterbury, New Zealand, pp. 652-665.

[2] Mariappan T. (2016) *Recent developments of intumescent fire protection coatings for structural steel: A review.* Journal of Fire Sciences, vol. 34, no. 2, pp. 1-44.

[3] Wang L., Dong Y., Zhan D. and Zhang C. (2015) *Experimental study of heat transfer in intumescent coatings exposed to non-standard furnace curves.* Fire Technology, vol. 51, no. 1, pp. 627-643.

[4] Comité Europeén de Normalization (CEN) (2013) *EN 13381-8:2013 Test methods for determining the contribution to the fire resistance of structural members - Part 8: Applied reactive protection to steel members.* Brussels, Belgium.

[5] Comité Europeén de Normalization (CEN) (2005) *EN 1993-1-2:2005 Eurocode 3: Design of steel structures - Part 1-2: General rules - Structural fire design.* Brussels, Belgium.

[6] Zhang Y., Wang Y.C., Bailey C.G. and Taylor A.P. (2012) *Global modelling of fire protection performance of intumescent coating under cone calorimeter heating conditions.* Fire Safety Journal, vol. 50, pp. 51-62.

[7] Li G.Q., Lou G.B., Zhang C., Wang L. and Wang Y. (2012) *Assess the fire resistance of intumescent coatings by equivalent constant thermal resistance.* Fire Technology, vol. 48, pp. 529-546.

[8] Lucherini A., Giuliani L. and Jomaas G. (2018) *Experimental study of the performance of intumescent coatings exposed to standard and non-standard fire conditions.* Fire Safety Journal, vol. 95, pp. 42-50.

[9] Bourbigot S., Duquesne S. and Leroy J.-M. (1999) *Modeling of heat transfer of a polypropylene-based intumescent system during combustion.* Journal of Fire Sciences, vol. 17, pp. 42-56.

[10] Anna P., Marosi Gy., Bourbigot S., Le Bras M. and Delobel R. (2002) *Intumescent flame retardant systems of modified rheology.* Polymer Degradation and Stability, vol. 77, pp. 243-247.

[11] Jimenez M., Duquesne S. and Bourbigot S. (2006) *Intumescent fire protective coating: toward a better understanding of their mechanism of action.* Thermochimica Acta, vol. 449, pp. 16-26.

[12] Maluk C., Bisby L., Krajcovic M. and Torero J.L. (2016) *A Heat-Transfer Inducing System (H-TRIS) Test Method.* Fire Safety Journal, Article in press.

[13] Lucherini A. and Maluk C. (2017) *Novel test methods for studying the fire performance of thin intumescent coatings.* Proceedings of 2nd International Fire Safety Symposium (IFireSS), Napoli, Italy, pp. 565-572.

[14] Emmons H.W. (1943) *The numerical solution of heat conduction problems.* Transactions of American Society of Mechanical Engineers (ASME), vol. 65(6), pp. 607-615.

[15] Dusinberre G.M. (1961) *Heat transfer calculations by finite differences.* International Textbook Company, Pennsylvania, USA, 293 pp., 1961.

Determination of mechanical properties of normal strength limestone concrete after exposure to elevated temperatures

Urška DOLINAR[1], Gregor TRTNIK[1,2] and Tomaž HOZJAN[1]

[1]*Faculty of Civil and Geodetic Engineering, University of Ljubljana, Slovenia*
[2]*Department for concrete, IGMAT Building Materials Institute, Slovenia*

urska.dolinar@fgg.uni-lj.si

ABSTRACT

Fire resistance is one of the essential requirements that has to be considered in the design process of any structure. Methods for ensuring fire resistance of new structures are therefore, relatively well developed. Thus, loss of bearing capacity of a concrete structure during fire is relatively well known and documented, while determination of remaining bearing capacity of concrete structure after fire is less known. This can mainly be attributed to the lack of adequate knowledge of mechanical properties of concrete after fire, duration of fire exposure, heating rate, and maximum temperatures reached in the structure during fire. These factors have to be taken into account when making a decision on the rationality of the structure's restoration. Estimation of bearing capacity of concrete structures after fire depends on precise but difficult determination of temperature dependent changes in material. Bearing capacity of structures after fire is in general, often assessed with simplified procedures and experimental investigation. These mainly covers visual inspection of the damaged structure and the execution of some standard, mostly destructive tests. Such tests are precise, but often difficult to implement, expensive, time-consuming, limited to certain pre-selected locations and can somehow weaken the individual structural element. Therefore, in order to determine bearing capacity of concrete structure after fire, various non-destructive methods are being developed. Among these, different ultrasonic (US) tests are frequently used and well described in the literature. These results indicate good connection between US pulse velocity and concrete compressive strength after exposure to elevated temperatures.

This article presents the first part of the ongoing research. Main goal of the research is to develop a numerical method based on artificial neural network approach to estimate concrete properties after fire using non-destructive tests. In this paper, results of destructive and non-destructive tests are presented and analyzed to determine changes in basic mechanical properties of standard concrete mixture after exposure to elevated temperatures, namely 200 °C, 400 °C, 600 °C, or 800 °C. With the aim of separating 'material' and 'structural' effects, temperature gradients within the concrete specimens, were minimized, as possible. Prior experimental testing, the damage to the concrete was roughly detected by observing the specimen's surface. Compressive strength, tensile strength, surface hardness, dynamic elastic modulus, and shear modulus are determined. It is noticed that crack formation at 400 °C has an influence on reduction of the US pulse velocity, tensile strength, dynamic elastic, and shear modulus. Analysis of variance is used for assessing the experimental results sensitivity on temperature elevation. It is noticed that US method detects statistically significant impact of elevated temperatures on US pulse velocity. On the other hand, rebound hammer technique indicates statistically significant impact of elevated temperatures on concrete surface hardness between 400 °C and 600 °C.

KEYWORDS: *concrete, elevated temperatures, compressive strength, non-destructive tests.*

IOP Conf. Series: Journal of Physics: Conf. Series **1107** (2018) 032018 doi:10.1088/1742-6596/1107/3/032018

INTRODUCTION

Concrete is a widespread construction material due to its strength, durability, and affordability. It is a composite material consisting of aggregate, cement, and water where the choice of aggregate has a significant impact, since it represents around 70 % of the total volume of concrete [1]. During fire exposure different chemical, physical, and mechanical changes of the material occur. Various experiments were made and presented in the literature, where limestone [2-6], granite [2, 5-6], silicate [2-3], quartz [6], and marine gravel [2] were used as an aggregate to investigate the concrete's mechanical properties after exposure to elevated temperatures. In general, silicate concrete has lower mechanical properties after exposure to elevated temperatures compared to limestone concrete [2], due to limestone's higher degradation temperature.

Mechanical properties of concrete after fire exposure are commonly determined using destructive techniques like compressive test, bending test, and test of elastic modulus determination [3, 7-8]. On the other hand, results of non-destructive tests are used as relation to destructive ones [9-10]. Different US tests based on measuring US pulse transition are frequently described in literature. This has been proved to be one of the most appropriate non-destructive tests for estimating mechanical properties of concrete at ambient temperature [11-12] and determining internal damage of concrete [13-14], due to its clear physical foundation. US method is very sensitive to the changes of microstructure inside the concrete and related mechanical properties [15-16]. With results of before mentioned non-destructive tests an advanced and effective method for estimating concrete compressive strength, at ambient temperatures, was developed using artificial neural networks [17]. As literature reveals, advanced non-destructive methods for assessing the mechanical properties of fresh and hardened concrete are commonly used at ambient temperature, but not often after fire testing. In case of non-destructive experimental research of silicate concrete after fire exposure, the correlation between US pulse velocity and concrete compressive strength was proposed [10]. During the experiment [10] it was found that the specimen size had no influence on the measured US pulse velocity and that concrete with silicate aggregate was damaged to that level that unrealistic results were obtained around temperature 600 °C. In the last year a relationship between measured concrete compressive strength after fire and results of nonlinear US methods were presented [9].

The article presents experimental materials, test methods, and results obtained from the concrete specimens. Experimental tests were carried out at ambient temperature and after exposure to elevated temperatures up to 800 °C. During heating and cooling down, temperature gradients within the concrete specimens were minimized to obtain isothermal conditions. Before experimental testing, the specimens were observed for crack propagation and spalling. The compressive and tensile strengths were determined with standard destructive tests and different non-destructive tests. Description of experimental work is followed by presentation of elevated temperatures influence on the concrete mechanical properties, after exposure to elevated temperatures.

MATERIALS AND METHODS

Specimens were made of normal strength concrete with high initial strength Portland cement (CEM I 52.5 R), limestone aggregate, water, and plasticizer, presented in Table 1. The maximum nominal aggregate size was 16 mm.

Table 1. Concrete mixture recipe.

Material	Type	Amount [kg]
Cement	Cem I 52.5 R	15.65
Water	Tap	7.36
Hyper plasticizer		0.09
Limestone aggregate	0-4 mm	40.50
Limestone aggregate	4-8 mm	12.15
Limestone aggregate	8-16 mm	28.35

Mixing was performed in a pan mixer. Specimens were either 40 x 40 x 160 mm or 100 x 100 x 400 mm concrete prisms, shown in Fig. 1 (left). The small prisms were used in standard bending tests, while the big prisms were divided into four cubes of 100 x 100 x 100 mm before used for experimental testing. One of such cubic specimens had two embedded temperature resistant thermocouples for monitoring the temperature development

5 millimeters below the surface and in the center of the cube. After a period of 28 days of moist curing (20 ± 2 °C, RH ≥ 95 %) the specimens were cured under laboratory conditions (20 ± 2 °C, RH 60 ± 5 %) for 5 days. For every tested temperature, there were 3 concrete cubes and 3 prisms available, i.e. 15 cubes and 15 prisms all together.

Before heating, specimens were tested with US and resonant frequency method. Time of the US pulse traveling through the specimen was measured with US method and US velocity was calculated according to EN 12504 standard [18]. The US tests were carried out with Proceq Pundit PL-200 instrument. The wave induction caused by mechanical impulse was measured with resonant frequency method. Based on these measurements, the dynamic elastic and shear modulus were determined as proposed in ASTM standard [19]. After preliminary testing, specimens were heated in an electric furnace (Fig. 1 (right)). Average heating rate in the center of the cube was between 0.78 °C/min and 1.61 °C/min for heating up to the maximum temperature of 200 °C and 800 °C, respectively. During the heating, temperatures within concrete cubes were continuously recorded with two thermocouples, one below the surface and one in the center of the cube. The isothermal conditions at 200 °C, inside the cube, was reached after 4.3 hours, and it took 8.3 hours to reach isothermal conditions at 800 °C.

Fig. 1. Fresh concrete specimens (left) and heating in the electric furnace (right).

After the maximum temperature inside the concrete cube was reached, it was maintained for a short period before the specimens were slowly cooled down to the ambient temperature. In Fig. 2 temperature development inside concrete cube is depicted. Presented experimental work was derived from standardized tests and thus temperature gradients within the concrete specimens were intentionally minimized, with the aim of separating 'material' and 'structural' effects, as far as possible. It is worth noting that during heating internal thermal gradients have an influence on structure's bearing capacity to some extent [20, 21], but this influence will be investigated in future research.

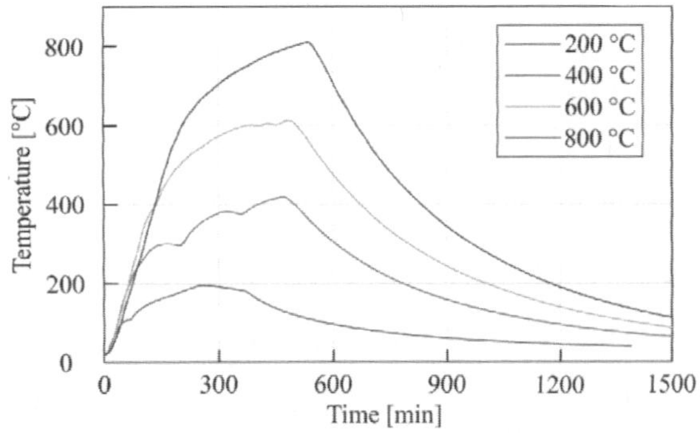

Fig. 2. Temperature development inside the concrete cube at different maximum temperatures.

After cooling and crack observation various non-destructive methods such as US, resonant frequency method, rebound hammer, and destructive ones such as compressive and bending tests were carried out to determine the mechanical properties of concrete. Schmidt rebound hammer was used for determination of surface hardness of concrete according to EN 12504 standard [22]. Compressive and tensile strength were measured by standard compressive and bending tests, according to EN 12390 standards [23-24]. Finally, analysis of variance [25] was used for assessing experimental results' sensitivity on temperature elevation.

EXPERIMENTAL RESULTS

Non-destructive tests

By observing the concrete surface after being subjected to elevated temperatures the damage of the concrete can be roughly detected. Assessment of fire-damaged concrete usually starts with visual observation of cracking, spalling and color change of concrete surface. Fig. 3 illustrates the concrete surfaces after exposure to elevated temperatures. There is no visible effect on the surface of the specimens heated up to 200 °C. The concrete starts to crack at temperatures above 400 °C but the effect is not significant at that temperature level. The cracks become pronounce at 600 °C and 800 °C, where the spalling of the specimens due to excessive cracking is not observed. In comparison to limestone concrete surface observation in [4], the cracks here appear at lower temperatures.

Fig. 3. Concrete prisms after exposure to elevated temperatures, from left to right: 20 °C, 200 °C, 400 °C, 600 °C, and 800 °C.

Results from non-destructive tests are shown in Fig. 3, where minimum, average and maximum values from three measurements are presented. US pulse velocity and surface hardness, shown in Fig. 3 (a, b), were determined on concrete cubes. Dynamic elastic and shear modulus, shown in Fig. 3 (c, d), were determined on concrete prisms.

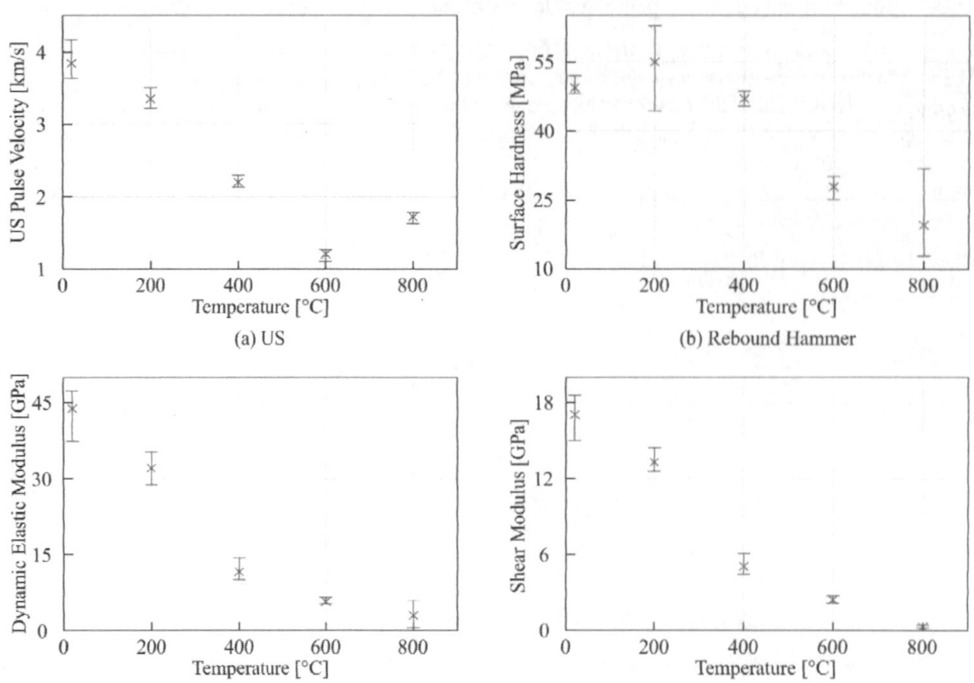

Fig. 4. US pulse velocity in concrete cubes (a), surface hardness of concrete (b), dynamic elastic modulus (c), and shear modulus (d) after cooling down.

Experimental results show that average US pulse velocity is decreasing with increasing temperature up to 600 °C. Increase at 800 °C may result from measurement error or other errors, therefore more tests at this temperature are required to prove or omit the observed increase of US pulse velocity. Average surface hardness at ambient temperature is lower than at 200 °C. Above 200 °C the average hardness is decreasing with increasing temperature. Great dispersion of results for surface hardness of concrete is observed at 200 °C and 800 °C, due to this, more tests at these temperatures are required to prove or omit the observed increase. Average dynamic elastic and shear modulus are decreasing with increasing temperature, where larger reductions of both values are observed up to 400 °C. At 400 °C, also larger reductions in US pulse velocity, is observed and concrete starts to crack.

Analysis of variance (Anova) [25] is used for observing sensitivity of used experimental tests at elevated temperatures. It is noticed that the US test detects statistically significant impact of elevated temperatures on US pulse velocity at all tested temperatures. On the other hand, rebound hammer detects statistically significant impact of elevated temperatures on concrete surface hardness between 400 °C and 600 °C. With resonant frequency test statistically significant impact of elevated temperatures on dynamic elastic and shear modulus is observed up to 600 °C and 800 °C, respectively.

Destructive tests

Results from destructive tests are shown in Fig. 4, where minimum, average and maximum values from three measurements are presented. Compressive and tensile strengths of concrete after cooling down are determined on concrete cubes and prisms, respectively.

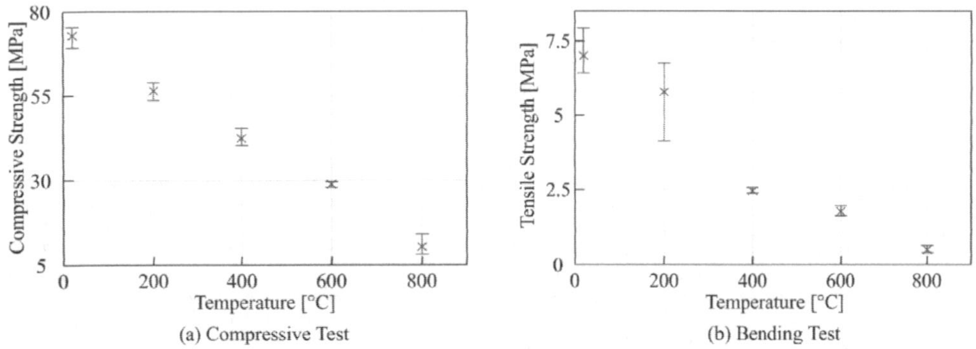

Fig. 5. Compressive (a) and tensile strength (b) of concrete after cooling down.

Experimental results show that average compressive strength and tensile strength of concrete after cooling down are decreasing with increasing temperature. It is observed that mechanical properties at selected temperatures are close together except at 200 °C where tensile strength results are more scattered than the other ones. Average compressive strength is 77.4 %, 52.5 %, 39.5 % and 14.1 % of the ambient temperature one at 200 °C, 400 °C, 600 °C and 800 °C, respectively. It is noticed that compressive strength is decreasing linearly, when tensile strength experience larger drop in strength at 400 °C, when also concrete starts to crack.

Statistically significant impact of elevated temperatures on the compressive strength is noticed at all tested temperatures, while for bending test this range is between 200 °C and 800 °C.

CONCLUSIONS

Presented experimental research was conducted on limestone concrete after exposure to elevated temperatures. Before testing concrete cubes and prisms were exposed to maximum temperature of 200 °C, 400 °C, 600 °C, or 800 °C and naturally cooled down. During heating and cooling, temperature gradients within the concrete specimens were minimized as possible, to obtain isothermal conditions. With destructive and non-destructive tests concrete compressive strength, tensile strength, surface hardness, dynamic elastic, and shear modulus were determined. The greatest impact of elevated temperature on decreasing mechanical properties was observed at dynamic elastic modulus. US pulse velocity and concrete compressive strength were observed to have similar decrease in case of exposure to 400 °C, around 42 % in comparison with properties at ambient temperature. In case of other temperatures, the influence of elevated temperature was not directly comparable. At 400 °C it was noticed that concrete started to crack at the same temperature the US pulse velocity, tensile strength, dynamic elastic, and shear modulus experienced larger drop in value. The cracking seems not to affect the concrete compressive strength and surface hardness to the same extent as the rest of the mechanical properties. When considering method's ability to detect the elevated temperature's impact on the mechanical properties US method, resonant frequency for determining shear modulus, and standard compressive test are able to detect statistically significant impact at all tested temperatures. Thus, obtained experimental results from non-destructive and destructive tests will be used in future research. Next step is to develop a numerical model for estimating different concrete's mechanical properties after fire with non-destructive measurements where artificial neural network approach will be used.

ACKNOWLEDGMENT

The work of U. Dolinar was financially supported by the Slovenian Research Agency with decision No. 802-7/2016-215. The work of G. Trtnik and T. Hozjan was also supported by the Slovenian Research Agency through the research core funding No. P2-0260. The support is gratefully acknowledged.

REFERENCES

[1] Hager, I., Tracz, T., Sliwinski, J., Krzeminen, K. (2015) The Influence of Aggregate Type on the Physical and Mechanical Properties of High-Performance Concrete Subjected to High Temperature, *Fire and Materials* 40:668-682, https://doi.org/10.1002/fam.2318.

[2] Hertz, K. D. (2005) Concrete Strength for Fire Safety Design, *Magazine of Concrete Research* 57: 445-453, https://doi.org/10.1680/macr.2005.57.8.445.

[3] Savva, A., Manita, P., Sideris, K. K. (2005) Influence of Elevated Temperatures on the Mechanical Properties of Blended Cement Concretes Prepared with Limestone and Siliceous Aggregates, *Cement & Concrete Composites* 27:239-248, https://doi.org/10.1016/j.cemconcomp.2004.02.013.

[4] Arioz, O. (2007) Effects of Elevated Temperatures on Properties of Concrete, *Fire Safety Journal* 42: 516-522, https://doi.org/10.1016/j.firesaf.2007.01.003.

[5] dos Santos, C. C., Rodrigues, J. P. C. (2016) Calcereous and Granite Aggregate Concrete After Fire, *Journal of Building Engineering* 8:231-242, https://doi.org/10.1016/j.jobe.2016.09.009.

[6] Tufail, M., Shahzada, K., Gencturk, B., Wei, J. (2017) Effect of Elevated Temperature on Mechanical Properties of Limestone, Quartzite and Granite Concrete, *International Journal of Concrete Structures and Materials* 11:17-28, https://doi.org/10.1007/s40069-016-0175-2.

[7] Yüzer, N., Aköz, F., Öztürk, L. D. (2004) Compressive Strength-Color Change Relation in Mortars at High Temperature, *Cement and Concrete Research* 34:1803-1807, https://doi.org/10.1016/j.cemconres.2004.01.015.

[8] Huang, Z., Liew, R. J. Y., Li, W. (2017) Evaluation of Compressive Behavior of Ultra-Lightweight Cement Composite after Elevated Temperatures Exposure, Construction and Building Materials 148:579-589, http://dx.doi.org/10.1016/j.conbuildmat.2017.04.121.

[9] Park, G. K., Yim, H. J. (2017) Evaluation of Fire-Damaged Concrete: An Experimental Analysis based on Destructive and Nondestructive Methods, *International Journal of Concrete Structures and Materials* 11:447-457, https://doi.org/10.1007/s40069-017-0211-x.

[10] Yaqub, M., Bailey, C. G. (2016) Non-destructive Evaluation of Residual Compressive Strength of Post-Heated Reinforced Concrete Columns, *Construction and Building Materials* 120:482-493, http://dx.org/10.1016/j.conbuildmat.2016.05.022.

[11] Popovics, S., Rose, J. L., Popovics, J. S. (1990) The Behavior of Ultrasonic Pulses in Concrete, *Cement and Concrete Research* 20:259-270, https://doi.org/10.1016/0008-8846(90)90079-D.

[12] Jacobs, L. J., Owino, J. O. (2000) Effect of Aggregate Size on Attenuation of Rayleigh Surface Waves in Cement-based Materials, *Journal of Engineering Mechanics-Asce* 126:1124-1130, https://doi.org/10.1061/(ASCE)0733-9399(2000)126:11(1124).

[13] Selleck, S. F., Landis, E. N., Peterson, M. L., Shah, S. P., Achenbach, J. D. (1998) Ultrasonic Investigation of Concrete with Distributed Damage, *Aci Materials Journal* 95: 27-36.

[14] Punurai, W., Jarzynski, J., Qu, J. M., Kurtis, K. E., Jacobs, L. J. (2006) Characterization of Entrained Air Voids in Cement Paste with Scattered Ultrasound, *Ndt & E International* 39: 514-524, https://doi.org/10.1016/j.ndteint.2006.02.001.

[15] Kim, B. C., Kim, J. Y. (2009) Characterization of Ultrasonic Properties of Concrete, *Mechanics Research Communications* 36: 207-214, https://doi.org/10.1016/j.mechrescom.2008.07.003.

[16] Trtnik, G., Gams, M. (2014) Recent Advances of Ultrasonic Testing of Cement Based Materials at Early Ages, *Ultrasonics* 54: 66-75, https://doi.org/10.1016/j.ultras.2013.07.010.

[17] Trtnik, G., Kavcic, F., Turk, G. (2009) Prediction of Concrete Strength Using Ultrasonic Pulse Velocity and Artificial Neural Networks, *Ultrasonics* 49:53-60, https://doi.org/10.1016/j.ultras.2008.05.001.

[18] SIST EN 12504-4 (2004) Testing Concrete – Part 4: Determination of Ultrasonic Pulse Velocity, European Committee for Standardization, Brussels: 14.

[19] ASTM E1876-99 (2001) Standard Test Method for Dynamic Young's Modulus, Shear Modulus, and Poisson's Ratio by Impulse Excitation of Vibration, ASTM International, West Conshohocken, PA, www.astm.org.

[20] Le, Q. X., Dao, V. T. N., Torero, J. L., Maluk, C., Bisby, L. (2018) Effects of Temperature and Temperature Gradient on Concrete Performance at Elevated Temperatures, *Advances in Structural Engineering* 21(8):1223-1233, https://doi.org/10.1177/1369433217746347.

[21] Torero, J. L., Law, A., Maluk, C. (2017) Defining the Thermal Boundary Condition for Protective Structures in Fire, *Engineering Structures* 149:104-112, http://dx.doi.org/10.1016/j.engstruct.2016.11.015.

[22] SIST EN 12504-2 (2002) Testing concrete in structures – Part 2: Non-destructive testing – Determination of rebound number, European Committee for Standardization, Brussels: 5.

[23] SIST EN 12390-3 (2009) Testing hardened concrete – Part 3: Compressive strength of test specimens, European Committee for Standardization, Brussels: 19.

[24] SIST EN 12390-5 (2009) Testing hardened concrete – Part 5: Flexural strength of test specimens, European Committee for Standardization, Brussels: 11.

[25] Scheffé, H., *The Analysis of Variance*, John Wiley and Sons, United States of America, 1999: 477.

IOP Conf. Series: Journal of Physics: Conf. Series **1107** (2018) 032019 doi:10.1088/1742-6596/1107/3/032019

Application of the Kragten method in order to evaluate the uncertainty of the heat release rate determination using of the cone calorimeter

Pierre CREMONA[1,2], Thomas ROGAUME[1], Franck RICHARD[1] and Benjamin BATIOT[1]

[1] Institut Pprime, UPR 3346 CNRS, Université de Potiers, ISAE-ENSMA, France
[2] Laboratoire Ceric, Poujoulat SA, France

p.cremona@laboratoire-ceric.com

ABSTRACT

Assigning uncertainties to measurement results is primordial in a certification framework using fire benches as the cone calorimeter [1] developed by NIST [2]. Indeed, this process allows in the first place to validate the measurement method accuracy and in second place to provide proper results to compare obtained between different labs (reproducibility).

However, even if numerous methods enable uncertainties determination, some can become complex when the data of interest is mathematical formulation powered by many measured parameters aimed to use in a model such as the Heat Release Rate (HRR).

This article aims to propose an intermediate uncertainty calculation method whose difficulty of application is sized between the GUM method [3] and the Monte Carlo method [4]: the Kragten one [5]. In order to validate the opportunity of using such a method, the three approaches were used to assess the measurement uncertainty on the HRR parameter in the specific case of small scale fire tests using cone calorimeter.

The results obtained show that the Kragten method allows to evaluate more easily and faster the HRR uncertainties measured at the Cone Calorimeter than the Monte Carlo method. Furthermore, the Kragten method is more reliable and reduces the risk of miscalculations than the GUM approach.

KEYWORDS

Suggested Keywords: Cone Calorimeter, Heat Release Rate, numerical method of Monte Carlo, GUM approach, Kragten approach, uncertainty.

INTRODUCTION

Uncertainty measurement is often an important issue dealing with experimental data processing. Although a proper uncertainties measurement is primordial to present reliable results in taking into account the factors affecting the outcome of the measurement as means, reliability and external parameters (*i.e.* ambient conditions).

Evaluation of experimental data uncertainty depends strongly on the mathematical model of the measurand or of the measurement system considered. Indeed, the measurand can be determined by direct or by multivariable measurements (equation of calculation). The measurement system corresponds itself to the integration of the input (*i.e.* sensor) at the output of the system (*i.e.* data processor or readout device) including the intermediate modules (*i.e.* amplifier, A/D converter, multiple interfaces). NASA [6] proposes procedures to evaluate uncertainty according to the measurement system considered (single or multi-variable and complete system). For the case of a direct measurement (mono variable), the uncertainty depends only on the Type A (statistical analysis) and Type B (resolution of instrumentation, calibration certificates, manufacturers' specifications, previous measurement data and other relevant information) uncertainty components of the measuring device [7]. When using multiple variables to determine the data of interest (*i.e.* HRR), the definition of the mathematical relationship between the measurand and its component variables is important [6],especially considering the notion of linearity and dependence between variables, in particular when the different terms of the mathematical model contain the same variables [5].

Linearity and variable dependences are the parameters allowing to choose the most adapted uncertainties quantification method. Past study [8] highlighted characterization of HRR uncertainty, but at large scale. At small scale however, there is no available studies that explores different method to determine uncertainties on the HRR measurement (commonly evaluated with cone calorimeter [1]). The present study thus proposes to apply GUM, Monte Carlo Simulation and Kragten methods in order to determine the heat release rate measurement uncertainties. Besides in this study, are considered the parameters which are meant to be used for the HRR calculation and not the direct measurements from the sensors as in [8]. Thus, the proposed method also takes into account the uncertainties associated with transposition of measured data (signal conversion). The results obtained allows to highlight the advantages of the Kragten method comparing to other aforementioned approaches.

PRESENTATION OF THE 3 METHODS

Our study focuses on 3 main methods. The first is the law of propagation of variances or GUM (the most used) [3]. The uncertainty of the result $u(Y)$ is evaluated by combining the uncertainties of each variable x_i and the covariance $cov(x_i,x_j)$ between each correlated variable, according to Eq. 1 [3]. When all the variables are independent, the covariance term is zero ($cov(x_i,x_j) = 0$). Nevertheless, the GUM approach encounters some limitations, detailed by [7] and [8]: especially the non-linearity of the mathematical relation. Furthermore, if the mathematical formulation is complex and that the variables are correlated, the partial derivations of Y according to x_i (c_i or $\partial Y/\partial x_i$) requires advanced mathematical skills and generate a risk of calculation errors.

$$u^2(Y) = \sum_{i=1}^{n} \left(\frac{\partial Y}{\partial x_i}\right)^2 u^2(x_i) + 2\sum_{i=1}^{n-1}\sum_{j=i+1}^{n}\left(\frac{\partial Y}{\partial x_i}\frac{\partial Y}{\partial x_j}\right) cov(x_i, x_j) \quad (1)$$

The second is the Monte Carlo method (MCS) [4]. It is a numerical approach for solving mathematical problems by means of the simulation of random variable [8]. The principle is to calculate a result Y from the generation of a random number i for each input variables according to the assigned probability distribution function. Each simulation must generate at least 100,000 random numbers and this simulation must be repeated at least 10 times [9]. The uncertainty is then determined by the average of the uncertainties of these n simulations, described by Eq. 2. This approach is favoured for complex models (correlated variables, significant uncertainties of the input variable or model involving many variables) or non-linear models [7, 9, 10]. Nevertheless, the calculation times can be long in complex cases [9]. Another difficulty consists in selecting the probability distribution functions of input variables [9, 11].

$$u^2(Y) = \sum_{1}^{n}\left(\frac{1}{i-1}\sum_{k=1}^{i}\left(Y^{(k)} - Y\right)\right)/n \quad (2)$$

The third is the Kragten numerical method (KM) [5, 11]. This approach implies a modification of the propagation law of uncertainties concerning the sensitivity coefficient c_i [11]. Thus, each uncertainty contribution is calculated between two values of Y: one corresponding to the measured value and the other corresponding to the measured value corrected by a term which is the uncertainty of this variable. So, this approach represents a gradient method similar to the first order finite difference approach. Instead of having Eq. 1 to determine the uncertainty of the measurand, Eq. 3 is used. Besides using an approximative numerical method of differentiation, the advantage of this method is that it only requires the mathematical formula used to obtain the measurand and the numerical values of the various variables and their uncertainties [11]. Moreover, according to Kragten [5], this method takes into account dependences of the variables within the mathematical formulation, reduces the risks of calculation errors and allows the calculation of uncertainty very quickly. Indeed, Kragten has developed a spreadsheet method that is adaptable to all error propagation calculations. The design of this sheet is clearly explained by [5] and an example of its use is presented in Table 1.

$$u^2(Y) = \sum_{i=1}^{n} u^2(Y, x_i) + 2\sum_{i=1}^{n-1}\sum_{j=i+1}^{n} \left(u(Y, x_i) \cdot u(Y, x_j)\right) cov(x_i, x_j) \tag{3}$$

THE HEAT RELEASE RATE

The Heat Release Rate (HRR) produced during the combustion of materials subjected to a radiative heat flux can be determined in different ways depending on the variables taken into account (*i.e.* with additional gas analysers, such as CO_2, CO and H_2O) [12]. The most used mathematical formulation of HRR is the one given in the ISO 5660-1standard [1]. It is expressed by Eq. 4 and depends on 6 variables.

$$\dot{q}(t) = \left(\frac{\Delta H_c}{r_o}\right) \cdot (1.10) \cdot C \cdot \sqrt{\frac{\Delta P}{T_e}} \cdot \frac{x^0_{O_2} - x_{O_2}}{\alpha - 1.5 x_{O_2}} \qquad (kW) \tag{4}$$

With ΔH_c the heat of combustion (MJ/kg), r_o the stoichiometric demand in oxygen, C the orifice flow meter calibration constant (named hereafter C-factor), ΔP the orifice meter pressure differential (Pa), T_e the absolute temperature of gas at the orifice meter (K), $x^0_{O_2}$ and x_{O_2} respectively the initial concentration of oxygen and actual oxygen analyser reading (mol/mol), and α the expansion factor equal to 1.105.

Several studies deal with calculating HRR uncertainties [13, 14, 15] or one of the variables of the HRR [16] according to the GUM method. *Enright et al.* [14] highlight that the HRR calculation variables are dependent of each other. This especially true for the gas-analysis terms and flow rate term both associated with the oxygen depletion factor [14] and for the C-factor which is calculated using variables also used to determine HRR (especially the temperature of the exhaust duct) [13, 14]. Nevertheless, *Guillaume et al.* [13] and *Enright et al.* [14] consider the effect insignificant. The results of the partial derivatives are detailed in [13] and [14].

Since the mathematical formulation is a combination of division and multiplications, Guillaume *et al.* [13] express the relative uncertainty of the HRR in order to simplify the calculation (Eq. 5).

$$u^2(\dot{q}(t)) = (\dot{q}(t))^2 \cdot \left(\frac{u^2(C)}{C^2} + \frac{u^2(\Delta P)}{\Delta P^2} + \frac{u^2(T_e)}{T_e^{\,2}} + \frac{u^2(x_{O_2})}{(x_{O_2})^2} + \frac{u^2(x^0_{O_2})}{(x^0_{O_2})^2} + \frac{u^2(\Delta H_c/r_o)}{(\Delta H_c/r_o)^2}\right) \quad (kW^2) \tag{5}$$

Given the number of variables involved, the complexity of the partial derivatives that can lead to calculation errors and the presence of dependence between variables, it seems interesting to use alternative methods to the GUM to estimate the uncertainties of the HRR: those of Kragten and Monte Carlo.

HRR UNCERTAINTY WITH KRAGTEN'S METHOD

Bryant and Mulholland [8] have already shown that the Kragten method allows to calculate the HRR uncertainty for full-scale fire tests with the same precision as the GUM method. However, in order to avoid dependencies between variables, the calculation of the uncertainties mainly concerns the basic measurement inputs such as instrument voltages, thermocouple temperatures and constant parameters. In our study, we use output data from measuring devices. So, the uncertainties of the variables (ΔP, T_e, C, $x_{O_2}$and $x^0_{O_2}$) are determined according to the evaluation methods proposed by [13]. Finally, we also choose the uncertainty proposed by [14] for the effective heat of combustion ($\Delta H_c/r_o$ or *EHC*) and we neglect the uncertainty expansion factor as proposed by Brohez [15].

The HRR data for this sensibility study are related to a pine tar which have been tested in a standard cone calorimeter under a radiant flux of 40 kW/m² [17]. The calculated HRR curve (according to Eq. 4) is illustrated in Fig. 1, which shows the maximum, the average and low (between 0,07 and 0,40 kW) values of HRR taken into account for this study (red squares).

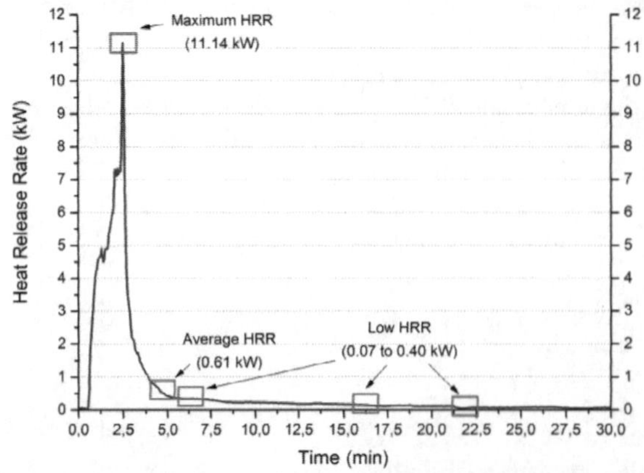

Fig. 1. Evolution of HRR as a function of time - Pine tar - 40 kW/m².

Table 1 presents the spreadsheet to compute the maximum HRR uncertainty. In the first column, each parameter x_i measured, equivalent to the maximum HRR, is entered (*i.e.* 391.05 K for the temperature). Then, is entered the absolute uncertainties of each variable u(x_i) in the second row (*i.e.* 1.44 K for the temperature). These uncertainties are added to each corresponding x_i along the diagonal. For example (second column), the temperature uncertainty is added only at the temperature. Thus, the measured temperature of 391.05 K becomes 392.49 K and the values of the other parameters remain unchanged. For the third column, only the orifice meter pressure differential is corrected. In the 9th line, the mathematical formulation for $\dot{q}(t)$ is entered and copied to all right columns on the same line. We can observe a difference between the measured result equal to 11.143 kW (first column) and the corrected results (second to 7th column). The sum of the differences observed between the corrected HRR value and the measured HRR value make it possible to calculate the uncertainty of 0.647 kW (penultimate row). The expanded uncertainty of the HRR, equal to 1.294 kW, is obtained by applicant the enlargement factor *k* (last line).

Table 1. Example of Kragten spreadsheet using maximum HRR values.

Base computation of $\dot{q}(t)$		**Numerical differentiation by change of the diagonal elements with the absolute uncertainty**					
		u(T_e) = 1.44	u(ΔP) = 0.60	u($x^0_{O_2}$)= 0.002	u(x_{O_2})= 0.007	u(*C-factor*) = 0.00126	u(*EHC*) = 655
T_e (K) =	391.05	392.49	391.05	391.05	391.05	391.05	391.05
ΔP (Pa) =	99.77	99.77	100.37	99.77	99.77	99.77	99.77
$x^0_{O_2}$(mol/mol) =	20.954	20.954	20.954	20.956	20.954	20.954	20.954
x_{O_2}(mol/mol) =	17.986	17.986	17.986	17.986	17.993	17.986	17.986
C-factor (-) =	0.04308	0.04308	0.04308	0.04308	0.04308	0.04434	0.04308
EHC (kJ/kg) =	13 100	13 100	13 100	13 100	13 100	13 100	13 755
$\dot{q}(t)$ (kW) =	11.143	11.122	11.176	11.151	11.118	11.468	11.700
u$_c$($\dot{q}(t)$) =	$\sqrt{(11.143-11.122)^2+(11.143-11.1176)^2+\cdots+(11.143-11.700)^2}=0.647\ kW$						
U($\dot{q}(t)$) =	$u_c(\dot{q}(t))\cdot k = 0.647\cdot 2 = 1{,}294\ kW$						

Like the GUM method, the Kragten one can take into account the uncertainties of measuring devices, but also those related to ambient conditions, operators, materials or other sources of influence. In order to, introduce these parameters a parameter representing a global source of error (repeatability, accuracy and reproducibility) have to be added in Table 1, whose value is always equal to 1 and whose value of its uncertainty corresponds to the influence of the parameter (*i.e.* the standard deviation for a repeatability).

COMPARISON WITH OTHERS METHODS

Table 2 compares the results obtained with the GUM, the Kragten and the Monte Carlo approach. Results obtained with Kragten method are very close to the ones obtained with Monte Carlo approach. However, the calculation time of the Monte Carlo method is close to 2h for each value of HRR (20 runs of $1e^6$ random number) whereas the results are obtained several minutes using Kragten's approach.

Table 2. Uncertainty of HRR according to the GUM, Kragten and Monte Carlo (MCS) method.

Value of HRR (kW)	Uncertainty of HRR (kW)		
	GUM	Kragten	MCS
11.143	0.650	0.647	0.642
0.610	0.036	0.047	0.047
0.404	0.024	0.039	0.035
0.151	0.009	0.034	0.033
0.066	0.004	0.033	0.032
Value of HRR (kW)	**Relative Uncertainty of HRR with *k=2* (%)**		
11.143	11.7	11.6	11.3
0.610	11.6	15.5	15.4
0.404	11.6	19.4	16.7
0.151	11.6	43.8	43.7
0.066	11.6	99.7	97.0

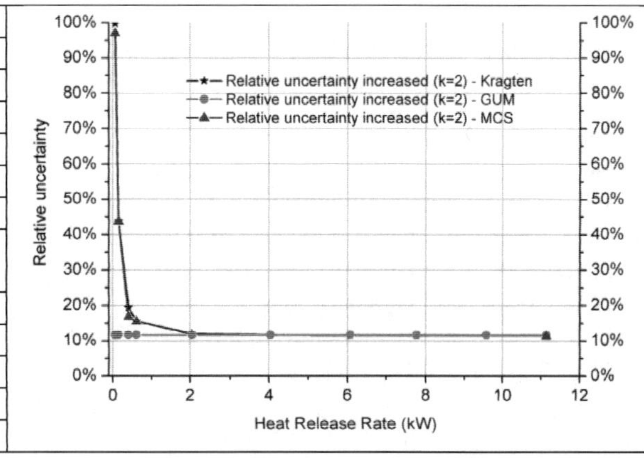

Figure 2 shows that of the relative contribution of each variable of the uncertainty calculated using Kragten's method or GUM (*Guillaume et al.* [13]) are in accordance. For HHR less than 2 kW, the variable of oxygen concentration pilots the uncertainty while above it is the C-factor. These observations had already been proven for full-scale fire tests [8]. In this specific case, the GUM method does not capture the influence of the oxygen concentration as the two other approaches. The error is caused by the covariance of several parameters and the non-linearity because the term x_{O_2} appears twice in the Eq. 4.

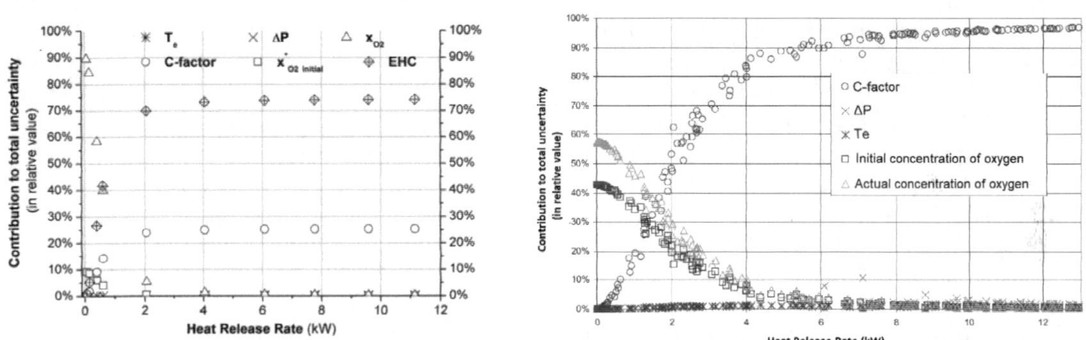

Fig. 2. Relative evolution of the different components uncertainty according to the HRR. Left: this study – Right: *Guillaume et al.* [13].

Then, the relative uncertainty changes as a function of the HRR decay for the Kragten and Monte Carlo methods, whereas it remains constant using the GUM approach (Table 2). This last point is contrary to the observations of *Guillaume et al.* [13] which specifies that above a 2 kW HRR, the relative uncertainty (according to the GUM method) increases very rapidly. Above 12 kW, the relative uncertainty of HRR remains constant and slightly less than 10% for the Kragten and the Monte Carlo methods, as shown in Table 2.

CONCLUSION

When the mathematical formulation of the measurand becomes complex or involves many parameters, it is shown that the Kragten method is a good alternative to the GUM or the Monte Carlo approaches. The calculation of the uncertainty is easier because of the absence of the coefficients of sensitivity. Besides Kragten method is simple and fast to use (a few minutes of treatment), because only the uncertainty of each variables, the value of the variables and the mathematical formulation to determine the HRR has to be specified (Eq. 4).

In the end, the precision of the uncertainties is very close to the Monte Carlo method whereas the computation time is extremely fast. The Kragten method is thus the most accurate when variables are interdependent and when covariance is unknown or poorly controlled, which is exactly the case concerning the heat release rate determination.

Moreover, the Kragten method is applicable to all results of the cone calorimeter determined by a multivariable measure. This concerns the mass loss rate, specific mass loss rate and total heat release rate.

ACKNOWLEDGMENTS

The authors wish to thank Poujoulat for its financial support and the "Institut des Risques Industriels Assurantiels et Financiers" for its technical and human support into the HESTIA platform.

REFERENCES

[1] International Standard Organisation, "ISO 5660-1 Reaction-to-fire tests -- Heat release, smoke production and mass loss rate - Part 1: Heat release rate (cone calorimeter method) and smoke production rate (dynamic measurement)", Genève: ISO, 2002, p. 45, https://doi.org/10.3403/30255889.

[2] Babrauskas, V., "Development of the cone calorimeter – A bench-scale heat release rate apparatus based on oxygen," *Fire and Materials*, vol. 8, pp. 81-95, 1984, https://doi.org/10.6028/nbs.ir.82-2611.

[3] BIPM, IEC, IFCC, ISO, IUPAC, IUPAP and OIML, "Guide to the expression of Uncertainty in Measurement", 2nd ed., JCGM, 2008, p. 134, https://doi.org/10.1017/cbo9781139135085.018.

[4] BIPM, IEC, IFCC, ISO, IUPAC, IUPAP and OIML, "Evaluation of measurement data - Supplement 1 to the guide to the expression of Uncertainty in Measurement-Propagation of distributions using a Monte Carlo method", JGGM, 2008, p. 90.

[5] Kragten, J., "Calculating Standard Deviations and Confidence Intervals with a Universally Applicable Spreadsheet Technique," *Analyst*, vol. 119, pp. 2161-2165, 1994, https://doi.org/10.1039/an9941902161.

[6] NASA, "Measurement Uncertainty Analysis Principles and Methods, NASA Measurement Quality Assurance Handbook - Annex 3," NASA Handbook, Washington, 2010.

[7] Chen, A. and Chen, C., "Comparison of GUM and Monte Carlo methods for evaluating measurement uncertainty of prespiration measurement systems," *Measurement*, vol. 87, pp. 27-37, 2016, https://doi.org/10.1016/j.measurement.2016.03.007.

[8] Bryan, R. A., Mulholland, G. W., "A guide to characterizing heat release rate measurement uncertainty for full-scale fire tests," *Fire and Materials*, vol. 32, pp.121-139, 2008, https://doi.org/10.1002/fam.959.

[9] Herrador, M. and Gonzales, A., "Evaluation of measurement uncertainty in analytical assays by means of Monte-Carlo simulation," *Talanta*, vol. 64, pp. 415-422, 2004, https://doi.org/10.1016/j.talanta.2004.03.011.

[10] Farrance, I. and Frenkel, R., "Uncertainty in measurement: A review of Monte Carlo Simulation Using Microsoft Excel for the Calculation of Uncertainties Through Functional Relationships, Including Uncertainties in Empiririically Derived Constants," *Clin Biochem Rev*, vol. 35, no. 1, pp. 37-61, 2014.

[11] Theodorou, D., Zannikou, Y. and Zannikos, F., "Estimation of the standard uncertainty of a calibration curve:application to sulfur mass concentration determination in fuels," *Accred Qual Assur*, vol. 17, pp. 275-281, 2012, https://doi.org/10.1007/s00769-011-0852-4.

[12] Janssens, M., "Measuring Rate of Heat Release by Oxygen Consumption," *Fire Technology*, pp. 234-249, 1991, https://doi.org/10.1007/bf01038449.

[13] Guillaume, E., Marquis, D., Saragoza, L. and Yardin, C., "Uncertainty on heat release rate measurement with cone calorimeter during the combustion of a material," *Revue française de métrologie*, Vols. 2012-3, no. 31, pp. 3-11, 2012.

[14] Enright, P. and Fleischmann, C., "Uncertainty of Heat Release Rate Calculation of the ISO5660-1 Cone Calorimeter Standard Test Method," *Fire Technology*, vol. 35, no. 2, pp. 153-169, 1999.

[15] Brohez, S., "Comments to the paper Uncertainty of HEat Release Rate Calculation of the ISO5660-1 Cone Calorimeter Standard Test Method," *Fire Technology*, vol. 45, pp. 381-384, 2009, https://doi.org/10.1007/s10694-008-0050-z.

[16] Guillaume, E., Marquis, D. and Saragoza, L., "Calibration of flow rate in cone calorimeter tests," *Fire and Materials*, vol. 38, no. 2, 2012, https://doi.org/10.1002/fam.2174.

[17] Cremona, P., "Caractérisation expérimentale et numérique des scénarii de feu impliquant un conduit de fumée d'appareils de combustion bois," Master Thesis, Université de Poitiers, Poitiers, 2017.

Influence of a misalignment of the cone or the sample holder on the view factor with a Monte Carlo approach

Pierre CREMONA[1,2], Yan BILLAUD[1], Thomas ROGAUME[1], Franck RICHARD[1] and Benjamin BATIOT[1]

[1] *Institut Pprime, UPR 3346 CNRS, Université de Poitiers, ISAE-ENSMA, France*
[2] *Laboratoire Ceric, Poujoulat SA, France*

p.cremona@laboratoire-ceric.com

ABSTRACT

The cone calorimeter is a fire test internationally standardized as ISO 5660-1 [1], developed at The National Institute of Standards and Technology in the 1980's [2]. It allows to study different flammability and combustibility parameters, as the ignition time, mass loss, mass loss rate, gaseous emissions and the rate of heat released by materials exposed to radiant heat flux. However, the results obtained with this set-up are clearly dependent of several critical points, especially the aeraulic conditions around the material and the radiation thermal stress imposed to the material. It appears important to properly characterize this last one to ensure that the parameters obtained experimentally are correlated to the true experimental conditions.

Among the critical points, the incident flux received by the sample holder through the radiant spirals of the cone calorimeter is identified as being the most significant. This incident flux depends on the temperature and the radiative properties of the spirals, and on the view factor between the spirals and the sample. This work focuses on the latter, and most specifically on the sensitivity evaluation of the geometrical parameters on the incident radiative flux. In order to determine the view factor, two distinct strategies may be applied. The first one is the algebraic derivations of geometric view factor that may be fastidious when the geometrical configuration increase in complexity. The second one consists in a rigorous experimental study by measuring the incident flux at different locations of the sample holder using flux meter. However, this approach is time-consuming and concedes a lot of measurement uncertainties. This article presents an intermediate method, easy, precise and allowing a parametrical study (axiality, horizontality, radiation coefficient, dimensions *etc.*) by means of a numerical Monte Carlo method.

KEYWORDS

Suggested Keywords: cone calorimeter, numerical method of Monte Carlo, view factor, uncertainty, experimental error, homogeneity of the radiative flux.

INTRODUCTION

The Construction Products Regulation [3] requires the qualification of the fire reaction of building materials. This is why the cone calorimeter is very widely used. However, this type of device is also used in the field of research and development, in order to compare newly created materials, as well as for the development and validation of numerical models.

As stated by *Guillaume et al.* [4], it is essential during experimental investigations to determine test bench uncertainties. In fire safety science, at small scale, the experimental uncertainties depend both on the heterogeneity of materials (especially for polymers and natural materials) and the accuracy of the different measurement sensors (temperature, oxygen concentration, differential pressure...) [5, 6, 7, 8, 9]. Other parameters (*i.e.* the positioning of the igniter, the radiant spiral and finally the sample holder) that are difficult to model mathematically are taken into account in the calculation of uncertainty by repeatability and reproducibility [4]. However, it is interesting to know the numerical impact of these parameters.

This study can be conducted experimentally with a flow meter like *Wilson et al.* [10] and *Kang et al.* [11], but the experimental errors are too important and the measurement not fine enough. This observation motivated us to develop a numerical method in order to study the biases with regard to the positioning of the different devices. To reach this aim, a radiation transport model based on a Monte Carlo approach is developed following the concept described in the reference book of Modest [12].

The interest of this paper thus relies in an explorative research of the uncertainties associated with the relative positioning of the sample holder and the radiative spiral. Indeed, the sources of errors related to the positioning of these two devices consist of (as illustrated on Fig. 1): i) the axiality of the sample holder with respect to the radiant spiral; ii) the horizontality of the sample holder and the radiant spiral; and iii) the distance between the sample holder and the radiant spiral. To determine the effect of the three parameters considered, we compare the results obtained with the ideal configuration (according to ISO 5660-1 [1]).

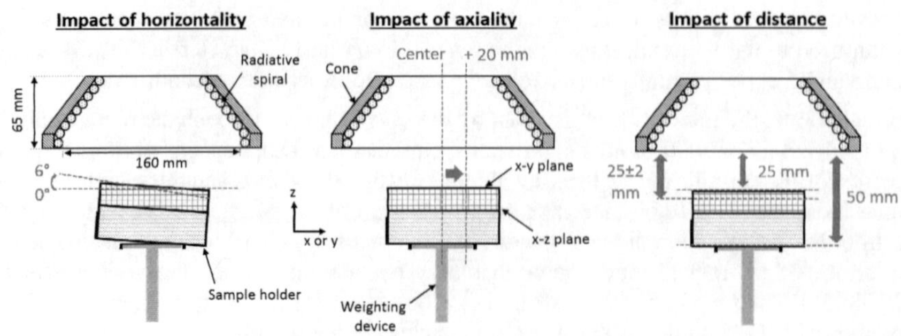

Fig. 1. Schematic diagram of the main errors of implementation and dimensions of the cone.

It should be noticed that the ISO 5660-1 [1] requires the control of axiality and horizontality in the context of regulatory tests. Therefore, this study is only relevant for fundamental and applied research trials, especially when specific sample holders are used.

MONTE CARLO RADIATIVE APPROACH TO DETERMINE THE VIEW FACTOR

The Monte Carlo method is a probabilistic numerical approach that consists in simulating the trajectory of quanta (*i.e.* bundles of energy), in a ballistic way, from an emitting sub-surface to a receiving one. For more information on this method you may refer to Modest [12].

In this study, the radiative scene is composed of the elements described previously, namely: the cone calorimeter and the sample holder, each of them being discretized into squared cells.

The sum of energy (proportional to the number of quanta) received by an elementary cell of the target, divided by the incident heat flux emitted from the radiative source, allows to calculate, thanks to Eq. 1, the view factor of the element.

$$F_{v_e} = \frac{\dot{q}_r''}{\dot{q}_e''} = \frac{N_r/S_r}{N_e/S_e} \qquad (1)$$

Where

F_{v_e} : View factor of each discretized element

$\dot{q}_r''$: Received heat flux (sum of specific energy of a quanta) (kW.m^{-2})
$\dot{q}_e''$: Incident heat flux from cone calorimeter (kW.m^{-2})
N_r, N_e: Respectively the number of received and emitted quanta (kW)
S_r, S_e: Respectively the surface of reception and emission (m²)

It is well-known that stochastic methods need a certain amount of quanta in order to produce an accurate resolution, in this case a view factor spatial distribution. The higher the number of photons shot, the better the result. One can notice that the calculation time is directly proportional to the number of quanta used. There is no specific rule to estimate the number of quanta needed to solve a radiative problem. In our case, we choose to perform 10 simulations with a number of quanta between 1e^{10} and 2e^{12} to determine when the standard deviation stabilizes.

To illustrate this point, Fig. 2 presents the evolution of the converges with respect to the number of quanta simulated. When 2e^{10} quanta are drawn (1.25e^{5} quanta shot per surface), the CPU time with a single Intel Core I7-5500 U processor (2.4 GHz) and 8 Go RAM is about 4.5 min. It is 43.9 min and 475.5min respectively for 2e^{11} and 2e^{12} quanta shot (1.25e^{6} and 1.25e^{7} quanta shot per surface).

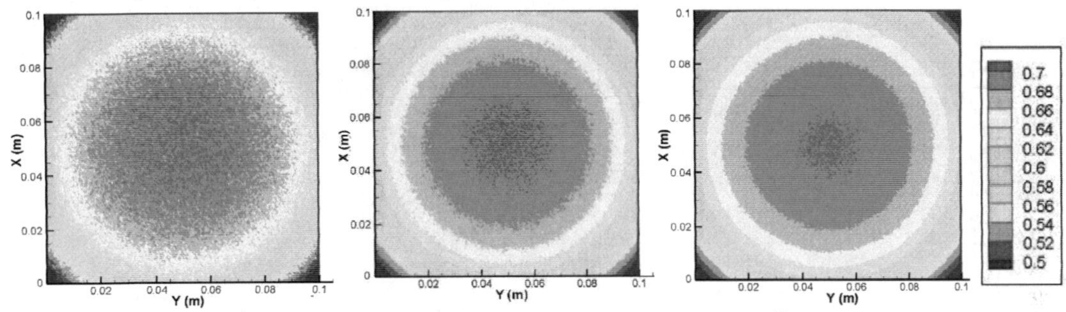

a. 2e^{10} quanta shot b. 2e^{11} quanta shot c. 1e^{12} quanta shot

Fig. 2. Simulation results of Monte Carlo radiative evaluation of the impact of the cone calorimeter on a plate plane using different number of quanta.

With a fine mesh of the sample holder (0.5 mm by 0.5 mm), Fig. 2b shows that a short calculation time (45 min equivalent to 2e^{11} quanta shot) is enough to obtain a relative standard deviation that starts to stabilize at 6,94%. Below this value, the quantity of quanta is insufficient to obtain a satisfactory solution (considering homogeneity) especially when the number of quanta is about 2e^{10} (Fig. 2a). To obtain a precise result (net outline of the different zones), it is necessary to shot at least 1e^{12} quanta above which the value of the relative standard deviation is constant at 6,93% (Fig. 2c).

VALIDATION OF THE MONTE CARLO RADIATIVE APPROACH

In order to validate the present model, the view factor obtained, using 1e^{12} quanta, are compared with values from literature [11] in Table 1. The comparison is performed with values at the centre of a horizontal plane for different altitude of the cone.

Table 1. View factor of *x-y* plane of the sample holder (exposed surface).

Z (mm)	Present study	Ref. [11]	Relative error (%)
15	0.7650	0.7660	0.13
20	0.7586	0.7599	0.17
25	0.7452	0.7461	0.12
30	0.7255	0.7261	0.08
35	0.7008	0.7014	0.08

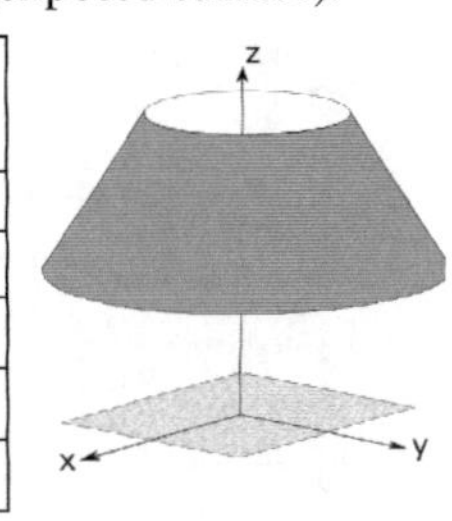

Table 1 shows that the results obtained are in perfect agreement with other studies. The relative errors calculated for all values are very low and the maximum error is reached for the height of 20 mm with only 0.17 %. This analysis confirms the feasibility to determine the view factor using the present Monte Carlo approach.

SENSITIVITY ANALYSIS

Distance between the cone and the sample holder

The positioning error between the cone and the sample holder, due to an incorrect installation of the sample holder for example, may affect the homogeneity of the incident flux. According to ISO 5660-1 [1], intumescent materials are positioned at 50 mm instead of 25 mm from the cone.
To study the effect of the distance on the incident radiative heat flux, simulations are performed with the distance between the cone and the sample holder ranging from 10 mm to 50 mm along the *z* axis. Firstly, the variation of the view factor along the z-axis is studied by generation the view factor distribution in the *x-z* plane at *y*=0. Then, the view factor distribution on the exposed surface (*x-y* plane) at different axis (*z*=23, 25, 27 and 50 mm) from the cone is interpreted.

Table 2 presents the average, the minimum and the maximum values of the view factor in the x-y plane at different heights. The standard deviation is calculated over the entire mesh of the exposed surface (200 × 200 values). The variation of the distance between the cone and the sample holder impacts particularly the maximum values and the homogeneity of the view factor on the exposed surface.

Table 2. Minimum, maximum and standard deviation of view factor of *x-z* plane of the sample holder.

Z (mm)	View factor				
	Average	**Min.**	**Max.**	**SD.**	**RSD. (%)**
10	0.7937	0.7675	0.8044	0.0093	1.2
20	0.6442	0.5931	0.6678	0.0216	3.3
25	0.5769	0.5196	0.6051	0.0248	4.3
30	0.5146	0.4528	0.5440	0.0263	5.1
40	0.4078	0.3503	0.4377	0.0255	6.2
50	0.3221	0.2723	0.3493	0.0221	6.8

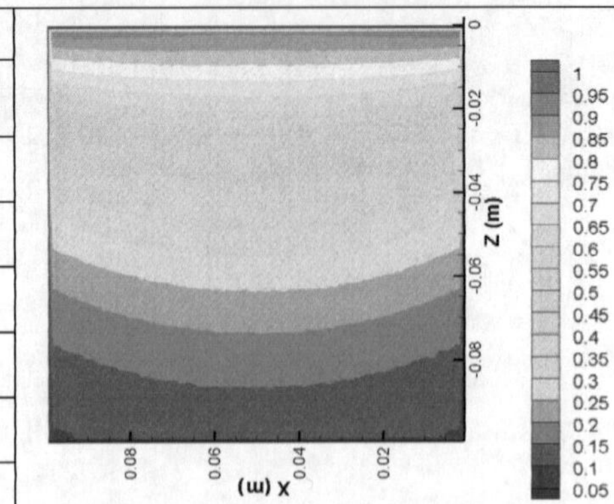

Table 2 shows that the further the sample holder is from the cone, the more homogeneous and weak the factor of view is. Indeed, less than 10 mm below the cone, the standard deviation is very low. On the contrary, the standard deviation for distances between 20 and 50 mm below the cone is stable and at least twice as large as at 10 mm. The relative standard deviation continues to increase depending on the distance of the cone. Even if the incident flux is theoretically identical at the center of the sample at 25 and 50 mm, the thermal stress conditions of all the exposed surface vary strongly between these two positions. Therefore, it is difficult to compare the fire behaviour of two materials at 25 and 50 mm (*i.e.* non-intumescent and intumescent materials). This observation is all the truer that the aeraulic conditions are also different.

On the other hand, Table 3 shows that an incorrect installation of the sample holder of ± 2 mm (25±2 mm from the cone) does not significantly impact the view factor and so the incident flux received by the exposed surface. Thus, the requirement of the standard on this point is relevant.

Table 3. Minimum, maximum and standard deviation of view factor of *x-y* plane of the sample holder (exposed surface) as a function of the distance.

Z (mm)	Average view factor	Minimum view factor	Maximum view factor	SD.	RSD. (%)
23	0.7360	0.5502	0.7576	0.0295	4.0
25	0.7221	0.5289	0.7452	0.0331	4.6
27	0.7078	0.5140	0.7382	0.0364	5.1
50	0.5430	0.3744	0.6129	0.0481	8.9

Axiality of the sample holder

A bad axiality is caused by a misalignment of the sample holder on the weighting device and/or the radiant spiral in the cone.

IOP Conf. Series: Journal of Physics: Conf. Series **1107** (2018) 032020 doi:10.1088/1742-6596/1107/3/032020

To study the impact of axiality or eccentricity, it is chosen to decenter the sample holder from 0 mm (reference case) to 20 mm along one axis (*x* or *y*) and from 0 mm to 10 mm along two axes (*x* and *y*). The sample holder is positioned at 25 mm under the cone.

Table 4. Minimum, maximum and standard deviation of view factor of x-y *plane* of the sample holder (exposed surface) as a function of the axiality.

Decentred value		**View factor**				
X (mm)	**Y (mm)**	**Average**	**Minimum**	**Maximum**	**SD.**	**RSD. (%)**
0	0	0.7233	0.5480	0.7452	0.0311	4.3
1	0	0.7232	0.5386	0.7453	0.0312	4.3
5	0	0.7218	0.4949	0.7453	0.0347	4.8
10	0	0.7171	0.4301	0.7457	0.0444	6.2
20	0	0.6972	0.2877	0.7457	0.0779	11.2
1	1	0.7232	0.5292	0.7416	0.0314	4.3
5	5	0.7203	0.4365	0.7440	0.0380	5.3
10	10	0.7108	0.3106	0.7419	0.0562	7.9

Table 4 shows that a small centering error (up to 5 mm), when placing the sample holder or the radiant spiral, has a small impact on the exposed surface. It is necessary that the decentering must be strong to have a significant impact on the factor of view: at least 10 mm. This configuration never occurs when the sample holder is in position unless decentering is voluntary. On the other hand, the risk is higher when repositioning the spiral. However, a displacement less than 5 mm may be considered as inconsequential on view factor estimation.

Horizontality of the sample holder or the cone

A bad horizontality is caused by a misalignment of the sample holder or the cone. To study the influence of such defect, it is chosen to vary the tilted angle from the horizontal axis of the sample from 0° (reference case) to 6°. In other words, a 10 mm offset from one side to another when the dimensions of the sample holder are 100 x 100 mm² (length x width). The sample holder is positioned at 25 mm from the cone.

Table 5. Minimum, maximum and standard deviation of view factor of *x-y* plane of the sample holder (exposed surface) as a function of horizontality.

Tilt angle (°)	**Average view factor**	**Minimum view factor**	**Maximum view factor**	**SD.**	**RSD. (%)**
0	0.7203	0.5416	0.7459	0.0301	4.2
1	0.7203	0.5333	0.7474	0.0304	4.2
2	0.7203	0.5248	0.7492	0.0313	4.3
3	0.7203	0.5163	0.7523	0.0327	4.5
4	0.7203	0.5077	0.7554	0.0345	4.8
5	0.7203	0.4992	0.7593	0.0368	5.1
6	0.7203	0.4906	0.7646	0.0393	5.5

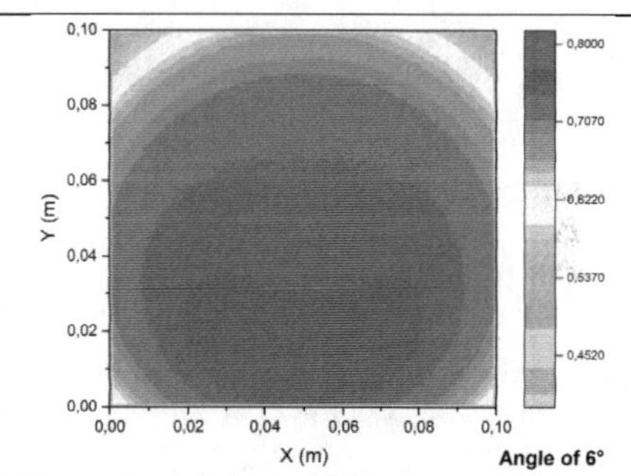

An error of horizontality, as important as it may be, does not seem to affect the average view factor of the exposed surface. On the other hand, the distribution of the view factor at the surface is impacted. For a tilt angle value inferior to 2°, the relative standard deviation is not affected and is about 4.2%. For higher angle values, the relative standard deviation increases slightly until it reaches nearly 5.5% with a tilt angle of 6° (figure of Table 5). The difference between the maximum and minimum view factors increases more and more with angle, leading to an increase in heterogeneity of the radiation exposure.

CONCLUSION

The results of this numerical study show that:

- It is also shown that the positioning of the cone or the sample holder (distance or misalignment) strongly impact the homogeneity of the flux received by the sample. Therefore, depending on the physical parameter studied, not all materials can be compared with each other, especially with regard to intumescent materials. Moreover, a misalignment of a few millimeters is negligible on the distribution of the factor of view.

However, the incident flux becomes more and more homogeneous when the sample holder moves away from the cone.

- The impact of axiality on the distribution of the view factor on the exposed surface has little effect for an eccentricity inferior to 10 mm. Beyond, the impact is significantly large.
- The impact of axiality by a few millimeters is negligible on the distribution of the factor of view.
- The impact of a horizontality error is negligible as long as the angle of inclination is lower than 3°. The minimum and maximum view factors are impacted.
- This study can be extended to other normative benches such as the IMO-LIFT. (ISO 5658 [13]).
- The Monte Carlo numerical approach makes it possible to propose an alternative method to experimental method which involves introducing new uncertainties due to the properties of the sensor and its positioning.

Knowing the view factor, the emissivity of the material, the temperature of the radiating spirals and the Stefan-Boltzmann constant, the incident flux on the exposed surface can be determined. Thus, it is possible to determine the energy loss or gain generated by these errors.

ACKNOWLEDGMENTS

The authors wish to thank Poujoulat for its financial support and the “Institut des Risques Industriels Assurantiels et Financiers” for its technical and human support into the HESTIA platform.

REFERENCES

[1] International Standard Organistion, “ISO 5660-1 Reaction-to-fire tests -- Heat release, smoke production and mass loss rate - Part 1: Heat release rate (cone calorimeter method) and smoke production rate (dynamic measurement)”, Genève: ISO, 2002, p. 45, https://doi.org/10.3403/30255889.

[2] Babrauskas, V., “Development of the cone calorimeter – A bench-scale heat release rate apparatus based on oxygen,” *Fire and Materials*, vol. 8, pp. 81-95, 1984, https://doi.org/10.6028/nbs.ir.82-2611.

[3] The European Parliament, “Regulation (EU) No 305/2011 of the European Parliament and of the council of 9 March 2011,” *Official Journal of European Union*, vol. L 88, pp. 5-43, 2011.

[4] Guillaume, E., Marquis, D., Saragoza, L. and Yardin, C., “Uncertainty on heat release rate measurement with cone calorimeter during the combustion of a material,” *Revue Française de Métrologie*, Vols. 2012-3, no. 31, pp. 3-11, 2012.

[5] Enright, P. and Fleischmann, C., “Uncertainty of Heat Release Rate Calculation of the ISO5660-1 Cone Calorimeter Standard Test Method,” *Fire Technology*, vol. 35, no. 2, pp. 153-169, 1999.

[6] Zhao, L., “Bench scale apparatus measurement uncertainty and uncertainty effects on measurement of fire characteristics of material system,” Worcester polytechnic Institute, 2005.

[7] Zhao, L. and Dembsey, N., “Measurement uncertainty analysis for calorimetry apparatuses,” *Fire and Materials*, vol. 32, no. 1, pp. 1-26, 2008, https://doi.org/10.1002/fam.947.

[8] Brohez, S., “Comments to the paper uncertainty of heat release rate calculation of the ISO5660-1 cone calorimeter standard test method,” *Fire Technology*, vol. 45, no. 4, pp. 381-384, 2009, https://doi.org/10.1007/s10694-008-0050-z.

[9] Guillaume, E., Marquis, D. and Saragoza, L., “Calibration of flow rate in cone calorimeter tests,” *Fire and Materials*, vol. 38, no. 2, 2012, https://doi.org/10.1002/fam.2174.

[10] Wilson, M., Dlugogorski, B. and Kennedy, E., “Uniformity of radiant heat fluxes in cone calorimeter,” *Fire Safety Science,* vol. 7, pp. 815-827, 2003, https://doi.org/10.3801/iafss.fss.7-815.

[11] Kang, S., Choi, S. and Choi, J., “View factor in cone calorimeter testing,” *International Journal of Heat and Mass Transfer*, vol. 93, pp. 217-227, 2016, https://doi.org/10.1016/j.ijheatmasstransfer.2015.09.067.

[12] Modest, M. F., “Radiative Heat Transfer (Third Edition),” Academic Press, Amsterdam, 2013, p. 904, https://doi.org/10.1016/B978-0-12-386944-9.50026-1.

[13] International Organization for Standardization, “ISO 5658-2:2006+A1:2011 - Reaction to fire tests - Spread of flame - Part 2: Lateral spread on building and transport products in vertical configuration,” International Organization for Standardization, Genève, 2006, p. 35, https://doi.org/10.3403/30090264.

Heat release and flame spread assessment of insulation in External Thermal Insulation Composite System (ETICS) façades

Martyn MCLAGGAN[1,2], Konrad WILKENS FLECKNOE-BROWN[1,2], Anders DRAGSTED[2] and Patrick VAN HEES[1]

[1]*Division of Fire Safety Engineering, Lund Technical University, Sweden*
[2]*Advanced Services, DBI – Danish Institute of Fire & Security Technology, Denmark*

martyn.scott_mclaggan@brand.lth.se

ABSTRACT

External Thermal Insulation Composite Systems (ETICS) represent a popular modern façade for reducing energy consumption, particularly in retrofit applications. Insulation – typically in the form of either EPS (expanded polystyrene) or mineral wool – is applied to the exterior of buildings using mortar adhesive, and then coated with a final layer of sand-cement render and glass fibre reinforcement. The materials are relatively cheap, and the labour can be easy and inexpensive compared to other solutions.

The system does however mean that highly flammable EPS insulation is protected by only a thin layer of render, normally 3–8mm. The addition of this combustible material to the façade of the building represents a change in the fire risk, as the typical fire safety strategy involving compartmentation does not anticipate vertical fire spread on the exterior of a building. It is therefore necessary to characterise the fire risk, and ensure that an optimal fire barrier – namely, the render layer – is defined adequately.

In this work, the insulation component of an EPS ETICS façade is tested using micro- to small-scale methodologies. The objective of this is to characterise some of the fundamental thermal material properties, ignitability, heat release, and flame spread of the specific components. The aim is to then extend this to a larger-scale methodology for testing of complete systems, which will be complemented by numerical modelling to enable scaling. Previous attempts at testing ETICS in a small-scale have been largely unsuccessful, and efforts have pointed to large-scale apparatuses – generally 4–8m high – as the only viable solution at present. These are however expensive and time consuming, and deliver little information on how the façade performs. Testing is especially problematic for EPS, which melts and shrinks at low temperatures, leading to difficulties in attempting to extract global flammability properties.

MCC (Microscale Combustion Calorimeter) testing has been performed to benchmark the different EPS insulation that can be used in buildings. It is found that black EPS containing expanded graphite used in façades has substantially improved performance compared to ordinary white EPS used as building insulation. The peak heat release was reduced from 1160 W g^{-1} down to 490–740 W g^{-1}, occurring at a temperature of 430–440 °C. An extra additive to improve the moisture performance of the system also enhances the fire performance slightly by reducing the heat release further.

Previous attempts of testing ETICS and EPS using LIFT (Lateral Ignition and Flame spread Test) have in many cases been unproductive. For this experimental series, a sheet of paper has been applied to the surface of the first 100 mm of EPS in order to force ignition. Results for the flame spread velocity near to ignition have then been discarded to minimise the influence of this method. Flame spread results for the white EPS were not satisfactory due to the fact that the flame spread velocity had a weak dependence on the incident heat flux, and that that minimum critical heat flux for flame spread was 0.0 kW m^{-2}, that is, no external energy was required to sustain burning. For black EPS specimens, the addition of the expanded graphite was effective in improving the flame spread performance. The minimum heat flux for flame spread was increased to 0.75–1.09 kW m^{-2}, and the velocity was sufficiently moderate that a reasonable value for the flame spread modulus could be obtained.

In future, attempts will be made to correlate the flame spread results of these individual EPS products to the flame spread as part of a complete ETICS façade. This would then provide a more meaningful method to compare different ETICS solutions as a first step without the need for expensive large-scale testing during the main development phase of systems.

KEYWORDS:

Keywords: ignition, heat release, flammability, flame spread, facades; performance-based design, ETICS.

IOP Conf. Series: Journal of Physics: Conf. Series **1107** (2018) 032021 doi:10.1088/1742-6596/1107/3/032021

INTRODUCTION

ETICS (External Thermal Insulation Composite Systems) are a façade kit where insulation is attached to a solid wall, such as concrete or masonry, using mortar adhesive. An additional layer of mortar is applied to the insulation, before a sheet of glass fibre reinforcement and a thin finishing layer of render and paint. The system requires time to dry and cure due to the presence of the sand-cement render/mortar, but the application of the kit is otherwise rapid and relatively straightforward compared to other façade solutions [1].

The two most common insulation materials used in ETICS are either mineral wool or polystyrene in the form of EPS (expanded polystyrene) or XPS (extruded polystyrene). In some cases PUR/PIR (polyurethane/polyisocyanurate) or phenolic foam are used but at present these are less popular. EPS has good insulation properties at low cost, and is light and easily workable. However, its flammability is extremely poor. The material has a low thermal inertia which causes rapid surface heating leading to short ignition times [2]. In addition, the material melts and drips prior to its pyrolysis temperature which enhances flame spread. Upon ignition, the material releases very large amounts of energy owing to its high heat of combustion. Its use in buildings therefore requires careful consideration and proper fire risk mitigation strategies to be in place to enable its safe usage. Furthermore, the scaling up of results in the field of fire science is difficult, especially in the case of façades where the interaction between components is highly relevant [3]. In many cases the use of the material can be limited, which acts as a barrier to innovation. This overall aim of this work seeks to optimise the insulation to achieve maximum energy savings for quantified fire performance. This is a first step a complicated process, but utilises performance-based methods to rank material formulations to predict their behaviour in large-scale tests, with later applicability to predicting the performance of actual buildings.

MATERIALS

There were three types of EPS chosen for study. The first was a typical white EPS product used for insulation in walls and can be easily purchased on the market. This type of insulation is not typically used in façades, but acts as a useful benchmark for comparing results to literature. The other two products were both black EPS materials as used in ETICS façades. The blackened nature of the EPS indicates that it includes expanded graphite in order to improve its thermal performance but also results in a slight increase in cost. Both of these products were also available on the market. The second of the two black EPS products further contains an extra additive to improve the hygrothermal performance of a façade. This is due to the fact that EPS ETICS tend to suffer the effects of high moisture [4]. A crude estimation based on visual observation is that approximately one in 400 beads contain the additive – indicated by yellow beads – for the latter EPS product.

IGNITABILITY AND HEAT RELEASE METHODOLOGY

Microscale Combustion Calorimeter (MCC)

The microscale measurements were performed in an MCC produced by the FAA (Federal Aviation Authority, New Jersey). The methodology follows the standard defined in ASTM D7309 [5]. Experiments for all materials were performed in nitrogen and some additional experiments were performed in air. The flow rate of gas was 0.1 L min^{-1} and the heating rate in all cases was 1.0 °K s^{-1}. Sample weight across all experiments was within the range of 2.1–2.5 mg, and each specimen was prepared using a scalpel.

Bench-scale calorimetry

Flammability and flame spread assessments generally require accurate determination of the critical heat flux of the material, which in turn necessitates a large number of tests. Previous work on the flammability of EPS has identified the difficulties in obtaining a representative value of the critical heat flux [6,7]. This is due to the fact that the material softens at a temperature much lower than its pyrolysis and exothermic oxidation reactions. As a result, the surface of the material rapidly regresses and ignition in many cases either does not occur or is delayed. Additionally, the incident heat flux upon ignition will be lower due to the greater distance from the cone heater. For these reasons, the bench-scale ignitability and heat release results are instead taken from the highly detailed studies by Hidalgo *et al.* [6] and Cleary & Quintiere [7]. In future, it will be necessary to obtain specific and accurate critical heat fluxes for the actual materials tested in this project.

FLAME SPREAD METHODOLOGY

The LIFT (Lateral Ignition and Flame spread Test) apparatus is designed to extract an experimentally determined empirical correlation for the flame spread of materials. It was developed over a number of years by Quintiere and colleagues [8–10] using a radiant panel positioned at a set angle from a 155 mm high by 800 mm long specimen. Flame spread is evaluated in opposed flow conditions but using dimensionless parameters it can be adapted for upward, wind-aided or forward flow conditions [11]. The apparatus in this study was in the ISO 5658 configuration [12], where a strong pilot is positioned vertically and partially impinges on the surface of the sample. Calculations of the heat release based on oxygen depletion [13] in the hood indicated that the pilot flame released 1.0–2.0 kW of energy. This provides additional heat flux to the surface which reduces the time to ignition. Other difficulties of this pilot position have been noted in the literature [14].

Radiant panel gas

The burner as specified by the standard should be fuelled by methane and provide a standardised heat flux distribution across the length of the specimen. However, practicalities in the laboratory at Lund University dictated that propane was a more convenient gas to use.

Peak heat flux modification

In the original development of the LIFT apparatus there was a methodology which outlined a different peak heat flux. In current standards there is only a single heat flux specified which is intended to reasonably ignite all materials, which is 50 kW m^{-2}. This is done for the sake of simplicity, especially for complex materials, and so that it is not necessary to perform ignitability tests before running LIFT. The original method set the peak heat flux to instead be approximately 5 kW m^{-2} above the critical heat flux of the material This value could be obtained either through bench-scale testing in the cone calorimeter or by using the LIFT apparatus in its ignition configuration. A range of 15.5–70 kW m^{-2} was also specified based on the limitations of the radiant panel [9].

In this research, the latter method has been used. The determination of an accurate critical heat flux in EPS proves difficult, as previously noted. Values in the literature indicate nonetheless that the range may be between 6–15 kW m^{-2}, and a maximum of 20 kW m^{-2}. Therefore the peak heat flux of the panel was set to be 25.0 kW m^{-2} and measured before every experiment.

Preliminary testing

A set of preliminary testing was performed to ensure the suitability of these changes and that the results were still in-line with standardised testing. Tests were performed on plywood, MDF (Medium Density Fibreboard) and particle board (Swedish spånskiva). A total of twenty tests were performed with the propane fuelled radiant panel, and four tests were performed with the reduced radiant peak heat flux. The calculated flame spread modulus agreed with those found in the literature. Additionally, testing of EPS in the traditional unaltered configuration was performed. In this case, the material failed to ignite and instead melted along the length of the material. This matches the conclusions reached in the literature [15].

RESULTS AND DISCUSSION

Ignitability and heat release

The heat release for both of the black FR (fire retarded) EPS materials (EPS-A and EPS-B) are both reduced when compared to white non-FR EPS-C (Table 1). The heat release capacity, η_c, is reduced from 1160 J g^{-1} K^{-1} for EPS-C down to 740 J g^{-1} K^{-1} for EPS-B and down to 490–640 J g^{-1} K^{-1} for EPS-A when pyrolysed in an inert atmosphere (Fig. 1, left). It is therefore clear that the addition of expanded graphite is effective in reducing the heat release of the polystyrene product. Additionally, the yellow beads of EPS-A had the lowest heat release capacity and overall had a slightly lower heat release (14–35 % reduction) when compared to EPS-B. However, the temperature for the peak heat release, T_p, was reduced from 430–440 °C to 405°C which indicates that the material would ignite earlier. The reduction in the heat release for the yellow beads may not be sufficient to offset the decreased onset of ignition. Nonetheless, there is still substantial improvement over the non-FR white EPS-C.

The apparent heat of combustion, ΔH_c, is substantially improved from 45 kJ g^{-1} for EPS-C down to a consistent value of 30 kJ g^{-1} for EPS-B and EPS-A. The calculation for heat of combustion uses the original mass, m_0, and so the presence of a char yield for the black EPS products will provide some improvement in the calculated heat of combustion. Even if corrected only for the mass lost, there is still some improvement in the heat of combustion. This may be due to the expanded graphite reacting and reducing heat transfer to the polystyrene. This however would violate some of the principles of microscale calorimetry, namely that there should not be temperature gradients within the specimen, and would therefore require slower heating rates to ensure the conclusions are correct.

The results of EPS-A in an oxidative environment indicate much lower onset and peak reaction rate temperatures (Fig. 1, right). The temperature for peak heat release is reduced from 435 and 405 °C down to

345 and 365 °C for black and yellow beads respectively. Additionally, the pyrolysis occurs over a much wider range (approximately 130 °C compared to 100 °C) which leads to a comparable value of apparent heat of combustion for the black EPS-A beads. In the oxidative environment, the yellow beads are found to be more reactive. Whilst there is a marginal increase in the pyrolysis onset temperature, this is offset by higher heat release and higher heat of combustion. This leads to a higher value of the heat release capacity, η_c, and would suggest that it has the poorest performance between the two EPS-A beads.

Table 1. Summary of micro-scale calorimetry results for each of the EPS materials tested.

ID	Colour	Gas	η_c	$\dot{q}_p$	ΔH_c	T_p	Y
-	-	-	J g^{-1} K^{-1}	W g^{-1}	kJ g^{-1}	°C	%
EPS-A	Black	N_2	639.60	639.78	30.81	432.98	26.09
		Air	1120.71	376.60	30.49	346.56	4.35
	Yellow	N_2	487.50	486.32	30.40	402.92	12.00
		Air	1438.89	479.05	38.37	363.32	4.17
EPS-B	Black	N_2	741.75	741.36	31.65	439.13	14.29
EPS-C	White	N_2	1163.22	1163.10	45.52	436.61	0.00

Some differences in the behaviour of the materials between the inert and oxidative environments reveal different reaction pathways. Whilst the char yields, *Y*, for the black and yellow beads of EPS-A are quite different in an inert atmosphere, 26 and 12% respectively, these are found to both be 4% in air. For black EPS-A, the heat of combustion between the two atmospheres is found to be the same. However, for yellow EPS-A the heat of combustion is higher in air which suggests that the oxidative pathway is more exothermic than the pathway in an inert atmosphere.

Lyon *et al.* [16] has studied six white EPS products previously. Comparison with EPS-C shows that the temperature of peak reactions, T_p, is given as 445±1 °C which is within ±5 % of the results presented here. The ±5 % is the anticipated accuracy of the MCC apparatus, as detailed in the same report. The result for heat of combustion is however somewhat higher in this study; 46 kJ g^{-1} compared to 40 kJ g^{-1} in the results of Lyon *et al.* [16]. This is also higher than values quoted for styrene, 39.4–42.2 kJ g^{-1} [17], and so may require additional testing or modification to the calculation.

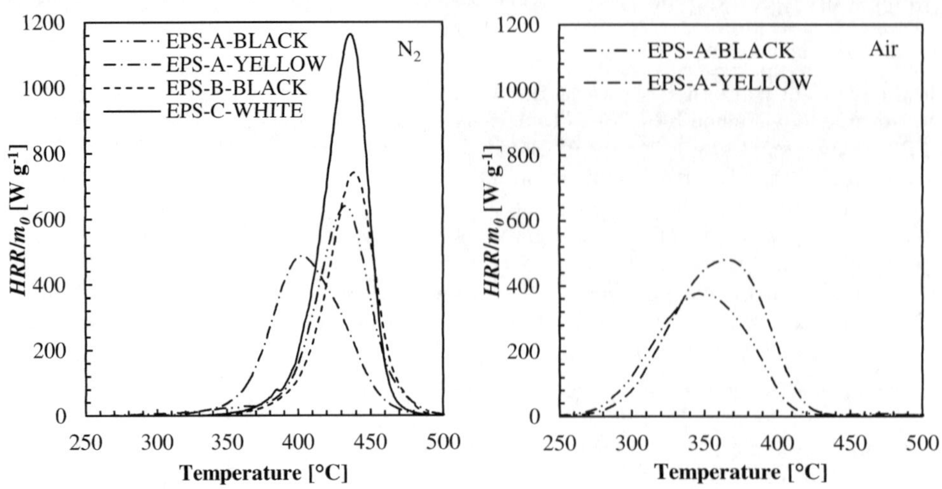

Fig. 1. Results from MCC testing for all three EPS materials in nitrogen (left) and only EPS-A in air (right).

Flame spread

In order to achieve ignition, a piece of standard 80 g m^{-2} paper (100 mm wide by 150 mm high) was attached to the surface of 47.5 mm thick EPS. This caused ignition after 4–8s for white EPS and 5s for black EPS. This is more rapid than the ignition times of EPS obtained from the literature in bench-scale calorimetry and so reduces the preheating time of the upstream material before the flame front arrives. Analysis by Quintiere [8] indicates that a longer preheating time is preferable when testing a material for research purposes but for polystyrene this is not the case due to the rapid melting causing loss of material.

Despite the forced ignition, the results for non-FR white EPS-C are still somewhat inadequate and the scatter is large. The flame spread velocity is found to have a weak dependence on the incident heat flux (Fig. 2, left), and the minimum heat flux for flame spread, $\dot{q}_{0,s}''$, is found to be 0.0 kW m^{-2} due to the fact that all material was consumed. Nonetheless, a flame front was still established which spread along the length of the material. There was generally not significant generation of flaming droplets – although sufficient to still fail the standardised test – and molten styrene was instead contained within the holder.

A substantial improvement was found in the black EPS specimens containing expanded graphite. As indicated by the MCC results previously, the introduction of the graphite increases the char yield of the product. On exposure to heat, this rapidly expands and inhibits some heat transfer to the polystyrene. This in turn increases the time required to pyrolyse the EPS upstream and reduces the flame spread velocity. A stronger correlation between the incident heat flux and the flame spread velocity is then found (Fig. 2, right) and the minimum heat flux for flame spread in this case is found to be 0.75–1.09 kW m^{-2}. This value is still extremely low but the product would nonetheless require some external heat source to sustain burning. At present, four flame spread tests of EPS-A and three tests of EPS-C have been performed. Specimens of EPS-B are currently being tested, along with complete ETICS kits with the mortar render in place. Once the flame spread is understood to some extent then suitable experiments to evaluate the integrity of the render layer can be performed.

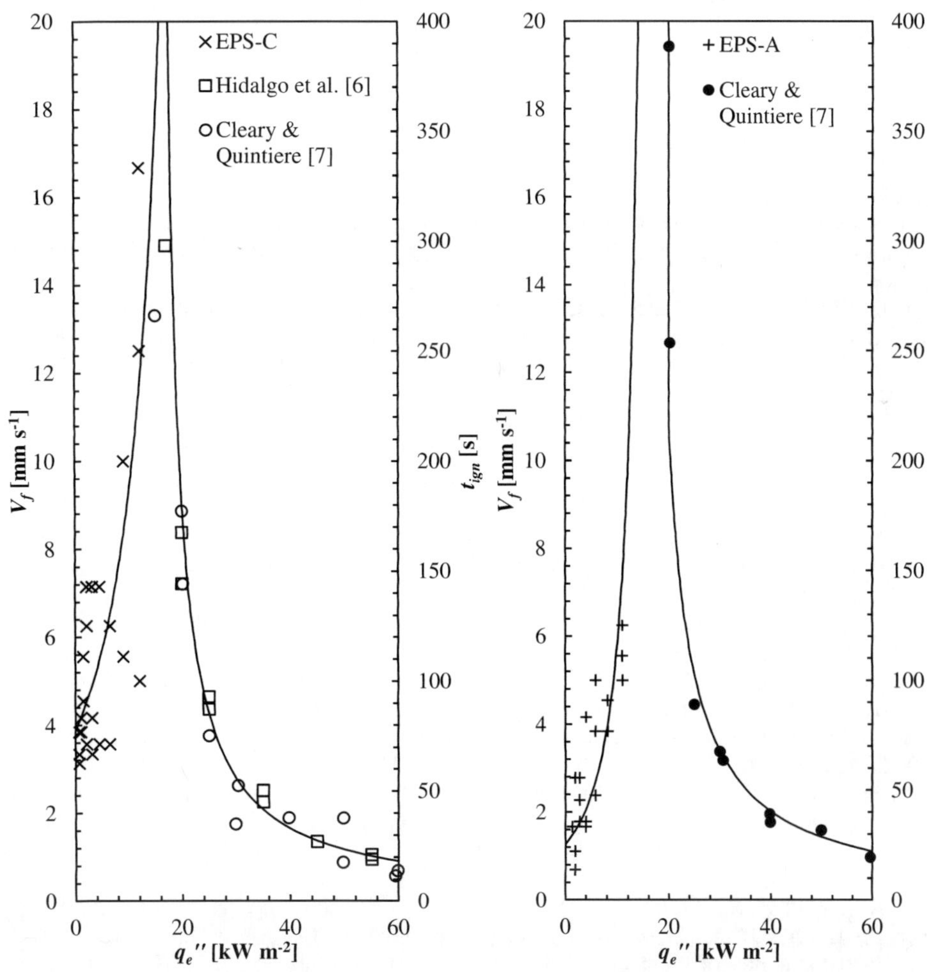

Fig. 2. Flammability diagrams for white EPS (EPS-C, left) and black EPS (EPS-A, right). Ignition data is taken from Hidalgo *et al.* [6] for 50 mm non-FR EPS, and Cleary & Quintiere [7] for both FR and non-FR 50 mm EPS samples.

CONCLUDING REMARKS

A successful method for extracting apparent flame spread properties from flame retarded polystyrenes has been outlined. This has shown to be effective for FR EPS containing expanded graphite where the minimum critical heat flux is greater than 0 kW m^{-2}. For ordinary white EPS, the flame spread velocity is too high and no external energy is required to sustain flame spread and thus a flame spread parameter cannot adequately be determined. These results are in agreement with the literature obtained using a different methodology. MCC results quantify the char yield in micro-scale and show that the black EPS products have lower heat release. Additional work will extend this method to EPS with render in place to simulate ETICS kits, and attempt to determine relevant flame spread properties for modelling and predicting full-scale tests. From this, product development to define optimal thermal properties and thickness of the insulation and render can be performed.

ACKNOWLEDGEMENTS

The Danish Innovation Fund (Innovationsfonden) are gratefully acknowledged for their support through an Industrial Postdoc award (5190-00020B). Part of this work was also funded by PROFIL (Prototype Fire Laboratory) from DBI in agreement with the Danish Agency for Research and Innovation. Thomas Krogsgaard on behalf of Sto Denmark is thanked for providing samples. Dr Stefan Svensson and Dan Madsen (Lund Technical University, Sweden) are thanked both for their help in the lab and for fruitful discussions.

REFERENCES

[1] P. Nelson, R. Kroll, *ASTM STP1269 – Exterior Insulation Finish Systems (EIFS): Materials, Properties, and Performance*, ASTM International, West Conshohocken, PA, 1996. doi:10.1520/STP1269-EB.
[2] D.D. Drysdale, Fundamentals of the Fire Behaviour of Cellular Polymers, *Cell. Polym.* 4 (1985) 405–419. doi:10.1007/978-94-009-3443-6_4.
[3] V. Babrauskas, Facade fire tests: Towards an international test standard, *Fire Technol.* 32 (1996) 219–230. doi:10.1007/BF01040215.
[4] E. Barreira, V.P. de Freitas, Experimental study of the hygrothermal behaviour of External Thermal Insulation Composite Systems (ETICS), *Build. Environ.* 63 (2013) 31–39. doi:10.1016/j.buildenv.2013.02.001.
[5] ASTM International, *ASTM D7309-13: Standard Test Method for Determining Flammability Characteristics of Plastics and Other Solid Materials Using Microscale Combustion Calorimetry,* (2013) 11pp. doi:10.1520/D7309.
[6] J.P. Hidalgo, J.L. Torero, S. Welch, Fire performance of charring closed-cell polymeric insulation materials: Polyisocyanurate and phenolic foam, *Fire Mater.* (2018) 1–16. doi:10.1002/fam.2501.
[7] T.G. Cleary, J.G. Quintiere, *Flammability Characterization of Foam Plastics.* NISTIR 4664, Gaithersburg, MD, 1991. doi:10.6028/NIST.IR.4664.
[8] J. Quintiere, A simplified theory for generalizing results from a radiant panel rate of flame spread apparatus, *Fire Mater.* 5 (1981) 52–60. doi:10.1002/fam.810050204.
[9] J. Quintiere, M. Harkleroad, D. Walton, Measurement of Material Flame Spread Properties, *Combust. Sci. Technol.* 32 (1983) 67–89. doi:10.1080/00102208308923653.
[10] J.G. Quintiere, M.T. Harkleroad, New Concepts for Measuring Flame Spread Properties, in: T.Z. Harmathy (Ed.), *Fire Saf. Sci. Eng. ASTM STP 882*, Philadelphia, PA, 1985: pp. 239–267. doi:10.1520/STP35300S.
[11] T. Cleary, J. Quintiere, A Framework For Utilizing Fire Property Tests, *Fire Saf. Sci. 3* (1991) 647–656. doi:10.3801/IAFSS.FSS.3-647.
[12] British Standards Institution, *BS ISO 5658-2:2006 Reaction to fire tests -- Spread of flame -- Part 2: Lateral spread on building and transport products in vertical configuration*, 2006.
[13] M.L. Janssens, Measuring rate of heat release by oxygen consumption, *Fire Technol.* 27 (1991) 234–249. doi:10.1007/BF01038449.
[14] M.S. McLaggan, *Novel fire testing frameworks for Phase Change Materials and hemp-lime insulation*, Ph. D. The University of Edinburgh, 2016. doi:http://hdl.handle.net/1842/15896.
[15] I. Oleszkiewicz, Fire exposure to exterior walls and flame spread on combustible cladding, *Fire Technol.* 26 (1990) 357–375. doi:10.1007/BF01293079.
[16] R.E. Lyon, R.N. Walters, S.I. Stoliarov, N. Safronava, *Principles and practice of microscale combustion calorimetry*, DOT/FAA/TC-12/53, R1, Atlantic City Airport, USA, 2014.
[17] A. Tewarson, Flammability of Polymers, in: A.L. Andrady (Ed.), *Plast. Environ.*, 1st ed., John Wiley & Sons, Hoboken, USA, 2003: pp. 403–489. doi:10.1002/0471721557.ch11.

IOP Conf. Series: Journal of Physics: Conf. Series **1107** (2018) 032022 doi:10.1088/1742-6596/1107/3/032022

Experimental tools applied to the ignition study of spruce wood under cone calorimeter

Lucas TERREI[1], Zoubir ACEM[1], Véronique GEORGES[2], Paul LARDET[2], Pascal BOULET[1] and Gilles PARENT[1]

[1] *LEMTA, UMR CNRS 7563, Université de Lorraine, France*
[2] *Université Paris-Est, Centre Scientifique et Technique du Bâtiment (CSTB), France*

lucas.terrei@univ-lorraine.fr

ABSTRACT

Wood-based materials, which are more and more used in building construction, are often characterized by normative tests using the cone calorimeter. The ignition time and temperature, the mass loss rate or heat release rate are the most current measurements performed. The sample ignition can vary according to a large number of variables (humidity, density, scale sample, experimental conditions). The auto ignition phenomenon has been studied so far but opinions concerning the involved physical processes remain controversial. The purpose of the present work is to improve the study of thermal degradation of wood under cone calorimeter thanks to a dedicated experimental set up. This set up includes mainly, an infrared camera, enabling an accurate surface temperature measurement and a fast visible camera (3000 fps), used to highlight the "auto" ignition phenomenon. Surface temperature and mass loss rate evolution are presented for different heat flux expositions. The fast visible camera pictures showed an ignition piloted by the coil. The ignition occurs only above 55 kW.m^{-2}. When this external heat flux was sufficiently low, it was possible to observe the transition between pyrolysis solely and smouldering combustion for a sample surface temperature around 400 ° C.

KEYWORDS

Cone calorimeter; ignition; surface temperature; flame spread.

IOP Conf. Series: Journal of Physics: Conf. Series **1107** (2018) 032022 doi:10.1088/1742-6596/1107/3/032022

INTRODUCTION

With the ecological transition and the modernisation of urban planning, uses of wood-based material as construction structures or facades becomes increasingly popular. Among available solutions, wood appears to be an interesting material because of its mechanical properties. For fire safety concerns in construction domain, the study of fire behaviour and flame propagation in wood structures is important due to its flammability. Nevertheless, some previous studies have shown that a good choice of wood exhibits encouraging resistance to fire [1, 2], especially a slow in-depth propagation and an ability to self-extinguishment.

The fire behaviour of wood has been widely studied for several years, mainly the ignition criteria [3]. When the sample surface is exposed to an external heat flux or an external flame, pyrolysis occurs within the sample releasing pyrolysis gases. Pyrolysis process depends on several parameters such as moisture, density, heat flux, experimental conditions and so on, and it is consequently very difficult to clearly identify and characterise "the ignition conditions" [4, 5, 6], even if it is well known that ignition occurs when the pyrolysis products, at a sufficient concentration, reach their ignition point. The most difficult part remains to experimentally determine this ignition point. Moreover, after ignition, the additional heat provided by flames on the surface helps to sustain or even to hasten the pyrolysis process. Before their usage, building materials are subjected to different fire normative tests. The cone calorimeter test is one of them. It is a standard test (ASTM and ISO) useful for measuring the Heat Release Rate (HRR) of materials as a function of calibrated incident fluxes. Many fire models need to link this HRR to the net heat flux into the surface in order to best describe the degradation process.

This net heat flux can be calculated provided the measure of surface and in-depth temperature during the test [7]. This test, at a small scale, can be done in two orientations (vertically or horizontally) even if the horizontally orientation was often preferred for practical reasons. However, studies showed that vertical orientation seems more critical than the horizontal one [888, 9]. As stated by Babrauskas [3], physical data such as temperature and time to ignition seemed to give a wide variety of results. It is getting worse if we also consider the time to incandescence [10]. This time could be confused with the time of ignition when the heat flux is high. The surface temperature measured at this time is often considered to be the ignition temperature.

The aim of the present work is to present some experimental tools applied to the study of the auto-ignition of wood under exposure of an external heat flux provided by a vertically oriented cone calorimeter. The ignition of the wood will be highlighted using a fast visible camera. A precision scale is used in order to record the mass loss during the test. In addition, an infrared camera is used to observe the radiation from the surface of wood samples during thermal degradation experiments. Infrared images are post-processed involving a subtraction method between successive images, to withdraw the high incident flux from the cone calorimeter partly reflected by the sample. Then, an identification step was implemented linking the rise in intensity between two time steps with the increase in sample surface temperature.

MATERIALS AND METHODS

Samples and cone calorimeter test

The selected wood material was spruce. The sample sizes were 100 × 100 mm with a 50 mm thickness. The sample average density was about 480 kg m^{-3}, given with an average moisture content around 9 %. As specified in the standard ISO 5660-1, the samples were wrapped with two layers of aluminium foil, apart from the top side exposed to the radiative flux. The distance between the sample and the heater was 25 mm. The samples were exposed vertically to an external constant heat flux between 15 and 75 kW.m^{-2}. This radiative heat flux emitted by the cone was controlled before each test thanks to a Schmidt-Boelter fluxmeter (Medtherm). Cone emission could be assumed to be close to those of a blackbody [11]. The surface temperature was measured as explained below. A sample was considered ignited when a visible flame was observed.

Temperature measurement

Surface or ignition temperatures are usually measured using thermocouples. However, these measurements are punctual and could be false due to the heat sink along the thermocouple wire or to bad contact between the thermocouple and the sample. This contact seems all the more difficult for the wood that is degrading, produces char and cracks [12]. In this work, surface temperatures will be measured using a multispectral infrared camera (Orion SC7000 by FLIR). This non-intrusive measurement allows studying the temperature field evolution on a large surface. The contact problem is also eliminated. During infrared measurements, the flame between the infrared camera and the sample surface may lead to a mismeasurement of the temperature.

Some studies have shown that the emission by flames is observed to be predominant in specific wavelengths where major combustion gases (CO_2 and H_2O) can emit [13]. Outside these bands, only soot can emit, and since the optical thickness for the flames that are involved in the cone calorimeter experiments is very small, the flame can be considered as almost transparent. The camera is equipped with a filter wheel to select wavelengths. In the present paper, the selected wavelength was 3.9 μm (2564 cm^{-1}), which is outside the emission bands of combustion gases. This type of measurement was already successfully performed by our team on other materials [14]. Once the camera signal is acquired, it is converted to a spectral intensity, thanks to a preliminary calibration of the camera involving a blackbody, and the temperature is calculated using the inversion of the Planck's law. The emissivity for wood was studied in [15]. It is shown in this paper that the emissivity of wood at 2564 cm^{-1} is close to 0.9 and that it doesn't vary more than ± 5% around this value when the wood deteriorates under a heat flux. For Chaos et al, a misestimate of the emissivity in 0.8-1 range can lead to an error of 8% in the surface temperature [16]. Fig 1 shows the temperature field obtained for two different tests (20 and 55 kW m^{-2}) at the same time. The white circle is the study area where the mean surface temperature was calculated.

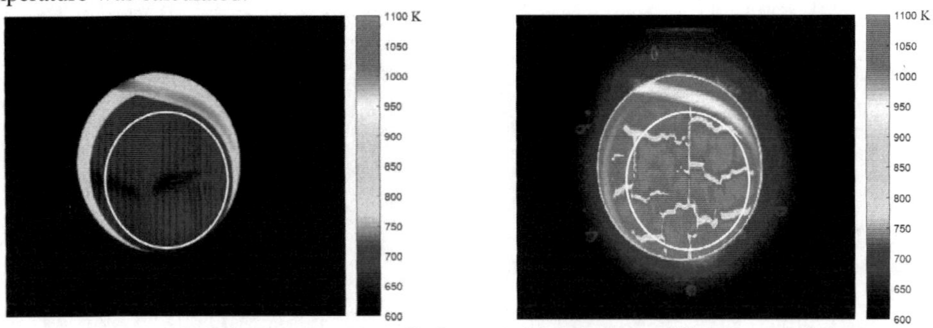

Fig. 1. Temperature field in Kelvin and studied area (white circle) at 120 s for two different external heat fluxes (20 $kW.m^{-2}$ (left) and 55 $kW.m^{-2}$ (right)).

The time evolution of the surface temperature obtained from the IR camera and a K-type thermocouple (1 mm diameter) are plotted in Fig 2. In a general way, the surface temperature increases more or less quickly before reaching a plateau. Until 1500 s, the temperature given by the thermocouple and the infrared camera are in pretty good agreement, even if the thermocouple underestimates the surface temperature. The discrepancy between the two measurements increases for large time. Given that the sample surface burns, the contact between the thermocouple and the surface can no longer be ensured, resulting in a decrease of the temperature. In addition, the thermocouple signal becomes more and more "noisy" since it is subjected to the temperature variation of surrounding, caused by the flame notably. Although widely used, we show here that surface temperature measurement by thermocouple can lead to great uncertainty. For the sake of accuracy, surface temperature measurement by IR camera should be preferred. This, of course, means that radiative properties have to be well known, especially the sample emissivity, which may vary during the degradation process. A parametric study of the sample emissivity was also done. Three values of emissivity (0.88, 0.90 and 0.95) were chosen in the typical range of variation of wood emissivity during its thermal degradation, as reported by Boulet et al. [15]. For the two tests at 20 and 55 $kW.m^{-2}$, the temperature range due to the emissivity uncertainty holds within 20 °C in the worst case, which is very reasonable. For future work we plan to use a multispectral thermography in order to estimate both temperature and emissivity.

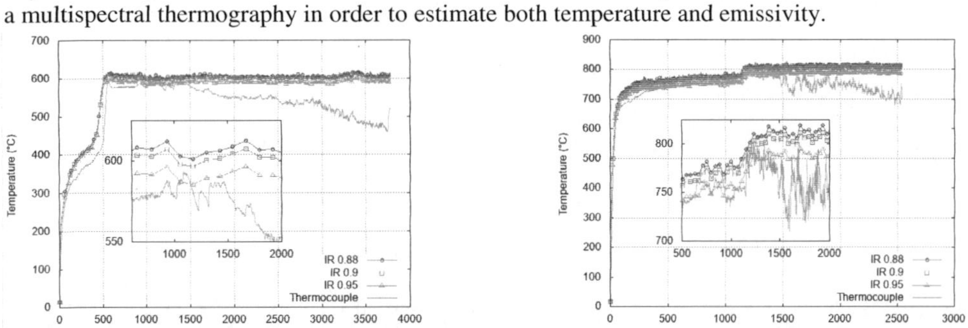

Fig. 2. Time evolution of the surface temperature for two different external heat fluxes (20 $kW.m^{-2}$ (left) and 55 kW. m^{-2} (right)) calculated with the infrared camera and measured with a thermocouple located at the center of the surface .

RESULTS AND DISCUSSION

Samples were exposed to five external heat fluxes (15, 20, 45, 55, 75 kW.m^{-2}). Flame ignition was only present for the tests at 55 and 75 kW.m^{-2}.

Samples with smouldering combustion (no flame)

The cone coil temperature was set to 770 °C for obtaining a desired radiative heat flux of 45 kW.m^{-2}, to 570 °C for 20 kW.m^{-2}, and to 520 °C for 15 kW.m^{-2}. For tests with externals heat fluxes of 15 and 20 kW.m^{-2}, the total exposition time was respectively 8000 s and 3600 s. During these tests, no flame appeared but smouldering combustion phenomenon was observed as discussed below. Figure 3 shows the Mass Loss Rate (MLR) and surface temperature evolution provided by IR camera acquisitions. Since the MLR and the surface temperature remain constant after 2000 s, the curves are plotted only from 0 to 3600 s.

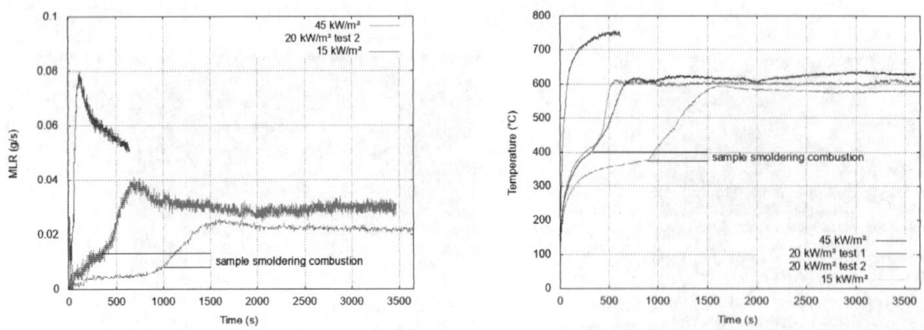

Fig. 3. MLR and surface temperature for tests without sample ignition (15, 20 and 45 kW.m^{-2}).

In a general way the MLR first increases until a maximum. Then the MLR decreases, as the char plays the role of an insulator layer that reduce heat transfer within the sample, before reaching a plateau. For the test at 15 and 20 kW.m^{-2}, an acceleration of the MLR is observed at 900 s and 345 s respectively. This acceleration is also observed for the temperature for which the transition occurs around 370 and 410 °C respectively. This phenomenon corresponds to the beginning of the smouldering combustion of the char that leads to an additional amount of energy release. The lower the external heat flux is, the more the transition between the pyrolysis without char combustion and pyrolysis accompanied by smouldering combustion can be remarked. For the test at 45 kW.m^{-2} this transition is hardly visible because the MLR and temperature increase is too fast.

Samples with presence of flame

Tests were also carried out with higher radiative heat fluxes for which the ignition occurs. Fig 4 shows the MLR (left) and the surface temperature (right) evolutions for tests under 55 and 75 kW.m^{-2}. The cone coil temperature was set to 825 °C and 920 °C respectively.

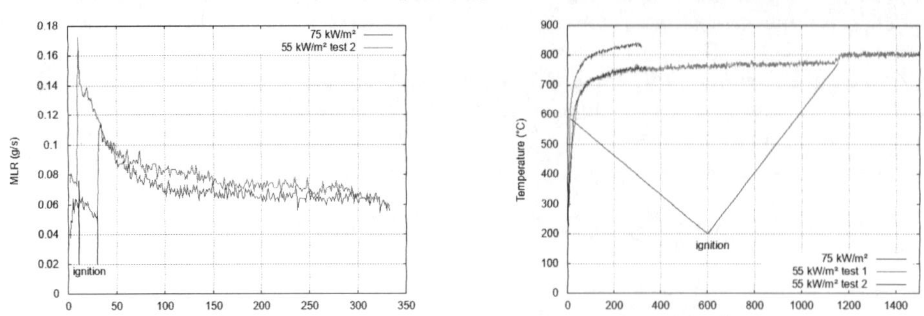

Fig. 4. MLR and surface temperature for tests with sample ignition (55 and 75 kW.m^{-2}).

Similar overall trends as previous tests without ignition are observed. The main difference is a drastic increase in the MLR and the surface temperature when the ignition occurs. This is clearly pointed out in the MLR curves of the figure 4. This increase is all the more pronounced that the flux is high. For the test 75 kW.m^{-2}, the MLR is doubled from 0.08 to 0.17 g s^{-1}. The ignition occurs at 11 s for the test at 75 kW.m^{-2} and at 30 s for the test at 55 kW.m^{-2}. The corresponding surface temperature were estimated to 609 °C for 75 kW m^{-2} and to

776 °C and 618 °C for the two tests at 55 kW m^{-2}. Besides some heterogeneities of each sample, the experiment conditions, in particular the flame origin can affect results. After the ignition, the MLR decreases due to the protecting role plays by the char, as discussed previously.

Ignition of the sample

As explained above, the ignition time corresponds to the moment when a flame appears on the sample surface. At this time, the surface temperature is often called ignition temperature. However, it is difficult to explain the conditions which lead to the ignition. We planned to find out if ignition corresponds to a real auto ignition as it is often claimed or if it is rather piloted by the fact that pyrolysis gases encounter the hot cone coil. To answer this question, a fast camera was placed to observe the sample ignition. The camera was set with 3000 frames per second, which allowed to record 1.3 seconds. The recorded pictures during ignition are presented in figure 5 and 6 for two tests, performed respectively at 55 and 75 kW.m^{-2}. As suspected, the gases ignition is well caused by the cone coil and not by the sample itself. Then, the corresponding surface temperature can no longer be really considered as the ignition temperature.

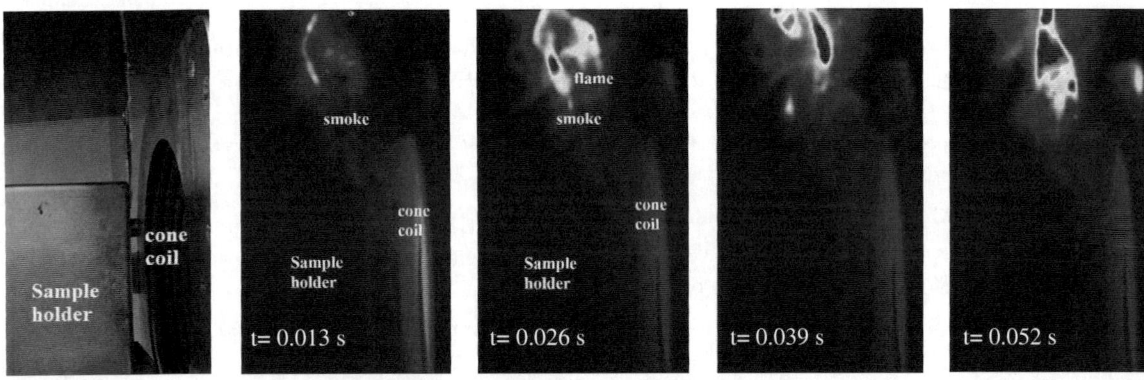

Fig. 5. Ignition at 55 kW.m^{-2}.

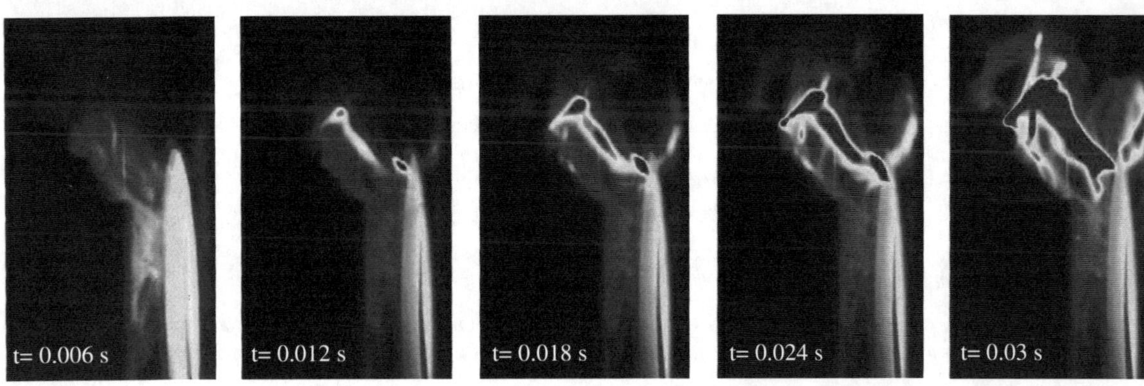

Fig. 6. Ignition at 75 kW.m^{-2}.

CONCLUSION

A dedicated experimental set up was used in order to enhance the thermal degradation study of spruce wood under cone calorimeter. A fast visible camera was used in order to best describe the ignition phenomenon. A significant progress was made by stating that the ignition occurs when the pyrolysis gases impacting the hot cone coil. The surface temperature was also measured with an infrared camera. This temperature is determined using a filter to overcome the presence of the flame. The calculation was done by inverting Planck's law, fixing the emissivity. In future tests, more filters at different wavelengths will be used in order to improve the accuracy of the surface temperature calculation. This step will also make it possible to take into account the emissivity evolution over time.

Thermocouples will be then placed inside the sample to deduce the char presence and to calculate the conductive flux. The char layer prediction is crucial for the design of wood structure. In first tests, pyrolysis

IOP Conf. Series: Journal of Physics: Conf. Series **1107** (2018) 032022 doi:10.1088/1742-6596/1107/3/032022

was immediately stopped with liquid nitrogen in order to get an idea of the proportion of char layer. Then, the proportion of wood consuming by smouldering could be studied.

Some studies had shown that the sample moisture had a significant influence on the temperature, time to ignition, and pyrolysis. In future tests the wood moisture will be increased up to 30 %. Finally, the set up will be further improved in order to study the extinction phenomenon too.

REFERENCES

[1] Emberley, R., Inghelbrecht, A., Yu, Z., & Torero, J. L. (2017). Self-extinction of timber. *Proceedings of the Combustion Institute*, *36*(2), 3055-3062.

[2] Emberley, R., Do, T., Yim, J., & Torero, J. L. (2017). Critical heat flux and mass loss rate for extinction of flaming combustion of timber. *Fire Safety Journal*, *91*, 252-258.

[3] Babrauskas, V. Ignition of wood: a review of the state of the art. *Journal of Fire Protection Engineering*, 2002, vol. 12, no 3, p. 163-189.

[4] Friquin, K. L. (2011). Material properties and external factors influencing the charring rate of solid wood and glue laminated timber. *Fire and materials*, *35*(5), 303-327.

[5] Friquin, K. L., Grimsbu, M., & Hovde, P. J. (2010). Charring rates for cross-laminated timber panels exposed to standard and parametric fires. In *Proceedings of the 11th World Conference on Timber Engineering*.

[6] Janssens, M. (1991). Piloted ignition of wood: a review. *Fire and Materials*, *15*(4), 151-167.

[7] Urbas, J., & Parker, W. J. (1993). Surface temperature measurements on burning wood specimens in the cone calorimeter and the effect of grain orientation. *Fire and materials*, *17*(5), 205-208.

[8] Boonmee, N., & Quintiere, J. G. (2002). Glowing and flaming autoignition of wood. *Proceedings of the combustion institute*, *29*(1), 289-296.

[9] Tsai, K. C. (2009). Orientation effect on cone calorimeter test results to assess fire hazard of materials. Journal of hazardous materials, 172(2-3), 763-772.

[10] Crémona, P. (2017). *Caractérisation expérimentale et numérique des scenarii de feu impliquant un conduit de fumée d'appareils de combustion bois* (Doctoral dissertation, ISAE-ENSMA Ecole Nationale Supérieure de Mécanique et d'Aérotechique-Poitiers).

[11] Boulet, P., Parent, G., Acem, Z., Rogaume, T., Fateh, T., Zaida, J., & Richard, F. (2012). Characterization of the radiative exchanges when using a cone calorimeter for the study of the plywood pyrolysis. *Fire safety journal*, *51*, 53-60.

[12] Urbas, J., Parker, W. J., & Luebbers, G. E. (2004). Surface temperature measurements on burning materials using an infrared pyrometer: accounting for emissivity and reflection of external radiation. *Fire and materials*, *28*(1), 33-53.

[13] Parent, G., Acem, Z., Lechêne, S., & Boulet, P. (2010). Measurement of infrared radiation emitted by the flame of a vegetation fire. *International Journal of Thermal Sciences*, *49*(3), 555-562.

[14] Acem, Z., Brissinger, D., Collin, A., Parent, G., Boulet, P., Quach, T. H. Y., Batiot, B., Richard, F., & Rogaume, T. (2017). Surface temperature of carbon composite samples during thermal degradation. *International Journal of Thermal Sciences*, *112*, 427-438.

[15] Boulet, P., Brissinger, D., Collin, A., Acem, Z., & Parent, G. (2015). On the influence of the sample absorptivity when studying the thermal degradation of materials. *Materials*, *8*(8), 5398-5413.

[16] Chaos, M. (2014). Spectral aspects of bench-scale flammability testing: application to hardwood pyrolysis. *Fire Safety Science*, *11*, 165-178.

Experimental investigation of the fire behaviour of a carbon-PEKK composite used for aeronautical applications

Nathan GRANGE[1], **Brady MANESCAU[2]**, Khaled CHETEHOUNA, Nicolas GASCOIN, Samuel SENAVE and Isabelle REYNAUD[3]

[1]*INSA Centre Val de Loire, Univ Orléans, France*
[2]*DAHER, France*

brady.manescau@insa-cvl.fr

ABSTRACT

Composite materials used for aerospace applications are vulnerable when exposed to high temperatures. Therefore, the development and the using of these materials are strictly framed by aviation regulations where severe fireproof requirements must be accomplished. In the light of this, experimental research is of crucial importance in the understanding of the thermal behaviour of composites materials for aircraft parts, especially when they might be exposed to high-temperature or fire conditions. In this context, thermal degradation of the thermoplastic resin polyether-ketone-ketone (PEKK) with carbon reinforcement is studied. This composite belongs to the polyaryletherketone (PAEK) family, which is known for its high performances. The carbon-PEKK thermal behaviour is evaluated from small scale to medium by scale by means of thermogravimetric analysis, Differential scanning calorimetry, Py-GC-MS and cone calorimeter. The different experiments demonstrate the high thermal performance of the carbon-PEKK composite. Indeed, the latter does not present any decomposition below 500°C, as well as a low amount of volatiles emitted during the resin decomposition. The cone calorimeter experiments confirm the results obtained at small scale and demonstrate that the low yield of volatiles induces a low amount of heat release when exposed to a radiative heat flux. Moreover, the samples present a reduced temperature on the backward face. Finally, these results suggest that the carbon-PEKK might ensure the confinement of a fire as well as a limitation of the fire spread.

KEYWORDS:

Aircraft Composite material, Thermal degradation, cone calorimeter, py-GC-MS

INTRODUCTION

Nowadays, composites materials are widely employed in several engineering fields and especially in aerospace industry. They represent more than half of the constituent elements an aeronautical vehicle. For instance, approximately 50 percent in mass of the Boeing 787 Dreamliner is made by carbon-reinforced composites; in this sense, another example is the Airbus A350-XWB with 52 percent in mass **[1]**, as well as the Bombardier CSeries with a percentage of 47% **[2]**. Besides the high structural performance requirements, the using of these composite materials is based on the respect of a high-level standard used for the aviation safety especially for fire safety. In 2017, according to the Annual Safety Review, edited by the European Aviation Safety Agency, the second most frequent cause of fatal events, involving airplanes, was the post-impact fire/smoke **[4]**. This highlights the crucial need to understand the thermal behaviour of composites exposed to high temperature and hazardous fire aggression, to improve the ability to predict their thermo-structural behaviour, when employed as system or engine components.
The present study aims at characterizing the thermal degradation of the carbon-reinforced thermoplastic resin Poly-ether-ketone-ketone (PEKK), originally developed for automotive and aerospace applications in particular for exterior structures and cabin interiors. Carbon-PEKK belongs to the Polyaryl-Ether-Ketone (PAEK) family, known for their high mechanical performance and high temperature stability as demonstrated by Riviere et al. **[5].** This composite is receiving a particular attention for its multiple applications. Indeed, PEKK-based composites show due to their low flammability and limited smoke emissions when subjected to fire **[6]**. Thus in order to evaluate the thermal behaviour of carbon-PEKK composite, different characterizations have been performed from small scale up to medium scale with samples directly cut from large panel. At small scale, Thermogravimetric analysis has been performed under both inert (argon) and oxidative atmosphere (air) up to 1000°C to evaluate the mass loss and thermal stability. The pyrolysis gases are then assessed at small scale using a Py-GC-MS apparatus for different temperatures defined from the previously obtained TG curves. Finally, fire behaviour of the composite is evaluated at medium scale with a cone calorimeter apparatus at a fixed heat flux of 100 $kW.m^{-2}$. The latter value corresponds to the heat flux encountered on large-scale samples (116 kW/m^2) during the aeronautical standard fire tests **[3]**. These experiments aims to provide physical quantities such as Heat Release Rate (HRR), surface temperature and Mass Loss Rate (MLR), In particular, MLR which is one of the most important parameter as it is directly linked to the pyrolysis gas flow rate and represents the initial factor of the combustion process **[7]**.

MATERIALS AND METHOD

Materials

The carbon-reinforced (fibres of approximately 7 μm diameter) polymer resins considered is a thermoplastic PEKK (66 wt% fibre ratio) provided by Daher. PEKK is a semi-crystalline thermoplastic material belonging to the family of PAEK, known for their high mechanical performance and high temperature stability **[5]**. PEKK-based composites find their main use in aerospace applications, especially for exterior structures and cabin interiors, due to their low flammability and smoke emission when subjected to fire **[6]**. The PEKK monomer is characterized by three aromatic groups bonded together by ether (R–O–R') and ketone (RC(=O) R') links (see Figure 1-left). The glass transition temperature is around 156 °C and the melting point 338 °C **[8]**.

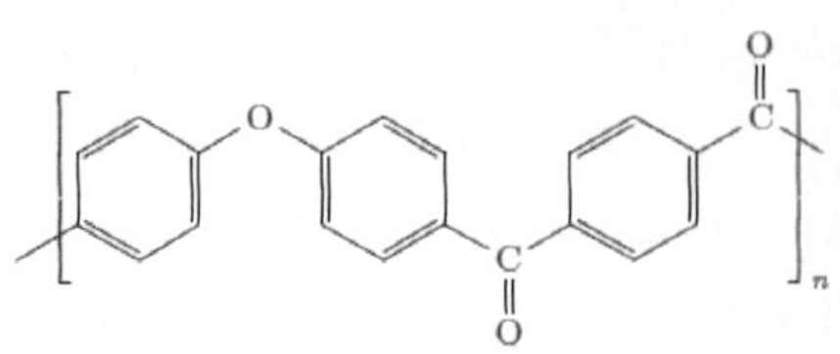

Fig. 1. Overview of the PEKK Monomer (left) and carbon-PEKK sample (right).

Experimental set-up

At small scale, thermogravimetric (TG) measures were performed with a SETARAM Setsys-16/18 TG, under both inert (argon) and oxidative (air) conditions at atmospheric pressure. This device consisted of a vertical-type TG where the Al_2O_3 crucible was hung to the electronically controlled balance. All the inert experiments were carried out with an argon flow rate of 16.6 $ml.min^{-1}$. Three non-isothermal heating rate of 5, 15 and 25°$C.min^{-1}$ and a final temperature of 1000°C were fixed. Samples were prepared in form of particulate of masses of 7–9 mg directly cut from virgin plate representative of the industrial material, in order to limit the temperature gradient inside the samples. DSC measurements are also performed at small scale on same sample type. The experiments have been carried out under nitrogen atmosphere (volume flow rate of 40 ml.min^{-1}) and atmospheric pressure using a Nietzsch STA 449 F3 DSC from ambient to 1000°C, with a heating rate of 15 °C/min. Two tests have been carried out on the two samples extracted from two different locations of the plate. Pyrolysis gases of the sample were estimated with a Pyroprobe 5000 (CDS) for three pyrolysis temperatures defined along the decomposition of the composite. A solid composite sample of 1–2 mg, directly cut from large panel was placed inside a quartz tube which was then introduced into the pyrolyser and heated up at 5 °C.ms^{-1} to the desired temperature and maintained 10 s at this temperature. The pyrolysis products are transported to the GC-MS by a helium flow via a heated transfer line. GC separations were performed with a Trace Ultra GC (Thermo) equipped with a SLB-5ms capillary column (30 m long, 0.25 mm i.d., 0.25µm film thickness). The column temperature was programmed from 60°C to 300°C at a rate of 5 °C.min^{-1} and finally held for 1 min at 300 °C. Mass spectra were recorded with a Thermo DSQ II quadrupole in the electron ionization mode at 70 eV. The identification of the pyrolysis products has been based on the comparison of their mass spectra with the NIST mass spectral library.

Cone calorimeter experiments are finally carried out according to the ISO 5660-1: 2002 standard **[9]** on medium scale samples representative of structural parts materials. Samples with an exposed surface of 88.4 cm² are located at 35 mm from a heat source (to assess a similar exposition between the different configurations) producing a constant heat flux of 100 kW/m². Samples are placed successively in horizontal (standard cone calorimeter sample holder) and vertical configurations (hang into a support and isolated on the side of the samples) as presented in Figure 2. This heat flux level correspond to the requirements of the aircraft fireproof certification standards tests on large-scale samples. Two samples per configuration (Experiment 1& 2) have been tested to ensure the repeatability of the measurements with a thickness of 1.68±0.05 mm (12 plies) and their average masses are about 26±0.3 g. The vertical configuration allows measuring the backward temperature using a Flir A600 infrared thermography apparatus. The Camera measures the temperature of a zone located at the centre of the backward face of the sample covered by a graphite layer (with a known emissivity of 0.95). The horizontal configuration allows measuring the mass loss (i.e. MLR) during the tests thanks to a load cell situated under the sample holder. The HRR is determined in both configurations using the oxygen depletion principle based on the concentration of O2, CO and CO2 **[10]**. To this end, a smoke absorption is performed at 24 l/s with a gas sampling of 58.3 ml/s.

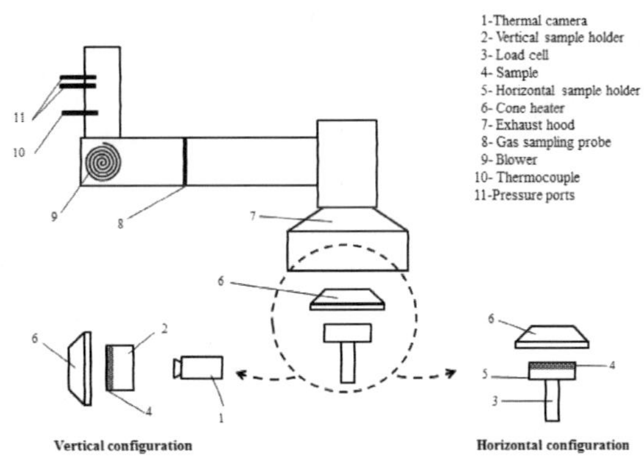

Fig. 2. Overview of the cone calorimeter experiments

RESULTS AND DISCUSSION

Thermogravimetric analysis

In figure 3, the experimental mass loss curve and DTG curves obtained under inert and oxidative atmosphere are presented for the three non-isothermal heating rates. Under inert atmosphere, the PEKK resin is characterized by a mass release of about 20% and does not present any mass variation up to about 515°C; then a faster degradation process occurs, highlighting one global reaction, which comes to end at about 900°C. Similar behaviour is reported in the literature for composites with Poly-ether-ether-ketone (PEEK) resin, which also belongs to the PEAK family **[6]**. The identified reaction lies between about 510-710 °C and it might be associated with the random scission of ether bonds **[7]**. In particular, the decomposition process is expected to begin with the rupture of weakest bonds between the aromatic groups, whereas the major volatiles release should consist of phenol **[6]**. The onset of this decomposition occurs after the transition phase of PEKK resin, typically observed by DSC measurements around 340 °C **[8]**.

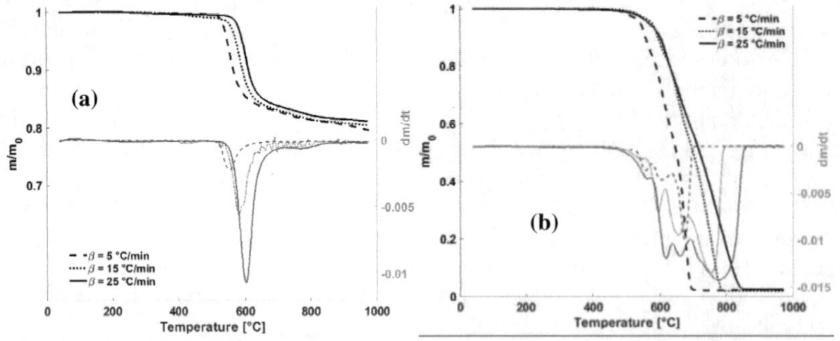

Fig. 3. Experimental mass loss and DTG Curves obtained under inert (a) and oxidative (b) atmosphere.

Under oxidative atmosphere, the samples have been almost completely consumed (no residual mass obtained). As for inert atmosphere, the degradation begins at about 500 °C and concludes between 700-850 °C, depending on the heating rate of the test. Three inflection points are visible in the TG curve correspondinf to the three peaks on the DTG curves. This underlining the occurring of a complex reaction pattern. The first inflection peak is associated with resin decomposition, in correspondence of the same temperatures observed in inert conditions, between 550-615 °C. The second appear between 575-689°C while the third follows between 633-862°C. These peaks might refer to the combustion of carbon fibers, whose onset is expected between 600-700 °Cwhich develops simultaneously with resin decomposition and combustion, as well as that of char **[6,11].** It can also be seen that, befoe the first peak, especially for β=15 and 25°C.min^{-1}, a less intense peak is reveled on the DTG curves. This peak might eventually due to the oxydation of the pyrolysis products release by the early scission of the weakest bond **[6]**. Moreover,by increasing the temperature, also carbon-PEKK shows several further overlapped peaks, revealing a complex kinetic mechanism, driven by the occurring of multi-step reactions.

DSC measurements

In Figure 4, one can see the heat flow (corrected as function of mass loss) measured for carbon-PEKK under inert atmosphere. It reveals a major endothermic peak at about 600°C which corresponds to the single-step decomposition reaction observed previously in TG experiments. Moreover, the DSC analysis highlights the importance of the melting phase observed between 350°C and 550°C. Indeed, a wide endothermic peak is visible on the DSC curves in this interval. It is important to note that this phenomenon could influence the heat transfer in the composite sample while exposed to fire.

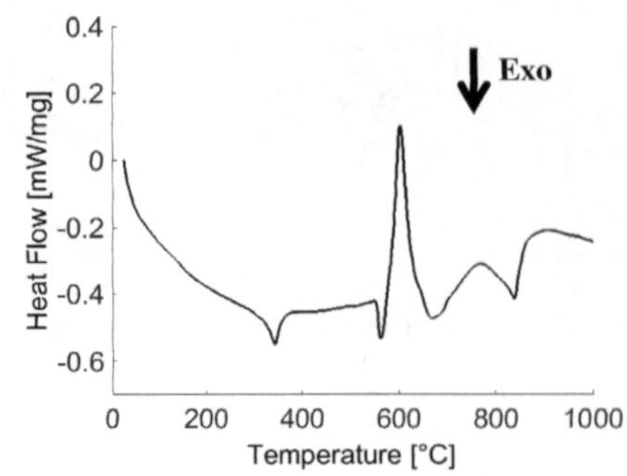

Fig. 4. DSC curve of the carbon-PEKK sample

Volatiles Emission

From the TG curves, the pyrolysis temperatures have been chosen to be 450, 590 and 750°C. Looking in details at the pyrolysis products (figure 5), one can see that no compounds are detected at 450°C which is just before the decomposition temperature. At 590 °C, once the thermal degradation started, three hydrocarbons are mainly identified: butane, benzene and toluene. They might be released by the cleavage of the weakest bond in the PEKK monomer (ether bond). At higher temperature (750°C), when the decomposition of the resin is finished an important release of phenol is observed (46 %), probably associated with the cleavage of the ether bond producing phenolic radical which leads to phenol after hydrogen loss, as observed on PEEK resin **[6]**. Low amount of aromatic compounds and ketones such as pentanone and p-benzoquinone are also suspected are also released by the cleavage of ketone bond at this temperature (7%). This low amount of volatiles (only 6 compounds at 750°C) might be a great benefit for the carbon-PEKK compared to other composites. Indeed, this induces a limited combustion on the surface leading to a small heat release when exposed to fire. Such behaviour might be interesting for aircraft safety. In fact, the fire spread would be limited and the smoke would be reduced lowering the potential risk of intoxication and ignition.

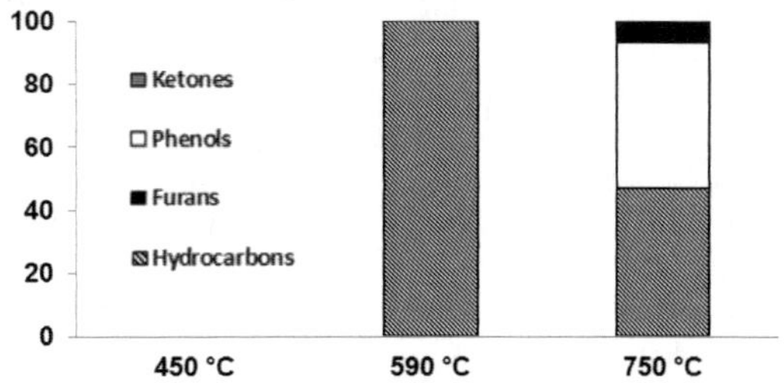

Fig. 5. Volatiles emission during carbon-PEKK sample pyrolysis

Cone calorimeter experiments

The figure 6 presents the different results obtained for the cone calorimeter experiments. The two experimental mass loss curves (fig 6.a) highlight a good repeatability of the two tests. Both curves are characterized by two different phases. First, a quasi-constant mass up to 30 s corresponding approximately to the ignition time. Then, the latter is followed by a monotonous behaviour of the mass loss until 160 s where the mass start to stabilize. The mass loss at the end of the experiment represents 15 % of the initial mass of the sample. The thermoplastics PEKK resin has a good thermal stability and high melting point **[8]**. Moreover, the resin decomposition is characterized by a high yield of stable carbonaceous char (approximately 60 % of the original mass) leading to a superior flammability resistance **[6]**.

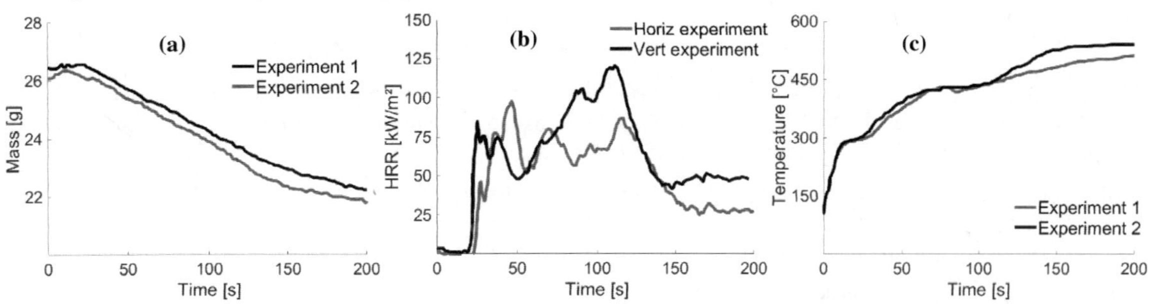

Fig. 6. Experimental mass loss (a), HRR (b) and backward face temperature (c) obtained with cone calorimeter experiments.

Experimentally and for both configurations, the HRR (Fig 6.b) presents a rapid increase before reaching a maximum value of approximatively 100 kW/m², which is limited compare to other composite materials **[12-13]**. The HRR values are oscillating within a range of 20 kW/m², which might correspond to the flame oscillation on the surface of the sample. Then, the HRR decreases down to 35 kW/m² at 160 s and seems to remain constant. After 160 s, experimentally, no more flame is visible on the surface of the carbon-PEKK composite, which indicates that internal heat generation process occurs, probably associated with the resin decomposition in the internal plies or the carbon fibre combustion on the surface of the sample. Moreover, Horizontal and vertical configurations present also similar ignition times due to the high value of the cone calorimeter heat flux reducing the convection role in the vertical configuration. However, in the vertical configuration, the buoyancy will generate a boundary layer on the surface of the sample and the bulk of the flow will be pulled up, conducting to a difference of temperature between the top and the bottom of the sample leading to a change in the HRR behaviour. The general trend of the backward face temperature (Fig 6.c) can be divided into two phases, a transient one with a quick increase of the temperature followed by a slow and constant increase of the temperature. The transient phase can be separated into two phases. The first phase presents a fast increase of the temperature between 0 and 30 s followed by a short stabilization at about 300 °C before a second temperature increase of approximatively 100 °C between 30 s and 90 s. A last increase of 100°C between 90 s and 180 s is observable at the end and can be associated with the stable phase. The first stabilization at 300°C might be explained by the high melting temperature of the PEKK resin (T_m = 338 °C **[8]**) leading in an absorption of energy when reaching this temperature reducing the increase of temperature on the backward face.

CONCLUSION

In this work, the thermal degradation of the thermoplastic carbon-PEKK composite is evaluated in order to contribute to the understanding of its fire behaviour. The different results demonstrate the important thermal stability of the carbon-PEKK, characterized by high temperature resistance, starting its decomposition at about 500 °C. This high performance can be attributed to the chemical structure of the monomer make it as a promising material for aircraft components, especially where severe fire safety requirements must be accomplished. Moreover, the low amount of volatiles emitted during the pyrolysis of the resin limits the heat release and the fire spread risks. The latter is confirmed with the cone calorimeter experiments on medium scale samples. Cone calorimeter experiments also highlight the low mass loss and the moderate temperature on the backward face when sample are exposed to a large heat flux. The achieved results highlight the correlation between the data obtained at small scale and the one obtained at medium scale. Moreover, the different results demonstrate the interesting behaviour of the carbon-PEKK when exposed to fire or large heat flux. Thus, such composite might be interesting as firewall of fire protection for aircraft parts. In the light of this, the next stage of this study will be focused on the implementation of large-scale fire tests with sample of approximately 500 x 500 mm, sample representative of a full scale aircraft part. The latter will be exposed to a kerosene flame as required in the aeronautical fire standards to confirm the carbon-PEKK fire behaviour.

REFERENCES

[1] Marsh, G. (2007). Airbus takes on Boeing with reinforced plastic A350 XWB. Reinforced plastics, 51(11), 26-29.

[2] Marsh, G. (2011). Bombardier throws down the gauntlet with CSeries airliner. Reinforced Plastics, 55(6), 22-26.

[3] International Standards (1998). Aircraft Environmental Test Procedure for Airborne equipment Resistance to Fire in Designated Fire Zones

[4] European Aviation Safety Agency (2017). Annual Safety Review. Doi: 10.2822/26228

[5] Riviere, L., Causse, N., Lonjon, A., Dantras, E., & Lacabanne, C. (2016). Specific heat capacity and thermal conductivity of PEEK/Ag nanoparticles composites determined by Modulated-Temperature Differential Scanning Calorimetry. Polymer Degradation and Stability, 127, 98-104.

[6] Patel, P., Hull, T. R., McCabe, R. W., Flath, D., Grasmeder, J., & Percy, M. (2010). Mechanism of thermal decomposition of poly (ether ether ketone) (PEEK) from a review of decomposition studies. Polymer Degradation and Stability, 95(5), 709-718.

[7] Mouritz, A. P., & Gibson, A. G. (2007). Fire properties of polymer composite materials (Vol. 143). Springer Science & Business Media.

[8] Chang, I. Y., & Lees, J. K. (1988). Recent development in thermoplastic composites: a review of matrix systems and processing methods. Journal of thermoplastic composite materials, 1(3), 277-296.

[9] ISO 5660-1: 2002. (2002). Reaction-to-fire tests–Heat release, smoke production and mass loss rate–Part 1: Heat release rate (Cone calorimeter method).

[10] Babrauskas, V. (1982). Development of the cone calorimeter, a bench scale heat release rate apparatus based on oxygen consumption. NASA STI/Recon Technical Report N, 83.

[11] Feih, S., & Mouritz, A. P. (2012). Tensile properties of carbon fibres and carbon fibre–polymer composites in fire. Composites Part A: Applied Science and Manufacturing, 43(5), 765-772.

[12] Grange, N., Chetehouna, K., Gascoin, N., Coppalle, A., Reynaud, I., & Senave, S. (2018). One-dimensional pyrolysis of carbon based composite materials using FireFOAM. Fire Safety Journal, 97, 66-75.

[13] Toldy, A., Szolnoki, B., & Marosi, G. (2011). Flame retardancy of fibre-reinforced epoxy resin composites for aerospace applications. Polymer Degradation and Stability, 96(3), 371-376.

IOP Conf. Series: Journal of Physics: Conf. Series **1107** (2018) 032024 doi:10.1088/1742-6596/1107/3/032024

Effect of Back Boundary Condition on Pyrolysis of Charring and Non-charring Materials

Sabarilal S[1], Amit Kumar[2]

National Centre for Combustion Research & Development & Department of Aerospace Engineering, Indian Institute of Technology Madras, Chennai, Tamil Nadu, India

sabarilals@gmail.com

ABSTRACT

Predicting pyrolysate fuel generation rate form a condensed fuel is an important component of fire models. For validating or tuning these subgrid models, mass loss rate (MLR) of small samples exposed to known radiant heat flux is measured in standard devices such as the cone calorimeter and Fire Propagation Apparatus (FPA). In order to accurately predict the MLR, pyrolysis model input parameters should represent actual fuel degradation process. The pyrolysis process is intricately coupled to heat transfer in the fuel. A few experimental studies on non-charring polymers [1][2] have shown that these results are sensitive to back boundary, therefore, special care should be taken in interpreting results from these bench-scale instruments. To elucidate the effect of back boundary on the pyrolysis process, here a numerical investigation is carried out for both charring and non-charring fuel material. In this work heat transfer process into the fuel is determined to understand the pyrolysis process. Heat transfer process in the two categories of fuel are quite different. For each fuel category study is carried out for 1 mm and 10 mm thick fuels exposed to low (20kW) and high (100 kW) fluxes. This work aims improve the understanding of burning process in these bench-scale instruments especially for charring substances which have not been addressed in literature in this regard. The study shed important insight for proper use bench tests as a versatile tool for fire safety engineers.

Keywords: *Cone calorimeter, FPA, back boundary, fire behaviour, pyrolysis, charring material, non-charring material, thin and thick fuel, burning process.*

NOMENCLATURE LISTING

A	Pre-exponential factor (1/s)	**Greek**	
Cp	Specific heat (J/kg/K)	κ	Absorption coefficient (m^{-1})
dz	Cell thickness (mm)	η	Surface transmissivity
E_a	Activation energy (J/mol)	σ	Stefan-Boltzmann constant ($W/m^2/K^4$)
h	Convective heat transfer coefficient ($W/m^2/K$)	ρ	Density (kg/m^3)
H_p	Heat of pyrolysis (s^{-1})	ε	Emissivity
k	Thermal conductivity (W/m/K)	**Subscripts**	
n	Order of reaction	v	Virgin fuel
T	Temperature (K)	c	Char
		i	i^{th} computational cell

INTRODUCTION

In the recent years a number of pyrolysis models are developed to predict the production rate of gaseous fuel from condense phase fuel subjected to external heat flux[3][4].Pyrolysis modeling is complex since many of the input parameters are not known with high accuracy. Also current flammability test fails to provide all input parameter required for modelling[4]. An important drawback of current bench scale (FPA and cone calorimeter) test methods is that it fails to exactly define the thermal boundary condition at the rear face of the sample. To understand the pyrolysis process in detail the heat transfer across the entire fuel is essential. As per standard test the rear surface of the sample is insulated. To understand the heat transfer phenomena in standard configuration we need to consider entire assembly of sample and insulation also need to account for thermal contact resistance at interface along with this variation of insulation properties which make this job extremely complex and uncertain.[1] Suggests a well-defined back boundary condition by placing highly conductive inert material (aluminum block) and by introducing thermocouple into the block heat transfer across the back boundary during the test can be estimated. In this study same approach is adopted by introducing aluminum block at the back side of specimen and the variation in material behavior against standard configuration (insulated boundary condition) is investigated numerically. This work fuel of two thicknesses 1 mm and 10 mm is studied for both charring and non-charring fuels. The fuel is considered exposed to two heat fluxes, high (100 kW) and low (20 kW).

NUMERICAL MODELING

A simplified 1D pyrolysis model was developed based on principle of conservation of mass and energy similar to the way described in [4] is used to predict the pyrolysis of charring and non-charring material. A control volume approach is employed and the governing mass and energy conservation equations are solved numerically using a fully implicit scheme. In this present study 3 species (virgin solid, char, and pyrolysate) are considered and for charring and 2 species (virgin solid and pyrolysate) for non-charring material. The decomposition of virgin solid to pyrolysate takes place through a single step nth order Arrhenius type endothermic reaction. Pyrolysis gas is assumed to be in thermal equilibrium with the solid and escape immediately once it is formed so that no pressure buildup within the solid.

Thermal Balance and Governing Equations

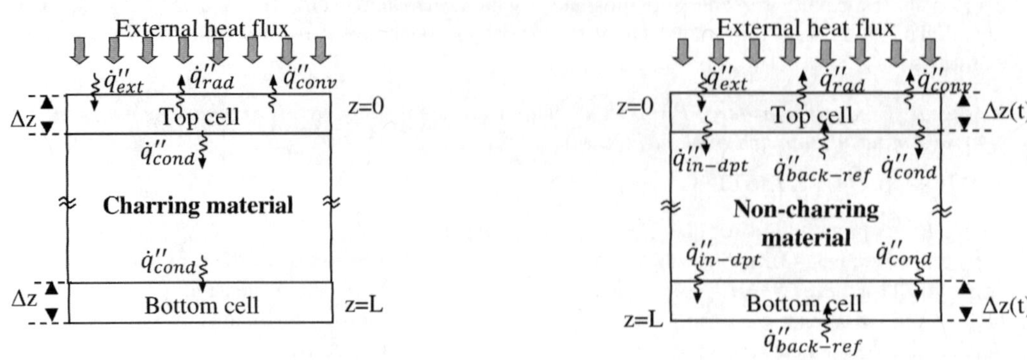

Fig. 1 Schematic of thermal boundary condition for charring and non-charring material.

$$\frac{\partial(\rho C_p T)}{\partial t} = \frac{\partial}{\partial z}\left(k\frac{\partial T}{\partial z}\right) - \dot{m}'''H_p \quad (1)$$

$$\frac{\partial(\rho C_p T)}{\partial t} = \frac{\partial}{\partial z}\left(k\frac{\partial T}{\partial z}\right) + \frac{\partial \dot{q}''_r}{\partial z} - \dot{m}'''H_p \quad (2)$$

Fig 1 shows schematic of energy balance for charring and non-charring material. In case of charring material the surface is assumed to be completely opaque so that external heat flux ($\dot{q}''_{ext}$) is first absorbed at the surface and energy is then conducted (q_{cond}) into the fuel. The radiation absorbed will be $\varepsilon\dot{q}''_{ext}$ since emissivity and absorptivity are considered same. Were ε is effective emissivity, which is evaluated using volume fractions of virgin material and char present at the surface. Similarly approach is adopted for evaluating effective thermal conductivity (k) and density (ρ) but effective specific heat (C_p) is evaluated using mass fractions. $\dot{q}''_{rad}$ and $\dot{q}''_{conv}$ represent the radiation and convection losses from surface.

Special consideration is taken for boundary condition of the non-charring semi-transparent material. For these kinds of materials a fraction of the applied external radiation ($\dot{q}''_{ext}$) is absorbed in-depth ($\dot{q}''_{in-dpt}$) according to its surface transmissivity and absorption coefficient. Unlike in charring material the energy is transported in semi-transparent material by in-depth absorption and conduction. As compared to other pyrolysis models for non-charring material this work considers in-depth emission along with in-depth absorption by solving RTE equation this additional term is given as $\frac{\partial \dot{q}''_r}{\partial z}$ in equation(2). The transmitted radiation which reaches the back boundary is reflected back ($\dot{q}''_{back-ref}$) based on the reflectivity of back surface. Further this reflected radiation is again absorbed and finally leaves the top surface of specimen. After considering in-depth emission and absorption, radiation leaving the cell $E(i)$ can be evaluated using equation (3), $E(i-1)$ is radiation entering the cell.

$$E(i) = E(i-1)e^{-\kappa dz(i)} - \sigma T_i^4\left(1 - e^{-\kappa dz(i)}\right) \qquad (3)$$

NUMERICAL MODEL VALIDATION

The pyrolysis model developed is first validated against bench-scale experiments (FPA) of [5]. All the experiments in [5] were performed in 100% Nitrogen environment so that the thermal decomposition of fuel takes place in the absence of flame. Charring and non-charring pyrolysis models are validated against experiments conducted on chlorinated polyvinyl chloride (CPVC) and poly methyl methacrylate, PMMA respectively as shown in Fig. 2. Model input parameters for CPVC and black PMMA values are presented in Table 1.

Table 1 Model input parameters

	CPVC	PMMA		CPVC	PMMA		CPVC	PMMA
k_v	0.184 [5]	0.188 [6]	ρ_c	589.7 [5]	-	E_a	2.5*10^5 [5]	1.64*10^5[6]
ρ_v	1908.7 [5]	1190 [6]	Cp_c	1021.6 [5]	-	A	7.9*10^20 [5]	2.5*10^11 [6]
Cp_v	1933.8 [5]	1465 [6]	ε_c	1 [5]	-	H_p	7.8*10^5 [5]	6.37*10^5
ε_v	0.63 [5]	-	η	-	0.93 [6]	n	1.93 [5]	1
k_c	0.038 [5]	-	κ	-	960.5 [6]	h	10 [6]	10 [6]

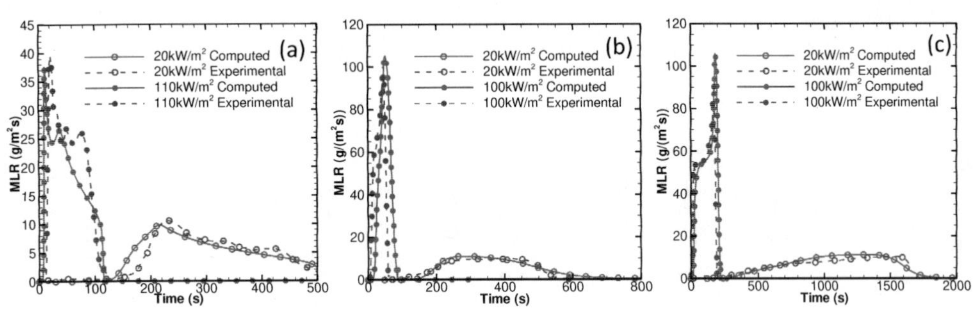

Fig. 2 Comparison between experimental [5] and numerical predictions. (a) 1.58 mm thick charring material (CPVC). (b) 3.18 mm thick non-charring material (PMMA). (c) 9.53 mm thick non-charring (PMMA) material.

IOP Conf. Series: Journal of Physics: Conf. Series **1107** (2018) 032024 doi:10.1088/1742-6596/1107/3/032024

MODIFIED BACK BOUNDARY CONDITION

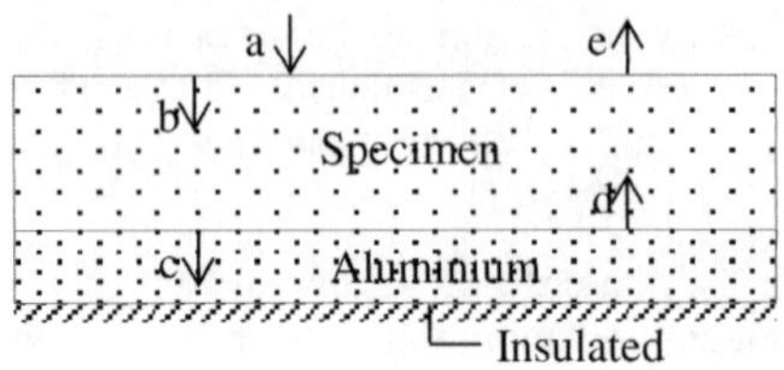

a	Incoming radiation
b	Radiation getting into specimen
c	Energy transferred to aluminium block
d	Back reflected radiation (only for translucent fuels)
e	Radiation leaving surface

Fig. 3 Schematic of modified boundary condition.

As mentioned before, one of the limitations of standard flammability test analysis (insulated boundary condition) is that it neglects the effect associated to back boundary of the sample. To understand the flammability nature of different material in more detail, understanding of energy transfer across the specimen especially the back surface is necessary. In this study a material of known thermal properties (Aluminum (Al) block) is introduced on the back side of the specimen as shown in Fig. 3 by doing this heat transfer to the block can be estimated and more robust and concrete pyrolysis model can be developed. For simulation aluminum block of 5mm thickness is considered with adiabatic back boundary and zero thermal contact resistance is assumed at the interface of specimen and Al block. For Al block thermal conductivity, specific heat and density are taken values 244 W/(mK), 921 J/(kg/K) and 2700 kg/m^3 respectively. Surface emissivity of 0.2 is taken for the block which is important in case of non-charring translucent fuel were in-depth radiation is considered.

RESULTS AND DISCUSSION

In this section the effect of back boundary on eight different cases are discussed, these include 1 mm and 10 mm thick fuel exposed to low (20 kW) and high (100 kW) heat flux each for charring and non-charring material. The cases of charring material are discussed first followed by cases for non-charring material.

Effect of back boundary on pyrolysis of charring material

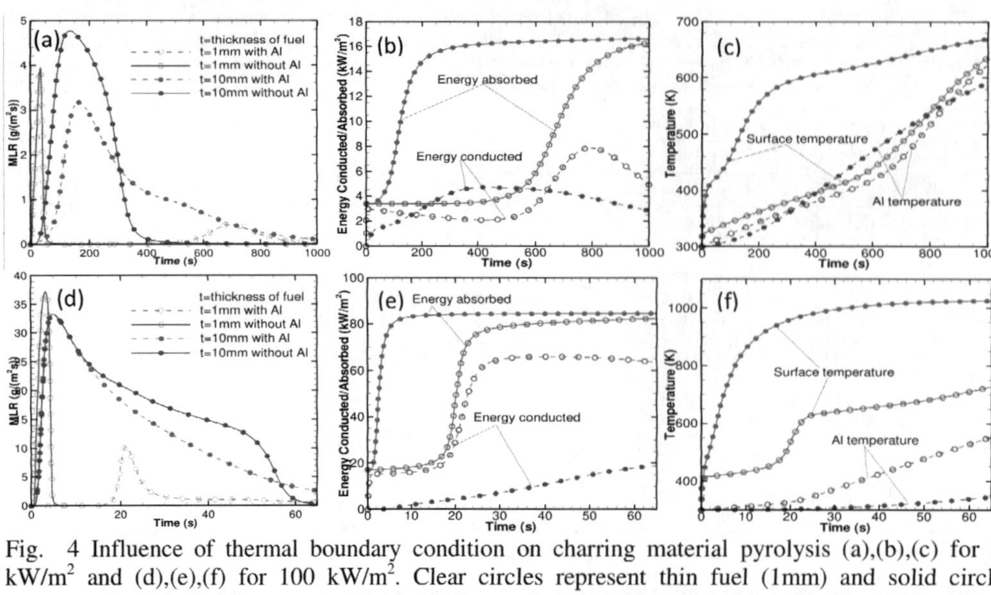

Fig. 4 Influence of thermal boundary condition on charring material pyrolysis (a),(b),(c) for 20 kW/m^2 and (d),(e),(f) for 100 kW/m^2. Clear circles represent thin fuel (1mm) and solid circles represent thick fuel (10mm). (a) and (d) MLR comparison with and without Al block. (b) and (e) Energy absorbed (solid line) and conducted (dashed line) to Al block during pyrolysis. (c) and (f) surface (solid line) and Al block (dashed line) temperature variation.

Fig. 4 shows influence of aluminum block on mass loss rate of charring material, as seen from Fig. 4(a) and (d) for 1 mm fuel, significant ignition delay was observed along with reduction in peak MLR. Because for 1 mm thick fuels the energy conducted to the aluminum block is high in the initial stages as seen in Fig. 4(b) and (e) and therefore, less energy will be available for sensible heating of the fuel and start the pyrolysis process. Once the specimen attains required temperature the pyrolysis process begins. During pyrolysis a char layer is formed on the specimen surface which increases the energy absorbed this is due to high surface absorptivity of char. As a result a sudden increase in absorbed energy and thereby an increase in energy conducted can be seen in Fig. 4 (b) and (e). For the four cases presented here, except for 100kw/m^2 on 10 mm thick fuel case all other show a delay in ignition. This is due to thermally thick and thermally thin nature of the fuel. In case of low heat flux (20kW/m^2) both 1mm and 10 mm act as thermally thin fuel. 1 mm fuel exposed to high heat flux (100kW) also behaves as thermally thin, since the thermal thickness is greater than material thickness prior to ignition.For the three thermally thin cases mentioned above from Fig. 4 (c) and (f) it is evident that the aluminum block temperature increases prior to MLR begins, this is due to the fact that the thermal wave travels faster than the pyrolysis. In case of 10 mm fuel exposed to high heat flux the heating rate is very high and due to higher material thickness the thermal wave takes more time to reach the back boundary that is why in Fig. 4 (f) the aluminum block temperature will not show an initial increase. As a result 10 mm fuel exposed to high heat flux act as thermally thick and will not show an ignition delay.

Effect of back boundary on pyrolysis of non-charring material

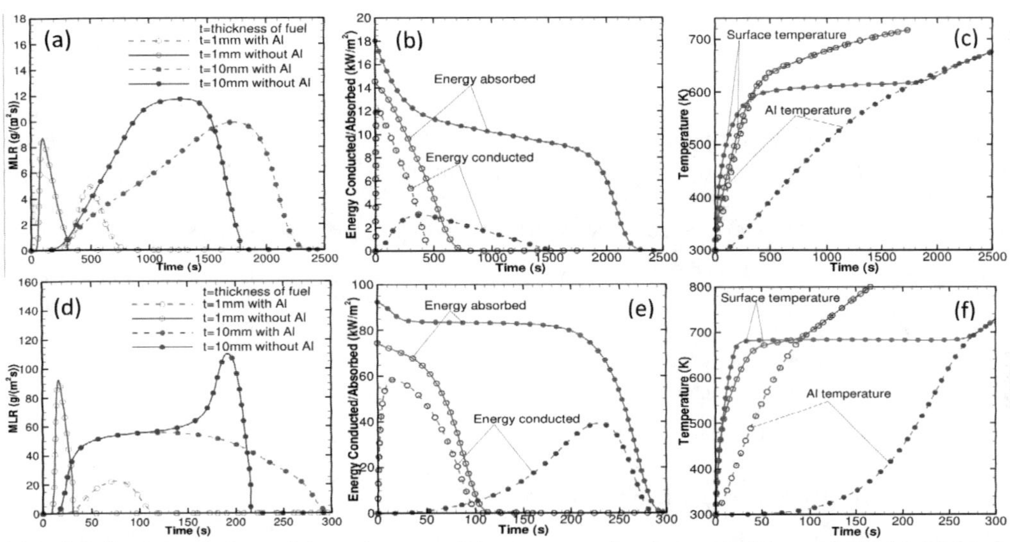

Fig. 5 Influence of thermal boundary condition on non-charring material pyrolysis (a),(b),(c) for 20kW/m^2 and (d),(e),(f) for 100kW/m^2. Open circles represent thin fuel (1mm) and solid circles represent thick fuel (10mm). (a) and (d) MLR comparison with and without Al block. (b) and (e) Energy absorbed (solid line) and conducted (dashed line) to Al block during pyrolysis. (c) and (f) surface (solid line) and Al block (dashed line) temperature variation.

Fig. 5 shows influence of aluminum block on mass loss rate of non-charring material, similar to charring material in Fig.4. One can note that pyrolysis in 1 mm thick fuel is more sensitive to back boundary conditions. One can see from Fig. 5 (a) for 1 mm fuel exposed to low heat flux (20 kW) significant ignition delay is observed. In charring material the only mode of heat transfer is by conduction as compared to non-charring translucent material where in-depth radiation also plays a role. This probably is the reason for higher sensitivity of 1 mm thick charring material to back boundary condition compared to 1 mm thick non-charring fuel. For all four cases the energy conducted to the aluminum block shows the same trend (initially increases and then decreases) during initial period the specimen temperature will be higher so heat conduction to aluminum increases, once the pyrolysis starts specimen thickness decreases and more radiation reaches aluminum block and its temperature increases results in a decrease in heat conduction as seen in Fig. 5 (b) and (e). As time progresses the specimen thickness decreases and energy absorbed also decreases but in case of charring material energy absorbed increases with time. In this case, the aluminum block temperature rise is due to two reasons heat conduction from specimen along with in-depth radiation which absorbed by aluminum surface, so toward the end of pyrolysis process (when specimen thickness is very less), the block temperature can be greater than specimen temperature as seen in Fig. 5 (c). Here also 1 mm and 10 mm thick fuel samples exposed to low heat flux and 1 mm thick fuel exposed to high heat flux behaves as thermally thin fuel and 10 mm thick fuel sample exposed to high heat flux behaves as thermally thick which is clear from Fig. 5 (e) and (f). Another interesting result observed in thick fuel exposed to high heat flux Fig. 5 (d) is that due to the introduction of aluminum block the peak in MLR disappears which was observed in insulated boundary condition. Similar experimental observation is reported for PA6 and HIPS[1][2] . When using insulated boundary condition this peak in MLR at the end of burning is caused due to the reduction in conduction heat flux to the specimen when the pyrolysis zone approaches the insulation back boundary. By introducing aluminum at the back boundary heat is transferred to the block even if the pyrolysis front reaches back boundary and peak disappears.

CONCLUSION

A 1D pyrolysis model validated against bench-scale tests is used to analyze the effect of back boundary on pyrolysis process of charring and non-charring material. By introducing an aluminum block, a delay in pyrolysis process is observed especially for thin fuels (here 1mm thick). In case of charring material and non-charring material 1 mm and 10 mm thick fuel samples exposed to low heat flux (20 kW/m^2) and 1 mm thick fuel exposed to high heat flux (100 kW/m^2) behaves as thermally thin and 10 mm thick fuel sample exposed to high heat flux (100 kW/m^2) behaves as thermally thick fuel. When 10 mm thick non-charring material exposed to higher heat flux a sudden peak in MLR is observed towards the end of pyrolysis is not a material property, it is an after effect of insulated back boundary condition. The main intention of introducing the aluminum block at the back side of the sample is to quantify the heat transfer across the rear side and those experimental data can be better fitted with numerical models thereby a better understanding of pyrolysis process.

REFERENCES

[1] R. Carvel, T. Steinhaus, G. Rein, and J. L. Torero, "Determination of the flammability properties of polymeric materials: A novel method," *Polym. Degrad. Stab.*, vol. 96, no. 3, pp. 314–319, 2011, 10.1016/j.polymdegradstab.2010.08.010

[2] B. Schartel, M. Bartholmai, and U. Knoll, "Some comments on the use of cone calorimeter data," *Polymer Degradaton and Stability.*, vol. 88, no. 3, pp. 540–547, 2005, 10.1016/j.polymdegradstab.2004.12.016

[3] S. I. Stoliarov, S. Crowley, R. N. Walters, and R. E. Lyon, "Prediction of the burning rates of charring polymers," *Combustion and Flame*, vol. 157, no. 11, pp. 2024–2034, 2010, 10.1016/j.combustflame.2010.03.011

[4] C. Lautenberger and C. Fernandez-pello, A Generalized Pyrolysis Model for Combustible Solids, no. April. *Fire Safety Journal*, 2007, 10.1016/j.firesaf.2009.03.011

[5] M. Chaos, M. M. Khan, N. Krishnamoorthy, J. L. De Ris, and S. B. Dorofeev, "Evaluation of optimization schemes and determination of solid fuel properties for CFD fire models using bench-scale pyrolysis tests," *Proceedings of the Combustion Institute.*, vol. 33, no. 2, pp. 2599–2606, 2011, 10.1016/j.proci.2010.07.018

[6] N. Bal and G. Rein, "Numerical investigation of the ignition delay time of a translucent solid at high radiant heat fluxes," *Combustion and Flame*, vol. 158, no. 6, pp. 1109–1116, 2011, 10.1016/j.combustflame.2010.10.014

Numerical simulation of EN 50399 standard test – Calibration and validation of a numerical virtual test

Bertrand GIRARDIN[1], Damien FLAMMIER[2] and Gildas AUGUIN[1]

[1] *Efectis, Research & Innovation Direction, France*
[2] *Efectis, Laboratory, Fire resistance Testing Division, France*

bertrand.girardin@efectis.com

ABSTRACT

Wires and cables have recently been included in the EN 13501-6 classification standard within the Construction Products Regulation (CPR). Thus they need to be tested according to the EN 50399 standard. It consists in a large scale test for classification of vertically mounted cables. This test represents a limit to the development of new flame retarded cable formulations because of time and material consuming. Moreover, all the developed formulations cannot be tested. A solution for the development and orientation tests phase could be to predict the fire behavior of cables inside the EN 50399 apparatus, thanks to CFD simulations carried out with the computational fluid dynamics code Fire Dynamics Simulator (FDS). Nevertheless, due to their particular geometries and the complex phenomenon cables undergo, it is difficult to simulate and predict how they perform in such large scale test.

So a first step is to evaluate the ability of FDS to reproduce quantitative results in terms of gas temperatures, velocity and heat fluxes distribution inside the test apparatus and in the cables environment for further evaluation of their fire performances. A numerical model of the test facility without burning cables is developed and assessed.

When experimental results are compared with numerical calculations, good agreement is found for gas temperature and heat flux. Temperatures with order of magnitude up to 1000 °C in front of the burner and up to 200°C are observed between 400 and 500 mm above the burner. For the thermocouples located 1.3 m and 2.6 m above the burner, the maximum gas temperature achieved are close to 100 °C. The proposed numerical model gives correct thermal loads in the vicinity of the ladder and the cables during calibration tests and can be used for further evaluation of combustible material on cables.

KEYWORDS

Modeling; EN50399; Cables; CFD; Validation; Flame spread

INTRODUCTION

Cables transmit energy and an ever growing amount of data. Numerous cables are installed to connect rooms and floors in buildings and installations. Low up to medium voltage cables, telephone and LAN cables are then part of all residential, industrial or public building. Independent from the conductor (mainly conductive copper wires), those cables generally contain polymer-insulation, sheathing or bedding. All types of cables can then produce a significant amount of fuel that can contribute to fires growth or propagation. As they need to pass through fire walls and are frequently located in horizontal and/or vertical ducting, they have significant potential to increase the fire hazard in a building spreading of fires.

However, a harmonization of the European fire safety test requirements for different building products, including cables have taken place in the frame of the European Construction Products Regulation (CPR) [1] since the last decades and cables now need to be CE-marked. The purpose is to ensure the free movement of all construction products within the European Union by harmonizing national laws with respect to the essential requirements applicable to these products in terms of health and safety. Cables are then classified in the so-called "Euroclass". The EN 13501 [2] standard summarizes the criteria and the tests to be performed for the Euroclass of a cable. The Euroclasses are defined according to the results of either ISO 1716 [3], EN 50399 [4] and EN 60332-1-2 [5] tests.

One of the limitations when developing new materials for cables is the relevance of the tests performed for evaluating the fire performances of materials. Indeed, small-scale standard tests are generally not discriminant and large-scale tests are time and materials consuming. The development of new materials for cables implies then to test the materials themselves rather than the cables. These tests on the materials have then to be representative of the fire scenario involved when cables burn.

In that context, cone calorimetry is widely used even if it does not permit to examine flame spread. Some studies tried to investigate the lateral [6] or vertical [7] flame spread with a modified and/or inclined cone calorimeter with more or less success. Indeed, these studies considered forced and radiative burning conditions when the specimen are ignited with a flame in the EN 50399 apparatus or IEC 60332-1 apparatus. Moreover, the size of the ignition source (the heating cone) used in these studies is similar to that of the sample size (the size of the cone plaque). On the contrary, in the previously mentioned tests, the height of specimen is much larger than the one of the ignition source and thus allows studying flame spread.

In order to avoid massive cables production and time/materials consuming test that can't be performed in the new materials development phase, one could use numerical modeling and virtual testing. Indeed, based on limited inputs and using such model could permit to predict the fire performance even in terms of flame spread in complex tests such as the EN 50399. To do so, a first step would be to simulate experiments performed without cable and with incombustible cable (flow obstruction) in order to validate the fire model in terms of gas temperatures, gas velocity and/or heat fluxes.

The goal of this study is thus to present and validate the virtual model developed for the EN 50399 based on experimental data [8]. In a first part, the implementation of the EN 50399 apparatus using FDS v6 will be detailed, then critical aspects of the models such as the burner and the way cables can be implemented will be discussed. Special attention will be given in the temperature fields and impacting heat flux in the apparatus.

DESCRIPTION OF THE MODEL

Implementation of the EN 50399 cable test apparatus

The cable test apparatus was implemented according to the requirements in the standard EN 50399 (see Fig. 1). The inner dimensions of the apparatus are 1 x 2 x 4 m^3 (width x depth x height). The inlet air is applied as volumetric flow rate of 0.133 m^3/s injected vertically upward from a 0.8 x 0.4 m^2 opening in the floor. The extraction hood (0.5x1 m^2) is placed 400 mm above the outlet of the cabin and a constant outlet flow rate of 1 m^3/s is imposed.

After performing a mesh sensibility analysis, it was found that 25 x 25 x 50 mm^3 mesh size was an appropriate cell size for the part of the test apparatus where the cable ladder is mounted. The size of the domain is extended to 4.5 m in height to take into account for the extraction hood. There is however a large volume of the test apparatus where there are no flames and where the cell size can be larger (50 x 50 x 50 mm^3).

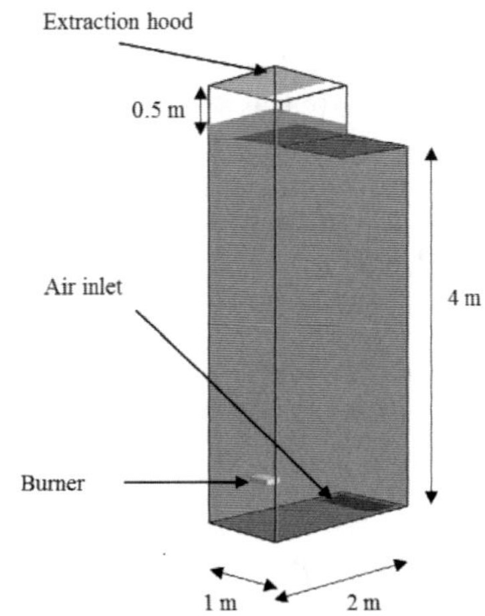

Fig. 1. Implementation of the EN 50399 apparatus in FDS.

Implementation of the burner

As for the cables, the mesh size that can be used for CFD calculations is coarser than the burner size. Indeed, the burner in the EN 50399 apparatus consists of 242 holes of 1.32 mm diameter each throwing 1550 mg/s of air and 442 mg/s of propane. The total surface throwing gases is thus of around 331 mm² and the imposed mass flux are thus of 4.68 kg/m²/s of air and 1.33 kg/m²/s of propane. However, in FDS, the burner is described by a unique obstacle than cannot be larger than at least one grid cell (25 mm). When the mass flow of premixed gas leaves the holes of the burner the velocity is relatively high giving a horizontal flame that hits the rear side of the test apparatus. On the contrary, in the simulations the same mass flow leaving the 12500 mm^2 hole gives a much lower velocity and the flame becomes vertical. This obviously has implications on the heat transfer from the burner flame to the cables.

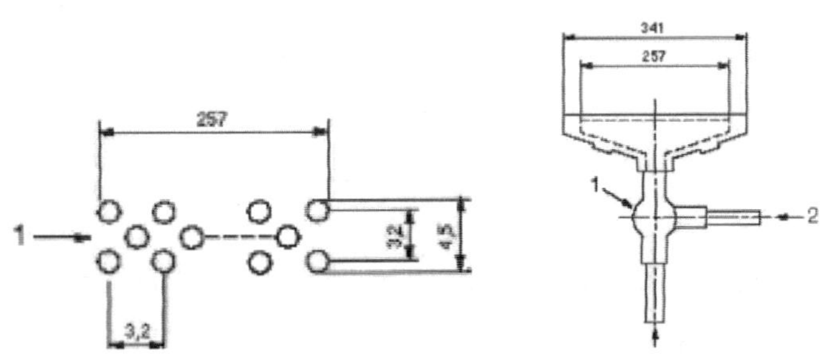

Fig. 2. Burner description in the EN50399 apparatus [9].

Different methods to numerically implement the burner are possible with FDS:

-Method 1: Adapt the mass flow per unit area of air and propane to the actual surface of the burner in the model so the burner power remains at 20.5 kW.

-Method 2: Maintain the total mass flux (6.01 kg/m²/s of gas) but decrease the mass flux of propane so the burner power remains at 20.5 kW.

-Method 3: Adapt mass flow of air and propane to the surface so the burner power remains at 20.5 kW but use the feature of FDS to "patch" the velocity and impose a horizontal velocity of 2 m/s in the vicinity of the burner.

-Method 4: Implement 242 particles with a total debiting surface of 331 mm² as for the real burner.

IOP Conf. Series: Journal of Physics: Conf. Series **1107** (2018) 032025 doi:10.1088/1742-6596/1107/3/032025

Implementation of the cables

One of the major limitations of FDS is the fact that the meshing should be lower than the diameter of the cables (10 or 20 mm for instance). This implies to have very long computational times. However, FDS code introduced the so called fixed cylindrical Lagrangian particle [10][11]. Using these features, sub-grid modelling of cylindrically symmetric cables can be performed on a coarser mesh. This allows detailed multi-scale modelling of fine objects, such as cables, in a much larger computational domain.

VALIDATION OF THE MODEL - COMPARISON OF SIMULATED TO EXPERIMENTAL DATA WITHOUT CABLES

In [8], Försth *et al.* conducted experiments in a cable test apparatus with a ladder without cable. Instead, thermocouples and plate thermometers (PT) were implemented on the ladder, around 430 mm, 1300 mm and 2600 mm above the burner location. At each location one thermocouple for gas temperature measurements and two plate thermometers were implemented (see Fig. 3). One of the PT was facing the burner and the front side of the cable test apparatus (Front) and one facing the rear side of the apparatus (Rear). The experiments were simulated using the four different methods. The numerical results in term of gas temperature and PT temperature are compared to the experimental measurements.

At the exception of the method 2, the results in terms of gas and PT temperature are similar to the experimental ones. Indeed, the experimental gas temperatures 430 mm, 1300 mm and 2600 mm above the burner are experimentally measured at 273±10 °C, 118±108 °C and 91±10 °C while they are predicted at 247±15 °C, 132±13 °C and 109±12 °C respectively with the method 3 for instance.

Table 1: Experimental and simulated gas and PT temperature.

	430mm	1300mm	2600mm		430mm	1300mm	2600mm
Experimental	279 ± 10°C	118 ± 10°C	96 ± 10°C	Front	185 ± 15°C	99 ± 15°C	81 ± 15°C
				Rear	224 ± 15°C	107 ± 15°C	88 ± 15°C
Method 1	259 ± 15°C	151 ± 15°C	122 ± 14°C	Front	225 ± 21°C	117 ± 19°C	101 ± 19°C
				Rear	224 ± 18°C	123 ± 18°C	102 ± 18°C
Method 2	74 ± 4°C	83 ± 8°C	94 ± 11°C	Front	70 ± 7°C	76 ± 12°C	82 ± 126°C
				Rear	125 ± 13°C	89 ± 13°C	85 ± 15°C
Method 3	247 ± 15°C	132 ± 13°C	109 ± 12°C	Front	197 ± 18°C	105 ± 17°C	91 ± 17°C
				Rear	272± ± 18°C	121 ± 18°C	95 ± 17°C
Method 4	223 ± 13°C	150 ± 15°C	119 ± 14°C	Front	229± ± 21°C	119 ± 19°C	101 ± 19°C
				Rear	206 ± 18°C	122 ± 18°C	100 ± 18°C

After performing such sensibility analysis, the experimental results were reproduced with the minimal errors using the Method 3 ("velocity patch"). Indeed, the model slightly over-predicts the PT temperature especially for the PT facing the rear side of the apparatus and 430 mm above the burner (260 °C with the model against 225 °C for the experiment). Apart for this point, the PT temperature prediction lies within ±15 °C. A visualization of the temperature and velocity field can be seen in Fig. 3 Gas temperatures up to 1000 °C are obtained close the burner, thus validating the measurements of Försth *et al.* [8] in front of the burner.

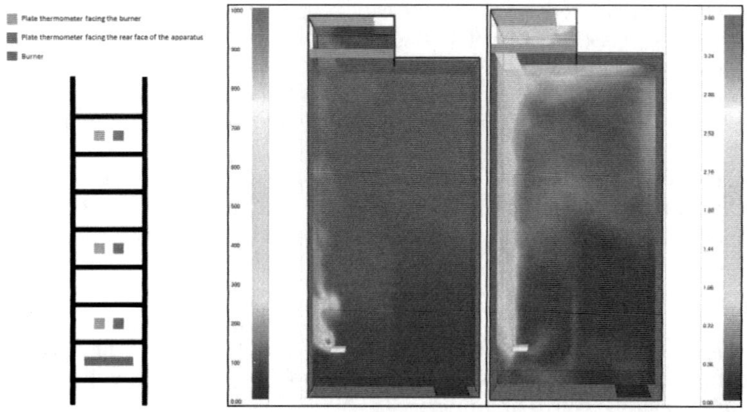

Fig. 3. Plate thermometer location [8] (left), simulated temperature (middle) and air velocity (right) in the cable test apparatus at the steady state.

IMPACT OF THE BURNER ON THE NUMERICAL BEHAVIOR OF THE CABLES

As shown in Table 1, the PT temperatures are different for the PT facing the burner than for those facing the rear side of the tests apparatus meaning that the impacting heat flux are different. Cables were numerically implemented to investigate the impact of the burner implementation on the ignition and heat release rate response of the model. The THIEF model [11][12] was used to model the heat transfer inside the cables. Namely, cables of 12 mm diameter were implemented and the results in terms of cable surface temperature after 1000 s of exposure are presented in Fig. 4.

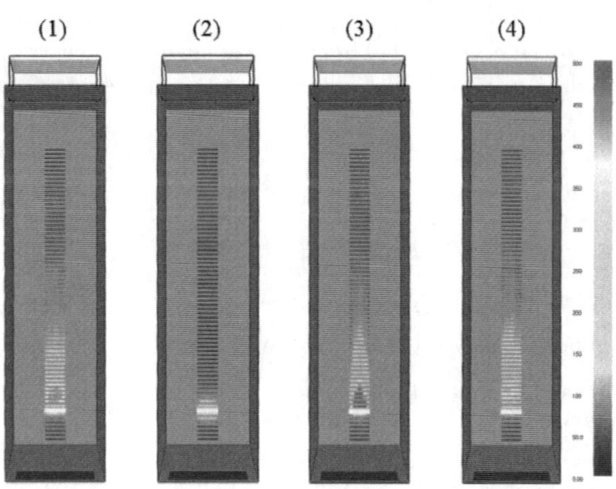

Fig. 4: Temperature distribution along cables without cable ignition after 1000 s exposure to the burner.

It can be seen that the area where the temperature increases have different shapes depending on the burner type. Namely, the temperature increase is limited to around 400 °C on all the width of the ladder for the Method 1 when similar temperatures are obtained with the Method 4 but in a smaller area. The highest temperature (around 500 °C) is obtained for the Method 3 as well with an inversed V-shape. This behavior is typical of the EN 50399 testing where the highest damaged lengths are generally obtained at the middle of the ladder width.

IOP Conf. Series: Journal of Physics: Conf. Series **1107** (2018) 032025 doi:10.1088/1742-6596/1107/3/032025

CONCLUSION

The purpose of this study was to develop a numerical model to simulate the fire environment of the EN 50399 test facility. A comparison of experimental and simulated data using FDS v.6 showed that it is possible to predict the gas temperature and PT thermometer measured at different location of the EN 50399 apparatus. The optimum results were obtained when "patching" the velocity field in front of the burner. Furthermore, the anisotropy shown by Försth *et al.* [8] is well considered by the model. This validates the model developed in this study.
Further investigations on how to obtain reliable and meaningful data to be used as input data for numerical modelling of the flame spread along cables have to be carried out. Moreover, the model can be further validated by measuring the temperature field and also the velocity field inside the test apparatus.

REFERENCES

[1] European Union, 2011, Regulation (EU) No 305/2011 of the European Parliament and of the Council of 9 March 2011 laying down harmonized conditions for the marketing of construction products and repealing Council Directive 89/106/EEC.

[2] EN 13501-2 Fire classification of construction products and building elements. 2007.

[3] ISO 1716, Reaction to fire tests for products - Determination of the gross heat of combustion (calorific value). 2010.

[4] EN 50399 Common test methods for cables under fire conditions. Heat release and smoke production measurement on cables during flame spread test. Test apparatus, procedures, results. 2011.

[5] EN 60332-1-2 Tests on electric and optical fibre cables under fire conditions - Part 1-2: Test for vertical flame propagation for a single insulated wire or cable - Procedure for 1 kW pre-mixed flame. 2004.

[6] Tsai, K.-C. Using cone calorimeter data for the prediction of upward flame spread rate. *J. Therm. Anal. Calorim.* **2012**, *112*, 1601–1606.

[7] Merryweather, G.; Spearpoint, M. J. Flame spread measurements on wood products using the ASTM E 1321 LIFT apparatus and a reduced scale adaptation of the cone calorimeter. *Fire Mater.* **2009**, *34*, 109–136.

[8] Försth, M., Sjöström, J., Wickström, U., Andersson, P., & Girardin, B. (2015)"Characterization of the thermal exposure in the en 50399 cable test apparatus." *International Conference and Exhibition on Fire and Materials: 02/02/2015-04/02/2015.* (pp. 23-37), Interscience Communications, 2015.

[9] EN 60332-3-10: 2010. Tests on electric and optical fibre cables under fire conditions - Part 3-10 : test for vertical flame spread of vertically-mounted bunched wires or cables - Apparatus

[10] Fire Dynamics Simulator FDS, version 6.5.2, www.nist.gov

[11] McGrattan, K., McDermott, R., Hostikka, S.,Floyd, J. & Overholt, K. (2017). Fire dynamics simulator (version 6), user's guide. NIST special publication, 1019(6), 1-296.

[12] McGrattan, K. B., & Dreisbach, J. (2008). Cable Response to Live Fire (CAROLFIRE) Volume 3: Thermally-Induced Electrical Failure (THIEF) Model (No. NIST Interagency/Internal Report (NISTIR)-7472).

IOP Conf. Series: Journal of Physics: Conf. Series **1107** (2018) 042001 doi:10.1088/1742-6596/1107/4/042001

Experimental Study about the Influence of Heat Tightness of an Enclosure Fire on Ignition Risk of Unburnt Gases in a Connected Exhaust System

Brady MANESCAU, **Hui-Ying WANG**, Bruno COUDOUR and Jean Pierre GARO

Institut P', Fluides-Thermique-Combustion, CNRS, ENSMA, Université de Poitiers, France

wang@ensma.fr

ABSTRACT

This experimental study was performed from a reduced scale enclosure with a length/height and width of 2 m. In this work, we investigate whether heat tightness of enclosure can change the compartment fire dynamics and pose ignition risk in exhaust system. An external ventilation system provides an air supply rate ranging from 24 to 40 m^3/h, corresponding respectively to 3 and 5 Air Change Per Hour (ACPH). A circular dodecane pan with a diameter of 40 cm is placed in the middle of the enclosure. Two situations with and without insulation of the enclosure are compared for several ACPH. The results show that heat tightness of enclosure leads to faster fire growth implying more important peak in heat release rate, and thus more dangerous fire with regarding the ignition risk. Heat tightness of enclosure makes the depression level of the compartment decrease, and consequently the air inlet supply rate decrease too, but the mass loss rate of liquid fuel increase. With heat tightness of enclosure, the fire becomes very-under-ventilated, and the vaporized fuel does not completely contribute to the heat release due to a significant formation of the unburnt gases as hydrocarbons, CO and H_2. It is found that ignition of unburnt volatiles near the exhaust system occurs more easily when the compartment is more heat-tight. Oscillating flames in under-ventilated conditions were observed experimentally due to the ignition/extinction of the liquid fuel pan in a vitiated air enclosure. After the fire extinction, the remaining flames as ghosting flame in the vicinity of the enclosure ceiling were not observed under lowered oxygen vitiation with a mass fraction of around 7%. With a time delay of about 15 min in the current situation, the energy released per mass of oxygen consumed allows to raise the temperature of fuel-air mixture to about 300°C high enough for ignition of the unburnt gases. When the molar fraction of unburnt fuels is above the low flammability limit, any return of air from dilution duct after the fire extinction in a vitiated air enclosure could lead to spontaneous ignition near the extraction duct without needing a pilot flame. Note that heat-tightness of enclosure contributes to a reduction in the ignition delay time at entrance of the connected extraction duct due to a quick accumulation of unburnt fuel gases there. The experimental results from such reduced scale enclosure can be extended to a full scale one by imposing a preservation of the ratio ($\dot{Q}^2/L^5$) between heat release rate ($\dot{Q}$) and characteristic length (L). This relationship is derived from dimensionless variable of the energy equation.

KEYWORDS:
risk assessment; ignition risk; fire growth; compartment fires; liquid fuel; heat tightness; extraction duct; dilution; depression

INTRODUCTION

Ignition has gained increasing attention in fire science mainly in industrial scenarios, though knowledge is limited for explosions or gas deflagrations through ventilation system. This phenomenon generally occurs when the fire is ventilation-controlled or during its extinction phase, typically when a compartment containing a fire has a very limited fresh air supply. Liquid pool fires usually generate a significant hazard in an enclosure due to the amount of unburnt volatile gases. The liquids could originate e.g. from leaks from transformers generators or other machinery. Knowing the liquid fire burning rate is the starting point of any fire safety analysis. The factors affecting the burning rates of liquid pool fires in open atmosphere are well known for a wide variety of liquids [1, 2]. However, many liquid pool fire scenarios involve fires in highly confined spaces, especially encountered in the nuclear facilities but also in the current constructions. The burning rate can be significantly different from the ones measured in open atmosphere [3]. These differences are caused essentially by the influence of air ventilation, heat radiation from hot walls and hot gas layer.

In order to contain the potential release of radioactive material and avoid dispersion to the outside, nuclear installations are rapidly moving towards building envelopes more air and heat-tight [4]. Intensive research has been carried over decades on the oscillatory phenomena of unstable flame under confined conditions with mechanically ventilation system [4-6]. The low-frequency oscillating flames were observed [4, 5] with a drastic change in flame behavior in a vitiated air enclosure. There is still a lack of knowledge for complete understanding of an oscillatory behavior but conjectured that issues involved are ignition/extinction of condensed fuel, phenomena leading to a reverse flow in the supply ventilation system and successive combustion of unburnt gases [4, 5]. Only a small proportion of the work has looked specifically the effect of heat tightness of enclosure on the fire behavior, notably the thermal stratification of the unburnt fuel [7].

In a typical ignition scenario, the atmosphere in the extraction duct consists of large un-combusted gases with a concentration generally above the lower flammability limit at elevated temperature. The experimental study of Lassus [8] showed that in a highly confined setting with interconnecting ventilation ducts, an extraction duct is an important pathway for the onset of ignition of a flammable mixture of pyrolysis gases and vitiated air with presence of a dilution duct. Indeed, with a compartment well heat-tight, heat feedback to the surface of liquid fuel is more important in comparison with that less heat-tight [9]. We can expect that the ignition-related risks may become more significant when the enclosure is more heat-tight. In the current study, we verified theoretically and experimentally the influence of heat tightness of enclosure on the compartment fire dynamics and the ignition delay time at entrance of the extraction duct.

EXPERIMENTAL SET-UP

The experimental facility with a length/height and width of 2 m and its geometry are shown in Figure 1(a, b). An external ventilation system consisting of two intake and an extraction ducts, provides an air supply rate ranging from 24 to 40 m^3/h, corresponding respectively to 3 and 5 Air Change Per Hour (ACPH). Ducts have a square section of 0.2x0.2 m^2. The intake was placed closer to the floor at a height of 0.3 m and the exhaust was placed near the ceiling at a height of 1.7 m. Walls of the enclosure are made of concrete (thermal conductivity: λ =1 W/m.K, density: ρ =2100 kg/m^3, specific heat: C_p=0.88 kJ/kg.K) with a thickness of 0.2 m; there is always a conductive heat loss through the walls during the fire. The enclosure is considered as perfectly heat tight when the wall and ceiling are recovered by an insulating material as Promatech H (λ = 0.175 W/m.K, ρ =400 kg/m^3, C_p=0.92 kJ/kg.K) with a thickness of about 0.035 m.

The stainless-steel circular pan with a diameter of 40 cm was placed in the middle of the enclosure, slightly elevated at a height of 0.3 m. Dodecane is typically used as the fuel load, and a good repeatability was obtained with a difference below 5% for each fire tests. The fuel mass was continuously monitored using a load cell, installed under the pan. Measurements performed include pressure, temperature and chemical species concentrations of gases filling the enclosure. As shown in Fig.1b), the gas temperature was analyzed from seven vertical profiles inside the enclosure, at the extraction duct and the dilution one. Each of the seven profiles comprises eight thermocouples positioned at regular intervals of 0.2 m over the height of the enclosure.

A mechanical ventilation network is used for air supply by using a centrifuge fan at the end of the dilution duct with a fixed speed (cf. Fig.1a). Before activating the fire, a period of 10 minute was included for establishing a depression below atmospheric pressure inside the enclosure. The depression level is a function of the Air Change Per Hour (ACPH) of the enclosure, depending on the resulting outlet volume flow rate of the fan at the end of the extraction duct. After activating the fire, the velocity, the temperature and the pressure at both the intake and

IOP Conf. Series: Journal of Physics: Conf. Series **1107** (2018) 042001 doi:10.1088/1742-6596/1107/4/042001

extraction ducts are connected to the fire dynamic. Under the depression condition, the experimental facility is not perfectly airtight mainly due to passage of the thermocouples through the wall. Based on the mass conservation at the admission and extraction ducts, the leakage area around the enclosure is estimated in the order of 14 cm^2.

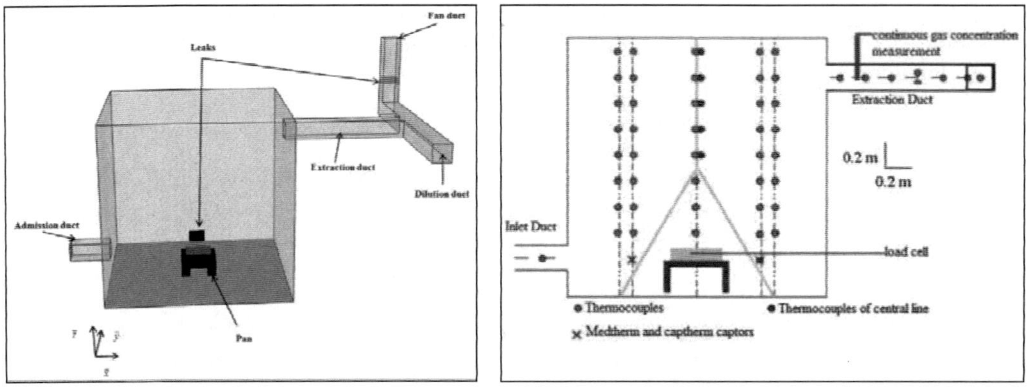

a) 3D view of the enclosure b) Measurement locations from the side (x, z) view

Fig. 1. Confined and mechanically ventilated enclosure fire connected to an extraction duct

RESULTS AND DISCUSSION

Histories of the relative pressure *(P_0-P)* and the inlet velocity for 5 ACPH at the admission duct are illustrated in Figure 2. Here, P_0 denotes the atmospheric pressure, and *P* the enclosure one. The fire-induced pressure can be said to follow three main stages : 1) activating the fire causes an overpressure inside the compartment which leads to a significant reduction in the inlet flow velocity at the admission duct; 2) with some time delay, the fully developed fire becomes ventilation controlled due to the fire growth, resulting in large fluctuations of the pressure and oscillations of the inlet velocity; 3) during the decay phase, extinction induces a depression peak, followed by a significant increase in velocity of fresh air supplied by the admission duct. Heat-tightness of the enclosure can sensitively reduce the depression level (Fig.2b), and consequently, the air inlet velocity from the admission duct. Therefore, heat-tightness of an enclosure has a significant impact on the dynamic confinement which prevent from hazardous gas leaks.

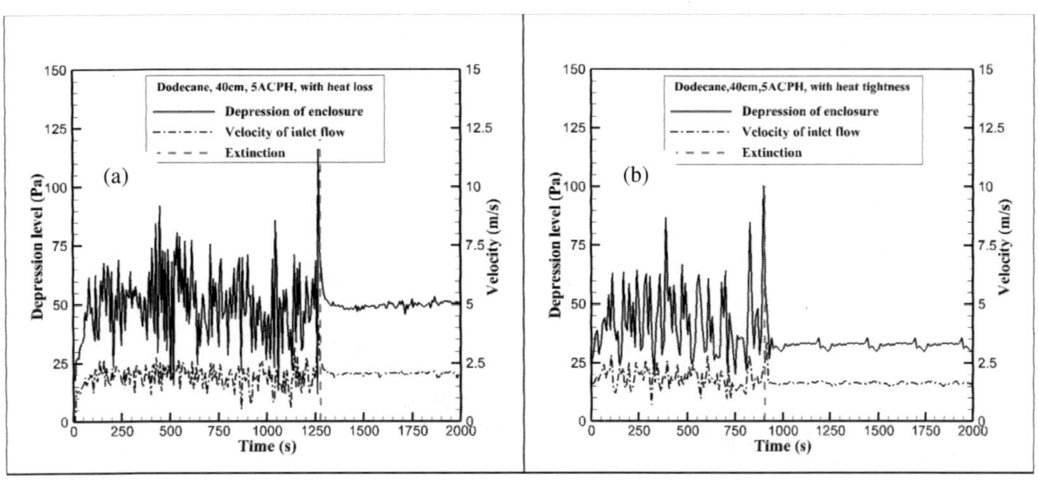

Fig. 2. Evolution of the pressure and inlet velocity at 5 ACPH. a) with heat loss; b) with heat tightness

A typical evolution of the mass loss rate and the associated HRR at 3 and 5 ACPH with heat loss and tightness is illustrated in Fig.3(a-d). Note that we lack the ability to accurately quantify the HRR by using oxygen calorimetry in such confined facility. Based on the measured mass loss rate $\dot{m}_F$, the theoretical HRR is derived as $\dot{Q} = \dot{m}_F \Delta H_c$. The effective HRR is approximately determined from the variation of the oxygen concentration

IOP Conf. Series: Journal of Physics: Conf. Series **1107** (2018) 042001 doi:10.1088/1742-6596/1107/4/042001

(ΔY_{O2}) at the entrance of extraction duct and the inlet air rate ($\dot{m}_A$) as: $\dot{Q} = (Y_{O2,\infty}\dot{m}_A)\Delta Y_{O2}\Delta H_o$. Here, ΔH_c and ΔH_o are the energy released per kilogram of fuel and oxygen consumed, respectively. With heat loss of enclosure (cf. Fig.3a, c), an increase of ACPH from 3 to 5 contributes to an increase of the mass loss rate due to the higher ventilation flow rate over the liquid surface, which enhances liquid vaporization rate via convection. With heat tightness of enclosure (cf. Fig.3b, d), the hotter smoke spreads rapidly towards the liquid base, resulting in a significant increase of the mass loss rate due to the enhanced radiative heat feedback to the liquid surface. The mass loss rate is practically insensitive to an increase of ACPH under the heat-tight conditions due to the dominant radiative heat transfer. In the initial growth phase, burning is fuel-controlled with a stable flame. Followed by the fully developed, post-flashover phase, in which burning is ventilation controlled with appearance of an unstable flame. The lowed oxygen vitiation induces oscillations of the HRR following the trend of the mass loss rate due to an unstable flame. The oscillatory phenomena of the HRR is attributed to the coupling process between the fuel evaporation rate and the inlet flow rate of air, both attached to heat tightness of enclosure. With heat loss of enclosure (cf. Fig.3a, c), the HRR curves show a gradual growing trend with a moderate peak of about 90 kW. The oxygen concentration initially present in the enclosure and the one supplied by mechanical ventilation entrain a sufficiently ventilated fire. The effective HRR corresponds approximately to the theoretical one derived from the mass loss rate because the reaction is almost complete. It is observed that a heat tightness of enclosure leads to the ultra-fast fire growth (cf. Fig.3b, d), and flashover is clearly identified by a rapid increase of HRR from 40 to 140 kW. The ultra-fast growing fire in a very consistent manner becomes progressively oxygen limited, that is to say ventilation-controlled, and the theoretical HRR is higher than the effective one. The extinction due to unstable combustion in a vitiated air enclosure happened early when the wall envelope is heat-tight.

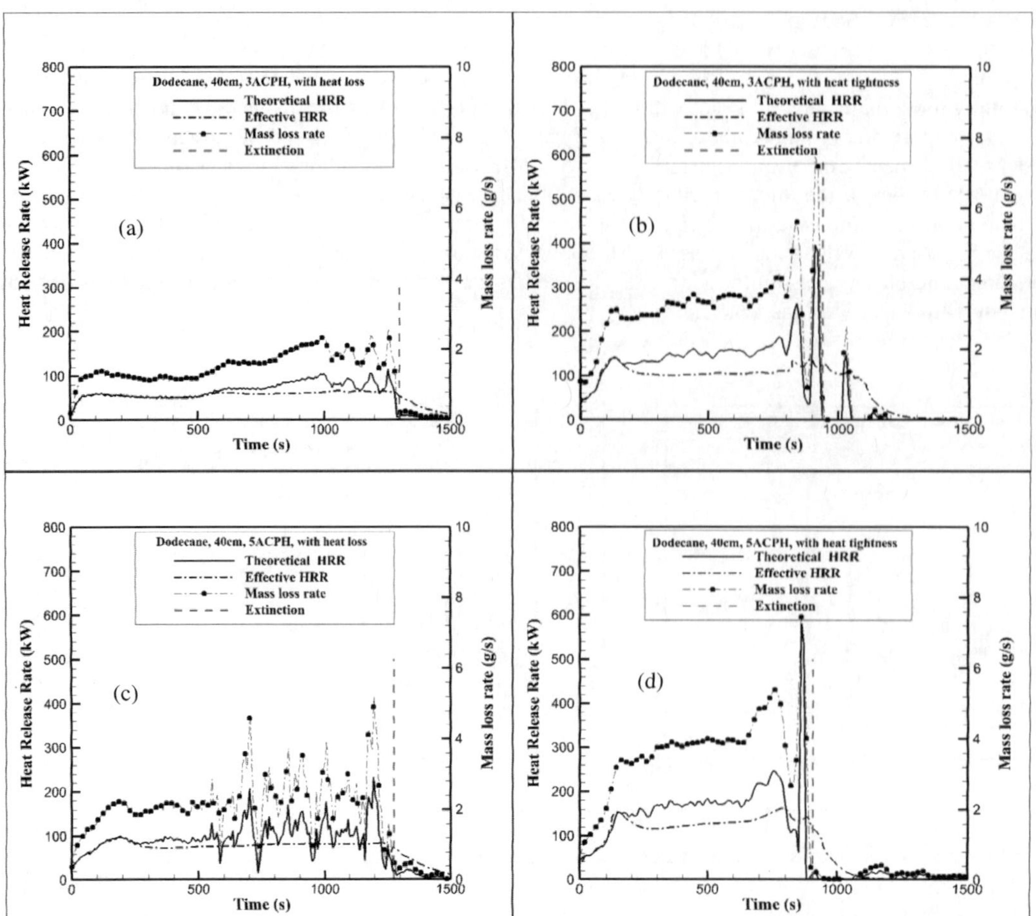

Fig. 3. Evolution of the mass loss rate and HRR. a) Heat loss at 3 ACPH with extinction due to burn out; b) Heat tightness at 3 ACPH with extinction due to unstable combustion; c) Heat loss at 5 ACPH with extinction due to burn out; d) Heat tightness at 5 ACPH with extinction due to unstable combustion

In such reduced scale enclosure, the Dodecane pan of 40 cm in diameter leads to under-ventilated fire, thus generating a large amount of unburnt volatiles as hydrocarbons, CO and H_2. The molar fraction of such unburnt volatiles near the extraction duct is presented in Figure 4(a-d). The aim of this study is to verify if the auto-inflammation risk occurs when the unburnt fuel concentration is beyond the Low Flammability Limit (LFL) at the point of Auto-Ignition Temperature (AIT) [10]. The LFL_i of each fuel and the LFL of gas mixtures as a function of the fuel molar fraction X_i of each fuel composing it are determined as well as a function of the gas temperature as follows :

$$LFL_i(T) = LFL_i(T_0)\left[1-\frac{T-T_0}{1300-T_0}\right] \quad \text{and} \quad LFL(T) = 100\times\left[\sum_i \frac{X_i}{LFL_i(T)}\right]^{-1}$$

The values of the AIT and LFL at T_0=25°C [10] for three fuel species are summarized in Table 1.

Table 1. Values of the AIT and LFL at T_0=25°C for the three major fuel species

Parameter	CO	H_2	Dodecane
LFL (T_0=25°C)	12.5%	4%	0.6%
AIT (°C)	588	520	204

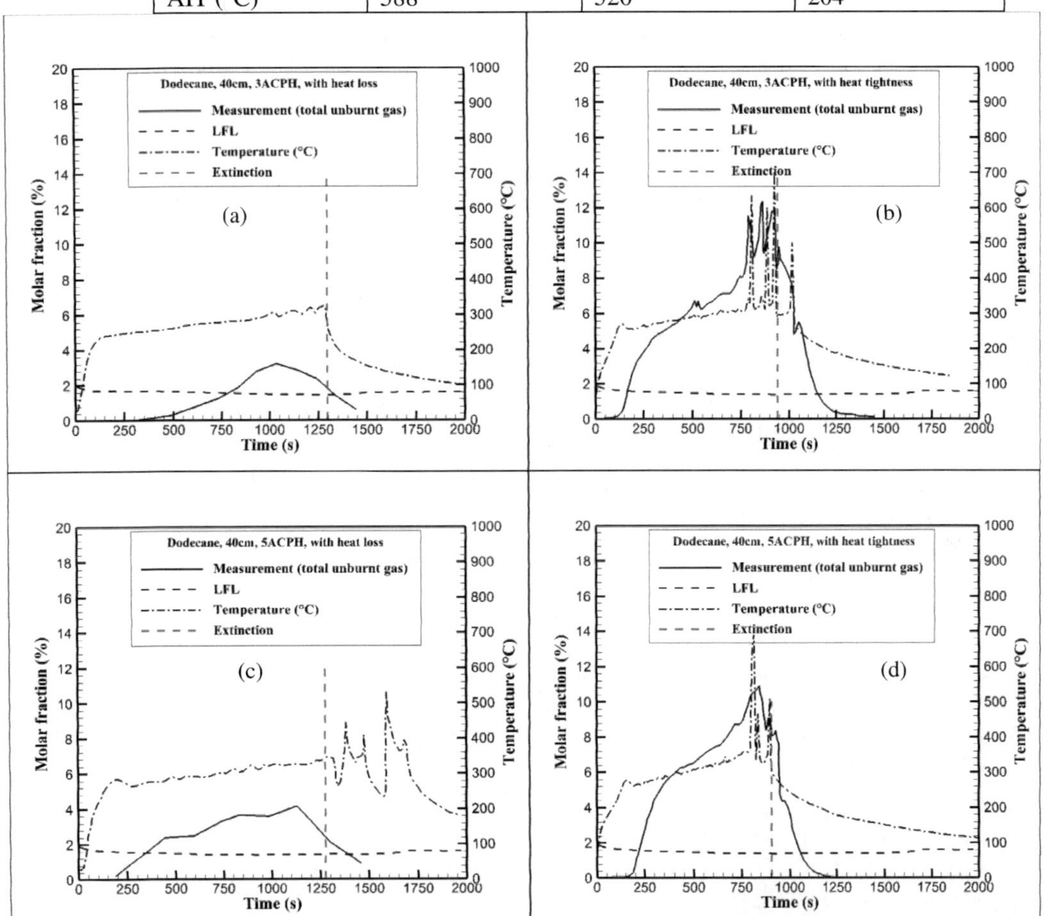

Fig. 4. Risk assessment of unburnt gas ignition at entrance of the extraction duct. a) with heat loss at 3 ACPH; b) with heat tightness at 3 ACPH; c) with heat loss at 5 ACPH; d) with heat tightness at 5 ACPH

With heat loss of the enclosure at 3 ACPH (cf. Fig.4a), ignition was not observed experimentally due to less important formation of the stratified hotter unburnt fuels layer even the gas temperature is above 300°C high enough for ignition. It is clear that the ignition is not merely dependent on the existence of the fuel-air mixture with a concentration beyond the LFL at the AIT points near the entrance of the extraction duct. As illustrated in Fig.4b, we can see that with heat tightness of the enclosure, starting from 250 s, the molar fraction of unburnt fuels is above the low flammability limit (LFL) at the entrance of the extraction duct. For many minutes, the

IOP Conf. Series: Journal of Physics: Conf. Series **1107** (2018) 042001 doi:10.1088/1742-6596/1107/4/042001

energy released per mass of oxygen consumed allows to raise the temperature of fuel-air mixture above the AIT near the ceiling. With a time delay of about 900 s, the auto-ignition would be caused near the extraction duct due to a return of air from dilution duct after the fire extinction at the gas temperature of about 400°C high enough for ignition. The phenomena with ignition near the extraction duct can be characterized by a rapid decrease of unburnt fuel concentration towards a stoichiometric fuel-air mixture and a sharp increase of temperature with a peak reaching a typical flame temperature of 550°C there. It seems that the global equivalent ratio, $\phi = \dot{m}_F s / \dot{m}_A$, calculated from the fuel supply rate $\dot{m}_F$, air inflow rate $\dot{m}_A$ and stoichiometric coefficient s, higher than 1.5 represents a set of dangerous conditions, as any return of air from dilution duct after the fire extinction could lead to spontaneous ignition near the extraction duct. In Fig.4c with heat loss of the enclosure, occurrence of auto-ignition near the extraction duct was consistently observed at 5 ACPH with a longer time delay of about 1320 s after the extinction at 1250 s with presence of a return of air from dilution duct. As illustrated in Fig.4d, heat tightness of the enclosure contributes to a significant reduction in the ignition delay time to about 780 s due to a rapid formation of the hotter unburnt fuels layer near the ceiling. This is mainly attributed to a sharp increase of the mass loss rate at the initial fire growth stage up to 200 s (cf. Fig.3b, d).

CONCLUSION

In summary, heat-tightness of enclosure can sensitively increase the mass loss rate, and reduce the inlet air flow rate due to a lower depression level inside enclosure, both decreasing the oxygen concentration. This implies that burning is quickly ventilation controlled with heat-tightness of enclosure, forming significant amount of unburnt volatiles. If the concentration of unburnt volatiles is above LFL, and the fire has been established for sufficiently long (in the current study, more than 10 min), there is significant risk of ignition inside the extraction duct without needing a pilot flame. Occurrence of an ignition of the hotter concentrated unburnt fuels at entrance of the extraction duct is attributed to a return of air from dilution duct at the stage of fire extinction which induces a sharp increase of depression level inside enclosure. The experimental results show clearly that ignition can happen early near the extraction duct, as long as the enclosure remains sufficiently heat tight. With heat loss of the enclosure, occurrence of ignition is also identified but with a longer time delay. The current study can be applied to a full scale enclosure fire with a volume of approximately 100 m^3. This implies that ignition of unburnt fuel gases may occur in an exhaust system when the HRR is beyond 1.2 MW, corresponding approximately to a global equivalence ratio beyond 1.5. The work is continuing with the aim of improving the evaluation of auto-ignition near the extraction duct by measuring more detailed chemical species and heat flux over the liquid surface.

REFERENCES

[1] Peatross, M.J. and Beyler, C.L. (1997) Ventilation effects on compartment fire characterization, Fire Safety Science –Proceedings of the Fifth International Symposium, 5:403-414, doi:10.3801/IAFSS.FSS.5-403.

[2] Beaulieu, P. and Dembsey, N. (2007) Effect of oxygen on flame heat flux in horizontal and vertical orientations. Fire Safety, 43:410-428, doi:10.1016/j.firesaf.2007.11.008.

[3] Vilfayeau, S., Ren, N., Wang Y. and Trouvé A. (2015) Numerical simulation of under-ventilated liquid-fueled compartment fires with flame extinction and thermally-driven fuel evaporation. Proceedings of the Combustion Institute, 35:2563–2571, doi.org/10.1016/j.proci.2014.05.072.

[4] Prétrel, H., le Saux W. and Audouin L. (2012) Pressure variations induced by a pool fire in well-confined and force-ventilated compartment, Fire Safety, 52:11-24, doi:10.1016/j.firesaf.2012.04.005.

[5] Zavaleta, P. and Audouin, L. (2018) Cable tray fire tests in a confined and mechanically ventilated facility, Fire and Materials, 42:28-43, doi:10.1002/fam.2454.

[6] Wahlqvist, J. and Van Hees, P. (2013) Validation of FDS for large-scale well-confined mechanically ventilated fire scenarios with emphasis on predicting ventilation system behavior, Fire Safety, 62:102-114, doi.org/10.1016/j.firesaf.2013.07.007.

[7] Sinai, Y.L. (1999) Comments on the role of leakages in field modelling of under-ventilated compartment fires, Fire Safety Science, 35-1:11–20, doi.org/10.1016/S0379-7112(99)00005-3.

[8] Lassus, J., Studer, E., Garo, J.P., Vantelon, J.P., Jourda P. and Aine, P. (2010) Influence of ventilation on ignition risk of unburnt gases in the extraction duct of under ventilated compartment fire. Combust. Sci and Tech, 182:517-528, doi.org/10.1016/j.ijthermalsci.2013.07.015.

[9] Nasr, A., Suard, S., El-Rabii, H., Garo, J.P., Gay L. and Rigollet, L. (2013) Heat feedback to the fuel surface of a pool fire in an enclosure, Fire Safety, 60:56–63, doi:10.1016/j.firesaf.2012.12.005.

[10] Babrauskas, V. (2003) Ignition Handbook, Fire Science Publishers, Chapter 4:41-140.

Calculation of radiation transmittance through n-heptane in a pool fire

Teemu ISOJÄRVI[1], Hadi BORDBAR[1] and Simo HOSTIKKA[1]

[1]*Department of Civil Engineering, Aalto University, Finland*

teemu.isojarvi@aalto.fi

ABSTRACT

The radiative transfer mechanisms in a volatile hydrocarbon pool fire were investigated by solving the transmittance of infrared radiation through fuel (n-heptane) layers of different depths. The incoming radiation was assumed to have the same relative amounts of different wavelengths as a spectrum obtained experimentally for a 2 meter pool fire, and the transmittances were calculated by integrating either the single-ray Lambert-Beer formula, the two-flux (Schuster-Schwarzschild) method solution or the analytical plane-parallel monochromatic/gray solution of the radiative transfer equation over wavelength, using the liquid absorption coefficients from several sources. The obtained transmittances were compared to earlier heat flux measurements, and the possibility of calculating them with significantly less computation time by using a k-distribution method was investigated. The results managed to replicate the measured heat flux values at depths over 1 mm in the liquid with reasonable accuracy, and the use of the k-distribution, more known and used in gas phase thermal radiation calculations, significantly speeds up the calculations.

KEYWORDS:

pool fire; thermal radiation; hydrocarbon; transmittance; infrared spectra

IOP Conf. Series: Journal of Physics: Conf. Series **1107** (2018) 042002 doi:10.1088/1742-6596/1107/4/042002

INTRODUCTION

A pool fire, where an evaporating liquid undergoes combustion, is an important scenario in fire safety science due to large amounts of liquid fuels being stored in industrial facilities and carried between locations both by land and sea routes. A large pool fire can generate dangerous radiative heat fluxes at distances of tens of meters or more from the burning site, making it a large risk to surroundings. In addition to experiments on liquid pools, information about the relevant physics can be obtained by numerical simulations, which represent a complicated multiphysics problem with chemical reaction, hydrodynamics and radiation. The radiative heat transfer back to the pool from the flames above is the most important heat transfer mechanism for large scale pool fires, and it controls the mass loss and heat generation rate in combustion through the evaporation rate on the pool surface. In principle, the physics of this radiative transfer can be calculated by solving the radiative transfer equation (RTE), coupled to the hydrodynamic and reaction rate equations, by using FVM or some other numerical method.

It is, however, much simpler to measure empirical infrared flame spectra for pool fires of different surface areas and use this as the input data in an RTE calculation of radiative transfer downwards from the pool surface (using measured infrared absorption spectra of the liquid-phase fuels, from, e.g. Refs. [1,2,3]). The motivation for doing this is the attempt to find a radiative transfer model, usable in pool fire simulations, which would require less processing power through use of acceptable simplifications and empirical data instead of a full solution of the RTE. Earlier publications about radiation in pool fires, made by other authors, include Refs. [3,4,5], and some other sources discuss the more general problem of evaporation, heat transfer and flows in volatile liquid pools, which may contain mixtures of liquids instead of a single substance [7].

METHODS

Flame emissivity

As experimental flame spectra from different sources have been measured from pool fires of different sizes, there is a need to approximate the spectrum of a small pool fire by scaling the measured spectrum of a larger one. The simplest way to do this is to assume that the relative amounts of different wavelengths in the spectrum depends only on the fuel, and that the effect of pool size can be taken into account with a simple constant multiplicative factor, as in the commonly used equation of flame emissivity

$$\varepsilon = 1 - exp(-KL), \tag{1}$$

where the constant K depends on the fuel and L is a characteristic length of the flame. This, of course, is not necessarily a justified assumption, and it is possible that the ε in Eq. 1 is actually a function of the wavelength λ and affects the relative intensities at different wavelengths. In this work it is assumed that it is not a function of λ and the validity of this assumption is tested.

Absorption in a uniform medium with a simple geometry

The inward radiation flux at depth s in liquid can, in principle, be approximately calculated by integrating the monochromatic (or gray-gas) plane-parallel RTE solution for non-scattering and non-emitting medium [8] over the relevant wavelength interval, using the incoming flux as a weight function.

$$q^+(s) = 2\int_0^\infty q_0^+(\lambda)E_3(\kappa(\lambda)s)d\lambda \tag{2}$$

The function E_3 here is the third exponential integral function. An approximate version of this is the single-ray approximation implementing the Lambert-Beer law that predicts exponential decay of flux/intensity.

$$q^+(s) = \int_0^\infty q_0^+(\lambda)exp(-\kappa(\lambda)s)d\lambda \tag{3}$$

A more accurate approximation is the two-flux model [8], which predicts an inward radiative flux corresponding to a 2 times larger absorption coefficient than in the Lambert-Beer result:

$$q^+(s) = \int_0^\infty q_0^+(\lambda)exp(-2\kappa(\lambda)s)d\lambda. \tag{4}$$

The wavelength dependent quantity $q_0^+(\lambda)$ is obtained from experimental data by multiplying with a flame size parameter equivalent to the ε in Eq. 1. To obtain an intensity value, the corresponding radiative flux is divided by π steradians.

Significance of medium emission

Adding a term corresponding to radiative emission by liquid material to Eqs. 2 and 4 would not be difficult, if temperature distribution as a function of depth were known. However, in this work it is assumed that the significance of this can be ignored. At the boiling point of n-heptane (about 371 K), the blackbody emissive power σT^4 is about 1.1 kW/m^2, and at 50 °C it is about 0.62 kW/m^2, so it can be expected that medium emission is not among the most significant considerations. In the more accurate Eq. 2, the inclusion of emission in the liquid phase would mean having a term with an integration over the 2nd exponential integral function E_2 and in Eq. 4 there would be growing exponential term in the integrand in addition to the decreasing one. In pool fires with less volatile fuels, such as mineral spirits, with boiling temperature of 160-190 °C the thermal radiation from the liquid probably has to be considered.

The k-distribution method

The integrations in Eqs. 2, 3 and 4 can consume a lot of time if the flame spectrum and absorption spectrum have to be known in high resolution. A faster k-distribution method for calculating the fractional transmittances is based on converting the absorption coefficient function $\kappa(\lambda)$ into a monotone increasing function $g(k)$, which tells how large fraction of incoming radiation is on the wavelengths where the spectral absorption coefficient $\kappa(\lambda)$ is smaller than the variable k [8]. To find this function, we first take the function $q_0^+(\lambda)$, and multiply it with a constant that normalizes its integral to unity. If this normalized function is called $B_{\text{fl,n}}(\lambda)$, we can calculate the function $g(k)$ as an integral

$$g(k) = \int_{D(k)} B_{\text{fl,n}}(\lambda) d\lambda, \tag{5}$$

where $D(k)$ is defined as the set of wavelengths where the absorption coefficient is smaller than k:

$$D(k) = \{\lambda \in \mathbb{R} \mid \kappa(\lambda) < k\}. \tag{6}$$

Next the function $g(k)$ is converted to the inverse function $k(g)$ and finally the transmittance through a path length of s (assuming Lambert-Beer law) is found by integrating

$$\tau(s) = \int_0^1 exp(-k(g)s) dg, \tag{7}$$

or equivalently with the exponential integral decay law

$$\tau(s) = 2\int_0^1 E_3(k(g)s) dg. \tag{8}$$

An advantage of this kind of calculation is that, as the functions $g(k)$ and $k(g)$ are monotone increasing and therefore do not contain any rapid oscillations, the integral can be very accurately approximated with a Gauss-Legendre quadrature, using only a few quadrature points.

Data measured by other researchers

Data of the flame spectrum of a hydrocarbon (n-heptane) fire and the absorption spectrum of the liquid were found and combined from references [1,2,3]. The experimental data of the radiation penetration under the n-heptane liquid surface were obtained from [4].

The n-heptane IR absorption spectrum, combined from [1] and [3] is shown in Fig. 1.

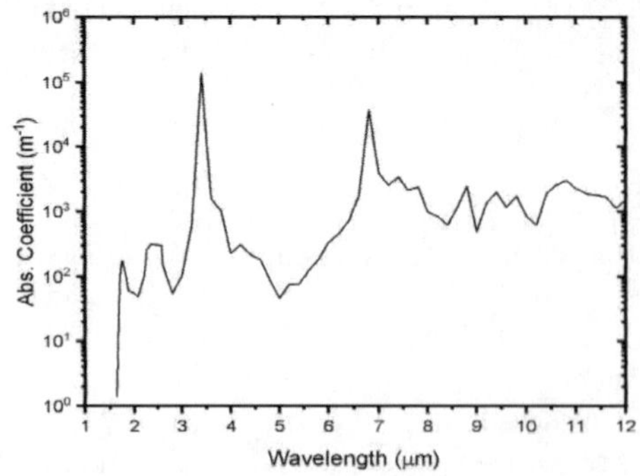

Fig. 1. The infrared absorption coefficient of liquid n-heptane, combined from data in two sources ([1, 3]).

The data in Suo-Anttila's work [3] was in the form of transmittances through a fixed path length (3.175 mm) of liquid, and the absorption coefficients in the Figure 1 are calculated by assuming Lambert-Beer law (exponential decay of intensity with increasing path length). The IR absorbance data in the NIST Chemistry Webbook for the range 2.65-20 μm has higher resolution than the data in Tuntomo's article [1], but as the former is from 1960 and the latter from 1992, the Tuntomo's results are used in these calculations. The measured IR emission spectrum of an n-heptane flame in the 2-meter pool fire of Suo-Anttila is in Fig 2., digitized from the original reported curve [3]. The article claims that the nearest blackbody equivalent of this spectrum has an absolute temperature of 1450 Kelvin and an emissivity of 0.4, and we assume that the spectrum outside this measured range of about 1.3-5.0 micron can be approximated with the corresponding Planck distribution of longer wavelengths (Fig. 2).

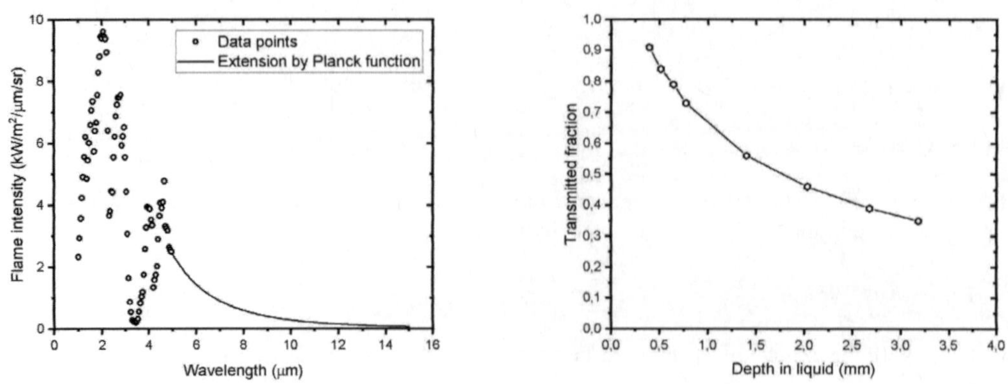

Fig 2. Left: the experimental intensity spectrum of an n-heptane flame measured in [3] continued to longer wavelengths with the spectral intensity of a blackbody with emissivity 0.4 and the absolute temperature of 1450 Kelvin. Right: transmitted fraction of flame radiation reaching different depths in the pool, calculated in [3].

RESULTS

Zhao et al. [4] fitted simple curves to their measured heat fluxes as a function of depth, valid for the 10 cm, 20 cm and 30 cm diameter pools. Similarly, a function of form $y = y_0 + A exp(-kx)$, in which y_0, A and k are fitting parameters, fits well to the Sandia laboratories transmittance data (the graph on the right in Fig. 2) and the parameter k (describing the decay rate of intensity and not to be confused with the k in function $g(k)$) appears to have a value of about 0.83 mm^{-1}, which is comparable to the values in the order of 0.5 mm^{-1} obtained in [4] when the very first datapoints with the lowest depth were ignored in the fitting by the authors of that article.

In Fig. 3, the results of single-ray, two-flux and exponential integral calculations are compared to the 20 cm and 30 cm curve fits given in [4]. When normalized to transmittances beginning from 1 at 0 mm, the curve fits describing the heat flux in 10, 20 and 30 cm pool fires are almost identical but distinguishable, which makes the two plots in Fig. 3 practically the same and is an indication that the relative amounts of wavelengths in the flame spectrum remain almost the same when the pool radius is multiplied by a small integer.

IOP Conf. Series: Journal of Physics: Conf. Series **1107** (2018) 042002 doi:10.1088/1742-6596/1107/4/042002

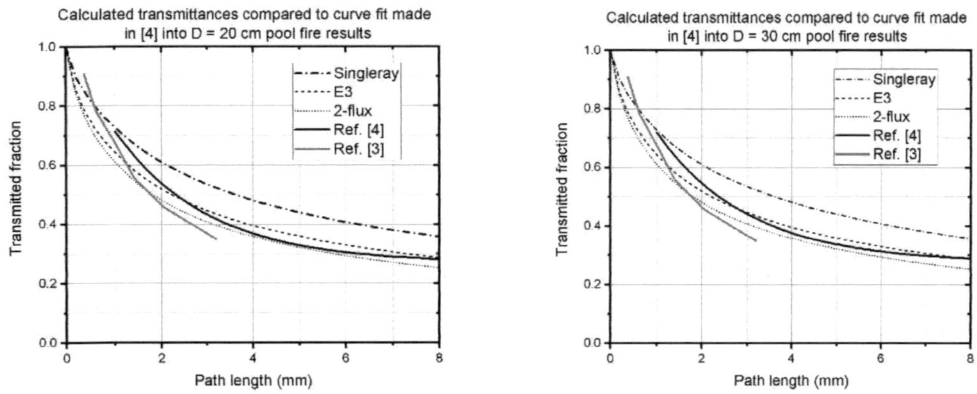

Fig 3: the calculated thermal radiation transmittances at different depths under the surface of a burning n-heptane pool, compared to curve fits (solid line) made in [4] to experimental heat flux data and to the results of calculation in [3].

When the k-distribution method of Eq. 7 is used for the calculation with single-ray Lambert-Beer model, the result is very similar to the result of Eq. 3 as seen in Fig. 4, and is expected to become identical when the k-distribution function $g(k)$ is calculated with high enough resolution in the numerical integration. Also shown in Fig. 4 is the result of the integration when done with a Gauss-Legendre quadrature using only 3 quadrature points instead of adding quadrature points until convergence.

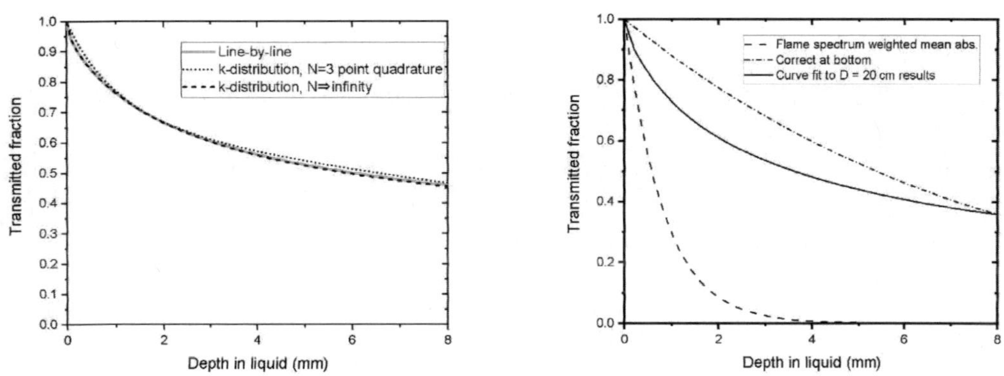

Fig 4: Left: the results of calculating the single-ray transmittance of flame radiation through n-heptane, done both with k-distribution and Eq. 3 ("Line by line"). The result of a Gauss-Legendre quadrature with only 3 nodes is also included.
Right: Experimental results in [4] compared to two different mean absorption coefficient calculations.

Finally, the calculations were performed using two alternative mean absorption coefficients, revealing the problems of grey-medium calculations. The first coefficient is the flame spectrum weighted mean absorption coefficient of n-heptane, and the result of Eq. 3 was compared to a simple exponential decay using that absorption coefficient. Numerical integration gave a value of k_{mean} = 1230 m^{-1}, and the corresponding graph is shown with dashed line in the right side of Fig. 4. From the image it is apparent that the absorption by the liquid cannot be described by a single absorption coefficient, and the result of such an approximation differs from the accurate result by a factor of 10 or more. Another way to define a mean absorption coefficient is to choose the k in the function exp($-ks$) so that it produces correct transmittance at the bottom (s = 8 mm) of the pool. This kind of calculation gives the coefficient a value of about 158 m^{-1}, and the transmitted fraction decreases almost linearly when nearing the bottom as seen drawn with a dashdot line on the right side of Fig. 4, which is clearly unrealistic.

CONCLUSIONS

In the results presented in this article, it is observed that the transmitted fraction of heat flux at different depths in the liquid phase of a pool fire can be calculated with reasonable accuracy from empirical flame spectrum and absorption coefficient data, and the effect of the size of the fire seems to be quite accurately taken in account by a simple multiplicative constant in the incoming intensity/flux. The calculations done with the two-flux method or with the exact plane parallel RTE solution are expectedly more similar to the experimental results than the single-ray calculation.

The most useful finding presented in this paper is that the results can also be obtained with similar accuracy by a k-distribution method in a similar manner to what is used for gas phase radiation calculations, and that the integrations in a k-distribution calculation can be done accurately with numerical quadrature involving a very small number of arithmetic operations, leading to fast computations. This kind of model can be implemented in fire simulation models attempting to predict liquid pool temperature distributions and evaporation.

REFERENCES

[1] A. Tuntomo, C. L. Tien, S. H. Park, "Optical constants of liquid hydrocarbon fuels", Combust, Sci. and Tech., 1992, Vol. 84, pp. 133-140. DOI: 10.1080/00102209208951849

[2] NIST Chemistry WebBook, SRD 69, url: http://webbook.nist.gov/cgi?ID=C142825&Type=IR-SPEC&Index=2#IR-SPEC (11 Jan 2018).

[3] J. M. Suo-Anttila, T. K. Blanchat, A. J. Ricks, A. L. Brown, "Characterization of thermal radiation spectra in 2 m pool fires", Proceedings of the Combustion Institute 32 (2009) 2567-2574. DOI: 10.1016/j.proci.2008.06.044

[4] J. Zhao, H. Huang, H. Wang, Z. Zhao, Q. Liu, Y. Li, "Experimental study on burning behaviors and thermal radiative penetration of thin-layer burning", J. Therm. Anal. Calorim. 130 (2017) 1153-1162. DOI: 10.1007/s10973-017-6367-4

[5] T. Sikanen, S. Hostikka, "Modeling and simulation of liquid pool fires with in-depth radiation absorption and heat transfer", Fire Safety Journal 80 (2016) 95-109. DOI: 10.1016/j.firesaf.2016.01.002

[6] D. Drysdale, *An introduction to fire dynamics, 3rd ed.*, (2011), pp. 184-187. DOI: 10.1002/9781119975465

[7] M. I. Fernandez, "Modelling spreading, vaporization and dissolution of multi-component pools", Ph.D. thesis, University College London (2013).

[8] M. F. Modest, "Radiative Heat Transfer", Elsevier Science, 2003, pp. 454-479 and 581-637.

Towards the prediction of the convective and the radiative heat fluxes in turbulent buoyant flames

Germain BOYER[1]
[1]*PSN-RES/SA2I, Institut de Radioprotection et de Sûreté Nucléaire (IRSN), France*
germain.boyer@irsn.fr

ABSTRACT

The present paper aims at describing the flow dynamics of turbulent buoyant flames encountered in gaseous or liquid pool fires or horizontally oriented solid materials. Indeed, the predictive simulations of liquid pools or solid fuel fires requires the prediction of the thermal stress received by the evaporating liquid or the pyrolysing material, and therefore a fine description of the surrounding flame. Yet, reactive Large Eddy Simulation is a powerful tool in fluid mechanics to correctly simulate turbulent flows such as large-scale buoyant flames, so long as the discretization and the numerical schemes allow to resolve the large scale turbulent structures which determine the global flame shape.

Thus, this study is divided into two steps. In the first step, Large Eddy Simulations of the SANDIA FLAME methane pool fire are performed using the CALIF^{3}S-ISIS software, for different grid refinements and convective schemes. The ability of the different numerical choices in correctly describing this flame is compared in terms of velocity variance-to-average ratio, puffing frequency and turbulent structures resolution. In the second step, the flame above a 40×40 cm^2 PMMA slab is simulated with the best numerical parameters found in the first step. The simulation of the flame is decoupled from pyrolysis aspects assuming by imposing the fuel mass flow rate observed at a particular stage of the experiment. The average temperature axial profile and the transverse profile of convective, radiative and total heat flux at the sample surface are compared to their experimental counterparts.

It is shown that, at least for large scale pool fires such as the SANDIA methane flame, the increasing the mesh refinement permits to reach a good balance between the average velocity and the turbulent kinetic energy and to correctly recover the characteristic flame puffing frequency, whereas using non-dissipative centred schemes allows to catch a wider range of turbulent scales in the flowfield. Unfortunately, for smaller fuel sources such as the 40×40 cm^2 PMMA slab, the present Large Eddy Simulations do not exhibit fully developed turbulence in most of the combustion zone, in spite of the use of centred schemes and of a fine grid. This uncertainty on the behaviour of the turbulence in such a configuration may be responsible for the observed anomalies, namely the overprediction on the average temperature and the radiative-to-convective heat flux ratio. Further experimental investigations would be useful to better understand the behaviour of turbulence on such mid-scale fires and prescribe relevant numerical model modifications.

KEYWORDS:

Fluid dynamics; Large Eddy Simulation; heat transfer; turbulent buoyant flames; cell size convergence; convective schemes; radiative/convective heat transfer balance

INTRODUCTION

In computational fire dynamics, the fuel load mass loss rate and therefore the total heat release rate are often provided by correlative approaches [1]. However, for more complex fuels, in particular organic materials, predictive approaches are required and two main issues arise: the prediction of the material thermal degradation rate, and the assessment of the thermal stress received by the material surface [2]. The present paper deals with this second issue. More precisely, the assessment of the feedback flux applied by the flame on the fuel load includes radiative and convective effects. The radiative effects mainly depend on the steady flame structure (height, temperature average and variance, optical flame properties). In this respect, the radiative losses can be directly fixed by imposing a flame radiative fraction in the heat transfer equation [3]. But in a predictive approach, special attention must be paid on the combustion model (including soot generation and oxidation) and the resolution of the Radiative Transfer Equation (including the description of the radiative absorption). In the same way, the convective heat exchange between the flame and the fuel interface is often modelled using a heat exchange coefficient depending on the temperature gap between the flame and the surface. But in a predictive approach, the correct evaluation of the convective exchange depends on the relevance of the turbulence modelling and/or simulation. In particular, numerical and experimental studies of the flow dynamics in buoyant turbulent flames [4,5] have highlighted the interaction between several instability mechanisms which trigger the turbulence in these buoyant flows, namely a main puffing phenomenon of well characterized frequency [6] due to the alternate accumulation and ejection of the light injected fluid, and Rayleigh-Taylor and Kelvin-Helmholtz instabilities occurring very close to the light fluid injection which favours the turbulent dissipation and mixing in the flame and the plume. All these phenomena are unlikely to be simulated by RANS approaches even if they have often be used for industrial applications [7]. Conversely, Direct Numerical Simulations of such large-scale configurations is out of reach in terms of CPU capabilities. In this respect, similarly with previous attempts [6,8], Large Eddy Simulations (LES) of these flows can be tentatively performed to accurately evaluate the flame structure related to a restricted period of a fire scenario. Implicitly, one assumes that the high-frequency dynamics of the flame is decoupled to the long-term variations of the flow conditions. Simulations with prescribed, constant injection rate can therefore be performed during a physical duration that is large enough to reach statistic convergence, but that remains smaller than the characteristic time of the fire scenario. Two experimental cases will be simulated with the CALIF^{3}S-ISIS fire simulation software developed by French "Institut de Radioprotection et de Sûreté Nucléaire" (IRSN) [9]: the 1 m diameter methane pool fire performed in the SANDIA FLAME facility [5], which will be used for both physical and numerical model validation; and the pyrolysis experiments performed on 40x40 cm^2 PMMA slabs in open atmosphere by Kacem *et. al.* [10]. Particular attention will be paid on the flame temperature profiles and the convective and radiative contributions to the flame feedback flux.

CFD MODEL DESCRIPTION

Physical model

The detail of the models used in the present simulations is available in the CALIF^{3}S-ISIS physical documentation [9]. For both simulated experimental cases, the Favre-filtered mass ($\bar{\rho}$), momentum ($\bar{\rho}\tilde{\boldsymbol{u}}$), mixture fraction $\tilde{Z}$, fuel mass fraction $\tilde{Y}_F$ and total enthalpy $\tilde{h}$ conservation equations are solved for. The momentum conservation equation is written in Low-Mach number formulation, which allows to uncouple the thermodynamic pressure (which is constant in open flows) and the dynamic pressure, and to only consider density variations related to temperature and mass fractions variations. The subgrid scale momentum stress tensor $\boldsymbol{\tau}_{sgs}$which appears in the filtered Navier-Stokes equations is modelled using a Boussinesq assumption and depends on the filtered strain rate tensor and the subgrid scale turbulent viscosity μ_s. μ_s is determined by a Smagorinsky model which general formulation is $\mu_s = \bar{\rho}(C_s\,\Delta)^2||\tilde{S}||$, where Δ is the characteristic cell size and C_s a non-dimensional parameter. Instead of setting a constant value of C_s as in the original Smagorinsky model, the dynamic version based on the Germano identity [11] is used to allow the decay of C_s in laminar and transitional zones, which is particularly important near the injection of fuel. Note that the maximum allowed value of C_s is 0.12. The subgrid scale scalar turbulent fluxes $\boldsymbol{\tau}_{sgs,\phi}$ are also modelled using a Boussinesq assumption and a simple gradient diffusion hypothesis. The ratio between dynamic and scalar mixing is fixed using a turbulent Prandtl or Schmidt number σ_ϕ, which is set equal to 0.5.

The combustion of a single fuel, namely methane or Methyl Metacrylate (MMA) is modelled using a single step combustion reaction which reads $F + \nu\, O_2 \rightarrow \alpha\, CO_2 + \beta\, H_2O + \sigma\, S$ in mole formulation. Here ν, α, β

and σ are the molar stoichiometric coefficients. In particular, the soot production is accounted for here; the mole soot yield σ is related to the mass soot yield $y_s = \sigma W_C / W_F$ characterized for a large range of fuels in Ref. [12]. The multicomponent system including the transport of fuel, oxidizer and combustion products mass fractions is reduced to the mixture fraction/fuel mass fraction formulation the under the unit Lewis number assumption. The combustion rate $\dot{\omega}_f$ is described by the EDM approach [13]: $\dot{\omega}_f = -\frac{\mu_s}{\sigma_t C_{LES} \Delta^2} \min\left(\widetilde{Y_F}, \frac{Y_{O_2}}{s}\right)$, where σ_t is the turbulent Schmidt number, C_{LES} is a constant which default value is 0.1 and $s = \nu W_{O_2}/W_F$ is the mass stoichiometric ratio (4 for methane and 8.2 for MMA).

The radiative losses are accounted for in the enthalpy conservation equation by solving the Radiative Transfer Equation (RTE). A Weighted Sum of Grey Gases Model (WSGGM) is used for the gas phase radiative absorption κ_G. Two methods are proposed: for the SANDIA methane flame, the P1 approximation of the RTE is chosen because the radiative transfer is not expected to have a major influence on the flame dynamics; the low computational cost of this method therefore allows using relatively large meshes. For the case of PMMA slabs, the Finite Volume Method is used to solve for the RTE, so as to correctly evaluate the radiative transfer, in particular near the sample surface. Considering the large computational cost of this method, it is solved every 20 time steps. Soot formation is accounted for in the flame radiative properties: the molar soot yield is $\sigma = 0.185$, and the additional radiative absorption κ_S due to the soot reads $\kappa_S = A\tilde{T} f_v$ where f_v is the soot volume fraction and $A = 1264$ m^{-1}K^{-1}.

Numerical model

The integration of the system of equations depicted above is performed using a fractional time step algorithm [14]: the scalar equations are solved for with a backward Euler method as time discretization. Then, the density is updated accounting for its dependence with respect to the updated scalars. Finally, the Navier-Stokes equations are solved for using a pressure correction technique and a second order Crank-Nicolson time discretization. All the above conservation equations are discretized using a Marker-And-Cell staggered discretization for the Navier-Stokes problem and the scalar transport equations. Two couples of space discretisations for scalar/Navier-Stokes equations are considered: (Sch1) MUSCL/centred schemes and (Sch2) MUSCL implicit/QUICK implicit schemes [15]. For (Sch1), the CFL number based on a characteristic velocity $U_{max} = 5\ m/s$ and the minimum cell size δ_{min} must remain lower than 0.1 because of the instability of the explicit MUSCL scheme and the centered scheme. For (Sch2), the constraint CFL=1 is imposed even if the schemes are implicit and thus stable. The second-order centred scheme used in (Sch1) is assumed to be conservative with respect of the kinetic energy balance for the Navier-Stokes equations, contrary to the upwind third order QUICK implicit scheme used in (Sch2) which is dissipative.

SIMULATION OF THE SANDIA METHANE POOL FIRE

The first experimental test case considered here is the 1 m diameter methane pool fire carried out in the SANDIA FLAME facility [5]. The computational domain represents the complete 6.2 m cubic central chamber, including the 2 m diameter flat plate located 2.45 m above the floor, the cylindrical shield surrounding an annular air inlet, the trapezoidal ceiling and the chimney outlet. The cells of the first considered cuboid MAC meshing (M1) are $\delta_{min} = 2.5$ cm cubic cells near the fuel inlet and are enlarged in the three directions with a maximum cell size of $\delta_{max} = 20$ cm near the external walls, the air supply inlet, the roof and the chimney outlet. A second refined mesh (M2) with $\delta_{min} = 1.25$ cm has also been used. In these conditions, the time step is of 8.33×10^{-4}s for (M1/Sch1), 8.33×10^{-3}s for (M1/Sch2), 4.16×10^{-4}s for (M2/Sch1), 4.16×10^{-3}s for (M2/Sch2). For the boundary conditions, a surface inlet fuel mass flux of 0.053 kg/m^2/s is imposed, which corresponds to a heat release rate of 2.07 MW. The corresponding injection velocity, equal to 0.097 m/s, is not disturbed by synthetic turbulence injection. The inlet temperature is of 286 K. Pure methane is injected with $Y_F = 1$. Note that for this fuel the soot conversion factor can be set equal to 0. Fresh air is injected at the annular inlet with a fixed velocity of 0.327 m/s at the temperature of 290 K. The walls temperature is fixed to an average value of 302 K.

The axial velocity and kinetic energy radial profiles depicted in Fig. 1 show that mesh M2 and convective schemes Sch2 always overestimate the average velocity and the turbulent kinetic energy compared to mesh M1 and schemes Sch1, respectively. For the axial velocity, the gap to experimental data remains in the uncertainty range of 20% [5] for all the simulations. For the turbulent kinetic energy, the use of mesh M2 is required to reach radial profiles which remain within the 30% experimental uncertainty range.

IOP Conf. Series: Journal of Physics: Conf. Series **1107** (2018) 042003 doi:10.1088/1742-6596/1107/4/042003

The main "puffing" flame oscillatory behaviour [5] is recovered in these simulations. Quantitatively, considering the High Resolution analysis [16] of the axial velocity recordings at a 0.6 m height on the centreline, the observed puffing frequencies are of 1.25 Hz, 1.3 Hz, 1.45 Hz and 1.5 Hz for M1/Sch1, M1/Sch2, M2/Sch1 and M2/Sch2 respectively, with 20% variations around the average value. These values are underestimated for mesh M1 and are closer from the theoretical frequency of 1.5 Hz for M2; they are slightly underestimated compared to the experimental value of 1.65 Hz. In this respect, the minimal cell size has a larger influence on the puffing frequency than the convective schemes combination. On the contrary, the snapshots of instantaneous temperature depicted in Fig. 2 show that schemes combinations Sch1 and Sch2 exhibit different turbulent structures: Sch2 shows larger, laminar-like fluctuations due to large-scale, toroidal instabilities, whereas Sch1 shows fully turbulent fluctuations which partially hide these characteristic toroidal structures.

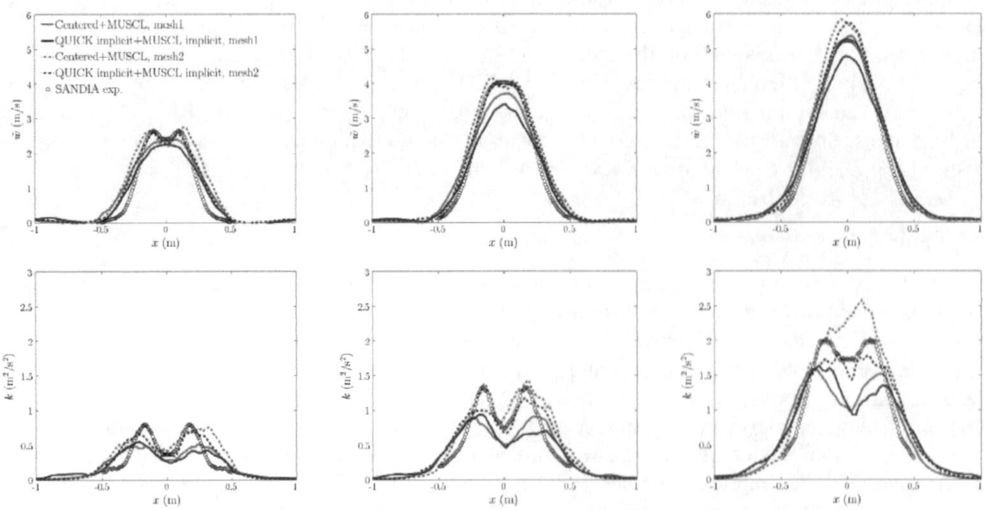

Fig. 1. Average axial velocity and resolved turbulent kinetic energy radial profiles at heights 0.3 m (left), 0.5 m (center) and 0.9 m (right) : comparison between simulations (cases M1/Sch1, M1/Sch2, M2/Sch1 and M2/Sch2) and the SANDIA FLAME experimental data.

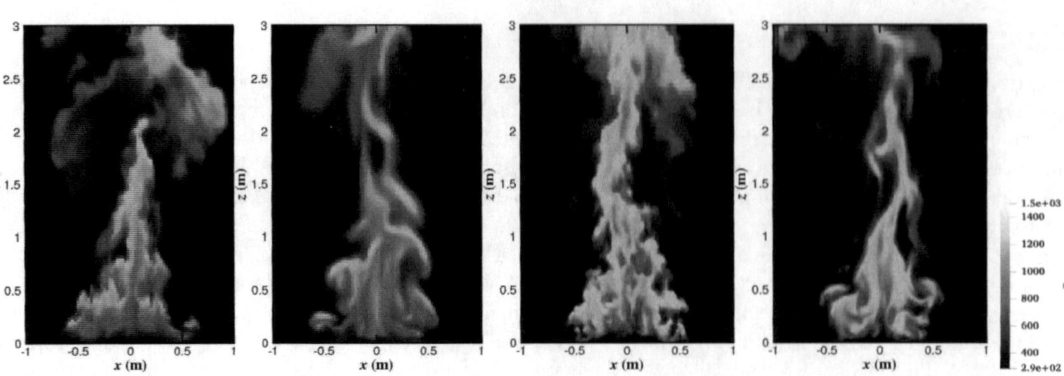

Fig. 2. Temperature fields in the plane $y = 0$ for the simulation cases M1/Sch1 (t = 10.62 s), M1/Sch2 (t = 10.25 s), M2/Sch1 (t = 10.87 s) and M2/Sch2 (t = 14s) - from left to right.

SIMULATION OF THE FLAME ABOVE A PMMA SLAB

Here the simulation of the flame above the square 40 × 40 cm^2 PMMA slab studied by Kacem *et. al.* [8] is presented. Convective schemes (Sch1) are used, the minimal and maximal cell sizes are $\delta_{min} = 0.5$ cm and $\delta_{max} = 10$ cm. A similar mesh as depicted above is used, but this time the external faces are open-air boundaries. The time step is chosen to ensure that the maximum average velocity-based CFL number remains lower than 0.05. A fixed mass flow rate of 0.01 $kg/m^2/s$ is imposed at the fuel inlet, which corresponds to the

IOP Conf. Series: Journal of Physics: Conf. Series **1107** (2018) 042003 doi:10.1088/1742-6596/1107/4/042003

lower limit of the experimentally measured value. This time, 200 random Fourier modes [17] are introduced as inlet velocity disturbances to trigger turbulence near the fuel injection. The simulation is carried out up to 15 s of physical time; statistic averages are performed on the time interval $[5, 15]$ s.

In Fig. 3, the axial average temperature profile is compared to the experimental thermocouple measurements and the spatial evolution of the received convective, radiative and total heat flux over the sample width are depicted. Obviously the simulation overpredicts the temperature in the combustion zone and recovers the correct far field decay after 1 m height. Accordingly, the radiative and total heat flux are underestimated compared to the experimental measurements at the middle of the sample (40 kW/m^2 and 57 kW/m^2 numerically against 50±6 kW/m^2 and 63 ±4.5 kW/m^2). The convective part, defined in the simulations as the sum of the conductive and the turbulent heat flux $\left(\lambda + \frac{\mu_t}{\mathrm{Pr}_t}\right)\nabla\tilde{T} \cdot \boldsymbol{n}$, is overestimated (17 kW/m^2 against 12 kW/m^2). Yet, it can be shown that the buoyant flame is characterized by a puffing frequency ranging between 1.5 and 1.9 Hz. Accordingly, the description of the flame puffing over a whole period of Fig. 4 shows that the hot gases accumulation which leads to the flame puffing is triggered by a pure Rayleigh-Taylor-like toroidal structure. This structure, which is more or less strongly dissipated by developed turbulence in the case of the SANDIA methane flame, remains coherent up to z =0.3 m. This shows that, in spite of the introduction of inlet random Fourier modes on the injected velocity, turbulence is not triggered in this zone. However, further experimental data (high-frequency velocity measurements or flame visualisations) would be necessary to know the actual turbulence behaviour in this zone. This lack of turbulent mixing possibly implies that combustion occurs too far from the sample and that the flame height is too large. Yet, a smaller flame height may induce a larger radiative heat flux received by the sample. It may also induce a lack of mixing with the external fresh air, and therefore explain the overestimation of the convective flux.

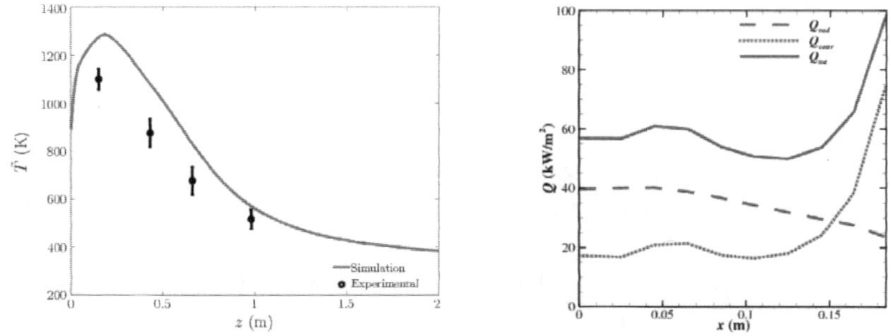

Fig. 3. Left: Axial profile of the simulated average temperature and comparison to the experimental measures. Right: Transverse profile of the simulated convective, radiative and total flux received by the sample surface

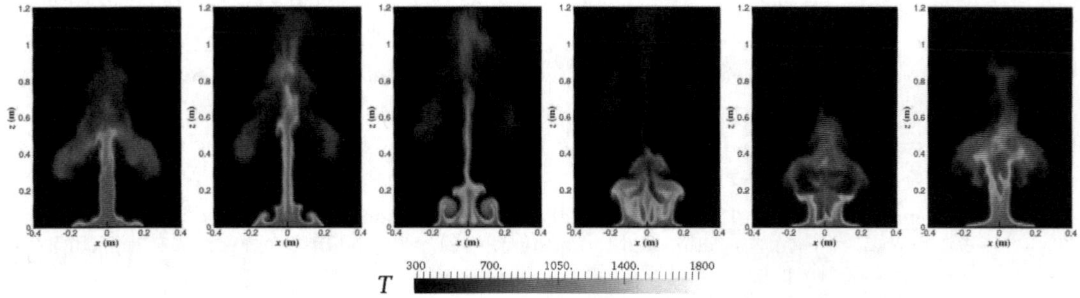

Fig. 4. Temperature fields in the plane $y = 0$ from $t = 9.9$ s (left) to $t = 10.4$ s

CONCLUSION

The Large Eddy Simulations presented in this paper have shown the ability of the CALIF^{3}S-ISIS CFD software to catch most of the phenomena occurring in turbulent buoyant flames, namely the balance between mean and turbulent kinetic energy, the flame puffing and the characteristic instabilities of such flows. In particular, for large pool fires such as the SANDIA methane flame, using upwind high-order schemes allows catching these phenomena, but the non-dissipative centred, second-order centred convective scheme for the Navier-Stokes equations is a better choice to catch more turbulent structures. If the average flame structure is recovered with 40 points clustered in the pool diameter, twice more points are required to correctly predict the turbulent kinetic energy. For smaller buoyant flames such as the one occurring above a square 40 × 40 cm^2 PMMA slab, using both 80 points in the slab width and non-dissipative schemes does not allow recovering a fully turbulent flame, which induces an average temperature and a slight radiative heat flux underprediction, and a slight convective flux overprediction. As a consequence, the accurate evaluation of the flame flux received by a pyrolysing sample requires more attention on the behaviour of the turbulence in the near-sample flame zone. From a numerical point of view, imposing inlet temperature fluctuations instead of the velocity fluctuations may be a better mean to trigger turbulence in this zone. Including buoyancy effects in the subgrid scale viscosity model may also help recovering the correct flame shape. But, so as to validate such models, accurate experimental data on smaller flames than those tested in the SANDIA FLAME facility would be useful.

REFERENCES

[1] V. Babrauskas. Estimating large pool fire burning rates. Fire Technology, 19:251–261, 1983.

[2] G. Boyer. Fully coupled CFD simulation of the pyrolysis of non-charring polymers: A predictive approach. International Symposium of Fire Safety Science, volume 12, pages 208–217, 2017.

[3] G. H. Markstein. Scaling of radiative characteristics of turbulent diffusion flames. Symposium International on Combustion, volume 16, pages 1407–1419, 1976.

[4] E. J. Weckman and Sobiesiak. The oscillatory behaviour of medium-scale pool fires. In Combustion Institute, editor, Symp. (Int) Combust., volume 22, pages 1299–1310, 1988.

[5] S. R. Tieszen, T. J. O' Hern, R. W. Schefer, E. J. Weckman, and T. K. Blanchat. Experimental study of the flow field in and around a meter diameter methane fire. Combust. Flame, 129: 378–391, 2002.

[6] B. M. Cetegen and K. D. Kasper. Experiments on the oscillatory behavior of buoyant plumes of helium and helium-air mixtures. Phys. Fluids, 8(11):2974–2964, 1996.

[7] S. Suard, C. Lapuerta, A. Kaiss and B. Porterie, Sensitivity Analysis of a Fire Field Model in the Case of a Large-Scale Compartment Fire Scenario, Numerical Heat Transfer Applications, 63(12):879-905 June 2013

[8] G. Maragkos and B. Merci, Large Eddy Simulations of CH4 Fire Plumes, Flow Turbulence Combust. 99(1), February 2017

[9] ISIS development team. ISIS 5.0.0: Physical modelling. Technical report, IRSN/PSN-RES/SA2I/LIE, 2017. URL https://gforge.irsn.fr/gf/project/isis/docman/Physicalmodelling.

[10] A. Kacem, M. Mense, Y. Pizzo, G. Boyer, S. Suard, P. Boulet, G. Parent, and B. Porterie. A fully coupled fluid/solid model for open air combustion of horizontally-oriented PMMA samples. Combust. Flame, 170:135–147, 2016.

[11] D. K. Lilly. A proposed modification of the Germano subgrid-scale closure method. Phy. Fluids, 4(3):633–635, 1992.

[12] A. Tewarson. Generation of Heat and Chemical Compounds in Fires. National fire Proctection Association, fourth edition, 2008.

[13] B. F. Magnussen and B. H. Hjertager. On mathematical modeling of turbulent combustion with special emphasis on soot formation and combustion. In Combustion Institute, editor, Symp. (Int) Combust, volume 16, pages 719–729, 1976.

[14] F. Boyer, F. Dardalhon, C. Lapuerta, and J.-C. Latché. Stability of a Crank-Nicolson pressure correction scheme based on staggered discretizations. . Int. J. Numer. Methods Fluids, 2013.

[15] L. Piar, F. Babik, R. Herbin, and J.-C. Latché. A formally second-order cell centred scheme for convection-diffusion equations on general grids. Int. J. Numer. Methods Fluids, 71(7):873–890, 2013.

[16] R. Badeau. Méthodes à haute résolution pour l'estimation et le et le suivi de sinusoïdes modulées. Application aux signaux de musique. PhD thesis, ENST, 2005.

[17] N. Jarrin, S. Benhamadouche, D. Laurence, and R. Prosser. A synthetic-eddy-method for generating inflow conditions for large-eddy simulations. Int. J. Heat Fluid Flow, 27:585–593, 2006.

Assessment of an insulating air layer model of façade external system: contribution to fire simulation of facade performance fire test

Virginie DREAN[1], Renaud SCHILLINGER[2] and Gildas AUGUIN[1]

[1] *Efectis, Research & Innovation Direction, France*
[2] *Efectis, Laboratory, Fire resistance Testing Division, France*

virginie.drean@efectis.com

ABSTRACT

Ventilated façades as a part of external thermal insulation are useful systems when applied to building design, especially in bioclimatic construction. In the context of the fire safety engineering, well taking into account the air layer arranged in such systems is of importance to assess the fire behavior of a building. In fire engineering studies, CFD codes as FDS (Fire Dynamics Simulator) are widely used. However, mesh sizes used in simulation are generally greater than 100 mm. Thus, the air layer cannot be modeled accurately because of typical thickness lower than mesh size.

In the present study, a method based on an analytical formulation of the thermal resistance for building components is evaluated to approach the local air layer behavior. Radiative, convective and conductive mechanisms in the layer are then accounted. The proposed method could be implemented in a numerical model when a ventilated layer must be considered through an equivalent thermal conductivity for the air layer.

CFD simulations are performed with FDS for fire exposed façade during LEPIR II performance tests where the ventilated façades with the larger air layer thicknesses are modeled. Ambient conditions in terms of flow temperature and velocity inside the air layer are evaluated in several locations of the façade.

These façade models are then modified to replace the air layer by an equivalent thermal conductivity analytically assessed. Its local internal temperature is then compared with the one predicted with CFD, using 2D thermal calculations with the FEM code SAFIR. Extrapolations for smaller thicknesses of air layers that cannot be model with FDS can then be assessed numerically.

This preliminary work shows encouraging application for air layer reserved in façades during a fire. Further development and validation are in progress to apply this method to other configurations of air layering.

KEYWORDS

Fire modeling; Performance-based design; Façade fires; Heat transfer; Fluid dynamics; CFD; thermal analysis FEM; External thermal insulation.

IOP Conf. Series: Journal of Physics: Conf. Series **1107** (2018) 042004 doi:10.1088/1742-6596/1107/4/042004

INTRODUCTION

Ventilated façades as a part of external thermal insulation are useful systems when applied to building design, especially in bioclimatic construction. Actual energetic situation encourages engineers and architects to install such a solution in buildings to potentially improve their energy performance. However, in the context of the fire safety engineering, quite well accounting the air layer reserved in such systems is important to assess the fire behavior of a building, and in particular, the fire spread through façades.
A ventilated façade is a complex multi-layered structural system. It consists, in a simplified way, of an external façade cladding, a frame structure generally made of steel, aluminum or wood profiles anchored to the wall of the building, an insulating material and an air gap between the cladding and the insulation. In a ventilated cavity, as defined in [1], with openings at top and bottom, there is a continuous stream of air through the cavity The air layer thickness is generally with a minimum of 20 mm to a maximum of 100 mm.

For combustible or weakly combustible insulation materials, further to its heating by radiation of the external cladding and by convection with the ambient conditions in the air layer, pyrolysis gases are released and can ignite, leading to a decrease of the system performances. The system with the lowest thickness of air layer is the most penalizing in case of a façade fire, the insulating material being closer of the cladding submitted to the flames. However, the low thickness of the air layer limits the inflammation of gases of pyrolysis released and transported in the layer. When the air layer presents a more important thickness, the insulating material is more distant from the cladding submitted to the external flames, and then less heated. However, the chimney effect, allowing the convection of hot gases, is more important for this configuration and the inflammation of released gases of pyrolysis is then favored. The temperatures in the air layer can also be high enough to reduce significantly the performances of the structure linking the cladding to the wall building. In particular, structural frame performances decrease at lower temperatures for aluminum or wood than for steel structures.

The building geometry is often modeled numerically in fire safety studies. CFD codes, as for example FDS [2], are widely used for this purpose in the French fire community. However, mesh sizes used in simulation are generally greater than 100 mm to ensure a reasonable calculation time. Thus, the air layer cannot be modeled accurately with only one or two cells. This study attempts to evaluate a method that could be implemented in numerical model when a ventilated layer as double-skin façade configuration must be considered. Based on an analytical formulation of thermal resistance for building components [1], the radiative, convective and conductive mechanisms in the layer are introduced through an equivalent thermal conductivity for air layer. CFD simulations are performed with FDS for fire exposed façade during LEPIR II tests [3]. Ventilated façades with larger thicknesses of 50 and 100 mm are modeled. Ambient conditions in terms of flow temperature and velocity inside the air layer are evaluated in several locations of the façade. These façade configurations are then modified to replace the air layer by the equivalent thermal conductivity assessed analytically. Its internal temperature is then compared with the one obtained with CFD using 2D calculations with the FEM thermal analysis code SAFIR [4]. Extrapolations for smaller thicknesses of air layers that cannot be model with FDS can then be assessed numerically.

EQUIVALENT MODEL FOR AIR LAYERS

Heat transfer across cavities in buildings is a complex process, made up of a combination of conduction in still air, convection from air movement within the cavity and radiation between the cavity sides. Some information is available on the air movement through wall cavities ([5] to [15]). For practical calculations of wall and air layer thermal transmittance, known as U coefficient, it is assumed in [1] that the heat transfer can be replaced by a standard thermal resistance, which depends on the direction of the heat flow. A similar method was used in [6] and [7]. The principle of the calculation method is to determine a thermal resistance R, expressed in m^2.K/W, for each thermally homogeneous layer of the system. For air layers, it is calculated in function of convective, conductive and radiative coefficients representative of flow condition in the cavity (Fig. 1). Then, these individual resistances are associated to determine the total thermal resistance of the system, expressed as a combination of resistors connected in series. The U coefficient is the inverse of the thermal resistance R. It is expressed in W/m^2.K and is representative of the thermal conductivity and the thickness of the layer.
The method applies for air gaps limited by two parallel faces, perpendicular to the direction of the heat flow, and whose emissivity are at least 0.8. The thickness of the air gap must be less than 300 mm and less than 10% of the two other dimensions (length and height) taken individually.
The formulas indicated in [1] are accounted directly to calculate the total transmittance of the air layer, without considering their derivation. However, the expression of the convective transmittance is easily

IOP Conf. Series: Journal of Physics: Conf. Series **1107** (2018) 042004 doi:10.1088/1742-6596/1107/4/042004

comparable with the one used in FDS or more generally expressed as (1) for the heat transfer coefficient depending on the local Nusselt number, or Reynolds and Prandtl numbers. The radiative transmittance (5) is close to the expression used to calculate the radiation between two parallel planes using the view factor.

$$h_{conv} = \frac{\lambda}{L} Nu = \frac{\lambda}{L} Re^{1/3} Pr^{0.8} \quad (1)$$

The conductive thermal transmittance U_{cond} is expressed in function of the air layer thickness e (m) and the air thermal conductivity λ_{air} (W/m.K) as given by the formula (3). The air layer is considered still.
Convection is practically non-existent in air layers less than 4 mm thick. Beyond 20 mm thickness for air layers, the convective regime changes from viscous to dynamic domain. The empirical formula (4) allows calculating the convective transmittance U_{conv} in this region.
The energy exchanged by radiation between two plane and parallel surfaces separated by a distance e small relative to the dimensions of the surface is independent of this distance as well as the presence of a gas (transparent medium). However, the radiative transmittance U_{ray} strongly depends on the average temperature and of the nature of the walls as indicated by the formula (5), with σ_B (W/m².K^4) the Boltzmann constant, ΔT (K) the average temperature of the walls delimiting the air layer, and ε their respective emissivity. The radiative transmittance is large compared to the conductive and convective ones.

In thin air layers, conduction, convection and radiation contribute to the total transmittance $U_{tot,air}$ (W/m².K) as expressed in (2). Thus, for a given cavity thickness e, the more $U_{tot,air}$ is high (inversely, the more $R_{tot,air}$ is low), the more the equivalent thermal conductivity of the layer is elevated.

$$U_{tot,air} = U_{cond} + U_{conv} + U_{ray}$$

$$R_{tot,air} = {}^{1}/_{U_{tot,air}} \quad (2)$$

$$U_{cond} = \frac{\lambda_{air}}{e} \quad (3)$$

$$U_{conv} = 0.73\Delta T^{1/3} \qquad (\Delta T > 5K) \quad (4)$$

$$U_{ray} = \frac{4\sigma_B\, \Delta T^3}{\frac{1}{\varepsilon_1} + \frac{1}{\varepsilon_2} - 1} \quad (5)$$

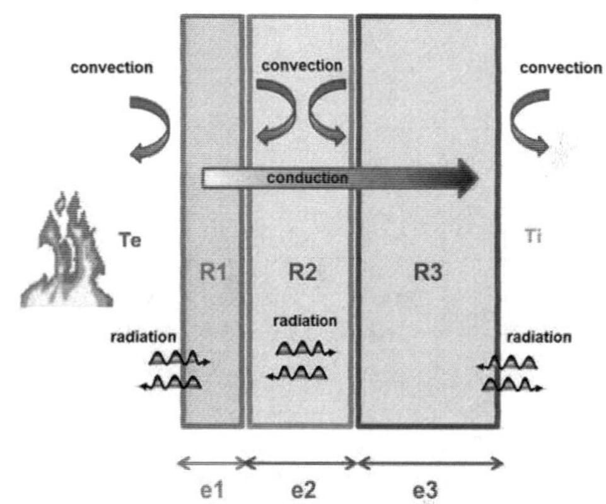

Fig. 1. Thermal resistance calculated in function of convective, conductive and radiative mechanisms

Calculating a local thermal transmittance $U_{tot,air}$ in the cavity is then representative of local flow conditions and captures the main heat transfer phenomena. In the specific case of a façade fire, the heat transfer inside the cavity is mainly driven by the external fire loads, inducing a chimney effect inside the air layer from the fire exposed bottom to the colder top. The external wind pressure coefficient is neglected in calculation. Despite the ascendant flow movement in the cavity, the cladding temperature exposed to fire and flame radiation is higher than the air gap temperature if there is no flame passage in the air cavity. Thus, the heat flow can be locally considered as horizontal.

For given temperatures in the cavity, representative of the parameter ΔT in (4), the air conductivity λ_{air} used in (2) is known. Thus, $U_{tot,air}$ can be evaluated for given flow condition assuming a horizontal heat flow. The equivalent $U_{tot,air}$ calculated for air layers with thicknesses e = 20, 50 and 100 mm are indicated in Fig. 2. The layer thickness shows reduced influence on the value of $U_{tot,air}$ in function of the air temperature.

An equivalent thermal conductivity of the air gap can be then calculated in function of the layer thickness and of the average temperature in the layer as (6).

$$\lambda_{air} = U_{tot,air} \times e \quad (6)$$

The implementation of this model in a heat transfer analysis is then possible, remembering that the thermal transmittance of the cladding is directly calculated using external thermal loads and material conductivity. The results for the equivalent thermal conductivity show that the value of λ_{air} for vertical air gaps increases, when the width *e* of the air gap increases. Same result is observed in [6], both analytically and experimentally. However, the reader should be careful in the analysis of this result and must remember that the equivalent conductivity is calculated analytically in function of a generic average temperature ΔT in the cavity. Thus, for larger air cavities, the local average temperature is generally smaller than for thin cavities, and so will be the equivalent thermal air conductivity.

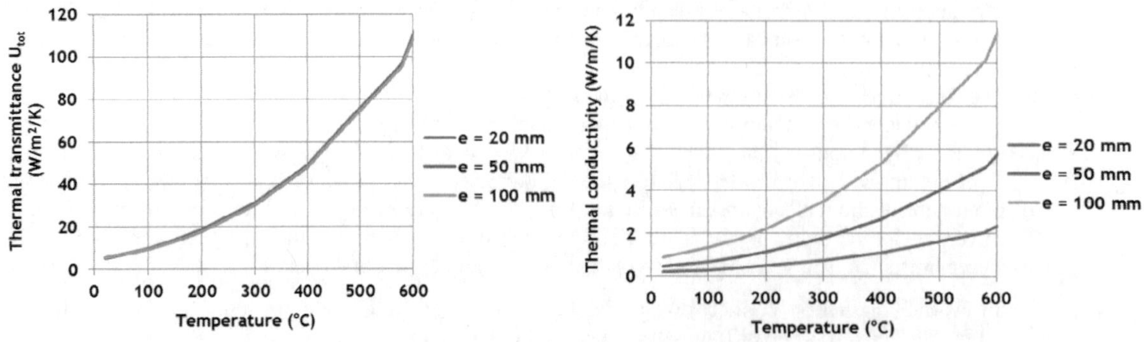

Fig. 2. Evaluation of the equivalent thermal transmittance and thermal conductivity of air gap with the average temperature for different air layer thickness and a height H of 1.5 m

NUMERICAL MODELING OF THE AIR LAYER

Thermal environment

In this study, fire exposed façade during LEPIR II [3] tests are addressed. A numerical model of the facility was validated [16] based on experimental results for two concrete calibration façades (Fig. 3 a and b). Thus, the fire source and the thermal loads imparted to the tested façade can be evaluated numerically with confidence. Simulations are performed with the CFD code FDS V6.5.2 [2]. In this study, the cladding consists in external panels made of cement fiber sheets of thickness e = 8 mm, a ventilated air layer, and a glass wool insulation with a thickness of e = 300 mm. Air layer thicknesses e = 50 mm and e = 100 mm are considered in simulation. In this paper, only results for e = 50 mm air gap are presented. In the vicinity of the façade, the mesh size is refined enough to capture the main flow phenomenon and cells with dimensions of 50 mm are used. Cell size of 12.5 mm is used for the air layer.

The thermal loads received at the exposed face of the system are then known and can be used locally to evaluate the heat transfer within the façade at different positions (Fig. 3 c, thermocouples in the air gap). Thermocouples TC 1 and TC 2 are located in the façade area directly exposed to flame outgoing from the windows of the fire room. Thermocouples TC 3 and TC 4 are rather located in the façade area exposed to radiation from external flames. Thermocouples TC 5 to TC 9 are in the façade area not impacted by flames.

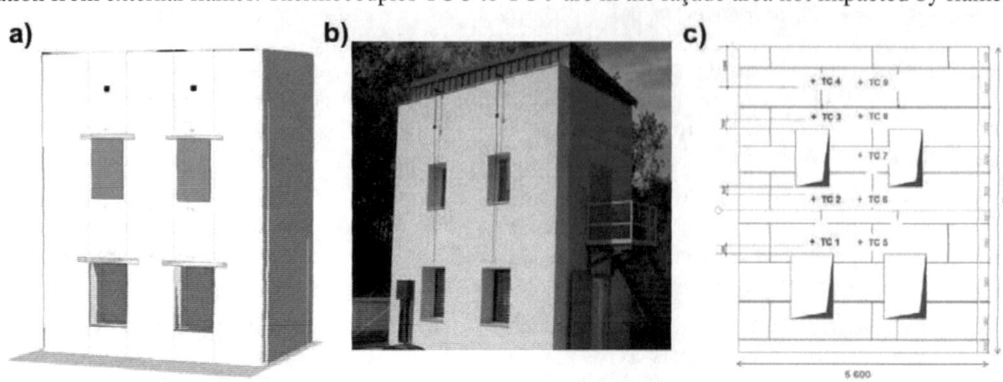

Fig. 3. a) Numerical and b) experimental LEPIR II facility – c) location of thermocouples

IOP Conf. Series: Journal of Physics: Conf. Series **1107** (2018) 042004 doi:10.1088/1742-6596/1107/4/042004

Results for the air layer

Thermal transfer calculations are performed using the FEM code SAFIR [4]. A local 2D model with 1.5 m high is considered. The façade configuration is then modified to replace the air layer by the equivalent "air material" with the equivalent thermal conductivity assessed analytically. Simulations are performed at thermocouples location using the imparted heat fluxes obtained with FDS as boundary conditions for the cladding. The temperature of the air gap is then compared with the FDS ones on Fig. 4. The temperatures evaluated with SAFIR are very close to the FDS ones for thermocouples above the fire room of the facility (TC 2, 3 and 4). However, some overestimation is observed for TC 2, located directly above the fire room: the thermal analysis model cannot account for the strong convective effects at this location. For TC locations at the center of the facility not directly exposed to flames (TC 6, 7 and 8), good agreement is observed for temperatures, excepting for TC 8. At this location, the air layer is heated by the hot air coming from below and can be hotter than the cladding, so the heat flow is no more horizontal. In the case studied, the temperatures in the air layer are locally high enough to reduce the performances of the structure linking the cladding to the wall building.

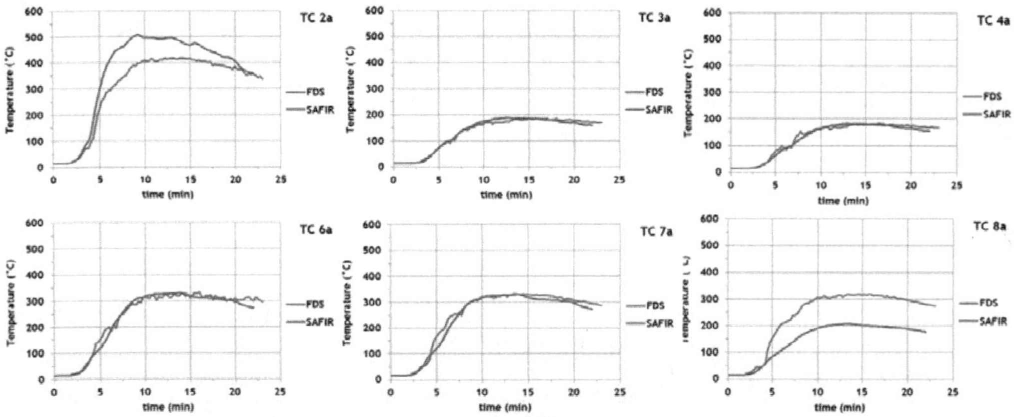

Fig. 4. Comparison of the e = 50 mm air gap temperature obtained with FDS and with SAFIR simulations with equivalent thermal conductivity of air at different TC location

Extrapolations to small air layers

Extrapolations for smaller thicknesses of air layers that cannot be model with FDS are then addressed. 2D simulation with the equivalent thermal conductivity of air are performed with SAFIR for thickness e = 20 mm and compared with temperatures obtained for air layer thickness of e = 50 mm in Fig. 5. We observe that temperatures are higher in the thinnest air layer at thermocouples location above the fire room (TC 1 and 2) and in the fire plume above the first level openings (TC 3 and 4). The system with the smallest thickness of air layer appears to be the most penalizing in the case studied, the insulating material being closer to the flames. When the air layer presents a more important thickness, the insulating material is more distant, then less heated. However, the chimney effect depending on temperature gradient is more important.

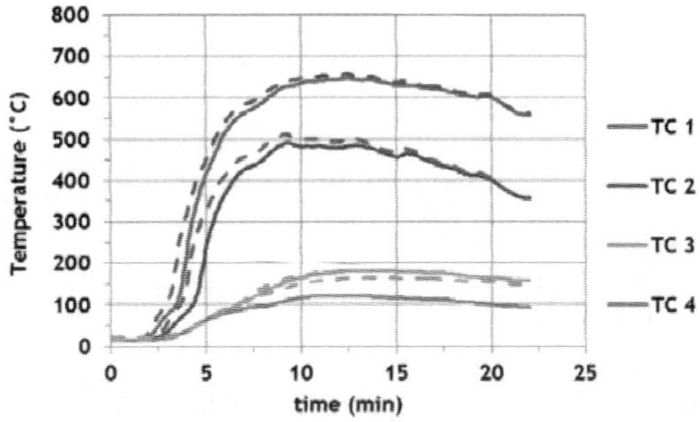

Fig. 5. Comparison of the e = 50 mm (straight line) and e = 20 mm (dotted line) air gap temperatures obtained with SAFIR simulations with equivalent thermal conductivity of air at different TC location

IOP Conf. Series: Journal of Physics: Conf. Series **1107** (2018) 042004 doi:10.1088/1742-6596/1107/4/042004

CONCLUSION

This preliminary work shows encouraging application for air layer reserved in façades during a fire. Good agreement is observed for air layer temperature evaluated with CFD simulations using FDS compared with simulations performed using the FEM thermal analysis code SAFIR using the equivalent air thermal conductivity model. This method will help to predict whether the critical temperatures for insulation and / or cladding frame are achieved or not, depending on the air gap configuration. Extrapolations to small air layers that cannot be modeled with FDS are then possible. Once fully validated, this method could be integrated in studies when ventilated systems exist and cannot be modeled accurately.

The analytical air layer model captures the main heat transfer phenomena in the cavity due mainly, in case of façade fire, to radiation between panels and convection by chimney effect. When applying the fire loads imparted to the cladding as boundary conditions, the heat transfer through the system will be representative of thermal gradients in the system and thus in the cavity. The local air temperature accounts for local chimney effect and radiation through the temperature evolution in the cavity, representative of the local convective and radiative transmittance. However, the model failed to predict the air layer temperature when strong convective effects occurred or when the heat flow is no more horizontal (ascendant hot gases hotter than cladding). The model is no more applicable if hot gases pass through the cladding, but can reasonably be used to assess local system performances in heat transfer analysis further to fire study.

Further development and validation are in progress for applying this method to other configurations of air layering. The model will be improved on more simple configurations to assess the thermal and fire behavior in ventilated systems. Specific small scale fire tests are necessary to appreciate the 3D convective phenomena in the air layer by measuring temperature, but also velocity. This will be part of the model validation process.

REFERENCES

[1] NF EN ISO 6946: 2008 - Building components and building elements – Thermal resistance and thermal transmittance – Calculation method, June 2008.
[2] Fire Dynamics Simulator, http://www.nist.gov.
[3] Arrêté du 24 mai 2010 portant approbation de diverses dispositions complétant et modifiant le règlement de sécurité contre les risques d'incendie et de panique dans les établissements recevant du public, Journal officiel de la République Française du 6 juillet 2010, texte 31.
[4] SAFIR, version 2011, Université de Liège, www.facsa.ulg.ac.be/cms/c_1584029/en/safir.
[5] Livkiss K. , Svensson S., Husted B., Van Hees P. (2018) Flame Heights and Heat Transfer in Façade System Ventilation Cavities, Fire Technology, 54, 689–713, 2018
[6] Šadauskienė J., Buska A., Burlingis A., Bliūdžius R., Gailius A. (2009) The Effect Of Vertical Air Gaps To Thermal Transmittance Of Horizontal Thermal Insulating Layer, Journal of Civil Engineering and Management 15(3):309-315, 2009.
[7] Katunská V., Bullová I., Špaková M. (2016) Analysis of Air Flow in the Ventilated Insulating Air Layer of the External Wall, SSP Journal Of Civil Engineering Vol. 11, Issue 2, 2016.
[8] Odewole G. A. (2011) Airflow Patterns in Ventilated Wall Cavities, Thesis, University of Manchester, 2011.
[9] Sanders C.H. (2005) Ventilation of the Cavities in Timber Framed Walls, Scottish Building Standards Agency 2005.
[10] Gudrum C. and Rode C. (2004) Moisture transport by convection in lightweight exterior facades. Proceedings of Conference 'Thermal 9' Clearwater beach, December 2004.
[11] Schumacher C., (2003), Ventilation drying of wall systems. Second International Building Physics Conference, Leuven 2003.
[12] Straube J. et al. (2004), Field studies of ventilation drying. Proceedings of Conference 'Thermal 9' Clearwater beach, December 2004.
[13] Kakela P. (2004), Drying rate of timber framed external wall structures in Nordic Climate Proceedings of Conference 'Thermal 9' Clearwater beach, December 2004.
[14] Eckert E. R., Carlson W. O. (1961), Natural convection in an air layer enclosed between two vertical plates with different temperatures, Int. J. Heat Mass Transfer, vol. 2, pp 106-120, 1961.
[15] Gloriant F. (2014) Caractérisation et modélisation d'une fenêtre pariéto-dynamique à 3 vitrages. Thèse de la Faculté des Sciences appliquées, Université d'Artois, 2014.
[16] Drean V., Schilinger R., Leborgne H., Auguin G., Guillaume E. (2018) Numerical Simulation of Fire Exposed Façades Using LEPIR II Testing Facility, Fire Technology Volume 50, Number 2, 2018.

IOP Conf. Series: Journal of Physics: Conf. Series **1107** (2018) 042005 doi:10.1088/1742-6596/1107/4/042005

Study on Smoke Behavior in Under-viaduct Stations by using Scale Reduced Model.

Hideki OZUMI[1], Susumu MAFUNE[1], Osamu SUZUKI[1]
Hyun woo PARK[2], Yoshifumi OHMIYA[2]

[1]*East Japan Railway Company, Japan*
[2]*Tokyo University of Science, Japan*

h-ozumi@jreast.co.jp

ABSTRACT

In the event of fire at concourse of rail stations under elevated platforms, smoke may emit through stairways leading from the concourse to platforms as well as standard natural smoke discharge outlets decreed by the Building Standards Law. However there is little knowledge about smoke discharge from the concourse via the stairway. In this study, Experiments regarding smoke flow in the case of fire in concourse of rail stations below elevated platforms were carried out compare volume of smoke emitted through stairway opening with that emitted through standard natural smoke discharge outlets as decreed by the Building Standards Law, with a simple scale reduced model which stairway was placed at right angle to the concourse. As a result, in these experiments, relationship of smoke discharge between stair openings and standard natural smoke discharge outlet was found.

KEYWORDS

smoke control, stairway, model, smoke exhaustion, under-viaduct station, concourse

IOP Conf. Series: Journal of Physics: Conf. Series **1107** (2018) 042005 doi:10.1088/1742-6596/1107/4/042005

1. INTRODUCTION

In the event of fire in a station, appropriate smoke discharge to the outside is extremely important for safe evacuation of passengers. There are two types of smoke discharge: standard natural smoke discharge where smoke is discharged from openings and mechanical smoke discharge where smoke is discharged from openings by force using ducts and fans.

In case of fire in a station concourse of an under-viaduct station, smoke is assumed to be discharged from stairway opening as shown in Fig. 2. Such smoke discharge from stairway openings can be regarded as standard natural smoke discharge; however, stairway openings are not considered smoke outlets by the Building Standards Law (hereinafter "BSL"). Moreover, there is actually little knowledge about smoke discharge from the concourse via the stairway.

In light of this, we carried out experiments on smoke flow using a scale reduced model of rail stations under elevated platforms with a stairway at a right angle to the concourse. From the experiment results, we obtained volumes of smoke discharge from the stairway opening and standard natural smoke discharge outlets decreed by the BSL, and we evaluated the volume of smoke discharge from the stairway opening by comparing the volumes obtained.

2. PREVIOUS STUDY

We have carried out experiments on smoke flow in fire using a model of rail stations under elevated platforms with a stairway at a right angle to the concourse. The experiment results confirmed that the smoke of fire in the concourse was naturally discharged from the stairway opening to the passenger stairway, that the smoke was divided into two layers of a smoke layer and a fresh air layer, and that the smoke was discharged in the direction of the shed. From this, we can consider that experiments using that model successfully reproduced actual smoke flow.

3. EXPERIMENT OVERVIEW

In this study, we carried out experiments to evaluate the volume of smoke discharge from the stairway opening using a scale reduced model of rail stations under elevated platforms with a stairway at a right angle to the concourse as done in the past studies. From the experiment results, we sought to obtain the volume of smoke discharge from the stairway opening and from the standard natural smoke discharge outlets, and we evaluated the volume of smoke discharge from the stairway opening by comparing the

Fig.1. Image of concourse at rail stations under elevated platforms

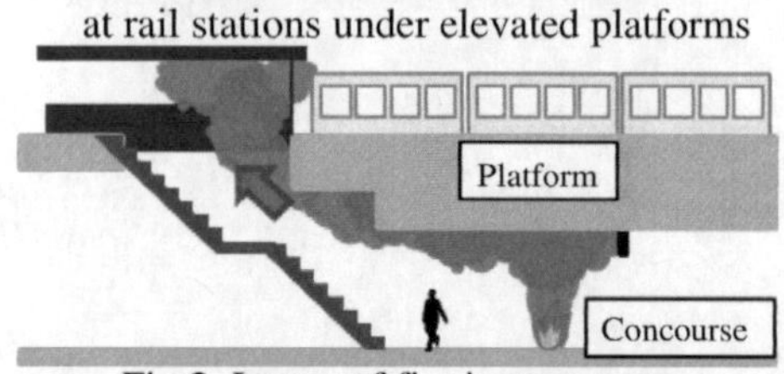

Fig.2. Image of fire in concourse at rail stations under elevated platforms

Fig.3. Smoke flow in passenger stairway (visually observed)

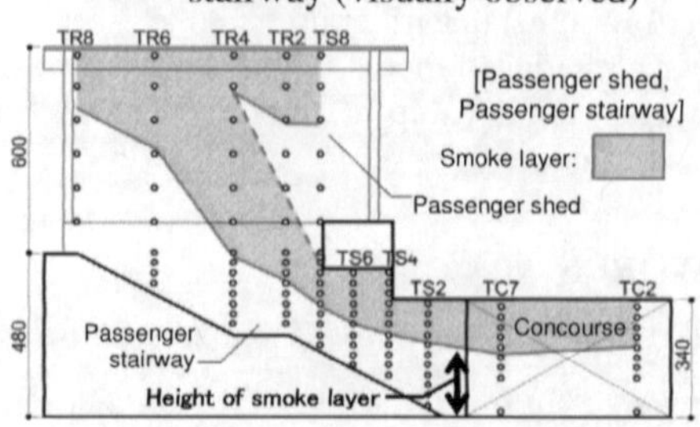

Fig.4. Smoke layer in passenger stairway calculated using heat rise value

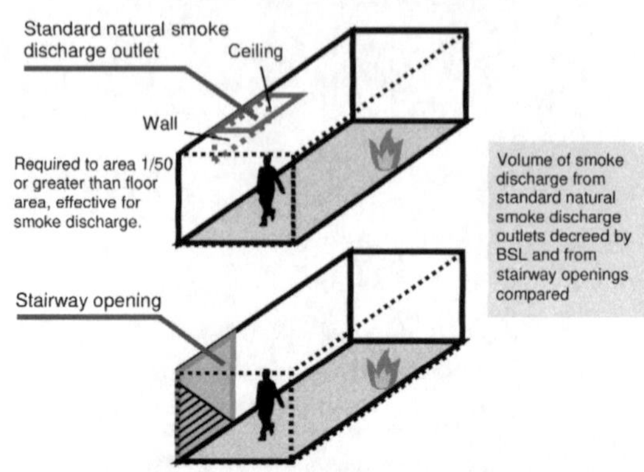

Fig.5. Verification of smoke discharge from standard natural smoke discharge outlets and stairway openings

volumes obtained.

3.1 Scaling Laws

In experiments using a model, the scaling laws need to be considered to make the behavior of the experiment subject analogous to that in the full-scale environment. In this study, we adopted a 1/10 scale model taking into account the space for the experiment and set experiment conditions to be as similar as possible in scaling laws.

Here, we determined the heat release rate, experiment time, and thermal inertia of model material as shown in the formulae 1) to 4) based on the scaling laws proposed by John H. Klote and James A. Milke in order to make the Froude numbers of the full-scale and the scale model match.

$$Q_m = Q_f(l_m/l_f)^{\frac{5}{2}} \quad \cdots ①$$

$$t_m = t_f\sqrt{l_m/l_f} \quad \cdots ②$$

$$(k\rho c)_{w,m} = (k\rho c)_{w,f}(l_m/l_f)^{\frac{9}{10}} \quad \cdots ③$$

$$T_m = T_f \quad \cdots ④$$

Here, Q: Heat release rate [W], l: Length [m], t: Time [s], T: Temperature [°C], k: Thermal conductivity [W/mK], ρ: Density [kg/m^3], c : Specific heat [J/kgK], $(k\rho c)w$: Thermal inertia [W^{2}s/m^{4}K^2], suffix f: Full-scale, suffix m: Scale model.

3.2 Experiment Models

For the experiments, we produced a model of a rail stations under elevated platforms with a stairway at a right angle to the concourse (Fig. 6a) and a model of a concourse with standard natural smoke discharge outlets and with no stairways (Fig. 6e, f) with an aim of evaluating the volume of smoke discharge from the standard natural smoke discharge outlets. As the objective of this experiment using models was to compare the volume of smoke discharge from standard natural smoke discharge outlets and the stairway opening, we closed both ends of the concourse, leaving the necessary air supply gaps to prevent smoke from being discharged from outlets other than the stairway opening and standard natural smoke discharge outlets.

3.2.1 Shape and Width of Stairway Ceiling

As the shape of the stairway ceiling in the models in the experiment, we adopted a stepped ceiling often used for actual stations (Fig. 6b) and a sloped ceiling (Fig. 6c) (Table 1).

As the minimum width of a station ceiling is stipulated as 1,500 mm or more by internal regulations, we produced four models with different stairway widths with the minimum being 150 mm.

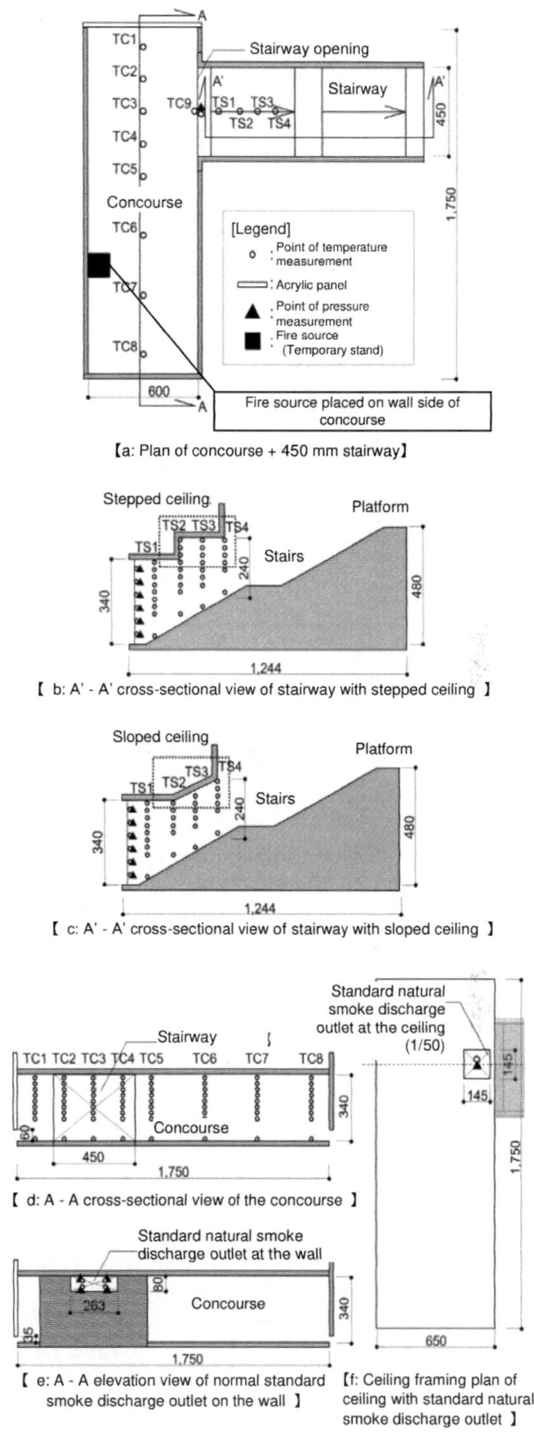

Fig.6. Diagrams of models with stairway at right angle to concourse (in mm)

3.2.2 Position and Size of Standard Natural Smoke Discharge Outlets

As the models of a concourse without stairways and with standard natural smoke discharge outlets, we produced models with the standard natural smoke discharge outlet in the ceiling and a model with that on the wall as stipulated by the BSL (Table 2). Those outlets were positioned to have almost the same distance between the stairway opening and the fire source. The standard natural smoke discharge outlet in the ceiling has a square shape. The shape of that on the wall was rectangular with the vertical height of 80 mm deemed to be effective by the BSL (800 mm in full-scale environment) and positioned to be adjacent to the ceiling (Fig. 8). As the ratio the outlet area to the floor area is stipulated to be 1/50 or more by the BSL, we set five types of ratios of standard natural smoke discharge outlet area to concourse floor area greater and less than 1/50.

3.3 Expected Fire Source

As the fire source, we assumed a temporary shop stall placed in the concourse as shown in Fig. 6. The heat release rate was set to 9.49 kW in the experiment using models based on scaling laws because the heat release rate of a stall of 1,200 mm^2 area in the full-scale environment is 3 MW. Furthermore, the duration of the fire was set to 6 minutes 19 seconds, also based on scaling laws, assuming firefighting to begin 20 minutes after the fire breakout.

3.4 Measurement Device

In this study, we measured the temperature and pressure in the models to find the volume of smoke discharge. The temperature in the models was measured using thermocouple trees that were attached to the positions TC 1 to 9 and TS 1 to 4 as shown in Fig. 6, and thermocouple trees had thermocouple thermometers at perpendicular intervals of 30 mm.

As the pressure in the models, differences in pressure on the concourse side and on the stairs side of the stairway opening were measured using a T-shape Pitot tube pressure gauge (Fig. 9). The pressure in the stairway opening was expected to vary at the upper and lower parts of the opening because high-temperature smoke was discharged along the opening ceiling. We thus decided to measure pressure difference at six points in the perpendicular direction. On the standard natural smoke discharge outlet on the wall, we attached a pressure gauge each on the concourse side and on the outside as with the stairway opening. Pressure difference in the standard natural smoke discharge outlet in the ceiling was measured using pressure gauges attached to the center of the outlet.

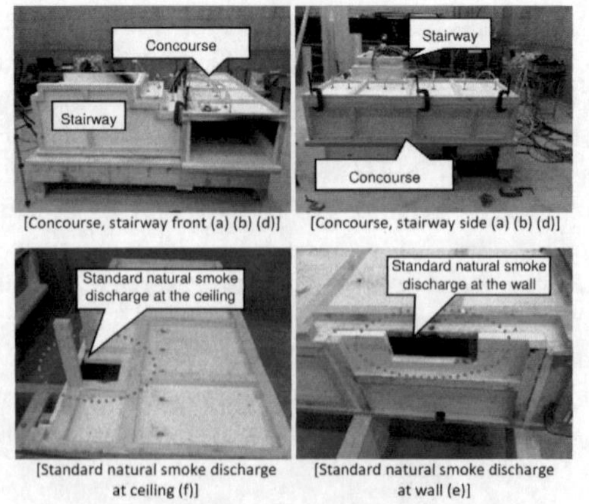

Fig. 7 1/10 Models of Concourse under Platform and Experiments

Table 1. Shapes of Models of Stairway Openings

No.	Shape of stairway ceiling	Stairway width [mm]	Ratio of opening area of stairway opening*
1	Stepped ceiling	150	1/87.5
2		300	1/43.8
3		450	1/29.2
4		600	1/21.9
5	Sloped ceiling	150	1/87.5
6		300	1/43.8
7		450	1/29.2
8		600	1/21.9

*80 mm downward from top considered to be effective.

Table 2. Shapes of Models of Natural Smoke Discharge Outlets

No.	Position of outlet	Dimensions of outlet [mm]	Ratio of opening area of outlet
9	Ceiling	123×123	1/70
10		133×133	1/60
11		145×145	1/50
12		163×163	1/40
13		188×188	1/30
14	Wall	188×80	1/70
15		219×80	1/60
16		263×80	1/50
17		329×80	1/40
18		438×80	1/30

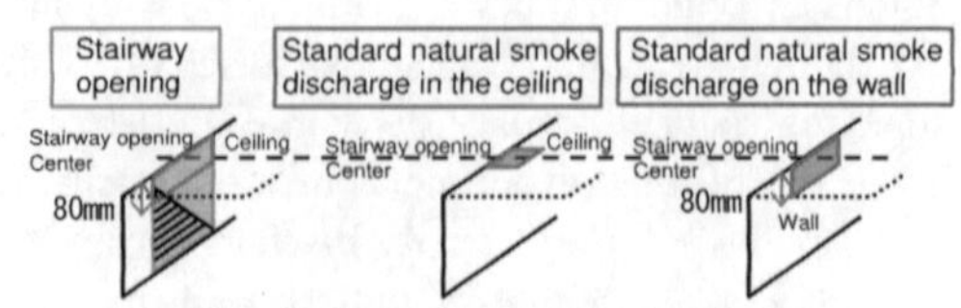

Fig.8. Comparison between stairway openings and standard natural smoke

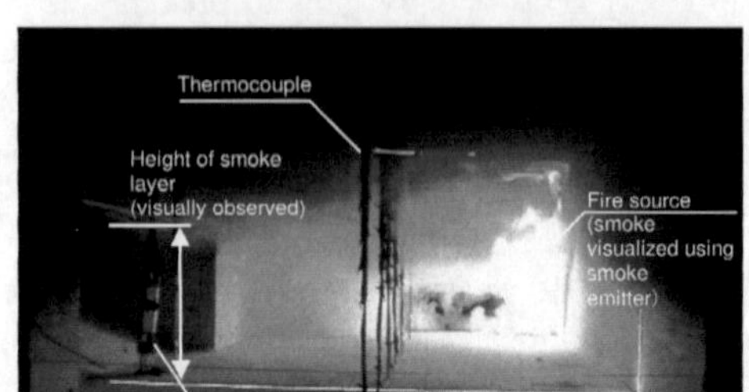

Fig.9. View of experiment (600 mm-wide stairway with stepped ceiling)

IOP Conf. Series: Journal of Physics: Conf. Series **1107** (2018) 042005 doi:10.1088/1742-6596/1107/4/042005

4. CALCULATION OF VOLUME OF SMOKE DISCHARGE FROM OPENINGS

(1) Stairway opening

The volume of smoke discharge from the stairway opening was calculated with formulae 5), 6), 7) by applying the temperature at the opening T_i and the pressure difference Δp_i. The experimental data was used the 60-second average of before the end of the experiment. The volume of smoke discharge from the stairway opening m_{mi} was calculated by dividing into six parts in the perpendicular direction, at the point of pressure and thermocouple measurement shown in Fig.6b with the sums of those defined as the individual volumes of smoke discharge m_m.

$$\rho_i = \frac{353}{(Ti+273)} \quad ⑤ \qquad v_i(z) = \sqrt{\frac{2\Delta p_i}{\rho_i}} \quad ⑥ \qquad m_{mi} = \alpha \rho_i v_i A \quad ⑦$$

v : Flow velocity [m/s] Δp : Pressure difference [Pa] ρ : Density [kg/m^3]

m_m : Volume of smoke discharge (Mass flow rate) [kg/s] A : Area of smoke discharge outlet [m^2]

α : Flow rate coefficient 0.7 T : Temperature [°C] Suffix i : Number of individual opening division

The BSL stipulates that openings in the upper zone 800 mm downward from the ceiling are effective for smoke discharge. We thus defined that the area effective for smoke discharge from the stairway opening to not be the whole area of the stairway opening facing to the concourse, rather that it be the product of the stairway width multiplied by the above-mentioned 80 mm (800 mm in the full-scale environment, Fig. 8).

(2) Standard natural smoke discharge outlet on the wall

The volume of smoke discharge from the standard natural smoke discharge outlet on the wall m_{mi} was calculated with formulae 5), 6) and 7) as done for that from the stairway opening. The volume of smoke discharge from the standard natural smoke discharge outlet on the wall was divided into two parts in the perpendicular direction (800 mm), at the point of pressure and thermocouple measurement shown in Fig.6e with the sums of those defined as the individual volumes of smoke discharge m_m.

(3) Standard natural smoke discharge outlet in the ceiling

The volume of smoke discharge from the standard natural smoke discharge outlet in the ceiling was calculated with formulae 5), 6) and 7) by applying the temperature measured at the center of the outlet T_i and the pressure difference Δp_i. (Fig.6f)

5. RESULTS AND DISCUSSION

We compared the volumes of smoke discharge measured at stairway openings and standard natural smoke discharge outlets. Fig. 10 shows the volumes of smoke discharge in the different conditions.

(1) Effect of the shape of stairway ceiling

Comparison between the volumes of smoke discharge from stairway openings with stepped ceiling and with sloped ceiling showed the volumes were almost equal to each other. Thus, we can say that the shape of the stairway ceiling does not have much effect on smoke discharge from stairway openings. Moreover, the larger the ratio of opening area was, the larger the volume of smoke discharge was.

(2) Effect of position of standard natural smoke discharge outlets

Comparison between the volumes of smoke discharge from standard natural smoke discharge outlets on the wall and in the ceiling revealed a tendency for the volume from outlets in the ceiling to be larger than that from outlets on walls. This is considered to be because the volume of smoke discharge from standard natural smoke discharge outlets in ceilings set upward of the smoke became larger as heated air raised smoke. Moreover, the larger the ratio of opening area was, the larger the volume of smoke discharge was.

(3) Comparison between volumes of smoke discharge from stairway opening and standard natural smoke discharge outlets

Comparison between the volumes of smoke discharge from stairway openings and standard natural smoke discharge outlets revealed a tendency for the volume from stairway openings to be larger than that from outlets. This is considered to be because the smoke layer in the stairway opening was lower than 80 mm downward from the ceiling (800 mm in the full-scale environment), the zone the BSL stipulates as being effective for smoke discharge, and the smoke was discharged from a further lower zone.

From this, we suppose that the part of the stairway opening lower than the zone 80 mm downward from the ceiling, which the BSL stipulates as being effective for smoke discharge, functioned as a smoke discharge outlet. As a result, the volume of smoke discharge from the stairway opening is expected to be greater than that from standard natural smoke discharge outlets if the ratio of the opening area is the same.

Further, from the temperature measurement results of the thermocouples in Fig. 6, the height of the smoke layer at each measurement point was calculated according to the He's law. We compared the visual results with the calculation results by the He's law and verified the validity of the calculation. The average smoke layer height in the concourse was almost the same between the case of the stairway opening and the case of the standard natural smoke discharge outlet. (Fig.11)

This experiment suggested that, when the area of the stairway opening effective for smoke discharge is the product of the stairway width and 80 mm downward from the ceiling (800 mm in the full-scale environment), the volume of smoke discharge from stairway openings with at least a 1/50 ratio of opening area will be larger than standard natural smoke discharge outlets with a 1/50 ratio of opening area required by the BSL.

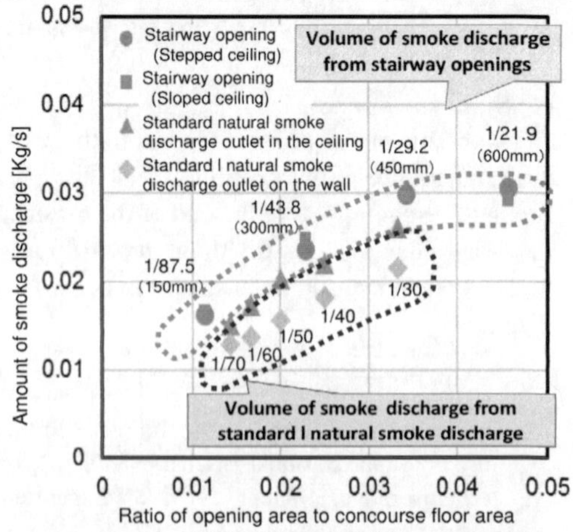

Fig.10. Comparison of volumes of smoke discharge from stairway openings and standard natural smoke discharge outlets

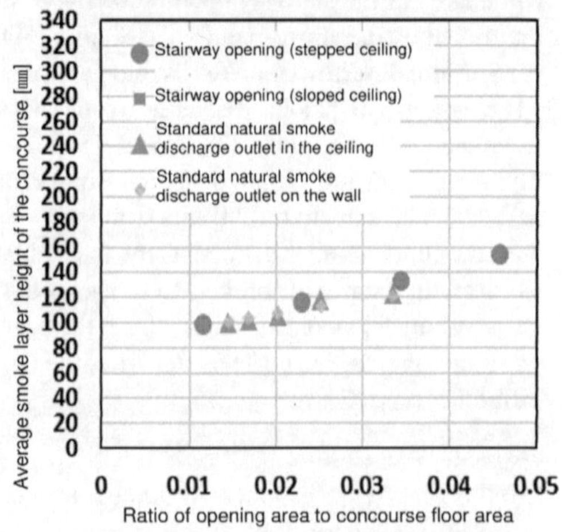

Fig.11. Comparison of the smoke layer height of the concourse, the case of the stairway opening and the case of the normal natural smoke discharge outlet.

6. CONCLUSIONS

We compared the volumes of smoke discharge from stairway openings and standard natural smoke discharge outlets in accordance with the BSL using models of a rail stations under elevated platforms with a stairway at a right angle to the concourse.

The models we used in the experiments in fires were of an under-viaduct station with a stairway at a right angle to the concourse only, although there are actually a variety of designs of rail stations under elevated platforms.

We are thus aware of the need to develop a method of identifying volume of smoke discharge more easily than experiments using models in order to evaluate volume of smoke discharge from stairway openings of rail stations under elevated platforms of different designs. For that method, we intend to investigate applicability of smoke flow simulation.

ACKNOWLEDGMENT

The authors would like to express the deepest appreciation to Mr. Daisuke Sato (Tokyo Fire Department) for assistance with this study.

REFERENCES

[1] Susumu Mafune, O Suzuki, D Sato, Y Ohmiya "Study on smoke behavior in the station underpass," Interflam --Proceedings of 14th International Fire Science and Engineering Conference, 2016, vol.1, pp 709-714, Poster Session

[2] John H.Klote and James A.Milke:"Principles of Smoke Management", American Society of Heating, Refrigerating and Air-Conditioning Engineers, Inc., pp.217-224, 2002

[3] Yaping He: "On Experimental Date Reduction for Zone Model Validation", Journal of Fire Sciences, pp144-161, March 1997

[4] Yaping He: "Determination of interface height from measured parameter profile in enclosure fire experiment", Fire Safety Journal, pp19-38, July 1998

IOP Conf. Series: Journal of Physics: Conf. Series **1107** (2018) 042006 doi:10.1088/1742-6596/1107/4/042006

A Large Eddy Simulation of LNG Pool Fire on Board a Chemical/Oil Tanker

Abdullah BOLEK and Selma ERGIN

Department of Naval Architecture and Marine Engineering, Istanbul Technical University, Turkey

abolek@itu.edu.tr

ABSTRACT

The air pollution from maritime transportation has become one of the major environmental concerns. The liquefied natural gas *(*LNG*)* has better environmental performance compared to conventional ship fuels. Therefore, the use of LNG as ship fuel has recently gained more attention in the maritime industry. On the other hand, LNG as a fuel can have high risks of explosion and fire on ships. Among the various LNG ship fires, the pool fire is the most common phenomenon. Therefore, this study focuses on the LNG pool fire on board a chemical/oil tanker with different pool aspect ratios. Also, the effects of wind speed on the flame characteristics are investigated. The LNG pool fire on board a chemical/oil tanker is studied by using large eddy simulation approach and Fire Dynamic Simulator (FDS) code, numerically. The results show that the flame characteristics are affected markedly by the pool aspect ratio and wind speed. While the aspect ratio increases, the mean flame height reduces, whereas the heat flux values increases. It is also found that the heat flux values increase with the wind speed.

KEYWORDS: *LNG pool fire; pool aspect ratio; wind speed; CFD; large eddy simulation.*

INTRODUCTION

International Maritime Organization (IMO), MARPOL Annex VI sets progressive stricter emission limits to control the emissions from ships, including NO_x and SO_x emissions which have adverse effects on both the environment and human health [1, 2]. LNG as a ship fuel can be an alternative to adapt the environmental regulations, since it eliminates almost all emissions of SO_x and particulate matter, reduce NO_x emissions by up to 80% and reduce CO_2 emissions by 15-25% [3-5]. However, handling or transporting this flammable and explosive substance, LNG especially under high pressures may lead severe fire accidents. Ergin and Sülüs investigated the engine room fire for a chemical tanker and presented the effects of the openings on the fire development [6]. The pool fire is the most common phenomenon among the various LNG ship fires such as jet fire, flash fire and fire ball [7, 8]. Therefore, the pool fire has become the subject of many papers.

The pool shape in the majority of the pool fire studies is circular, hence the existing correlations based on the circular pool assumptions. There are a limited number of studies which considers the effects of the pool aspect ratio on the fire characteristics in the literature. The experimental study on the asymmetric plume in a channel was conducted by Fan et al. [9] and it has been found that the heat flux increases with the aspect ratio due to the enhancement in the perimeter and a proportional increase in the air entrainment. Hu et al. [10] studied the burning behaviour of the square pool fires which are optical-thin under the cross-air flows. It has been reported that the burning rate changes with the pool diameter and the magnitude of the cross-air flows. Ji et al. [11] investigated experimentally the effects of the sidewall and aspect ratio on the flame characteristics and the burning rate of the n-heptane pool fires. The non-monotonic distribution of the burning rate was obtained by increasing the distance from the sidewall, but the burning rate increased with the pool aspect ratio. Jiang et al. [12] conducted several experiments to obtain a correlation for the burning rate of the rectangular pool fires. It has been shown that the correlation gives satisfactory results for n-heptane and gasoline pool fires and also reported that the small wind speed does not affect the burning rate significantly. Tang et al. [13] studied the radiative heat flux characteristics and mean flame height of the medium-scale rectangular pool fires and claimed that the atmospheric pressure has a low effect on the mean flame height, and also the slight difference was observed in flame radiation fraction. Tang et al. [14] scrutinized the flame pulsation characteristics of the rectangular pool fires for two different atmospheric pressures and different aspect ratios. The flame pulsation characteristics are slightly affected by the ambient pressure. However, it increases as the aspect ratio increases. Fan et al. [15] investigated the sidewall effects with various aspect ratios of the rectangular pool fires on the flame characteristics. It has been found that an increase in the aspect ratio causes an increase in the heat release rate due to the enlargement of the pool perimeter.

The present study investigates the flame characteristics of the LNG pool fire and heat fluxes on board a chemical/oil tanker for different pool aspect ratios and wind speeds, numerically. Four different pool aspect ratios, AR=1, 2, 4 and 8 and also a round pool (AR=0) are considered. The wind speed has been taken as 0 and 3 m/s. In order to calculate the flame characteristics and heat flux values of the fire on board a chemical/oil tanker, a genuine model is developed, and the governing equations are solved by using FDS code and the large eddy simulation approach. A brief explanation of the numerical model is given in the following section. The third section presents the findings of the study and discussions of the results. The last section, conclusions gives the major findings of the present study.

NUMERICAL MODEL

FDS code which is highly validated and widely used code to solve the low Mach number fire dynamic problems is also used in this study. FDS employs Large Eddy Simulation (LES) approach and the governing equations discretized by the second-order central difference method except for the thermal radiation [16]. In computations, Deardoff turbulence viscosity model is used. As a combustion model, the Mixing-Controlled Fast Chemistry model is used. The pool fires are classified as radiation (D>0.2 m) or conduction-controlled (D<0.2 m), since the pool fire which is investigated in the present study has the diameter larger than 0.2 m, it can be classified as radiation-controlled. Therefore, the radiation effects are also considered in the present study. FDS uses the radiative transport equation by considering the absorbing, emitting and scattering of the medium and the equation is discretized using finite volume method [17]. The computational domain of the study is shown in Fig. 1. The "stair stepping" method is used to generate the ship form, and the obstacles with LNG tanks on board ship are modelled as rectangular blocks. In order to obtain appropriate height of the computational domain, the correlation given by Heskestad [18] (Eq. 1) is used as an initial guess for the height of the domain to cover the complete flame.

$$L = 0.235\dot{Q}^{(2/5)} - 1.02D \tag{1}$$

where L is the mean flame height (m), $\dot{Q}$ is the energy release rate (kW/m^2) and D is the pool diameter (m).

The top and sides of the computational domain are set as open boundaries, only the bottom of the computational domain is the wall with no slip condition. The wind is blowing through the inlet without any oblique and leaves the domain from the outlet. The uniform distributions of the variables are used as the boundary conditions at the inlet and outlet of the domain. The ambient temperature and pressure are set to 20^0C and 1 atm, respectively. LNG includes mainly Methane (93.94%), Ethane (4.8%), Propane (0.81%), Nitrogen (0.25%) and Butane (0.2 %) and the heat of combustion value of LNG is 50 MJ/kg. The material of the ship, LNG tanks, and obstacles are steel with density 7850 kg/m^3, thermal conductivity 45.8 W/(m K) and specific heat 0.46 kJ/(kg K). The effects of four different AR values (AR=0.0, 1.0, 2.0, 4.0 and 8.0) and two different wind speeds (V_w=0 and 3 m/s) on the flame height, plume velocity, plume temperature and heat flux values are investigated.

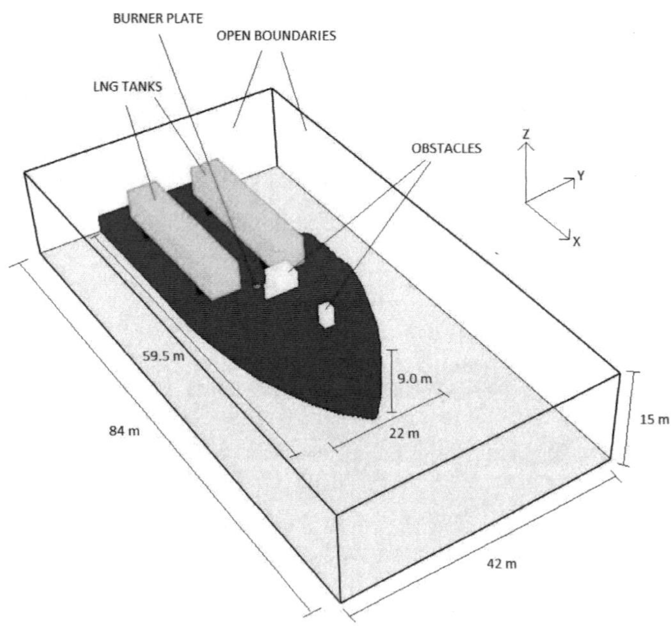

Fig. 1. Computational domain.

Three different grids are employed in the solution of the problem and the heat flux values are used to assess the desired grid resolution. Since LES intrinsically depends on the grid resolution, a final grid independent resolution does not exist. Therefore, the desired resolution is determined among three different grids. The resolved parts of Grid 2 (309x146x78) and Grid 3 (420x227x110) are found to be satisfactory to obtain the results, however, the resolution of Grid 1 (211x114x55) is found as inadequate. Therefore, Grid 2 is used in the solution.

RESULTS AND DISCUSSION

This section presents the calculated flame heights, vertical velocity values, temperature distributions and heat flux values for different pool aspect ratios and wind speeds. To obtain the fully developed results, the adequate computational time is specified as 30 s. The time step is determined by satisfying the CFL (Courant, Friedrichs, Lewy) condition, this parameter is set automatically by FDS [19]. Fig. 2 shows the instantaneous flame heights of the pool fire for different pool aspect ratios when t equals 30 s. The instantaneous flame heights for the round pool (AR=0) and the square pool fire (AR=1) are found similar. As can be clearly seen from Fig. 2, the flame height decreases while the aspect ratio increases.

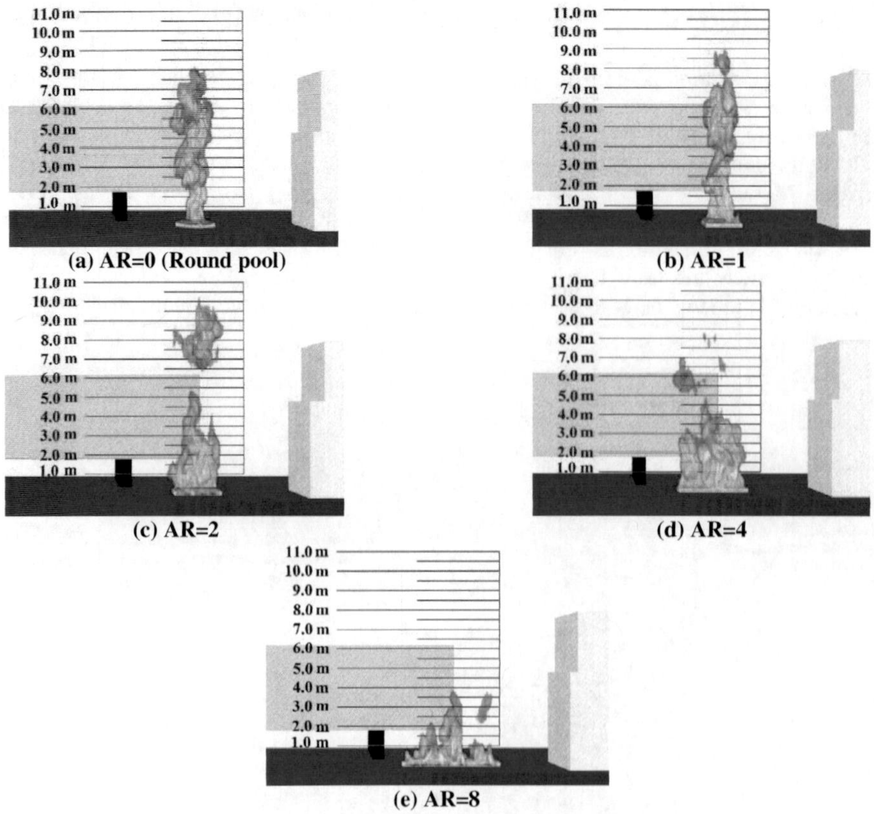

(a) AR=0 (Round pool) (b) AR=1 (c) AR=2 (d) AR=4 (e) AR=8

Fig. 2. Instantaneous flame heights of the pool fire at t=30 s. (a) AR=0, (b) AR=1, (c) AR=2, (d) AR=4 and (e) AR=8.

Fig. 3 shows the vertical plume velocity distributions at the centreline of the burner for different pool aspect ratios and wind speeds. As can be seen from Fig. 3 a, the velocity firstly increases, it reaches its maximum value, and then stays nearly constant along the centreline of the burner. The mean flame height decreases as the aspect ratio increases, since the flame height is significantly affected by the side entrainment of the pool shape. Hence the plume velocity decreases when the aspect ratio increases. The velocity distribution for the round pool (AR=0) is almost same as that for the pool aspect ratio, AR=2. The vertical velocity starts to oscillate at the certain distance above the burner. This distance becomes shorter when the aspect ratio increases. It can be seen from Fig. 3a that the amplitude of the velocity increases as the aspect ratio increases.

The flame characteristics are considerably affected by the wind speed. The flame is not only affected by the wind, but also by the obstacles, which are located in front of the pool. A recirculation region occurs behind the obstacles and it effects the vertical velocity and the temperature distributions. Since the flame height decreases with the increasing aspect ratio, relatively shorter flames will be less affected by the wind speed. Therefore, opposite trend is obtained for the wind velocity, V_w=3 m/s, the velocity values along the centreline of the burner increase as the pool aspect ratio increases (see, Fig. 3b). For the wind velocity, V_w=3 m/s, the velocity first increases sharply, reaches its maximum value and then decreases along the centreline of the burner (see, Fig. 3b). The maximum vertical velocity value decreases almost 50 % when the wind velocity increases from 0 to 3 m/s. The velocity values for the round pool are smaller than that of the square and rectangular pools.

Fig. 4 shows the plume temperature distribution along the centreline of the burner for different aspect ratios and wind velocities. For no wind (V_w=0 m/s) case, the plume temperature first increases sharply, reaches its maximum value and then decreases. The maximum temperature values are obtained at the persistent region of the flame as expected, and temperatures begin to decrease at the intermittent region of the flame as combustion reactions diminish and the cooling effect of the air became more dominant. The maximum temperature is almost same for all the aspect ratios, since the physical characteristics of the pool fire change with the size of the fire, and the temperature usually increases due to reduction in thermal radiation loss from the fire. The wind speed affects the temperature distribution significantly (see, Fig. 4ab). For the wind speed, V_w=3 m/s, the temperature

IOP Conf. Series: Journal of Physics: Conf. Series **1107** (2018) 042006 doi:10.1088/1742-6596/1107/4/042006

first rises very sharply, reaches its maximum value and then decreases sharply. On the other hand, the lowest temperatures occur for the case with round pool. The temperature distribution has a similar phenomenon as the vertical velocity distribution. The effect of wind speed is relatively small for the shortest flame length. Hence the plume temperatures decrease with decreasing pool aspect ratio.

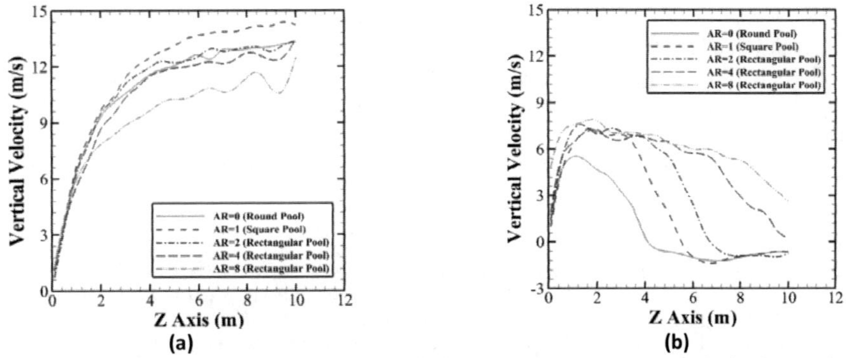

Fig. 3. Velocity distribution at the centreline of the burner. **(a)** V_w=0.0 m/s (no wind), **(b)** V_w=3 m/s

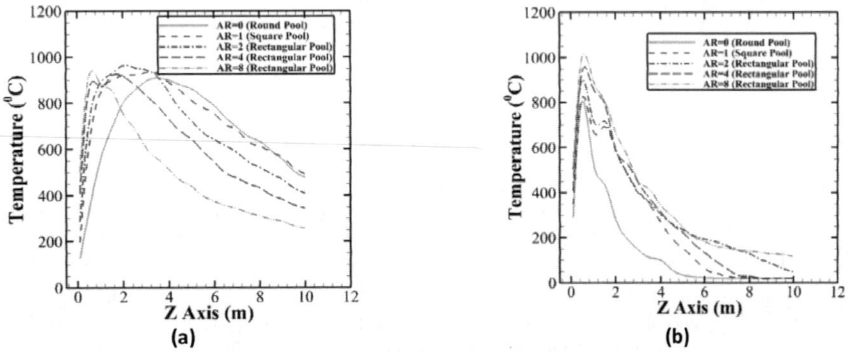

Fig. 4. Temperature distribution at the centreline of the burner. **(a)** V_w=0.0 m/s (no wind), **(b)** V_w=3 m/s

The sum of the heat fluxes by radiation, conduction and convection gives the total heat flux from flame to the fuel bed [20, 21] and the calculated heat flux values on the fuel bed for different aspect ratios and the wind speeds are presented in Fig. 5. Since the wind entrains the flame through the short side of the pool, the flame became more turbulent. Hence the heat flux values increase when the pool aspect ratio increases [22]. There is a sharp increase when the pool aspect ratio is small. Since the pool fire has the asymptotic burning rate for the large aspect ratios. On the other hand, the heat transfer characteristics, flame shape and turbulence mechanism are affected by the wind. The wind contributes the air-fuel mixing and that causes an improvement on the combustion efficiency and hence the burning rate [23]. Therefore, the heat flux values are increased with the wind speed.

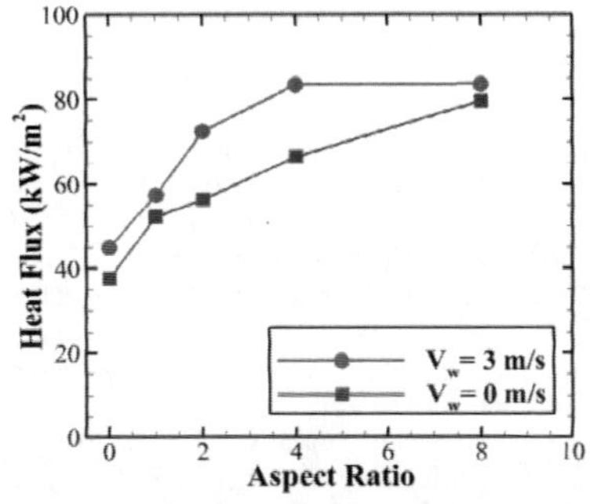

Fig. 5. Heat flux values on the burner plate for different aspect ratios.

CONCLUSIONS

This study investigates the LNG pool fire on board a chemical/oil tanker by using large eddy simulation approach. The effects of different pool aspect ratios and wind speeds on the flame characteristics, plume velocities, plume temperatures and heat fluxes are obtained, numerically. The solution of the governing equations is obtained by using Fire Dynamic Simulator (FDS) code. The major findings include:

- The flame height decreases while the pool aspect ratio and the heat flux values increase, since the flame height is significantly affected by the side entrainment of the pool shape.
- The pool fire that has a high aspect ratio is relatively less affected by the wind and recirculation region occurred due to the obstacles on board ship. Therefore, the temperature and vertical velocity values increase with the aspect ratio.
- The wind speed increases the combustion efficiency. As a result of this, the burning rate and heat flux values increase.

REFERENCES

[1] Kalender, S.S. and Ergin, S., (2017) Experimental Investigation into the Particulate Emissions from a Ferry Fuelled with Ultra-Low Sulfur Diesel, Journal of Marine Science and Technology 25(5);499-507, DOI 10.6119/JMST-017-0418-2

[2] Durmaz, M., Kalender, S.S. and Ergin, S., (2017) Experimental Study on the Effects of Ultra-Low Sulfur Diesel Fuel to the Exhaust Emissions of a Ferry, Fresenius Environmental Bulletin, 26(10):5833-5840.

[3] Paltrinieri N., Tugnoli, A., and Cozzani, V. (2015). Hazard identification for innovative LNG regasification technologies, Reliab. Eng. Syst. Saf. 137; 18–28. doi:10.1016/j.ress.2014.12.006.

[4] Noh, K. Y., Chang, Y. S., and Chang, D., (2014) Risk-Based Determination of Design Pressure of LNG Fuel Storage Tanks Based on Dynamic Process Simulation Combined with Monte Carlo Method, Reliab. Eng. Syst. Saf. 129; 76–82. doi:10.1016/j.ress.2014.04.018.

[5] Chu, B. and Chang, D., (2017), Effect of Full-Bore Natural Gas Release on Fire and Individual Risks: A Case Study for an LNG-Fueled Ship, J. Nat. Gas Sci. Eng. 37; 234–247. doi:10.1016/j.jngse.2016.11.043.

[6] Ergin, S., and Sulus, A., "The Effects of the Openings on the Fire Development in an Engine Room of a Chemical tanker". *Proc. Of 13th International Congress of the International Maritime Association of the Mediterranean*, IMAM 2009, Istanbul, Turkey, Vol. 1, pp: 343-349.

[7] Palazzi, E., Caviglione, C., Reverberi, A.P. and Fabiano, B. (2017), A Short-Cut Analytical Model of Hydrocarbon Pool Fire of Different Geometries, with Enhanced View Factor Evaluation, Process Saf. Environ. Prot. 110; 89–101. doi:10.1016/j.psep.2017.08.021.

[8] Planas-Cuchi, E., and Casal, J. (1998), Flame Temperature Distribution in a Pool-Fire, J. Hazard. Mater., 62;231–241. doi:10.1016/S0304-3894(98)00177-0.

[9] Fan, C.G., Zhang, J.Q., Zhu, K.J. and Li, K.Y., (2017), An experimental Study of Temperature and Heat Flux in a Channel with an Asymmetric Thermal Plume." Applied Thermal Engineering 113; 1128-1136.

[10] Hu, L., kuang, C., Zhong, X., Ren, F., Zhang, X., and Ding, H.(2017), An Experimental Study on Burning Rate and Flame Tilt of Optical-Thin Heptane Pool Fires in Cross Flows, Proceedings of the Combustion Institute 36.2;3089-3096.

[11] Ji, J., Fan, C.G., Li, Y.Z., Ingason, H., and Sun, J.H., (2015), Experimental Study of Non-Monotonous Sidewall Effect on Flame Characteristics and Burning Rate of N-Heptane Pool Fires, Fuel, 145;228-233.

[12] Jiang, C., Li, Y., Huang, H., Zhao, J., Wang, Z., and Zhang, J., (2016), Experimental Study of Burning Rate in Large-Scale Rectangular Pool Fire, Journal of Fire Sciences, 34.4;323-334.

[13] Tang, F., Zhu, K., Dong, M., and Shi, Q. (2015), Mean Flame Height and Radiative Heat Flux Characteristic of Medium Scale Rectangular Thermal Buoyancy Source with Different Aspect Ratios in a Sub-Atmospheric Pressure, International Journal of Heat and Mass Transfer 84;427-432.

[14] Tang, F., Hu, L., Wang, Q., and Ding, Z. (2014), Flame Pulsation Frequency of Conduction-Controlled Rectangular Hydrocarbon Pool Fires of Different Aspect Ratios in a Sub-Atmospheric Pressure, International Journal of Heat and Mass Transfer 76; 447-451.

[15] Fan, C.G., Ji, J., Li, Y.Z., Ingason, H., and Sun, J.H. (2017), Experimental Study of Sidewall Effect on Flame Characteristics of Heptane Pool Fires with Different Aspect Ratios and Orientations in a Channel, Proceedings of the Combustion Institute 36.2;3121-3129.

[16] Sikanen, T., and Hostikka, S., (2016), Modeling and Simulation of Liquid Pool Fires with In-Depth Radiation Absorption and Heat Transfer, Fire Saf. J. 80; 95–109. doi:10.1016/j.firesaf.2016.01.002.

[17] K. McGrattan, K., Hostikka, S., McDermott, R., Floyd, J., Vanella, M., Weinschenk, C., Owerholt, (2017), Fire Dynamics Simulator Techical reference guide Volume 1 : mathematical model.

[18] Heskestad, G., "Fire Plumes, Flame Height, and Air Entrainment" SFPE Handb. Fire Prot. Eng. Fifth Ed., Springer, New York, 2016: pp. 396–428.

[19] McGrattan, K., Hostikka, K., McDermott, S., Floyd, R., Vanella, J., Weinschenk, M., and Owerholt, C. (2017), Fire Dynamics Simulator User's Guide, NIST Special Publication 1019.

[20] Drysdale D. Pool fires. In: An introduction to fire dynamics. Chichester: John Willy & Sons Ltd.; 1999. p. 160–9.

[21] Hu, L., Tang, F., Wang, Q., & Qiu, Z. (2013), Burning characteristics of conduction-controlled rectangular hydrocarbon pool fires in a reduced pressure atmosphere at high altitude in Tibet, Fuel, 111, 298-304.

[22] Tang, F., Longhua, H., Zhang, X., Zhang, X. and Dong, M. (2015), Burning rate and flame tilt characteristics of radiation-controlled rectangular hydrocarbon pool fires with cross air flows in a reduced pressure, Fuel 139: 18-25.

[23] Jiang, P., & Lu, S. X. (2016), Pool fire mass burning rate and flame tilt angle under crosswind in open space, Procedia Engineering, 135, 261-274.

Fire Growth Rate Strategies in FDS

Francesco CIANI[1], Monica CAPOBELLI[1]

[1]*Fire Engineering Consultancy, Guidi & Partners s.r.l., Italy*

francescosaverio.ciani@gmail.com
monica.capobelli@gmail.com

ABSTRACT

The HRR curve is characterized by four phases: the incipient phase, the growth phase, the fully developed phase and the decay phase. A central parameter in the design of buildings and the provision of fire protection measures, is the rate at which the fire develops. Therefore the speed at which a space becomes untenable is mainly dependent on the fire growth rate, which is also closely related to the risk profile selection. To reproduce the fire growth phase, it is generally recommended the use of the t-squared fires [1].

Fire Dynamics Simulator (FDS) 6.6.0 [2] enables us to reproduce the t-squared fire in several ways, by using different "fire growth functions", i.e. RAMP-function and SPREAD_RATE function. These functions use different approaches. While the RAMP function associates the time history of the HRR curve with the entire surface, the SPREAD_RATE function is similar to a cellular automaton approach and it mimics a radially growing fire along the surface with a constant speed.

In this paper the behavior of the dimensionless heat release rate Q*, during the fire growth phase, has been studied for both the fire growth rate functions. Furthermore, a numerical study of the "growth function dependency" of some important outputs has been carried out, in order to quantify the implications on safety design. If on one hand, in accordance with the Swedish Best Practice [3], the SPREAD RATE should be preferred, seeing that the burning surface is kept small when the HRR is low and therefore it is possible to maintain applicable values of the $Q*$during the fire growth, on the other the case studies performed, using the RAMP function, have shown to be the worst case scenario analysing the gas temperature and the visibility in order to evaluate the Available Safe Egress Time.

KEYWORDS:

Modeling; performance-based design; Fire Safety Engineering, FDS 6.6.0; flame spread; fire growth; heat transfer; fluid dynamics; CFD; Froude number.

NOMENCLATURE LISTING

c_p	specific heat (J mol^{-1} K^{-1})	$\dot{Q}_C$	rate of heat release (kW)
D	Fire source diameter (m)	T_∞	ambient temperature (K)
Fr	Froude number	t	time (s)
g	gravity acceleration (m/s^2)	U	velocity of the fuel vapours
ΔHc	heat of comb. of the fuel vapour		**Greek**
HRRPUA	Heat Release Rate per unit area (Kw/m^2)	α	fire growth rate (kW/s^2)
L	Flame height (m)	ρ_∞	Fuel density (kg m^{-1}·s^{-1})

INTRODUCTION

In a life safety analysis that evaluates the ability of the occupants to escape from the enclosure of fire origin, it will normally only be the incipient and growth phases of the fire that are of relevance (after a flashover or full involvement, the possibility of escape can be discounted) [4].

For design purposes, fires are often assumed to grow proportionately to time squared (t-squared fires).
The t-squared fire growth can be reproduced by FDS using different approaches, but this paper is going to focus on SPREAD_RATE and RAMP functions. Although the RAMP function is widely used, the Swedish Best Practice guideline [3] recommends the use of the SPREAD_RATE function.

In this report the theoretical consequences of the use of these two methods have been explained, introducing the significant parameters related to the fire growth. Since the aim of this paper is to assess a responsible and informed use of these two approaches, the growth function dependency of some output parameters has been analysed for simple case studies, showing discrepancies and similarities

A SIGNIFICANT PARAMETER RELATED TO THE FIRE GROWTH

The surface area of the fuel source is a matter of great concern during the fire growth phase. A high HRR generated over a small surface area, will cause the fire plume to be driven by momentum instead of buoyance, similar to a jet flame. Jet flames are more structured than buoyance driven flames which usually occur in building fires and are also less affected by surrounding air flows [4]. If the surface area of the fire source is large with a low HRR, the flame will break up into smaller, separate, flames. This will not represent a "real" fire. Therefore the momentum (or inertia) and buoyancy in the flame will determine the type of fire.

The Froude number (Fr) may be used as a means of classification. It is a measure of the relative importance of inertia and buoyancy in the system, and is conveniently expressed as:

$$Fr = \frac{U^2}{gD} \tag{1}$$

A vast number of experiments have been carried out that relate flame heights to energy release rate and source diameter. The scientific community has found it convenient to express the data in terms of a non-dimensional energy release rate, denoted $\dot{Q}^*$, and given by the following expression

$$\dot{Q}^* = \frac{\dot{Q}_c}{\rho_\infty c_p T_\infty \sqrt{gD} D^2} \tag{2}$$

The $\dot{Q}^*$ represents the square root of the Froude number and has been found to be very important in controlling the geometry of fire plumes. So it is important to define an optimum value for the engineering applications [5]. Cox and Kumar [6] define that dimensionless HRR ($\dot{Q}^*$) should be in the range of 0.3 to 2.5 for natural fires in buildings. Using this range, along with the recommended design fires in the national guidelines, a range of applicable fire diameters can be calculated. With a known fuel surface diameter, the heat release rate per unit area (HRRPUA) can also be calculated.

T-SQUARED FIRE GROWTH

The growth phase begins when the fire exhibits a heat release rate that increases beyond a negligible level. Flame spread is the typical mechanism that governs fire growth. Flame spread is the phenomenon of flames moving across an object, igniting that which with they come into contact. When the growth stage is calculated analytically, it is done with the power low equation in engineering applications:

$$\dot{Q}_c = \alpha t^p \tag{3}$$

A fire growing outward radially at a constant rate should have an exponent of p=2 because the area of the involved circular surface grows with the square of the radius, where the radius is predicted to grow linearly in time. Testing has shown that this exponent is valid for a wide range of material and objects. These fires are commonly referred to as t-squared fires [7].
There are several ways to reproduce these t-squared fires in **FDS 6.6.0** using the simple pyrolysis model. This paper is going to discuss the following functions, characterized by two different approaches:

- RAMP function;
- SPREAD_RATE function (hereinafter reffered to as S_R);

It is worth mentioning that there is also a third method that achieves the S_R behavior, using the RAMP function with a control logic. Since the aim of this paper was to study these two "pure" approaches that are mainly used, this hybrid method has not been analysed.

The RAMP function controls the heat release rate of the burner, by specifying the time history of the RHR curve. Using this function, the fuel area is constant during the whole fire phenomenon. On the contrary the HRRPUA and $\dot{Q}^*$ vary during the fire growth phase. Fig. 1 shows the typical flame variation of a fire source modeled with the RAMP. It is clear that the flame behavior varies over time. Therefore the $\dot{Q}^*$ of the fire, during the first seconds, is outside of the above-mentioned range defined by Cox and Kumar [6], but then it increases untill it reaches optimum value.

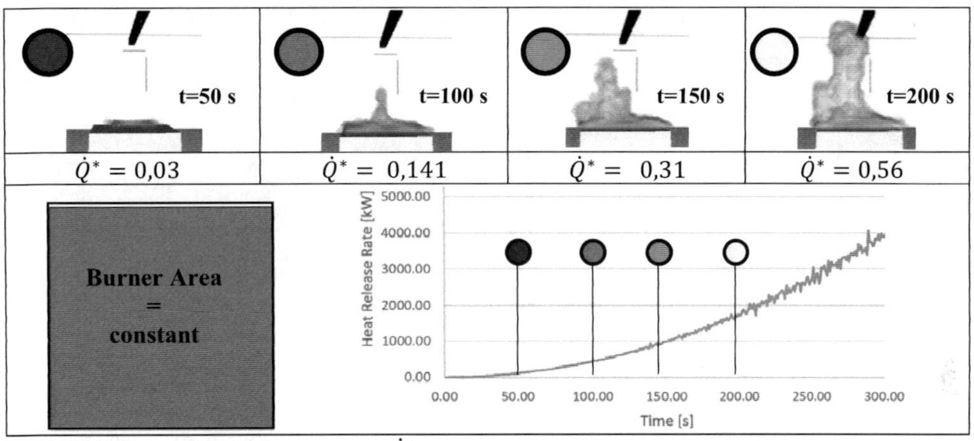

Fig. 1. Variation of $\dot{Q}^*$ in the growth phase using the RAMP

Instead the S_R function mimics a fire that is growing radially at a constant speed. In particular, the surface of the burner is discretized with a regular square array and flame spread occurs as a series of cell ignitions, using a sort of cellular automaton technique. The advantage of using the S_R function is that the burning surface is kept small when the HRR is low. Therefore it is possible to maintain applicable values of the dimensionless HRR ($\dot{Q}*$) during the fire growth. To calculate the S_R value, the following equation can be used [3]:

$$SPREAD\_RATE = \sqrt{\frac{\alpha}{\Pi} \cdot \frac{1}{HRRPUA}} \tag{4}$$

The disadvantage of using the above mentioned function is that the fire source needs to be properly resolved for the HRR to resemble a continuous function[3].

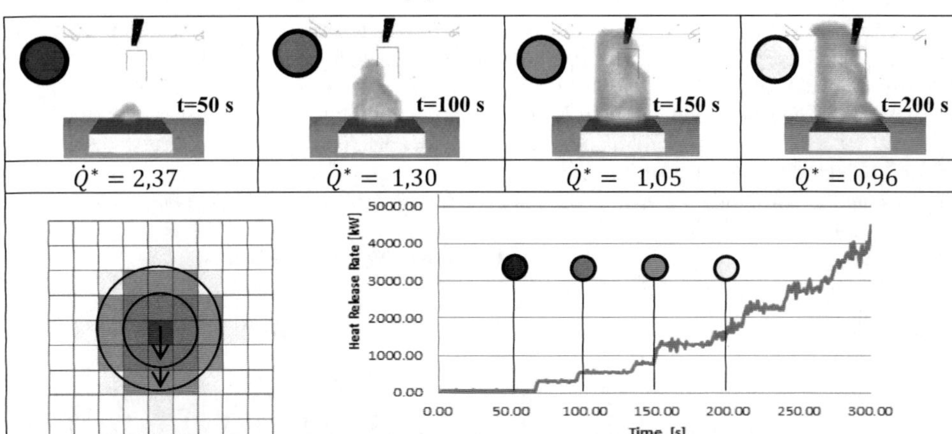

Fig. 2. Variation of $\dot{Q}^*$ in the growth phase using the S_R

Fig. 2 illustrates how the HRR curve is not smoothly varying when the S_R function is being used. The phenomenon is more obvious when larger grid cells are being used. Thus it is recommended that a square shaped fire source is modeled with an odd number of grid cells covering each side of the fire source. [3]

IOP Conf. Series: Journal of Physics: Conf. Series **1107** (2018) 042007 doi:10.1088/1742-6596/1107/4/042007

COMPARISON BETWEEN RAMP **AND** SPREAD_RATE **FUNCTIONS**

The S_R and RAMP functions have been tested on three different geometrical shapes (so six simulations have been performed). Table 1 shows the comparisons between three pairs of scenarios (S_01 – S_02, S_03 - S_04, S_05 - S_06), where each pair compared has the same inner geometry. Furthermore, each comparison differs from the others in size, in order to perform a sensitivity analysis. The aim of this sensitivity analysis is to assess if the differences, in terms of effect, between the two fire growth functions, are affected by building size and inner geometry.

Table 1. Different fire scenarios studied

Simulation	S_01	S_02	S_03	S_04	S_05	S_06
Function	S_R	RAMP	S_R	RAMP	SP_R	RAMP
Model Geometry	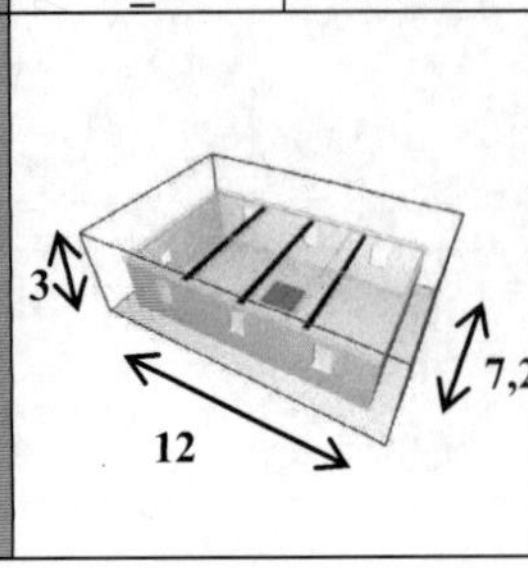		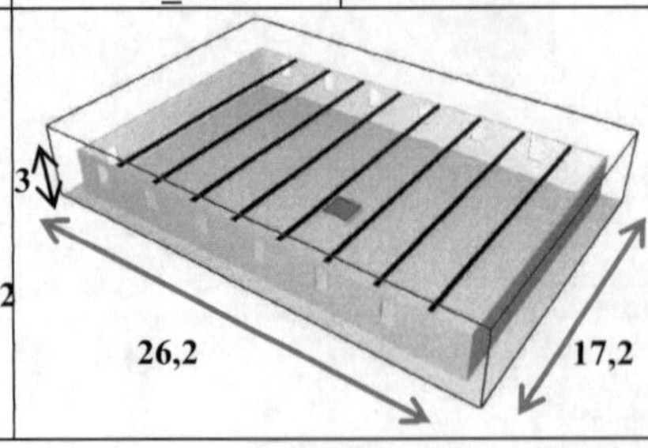		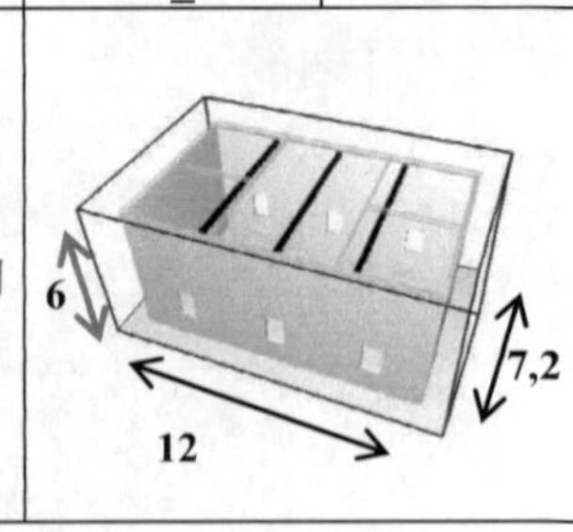	

In each one of the three comparison the fixed parameters, in addition to the enclosure size, are mesh size, number of ventilation openings, fire position and fire size. The output devices position is a fixed parameter as well in each one of the three comparison. Some fixed parameters, such as mesh size, were defined by performing sensitivity analysis (it has been ensured that all the ASET parameter are well resolved using this mesh size). Mesh size is 0.2×0.2×0.2 m^3. Regardless of the function used to perform the fire growth, the fire size was selected using the Italian Fire Code [8] that advices of the characteristics of the burners that can be applied to different occupancies. The occupancy selected was the "civil occupancy" and the characteristics of these fire case studies are presented in the following table.

Table 2. Fire Characteristics

Parameter	Civil occupancies
Fire Growth Rate t_α	150 s (fast)
RHR_{MAX}	5 MW
Fire Dimensions	1,8 x 1,8 m^2
$\dot{Q}^*$ (at RHR_{MAX})	0,8

The three comparisons between the HRR curves of the two fire growth function have been shown a similar behavior regardless of the enclosure geometry. In the first 50 seconds (highly significant for the ASET analysis) the HRR curves produced by the S_R show a *plateau* trend with higher values than the RAMP Curves (see figure 3, on the left), due to the first cell ignition that produces a flat Heat Release Rate trend. Furthermore the fire growth modeled with the S_R function has shown a discrepancy with the RAMP function as the fire HRR curve approaches the maximum HRR. This discrepancy occurs because the chosen fire source is square shaped whereas the fire spread over the fire source surface is radial. In order to avoid this discrepancy and to only study the growth phase of the fire, the following comparisons are evaluated during the first 300 seconds.

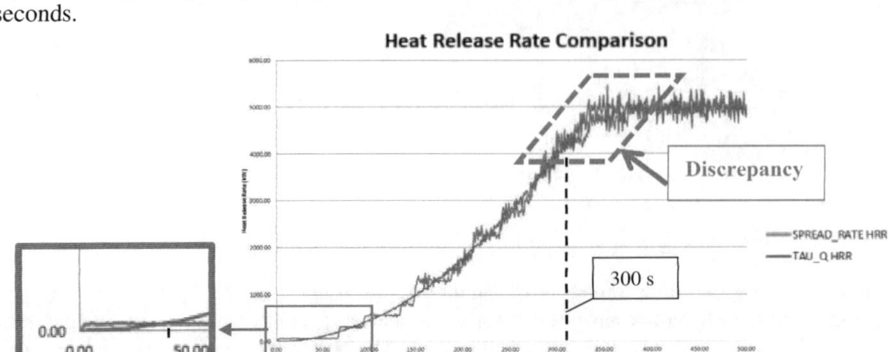

Fig. 3. HRR comparison between the SPREAD_RATE cases and RAMP cases

Seeing that the fire growth is a matter of concern for life safety, we analysed smoke layer height, visibility and gas temperature. The comparison of the graphs between the outputs of the smoke layer height devices (they were measured in multiple points and then averaged) of the two growth functions are represented below. The smoke layer for S1/S2 and S5/S6 drops below head height faster with S_R than without, but after 75 seconds, in all three comparisons, the layer heights of the S_R function cases are slightly higher than the ones of the RAMP function cases.

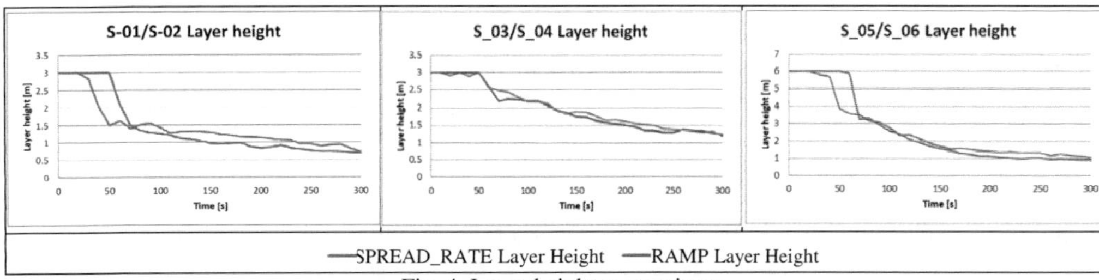

Fig. 4. Layer height comparisons

Furthermore the comparisons have been assessed for the visibility at 1.8 m off the ground. The scenarios that use S_R, after approximately 75 seconds, show a higher visibility than the ones that use RAMP. Fig. 5 shows the comparison between S_05 and S_06. This pair of scenarios has been chosen because the enclosure obscuration by the smoke is more gradual and the results are clearer than in the other comparisons. As a result it is clear that the visibility of the scenario S_05 (S_R) is always higher than the S_06 (RAMP).

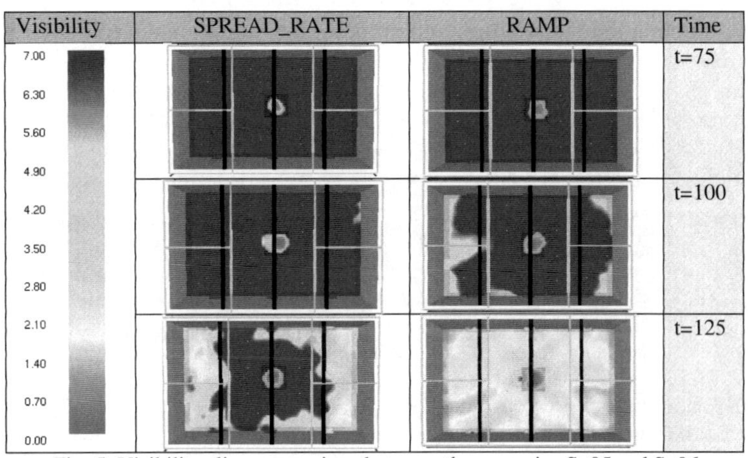

Fig. 5. Visibility slice comparison between the scenarios S_05 and S_06

Another important parameter related to life safety is gas temperature at 1.8 m off the ground. In all three comparisons the gas temperatures in the case studies that use the S_R are slightly lower than the ones that use the RAMP. This is clearly reflected in Fig. 5 that shows the gas temperature comparison between S_05 and S_06.

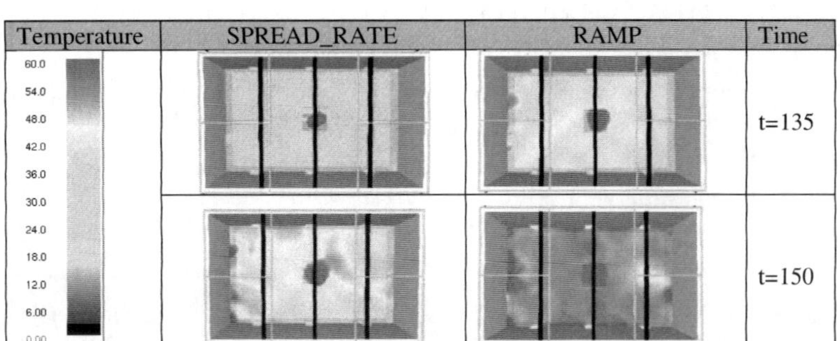

Temperature	SPREAD_RATE	RAMP	Time
See previous page	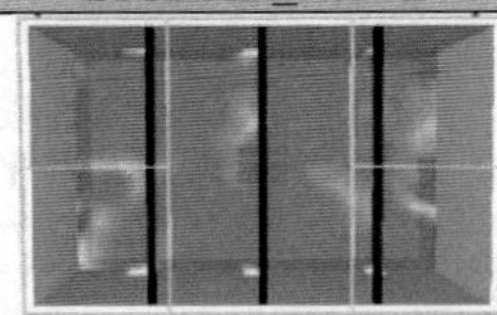	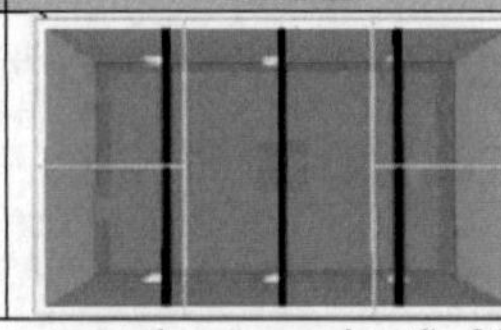	t=165

Fig. 6. Gas temperature slice comparison between the scenarios S_05 and S_06

Finally, the growth function dependency of the adiabatic surface temperature (AST) trend has been analysed to evaluate the fire resistance of structures. The curve trends have shown that there are no significant differences between the two functions of fire growth utilized.

CONCLUSION

Based on the information above, the following conclusions can be drawn.

- Since the S_R function associates a radial diffusion on a surface with a t-squared function of the HRR Curve, this behavior can be considered consistent with the real behavior of a fire. Moreover using the S_R function, the burning surface is kept small when the HRR is low and therefore it is possible to maintain applicable values of the $Q*$during the fire growth. On the contrary the HRR curve produced by this function has shown some discrepancies compared to the theoretical t-squared function, which have not been noticed in the RAMP function (excluding the small oscillations).
- As the S_R function, during the fire growth, generates a more structured flame than the RAMP-functions, the smoke produced is expected to layer in a more orderly manner, slowing down the fall of smoke level. This has been confirmed by the visibility and gas temperature analyses at 1.8 m off the ground. Therefore in all three comparisons the visibility and gas temperature slices have shown that the untenable conditions (related to this parameters) for occupants of the room occur before using the RAMP function using the S_R function.
- The smoke layer for S1/S2 and S5/S6 drops below head height faster with S_R than without, but after 75 seconds, in all three comparisons, the layer heights of the S_R function cases are slightly higher than the ones of the RAMP function cases. The phenomenon emerged during the first seconds could be due to the fact that, in the first 50 seconds, the HRR curves produced by the S_R show a *plateau* trend with higher values than the RAMP HRR Curves, where the trend is t-squared and start from 0 (see figure 3).
- The growth function dependency of the AST trend has been analysed to evaluate the fire resistance of structures, but the comparisons have not shown any significant differences between the two functions.

REFERENCES

[1] PD 7974-0:2002, *Application of Fire Safety Engineering Principles To The Design Of Buildings. Guide To Design Framework And Fire Safety Engineering Procedure,* British Standards Institution

[2] McGrattan K., Hostikka S., McDermott R., Floyd J., Weinschenk C., Overholt. K. "Fire Dynamics Simulator User's Guide. FDS Version 6.6.0", NIST, Gaithersburg, USA. 2017.

[3] Norén N., Rosberg D., "Developing a Swedish Best Practice Guideline for Proper use Of CFD-Models When Performing Aset-Analysis", Proceedings, Fire and Evacuation Modeling Technical Conference (FEMTC), 2014

[4] PD 7974-1:2003, *Application of fire safety engineering principles to the design of buildings. Initiation and development of fire within the enclosure of origin (Sub-system 1)* , British Standards Institution

[5] Drysdale, D., *An Introduction to Fire Dynamics*, John Wiley and Sons, Chichester, 1985, p. 136.

[6] Cox G., Sumar K., "Modeling enclosure Fires Using CFD," The SFPE Handbook of Fire Protection Engineering, 3rd edition, Society of Fire Protection Engineers, 2002, pp. 3-205 till 3-207.

[7] Hurley M. J., Rosenbaum E. R., *Performance-Based Fire Safety Design*, CRC Press Book, 2015. p18.

[8] Decree of the President of the Republic, August 1, 2011, no. 151. Regulations for simplified application of the discipline procedures relating to fire prevention activities (in Italian).

Design of fire safety equivalence of building open arcades to exterior spaces

Emanuele GISSI[1], Piergiacomo CANCELLIERE[2] and Mauro CACIOLAI[2]

[1]*Ministry of Interior, Italian National Fire Rescue and Service, CNVVF Torino, Italy*
[2]*Ministry of Interior, Italian National Fire Rescue and Service, Central Directorate for Fire Prevention and Technical Safety, Rome, Italy*

piergiacomo.cancelliere@vigilfuoco.it

ABSTRACT

An arcade is a succession of arches, each counter-thrusting the next, supported by columns, piers, or a covered walkway enclosed by a line of such arches on one or both sides. In warm or wet climates regions, exterior arcades provide shelter for pedestrians. A building open arcade is sometimes called a porch. Arcades usually border a covered section of a building and sometimes they support roofs. They are very old architectural elements that date back thousands of years.

Building open arcades (*"Portici"*) are traditionally employed in cities to provide shelter for pedestrians along public and private streets. Along the closed side of the arcades, commercial activities of all sort often offer their main entrance and exhibition spaces. Other activities as private dwellings or any other kind of civil activity may communicate directly through doors or other openings towards the covered space under the arcade.

According to current prescriptive fire safety regulations, open arcades cannot be considered equivalent to exterior open spaces, and all the activities that have openings towards these covered spaces should be presumed in mutual communication from a fire safety design point of view. In fact, it is hypothesized that closed enclosures allow for the faster spread of fire effluents between the communicating spaces.

Nevertheless, according to experiences and engineering judgments, under certain conditions and constraints, the open arcades allow for evacuation of smoke and heat towards the exterior spaces through the open side. When these conditions are met, all the activities that have openings towards these covered spaces could be assumed mutually independent from a fire safety design perspective, because, in case of fire, the fire effluents do not spread easily towards the adjoining activities.

In this paper a fire safety engineering calculation is performed to assess in which circumstances an open arcade can be considered as equivalent to an exterior space for deemed-to-satisfy solutions of the Italian Fire Code (IFC 2015).

The study proposes an arcade parametric fire model, whose pre-processor has been developed in Python, for assessing conditions, safety measures and geometric constraints that allow to recognize building open arcades as exterior safety places. The open arcade can be considered equivalent to exterior space for deemed-to-satisfy solutions when occupants can safely escape from the open arcade towards a safe place, occupants of other activities abutting in the same open arcade are not harmed by the fire effluents coming from the activity involved in the fire, and when fire and its effluents do not spread to other abutting activities. Furthermore, the paper assumes a simplified equivalent model structure of the building arcade where the parametric study conducted with FDS is aimed at setting some basic rules for the fire safety design of building open arcades.

The results of the parametric study show that under certain conditions and constraints, the open arcades allow for evacuation of smoke and heat towards the exterior spaces through the open side. Therefore, when these conditions are met, all the activities that have openings towards these covered spaces could be assumed mutually independent from a fire safety design perspective, since in case of an outbreak of a fire the fire effluents do not spread easily towards the adjoining activities.

KEYWORDS:

Open arcades, fire modelling, CFD, Python, exterior safety places

IOP Conf. Series: Journal of Physics: Conf. Series **1107** (2018) 042008 doi:10.1088/1742-6596/1107/4/042008

1. INTRODUCTION

To reduce risks and achieve acceptable levels of safety, fire codes and regulations play a fundamental role both in buildings and high hazard facilities. Most of the Italian activities subjected to fire inspection and listed in the annexe 1 of the Decree DPR 151/2011 [1], are subject to prescriptive fire safety codes. These codes are composed of some requirements which try to address all the different components and devices of the system to provide fire safety for a building or an industrial activity. Moreover, when fire safety targets must be satisfied in pre-existing buildings or in activities that are going to be modified into more complex ones, there are more constraints to cope with and it could be difficult to achieve an acceptable level of fire risk using prescriptive-based fire codes and regulations. For these reasons, on August 20th, 2015, the Italian Home Office released the Ministerial Decree of August 3rd, 2015, that contains a new approach to the fire safety design of activities subjected to fire inspection. The technical Ministerial Decree is titled "Approval of fire prevention technical standards, pursuant to Article 15 of Legislative Decree 139 of March 8, 2006" [2], but is commonly recognized among Italian fire officers and practitioners as the Italian Fire Code (IFC) [3].
Following the fire safety engineering principles, the IFC sees the process of the fire safety design considering the system as a whole by focusing on the safety targets whether they are life safety, property loss, business interruption, environmental damage or heritage preservation. The fire design method is very flexible: for each fire-safety project offers design solutions that are semi-performance based (the so-called "deemed-to-satisfy solutions"). These compliant solutions contain prescriptive examples of materials, products, design factors, construction and installation methods, which - if adopted - comply with the performance requirements of the IFC. If a deemed-to-satisfy solution could not be satisfied, the IFC offers performance-based solutions called "alternative solutions". The alternative solution is any solution that can meet the IFC performance requirements, other than a deemed-to-satisfy solution, using other international fire safety standards, innovative products and fire safety technology and last, but not least, applying the Fire Safety Engineering (FSE) design. In fact, the whole section M (*Methods*) of the IFC deals with the FSE performance-based approach. The methods M.1 describes the minimum content of the additional fire safety documentation in case of a performance-based approach must be delivered along with the usual design reports (structural, mechanical, etc.), while the sub-methods M.2 contains the guidelines for the identification and the selection of the design fire scenarios to be used for "quantitative analysis". The last method M.3 contains some details for the application of performance-based approach methods for life safety problems.
According to current prescriptive fire safety regulations, open arcades cannot be considered equivalent to exterior open spaces, and all the activities that have openings towards these covered spaces should be presumed in mutual communication from a fire safety design point of view. Fig. 1 depicts some example of historical building open arcades.

Fig. 1. Some examples of historical building open arcades

Applying the FSE design procedure of the IFC, this paper proposes an arcade parametric fire model for assessing conditions, safety measures and geometric constraints that allow to recognize open arcades as exterior safety places.

2. IDENTIFICATION OF THE ARCADE FIRE SAFETY GOALS.

The fire safety engineering methodology proposed in chapter M.3 of the IFC is followed. the scope of the following calculation consists in assessing the conditions and the geometric constraints that allow for considering open arcades equivalent to exterior space. The open arcade is to considered equivalent to exterior space for deemed-to-satisfy solutions when the following safety goals are met :occupants can safely escape from the open arcade towards a safe place, occupants of other activities abutting in the same open arcade are not harmed by the fire effluents coming from the activity involved in the fire and fire and its effluents do not spread to other abutting activities. These safety goals are deemed satisfied if the following performance criteria [4] are respected in any condition:

- the hot smoke layer in the open arcade remains a1t a height not lower than 2.50 m;
- the same hot smoke layer has an average temperature not higher than 200°C.

No requirement on fire or smoke heat radiation limit is developed, since the temperature of the hot smoke layer implies a radiation level not greater than 2.5 kW/m^2 and the radiating heat from the burning activity would not be different if the same event had happened in an open space instead of under an open arcade.
Fire resistance rated walls or enclosures are introduced in trial designs to prevent fire and smoke spread towards neighbouring activities through common border elements. The predefined fire scenario for civil activities described in chapter M.2, table M.2-2 of the IFC is employed.
Therefore, a constraint on the typology of the activities abutting in the studied open arcade shall be enforced: only civil activities having a design fire load $q_f \leq 1200$ MJ/m^2 are allowed abutting in the open arcade. The fire is presumed to have an instantaneous growth to the maximum heat release rate value and an infinite duration. For simplicity, the fire is located directly in the middle of the open arcade.

3. GEOMETRY OF THE GENERIC OPEN ARCADE MODEL

By definition, an open arcade is an enclosed space that has at least one of its two longer sides open towards the exterior. The geometry of the generic open arcade is described as in Figure 2, where:

- pz is the average height of the vault,
- ha is the average height of the opening towards the open space,
- py is the length of the shorter horizontal dimension of the arcade,
- px is the overall horizontal width of the front arcade openings towards the open space, measured at ha height, between the end of the arcades or any element that obstacles smoke spread along the open arcade (eg. fire or smoke curtains).

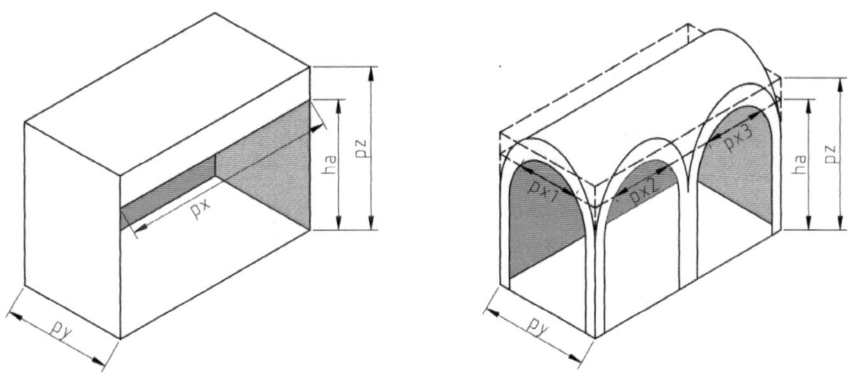

Fig. 2. Geometry of the generic open arcades

When multiple openings are present, as in Fig. 2, the value of px shall be obtained by summing the single opening widths, i.e. px=px1+px2+px3. No reduction of the horizontal width of the openings is allowed at heights lower than ha. The IFC defines the average height of an enclosed space as:

$$h_{avg} = \frac{\int_A h dA}{A} \qquad (1)$$

with A area of the enclosed space. The simplified geometrical model of arcades described above takes into account the main parameters that could affect the spread and the accumulation of the heat and smoke under the arcade, namely the average height of the vault (pz) and the average height of the opening towards the open space (ha). Moreover,no description of the occupants quantities, qualities and behaviour is required, because the safe condition for the occupants of the arcade is going to be conserved indefinitely.

4. DEVELOPMENT AND EVALUATION OF THE TRIAL DESIGN

Several parametric trial designs are proposed according to the geometry pictured in Fig. 2, and to the following geometric restrictions:

- $3\ \text{m} \leq py \leq 1.2\ pz$,
- $0.6\ pz \leq ha \leq pz$,

IOP Conf. Series: Journal of Physics: Conf. Series **1107** (2018) 042008 doi:10.1088/1742-6596/1107/4/042008

Several additional requirements should be applied to construction product elements an materials of other activities abutting in the same building open arcade of the designed activity to prevent fire and smoke spread towards neighbouring activities, as listed in detail in the next Section "5. SELECTION OF THE FINAL DESIGN".

The evaluation of trial designs is performed by using NIST FDS [5] and a custom parametric preprocessor written in Python [6]. The preprocessor allows the automatic variation of any quantity inserted by the analyst into the into FDS input file. The following variations of the parameters are simulated:

- $3 \text{ m} \leq pz \leq 9 \text{ m}$,
- $3 \text{ m} \leq ha \leq 9 \text{ m}$,
- $3 \text{ m} \leq py \leq 10.75 \text{ m}$,
- $12 \text{ m} \leq px \leq 200 \text{ m}$.

A sensitivity analysis of the model has been also done varying the simulation grid and fire size [7]. Due to the limited space, the results of the sensitivity analysis and numerical conditions are not reported. Trends are extrapolated over the upper limits of the variations. Around 300 numerical fire simulations have been completed in the ranges of variation of the parametric trial design. Some example results are shown in Fig. 4 a) and b).

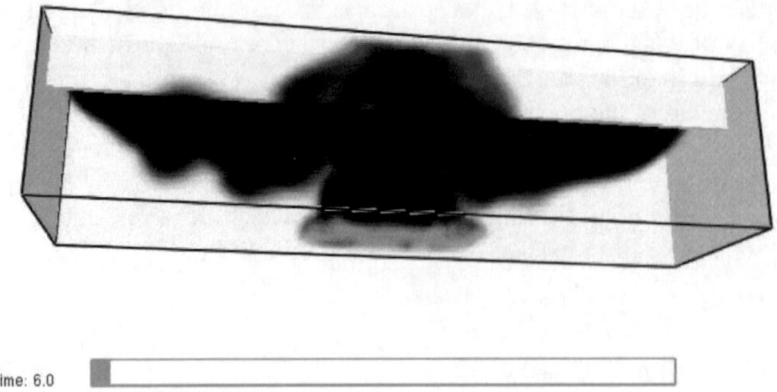

Fig. 3. An example of numerical fire simulation of the modelled open arcade.

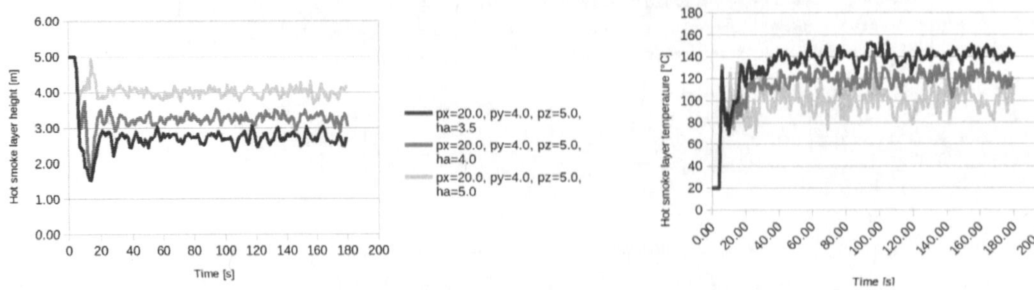

a) *b)*

Fig. 4. Hot smoke layer height results a) and hot smoke layer temperature result b).

Table 1 shows the results of the simulations of the hot smoke layer height for the parametric trial designs, considering px=20 m and py, pz and ha varying according to the trial design range. The fluctuating in time-height value is averaged after the studied system has reached stationary conditions at around 120 s of each simulation.

The empty cells represent out-of-allowed-bounds geometric conditions. It is important to underline that results displayed in italics are not respecting performance criteria: the hot smoke layer in the open arcade is lower than 2.50 m or it has an average temperature higher than 200°C.

The hot smoke layer temperatures for the parametric trial designs calculated during the simulation, as averaged after the studied system has reached stationary conditions at around 120 s of simulation, are all under 200°C thus respecting the assumed performance criteria.

Local temperature on structural elements in the vertical of the fire plume are significantly higher, and could lead to localized collapse of the open arcade, if not specifically designed as fire-resistant construction product elements.

Table 1. Hot smoke layer average height LH, after achieving stationary conditions at 120 s for px=20.00 m

			ha [m]					
px **[m]**	***py*** **[m]**	***pz*** **[m]**	**3.00**	**3.50**	**4.00**	**5.00**	**6.00**	**9.00**
20.00	3.00	3.00	*2.29*					
		3.50	*2.38*	2.72				
		4.00	*2.41*	2.82	3.25			
		5.00	*2.31*	2.78	3.35	4.22		
		6.00			3.27	4.25	5.27	
		9.00					5.39	8.38
	4.00	3.50	*2.27*	2.65				
		4.00	*2.37*	2.74	3.12			
		5.00	*2.35*	2.79	3.31	4.13		
		6.00			3.27	4.25	5.12	
		9.00					5.23	8.14
	6.00	5.00	*2.25*	2.72	3.30	4.04		
		6.00			3.22	4.23	5.00	
		9.00					5.06	7.98
	9.00	9.00					5.07	7.88
	10.75	9.00					5.03	7.72

5. SELECTION OF THE FINAL DESIGN

The final parametric designs are selected by comparing the numeric results to the performance criteria. Any open arcade respecting the hypothesis assumed for the fire scenarios selected in Section 3 and the prescriptions described in Section 4 shall be deemed equivalent to a safety open space when employing the deemed-to-satisfy solutions of the IFC. The following deemed-to-satisfy solution can be taken into account and shall be fulfilled for considering a building open arcade equivalent to an exterior safety spaces:

- Only civil activities having a design fire load $q_f \leq 1200$ MJ/m^2 are allowed abutting into the open arcade;
- The following geometric constraints for the open arcade apply: 3 m $\leq$ py $\leq$ 1.2 pz and 0.6 pz $\leq$ ha $\leq$ pz;
- The hot smoke layer shall be considered filling the following upper portion of the open arcade:
  - when px < 15 m, the entire height of the open arcade, from the ceiling to the floor;
  - when px $\geq$ 15 m, the upper volume of the open arcade from the ceiling to the ha - 1 m height;
- Requirements for construction elements:
  1. any vertical and horizontal separation (eg. walls and floors) between the designed activity and the rest of the building shall be classified at least EI 30 for fire resistance;
  2. any vertical and horizontal separation (eg. walls and floors) placed above the designed activity and abutting the open façade shall be classified at least EI 30 for fire resistance;
  3. any opening placed above the designed activity shall be protected by a closure classified at least E 30-Sa for fire resistance;
  4. the open arcade shall be classified at least R 30 fire resistance rated in front and in the vertical walls or partitions of the designed activity, where the thermal stress due to the fire is presumed significant and could cause local collapse.
  5. any opening that may become impinged by the hot smoke layer shall be protected by a smoke barrier classified at least D 30 according to EN 12101-1 [8];
- The effects of the fire and its effluents along the building open arcade are considered contained by:
  6. smoke barriers operating at high temperature classified at least DH 30 according to EN 12101-1 [8] , closing the entire py dimension of the open arcade, descending to the floor level, if located at a distance not greater than 2.5 m from the designed activity facade,

7. smoke barriers classified at least D 30 according to EN 12101-1 [8] , closing the entire py dimension of the open arcade and descending to the lower height of the hot smoke layer, in other cases.

Fig. 5 resumes the 1 to 7 aforementioned requirements for construction product elements of the building open arcades that have to be fulfilled for considering them as an open safety place.

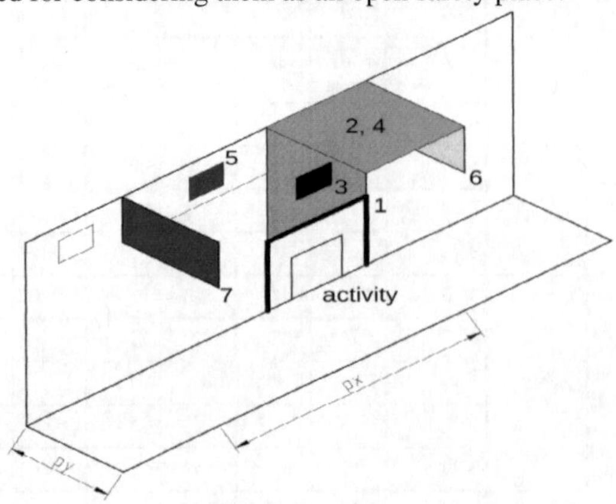

Fig. 5. Building open arcade construction product elements requirements for considering an open arcade as an open safety place.

CONCLUSION

According to current prescriptive Italian fire safety regulations, building open arcades cannot be considered equivalent to exterior open spaces, and all the activities that have openings towards these covered spaces should be presumed in mutual communication from a fire safety design point of view. The study proposes a building arcade parametric fire model, whose pre-processor has been developed in Python, that is used for assessing conditions, safety measures and geometric constraints that allow to recognize open arcades as exterior safety places. Under certain conditions and constraints, the open arcades allow for evacuation of smoke and heat towards the exterior spaces through the open side. When these conditions are met, all the activities that have openings towards these covered spaces could be assumed mutually independent from a fire safety design perspective, because, in case of an outbreak of a fire, the fire effluents do not spread easily towards the adjoining activities. Moreover, this paper is an example of a simple method to build deemed-to-satisfy solutions for inclusion in prescriptive fire safety codes that are fully justified by quantitative methods and are not based on "magic numbers" [9].

REFERENCES

[1] DECREE OF THE PRESIDENT OF THE REPUBLIC August 1, 2011, no. 151. Regulations for simplified application of the discipline procedures relating to fire prevention activities (in Italian).

[2] DECREE OF THE MINISTER OF THE INTERIOR 3 August 2015 "Approval of fire prevention technical standards, pursuant to Article 15 of Legislative Decree 139 of 8 March 2006" (in Italian)

[3] Cancelliere P. et al., *Italian hybrid Fire Prevention Code*, WIT Transactions on The Built Environment, Vol. 174, Pages 117-117, Published 2018

[4] Purser, D. A. (2008). Assessment of hazards to occupants from smoke, toxic gases and heat. In P. J. Di Nenno (Ed.), SFPE handbook of fire protection engineering. Chapter 2–6 (4th ed.,pp. 96–193). Quincy, MA: National Fire Protection Association.

[5] NIST SOFTWARE, FDS and Smokeview, https://www.nist.gov/services-resources/software/fds-and-smokeview

[6] *Python Software Foundation*, US, https://www.python.org/

[7] ISO 16730-1:2015 "Fire safety engineering — Procedures and requirements for verification and validation of calculation methods — Part 1: General

[8] EN 12101-1:2005+A1:2006 Smoke and heat control systems. Specification for smoke barriers

[9] Gissi E., Ronchi E., Purser D. A, . *Transparency vs magic numbers: The development of stair design requirements in the Italian Fire Safety Code*, Fire Safety Journal, Volume 91, July 2017, Pages 882-891

IOP Conf. Series: Journal of Physics: Conf. Series **1107** (2018) 042009 doi:10.1088/1742-6596/1107/4/042009

Simulations of a Turbulent Line Fire with a Steady Flamelet Combustion Model and Non-Gray Gas Radiation Models

Van Minh LE[1,2], Alexis MARCHAND[1], Salman VERMA[1], James WHITE[1], Andre MARSHALL[1], Thomas ROGAUME[2], Franck RICHARD[2], Jocelyn LUCHE[2] and Arnaud TROUVÉ[1]

[1]*Department of Fire Protection Engineering, University of Maryland, USA*
[2]*Institut Pprime (UPR 3346, CNRS), Université de Poitiers, France*

atrouve@umd.edu

ABSTRACT

The general objective of this project is to develop an accurate combustion and radiation modeling framework for high-fidelity large eddy simulations (LES) of well-controlled turbulent laboratory-scale fires for which the fuel composition and fuel oxidation chemistry are known. This modeling framework is aimed at providing a solid basis for the development and validation of engineering-level models used in simulations of real-world fire problems for which the sources of fuel are diverse, complex, and in many cases, poorly characterized. The combustion model features a library of flamelet solutions corresponding to one-dimensional, steady, laminar, counterflow diffusion flames simulated with specialized software, a chemical kinetic mechanism and an equi-diffusive molecular transport model (i.e., unity Lewis numbers). Two different flamelet libraries are considered here: a first library generated with a solver called libOpenSMOKE and a detailed chemical kinetic mechanism developed for C1-C3 combustion chemistry and a second library generated with a solver called FlameMaster and the GRI-Mech v3.0 chemical kinetic mechanism developed for methane combustion chemistry. The radiation model features a banded Weighted-Sum-of-Gray-Gases model but (so far) no description of subgrid-scale turbulence-radiation interactions (TRI). This modeling framework is incorporated into a LES solver developed by FM Global and called FireFOAM, and is evaluated in simulations of a two-dimensional, plane, buoyancy-driven, methane-air, turbulent diffusion flame experimentally studied at the University of Maryland. The configuration corresponds to an intermediate validation step in our model development strategy without the complications of flame extinction. The flame structure is characterized by new micro-thermocouple measurements of the temporal mean and root-mean-square gas temperatures. Comparisons between simulated and measured temperatures show significant discrepancies that are explained by the large values of the width of the presumed probability density function (PDF) representing subgrid-scale variations of mixture fraction and by the absence of a model for subgrid-scale TRI.

KEYWORDS:

Buoyant turbulent diffusion flame, Large Eddy Simulation, Flamelet combustion, Weighted-Sum-of-Gray-Gases radiation

IOP Conf. Series: Journal of Physics: Conf. Series **1107** (2018) 042009 doi:10.1088/1742-6596/1107/4/042009

INTRODUCTION

Despite recent progress made in simulations of fires using Computational Fluid Dynamics (CFD), CFD models for fire applications face several technical challenges, in particular, the need to provide accurate models for turbulent combustion [1–4]. The popular combustion model called the Eddy Dissipation Concept (EDC) model offers a cost-effective description of the location and intensity of heat release processes under a range of flow conditions but is known to have several limitations. First and foremost, EDC assumes that the rate of combustion is entirely determined by the rate of turbulent fuel-air mixing and therefore does not represent flame extinction/ignition phenomena, which also depend on combustion chemistry. Also, EDC has a domain of validity restricted to turbulent flames and does not correctly represent laminar combustion. In the present study, we consider two different flamelet libraries. A first library generated with a solver called libOpenSMOKE [5-7] and a detailed chemical kinetic mechanism developed for C1-C3 combustion chemistry [8] and a second library created with a solver called FlameMaster [9] and the GRI-Mech v3.0 chemical kinetic mechanism developed for methane combustion chemistry [10].

Another challenge in CFD simulations is the need to provide accurate models for thermal radiation [1–4]. In the presence of non-gray species, the emission and absorption terms in the Radiative Transfer Equation (RTE) depend on wavelength. However, the accurate description of the stiff spectral variations of radiation properties (as described by a Line by Line model for example) has a high computational cost that can easily dominate the overall cost in a CFD solver. A typical simplification to control cost is to assume a linear relationship between radiative power and heat release rate using a coefficient of proportionality called the global radiative loss fraction. The global radiative loss fraction is an empirical parameter that provides a convenient solution to the problem of representing changes in radiative emissions when considering different fuels, but that fails to represent changes in radiative emissions in response to evolving fire conditions. An alternative solution is the Weighted-Sum-of-Gray-Gases (WSGG) model which represents spectrally-resolved radiation at a reasonable cost.

The present study is aimed at addressing current limitations in combustion and radiation modeling for simulations of fires. In the following, we consider models that account for combustion chemistry and do not prescribe the global radiative loss fraction. Our modeling framework is incorporated into a Large Eddy Simulation (LES) solver developed by FM Global, called FireFOAM [11], and is part of an ongoing study corresponding to an experimental configuration called the turbulent line burner [12]. The turbulent line burner is representative of the buoyancy-driven turbulent flame conditions found in fires, albeit with a simplified fuel source: the fuel is gaseous methane and the fuel flow rate is controlled. The use of methane offers several benefits: methane is a well-studied fuel and there is detailed knowledge of its oxidation chemistry, which allows using state-of-the-art combustion models; and methane is a weakly-sooting fuel, which allows focusing on gas radiation. We consider in the following a flamelet model for combustion and a non-gray model for gas radiation. Comparisons with experimental data rely on new temperature measurements performed with micro-thermocouples. Our objective is to evaluate the performance of the flamelet combustion and non-gray radiation models in the simplified context of the turbulent line burner with the idea that these models will eventually provide a solid basis for the development and validation of engineering-level models used in real-world fire applications.

EXPERIMENTAL AND NUMERICAL CONFIGURATION

Experimental configuration

The configuration adopted in the present study is the turbulent line burner studied at the University of Maryland [12]. The configuration corresponds to a buoyant, turbulent, methane-fueled, slot-burner diffusion flame with a controlled coflow (Fig. 1). Flames are stabilized above a 5-cm-wide and 50-cm-deep Wolfhard-Parker burner surrounded by a small horizontal plate and an oxidizer coflow. The coflow is 50-cm-wide and 75-cm-deep. The methane velocity is 6 cm/s and the theoretical heat release rate of the flame is 50 kW. In the present study, the oxidizer coflow is made of pure air; the velocity of the air coflow is 25 cm/s. The flame structure is characterized through measurements of the temporal mean and root-mean-square (rms) gas temperature using 12.7-μm-diameter Type-S (Pt/10%Rh) thermocouples. The radiation loss from the micro-thermocouples is assumed small and is simply neglected.

Numerical configuration

Numerical simulations are performed using FireFOAM [11], a fire modeling solver developed by FM Global and based on an open-source, general-purpose, CFD software package called OpenFOAM [13]. FireFOAM is a second-order accurate, finite volume solver with implicit time integration; the solver features advanced meshing capabilities (structured/unstructured polyhedral mesh); it also features a massively parallel computing capability. FireFOAM uses a Favre-filtered, compressible-flow, LES formulation and provides a choice between several modeling options for the treatment of turbulence, combustion, and thermal radiation. In the current simulations: subgrid-scale (SGS) turbulence is described using the one-equation eddy viscosity model, which solves a transport equation for SGS turbulent kinetic energy. The computational domain is presented in Fig. 1 and is 2-m-wide in the cross-stream *x*-direction, 2-m-long in the spanwise *y*-direction and 2-m-high in the vertical *z*-direction. The grid is block-structured and features three levels of resolution. The level of resolution used in the flame region corresponds to a uniform grid with cubic cells of size $\Delta x_1 = 4.167$ mm (i.e., 12 grid cells across the burner width). The total number of cells is 2.36 million. All FireFOAM simulations are performed for a duration of 30 s. Turbulent statistics are collected for the final 20 s of each simulation after the flow and flame become statistically stationary and long enough for the statistics to be converged. A classical Courant-Friedrichs-Lewy (CFL) condition is used to control the time step: the time step is approximately equal to 0.25 ms. Each simulation is run using 192 processors on a large-scale Linux cluster with the adiabatic cases requiring 14,000 CPU-hours and the flamelet/non-gray model requiring 25,000 CPU-hours.

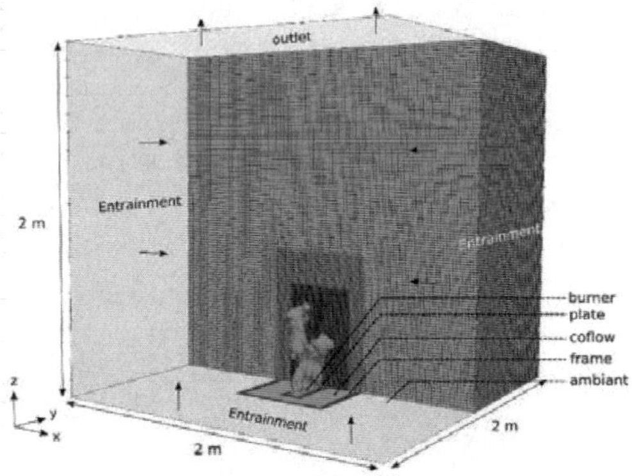

Fig. 1. Computational domain and multi-level mesh refinement. The flame height is approximately 50 cm.

Combustion modeling

As mentioned in the introduction, we use libOpenSMOKE and FlameMaster to generate laminar flame solutions that are first parametrized in terms of a reduced set of variables: the mixture fraction Z, the stoichiometric value of the scalar dissipation rate χ_{st}, and the enthalpy defect Δh ($\Delta h = 0$ for adiabatic conditions). These two flamelet libraries are then incorporated into FireFOAM using statistical averaging based on probability density functions (PDFs). The implementation of the flamelet libraries into FireFOAM is based on the following expression:

$$\tilde{q} = \int_{\Psi} q^{fl}(\Psi)\tilde{p}(\Psi)d(\Psi) \quad (1)$$

where $\tilde{q}$ designates the LES-filtered value of a quantity q (with q representing the gas temperature or a species mass fraction); where ψ represents the parametrization of the flamelet library, $\Psi = (Z, \chi_{st}, \Delta h)$; where q^{fl} is the value of q found in the flamelet library for a given set of values for ψ; and where $\tilde{p}$ is the joint (mass-weighted) PDF representing SGS variations of Z, χ_{st} and Δh. Assuming statistical independence, we write: $\tilde{p}(Z, \chi_{st}, \Delta h) = \tilde{p}(Z) \times \tilde{p}(\chi_{st}) \times \tilde{p}(\Delta h)$. In the present work, we adopt a classical presumed-PDF approach in which $\tilde{p}(Z)$ is a β-PDF, $\tilde{p}(\chi_{st})$ and $\tilde{p}(\Delta h)$ are Dirac δ-functions. $\tilde{p}(Z)$ is parameterized by the local LES-filtered mean and variance, $\tilde{Z}$ and $\widetilde{Z_v}$, with $Z_v = (Z - \tilde{Z})^2$; $\tilde{p}(\chi_{st})$ and $\tilde{p}(\Delta h)$ are parametrized by the local LES-filtered means $\widetilde{\chi_{st}}$ and $\widetilde{\Delta h}$. An efficient implementation of Eq. (1), consists of pre-tabulating $\tilde{q}$ in terms of the LES grid scale variables, $\tilde{Z}$, $\widetilde{Z_v}$, $\widetilde{\chi_{st}}$ and $\widetilde{\Delta h}$. Following standard modeling choices, $\tilde{Z}$ is obtained through the solution of turbulent transport equations, whereas $\widetilde{Z_v}$, $\widetilde{\chi_{st}}$ and $\widetilde{\Delta h}$ are obtained through algebraic closure models. In summary, the FireFOAM system of governing equations includes conservation equations for the LES-filtered mass density, flow velocity vector and mixture fraction; the flamelet model is informed by the values of

IOP Conf. Series: Journal of Physics: Conf. Series **1107** (2018) 042009 doi:10.1088/1742-6596/1107/4/042009

$(\tilde{Z}, \widetilde{Z_v}, \widetilde{\chi_{st}}, \widetilde{\Delta h})$ and provides in return the values of LES-filtered temperature and species mass fractions. In the following, we refer to the steady laminar flamelet models as "SLF/libOpenSMOKE" and "SLF/FlameMaster".

Radiation modeling

In the present study, we consider two alternative approaches to the prescribed global radiative loss fraction model. Both approaches rely on solving the RTE with full emission and absorption effects. The first approach corresponds to a gray model; the second corresponds to a more elaborate non-gray model [14]. The RTE equation for the gray model is:

$$\nabla \bar{I}.\vec{s} = \overline{\kappa I_b} - \overline{\kappa I} \quad (2)$$

where κ is the gas mean absorption coefficient and I_b the local blackbody radiation intensity, $I_b = (\sigma T^4/\pi)$, with σ the Stefan-Boltzmann constant and T the temperature. κ is obtained from the composition and temperature of the gas: $\kappa = p\,(\sum_k x_k a_k(T))$, where p is pressure, x_k the mole fraction of species k ($k = CO_2$, H_2O or CH_4) and a_k the Planck mean absorption coefficient for species k. In this description, the coefficients a_k are treated as known functions of temperature [15]. Note that in the present study, we ignore SGS turbulence-radiation interactions (TRI) and simply write: $\overline{\kappa I_b} = \bar{\kappa} \times (\sigma \tilde{T}^4/\pi)$; $\overline{\kappa I} = \bar{\kappa} \times \bar{I}$; and $\bar{\kappa} = p\,(\sum_k \widetilde{x_k} a_k(\tilde{T}))$. The assumption that SGS TRI effects can be neglected is questionable and will be re-visited in future work. In the following, we call this gray model "gray" and use $N_\Omega = 64$.

We now proceed to a brief discussion of the non-gray model. We consider the classical banded Weighted-Sum-of-Gray-Gases (WSGG) model and use the FireFOAM implementation originally proposed in Ref. [16]. The RTE equation is in that case:

$$\nabla \overline{I_j}.\vec{s} = \overline{\kappa_j a_j I_b} - \overline{\kappa_j I} \quad (3)$$

where the subscript j denotes a fictitious gray gas representing a certain spectral region; where κ_j is the mean absorption coefficient of the fictitious gas in that region; and where a_j is a model coefficient. In WSGG, the coefficients κ_j and a_j are calibrated for different values of the H_2O-to-CO_2 mole ratio using a spectrally-resolved radiation model (i.e. a Statistical Narrow Band model or a Line-by-Line model). In the present study, κ_j and a_j are described using the model proposed in Ref. [17] with the assumption that the H_2O-to-CO_2 mole ratio is equal to 2 (the maximum value allowed in Ref. [17]). In the following, we again choose to ignore SGS TRI, call this non-gray model "WSGG", and use 5 fictitious gray gases and $N_\Omega = 64$. Note that the computational cost of WSGG is 25,000 CPU-hours instead of 14,000 CPU-hours for the grey model.

RESULTS AND DISCUSSION

We focus on gas temperature results and present temporal mean and rms values, noted $\bar{T}$ and T_{rms}, as a function of elevation z along the flame centerline located at $x = 0$ (Fig. 2). $\bar{T}$ and T_{rms} are obtained by post-processing the LES solutions and correspond to the mean and amplitude of grid-resolved fluctuations (SGS contributions are not included). Experimental data in Fig. 2 show that $\bar{T}$ increases rapidly near the flame base, up to a peak value close to 1200 K, and then decreases at elevations larger than 0.3 m; T_{rms} also increases quickly near the flame base, up to values close to 400 K, and then decreases at elevations larger than 0.55 m. Numerical results in Fig. 2 show that: SLF/gray and SLF/WSGG do not correctly capture the shape of the z-variations of $\bar{T}$ and T_{rms} and significantly under-estimate (over-estimate) peak values of $\bar{T}$ (T_{rms}) by approximately 200 K (by more than 100 K). As discussed previously, in the flamelet solutions, grid-resolved temperatures depend on assumptions made in the shape of the subgrid-scale PDF of $\tilde{p}\,(\tilde{Z}, \widetilde{Z_v}, \widetilde{\chi_{st}}, \widetilde{\Delta h})$ (treated as the product of a β-PDF times two Dirac δ-functions) and on the controlling parameters $\tilde{Z}, \widetilde{Z_v}, \widetilde{\chi_{st}}, \widetilde{\Delta h}$. Errors in the assumed PDF shape and/or errors in predictions of the controlling parameters will lead to errors in simulated temperatures.

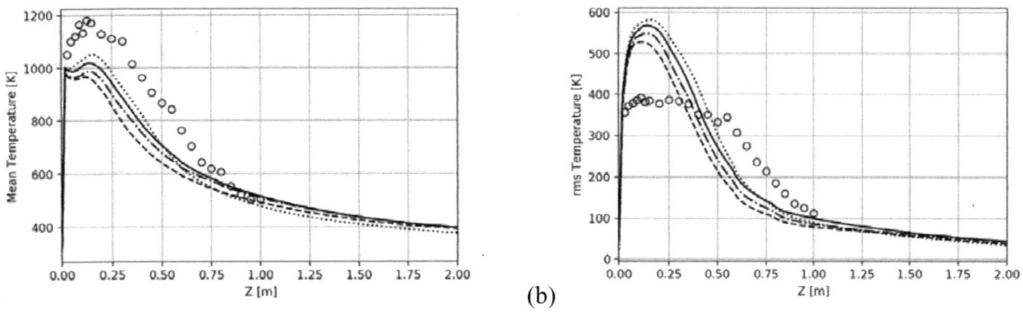

(a) (b)

Fig. 2. FireFOAM simulations of the turbulent line burner: vertical variations of (a) mean temperature $\overline{T}$ and (b) rms temperature T_{rms} along the flame centerline. Comparison between experimental data (symbols) and LES simulations performed with SLF/libOpenSMOKE - Adiabatic (solid line), SLF/FlameMaster – Adiabatic (dotted line), SLF/libOpenSMOKE - Gray (dashed line) and SLF/libOpenSMOKE - WSGG (dashed-dotted line).

We first consider the possibility that the under-estimated values of simulated temperatures are due to over-estimated levels of radiative heat losses. We proceed to calculate the global radiative power by spatially integrating the radiative source term that appears in the Δh-equation over the computational domain; we find: Q_{rad} = 7.6 kW for SLF/gray and Q_{rad} = 3.8 kW for SLF/WSGG. After dividing by the global heat release rate (50 kW), these results are found to correspond to χ_{rad} = 15% for SLF/gray and χ_{rad} = 7.6% for SLF/WSGG. These values are much lower than those measured experimentally: χ_{rad} = 23%. Thus, the low values of simulated temperatures are not explained by excessive values χ_{rad}; on the contrary, SLF/gray and SLF/WSGG significantly under-predict radiative emissions. Neglecting soot radiation does also contribute to reduce the total radiative emission (but the contribution of soot to radiant emission remains small).

Note that we have checked that the choice of the flamelet library (and of the chemical kinetic mechanism) did not affect the final results. Two adiabatic simulations were performed using SLF/libOpenSMOKE and SLF/FlameMaster. We find that both simulations provide similar results (see Fig. 2) and conclude that the solution does not depend on the choice of the flamelet library.

Finally, we consider the possibility that the underestimated values of simulated temperatures are due to over-estimated values of the SGS Z-variance, $\widetilde{Z_v}$. We confirm this possibility after checking that values of $\widetilde{Z_v}$ are relatively high in regions where $\tilde{Z} \approx Z_{st}$, we find that $(\widetilde{Z_v})^{1/2} \geq Z_{st}$ where Z_{st} = 0.055 is the stoichiometric value of mixture fraction. High values of $\widetilde{Z_v}$ correspond to a wide β-PDF in mixture fraction space, which in turn lead to reduced temperatures, in particular for $(Z_{st}/2) \leq \tilde{Z} \leq (2 \times Z_{st})$ (the reduction may be as high as several 100s K). Thus, the low values of simulated temperatures can be explained by excessive values of $\widetilde{Z_v}$ which are responsible for increasing the effect of statistical filtering associated with the SGS Z-PDF. These results suggest that the SGS Z-PDF model should be re-examined in future work in order to improve the performance of the flamelet model. On a separate but related note, the low values of $\tilde{T}$ result in low values of the radiation emission term $\overline{\kappa I_b}$ (recall that in the current formulation, $\overline{\kappa I_b}$ is treated without SGS TRI and is proportional to $\tilde{T}^4$), which explains the low values of radiative emissions obtained with SLF/gray and SLF/WSGG. These results suggest that the modeling of radiation emission should be re-examined in future work and account for SGS TRI to improve the performance of the gray and non-gray models.

CONCLUSION

This paper presents LES simulations of a buoyancy-driven methane-air turbulent diffusion flame with a new modeling framework that accounts for combustion chemistry and does not prescribe the global radiative loss fraction. The modeling framework is based on steady laminar flamelet combustion and nongray/WSGG radiation, and its performance is evaluated against new temperature measurements performed with micro-thermocouples. It is found that the flamelet model significantly under-predicts mean temperatures, a result that is explained by the large values of the modelled SGS variance of mixture fraction $\widetilde{Z_v}$; $\widetilde{Z_v}$ is an input parameter of the presumed SGS Z-PDF that controls the magnitude of the classical LES filtering effect. These results suggest that the modeling of both the shape of the SGS Z-PDF (described as a β-PDF) and the SGS Z-variance requires additional investigation. Besides, it is found that the radiation model significantly underpredicts global radiative emissions, a result that is explained by the low values of mean temperatures and the absence of a model for SGS turbulence-radiation interactions. These results suggest that the modeling of radiation requires additional investigation and should, in particular, take advantage of the subgrid-scale information provided by the flamelet model. Future work will also be focused on the implementation of an unsteady flamelet model based on the FlameMaster solver in order to describe flame extinction phenomena [18-20].

ACKNOWLEDGEMENT

This project is financially supported at the University of Maryland (UMD) by the U.S. National Science Foundation (NSF, CBET Program, Award #1604907) and by FM Global. It is also financially supported at Université de Poitiers by the French Government program "Investissements d'Avenir" (LABEX INTERACTIFS, reference ANR-11-LABX-0017-01). The project is also supported by supercomputing resources made available by UMD (http://www.it.umd.edu/hpcc) and by NSF (XSEDE Program, Grant #TG-CTS140046). Dr. James White contributed to this work while a PhD student at UMD and is now employed at FM Global. This work also benefitted from the assistance of Drs. I. Sikic and J. Wen who shared their FireFOAM implementation of their WSGG model and who are here gratefully acknowledged.

REFERENCES

[1] V. Novozhilov, Prog. Energy Combust. Sci. 27 (2001) 611–666.

[2] A. Trouvé, Y. Wang, Intl. J. Computational Fluid Dynamics 24 (2010) 449–466.

[3] B. Merci, Fire Safety Science, Proc. Eleventh International Symposium, International Association for Fire Safety Science 24 (2014) 46–65.

[4] K. McGrattan, S. Miles, SFPE Handbook of Fire Protection Engineering (5th Ed.), Society of Fire Protection Engineering (2016) 1034–1065.

[5] A. Cuoci, A. Frassoldati, T. Faravelli, E. Ranzi, Combust. Flame 156 (2009) 2010–2022.

[6] A. Cuoci, A. Frassoldati, T. Faravelli, E. Ranzi, Env. Eng. Sci. 25 (2008) 1407–1422.

[7] Flamelet model for OpenFOAM, developed by Holtzmann CFD, available at: https://holzmann-cfd.de/publications/flamelet-model, accessed: 13-03-2018.

[8] E. Ranzi, A. Frassoldati, R. Grana, A. Cuoci, T. Faravelli,A. Kelley, C. Law, Prog. Energy Combust. Sci. 38 (2012) 468–501.

[9] H. Pitsch, FlameMaster, A C++ Computer Program for 0D Combustion and 1D Laminar Flame Calculations, available at: https://www.itv.rwth-aachen.de/index.php?id=flamemaster&L=1, accessed: 13-03-2018.

[10] G.P. Smith, D.M. Golden, M. Frenklach, N.W. Moriarty, B. Eiteneer, M. Goldenberg, C.T. Bowman, R.K. Hanson, S. Song, W.C. Gardiner Jr., V.V. Lissianski, Z. Qin, GRI-Mech v3.0, available at: http://www.me.berkeley.edu/gri_mech/, accessed: 13-03-2018.

[11] FireFOAM, developed by FM Global, available at: https://github.com/fireFoam-dev, accessed: 13-03-2018.

[12] J.P. White, Measurement and Simulation of Suppression Effects in a Buoyant Turbulent Line Fire, Ph.D. thesis, University of Maryland, College Park, USA, 2016.

[13] OpenFOAM, developed by the OpenFOAM Foundation, available at: http://www.openfoam.org, accessed: 13-03-2018.

[14] M. F. Modest, Radiative Heat Transfer, Academic Press, 2013.

[15] Radiation sub-models, TNF Workshop, available at: http://public.ca.sandia.gov/TNF/radiation.html, accessed: 13-03-2018.

[16] I. Sikic, J. Wen, S. Dembele, FM Global Fire Modelling Workshop, Norwood MA, USA, 2015.

[17] R. Johansson, B. Leckner, K. Andersson, F. Johnsson, Combust. Flame 158 (2011) 893–901.

[18] M. Ihme, H. Pitsch, Phys. Fluids 20 (2008) 055110.

[19] M. Ihme, Y.C. See, Combust. Flame 157 (2010) 1850–1862.

[20] M.E. Mueller, H. Pitsch, Combust. Flame 159 (2012) 2166–2180.

An Experimental Study of on the Combustion Properties of Lobby Chairs in Open Space and in ISO Room Compartment

Jaeyoung LEE[1], Yoshifumi OHMIYA[1], Fumiaki SAITO[1], Yusuke SHINTANI[2], and Kazunori HARADA[3]

[1]*Department of Architecture, Faculty of Science and Technology, Tokyo University of Science, Japan*
[2] *Takenaka Research & Development Institute, Takenaka Corporation, Japan*
[3]*Department of Architecture and Architectural Engineering, Kyoto University, Japan*

jaeyoung.lee @rs.tus.ac.jp

ABSTRACT

Performance-based design should reflect initial fire phenomena properly. For this purpose, necessary data are being accumulated through combustion tests. It is reported that the maximum heat release rate and the fire growth rate are significantly increased by thermal radiation feedback effects from a smoke layer and a wall in a compartment in comparison with burning in an open environment [1-2]. In this study, a series of burning tests were carried out using lobby chairs to compare burning behavior in open environment and in a compartment. The measured items were heat release rate, mass loss rate, heat flux and vertical distribution of temperature in the compartment and above the center of combustibles. The cause of flame spread to adjacent chair was different between in an open space and in a compartment. In an open space, flame spread was mainly caused by direct contact of flame with adjacent chair. However, the main cause of the flame spread in a compartment was thermal radiation. As a result, heat release rates in open space and in a compartment significantly differ in shape. In a compartment, heat release rate slowly decreases after it reaches at the maximum level. In open space, heat release rate increased and decreased as flame spread to adjacent chairs successively. Heat flux showed similar patterns with that of heat release rate. Both maximum heat release rate and maximum heat flux increased as the number of chairs were increased. Measured values in a compartment are 2-4 times larger than that in an open space. Fire growth rate was increased by hot smoke layer. However, the relationship with the number of chairs was not clear. In addition, it was confirmed that the air temperature around chairs was higher than that in an open space.

KEYWORDS:

Burning in a Compartment; Flame spread; Fire growth; Lobby chairs

NOMENCLATURE LISTING

HF	Heat flux (kW/m^2)	**Greek**	
Q	Heat release rate, *HRR* (kW)	α	Fire growth rate (kW/s^2)
T	Temperature (°C)	**subscripts**	
t	Time (s)	grow	Growth
		max	maximum *HRR*
		0	incubation

IOP Conf. Series: Journal of Physics: Conf. Series **1107** (2018) 042010 doi:10.1088/1742-6596/1107/4/042010

INTRODUCTION

It is recommended that performance-based design consider the initial fire phenomena properly. For this purpose, basic data sets necessary to predict behavior of a fire are being accumulated through combustion tests. It is reported that the maximum heat release rate (Hereinafter, *HRR*) and the fire growth rate are significantly increased due to thermal radiation feedback from a smoke layer and from enclosure walls in case of fires in a compartment [1-2]. For example, the maximum *HRR* of upholstered furniture is increased by 20 % in a compartment compared to that in an open space [1]. According to a study conducted using polyurethane foam blocks, the maximum *HRR* is increased by 80 % in an ISO room corner compartment. At a corner, 10-20% of additional increase was observed [2]. However, basic data and quantitative research on fires spreading to adjacent combustibles are not sufficient yet. Therefore, the purpose of this research is to examine the behavior of fire spread to adjacent combustibles quantitatively. The experiments were carried out using lobby chairs both in an open space and in a compartment. The influence on the initial fire growth and maximum *HRR* was investigated.

EXPERIMENTAL METHODS

The dimension of a lobby chair was W600 x D550 x H680 mm. The weight of each lobby chair is in the range 10.23-10.80 kg. As a result of disassembling one lobby chair, main constituent materials were natural wood frame 7.635 kg, polyurethane foam 0.46 kg for cushion materials with thickness 5-25 mm, PVC 0.765 kg for covering material and stainless steel 1.26 kg for legs. The estimated total heat release (*THR*) was 157.7 MJ [3]. An iron plate with the thickness of 1.6 mm is installed between the cushion part and the leg part before conducting a combustion experiment.

Fig. 1 and Fig. 2 show experimental methods in open space and in a compartment. An experiment is separately conducted in an open space (OS1-OS3) and in a compartment (CS1-CS3). For experiments in a compartment, ISO9705 room corner test facility was used [4]. Lobby chairs were arranged side by side. The distance between chairs was 40 mm. An ignition point was located at the center of the cushion part. A piece of ceramic blanket, 30 x 30 mm, was absorbed with a small amount of ethanol and ignited. *HRR* was measured by an oxygen consumption method [5].

Vertical temperature distribution was measured by using type K thermocouples of diameter 0.32 mm. Total heat fluxes were measured using Circular-foil gauges (Gordon gauge). Mass loss was measured by a load cell. The mass loss was measured only in open space experiment. As for heat release rate measurements, delay time was adjusted by preliminary experiments using methanol pool fires.

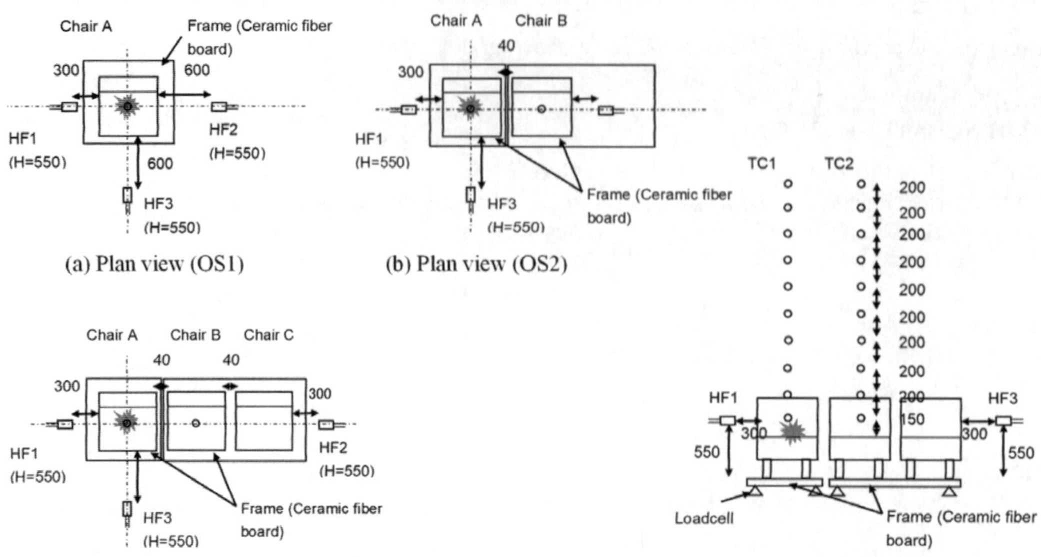

(a) Plan view (OS1) (b) Plan view (OS2) (c) Plan view (OS3) (d) cross sectional view (OS3)

Fig. 1. Combustion experiments of lobby chairs in an open space

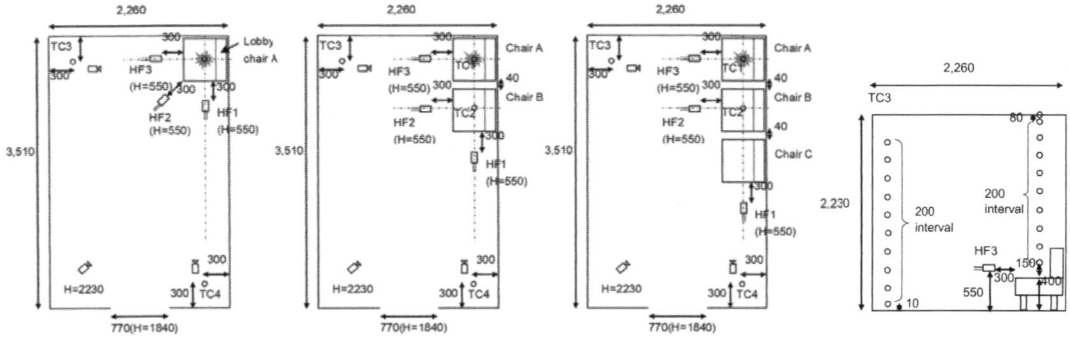

(a) Plan view (CS1) (c) Plan view (CS2) (c) Plan view (CS3) (d) Cross sectional view

Fig. 2. Combustion experiments of lobby chairs in a compartment

EXPERIMENTAL RESULTS

Combustion behaviors in an open space (OS1-OS3) and in a compartment (CS1-CS3) are shown in Fig. 3. In an open space, fire spread horizontally to the back of a chair in a circular shape. In a compartment, fire also spread to the back of a chair and a wall in a circular shape, but the spreading speed was much faster in the direction toward a wall. In addition, fire spread to the adjacent chair when chair were placed side by side. In an open space, flame spread to an adjacent chair by direct flame contact. In a compartment, flame spread to adjacent chair by thermal radiation without direct contact of flame. Flame spread time in a compartment is relatively shorter than that in an open space. After 15 minutes, burning was reduced because only the burning of wood frame continues.

Time	OS1	OS2	OS3	CS1	CS2	CS3
1'00"	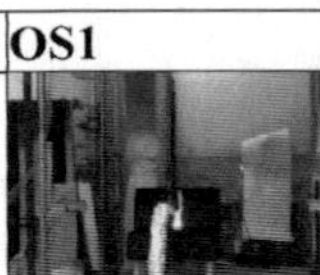	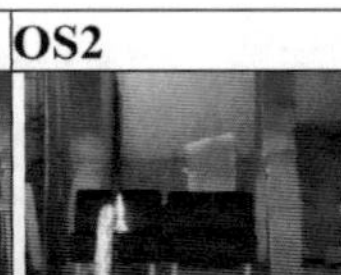	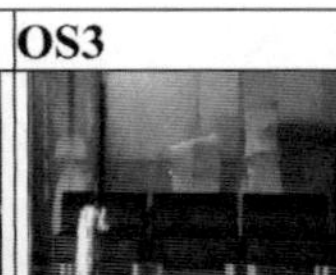	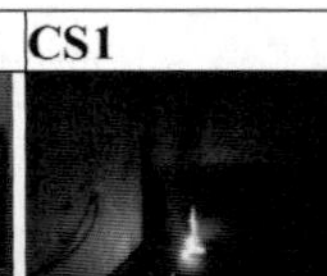	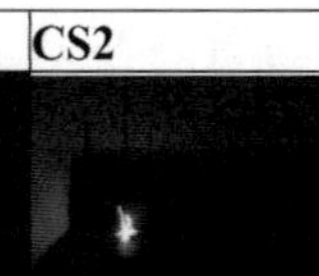	
3'30" (growth period)						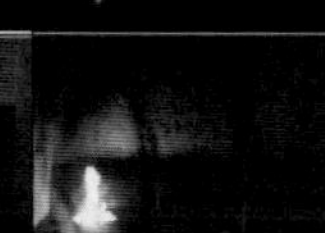
5'00"				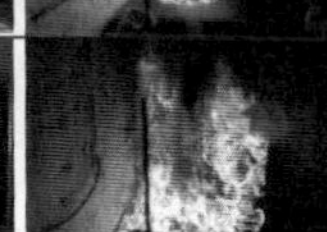	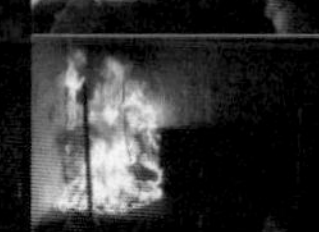	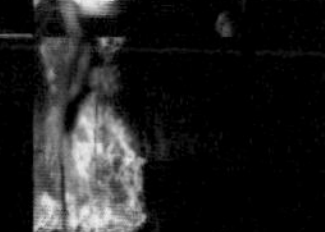
Flame spreadl A to B		 6'42" by contact of flame	 6'29" by contact of flame		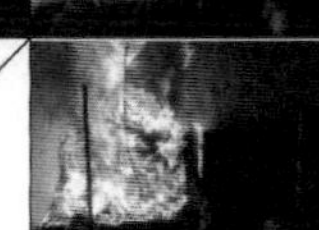 5'40" by radiation	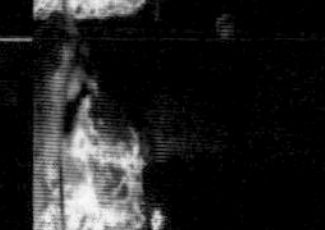 5'05" by radiation
Flame spread B to C			Not happened yet			 5'51" by radiation
10'00"			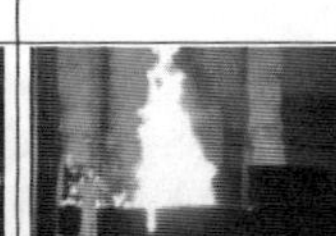			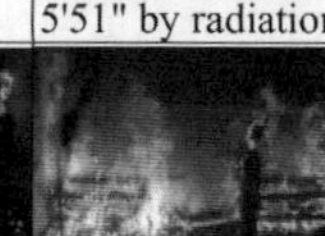
Flame spread B to C						already spread

Fig. 3. Burning behavior in each experiment

The results of *HRR* are shown in Fig.4. The maximum *HRR* was significantly increased as the number of chairs increased. *HRR's* in an open space and in a compartment differ considerably. In an open space, the number of *HRR* peaks is the same as the number of chair. The maximum *HRR* occurred at the second peak. In compartment, the maximum *HRR* is quickly increased to the maximum. Then, the *HRR* decreased gradually. Compared to CS1, the values of total heat release (hereafter, *THR*) of OS1 were smaller because wood frame did not burn completely.

The heat flux (hereinafter, *HF*) are shown in Fig.5. The change of *HF* was similar to that of *HRR*. *HF* in compartment is significantly larger than that in an open space.

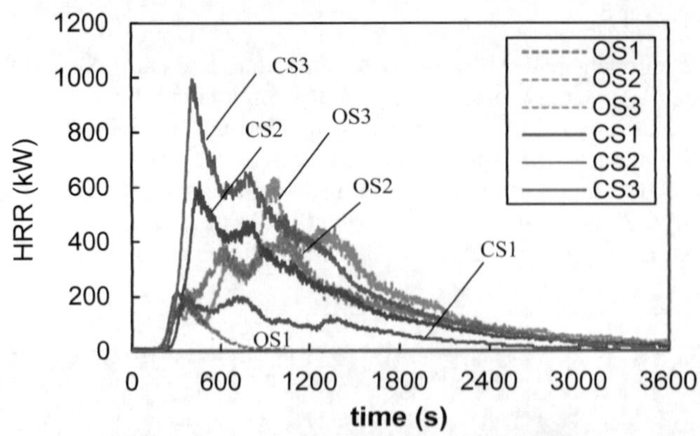

Fig. 4. *HRR* in open space and in a compartment

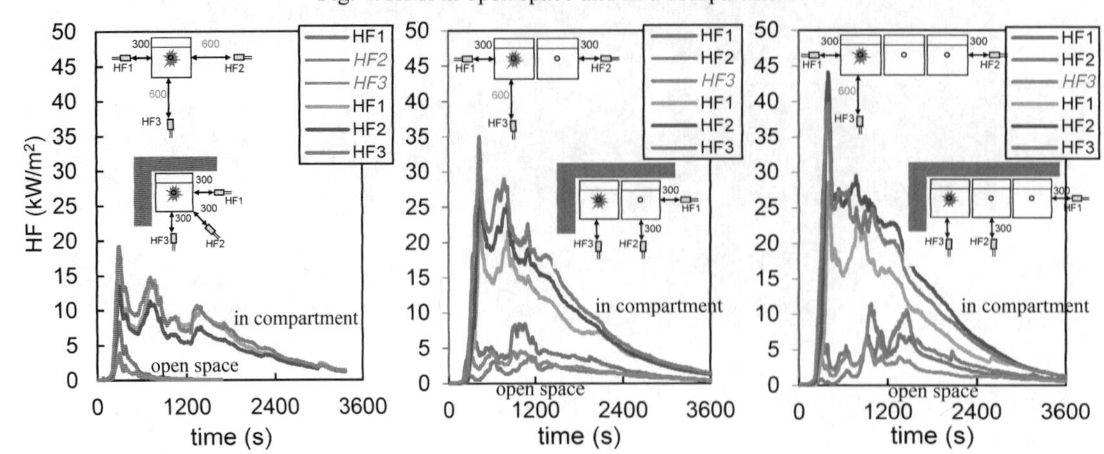

OS1 and CS1 OS2 and CS2 OS3 and CS3

Fig. 5. Measured heat flux

The height of a smoke layer was determined by using vertical temperature distribution measured by a thermocouple tree apart from flame (TC3). The smoke layer height was defined as the height where temperature increase was 10 % of the maximum temperature increase [6]. Change of the smoke layer height and the temperatures of smoke layer and air layer are shown in Fig. 6 (a). The smoke layer interface descends during 300 and 400 seconds but maintains stable position after 400 seconds. The height of the smoke layer was about 800-1000 mm from the bottom. The smoke layer height was lower as the number of chairs was increased. The smoke layer temperature was significantly increased when smoke layer height was low. The maximum temperature occurred at around 400 seconds. Thereafter, the temperature of the smoke layer gradually decreased.

The flame tip height was obtained from temperature distribution measured by a thermocouple tree installed above the chair (TC1). The flame tip height was defined as the point where the temperature decreased to 290°C [7]. Fig. 6 (b) indicates the measured results of OS1 and CS1. Comparing the flame size in an open space with that in a compartment, the flame grew more quickly in a compartment. The flame tip in a compartment was twice as high as that in an open space.

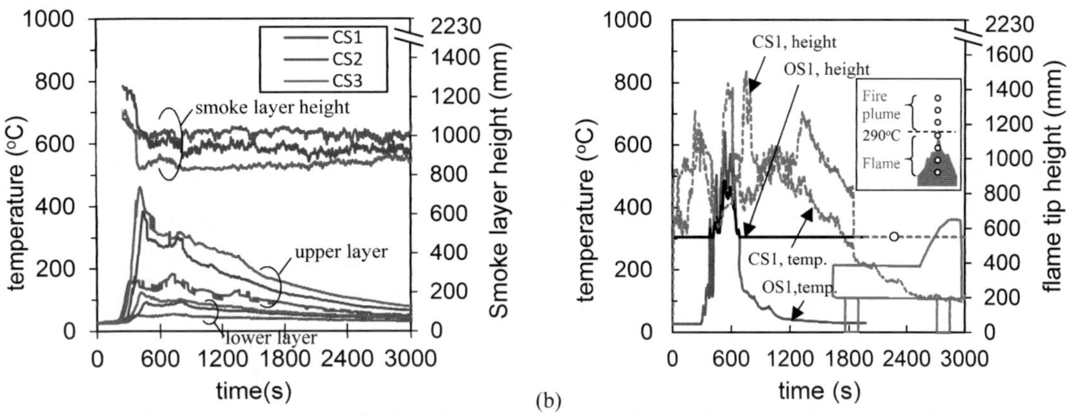

(a) (b)

Fig. 6. (a)Smoke layer temperature and height in compartment (at TC3), (b) Flame tip height and flame temperature upper chair in case of one lobby chair (OS1, CS1)

DISCUSSIONS

To compare between experiments, maximum *HRR* and time to maximum *HRR* were extracted from measured data and plotted in Fig 7. Both in an open space and in a compartment, the maximum *HRR* increases as the number of chair increases. However, it is revealed that the degree of increase is significant in a compartment than in an open space.

Maximum *HF*'s obtained by each test are shown in Fig. 8. *HF* in a compartment is larger than that in an open space. The *HF* in a compartment was larger than that in an open space, about 2 times in case of one chair burning, about 2.7 times in case of 2 chairs, about 4 times in case of 3 chairs.

The initial fire growth rate (hereinafter, fire growth rate) α_i (kW/s^2) was obtained from the result of *HRR* [7] using

$$\alpha_i = Q_{grow} / (t_{grow} - t_0)^2, \quad t_0 < t_{grow} < t_{max} \tag{1}$$

where Q_{grow} is the maximum *HRR* (kW) in the regression section, t_{grow} is the duration of the maximum *HRR* (s), t_0 is incubation time (s). Fig. 9 shows the relationship between fire growth rate and smoke layer temperatures. Here, the smoke temperature is the maximum temperature in the fire growth period. The graph also contains data on polyurethane foam blocks in ISO room corner test compartment [2]. According to the experiment results, the fire growth rate increases as the smoke layer temperature increases. The fire growth rates in a compartment are roughly twice as large as those in an open space. Compared to the results of experiments on corners in previous studies, it shows relatively larger fire growth rates overall. For a lobby chair, it is assumed that mutual thermal feedback between heated sides is a cause of increased fire growth rate because the cushion part and the back part of a chair burned simultaneously.

CONCLUSIONS

Burning experiments were conducted in an open space and in a compartment using lobby chairs. One to three specimens were set at a separation distance of 40 mm. The results were as follows;

- The cause of flame spread to an adjacent chair was the contact of flame in open space and thermal radiation in a compartment.
- The history of HRR was largely different between in an open space and in a compartment. In a compartment, the HRR gradually decreased after quickly reaching to the maximum value. In an open space, the increase and decrease of the HRR repeated due to successive spread to adjacent combustibles.
- History of HRR and HF were similar in shape. Maximum HRR and maximum HF increased with increasing number of chairs. The measured values in a compartment were about 2 to 4 times of those in an open space.
- The fire growth rate increased with the increase in the smoke layer temperature. In a compartment, the lower layer temperature around the chair also increased with the temperature increase in the smoke layer.

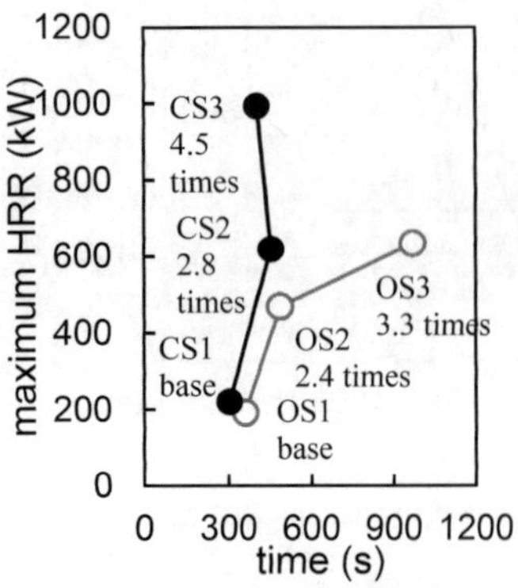

Fig. 7. maximum *HRR* and time to maximum *HRR*

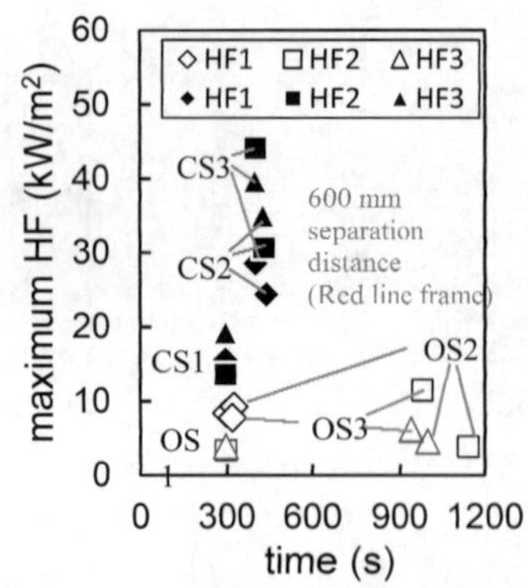

Fig. 8. Comparison of maximum Heat flux

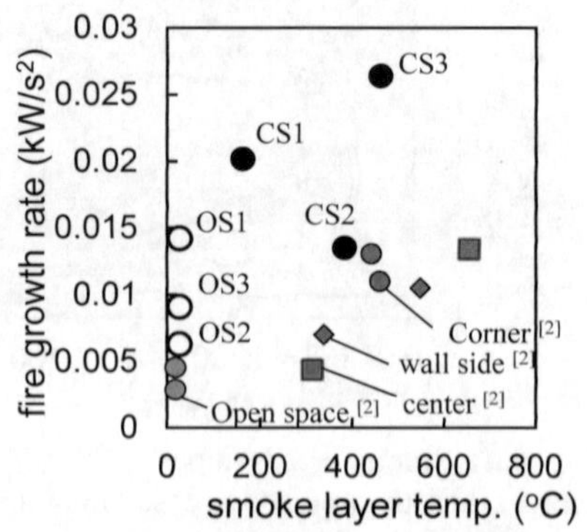

Fig. 9. Smoke layer temperature and fire growth rate [2]

ACKNOWLEDGEMENT

This work was supported by Grant-in-Aid for Scientific Research (C) by the Japan Society for the Promotion of Science (KAKENHI Grant Numbers 16K06620). This work was carried out in cooperation with members of the working group for Fire Load and Design Fire Behavior in Architectural Institute of Japan.

REFERENCES

[1] Predictions of room fire conditions, Chapter 6, Fire Safety of Upholstered Furniture, CBUF project final report, EUR 16477 EN, 1993

[2] Kazunori Harada, Ken Matsuyama, Kazuhiko Ido, Masaaki Noaki, Sungchan Lee, Jaeyoung Lee, Burning of Polyurethane Foam Block in ISO Room Compartment, IFireSS – International Fire Safety Symposium Coimbra, Portugal, 20-2, April 2015, pp.493-502

[3] Architectural Institute of Japan, Recommendations for design fire loads and fire actions in building, 2013

[4] ISO 9705, Fire tests – Full-scale room test for surface products, International Standardization Organization, 1993

[5] ISO 24473, Fire tests – Open Calorimetry – Measurement of the rate of production of heat and combustion products for fires of up to 40MW, International Standardization Organization, 2008

[6] Y. He, On Experimental Data Reduction for Zone Model Validation, Journal of Fire Sciences, Vol 15, Issue 2, 1997, pp. 144-161 https://doi.org/10.1177/073490419701500205

[7] Sizuo Yokoi, On the heights of flames from burning cribs, Bulletin of the Fire Prevention Society of Japan, Vol 13, No.1 Oct.1963, pp. 22-27

[8] Jaeyoung Lee, Yoshifumi Ohmiya, Kazunori Harada, Ken Matuyama, Yuhsiang Wang, Fumiaki Saito, Seiji Okinaga, Yusuke Shintani, Akihide Jo, Jun Kitahori, An experimental study on combustion characteristics of Printers depending on geometrical type and kind of plastics used, to be presented at 11th AOSFST 2018 – Asia-Oceania Symposium on Fire Science and Technology, Taipei, 21-25 October 2018

IOP Conf. Series: Journal of Physics: Conf. Series **1107** (2018) 042011 doi:10.1088/1742-6596/1107/4/042011

Performance and calibration of two subgrid extinction models for turbulent diffusion combustion in an under-ventilated enclosure fire

Alexander SNEGIREV[1], **Egor KUZNETSOV**[1], Ekaterina MARKUS[1],
John HARRIS[2], Bradford MORAVEC[2]

[1] *Department of Fluid Dynamics Combustion and Heat Transfer*
Peter the Great St-Petersburg Polytechnic University, Russia
[2] *The Boeing Company, USA*

kn0egor@gmail.com

ABSTRACT

The fast-chemistry eddy dissipation combustion model fails to predict abrupt flame self-extinguishment when the oxygen content falls below the critical threshold. To overcome this limitation without recourse to the highly resolved simulations with the detailed chemical mechanisms, the bespoke subgrid extinction model is required. In this work, two subgrid extinction models are implemented in ANSYS Fluent and applied. The first extinction model utilizes conventional critical flame temperature concept, and the second one is the Damköhler number-based approach. The model formulation is used as derived earlier from the perfectly stirred reactor model, with the finite-rate reaction kinetics and radiative heat losses from the unresolved reaction zone.

To evaluate model performance and to select optimum model parameters, the burner fire with the constant fuel supply rate (corresponding to the nominal heat release rate 400 kW in complete combustion) is considered in 4x6x4.5 m compartment with the 0.3 m diameter lower air inlet connected to the ambient atmosphere. Conjugate heat transfer at the compartment walls and transient thermal conductivity in the compartment structures are accounted for.

Provided that the model parameters are optimized, both extinction models enable prediction of flame self-extinguishing, and the predicted times for flame to extinguish are similar to those recorded in the experiments. For the CFT model, best agreement with the experimental time to flame extinguishment is obtained if the critical flame temperature is set to 1900 K. For the Damköhler number-based approach, the optimum value of 0.25 is found for the model constant in the proportionality relationship between the resolved strain rate and the residence time in the reaction zone, which is not resolved in the simulations.

In spite of the similar performance of the two extinction models in the experimental scenario with weakly strained flame, the Damköhler number-based model is expected to be applicable in a wider range of scenarios including highly strained flames such as those affected by strong jets, sprays, and cross-winds.

KEYWORDS:

Extinction, under-ventilated combustion, CFD fire modeling

IOP Conf. Series: Journal of Physics: Conf. Series **1107** (2018) 042011 doi:10.1088/1742-6596/1107/4/042011

INTRODUCTION

Extinction in under-ventilated fires is observed as soon as the oxygen concentration decreases below a critical threshold. Comprehensive modeling of extinction and ignition phenomena can only be performed by using the detailed chemical mechanisms. This methodology appears impractical in fire modeling since the reaction zones of the flamelets are not spatially resolved, and the chemical composition of fuels is often unknown. The eddy-dissipation combustion model (most widely used in fire modeling) implies infinitely fast chemistry and assumes that fuel burning rate is solely controlled by subgrid mixing. Combustion is predicted by this model if fuel and oxygen are available at any concentration, regardless of the temperature. Extinction cannot, therefore, be predicted by the fast chemistry and mixing-controlled combustion model. As such, additional submodel is required to allow for the abrupt flame extinction observed in experiments.

To date, two distinct approaches have been developed to model local flame extinction:

1. Critical flame temperature concept (see Refs. [1], [2] etc.) – very simple approach, good for vitiated flames at low strain. It is this approach which is used in the open source specialized CFD code FDS for fire modeling [3]. It has been, however, demonstrated in Refs. [4], that this approach fails to predict extinction at high strain (blow-off, suppression by sprays).
2. Damköhler number-based modeling (Refs. [5], [6], [4], [7] among others) – allows for finite-rate chemistry, kinetic parameters must be calibrated. Note, that the perfectly stirred reactor (PSR) subgrid model investigated in Ref. [7] also reveals applicability of the properly formulated Damköhler number as a unified extinction criterion (incorporating both high-strain blow-off and low-strain radiative limits). The advantage of this approach is that it is suitable for highly strained flames.

More recently, the reactive volume fraction approach has also been proposed (see Ref. [8]) based on the heat balance in an eddy containing the fuel-air mixture.

Existing CFD modeling experience of flame extinction covers simulations of unconfined flames [2] and enclosure fires. The latter were mainly considered at small spatial scale [1, 2], while simulations of large-scale compartments remain limited. No subgrid models of local flame extinction are available in ANSYS Fluent for joint use with the eddy dissipation model. In this work, we implement and compare two of the above-listed approaches (critical flame temperature and Damköhler number-based extinction models) by simulating self-extinguishment of the 400 kW methane flame in an under-ventilated large-scale enclosure [9].

SUBGRID EXTINCTION MODELING

Critical flame temperature extinction model

Critical flame temperature (CFT) extinction model utilizes the assumption that the flame can only exist if the temperature in the reaction zone, T_f, exceeds the critical value, $T_{f,cr}$. In this model, the finite rate reaction is not allowed, and the temperature in the reaction zone, $T_f = \tilde{T} + \tilde{Y}_{O_2}\,\Delta h_{c,O_2}/\hat{c}_P$, corresponds to that after complete burn-out of the stoichiometric mixture having oxygen mass fraction $\tilde{Y}_{O_2}$. In the above equality $\tilde{Y}_{O_2}$ and $\tilde{T}$ are the resolved oxygen mass fraction and temperature, $\Delta h_{c,O_2}$ is the heat of combustion per 1 kg O_2 consumed (known to be about 13 MJ/kg for many fuels), $\hat{c}_P$ is the mixture specific heat, which should allow for the specific heat of combustion products and heat consumed in its dissociation. In case of presence of inert diluents, its specific heat should also be accounted for in evaluating $\hat{c}_P$. Note, that proper assignment of the $\hat{c}_P$ value is an approximation to the formulation of the extinction condition in terms of enthalpies as is done in FDS 6. Thus, extinction is predicted if $\Delta h_{c,O_2}\tilde{Y}_{O_2} < \hat{c}_P\left(T_{f,cr} - \tilde{T}\right)$; the burning rate is set to zero in the grid cells where this inequality is obeyed. To obtain the best agreement with different experiments, different CFT values might be recommended. For example, in Ref. [10] best agreement of flame extinction by FDS were obtained with the critical flame temperature 1773 K.

The inherent assumption of the CFT extinction model is that the heat is released immediately, regardless of the residence time characteristic of the reactants in the reaction zone which is affected by the local velocity strain. In highly strained flows local residence time is too low, and the heat required to rise temperature up to T_f would not be completely released. The model is therefore suitable for the low-strain flames with a low probability of strain-induced blow-off.

Damköhler number-based extinction model

This model is designed to overcome the abovementioned limitation of the CFT extinction model. Here we apply the approach developed in Ref. [7] where perfectly stirred reactor model is utilized to predict the state of the reaction zone of the diffusion flamelet occupying a small fraction of grid cell volume. It has been shown in Ref. [7] that the extinction condition can be formulated as that of sufficiently low Damköhler number, $\mathrm{Da} < \mathrm{Da}_{cr}$, which is defined as a product,

$$\mathrm{Da} = \dot{r}_{fuel}\tau_{res}, \quad (1)$$

of the reaction rate, $\dot{r}_{fuel}$, and the residence time, τ_{res}, both are evaluated as explained below.

Evaluation of the reaction rate utilizes the single-step global reaction model,

$$\dot{r}_{fuel} = A\left(M_{fuel}/\rho\right)\left(\tilde{Y}_{fuel}\rho/M_{fuel}\right)^{n_{fuel}}\left(\tilde{Y}_{O_2}\rho/M_{O_2}\right)^{n_{O_2}}\exp\left(-\left(E_a/\mathcal{R}T_f\right)\right). \quad (2)$$

The temperature in the reaction zone, T_f, obeys the equation,

$$T_f = \tilde{T} + \left(\tilde{Y}_{O_2}\Delta h_{c,O_2}/\hat{c}_P\right)/\left(1+\tau_{res}/\tau_{loss}\left(T_f\right)\right), \quad (3)$$

which allows for the radiative heat losses, the rate of which is quantified by the characteristic time scale, $\tau_{loss}\left(T_f\right) = \hat{c}_P\rho/\left(4\kappa\sigma\left(T_f^2+\tilde{T}^2\right)\left(T_f+\tilde{T}\right)\right)$, where κ is the local effective absorption coefficient, $\sigma = 5.67\cdot10^{-8}$ W/(m$^2\cdot$K^4) is the Stefan-Boltzmann constant. Iterations are required to solve Eq. (3) for T_f.

Consistent with the CFT model, the assumption of stoichiometry, $\tilde{Y}_{fuel} = \tilde{Y}_{O_2}/\left(\nu_{O_2}M_{O_2}/M_{fuel}\right)$ (where ν_{O_2} is the stoichiometric mole oxygen to fuel ratio), is implied in evaluating temperature in the reaction zone and is also used to rewrite Eq. (2) as

$$\dot{r}_{fuel} = A\left(M_{fuel}/\left(\rho\nu_{O_2}^{n_{fuel}}\right)\right)\left(\tilde{Y}_{O_2}\rho/M_{O_2}\right)^{n_{fuel}+n_{O_2}}\exp\left(-\left(E_a/\mathcal{R}T_f\right)\right). \quad (4)$$

The residence time is evaluated as $\tau_{res} = C_{SGS}\tilde{S}^{-1}$, where $\tilde{S}$ is the resolved strain rate, and C_{SGS} is the model constant to be calibrated. We assume that the volume occupied by the reaction zone is much smaller than that of the grid cell, and therefore the resolved temperature $\tilde{T}$ and oxygen mass fraction $\tilde{Y}_{O_2}$ represent the environment for the reaction zone.

In the simulations presented below, we use the following kinetic parameters derived in Refs. [4, 7] for extinction of counter-flow methane-O_2-N_2 flames: E_a = 121 kJ/mol, $A = 0.883\cdot10^7$ (m, mol, s), n_{O_2} = 1.13, n_{fuel} = 0.56, Da_{cr} = 2. For methane fuel M_{fuel} = 0.016 kg/mol, $\Delta h_{c,O_2}$ = 12.54 MJ/kg, ν_{O_2} = 2. To obey the equilibrium adiabatic flame temperature T_{ad} = 2226 K in the stoichiometric methane-air mixture at initial temperature of 300 K, specific heat of combustion products is set $\hat{c}_P = Y_{fuel,st}\Delta h_c/\left(T_{ad}-T_0\right)$ = 1400 J/(kg·K).

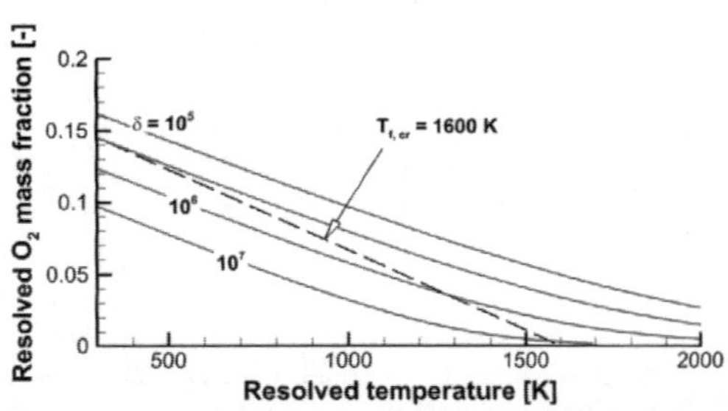

Fig. 1. The extinction domain predicted by the critical flame temperature model (at $T_{f,cr}$ = 1600 K) and by the Damköhler number-based model (n_{fuel} = 1, n_{O_2} = 0, κ = 0) at different values of δ

Comparison of the two extinction models is best illustrated for the special case when $n_{fuel} = 1$, $n_{O_2} = 0$, $\kappa = 0$ (first-order reaction, no radiative losses). Critical condition for extinction, $\mathrm{Da} = \mathrm{Da}_{cr}$, takes the form

$$\delta^{-1} = \left(\tilde{Y}_{O_2} / \left(\nu_{O_2} M_{O_2} / M_{fuel}\right)\right) \exp\left(-\left((E_a/\mathcal{R}) / \left(\tilde{T} + \tilde{Y}_{O_2} \Delta h_{c,O_2} / \hat{c}_P\right)\right)\right), \quad (5)$$

where $\delta = \tau_{res} A/\mathrm{Da}_{cr}$ is the dimensionless quantity inversely proportional to the local strain, $\tilde{S} \sim 1/\tau_{res}$. The dependence $\tilde{Y}_{O_2}(\tilde{T})$ at extinction boundary is shown in Fig. 1 for different values of δ, and therefore local strain. It can be seen that the higher is the strain rate, the larger is the extinction domain. Depending on the strain, it may be either larger or smaller than that predicted by the critical flame temperature model.

The Damköhler number-based model allows for the finite reaction rate in and the radiative heat losses from the reaction zone. As such, two distinct extinction mechanisms are captured, strain-induced blow-off and radiative quenching; both are affected by the oxygen and diluent availability.

CASE DESCRIPTION AND MODEL SETUP

Under-ventilated fire scenario

Enclosure fire tests performed in the Lawrence Livermore National Laboratory (LLNL) in 1986, see Ref. [9], are considered. In this series of tests, the rectangular enclosure with the horizontal dimensions 4x6 m, 4.5 height was utilized (see Fig. 2 and Fig. 4). Centrally located 0.57 m diameter circular burner elevated 0.23 m above the floor level is used to supply gaseous fuel (methane) at a constant flow rate. Air injection occurs through the lower of two air inlets (0.3 m diameter circular pipes). Walls, floor, and ceiling are 10 cm thick. The outlet duct designed for the fan-driven exhaust ventilation is closed in the scenario considered here.

Model setup

Simulations were performed using ANSYS Fluent 16.2. Large eddy simulation approach (LES) is applied to predict gaseous flow with the Smagorinsky-Lilly subgrid model. Eddy-dissipation model (EDM) is applied to allow for the turbulence-reaction interaction. The single-step irreversible reaction of complete fuel oxidation is considered. Soot and thermal radiation modeling are performed as explained in Ref. [11].

Block-structured hexahedral meshes are used both in fluid and solid domains. Simulation results shown below were obtained with the mesh, having the following amounts of grid cells: fluid 344 701, walls 99 888, floor and ceiling 36 296. The total amount of grid cells is 480 885. Cubic cells in the flame zone above the burner are of 3x3x3 cm size. The effect of mesh refinement was examined with both coarser and finer meshes in which the cell size above the burner was varied from 1.5 to 5 cm; simulation results were weakly affected. The coupled scheme was used for pressure-velocity coupling. Spatial discretization was set as follows: gradients – least squares cell based; pressure – second order; momentum – bounded central differences; energy, soot, species – second order upwind; transient formulation – bounded second order implicit.

The extinguished fuel is assumed to re-ignite as soon as the reactants in a grid cell do not obey the extinction condition. This implies overestimated ignitability of the reactants and might delay predicted flame extinguishment if the probability of the re-ignition is significant.

SIMULATION RESULTS

Here we consider enclosure configuration with no false ceiling, no fan-driven exhaust (the outlet duct is closed), open lower air inlet and closed upper air inlet (case TL-3 in Ref. [9]). Open air inlet prevents pressure growth as the temperature increases. Methane fuel is continuously supplied to the burner at the rate of 8 g/s corresponding to the nominal heat release rate of 400 kW. Conjugate heat transfer at the compartment walls and transient thermal conductivity in the compartment structures are accounted for.

In the experimental scenario replicated in this work flame was self-extinguished in 300 s after ignition. Dissimilar to the experimental observation steady burning with constant 400 kW heat release rate and slowly reduced peak temperature is predicted in the simulations without extinction model, at least for 500 s. The two-layer stratification is predicted, with the horizontal interface between the lower and upper layers located just above the burner. The air is entrained to the flame from the lower layer which is fed by the lower air inlet. The flow through the lower layer inlet oscillates in time, ejecting vitiated gas and sucking the fresh air. Combustion is predicted at any non-zero oxygen concentration.

Damköhler number-based extinction model

After the fuel supply is turned on, turbulent flame develops above the burner, and the fire plume impinges the ceiling. Simulations replicate formation of a hot upper layer with gradually reduced oxygen concentration, eventually causing flame extinguishment. Flame (which is visualized by 500 °C resolved temperature iso-surface) disappears shortly after 300 s, which agrees with the experimental observations. As soon as the flame disappears, heat release ceases, and temperature decreases due to the heat transfer to the structures. Simulated temperature and velocity distributions, oxygen mass fractions and reaction rates are presented in Fig. 2.

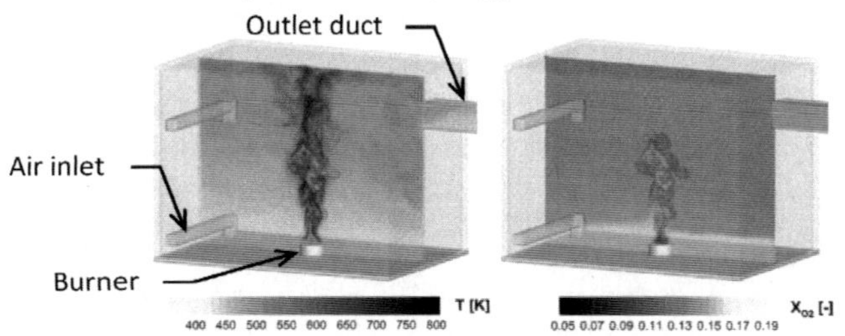

Fig. 2. Instantaneous resolved temperature and velocity, oxygen mass fraction and reaction rate distributions (left to right) in scenario 3-TL [9], time 250 s. Damköhler number-based extinction model, C_{SGS} = 0.25

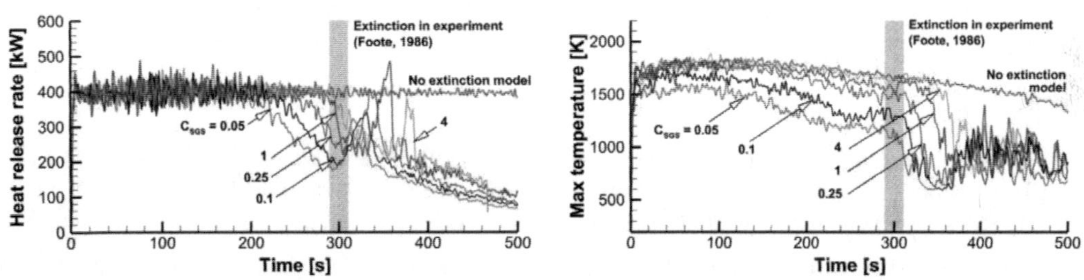

Fig. 3. The effect of the model constant C_{SGS} on predicted dynamics of integral heat release rate (left) and peak resolved temperature (right). Scenario 3-TL [9], Damköhler number-based extinction model

The effect of the model constant, C_{SGS} , on predicted dynamics of integral heat release rate and peak resolved temperature is shown in Fig. 3. It can be seen that the value of 0.25 enables little local extinction during the steady burning period and predicts flame extinguishment time close to that in the experiments.

Critical flame temperature extinction model

Flame development and self-extinguishment predicted with the CFT extinction model are similar to those with the Damköhler number-based model. Note, however, that some burning (although at a low reaction rate) is predicted at the horizontal interface as shown by arrows in Fig. 4 (250 and 300 s). When oxygen is low, the reaction zone shifts towards the air inflow (shown by the arrow at the right figure in Fig. 4, 350 s).

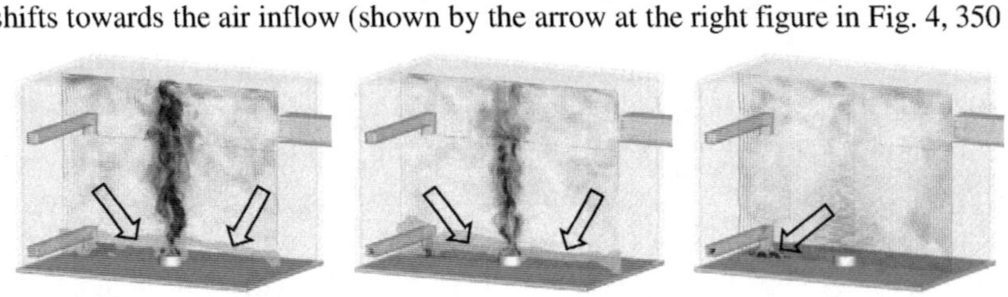

Fig. 4. Instantaneous resolved 500 °C temperature iso-surface and reaction rates distributions at 250, 300 and 350 s (left to right) in scenario 3-TL [9]. Critical flame temperature extinction model, $T_{f,cr}$ = 1900 K

The effect of critical flame temperature value on predicted dynamics of integral heat release rate and peak resolved temperature is shown in Fig. 5. Integral heat release rate exhibits slow and smooth decay, while the

IOP Conf. Series: Journal of Physics: Conf. Series **1107** (2018) 042011 doi:10.1088/1742-6596/1107/4/042011

peak flame temperature drops abruptly when the last pieces of flame disappear. With this criterion of flame self-extinguishment, Fig. 5 shows that the value of 1900 K is best suited to replicate the experimental time of extinction close to 300 s.

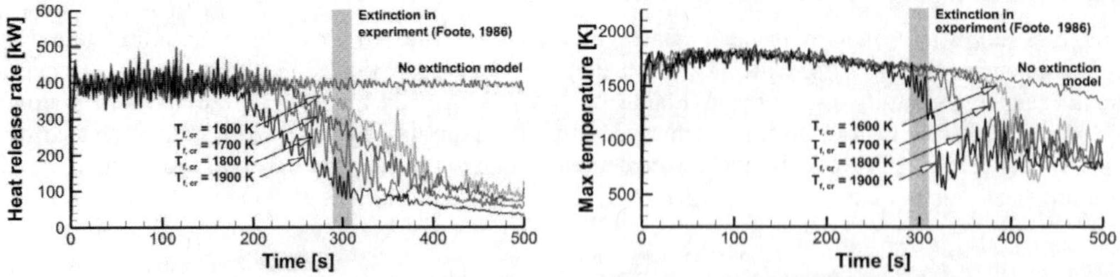

Fig. 5. The effect of critical flame temperature value on predicted dynamics of integral heat release rate (left) and peak resolved temperature (right). Scenario 3-TL [9], critical flame temperature extinction model

CONCLUSIONS

Provided that the model parameters are optimized, both extinction models enable prediction of flame self-extinguishing, and the predicted times for flame to extinguish are similar to those recorded in the experiments. The Damköhler number-based model is expected to be applicable in a wider range of scenarios including highly strained flames such as those affected by strong jets, sprays, and cross-winds.

Support by Russian Science Foundation (project 16-49-02017) and the Boeing company is gratefully acknowledged. Simulations were performed at the Supercomputer Center “Polytechnic”.

REFERENCES

[1] Hu, Z., Utiskul, Y., Quintiere, J.G. and Trouvé, A. (2007) Towards large eddy simulations of flame extinction and carbon monoxide emission in compartment fires, *Proc. Combust. Inst.*, 31: 2537-2545, https://doi.org/10.1016/j.proci.2006.08.053

[2] White, J.P., Vilfayeau, S., Marshall, A.W., Trouvé, A. and McDermott, R.J. (2017) Modeling flame extinction and reignition in large eddy simulations with fast chemistry. *Fire Saf. J.*, 90: 72-85, https://doi.org/10.1016/j.firesaf.2017.04.023

[3] Fire Dynamics Simulator (FDS) and Smokeview (SMV), https://pages.nist.gov/fds-smv/index.html

[4] Snegirev, A.Yu. and Tsoy, A.S. (2015) Treatment of local extinction in CFD fire modeling. *Proc. Combust. Inst.*, 35: 2519-2526, https://doi.org/10.1016/j.proci.2014.07.051

[5] Narayanan, P., Baum, H.R. and Trouvé, A. (2011) Effect of soot addition on extinction limits of luminous laminar counterflow diffusion flames. *Proc. Combust. Inst.*, 33: 2539-2546, https://doi.org/10.1016/j.proci.2010.07.003

[6] Vilfayeau, S., Ren, N., Wang, Y. and Trouve´, A. (2015) Numerical simulation of under-ventilated liquid-fueled compartment fires with flame extinction and thermally-driven fuel evaporation, *Proc. Combust. Inst.* 35: 2563–2571, https://doi.org/10.1016/j.proci.2014.05.072

[7] Snegirev, A.Yu. (2015) Perfectly stirred reactor model to evaluate extinction of diffusion flame, *Combust. Flame*, 162: 3622-3631, https://doi.org/10.1016/j.combustflame.2015.06.019

[8] Dorofeev, S.B. (2017) Thermal quenching of mixed eddies in non-premixed flames, *Proc. Combust. Inst.*, 36: 2947-2954, https://doi.org/10.1016/j.proci.2016.06.061

[9] Foote, K.L. 1986 LLNL enclosure fire tests data report. Technical Report UCID-21236, Lawrence Livermore National Lab., CA (USA), 1987

[10] Vaari, J., Hostikka, S., Sikanen, T. and Paajanen, A. (2012) Numerical simulations on the performance of water-based fire suppression systems. VTT Technology 54 http://www.vtt.fi/inf/pdf/technology/2012/T54.pdf

[11] Snegirev, A., Markus, E., Kuznetsov, E., Harris, J., Wu, T. (2017) On soot and radiation modeling in buoyant turbulent diffusion flames. *Heat Mass Trans.*, in press. https://doi.org/10.1007/s00231-017-2198-x

Application of a simplified pyrolysis model to predict fire development in rack storage facilities

Ekaterina MARKUS[1], Alexander SNEGIREV[1], Egor KUZNETSOV[1] and Leonid TANKLEVSKIY[1,2]

[1]*Peter the Great Saint-Petersburg Polytechnic University, Russia*
[2]*Gefest Company, Russia*

eksmarkus@gmail.com

ABSTRACT

The simplified (thermal) pyrolysis model is applied to simulate flame spread and fire growth in the rack storage facility using the FDS software. Pyrolysis in the combustible material is not explicitly considered, and the ignition temperature and the burning rate are used as the input parameters. Once the fuel surface temperature reaches the ignition temperature, the material ignites and burns at a prescribed burning rate. This approach is found to reasonably replicate both the transient development of the heat release rate and flame dynamics observed in the full-scale rack storage fire with 2x4x3 cardboard boxes in the rack, provided the model parameters (thermal properties of the fuel, ignition temperature, average heat release rate per unit area, burn-out time, and heat of gasification) are properly selected.

Distinct flame propagation regimes including buoyancy-driven upward flame spread, horizontal flame spread, and buoyancy-opposed downward flame propagation are observed in the simulations of the rack-storage fire development. Both measured and predicted HRR growth rates appear to be faster than that in a t-squared fire, mainly because of the developed combustible surface densely packed within the compact volume and due to availability of vertical gaps creating chimney-like effect and horizontal gaps providing permanent supply of fresh air.

It is shown how the full-scale test data enable selection of the model parameters. In particular, average value of the heat release rate per unit area is evaluated by joint consideration of the measured dynamics of the growing heat release rate and the surface area engulfed in fire. The burn-out time and the effective thickness of the fuel layer is estimated as the period from ignition to the time instant after which the measured heat release rate starts to decay. The heat of gasification is selected by fitting the predicted transient dependence of the heat release rate to that measured in the full-scale tests. Optimum values of the model parameters are consistent with the literature data for cardboard, and this indicates possibility of the simplified thermal pyrolysis model to predict rack storage fire dynamics with an alternative fire load, for which the full-scale test data may not be available. Further work is required to validate this approach for a wider range of full-scale fire tests.

KEYWORDS:

flame spread; ignition, rack-storage fire; fire growth; FDS; coupled simulations

IOP Conf. Series: Journal of Physics: Conf. Series **1107** (2018) 042012 doi:10.1088/1742-6596/1107/4/042012

INTRODUCTION

Rack storage fires are special because of the following issues greatly affecting the overall fire dynamics and growth rate: extremely high accumulation of fire load causing enormous peak heat release rates; developed surface area of the combustible items densely packed within the compartment volume; close face-to-face location of burning surfaces, high view factors, intensive radiative exchange causing rapid ignition; chimney-like effect facilitating intensive ventilation of the in-rack area. Obstructions hinder early detection of fire which often activates the suppression system only after engulfing considerable amount of combustibles. Design and efficiency assessment of fire detection and suppression systems could benefit from quantitative predictions, hence the efforts invested to develop comprehensive CFD models during the recent decade.

Time-dependent design fire selected as the input for engineering calculations of detection and alarm systems as well as smoke and heat venting routinely implies time-squared fire growth rate as recommended by NFPA 72, 92B, 204, among other codes. It is also used in the analysis of fire hazards in rack storages, for example, see Ref. [1]. The t-squared law was extensively validated for the performance of compact combustible items such as furnishings, and slow to ultra-fast fire types have been defined to allow for the variety of the growth rates. However, the different topology of fire load characteristic for high-rack storages would imply deviation from this law towards much higher growth rates. Indeed, early rack-storage fire tests conducted in Refs. [2, 3], indicated that the dependence on time for convective heat release rates produced in the initial fire growth period obey the third power; an even faster growth rate is possible due to the above-listed specific features of the rack storage fires, and this substantiates the need in detailed predictive simulations.

Simulations with most comprehensive modeling approaches applied to predict pyrolysis, charring and even exfoliation of practical combustibles burning in the rack-storage configurations have been recently undertaken in Refs. [4, 5, 6, 7], in which FireFOAM software tool with the bespoke pyrolysis model was applied. This series of works demonstrate that excellent replication of the heat release rates measured at the large-scale tests as well as of the realistic fire dynamics can be achieved as a result of the careful and long-term model refinement. Being tuned for a particular combustible material, such a comprehensive model might require re-calibration of a large number of model parameters if applied to alternative fire load. As such, a simplified approach incorporating as few as possible model parameters, yet capable of replicating the realistic heat release dynamics, should also be explored.

The simplified approach that describes ignition and burning of a flammable material in terms of the ignition temperature and prescribed burning rate (thereby avoiding consideration of the finite-rate pyrolysis) is applied here to predict the fire growth in the rack storage configuration. The examples of successful application of this approach in predicting upward flame spread over the vertical slab and in room-corner tests can be found in Refs. [8] and [9], respectively. It is of practical interest to understand whether this approach could potentially be applied in the rack storage occupied by the variety of materials for which pyrolysis kinetics may not be readily available. To explore this simplified approach by comparing the FDS predictions to the published measurements and to the results of more comprehensive simulations is the objective of this work.

It is worthy of note that the very few previously published studies used FDS to predict the rack-storage fires. The example of fully coupled simulations of this kind is Ref. [10] in which, however, an obsolete version 4 was applied. Taking into account all the improvements introduced in FDS submodels during recent years, it is necessary to validate it against a practical rack-storage fire scenario.

MODEL DESCRIPTION

Turbulence, combustion and thermal radiation

Fire Dynamics Simulator (FDS 6.6) numerically solves the Navier-Stokes equations for a multi-component reacting mixture in the low-Mach number limit. The large eddy simulation approach is used to simulate turbulent flow with the default Deardorff' sub-grid model. For turbulent combustion modeling, the eddy dissipation concept is utilized, and the single-step fast irreversible reaction is considered. Constant soot and CO yields are assumed. The rate of fuel consumption is set proportional to both the local limiting reactant concentration and the local rate of mixing, and the subgrid mixing time is evaluated taking subgrid kinetic energy, molecular diffusivity, and buoyancy into account.

Radiative transfer is simulated by solving the radiative transfer equation using the finite volume method with 104 discrete solid angles. Spectral properties of the gas-soot media are accounted for by the mean (gray) absorption coefficients, which are simulated by the RadCal procedure as a function of composition and

IOP Conf. Series: Journal of Physics: Conf. Series **1107** (2018) 042012 doi:10.1088/1742-6596/1107/4/042012

temperature. The ratio of the radiation emission in the flame to the chemical heat release is set equal to the pre-assumed radiative fraction. Unless otherwise stated, default heat transfer options of FDS 6.6 are used.

Two distinct approaches are applied in this and the earlier author's work to account for thermal decomposition of a solid combustible material.

Finite-rate pyrolysis model

This model employs the single step irreversible n-th order reaction of material decomposition. Temperature dependence of the reaction rate is described by Arrhenius type equation. Using this approach, the upward spread of the turbulent flame over the 5 m height PMMA slab is replicated in Ref. [11].

Simplified (thermal) pyrolysis model

This approach uses the ignition temperature, T_{ign}, and the mass loss rate, m''_{fuel}, as the input parameters. 1D heat conductivity equation without the reaction term is solved in the material layer of a constant thickness in the direction normal to the surface. The boundary condition at the surface takes into account conductive heat transfer to the material, absorbed radiation, reradiation, and the convective heat flux from the gas phase. Transparency of the material for the radiative flux is accounted for, and radiative transfer inside the material layer is solved by the two-flux method.

Once the surface temperature reaches the value of T_{ign}, the ignition occurs, a constant prescribed mass loss rate, m''_{fuel}, is set at the surface, and the net heat flux received by the surface is decreased by the value of $\Delta h_g m''_{fuel}$, where Δh_g is the effective heat of gasification. Note that in FDS 6.6 surface temperature is still calculated after ignition by solving the same heat conductivity equation with the corrected net heat flux at the boundary. This enables predicting considerable surface temperature growth in burning of charring materials. This default option is, therefore, retained in this work.

In order to account for the fuel burn-out, the corresponding heat release rate per unit area is ramped down to 0 kW/m^2 after the burn-out time, τ_b, is reached. The latter depends on burning rate and layer thickness and is defined by a user in the input data. The following estimate is used in this work for the burn-out time after ignition: $\tau_b = \rho\delta / m''_{fuel}$, where ρ and δ are the material density and the layer thickness.

Thus, it is possible to include heating up, ignition, developed and fire decay stages in the simulation in a simplified way, while still being able to capture most of the significant thermal feedback effects. The input parameters are as follows: thermal properties of the fuel (conductivity, specific heat, and heat of gasification), ignition temperature, heat release rate (or mass loss rate) per unit area of the burning material, burnout time.

RACK STORAGE FIRE SIMULATIONS

Simulation set-up

The experimental scenario studied in Ref. [6] is considered, with 2x4x3 combustible boxes in the rack. Each box has the dimensions of 1.07x1.07x1.07 m (see Fig. 1) and is simulated as a solid obstruction, for which a simplified pyrolysis model is employed as a boundary condition. The flues in the horizontal directions are 15 cm, in the vertical direction – 46 cm, the wooden pallets (often used in rack storages) are not considered. The first tier is elevated by 13 cm above the floor level. To initiate burning, four ignitors with constant heat release rate (22.5 kW each) were placed in the center of the rack so that 4 center bottom boxes are heated from different sides. Free-born scenario (no fire suppression) is considered.

Computational domain has the dimensions of 10.7x8.3x8 m (length, width, height). The grid is composed of 2.5, 5, and 10 cm size cubic cells. Nested rectangular zone with the finest grid surrounds the rack (see Fig. 1). The central area (4 central columns of boxes) is spanned by 2.5 cm cubic cells (6 cells across the horizontal flue). The cell size is increased to 5 cm in the vicinity of side columns and above the central columns where the main plume develops. At the periphery of the computational domain, 10 cm cells are used. The total number of cells is 3 103 220. It takes 125 hours of the wall clock time to compute 320 s of simulation time using 54 CPU cores (2 CPU Intel Xeon E5-2697 v3).

IOP Conf. Series: Journal of Physics: Conf. Series **1107** (2018) 042012 doi:10.1088/1742-6596/1107/4/042012

Model calibration

The following model parameters have to be defined: (i) material thermal properties and layer thickness, k, ρ, c, and Δh_c; (ii) ignition temperature, T_{ign}; (iii) heat release rate per unit area after ignition, which is coupled with the mass burning rate as $Q'' = m''_{fuel}\Delta h_c$; (iv) burn-out time, τ_b.

For the cardboard fuel considered in this work, we use the values of density, ρ = 184 kg/m^3, specific heat, c = 2700 J/(kg·K), and pyrolysate heat of combustion, Δh_c = 14.2 MJ/kg provided in Ref. [6]. For thermal conductivity, we adopt the value of k = 0.1 W/(m·K), which is closer to the literature data for cardboard (for example, see [13]) than the value of 0.42 W/(m·K) used in Ref. [6].

The ignition temperature, T_{ign}, is taken 360 °C as a representative value consistent with the literature data [14]. Note, that with the average net heat flux of 15 kW/m^2 time to ignition estimated in thermally thick limit is $t_{ign} = (\pi/4)k\rho c\left((T_{ign} - T_0)/q''_{net}\right)^2$ = 20 s. Consistent with the simulations performed in this work, during this time period the total measured heat release rate exceeds that provided by the igniters (90 kW) thereby indicating the beginning of sustained fire growth.

The HRRPUA value can be directly estimated from the HRR growth dynamics observed in Ref. [6], which shows that at time instant 85 s the HRR approaches 10 MW, and two central columns at the second and third tier are engulfed in fire. The latter implies that the burning area equals to the total surface area of 8 boxes, i.e. $1.07^2 \cdot 6 \cdot 8$ = 55 m^2, which results in Q'' = 10 000/55 = 182 kW/m^2. Also, at time instant 150 s 16 boxes are engulfed in fire, which indicates that the burning area roughly doubles (mainly due to the horizontal flame spread to the outside columns). The measured HRR at 150 s is about 20 to 23 MW, which indicates that the value of HRR per unit area remains nearly the same. Based on these estimates, we use the HRRPUA value of 200 kW/m^2. Recall that this is the average heat release rate per unit area of the box surface, which is much lower than the peak value of HRR per unit area of the horizontal cross-section of the rack.

The burn-out time, τ_b, determines the onset of the decay of the heat release rate. Transient variation of the HRR recorded in Ref. [6] shows that the burn-out time can be estimated as τ_b = 290 s. In fact, by assigning the burn-out time, we define the fuel mass per unit area, $m''_{fuel}\tau_b = \rho\delta$. For the given material density, mass burning rate and heat of combustion, it corresponds to the equivalent layer thickness $\delta = \tau_b m''_{fuel}/\rho = \tau_b Q''/(\rho\Delta h_c)$ = 0.022 m. This value is larger than the actual cardboard thickness (for example, 8.4 mm thick cardboard is used Ref. [12]), and this difference accommodates possible variation of the burning rate in time.

Finally, the heat of gasification, Δh_g, has to be estimated. In the model, the heat of gasification is subtracted from the net heat flux received by the material thereby reducing the predicted surface temperature. Thus, an underestimated value of Δh_g causes overestimated surface temperature, radiative flux emitted by the burning surface, and, therefore, radiative flux incident to the non-burning surfaces. In its turn, it causes earlier ignition of the non-burning surfaces and faster total heat release rate. This is shown in Fig. 2 where transient dependences of the predicted heat release rate are shown for three pre-assumed values of Δh_g. It can be seen that the value of Δh_g = 1 MJ/kg results in a reasonable agreement with the measurement.

Simulation results

Predicted fire development is visualized in Fig. 1. As highlighted in Ref. [7], the HRR growth rate is higher if flame propagates vertically upwards compared to that occurring if flame propagates horizontally; the growth rate, therefore, varies in time because of the varying direction of primary flame propagation. The following periods can be identified after the lowest boxes are ignited at time instant 16 s. Period 1 (16 – 50 s): flame spreads upwards over the vertical surfaces of the central boxes, and the flame tip appears above the upper tier. Period 2 (50 – 80 s): flame spreads horizontally covering the bottom surfaces of the central boxes in the upper tiers. The flame tip impinges the ceiling. Period 3 (80 – 120 s): flame spreads both vertically and horizontally: all vertical sides of the central boxes, 2nd and 3rd tier, burn. Period 4 (120 – 140 s): flame spreads horizontally towards the side boxes between 2nd and 3rd tier. Period 5 (140 – 190 s): flame spreads both vertically and horizontally. Note, that downward flame propagation over the external vertical surfaces of the side boxes at the second tier is also predicted. Finally, all vertical sides of the boxes in the 2nd and 3rd tiers burn, and the heat release rate attains its maximum. This fire dynamics corresponds well to experimental observations and simulation results presented in Refs. [6, 7], where a more complicated model was utilized.

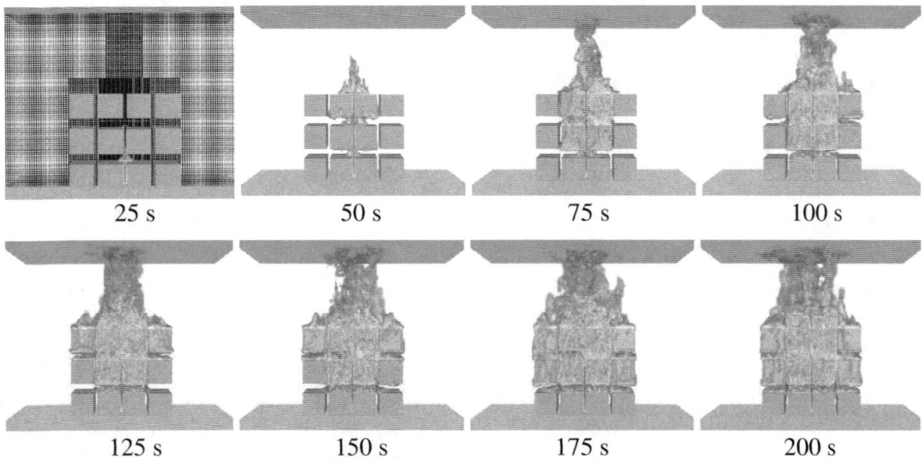

Fig. 1. Predicted flame snapshots (200 kW/m^2 heat release rate iso-surface) for the 2×4×3 rack-storage free-burn fire. The left upper figure shows the computational mesh in the central plane

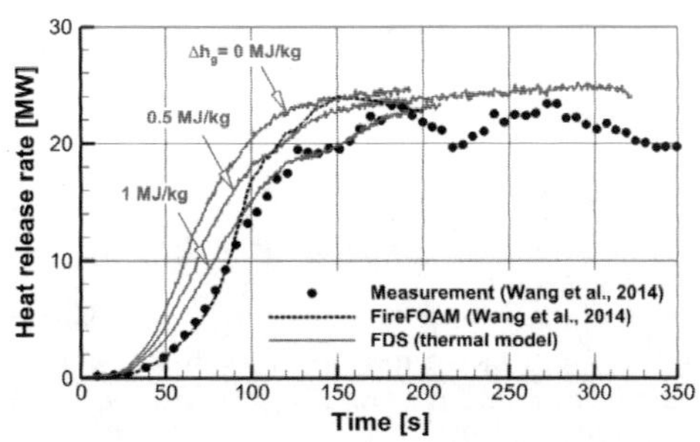

Fig. 2. Heat release rate history in the 2×4×3 rack-storage free-burn fire

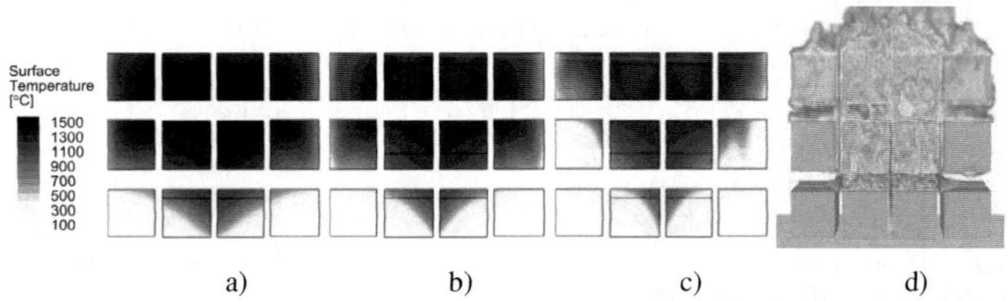

Fig. 3. The temperatures of the inner box surfaces (facing inside the rack) predicted with different values of the heat of gasification: a) – 0 MJ/kg; b) – 0.5 MJ/kg; c) – 1 MJ/kg at time instant 150 s and (d) the corresponding flame snapshot (200 kW/m^2 heat release rate iso-surface, the heat of gasification- 1 MJ/kg).

Predicted temperature distributions at the *inner* surfaces of the burning boxes is shown in Fig. 3. As expected, it appears to be much higher than that of the *outer* surfaces as presented in simulation results in Ref. [7]. This can also be attributed to the significantly different thermal conductivities used in these simulations.

CONCLUSIONS

A simplified (thermal) pyrolysis model enables approximate replication of the fire dynamics observed in full-scale fire tests of rack-storage fires. This can be achieved provided the model parameters (ignition temperature, HRRPUA, burn-out time, and heat of gasification) are properly selected for a given type of fuel load. In this work, thermal properties and the ignition temperature of the cardboard are selected based on the published measured values. The average HRRPUA value is evaluated by joint consideration of the measured dynamics of the growing heat release rate and the surface area engulfed in fire. The burn-out time and the effective thickness of the fuel layer is estimated as the period from ignition to the time instant after which the measured heat release rate starts to decay. The heat of gasification is selected by fitting the predicted transient dependence of the heat release rate to that measured in the full-scale tests. Optimum values of the model parameters are consistent with the literature data for cardboard. This implies potential capability of the simplified thermal pyrolysis model to predict rack storage fire dynamics with an alternative fire load, for which the full-scale test data may not be available.

This work was partially supported by the Russian Science Foundation (project 16-49-02017). Computational resources were provided by the Supercomputer Center "Polytechnic".

REFERENCES

[1] Trapp, A. C., Rangwala, A.S. (2015) Analyzing the Impact of In-Rack Sprinklers in a Warehouse Fire: A Demonstration of the Role Optimization has in mitigating damage, *Fire Safety Journal* 73: 55–62, https://doi.org/10.1016/j.firesaf.2015.03.002

[2] Yu, H.-Z. (1990). Transient Plume Influence in Measurement of Convective Heat Release Rates of Fast-Growing Fires Using a Large-Scale Fire Products Collector, *Journal of Heat Transfer* 112(1): 186-191, https://doi.org/10.1115/1.2910343

[3] Yu, H.-Z., Stavrianidis. "The Transient Ceiling Flows of Growing Rack Storage Fires", *Fire Safety Science – Proceedings of the Third International Symposium*, IAFSS, 1991, pp. 281-290, https://doi.org/10.3801/iafss.fss.3-281

[4] Chatterjee, P., Wang, Y., Chaos, M., Meredith, K.V., Zhou, X., and Dorofeev, S.B., "Numerical simulation of fire growth on corrugated cardboard commodities in three-tier-high rack storage arrays" – *Proc. 13th Int. Conf. Interflam*, UK, 2013, pp. 163-173.

[5] Meredith, K.V., Chatterjee, P., Wang, Y., Xin, Y., "Simulating Sprinkler Based Rack Storage Fire Suppression under Uniform Water Application," – *Proc. 7th Int. Seminar on Fire and Explosion Hazards*, 2013, pp. 511–520, http://dx.doi.org/10.3850/978-981-07-5936-0_07-08

[6] Wang, Yi., Meredith, K. V., Zhou, X., Chatterjee, P., Xin, Y., Chaos, M., Ren N., and Dorofeev, S. B., "Numerical Simulation of Sprinkler Suppression of Rack Storage Fires" – *Fire Safety Science – Proceedings of the Eleventh International Symposium*, IAFSS, 2014, pp. 1170-1183, https://doi.org/10.3801/iafss.fss.11-1170

[7] Ren, N., de Vries, J., Zhou, X., Chaos, M., Meredith, K. V., Wang, Y (2017). Large-Scale Fire Suppression Modeling of Corrugated Cardboard Boxes on Wood Pallets in Rack-Storage Configurations, *Fire Safety Journal* 91: 695–704, https://doi.org/10.1016/j.firesaf.2017.04.008

[8] Yan, Z., Holmstedt, G. CFD Simulation of Upward Flame Spread over Fuel Surface, *Fire Safety Science – Proceedings of the Fifth International Symposium*, IAFSS, 1997, pp. 345-356, https://doi.org/10.3801/iafss.fss.5-345

[9] Lewis, M. J., Rubini, PA., Moss, J. B. Field Modelling of Non-Charring Flame Spread, *Fire Safety Science – Proceedings of the Sixth International Symposium*, IAFSS, 2000, pp. 683-694, https://doi.org/10.3801/iafss.fss.6-683

[10] Guedri, K., Borjini, M. N., Jeguirim, M., Brilhac, J.-F., Said, R. (2011) Numerical Study of Radiative Heat Transfer Effects on a Complex Configuration of Rack Storage Fire, *Energy* 36: 2984-2996, https://doi.org/10.1016/j.energy.2011.02.042

[11] Markus, E., Kuznetsov, E., Snegirev, A. (2018) Buoyant turbulent diffusion flame attached to the vertical surface, *Combustion Explosion and Shock Waves*, accepted.

[12] Chaos, M., Khan, M. M., Dorofeev, S. B. (2013) Pyrolysis of corrugated cardboard in inert and oxidative environments, *Proceedings of the Combustion Institute* 34: 2583–2590, https://doi.org/10.1016/j.proci.2012.06.031

[13] Čekon, M., Struhala, K., Slávik, R. (2017) Cardboard-Based Packaging Materials as Renewable Thermal Insulation of Buildings: Thermal and Life-Cycle Performance, *Journal of Renewable Materials* 5 (suppl. 1): 84-93, https://doi.org/10.7569/JRM.2017.634135

[14] Quintiere, J.G., *Principles of Fire Behavior*, Delmar Publishers, NY, 1997, 258 p.

IOP Conf. Series: Journal of Physics: Conf. Series **1107** (2018) 042013 doi:10.1088/1742-6596/1107/4/042013

"Experimental assessment of separation distances of a load-bearing straw-bale construction"

Petr HEJTMÁNEK[1], Hana NAJMANOVÁ[2] and Tomáš VÁCHAL[3]

[1] *Department of Building Structures, Faculty of Civil Engineering, Czech Technical University in Prague, Czech Republic*
[2] *University Centre of Energy Efficient Buildings, Czech Technical University in Prague, Czech Republic*
[3] *Department of Construction Technology, Faculty of Civil Engineering, Czech Technical University in Prague, Czech Republic*

petr.hejtmanek@fsv.cvut.cz

ABSTRACT

With an increasingly dense urban structure and wider use of energy efficient and combustible materials, the requirements on fire separation distances are becoming an actual topic. In the Czech Republic, separation distances are determined according to "fire openness" categorisation of external walls. In line with the Czech Fire Standards, combustible external walls must be considered as unprotected areas (fire opened areas) until the opposite is proven. This classification might be a crucial element in building design and for location of an object on a plot. This paper introduces a project on fire safety evaluation of a load-bearing straw-bale construction and focuses on experimental determination of fire openness of combustible external walls with incombustible surface (A1/A2 reaction to fire class). To assess fire risk and a contribution to fire growth and flame spread of the walls, a full-scale fire test of an experimental straw-bale house was carried out in September 2017. The experimental object was designed as a mock-up of a residential house, single-storey, stand-alone building with ground floor dimensions of 4.0 × 6.0 m and a height of 3.5 m with two symmetrical connected rooms. The observations and measurements showed that the experimental wall composition, even though it was not certified as a construction with a specific fire resistance, did not contribute to the fire propagation and could be assumed as a fire closed area, where as unprotected areas are considered only windows. In the case of the evaluated object, this classification leads to a reduction in separation distances of almost 4 meters. According to the results can be concluded that a different approach to evaluate fire openness could be desirable to establish.

KEYWORDS

Large-scale fire test; fire; separation distance; heat flux; temperature; straw-bale construction; clay plaster; compartment fire

IOP Conf. Series: Journal of Physics: Conf. Series **1107** (2018) 042013 doi:10.1088/1742-6596/1107/4/042013

INTRODUCTION

In line with a growing importance placed on sustainable building, renewable resources and natural construction materials are becoming an inherent part of modern building design. Offering a new approach on the manner of designing buildings, this perspective challenges also other fields which at the first sight seemed to be in conflict with it, including fire safety design [1–2]. Regarding their combustibility, use of environmentally friendly materials may be questionable and strictly limited in national fire codes and standards. In context with an increasingly dense urban structure and a frequent application of these materials both in and on external walls, above all, the requirements on fire separation distances become an especially actual issue.

Among many possibilities of external fire spread, the radiative heat transfer is assumed as one of the main risks [3]. Considering this criterion, calculation of a separation distance is bounded with a line of critical radiative heat flux varying from 12.5 kW/m^2 for piloted ignition to 30.0 kW/m^2 for self-ignition [4]; in the Czech Republic, the critical value is 18.5 kW/m^2 [5]. Methodologies for determination of separation distances are usually given at national level by local standards and vary from simplified limit values based only on expected radiation up to performance-based evaluation considering variables such as configuration factor, heat release rate, emissivity, the thermal capacity of the affected structure, and others. A common aspect can be found in most of the methods: the amount of "unprotected areas" in the building enclosure. An unprotected area may be defined as any part of a wall which may contribute to radiation during the fire: parts without fire resistance (e.g. door, window) or parts that emit radiation energy; hence, combustible structures and structures with a combustible cladding must be also taken into account [6]. In the Czech Fire Standards, limit values for radiative heat flux released from a façade are laid down for estimation of "fire openness" of a structure: an external wall containing combustible materials is considered as a fire opened area (FOA) if the heat radiation flux at the outer face of the wall exceeds the value 60 kW/m^2, a range between 15 kW/m^2 and 60 kW/m^2 determines the wall as a partly fire opened area (PFOA), and an external wall is assessed as a fire closed area (FCA), only if the value of the radiation heat flux is less than 15 kW/m^2. Furthermore, regarding the Czech classification system for evaluation of structure's fire behaviour, combustible external walls must be considered as unprotected areas (fire opened areas), until the opposite is proven [5]. Calculation of separation distances are directly affected by the magnitude of a fire opened area, i.e. by openness categorisation of an external wall (FOA, PFOA, or FCA); as will be seen later in Fig. 7. Consequently, use of environmentally friendly and energy efficient materials in buildings may be strictly limited due to the large separation distances arisen from this classification.

To minimize separation distances of combustible external walls, i.e. to achieve other classification than FOA, "the opposite" must be proven. Regarding this aim, however, several problems are occurring. Firstly, although the limits of FOA, PFOA, and FCA are set down, no methodology how this characteristic can be obtained in practice is available. Since the set limits assess a contribution to fire of the surface, there is no simply way to measure the heat release without outer sources that may influence the results when it is tested physically. Furthermore, to comply current requirements when experimentally tested, both fire resistance tests from outside and inside must be conducted to get the knowledge about the external surface behaviour and to estimate surface temperatures on the unexposed side of the structure. Subsequently, from the maximum surface temperature on both sides, the heat flux is calculated using the Stefan-Boltzmann Law:

$$I = \varepsilon \cdot (T_{max} + 273)^4 \cdot 5{,}67 \cdot 10^{-11} \quad (1)$$

where I is heat flux [kW/m^2]; ε is emissivity [-], for fire $\varepsilon = 1$; T_{max} is the highest temperature on the unexposed side from both interior and exterior fire tests [°C]. Regarding this method is arguable, why surface temperature in the interior must be set and considered when an outer surface is assessed. Another way to estimate the value of heat radiation flux of a combustible wall is a summation of total heat release of all combustible layers of the structure; however, since an eventual incombustible surface (covering) layer and heat release of layers in time have no effect in the calculation, the real behaviour of the construction is neglected and may cause conservative results when using this method. This paper introduces a project on fire safety evaluation of a load-bearing straw-bale construction. With respect to the current scarcity of data describing fire characteristics and fire performance of straw as structural elements, the main goal of the project was to enhance the knowledge of fire behaviour of this material by acquiring new data in full-scale fire test of an experimental straw-bale house. **This contribution focuses on experimental determination of fire openness of combustible external walls with incombustible surface and aims to assess their fire risk and contribution to fire growth and flame spread.**

EXPERIMENTAL STRAW-BALE HOUSE

The experimental object was planned, constructed, tested, and burned during one-year period; besides the fire testing, other issues such as labour time dependency or thermal stability were investigated. It was designed as a single-storey, stand-alone building with ground floor dimensions of 4.0 × 6.0 m and a height of 3.5 m with two symmetrical connected rooms. Each room had two openings: a wooden door (0.7 × 2.0 m) on the shorter side of the building (western and eastern façade) and a plastic, double-glazed, window (0.8 × 1.4 m) on the southern façade. To compare influence of the openings, the northern façade was left without any opening, see Fig. 1.

To prevent moisture adsorption, the building was founded on concrete blocks. The walls were made of compressed straw bales having approximate dimensions of 0.60 × 0.40 × 0.35 m with initial density estimated of 130 kg/m^3. The straw bales were placed between a lower and upper wooden frame, where every row of bales was connected with the previous one with several reinforcement bars. Doors and windows were installed in wooden frames (cases). To provide structural stability, the whole construction was tied up with steel stripes. On the upper frame, wooden beams covered with OSB boards from both sides and water-proof insulation were laid as a roof construction, thermally insulated with mineral wool between the beams. The roof composition was designed according to Eurocode with REI 60, this evaluation was addressed in a parallel research. The straw bales were covered with an inner and outer plaster (approximately 50 mm in thickness), which due to research purposes varied in mixtures of clay (in-situ quarried clay, prefab clay mixture, or prefab lime-based clay), in reinforcement of the plaster (an inner plaster with reinforcement wire-mesh was used in one of the rooms), as well as in a way of its application: the plasters were cast either by professional worker, by trained students, or by amateurs under supervision of a professional. The window and door frames were left exposed.

The object was designed as a mock-up of a residential house. According to the Czech Fire Standards, load bearing external constructions must fulfil R 15 fire resistance in this case. Nevertheless, despite complying this requirement, an external wall is assumed as FOA, unless the contrary is approved by a certification.

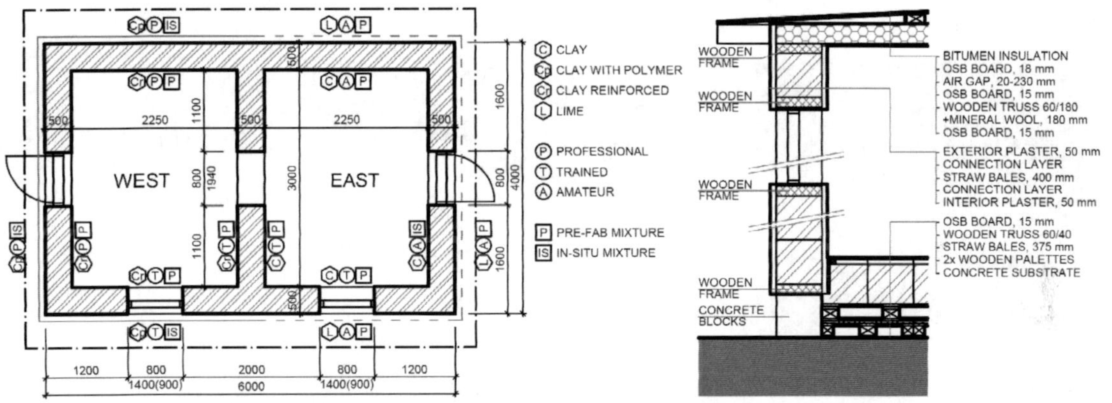

Fig. 1. Layout of the experimental object (left) and a cross-section (right)

LARGE-SCALE FIRE TEST

The fire experiment was carried out at the University Centre of Energy Efficient Buildings, Czech Technical University in Prague, in September 2017. Inside the building, wooden cribs, representing a residential house fire load of 40 kg/m^2, were deployed equally. Every crib consisted of wooden planks. To achieve the temperature development according to ISO 834 curve, every 4th row of the planks was replaced by OSB boards to slow down the burning rate. In total, there were 423 kg of wooden planks and 191 kg of OSB boards used in 12 cribs (6 cribs placed in each room). Windows were closed, doors were partly opened in 45° angle.

For this fire test, the façade with windows (southern) and the façade without openings (northern) were observed. On the southern façade, 28 surface thermocouples were arranged into the matrix 7 × 4 on the upper part of the wall, additional 2 thermocouples were put in the middle of the window overhead. On the northern façade, 12 surface thermocouples were installed (see Fig. 2). All thermocouples were K-type of 0,5 mm diameter ended with a copper disc sheet of 10 mm diameter, mechanically mounted to surface with a screw right next to the disc. Overall surface temperatures were also monitored with set of thermographic cameras. Heat flux meters (Schmidt-Boelter gauges, ranged up to 100 kW/m^2) were installed in the distance of the calculated separation distance only for windows. Unfortunately, the connection with the gauges was lost in the early minutes of the test and data describing heat flux were unavailable.

The fire test started at 10:50, two minutes after, all wooden cribs were simultaneously ignited with petrol channels. After 4 minutes, the first window broke and fire started to affect the southern façade. The second window broke after 8 minutes. 11 minutes after ignition, fire spread through openings onto the roof. In the 20th minute, the straw bales themselves were exposed to fire, when unprotected OSB frames of doors and windows fell off. The unexposed surface stood intact until the 25th minute, when the first small cracks emerged; since the 40th minute, the external plaster started to spall off. In the 55th minute, one of the window frames came down and since that moment, the straw bales located above were getting loosed. Eventually, the whole construction collapsed 64 minutes after the ignition.

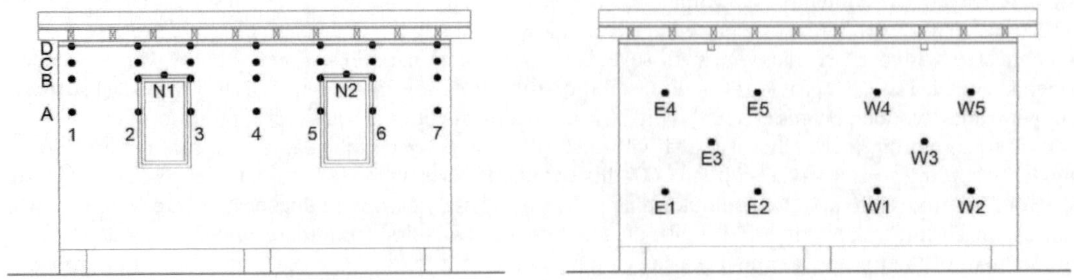

Fig. 2. Positions of thermocouples arranged on the northern (left) and southern façade (right).

During the fire test, the temperature rate in the object, monitored by one shielded thermocouple placed in the geometry centre of each room at a height of 2.0 m (2/3 of the room height), copied ISO 834 curve quite well; although the real values were higher: in the 30th minute, approximately 20% difference between ISO 834 and real temperatures was identified. Maximum temperatures reaching 1.100 °C were observed from the 25th to the 48th minute, followed by a gradual decline (see Fig. 3).

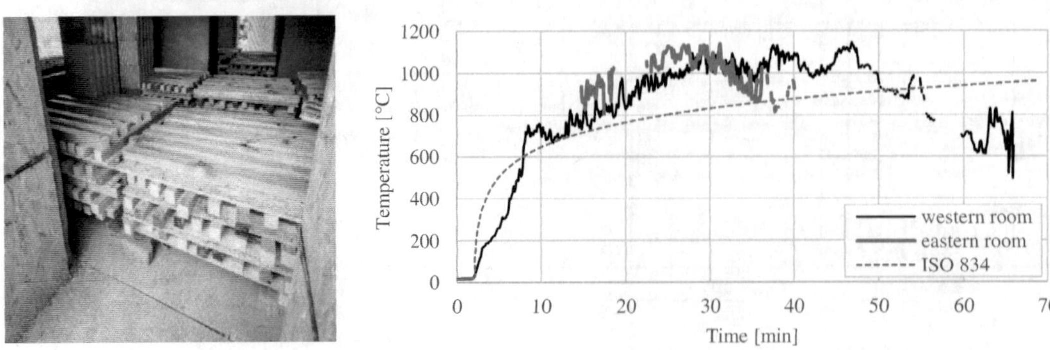

Fig. 3. Wooden cribs simulating residential house fire load (left). Development of temperature inside the tested object (right).

RESULTS

The observations showed that the straw-bale construction with an incombustible plaster resisted in fire conditions successfully. Even when exposed due to a fallen-off plaster or due to a burnt-away casing and when caught by fire on the surface, the straw showed satisfactory behaviour: thanks to a char layer formed on the surface, almost no contribution to the fire propagation was observed in the duration of the fire test. The good behaviour in fire can be illustrated in Fig. 4 comparing the southern and the northern façade just before the object's collapse. While the southern façade is damaged, the window overheads are completely fallen off and the straw bales are exposed to fire due to a plaster deflection, the northern façade without any opening seemed to be intact. During the fire test, the lime-based plaster started to drop off earlier – the first shard spalled in the 40th minute, whereas the clay plaster remained solid until the 63rd minute.

Since it was not intended to separate the heat flux of the fire and the eventual heat flux caused by the construction, a different approach was used to evaluate fire openness of the external walls. According to Eq. 1, the limit values for heat flux (15 kW/m^2 and 60 kW/m^2) were transformed into limits given by temperature, i.e. maximum temperature of 444 °C for classification as FCA and 741 °C for classification as PFOA (higher temperatures represent FOA). Using the thermocouple matrix, temperatures in defined points on the surface were measured and were later compared with the thresholds.

IOP Conf. Series: Journal of Physics: Conf. Series **1107** (2018) 042013 doi:10.1088/1742-6596/1107/4/042013

Fig. 4. Northern (left) and southern (right) façade in the 65th minute (just before the collapse).

The northern façade (without openings)

Thermocouples and thermal images on the northern façade had shown a very limited growth of temperature. In general, temperature did not exceed 150 °C, except sporadic peaks which lasted at most for minutes. Based on the observation and experience may be assumed that a measurement mistake occurred at those points, caused mainly by the wiring affected by high temperature. Even if the peaks corresponded to the real temperatures, because they lasted a very limited time, the heat released during these sequences would be negligible. Leaving out those peaks, the highest temperature was observed in the upper part of the wall (namely thermocouple E4 on the lime-based façade and W5 on the clay façade), see black dots in Fig. 5.

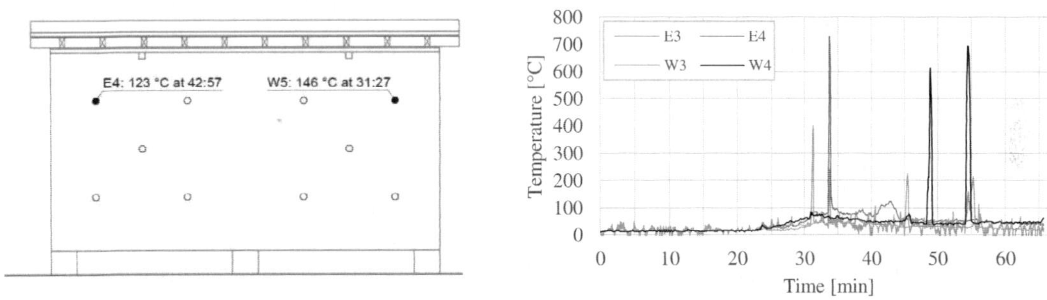

Fig. 5 The highest temperatures on the northern façade (left). Development of unexposed surface temperatures in selected points (right).

The southern façade (with windows)

As can be seen in Fig. 6, in time of the maximal fire intensity (around the 30th minute), temperatures exceeding the criterion of 444 °C can be found only around the windows and near the roof (dashed-line rim). The temperatures over 741 °C occurred only at one point just below the roof and at one point near the window (namely D2 and B2; solid-line rim). According to the visual observation, there was no crack or gap in the plaster at these points, elevated temperatures on the surface were caused by the flame only, either coming from the window or from the uncovered roof. On the remaining part of the wall only lower temperatures than 444 °C were measured. The highest temperature in each line of the thermocouple matrix was observed around the western window (namely A2, B2, C3, and D2), reaching almost 800 °C (indicated as black dots in Fig. 6).

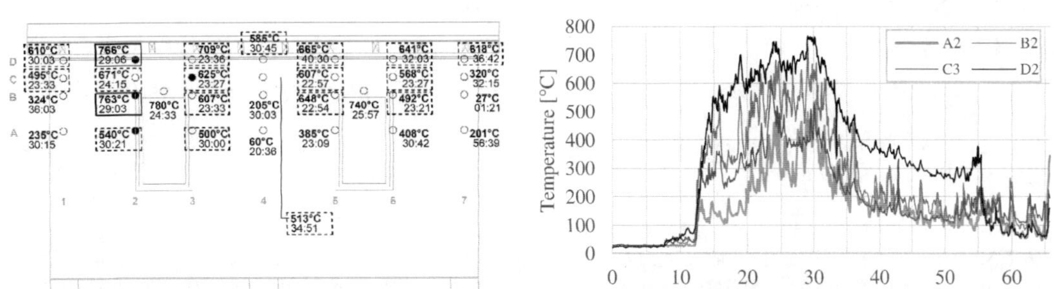

Fig. 6. The highest temperatures on the southern façade. Dashed-line rim: over 444 °C, solid-line rim: over 741 °C (left). Development of unexposed surface temperatures in selected points (right).

Separation distances evaluation

Since the openness of the tested straw-bale construction is not approved by a certification, according to the Czech Fire Standards, it must be treated as FOA. In this case, the separation distance is determined for the whole façade as d_{FOA} = 5.57 m. However, the fire test has shown that temperatures did not exceed the criterion for FCA at any point on the northern façade. In the case of the southern façade, this criterion was exceeded only around the windows and near the roof, what corresponded to the impact of present radiation heat flux of the fire. Thus, it can be inferred that the external walls did not emit any significant amount of heat, did not contribute to fire propagation and could be therefore assumed as FCA. The separation distance determined just for windows is calculated as d_{FOA} = 1.64 m, what is almost 4 meters less (for a comparison see Fig. 7). For further refinement, an FDS6 mathematical model with given heat flux in the openings corresponding the fire intensity equalling inside temperatures was used. Radiation heat flux was measured on a plane in the middle of windows, where the maximum separation distance defined by the Standard's critical value 18,5 kW/m^2 was determined as 1.81 m. Albeit slightly higher, this result is comparable with the value given by the Standard for FCA; however, a difference is observable on the edges of the area, where the Standard assumes broader radiation. Results for the southern façade are compared in Fig. 7. In case of the northern façade, the separation distance given by the Standard for FOA is the same 5.57 m; the wall assessed as FCA (with no openings) has no separation distance.

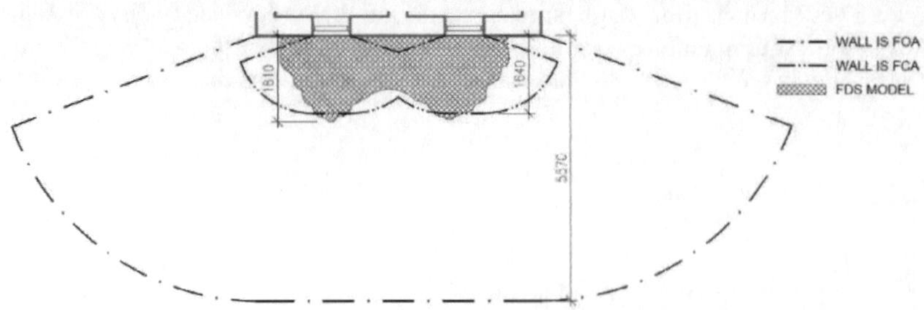

Fig. 7. Separation distances for southern façade.

CONCLUSIONS

In view of the fire openness categorisation of façades, determination of separation distances in the Czech Republic is quite strict and it limits use of environmental efficient (and combustible) materials and constructions. In this paper, contribution to fire growth of external load-bearing straw-bale walls with an incombustible surface was evaluated. The observations and measured values confirmed that the tested wall composition, even though it was not certified as a construction with a specific fire resistance, did not contribute to the fire propagation. Higher temperatures on the unexposed surface measured on the southern façade were caused by the flames and heat radiation of fire going only from the openings. Based on these results, the external walls could be therefore assumed as FCA, where separation distances are calculated only for windows. In the case of the evaluated object, the decision between FOA and FCA causes the difference in separation distances of almost 4 meters. If the FDS model is used to determine separation distances, a refined shape would be obtained lowering down the corner values. According to the executed fire test it is thought that a different approach to evaluate fire openness could be established. It can be assumed that a construction with fire resistance and solid incombustible surface is fire closed area unless it has openings installed. With openings present, the impact of such wall to separation distances is significantly lower (or even insignificant) than stated by the Czech Fire Standard.

REFERENCES

[1] B. Meacham, B. Poole, J. Echeverria, and R. Cheng, *Fire Safety Challenges of Green Buildings*. New York, NY: Springer New York, 2012.
[2] U. Krause, W. Grosshandler, and L. Gritzo, 'The International FORUM of Fire Research Directors: A position paper on sustainability and fire safety', *Fire Saf. J.*, vol. 49, pp. 79–81, Apr. 2012.
[3] C. R. Barnett, 'Fire Separation between External Walls of Buildings', in *Fire Safety Science - Proceedings of the 2nd International Symposium*, 1988, pp. 841–850.
[4] E. Carlsson, 'External Fire Spread to Adjoining Buildings - A review of fire safety design guidance and related research', Dept. of Fire Safety Engineering Lund University, Lund, Sweden, Report 5051, 1999.
[5] 'ČSN 73 0802 Požární bezpečnost staveb - Nevýrobní objekty [Czech Technical Standard 73 0802 Fire Protection of Buildings - Non-industrial Buildings]'. ÚNMZ, 2009.
[6] 'External fire spread: building separation and boundary distances', Building Research Establishment, Borehamwood, Herts, Building Research Establishment Report BR187, 1991.

IOP Conf. Series: Journal of Physics: Conf. Series **1107** (2018) 042014 doi:10.1088/1742-6596/1107/4/042014

CNG release from pressurized vessel through pressure relief safety device: isentropic flow approach versus CFD model

Lucie HASALOVÁ[1], Jiří IRA[2], Milan JAHODA[2] and Václav VYSTRČIL[1,2]

[1]*Technical Institute of Fire Protection in Prague, Fire Rescue Service of the Czech Republic, Czech Republic*
[2]*University of Chemistry and Technology Prague, Czech Republic*

lucie.hasalova@tupo.izscr.cz

ABSTRACT

This paper numerically studies flow characteristics of Compressed Natural Gas (CNG) from a fuel tank of a CNG passenger car through a 3 mm orifice of a temperature triggered pressure relief device (TPRD). The outflow velocity and the methane mass flow are of interest for modelling the gas dispersion both in the vicinity of the orifice as well as further from the leak source. It enables assessing the risk connected with possible accumulation of the explosive gas in enclosures or ignition of leaking gas.

An engineering approach of isentropic nozzle flow was compared to the results of a 2D CFD simulation over the whole range of the reservoir working pressure 200 bar (full fuel tank). The pressure in the reservoir was assumed constant. CNG was approximated as pure methane. Methane in the reservoir was set at ambient temperature as well as the surrounding air.

Both models predict the same trends both in the methane outflow velocity and mass flow through the orifice. Excellent correspondence between both models was found up to reservoir pressure 50 bar. Above this value, isentropic model predicts about 10 % higher outflow velocity, and even smaller difference was found in the predicted methane mass flow. Mass flow prediction of the isentropic model is, however, strongly dependent on the discharge coefficient.

For reservoir pressures between 50 to 200 bar, the Mach number predicted by CFD was in the range from 3.5 to 6 which corresponds to a transition region between super and hypersonic flow. Even at such high speed flow, CFD predictions were consistent with the isentropic nozzle flow theory and in general, the correspondence between both models is very satisfactory.

KEYWORDS: *CNG; gas dispersion; modelling; CFD; isentropic flow*

INTRODUCTION

During the last two decades, the amount of CNG vehicles steadily increase. CNG vehicles are nowadays commonly present in the everyday traffic. Presence of the pressure vessel on board the vehicle and the risk of CNG leak either from the high or low pressure pipeline, however, present a safety concern. To prevent the critical pressure buildup, each fuel tank has to be equipped with the temperature triggered pressure relief device (TPRD) with activation temperature of 110±10 °C. In case of the TPRD activation or leak from the high pressure pipeline, large amount of CNG is released through a small orifice or hole in a relatively short time to the ambient atmosphere. The outflow velocity and the CNG mass flow from the fuel system to the surroundings are of key importance for further analysis of gas dispersion, ignition probability analysis, design of the detection and ventilation systems etc.

Outflow of gas from pressurized vessel to ambient atmosphere

Working pressure in the passenger CNG vehicles is usually 200 bar. With such a high pressure difference, the resulting flow from a pressurized environment to the ambient well exceeds the speed of sound. Compressibility effects play significant role and must be considered in order to predict the flow. Isentropic models [1] are often used to describe the flow characteristics of both subsonic and supersonic compressible flows through a nozzle/orifice. Isentropic flow is any flow that is adiabatic and reversible. No friction or dissipative effects are considered. The outflow characteristics can be calculated based only on the stagnation state characteristics (reservoir pressure). The main assumption of isentropic flow is that it is a reversible process. When the gas that flows through the nozzle/orifice/hole is turned abruptly at the exit plane, shock waves emerge, and highly underexpanded jet is created [2]. The flow is irreversible, and isentropic relations are no longer valid. Another option to describe the flow characteristics of compressible fluid is to use CFD. The advantage of the CFD model is that compared to isentropic flow models it also provides the spatial information about the flow and structure of the underexpanded jet [3]. Friction and dissipative losses are accounted. The flow regimes can be distinguished based on the Mach number. Common CFD solvers are validated for supersonic flow applications typically characterised by the Mach number in the range 1.2 to 5. The outflow velocities from the full fuel tank are expected to locally exceed Mach 5 switching to hypersonic flow regime. At such flow speed, turbulence modelling becomes a significant concern, and the approximation of the gas thermodynamics behaviour as ideal becomes questionable. However, outflow velocity measurements at super or hypersonic speed is extremely difficult to perform. Validation of the CFD model compared to experimental data is thus rarely possible.
This work compares outflow velocity and gas mass flow of methane from pressurized reservoir through 3 mm orifice as predicted by the isentropic flow model and 2D CFD model. These characteristics are of significant importance serving as a boundary condition to CNG leak risk analysis. The fuel leak results in highly momentum driven flow. Jet fires may be observed and present significant safety concern. At the same time due to strong mixing the gas leaving the bottom region of the car looses big part of its momentum. From the long term view the proper estimate of the fuel mass flow is crucial for modelling gas dispersion further from the leak source.

ISENTROPIC FLOW MODEL

The outflow velocity v from the pressure vessel can be derived from the principle of energy conservation [4]. Assuming pressure and density of the isentropic flow are related as $p/\rho^{\gamma} = \text{constant}$, Bernoulli equation for ideal compressible fluid, where subscript zero denotes the conditions in the pressure reservoir (stagnation state), is obtained

$$\frac{{v_0}^2}{2} + \frac{\gamma}{\gamma-1}\frac{p_0}{\rho_0} + h_0 g = \frac{v^2}{2} + \frac{\gamma}{\gamma-1}\frac{p}{\rho} + hg\,. \tag{1}$$

Assuming the contribution of the hg term in Eq. 1 is negligible and the fluid velocity in the vessel is negligible compared to the outflow velocity ($v_0 \approx 0$), the outflow velocity can be calculated as

$$.\,v = \sqrt{\frac{2\gamma}{\gamma-1}\frac{p_0}{\rho_0}\left[1-\left(\frac{p}{p_0}\right)^{\frac{\gamma-1}{\gamma}}\right]}\,. \tag{2}$$

Equation 2 is known as Saint-Venant-Wantzel equation. γ is an isentropic exponent (isentropic expansion coefficient) which is a ratio of ideal gas isobaric and isochoric specific heat. In this work, γ for methane was set 1.304 (calculated at ambient conditions) independent of temperature.

Theoretical mass flow from the reservoir can be derived as $\dot{m} = \rho v A$. A is the outflow area. Substituting for ρ from the isentropic relation $p/p_0 = (\rho/\rho_0)^\gamma$, and multiplying by discharge coefficient C_d yields equation for the real mass flow in the subsonic region

$$\dot{m} = C_d A p_0 \sqrt{\frac{2\gamma}{\gamma-1}\frac{M}{RT_0}\left[\left(\frac{p}{p_0}\right)^{\frac{2}{\gamma}} - \left(\frac{p}{p_0}\right)^{\frac{\gamma+1}{\gamma}}\right]}. \quad (3)$$

After reaching the critical condition, the flow is choked. For calculating the mass loss in the supersonic region, velocity corresponding to the critical pressure ratio is substituted to the continuity equation resulting in relation for mass flow through orifice in the supersonic region as

$$\dot{m} = C_d A p_0 \sqrt{\frac{\gamma M}{RT_0}\left(\frac{2}{\gamma+1}\right)^{\frac{\gamma+1}{\gamma-1}}} \quad (4)$$

For the comparison with the mass flow predicted by the CFD model through the orifice depicted in Fig. 1., discharge coefficient 0.98 was used (recommended value for the rounded type orifices [5]).

2D CFD MODEL

A commercially available CFD solver ANSYS CFX 17.2 was used to perform all CFD simulations. Computational domain is shown in Fig. 1. Methane is supplied from the reservoir to the surroundings through the 3 mm orifice (real orifice size of one of the commercially used passenger vehicle TPRD). The created underexpanded jet is free and supposed to be axisymmetric. 2D simulation was thus used to save computational cost. Other possibility to save computational time is to create the case in cylindrical coordinates, which is close to the real reservoir and orifice geometry, and use axial symmetry and thus divide the mesh size by two. However, ANSYS CFX does not have a 2D-axisymmetric solver (as introduced in ANSYS FLUENT). Concerns about the capability of the axisymmetric solver to capture the behaviour of the supersonic to hypersonic flow on the axis of symmetry were also made. 2D Cartesian coordinates case was created although the curvature of the orifice is neglected.
The domain representing the ambient environment is 4x4 meters initially filled with air having 20 °C at ambient pressure 1.01325 bar. Pressure reservoir is not considered in the domain, only a section of the outflow path close to the orifice of the TPRD where the inlet boundary condition is defined (see the detail in Fig. 1.).
An unstructured hexahedral mesh with boundary layer on the walls was created for whole domain in ANSYS ICEM CFD 17.2. The first element height was equal $1 \cdot 10^{-5}$ m which proved to be sufficient for the SST k-ω turbulent model according to the y+ values on walls.

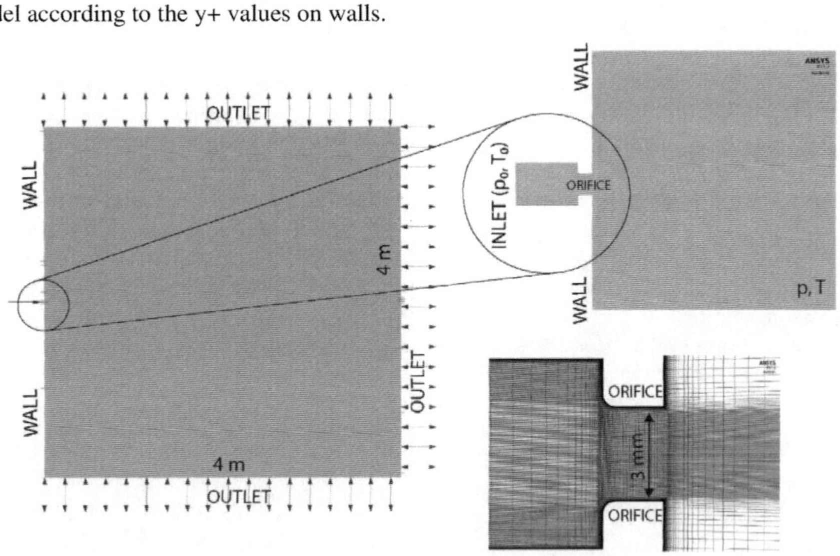

Fig. 1. Simplified computational domain representing the outflow from a pressure reservoir through 3 mm orifice to the surroundings.

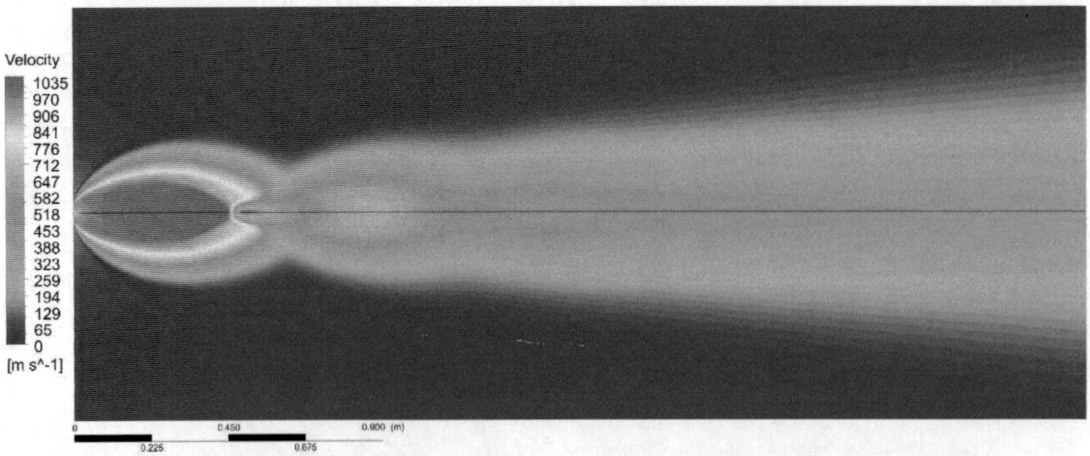

Fig. 2. Velocity field for the 200 bar inlet pressure simulation obtained from 2D CFD model.

Inlet boundary condition (which corresponds to reservoir conditions) is defined as a total pressure of pure methane in the reservoir having 20 °C (T_0). Pressure (p_0) is assumed to be constant. 10 cases were simulated: 1.5, 5, 10, 30, 50, 75, 100, 150 and 200 bar. Outlet boundary condition is of type opening with constant static pressure and ambient temperature 20 °C. Thermodynamic properties of both air and methane were governed by ideal gas law.

All performed simulations were steady state. The transient effects of mixing and turbulent structures cannot be predicted. However, for the purpose of this study, the global trend of the outflow velocity along the jet structure can be well captured by the steady state simulation and again significantly save computational cost.

The two-equation eddy-viscosity SST k-ω turbulent model with an automatic wall function was used for turbulence modelling. Second order accurate approximations for all terms in the governing equations were applied. Due to the convergence issues various expert parameters for linear solver and advanced options for solver control were tested.

RESULTS

For the comparison of the outflow velocities (Fig. 3), maximum velocity found along the jet structure predicted by CFD was used. Isentropic model predicts the characteristics of the flow, as a result of the pressure difference of reservoir and ambient conditions, with no spatial resolution. From this point of view, isentropic model is a 0D model.

Critical pressure for pure methane leaking to ambient atmosphere of 1.01325 bar with isentropic coefficient 1.304 is approximately 1.86 bar. Above this value, the flow is supersonic. The correspondence between two models is excellent up to 10 bar. The outflow velocity is rapidly increasing up to reservoir pressure 10 bar. After 50 bar, the increase in the outflow velocity is small, as predicted by both models. Isentropic model predicts about 10 % higher outflow velocities, but the overall trend is the same for both models.

Mach number is a strong function of temperature. Due to the gas expansion, the temperature of the gas drops significantly. It is thus difficult to compare the flow velocities as a function of reservoir pressure in terms of Mach number. However, for the 50 bar simulation, the maximum velocity predicted by CFD along the centreline of the jet corresponds to approximately 3.6 Mach at temperature -173 °C. Both models predicts the flow between 50 bar and maximum pressure of 200 bar between 3.5 to 6 Mach which corresponds to the transition between super and hypersonic region.

In terms of the mass flow, similar trends to velocity are observed (Fig. 4). The correspondence between models is excellent up to 50 bar. For higher reservoir pressure, there is a slight disagreement. The value of discharge coefficient 0.98 used in the isentropic model well represent the nozzle geometry used in the CFD model. As expected from Eq. 4., the mass flow above the critical conditions is a linear function of reservoir pressure.

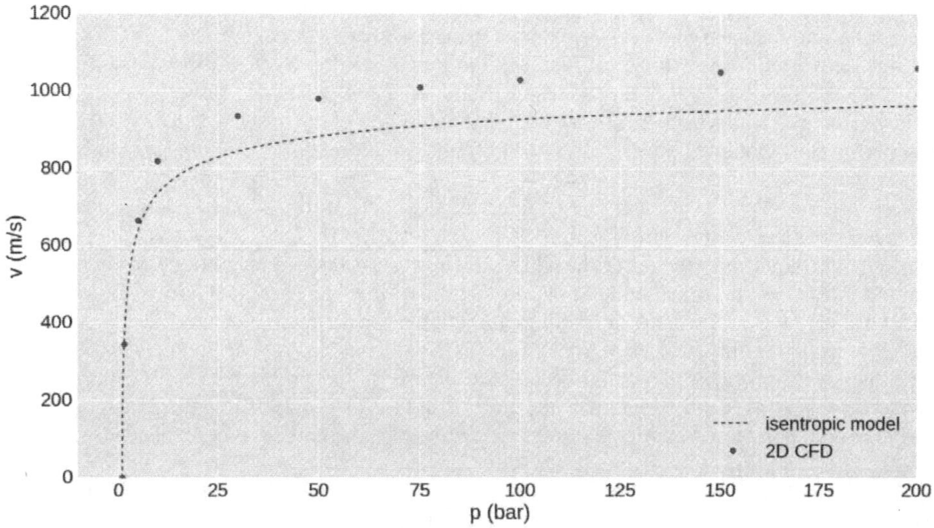

Fig. 3. Outflow velocity as a function of inlet/reservoir pressure predicted by isentropic model and 2D CFD model.

Maximum mass flow of methane at 200 bar through the orifice is approximately 10 g/s. It is necessary to point out that the outflow area is not the whole area of the orifice but area of the rectangle being 3 mm high and 1 computational cell thick as the simulation is only 2D. When scaled to the round orifice 3 mm in diameter, the mass flow is approximately 240 g/s. The same value was obtained from the isentropic model with discharge coefficient 0.98.

Experimental work measuring the mass flow through TPRD at 200 bar was performed as a part of the previous research. Experiments were transient. The pressure decrease is extremely fast (the pressure drops in five seconds about 50 bar). The estimate of the steady state mass loss rate to compare with the simulation is based on fitting the transient data and divided by 6 (number of TPRD orifices). The measured value of the mass loss rate through one 3 mm orifice at 200 bar pressure was experimentally estimated to be around 150 g/s. Considering, there will be significant pressure drop in the outflow path through the solenoid valve to TPRD and accounting for other nonideal factors that would be included in the discharge coefficient, mass flow predicted by both models corresponds in the order of magnitude with the experiments and literature values [6, 7].

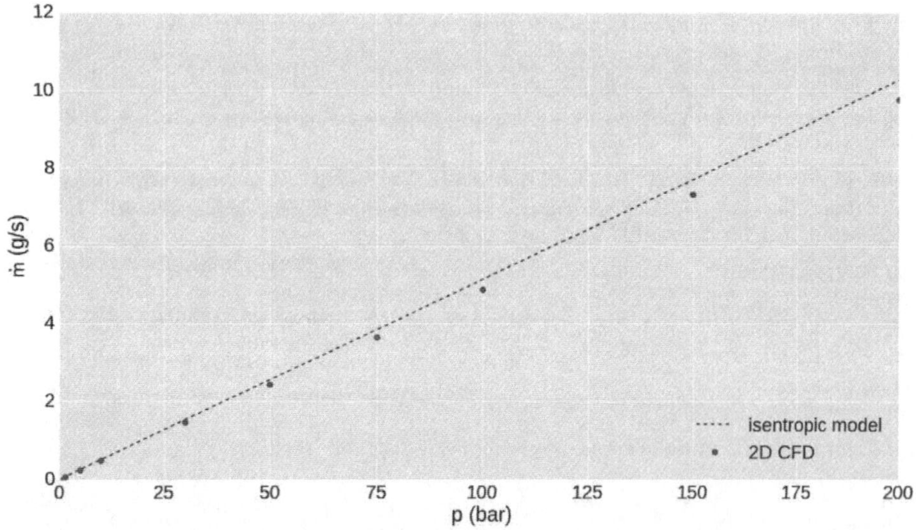

Fig. 4. Mass flow of methane as a function of inlet/reservoir pressure predicted using isentropic model and 2D CFD model.

IOP Conf. Series: Journal of Physics: Conf. Series **1107** (2018) 042014 doi:10.1088/1742-6596/1107/4/042014

CONCLUSIONS

The outflow velocity and methane mass flow predicted using isentropic model and 2D CFD model show very good correspondence over the whole range of fuel tank working pressure. Even at the hypersonic flow regime the CFD predictions seems consistent with the results obtained using isentropic flow theory of compressible fluid nozzle flow. However, it needs to be stressed out that validation of the outflow velocity over experimental data was not performed. Obtained mass flows were found to correlate with literature and our previous experiments.

Isentropic models provides easy to use and fast engineering estimates of the outflow characteristics of the gas leak from the pressurized vessel that can serve as a boundary condition for further analysis. The proper value of discharge coefficient is, however, crucial for the real mass flow estimates and may not be easily established for fuel leaks through TPRD or leak from the high pressure fuel line. However, it is generally recognized that using the idea of ideal mass flow is on the "safe side" predicting higher mass flow.

CFD modelling of compressible and supersonic flow is time consuming and requires experienced user. Convergence issues were repeatedly encountered as the nature of the problem is highly transient, but the simulation was performed as steady state. On the other hand CFD provides complex insight on the flow characteristics along the nozzle and further along the exit plane of the emerged underexpanded yet.

NOMENCLATURE LISTING

A	area (m^2)	T	temperature (K)
C_d	discharge coefficient	v	velocity (m/s)
g	gravitational constant ($m^3/kg/s^2$)	**Greek**	
h	elevation (m)	ρ	density (kg/m^3)
M	molar mass (kg/mol)	γ	isentropic exponent
p	pressure (Pa)	**subscripts**	
R	universal gas constant (J/mol/K)	0	reservoir – stagnation state

REFERENCES

[1] Franquet E., Perrier V., Gibout S., and Bruel P. (2015) Free underexpanded jets in quiescent medium: A review, *Progress in Aerospace Science*, 77:25-53, http://doi.org/10.1016/j.paerosci.2015.06.006

[2] Mack-Gardner A. and Santrock J. (2011) Analysis of Reservoir Pressure Decay, Velocity and Concentrations Fields of Natural Gas Venting from Pressurized Reservoir into the Atmosphere, *SAE International Journal of Passenger Cars - Mechanical Systems,* 4(1):216-230, https://doi.org/10.4271/2011-01-0252

[3] Perry R., *Perry´s chemical engineer handbook (7th ed)*, McGraw-Hill, 1997, Chapter 6 Fluid and particle dynamics.

[4] Sullivan D. A. (1981) Historical Review of Real-Fluid Isentropic Flow models, *Journal of Fluids Engineering* 103:258-267, http://doi.org/10.1115/1.3241728

[5] Dally J. W., Riley W. F., McConnell K. G., *Instrumentation for Engineering Measurements, (2nd ed)*, John Wiley and Sons,1993

[6] Hernandez M., Ma L., Huang Ch., Rosetto M, Martin J., Poisson D. (2013) *Safety investigation of CNG leaks in enclosed parking structures: computational fluid dynamics modelling and analysis*, National Research Council Canada, Technical report IFCI-OTHER-CTR-003

[7] Choi J., Hur N., Kang S., Lee E. D., Lee K. (2013) *A CFD simulation of hydrogen dispersion for the hydrogen leakage from a fuel cell vehicle in an underground parking garage*, International Journal of Hydrogen Energy, 38:8084-8091, http://dx.doi.org/10.1016/j.ijhydene.2013.02.018

ACKNOWLEDGEMENT

This work was supported by the Ministry of the Interior, Safety research programme of the Czech Republic 2015 – 2020 (BV lll/1 - VS) under number VI20172019077 - Accidental release of CNG from passenger vehicles.

Experimental and Numerical investigations of the flow characteristics in confined fires

Benjamin BETTING[1], Émilien VAREA[1], Carole GOBIN[1], Gilles GODARD[1], Bertrand LECORDIER[1] and Béatrice PATTE-ROULAND[1]

[1] Normandie Univ, INSA Rouen, UNIROUEN, CNRS, CORIA, France

benjamin.betting@coria.fr

ABSTRACT

Fire compartments are of major concern since they are responsible for fatal accidents. The decision and response times of rescue teams are mainly based on human decisions which result from empirical approaches. Since compartment fires are multi-physical and multi-scale problems, a perfect knowledge of the situation is impossible. A fundamental approach of fire compartments which involves fluid mechanics is of great interest. One of the main issues concerns the transition from located to generalized fire. The most important vehicle leading to generalized fires results in smoke transport. Indeed, temperatures are very high and many species are mixed. In this study, an experimental facility composed of a maritime container is set-up. In parallel, numerical simulations -Fire Dynamics Simulator (FDS) software- based on a LES approach, are used. This dual approach makes it possible to validate the FDS code in ventilated and under-ventilated conditions, considering both the temperature fields and the flow dynamics. The compartment fire allows power up to 1MW. Temperature fields are measured with thermocouples racks, and velocity fields are obtained by large scale Particle Image Velocimetry (PIV). The objectives of the paper are: I) compare and analyze temperature fields from both experimental and numerical data, II) introduce preliminary results on large scale PIV measurements located at the smoke exit. Data are also compared with simulations and III) check the numerical assumptions.

KEYWORDS:

Large scale PIV, enclosure fire, numerical simulations, FDS, smoke dynamics, under-ventilated conditions.

IOP Conf. Series: Journal of Physics: Conf. Series **1107** (2018) 042015 doi:10.1088/1742-6596/1107/4/042015

INTRODUCTION

Nowadays, the decision-making of the emergency services is done during the fire intervention and it is mainly based on a rapid analysis. This analysis comes from the experience gained over the years but also during the training given by the Firefighter fire simulators. However, this type of analysis does not allow for a complete knowledge of the disaster since it is mainly based on an empirical knowledge. The evolution of fire, the appearance of a secondary focus or some thermal phenomena, i.e. back-draft [1, 2] or flash-over, [3, 4, 5], are especially feared by the rescue services. Besides, they are difficult to predict with accuracy. All these factors may lead to fatal accidents.

In order to optimize and secure the intervention of rescue services, the knowledge of fires, dynamics and risks represent a major interest. One of the main aspects remains in determining the smoke dynamics at the openings where fresh air and hot fumes mix. Theoretical knowledge on this phenomenon will improve the understanding of smoke flows, particularly at the outlets, [6]. This particular phenomenon is the so called pulsating fire and has been extensively studied, [7-9]. The direct link between the mass of the flow and the air entrained by the plume justifies the interest in this work [10].

Fundamentally, analyzing temperature, velocity and pressure and their sensitivity in the development of a fire is of first importance, [11]. To study smoke dynamics with non-intrusive diagnostic, PIV in large scale can be used [12, 14]. However, few studies exist on the use of this technique for under-ventilated fires analysis. Indeed, the National Institute of Standards and Technology studied fire dynamics in enclosure at the doorway and in the axis of the fireplace, from the outside [15]. In this study, velocities were measured from the inside, for HRR until 1MW.

Besides, to complement this study, numerical simulations, using FDS code are done. The objectives of this work are to study smoke velocity and smoke dynamics and compare the evolution of the experimental and numerical temperature profiles.

EXPERIMENTAL AND NUMERICAL SET-UP

Experimental set-up

To investigate the smoke dynamics characteristics in real fires, an experimental test bench was set-up. The thermal phenomena studied in this experimental test bench are likely to be encountered by firefighters during their missions. This cell is composed of two maritime containers mounted perpendicularly, one is the measure cell and the other one is the test cell, see Fig 1.

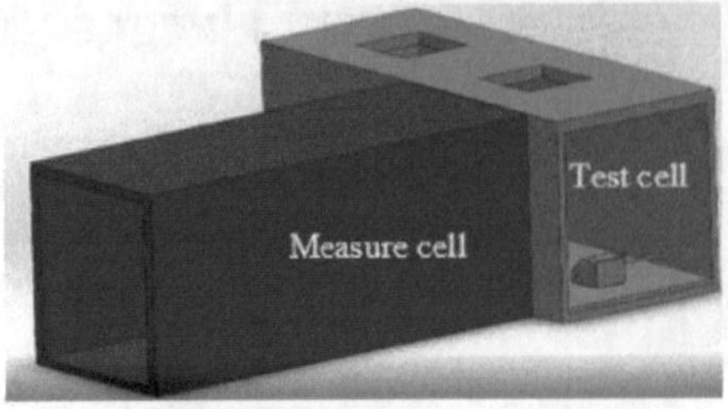

Fig. 1. Experimental set-up configuration.

The standard size of a maritime container is 6 x 2.59 x 2.45 m. These structures have emerged as training tools in rescue services. Moreover, the volume of these containers is close to that of a room of a standard flat, which is of 40 m^3, approximately.

The fireplace consists in 36 burners fueled with propane covering a 1 square meter surface. This configuration enables to reach a power of 1 MW and temperatures around 800 °C.

Propane was chosen because it produces little soot compared to a real compartment fire. Indeed, it has the advantage of being perfectly controlled and it guarantees the reproducibility of the experiments.

A large optical access of 1 m^2 enables both, the visualization and the possibility to implement advanced laser diagnostics such as Particle Image Velocity (PIV), [16-17]. This configuration is represented on Fig. 2.

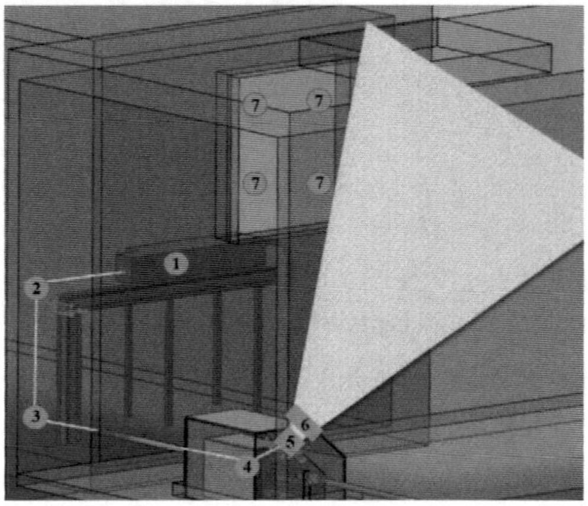

Fig. 2. View of PIV setup.

The laser is represented by number 1, the mirrors by 2, 3 and 4, the optical lenses in 5 and 6 and number 7 shows the position of cameras.

Large scale PIV is a complex task since a compromise between spatial resolution, and field of view exists. To overcome the issue, a special configuration is set-up. Instantaneous, two-dimensional velocity measurements relied on particle image velocimetry (PIV) in which alumina particles are used (~50μm diameter). These particles are injected trough an in-house air supply corona which is positioned below the burner. Special care is given to inject particles without modifying the fresh air injection dynamics. As low as 10% of air to reach stoichiometric condition is used to feed to burner with particles seeded air. The light source is a Nd:Yag laser (Quantel QSmart, 2*400 mJ/pulse) combined with a -20mm cylindrical lens to produce a 0.5mm thick laser sheet. Mie scattering from the particles is collected on 4 CCD (Charge-Coupled Device) cameras (JAIRM4200, 12 bits, 2048*2048 pix²) mounted with 50mm f/1.2 Zeiss lens. A 532 interferential filter reduces noise from ambient light sources. The optical arrangement yields a magnification of 5 pix/mm, calibrated using images of a precision reference grid. Data are recorded using Dantec Dynamics software. Raw images are post-processed with an in-house PIV algorithm developed by Dr. B. Lecordier. A short description is given below. The 4 images are combined on a same image using the reference grid. Distortion and image tilt are compensated. Then a preprocessing is achieved to reduce correlation on identified structures resulting from local soot generation or interfaces due to non-seeded fresh air coming from the opening. Then PIV processing was performed with a cross-correlation technique between pairs of successive images. The size of the PIV interrogation window is 64 pix² and a 50% overlap is processed. The time interval between two consecutive images is in the range of 1–2.5 ms.

For temperature measurements, 20 type-K thermocouples of a 1.5 mm diameter, mounted on two shafts of 10 thermocouples placed vertically and horizontally, are installed at the outlet.

Numerical set-up

Simulations are performed with the FDS code, version 5.5. The numerical domain is represented in Fig 3.

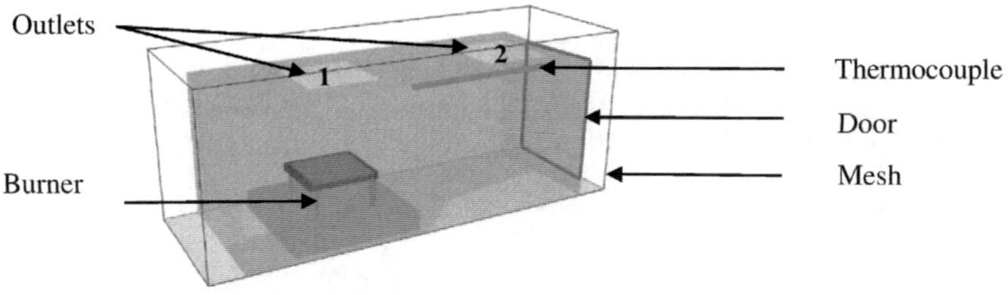

Fig. 3. FDS view of the setup.

In order to optimize the simulation, a convergence study has identified the compromise between resolution and calculation time for different mesh sizes. Simulations have been performed for a physical time of 600 s for four different cell sizes: 2.5 cm, 5 cm, 10 cm and 20 cm. In order to choose the ideal mesh, it was necessary to be interested in the computation times to obtain the best ratio between reliability and simulation time (not shown here). For these different reasons, the cell size of 5 cm was chosen.

To summarize, the initial conditions used in this work are:

- The enclosure is plunged into a bigger computational domain totally open to take the smoke exhaust and the intake of fresh air into account;
- Ramp fire very fast because the fuel is a propane;
- Consideration of insulation for heat losses (rockwool and Aerated concrete);
- Addition of leaks in order to feed the burner with air which is seeded by particles (20cm²).

RESULTS

Temperature fields

Simulations are performed for three different powers: 200 kW, 500 kW and 800 kW. The results of numerical and experimental temperature profiles are compared see Fig. 4.

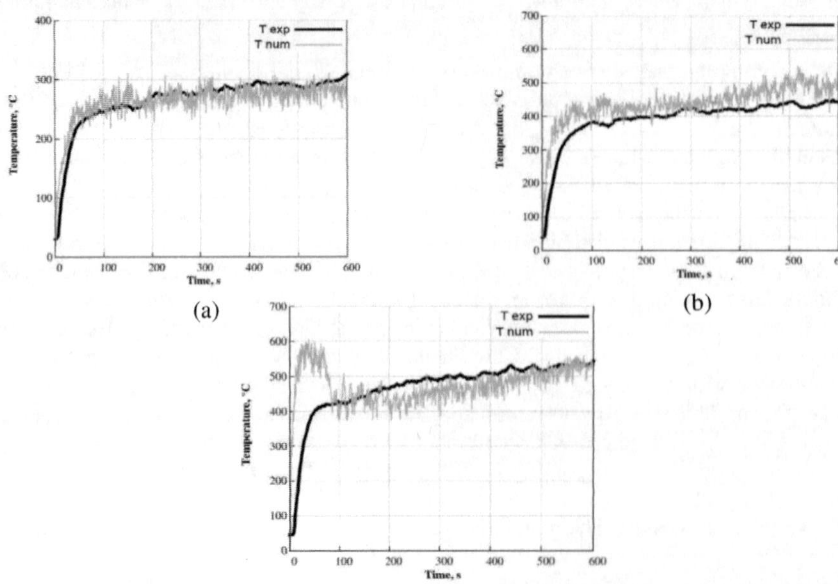

(a) (b) (c)

Fig. 4. Temperature profile comparison at 200 kW (a), 500 kW (b) and 800 kW (c).

In Fig. 4 (a), comparison is done for 200 kW. It is observed that the temperature evolutions with time for experimental and numerical results are very similar. The power rises are almost identical, as well as, the temperature values. In addition, the steady state is not reached.

In Fig. 4 (b), the temperature evolutions against time are plotted for a power of 500 kW. Numerical and experimental results have similar profiles. However, it is observed that a slight gap begins to appear between the two temperature evolutions. At 800 kW, see Fig. 4 (c), an important discrepancy between the experimental and numerical results appears at the beginning. Nevertheless, this difference is recovered after a few seconds. This deviation of the numerical data from the experimental results might be due to the range of validity of the CFD code. For all three cases, it can be seen that the numerical temperature profiles are much noisier than the experimental ones. This is due to the fact that, in the real case, the response time the thermocouples do not capture the high frequency fluctuations which results in a low pass filter. This comparison is done to validate numerical assumptions in simple cases. This made it possible to adjust the numerical parameters and thus improve the reliability of the simulations before the velocity measurements.

Velocity fields

In this study, two configurations are performed, see Fig. 3.

- In the first case, outlet N°2 is open. It's the only opening. Hot smokes have to go out through this opening. Fresh air necessary to the fire development mainly enters the volume by the same way. Indeed, a small amount of particle seeded air is introduced at the bottom of the burner, as described in the experimental part. This case allows studying confined fires. It's called configuration 1.
- In the second case, outlets are closed. The door is open. Hot gases go outside through the upper part of the volume. Fresh air enters the compartment by the lower part. This case is typical of well-ventilated fires. This case is called configuration 2.

In this paper, only instantaneous fields are performed for a power of 200kW. Comparison between numerical and experimental results is done. Results are plotted for configuration 1 and they are shown on Fig. 5.

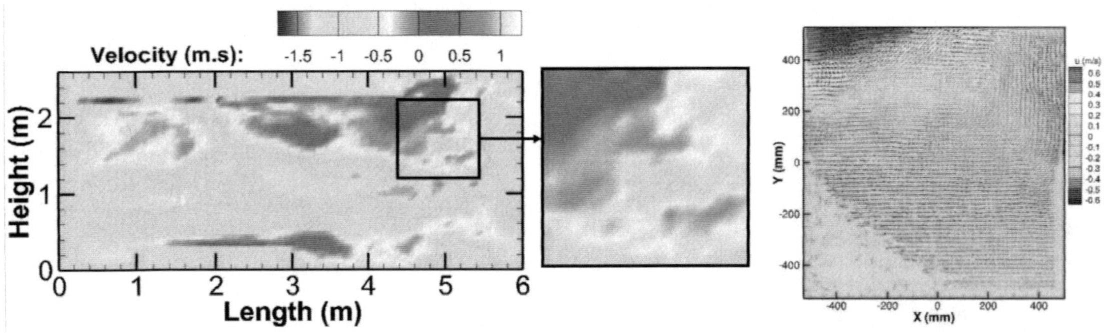

(a) (b)

Fig. 5 Velocity fields at 200 kW, numerical (a), experimental (b), configuration 1.

In Fig. 5 (a), we observe global velocity fields in the cell. A zoom is given for comparison with experimental results, visible on Fig. 5 (b). Although slightly overestimated the numerical case, velocity fields are similar. In fact, results of smoke dynamics and hot gas velocity obtained with PIV are correctly simulated with the FDS software. Considering the second configuration, Fig 6a, one can see that the velocity field is divided into two zones horizontally distributed. The first half part sees the smoke with positive velocity corresponding to smoke exit. The second part sees fresh air income. The same behavior is observed in the experimental field. Strong exit velocities are observed at the top left corner, while air inlet with negative velocities appears on the right side. In Fig. 5 (b); we can note the absence of vectors at the bottom left of the image which results from a non-total laser sheet. This point is being improved.

Fig. 6 (a) shows numerical results in configuration 2. As in the previous figure, a zoom is made to observe the same area as for the experimental tests, available on Fig. 6 (b).

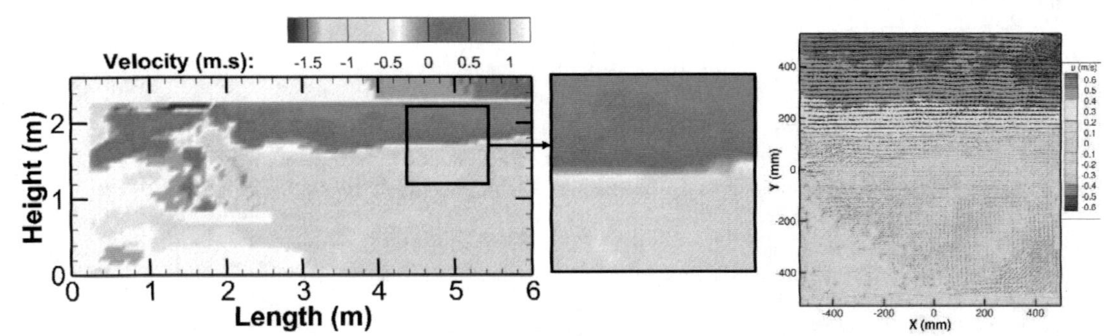

(a) (b)

Fig. 6. Velocity fields at 200 kW, numerical (a), experimental (b), configuration 2.

This case is ideal to make a comparison between numerical and experimental results. Indeed, the fireplace is well-ventilated. This allows studying fire with less soot (simpler for PIV measurements) and studying smoke flow with dynamics less complex (simpler for numerical simulations). In this configuration, the results are very close. In both cases, three different velocity fields are observables. They correspond to the hot gases are under ceiling, neutral plan and fresh air intake the enclosure. Velocity fields are still slightly overestimated in numerical case but faithfully represent the experimental flows.

CONCLUSION

This work introduces large scale PIV measurement as well as mean temperature profiles and comparison is made with numerical simulations.
Considering temperature evolutions, good agreement is observed for well ventilated conditions up to 500kW. Discrepancies exist for higher configurations where the fireplace is not well ventilated. The main issue may results from FDS code which is not yet fully validated for under ventilated conditions.
In this study, numerical and experimental velocity profiles were compared for powers up to 200kW. These tests are performed in representative large scale fire, in ventilated and confined configurations. Good agreement was observed for both cases. More data are being processed for analysis and comparison of mean and fluctuating velocity fields
This research makes it possible to observe the influence of the compartment on the development of fire. It also allows an experimental and numerical comparison and therefore, to observe the reliability of the FDS code in this configuration. Adjustments are needed, especially for laser lighting to obtain a fields of 1m². Otherwise, average fields of velocity are in progress.

ACKNOWLEDGEMENTS

Financial support from the French National Research Agency, under the project 'Firediag' is gratefully acknowledged. This project has been funded with the support from the European Union with the European Regional Development Fund (ERDF), FEDER MADNESS and from the Regional Council of Normandie. We thank the computing center CRIANN for the computation time.

REFERENCES

[1] Gottuk, D. T., Peatross, M. J., Farley, J. P., and Williams, F.W., "The development and mitigation of backdraft : a real-scale shipboard study", *Fire Safety Journal 33 (4) (1999) 261–282.*

[2] Bepoix, P., "Les phénomènes thermiques", *Editions Carlos Zaglia (2015)*

[3] J. Francis, J. and Chen, A., "Observable characteristics of flashover", *Fire Safety Journal*

[4] Thomas, P., Bullen, P., Quintiere, J. and McCaffrey, B., "Flashover and instabilities in fire behavior", *Combustion and Flame 38 (1980) 159–171*

[5] Bishop, S., Holborn, P., Beard, A.N. and Drysdale, D., "Non linear dynamics of flashover in compartment fires", *Fire Safety Journal 21 (1) (1993) 11–45*

[6] Morton, B.R., Taylor, G.I. and Turner, J.S., "Turbulent gravitational convection from maintained and instantaneous sources", *Proc. R. Soc. Lond. (1956) A234, 1 - 23.*

[7] Merci, B. and BEJI, T., "Fluid mechanics aspects of fire and smoke dynamics in enclosures", *CRC press (2016)*

[8] Bengston, L.G., "Enclosure fires, Swedish Rescue Service", (2001)

[9] Pretrel, H., Le Saux, W. and Audouin, L., "Pressure variations induced by a pool fire in a well-confined and force-ventilated compartment", *Fire Safety Journal (2012)*

[10] Heskestad, G., "Dynamics of the fire plume", The Royal Society (1998)

[11] You, D., Jourda, P., Hattenberger, J., Vantelon, J.P., Gaviot-Blanc, F., Joachim, A. and Mitanchez, G., "Study of the efficiency of the fire extinguishing systems used by fire fighters on the underventilated fires with 3D Firefighting and other application techniques", *Interschultz (2010)*

[12] Tieszen, S.R., O'hern, T.J., Schefer, R.W., Weckman, E.J. and Blanchat, T.K., "Experimental study of the flow field in and around a meter diameter methane fire", *Combustion and Flame, Vol.129, Issue 4, pp. 378-391 (2002)*

[13] Hou, X.C., Gore, J.P. and Baum, H.R., "Measurements and prediction of air entrainment rates of pool fires", *Symposium (International) on Combustion, Vol.26, Issue 1, 1453-1459 (1996)*

[14] Koched, A., Pretrel, H., Audouin, L., Vauquelin, O. and Candelier, F., „Application de la PIV sur un écoulement de fumée à un passage de porte induit par une source incendie", *CFTL (2012)*

[15] Bryant, R., "The application of stereoscopic PIV to measure the flow of air into an enclosure containing a fire", *Experiments in fluids (2009)*

[16] Koched, A. and al, "Utilisation de la stereo piv pour l'optimisation des mesures de vitesse de fumee d'incendie à un passage de porte par des sondes bidirectionnelles", *CFTL (2014)*

[17] Bryant, R.A., "The application of stereoscopic piv to measure the flow of air into an enclosure containing a fire". *Experiments in fluids, 47(2) :295–308, 2009.*

Capabilities and Limitations of the Fire Dynamics Simulator in the Simulation of Tunnel Fires with a Multiscale Approach

Matteo PACHERA[1], Xavier DECKERS [1,2] and Tarek BEJI[2]

[1] *Fire Engineered Solution Ghent, Ghent, Belgium*
[2] *Ghent University-UGent, Dept. Flow, Heat and Combustion Mechanics, Belgium*

mp@fesg.be

ABSTRACT

In tunnel fire safety engineering, long tunnels are generally simulated with one-dimensional fluid dynamics models (1D). Short tunnels can be simulated with a three-dimensional fluid dynamics model (3D). The 1D simulations are faster than 3D simulations, but they lack accuracy in the region where the flow is three dimensional, which is typically around the fire source. In this paper, we present a multiscale approach where large parts of the tunnel are simulated with the 1D approach, while a limited portion, close to the fire, is simulated with a 3D model. This allows to simulate with a good accuracy smoke dynamics and the induced flow field near the fire whilst large parts of the ventilation system and the tunnel's structure are simulated with 1D, reducing thus the computational costs.

The present paper explores the capabilities and limitations of the Fire Dynamics Simulator, FDS 6.6.0, in the simulation of tunnels with a multiscale approach. FDS has an embedded 1D model connected with the hydrodynamics solver (3D) which allows to implement the multiscale approach without making a link to external models.

In the first part of the paper, the fan-induced flow field is simulated with the 3D model in order to generate an operative map where the pressure rise induced by the fan is a function of its volume flow rate. The operative map is later implemented in the 1D model without the need to model explicitly jet fans in FDS.

The second part of the paper addresses the modelling of a fire in a tunnel. The fire has two effects on the response of the ventilation system: (1) local pressure losses near the fire section and, (2) higher pressure losses along the tunnel due to the high temperature of the combustion products. Very promising results were obtained with the multiscale approach. Nevertheless, the results show also that, in the 3D portion of the tunnel (where the fire is located), large velocity fluctuations may lead to unphysical pressure distributions, especially for large fires (above 3 MW) and when a fine mesh is used. Therefore, further analysis and investigation is required in order to be able to deal with higher fire loads such as a 100 MW HGV fire or a 200 MW tanker fire.

INTRODUCTION

A multiscale approach allows to couple models, with different accuracies and complexities, in order to reduce the computational cost and maintain a high quality of the calculation. In the field of fluid dynamics, the multiscale approach has been applied to complex networks where one-dimensional (1D) and three-dimensional (3D) models are linked together. For tunnel problems, different authors implemented the multiscale approach to study the flow field in cold and warm conditions [1-4].
Large parts of the tunnel are simulated with a 1D model due to its lower computational cost compared to a 3D model. However, 1D models may not be suitable to describe complex flows that are actually 2D or 3D. Therefore, the region of interest should be simulated with a different model. In case of a tunnel fire, the flow field near a fire is fully 3D as well as the flow near a ventilation device (e.g., Saccardo nozzle, Jet fan and supply or exhaust vents). When smoke movement in the tunnel is the main focus of the study, ventilation devices can be included in the 1D network, by including their operative map. This can be calculated by resorting to Computational Fluid Dynamics (CFD) simulations or by estimating an installation efficiency [5,6].

MULTISCALE MODELLING IN FDS

For the simulation of tunnels with a multiscale approach the Fire Dynamics Simulator (FDS 6.6) is chosen because it already contains a Heat Ventilation and Air Conditioning(HVAC) model that can be used to simulate the 1D portions of the tunnel. The HVAC model in FDS has some limitations: (1) there is no thermal model that takes into account the temperature losses along the tunnel, and (2) the pipes have a constant section. However, the direct coupling of the hydrodynamic solver with the HVAC solver allows to update the solution at every time step. The boundary conditions are calculated at the interface between the 1D and the 3D domains. The quantities calculated by the 1D model are imposed uniformly at the interface of the 3D domain. The quantities calculated in the 3D domain are averaged over the surface before being transferred to the HVAC model [7]. At the boundaries of the 3D domain, the mass flow and the velocity are imposed as boundary conditions, while at the boundaries of the 1D domain the pressure is imposed.

Figure 1: Scheme of a tunnel simulated with a multiscale approach. The red triangle in a circle refers to a jet fan.

Since the boundary conditions calculated by the 1D model are set uniformly at the interface of the 3D domain, the location of such interface is a key factor for the correct implementation of the multiscale approach. If the flow field in the 3D domain is not uniform over the section of the tunnel, the boundary conditions might affect the results of the simulation.

The tunnel under investigation has a square section of 25 m^2 and an overall length of 600 m. The tunnel is equipped with two jet fans with an area of 1 m^2 and a discharge velocity of 30 m/s, located 100 m from the inlet portal. The walls have a roughness of 0.003 m uniformly along the tunnel, the roughness is calculated considering a friction factor of 0.02, typical for road tunnels. The fire is located at 300 m from the inlet portal, in the middle of the tunnel. This location allows to freely move the interface between 1D and 3D domains without crossing the portals.

SIMULATIONS IN COLD FLOW

For the ventilation of the tunnel, two jet fans are installed in order to induce the longitudinal flow. The jet fans are first simulated with a 3D simulation and later, their operative map (i.e., pressure rise as function of volume flow rate) is implemented in the 1D model. The pressure rise induced by a jet fan in a tunnel is usually calculated according to Eq. 1 [5,6].

$$\Delta P = \rho \, n_F \, k \, \frac{A_F}{A_T} (u_F - u_T) u_F \qquad (1)$$

Where n_F is the number of jet fans, k is the installation efficiency, A_F is the area of the fan, A_T is the area of the tunnel and u_F and u_T are the discharge velocities of the fan and the mean velocity in the tunnel. It is important to note that Eq. 1 represents the performance map of the jet fans in the tunnel, which is different from the performance map of an 'isolated' jet fan (as shown in Figure 4 for example). In this study, the jet fan is assumed to operate with a constant discharge velocity.

In the numerical model, the jet fans have a square section and are directly attached to the ceiling. This simplification is required in order to avoid the difference between models when using different mesh resolutions.

Effect of the grid resolution

Before calculating the operative map of the jet fan in the tunnel section, a preliminary mesh sensitivity analysis is performed. The results from three different mesh resolutions (see Table 1) are compared in order to find a suitable mesh size. The jet fans are simulated in a 80 m-long portion of the tunnel, where the total pressure is imposed at the inlet equal 0 Pa and a velocity of 3 m/s is imposed at the outlet. The latter value is chosen based on an estimation of the critical velocity obtained with the equation of Li [8].

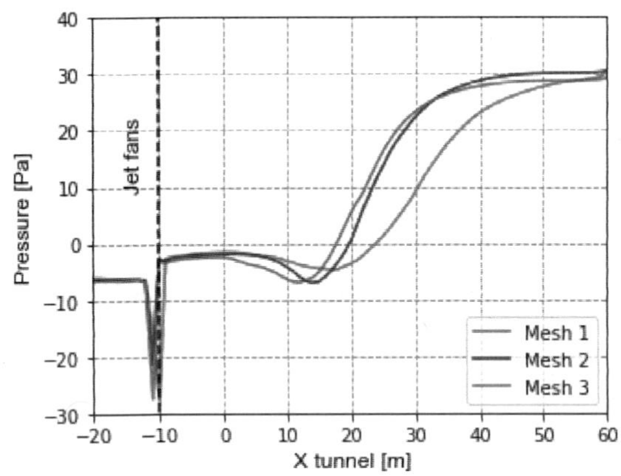

Figure 2: Pressure along the tunnel for different grid resolutions.

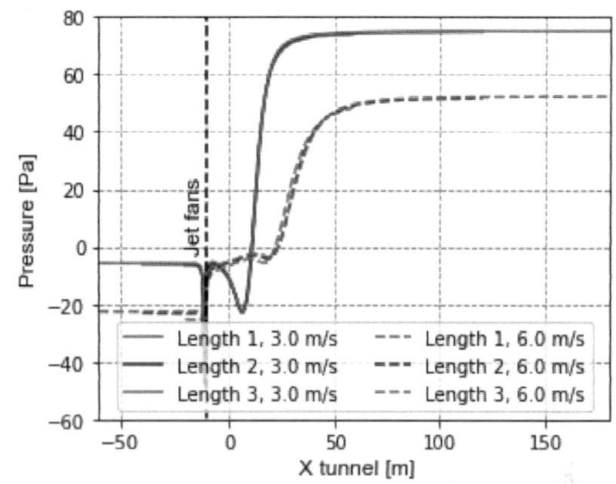

Figure 3: Pressure along the tunnel for different domain's lengths.

The pressure rise between the inlet and the outlet of the tunnel, obtained with different grid's resolutions, is similar, with a difference lower than 2 Pa (see Table 1). However, the pressure distribution predicted by Mesh 1 is different than the one obtained with Mesh 2 and 3 (see Figure 2). Therefore, Mesh 2 is chosen for the next simulations. The local pressure drop occurring at the fan's section is consequence of the flow acceleration at the inlet of the jet fans.

Effect of the length of the 3D domain for the simulation of jet fans

The location of the interface between the 1D and the 3D domains is fundamental for the correct calculation of the operative map of the jet fans. The flow field should be fully developed in the 3D domain (with a uniform flow over the cross section) before making the connection (i.e., interface) with the 1D domain. But at the same time, the 3D domain should be limited in length in order to reduce the computational cost of the simulation. A portion of the tunnel is simulated by imposing the pressure at the inlet and the velocity at the outlet. Two different velocities, i.e., 3 m/s and 6 m/s, are chosen because the decay length of the jet is affected by the velocity in the tunnel [7]. Higher values are not used because they are not common in 'real-life' applications. In this case, the tunnel is simulated without considering the friction losses at the walls, so the pressure distributions of the domains with different lengths can be compared together.

Table 1: Mesh sensitivity analysis in cold flow.

Case	El. Size	ΔP
Mesh 1	0.500 m	36.9 Pa
Mesh 2	0.250 m	36.7 Pa
Mesh 3	0.125 m	35.3 Pa

Table 2: Length independence study in cold flow.

Case	Length	$\Delta P_{3\ m/s}$	$\Delta P_{6\ m/s}$
Length 1	80 m	80.9 Pa	76.0 Pa
Length 2	160 m	80.6 Pa	75.0 Pa
Length 3	240 m	80.6 Pa	75.0 Pa

The pressure distribution for the different simulations is presented in Figure 3, while the pressure difference at the portals is summarized in Table 2. From the comparison, the simulation with Length 2 is chosen because it ensures that the high speed jet induced by the jet fans is fully decayed. The operative map of the jet fans in the tunnel is evaluated using the set-up of Mesh 2 and Length 2. The pressure at the outlet of the tunnel is changed in order to draw the operative map of the jet fans in the tunnel. In order to avoid possible uncertainties due to the friction losses the walls are modelled as free slip surfaces. The operative map of the jet fans is presented in Figure 4, together with the results of the Eq. 1 assuming an installation efficiency of 1.0.

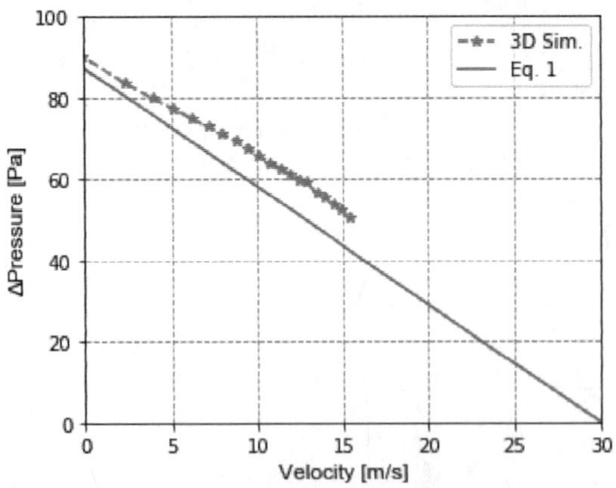

Figure 4: Operative map of the jet fans.

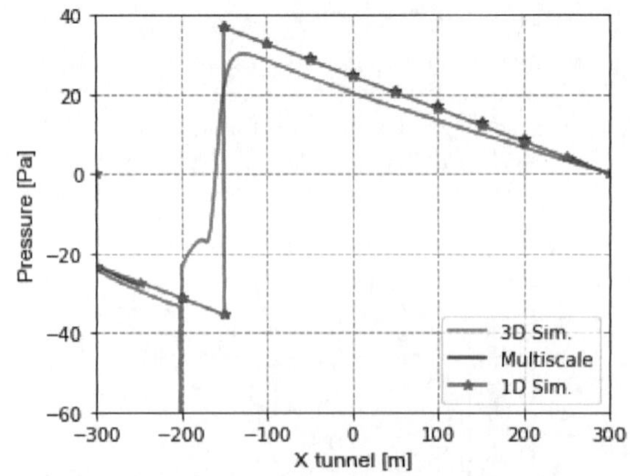

Figure 5: Pressure comparison in cold flow.

Comparison between multiscale, 1D and 3D simulations

The tunnel is now simulated embedding the operative map of the jet fans in the 1D model. The results obtained with 3D, 1D and multiscale simulations are compared together, Figure 5. For sake of simplicity the portals are not modelled and the pressure is imposed at the ends of the tunnel. This simplification is required to avoid uncertainties over the local pressure losses occurring due to the section's change. In the multiscale model only the first and the last 50 m of the tunnel are included. The results are presented in terms of pressure distribution along the tunnel and velocity (see Figure 5 and Table 3).

The results have a small error in the velocity, within 2%, so the model can be applied for the simulations of tunnels in cold conditions. The pressure rise induced by the fan is compared for the three models. The agreement between the 1D and the multiscale is better (with an absolute deviation of 0.2 Pa) than the one with the 3D simulation (with an absolute deviation of 2.5 Pa). However, the pressure rise induced by the fan is estimated by calculating the pressure losses along the tunnel, which might not be very accurate.

Table 3: Models' comparison in cold conditions.

Case	Velocity	ΔP_{Fan}
3D Sim.	6.30 m/s	74.8 Pa
Multiscale	6.17 m/s	72.3 Pa
1D Sim.	6.22 m/s	72.1 Pa

Table 4: Mesh sensitivity analysis in case of fire.

Case	El. Size	Velocity
Mesh 1	0.500 m	3.77 m/s
Mesh 2	0.250 m	3.71 m/s
Mesh 3	0.125 m	3.70 m/s

SIMULATIONS IN CASE OF FIRE

After the first simulations in cold conditions, the multiscale approach is tested with different fire loads and different lengths for the 3D domain. Also, in this case the tunnel is modelled without including the portals for sake of simplicity, but including a small 3D portion, 50 m long, near the ends of the tunnel. The fire is also simulated in a 3D domain, whose length varies as discussed later. The jet fans are not simulated in the 3D model but their operative map is embedded in the 1D network. Different fire loads are examined in order to show the applicability of the current model for the simulations of fire in tunnels.

Effect of the grid resolution.

The mesh sensitivity analysis is carried out with the same grid resolution tested previously, Table 4. The HRR in this case is 2.5 MW, which represents an intermediate value of the range of HRRs that will be simulated later, from 0.0 MW to 5.0 MW. A portion of tunnel, 100 m long, is simulated with a fire located at 20 m from the inlet portal. The total pressure, 20 Pa, is imposed at the inlet of the tunnel, while the static pressure, 0 Pa, is imposed at the outlet. In this case the pressure is imposed instead of the velocity because the latter numerical set-up yields some numerical issues occurring when a fire in a tunnel is simulated with FDS [9]. The temperature and pressure profiles are compared as a function of the grid resolution, Figure 6 and Figure 7.

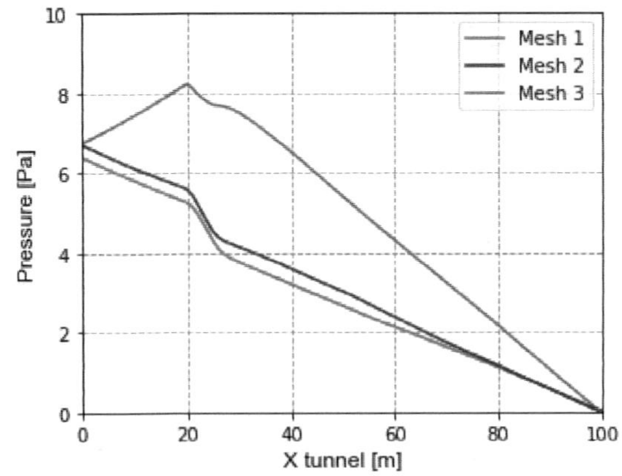

Figure 6: Pressure distribution for different mesh sizes

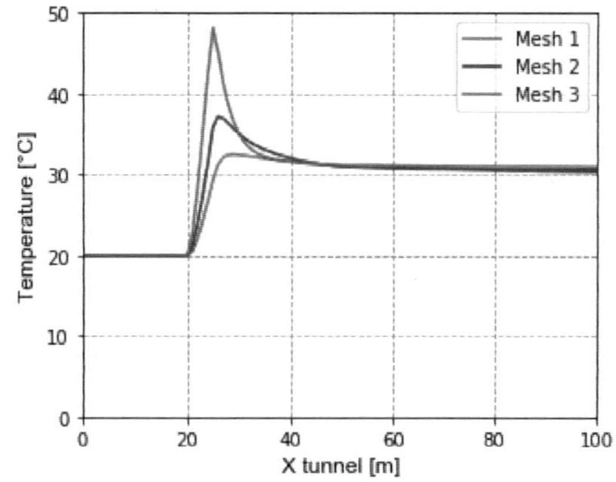

Figure 7: Temperature distribution for different mesh sizes

The longitudinal velocity upstream the fire for the given pressure difference at the portals is given in Table 4, showing that Mesh 2 and 3 predict the same longitudinal velocity. The temperature profiles along the tunnel are similar with the 3 grids. There are though some discrepancies in the fire region, which is out of the scope of the current study. The pressure shows an unexpected behavior for Mesh 3: the pressure gradient between the inlet portal and the fire is positive, while it is negative for Mesh 1 and 2. With a positive pressure gradient a negative velocity is expected, but the pressure distribution obtained with Mesh 3 is a consequence of the large velocity fluctuations in the fire region The pressure profile obtained with the finest mesh is not considered to be realistic,[9]. Therefore, the Mesh 2 is chosen for the next simulations where the length of the domain is investigated and the results are compared with a full 3D calculation.

Effect of the interface location between 1D and 3D domains

After comparing the results obtained with different grid resolutions, the size of the 3D domain is changed and the results are compared together for different values of the HRR, Table 5.

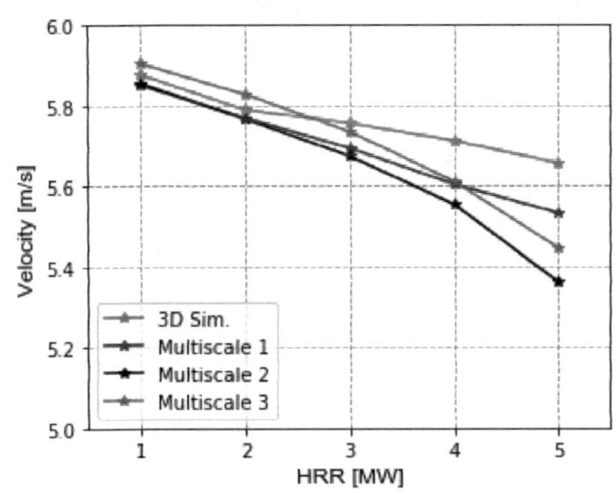

Figure 8: Velocity upstream the fire as function of the HRR.

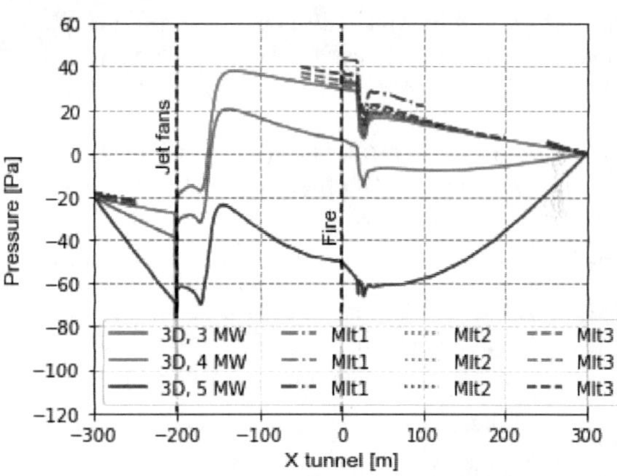

Figure 9: Pressure in the tunnel for different HRR.

The pressure distribution along the tunnel changes with the HRR, for low values of the HRR, 3.0 MW and below, the pressure decreases along the tunnel except for the region downstream the jet fans and near the fire section. When the HRR increases, 4.0 MW or higher, the pressure in the 3D simulation has a positive gradient in the region downstream the fire, where a negative pressure gradient is expected, Figure 9. The same effect is present, but less visible, also in the multiscale simulations, where the pressure rises in the fire section. This effect becomes more evident when the HRR increases, because the flow accelerations in the fire region have larger intensity. In a fully 3D simulation, even if the pressure distribution changes with the mesh resolution, the velocity upstream the fire is weakly affected, the difference between Mesh 2 and 3 was negligible. In the multiscale, due to the interfacing of different models, this is generally not true because the central domain of the tunnel is connected to the portals via 1D branches. The velocity along the tunnel have different trends increasing the HRR, showing that the multiscale is applicable only for relatively small fires, below 3.0 MW.

Table 5: Comparison of the longitudinal velocity for different domains' size and HRR.

Case	Length 3D	Vel. 1.0 MW	Vel. 2.0 MW	Vel. 3.0 MW	Vel. 4.0 MW	Vel. 5.0 MW
3D Sim.	500 m	5.88 m/s	5.79 m/s	5.76 m/s	5.71 m/s	5.66 m/s
Multiscale 1	100 m	5.85 m/s	5.77 m/s	5.69 m/s	5.60 m/s	5.53 m/s
Multiscale 2	150 m	5.85 m/s	5.77 m/s	5.67 m/s	5.55 m/s	5.36 m/s
Multiscale 3	250 m	5.90 m/s	5.83 m/s	5.73 m/s	5.61 m/s	5.45 m/s

CONCLUSIONS

The current paper describes the implementation of the multiscale simulation approach for tunnels in FDS. The tunnel was first simulated in cold conditions, without fire. A preliminary grid and length sensitivity study defined the appropriate numerical setup for the simulation of the jet fans in the tunnel. Based on this set up, the operative map of the jet fan was calculated with FDS and it was later embedded in the 1D model. When the tunnel is simulated in cold flow conditions, the velocity and the pressure distribution obtained with the different models are in good agreement. The pressure along the tunnel has a linear distribution as expected due to the friction losses along the walls. The comparison of 3D, 1D and multiscale simulations shows a satisfactory agreement between the numerical approaches. This result confirms the possibility of using FDS in a multiscale configuration to simulate the ventilation in cold conditions in tunnels.

The simulation of a fire in the tunnel induces an unexpected pressure distribution which is caused by the large flow accelerations in the fire region. The mesh sensitivity analysis shows that with a fine mesh, the pressure gradient is positive also in zones of the tunnel where the pressure is expected to decrease because of the friction losses. Therefore, a coarser mesh was chosen since the results of the two grids were similar in terms of velocity and temperature along the tunnel. When the fire HRR increases, some zones with positive pressure gradient appear also with the intermediate mesh. The level of agreement between the results obtained with the multiscale and with the 3D approaches decreases increasing the HRR. The current comparison shows that the pressure distribution along a tunnel is a key parameter for the feasibility of the multiscale approach. This is applicable only if the 3D and the 1D models can predict the same pressure distribution in the tunnel. This limits the possibility to currently use the multiscale approach for the simulations of fires in tunnels to HRRs below 3MW.

REFERENCES

[1] Ang, C. E., Rein, G., Peiro, J., & Harrison, R. (2016). Simulating longitudinal ventilation flows in long tunnels: Comparison of full CFD and multi-scale modelling approaches in FDS6. Tunnelling and Underground Space Technology, 119-125.

[2] Colella, F. (2010). Multiscale modelling of tunnel ventilation flows and fires. PhD thesis.

[3] Vermessi, I., Rein, G., Colella, F., Valkvist, M., & Jomaas, g. (2017). Reducing the computational requirements for simulating tunnel fires by combining multiscale modelling and multiple processor calculation. Tunnelling and Underground Space Technology, 146-153.

[4] Prince, J., (2015) "Coupled 1D-3D simulation of flow in subway transit networks.", *PhD thesis.*

[5] Tarada, F., & Brandt, R. (2009). Impulse ventilation for tunnels–a state of the art review. In Proceedings of 13th International Symposium on Aerodynamics and Ventilation of Vehicle Tunnels. BHRG. New Brunswick, New Jersey, USA (pp. 95-107).

[6] Beyer, M., Sturm, P. J., Saurwein, M., & Bacher, M. (2016, May). EVALUATION OF JET FAN PERFORMANCE IN TUNNELS. In 8th International Conference Tunnel Safety and Ventilation.

[7] McGrattan, K. B., McDermott, R. J., Weinschenk, C. G., & Forney, G. P. (2017). Fire dynamics simulator, technical reference guide. NIST Special Publication SP-1018.

[8] Li, Y. Z., Lei, B., & Ingason, H. (2010). Study of critical velocity and backlayering length in longitudinally ventilated tunnel fires. Fire Safety Journal, 45(6-8), 361-370.

[9] FDS user's group (2017), Discussion over 'Pressure rise induced by fires in tunnels' https://groups.google.com/d/topic/fds-smv/H5BL3j9j4aM/discussion

FireFOAM simulation of a localised fire in a gallery

Fabien SALMON[1], Delphine LACANETTE[1], Jean-Christophe MINDEGUIA[1], Colette SIRIEIX[1], Catherine FERRIER[2] and Jean-Claude LEBLANC[3]

[1] *UMR CNRS 5295 I2M, Bordeaux University, France*
[2] *UMR CNRS 5199 PACEA, Bordeaux University, France*
[3] *UMR CNRS 5608 TRACES, Toulouse University, France*

Fabien.Salmon@u-bordeaux.fr

ABSTRACT

Thermal marks were identified on the walls of the Chauvet-Pont d'Arc cave. In order to reproduce similar thermo-alterations, an experiment was carried out. The interests being archaeological, the chosen location was a natural gallery. Then, the combustion consisted in a localised wood fire within a quarry. The experimental site was instrumented and temperatures, velocities, gases concentrations and soot deposits were measured. A comparison between the experiment and a FireFOAM simulation is performed. Thanks to photogrammetry, the simulation takes place within the exact geometry of the quarry. FireFOAM manages to produce significant results that are largely in good agreement with experimental data. The encountered difficulties are also exposed as well as the corresponding solutions. The appropriate care taken to reach convenient results is detailed. Especially, a boundary condition expressing energy balance at a wall has been added to the software. Moreover, the existing thermocouple model has been modified in order to reach relevant corrected temperatures. In order to calculate the soot deposition, the Beresnev-Chernyak model has been implemented in FireFOAM. Finally, the hazard of toxicity, radiation and heat are also estimated by means of the fractional effective dose approach (FED).

KEYWORDS: *localised fire; gallery; FED; CFD; FireFOAM; rock art caves*

IOP Conf. Series: Journal of Physics: Conf. Series **1107** (2018) 042017 doi:10.1088/1742-6596/1107/4/042017

INTRODUCTION

For several years, numerical simulation has been on the rise in fire engineering. Especially, the computational fluid dynamics simulation (CFD) has been used to simulate fire. The increasing performance of computers enables simulations of sizeable combustions. Nonetheless, the coupling of complex physics is still a difficulty to manage accurate simulation. Many investigations try to simulate fire in different configurations thanks to CFD simulation. Compartment fires take a significant place in such researches. Besides, the LES modelling is used to compare this kind of experiment with simulation [1], [2]. Close to this configuration, this study falls within fires in geometries like galleries. This type of geometry is similar to tunnels and mines for instance. Furthermore, for safety reasons, a lot of investigations about tunnel fires simulation were conducted [3], [4].

The present research is driven by archaeological interests. High temperature caused alterations to the ceilings and walls of the Chauvet-Pont d'Arc cave (Ardèche, France) [5]. Some clues attest it. First, important spallings occurred in different locations within the cave. Secondly, some areas have a red colour, resulting from rubification (high-temperature chemical reaction). Eventually, soot is still on the walls of the cave. Therefore, it confirms that the walls were thermally marked, and more specifically by fires. Currently, the scales of these fires are unknown and the archaeologists would like to characterize their intensity to assume their function.

The first part of the presentation is the description of an experiment conducted with the aim of reproducing similar thermo-alterations to those in the Chauvet-Pont d'Arc cave. Wood was placed in the back of a former quarry. The quarry dimensions resemble to the Megaloceros Gallery ones (part of the Chauvet-Pont d'Arc cave). They are deeper described in the following section. A same fire was reproduced three times within three days to test the variability of the results. The fire consisted in a supplied wood hearth in a gallery. Thereby, the fire was localized during all the experiment. The experimental protocol is described thereafter.

The second part corresponds to the model description. The simulation is performed by the solver FireFOAM-4.0 [6], embedded in the software OpenFOAM [7]. The code structure and the models are outlined in this section. The set of general equations is presented and radiation, turbulence and combustion modellings are detailed. FireFOAM being an open-source software, contributions can be implemented. We add a boundary condition for temperature where the wall temperature is estimated from energy balance. This is similar to the default one in the Fire Dynamics Simulator (FDS [8]). In the same way, the simulation of soot deposit and the evaluation of toxicity, radiation and heat hazards are implemented to FireFOAM. Finally, a thermocouple correction is added as post-processing.

The last part is devoted to discussion about numerical results and comparison with experimental data.

EXPERIMENTAL PROTOCOL

Experiment description

The experiment was carried out within a former quarry in Lugasson (France). It is composed of two perpendicular galleries (Fig. 1). Each gallery is about 240 cm wide. A localized wood fire was placed at 80 cm from the back of the second gallery (250 cm high) and centered between walls. The hearth diameter was 80 cm. This fire was reproduced three times for three days (one per day) in order to evaluate its reproducibility. For archaeological reasons described in the introduction, the hearth was designed to impact the ceiling as much as possible in order to reach 250°C for 10 minutes in the rock ceiling (rubification threshold). According to preliminary tests, a tepee shape was chosen to maximize the flames height (Fig. 1). The fuel was pine sylvesteris because this is the burnt one in Chauvet-Pont d'Arc. The 80 cm branches were gathered in 4.5 kg bundles. At the ignition, four bundles were placed at the fireplace and a blowtorch started the fire. Then, at each supply time, two bundles were brought by firefighters with security equipment. The supply times were adopted to avoid any heat release rate (HRR) decrease during the first fire. The same times were kept for the two other fires. Globally, 30 bundles supplied the fire, corresponding to 135 kg.

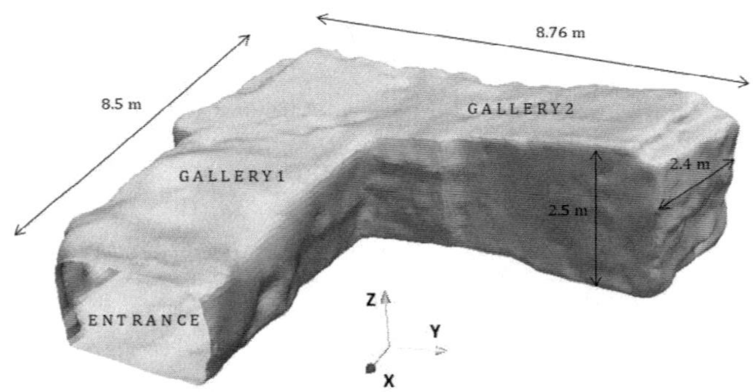

Fig. 1. On the left: 3D geometry of the quarry achieved by photogrammetry. On the right: Four bundles in a tepee configuration at the ignition

Instrumentation of the gallery

With the aim of comparing this experiment with the tool FireFOAM, the quarry was instrumented (Fig. 2). Sensors were positioned throughout the galleries to measure velocities, temperatures, soot deposits, gases and particles concentrations (oxygen, carbon dioxide and carbon monoxide). The type-K thermocouples trees were homogeneously placed to map the temperature during the fire. On each tree, several sensors allowed to measure the vertical gradient temperature. The gas concentration sensors were situated in the hot layer as well as the pegasor particle sensors (PPS), measuring particles concentrations. Two sensors measured the velocities of the cold and hot layers at the entrance of the quarry. In different locations, several plates were fixed to the wall in the purpose of collecting deposited soot at the end of the combustion. On the ground, a camera filmed the fire from the quarry corner. Eventually, a weighing scale was under the hearth, indirectly providing the HRR against time (considering that the heat of combustion remains constant).

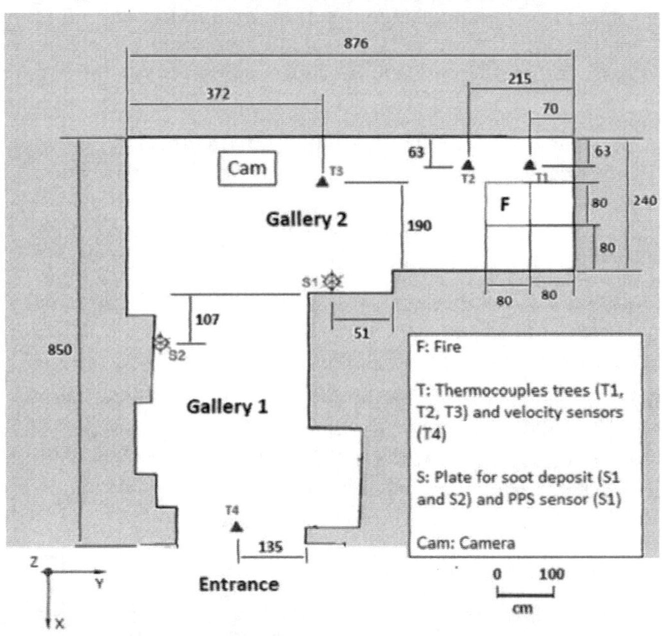

Fig. 2. Cut-away view of the quarry with the hearth location, the thermocouples trees locations and the soot deposit measures locations. A camera is placed on the floor, near the corner. Dimensions are in cm.

NUMERICAL MODEL

FireFOAM description

FireFOAM solves the Favre averaged aerothermochemistry equations. The LES modelling is chosen to simulate turbulence. The turbulent kinetic energy k (m².s^{-2}) is calculated thanks to the model [9] and the turbulent viscosity is then evaluated $\nu_t = C_k\sqrt{k}\Delta$ with $C_k = 0.094$ and Δ the filter size (m).

As regards radiation, the radiative transfer equation is solved with the underlying assumption of a grey medium. Moreover, the gases are supposed to be independent on wavelength. Scattering is not taken into account because of the low experimental soot concentrations. 30% of the combustion energy is assumed to be radiatively released [10]. To solve the radiation equation, the finite volume discrete ordinary method (fvDOM) is adopted. The weak particle quantity in smoke justifies the application of this model (optically thin medium).

Concerning combustion, the Magnussen model (Eddy-Dissipation Model) [11] turns out to be sufficient for the chemical reaction rate estimation:

$$\dot{\omega} = -\rho C\frac{\epsilon}{k}\min\left(Y_f, \frac{Y_o}{s}\right) \tag{1}$$

where the indices f and o correspond to the fuel and the oxidizer respectively, s is the stoichiometric coefficient of the oxidizer, ϵ is the turbulent kinetic energy dissipation (m².s^{-3}), k is the turbulent kinetic energy and C is a model constant.

Advances in FireFOAM simulation

First, the thick walls of the quarry compel the use of an adequate boundary condition corresponding to a heat flux balance:

$$q_r + h(T_g - T_w) = \lambda\,(T_w - T_x)/dx \tag{2}$$

with q_r the total radiative heat flux, h the convective heat transfer coefficient, λ the rock thermal conductivity and T the temperature. The indices g, w and x denote the gas outside the boundary layer, the wall surface and the center of the first cell in the wall respectively. The energy received by the wall is diffused through it. In order to estimate the heat conduction in the solid walls, the 1-D heat equation is solved for each surface cell of the geometry mesh. This boundary condition is implemented in FireFOAM that did not include it. The convective heat transfer coefficient is calculated from empirical formulations. This condition is similar to the default one in the Fire Dynamics Simulator [8].

Second, a consistent thermocouple model is implemented to the initial FireFOAM-4.0 in order to compare experimental data with numerical results. Indeed, a gap necessarily exists between the measurements of the sensor and the temperature of the surrounding gases. Then, based on energy balance, a post-processing calculation solving the Eq. 3 is added to the software:

$$\rho c_p d\frac{\partial T}{\partial t} = q_r + h_{conv}(T_g - T) \tag{3}$$

with ρ the thermocouple density (kg.m^{-3}), c_p its heat capacity (J.kg^{-1}.K^{-1}), d its diameter (m) and h_{conv} the convective heat transfer coefficient (W.m^{-2}.K^{-1}).

Third, the hazards of toxicity, heat and radiation are evaluated thanks to the fractional effective dose (FED). Some experiments were conducted on animals to estimate death rates in specified conditions. From them, mathematical correlations can value hazard [12], [13]. We add these estimations to FireFOAM. It worth noting that the gases toxicity correlations are implemented in FDS but not the radiation and heat ones. For instance, [14] deals with FED in FDS.

Eventually, by assuming that the main process of particles deposition is thermophoresis, the Beresnev-Chernyak model [15] is added to the global modelling in the purpose of evaluating the mass of the deposited soot. In such a case, particles mainly impact walls because of a temperature gradient. Consequently, this model highly depends on the convective heat transfer coefficient through the temperature gradient.

COMPUTATIONAL SETUP

The CAD file is the quarry morphology constructed from a photogrammetry survey. The mesh of the quarry is managed by the OpenFOAM tool "snappyHexMesh". The geometry mesh consists of about 250,000 cells (Fig. 3). 3 cm cells mesh the part above the hearth, 6 cm cells mesh the hot layer and other cells measure 12 cm. In addition to the geometry on the figure 3, a big box representing the outside is meshed with 96 cm cells. The mesh convergence has been performed.

At the boundaries, the velocity is fixed at zero. The wall temperature is computed thanks to (2). Outside, the box-wall temperature is fixed at the initial temperature 12°C. Each mass fraction gradient is set at zero at the walls.

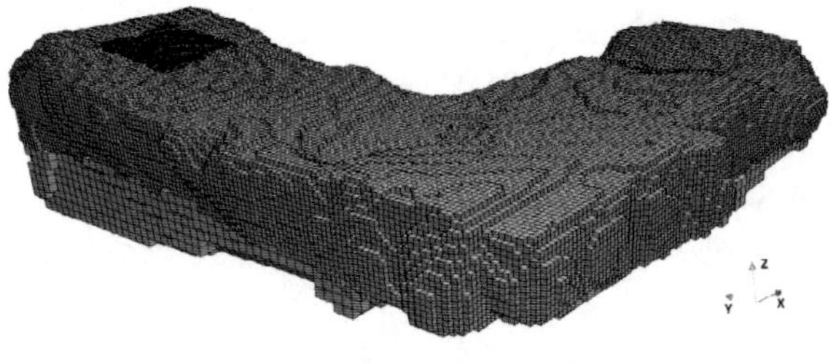

Fig. 3. Mesh of the quarry geometry

The HRR is dictated through a mass flow rate boundary condition. Therefore, the experimental measures of mass losses during the fire are prescribed at the burner. The fuel is then assumed to be a gas in the simulation instead of solid wood in reality. Thus, no pyrolysis model is used.
To solve the radiative transfer equation, a 32 solid angles discretization is adopted. The Courant number is compelled to remain inferior to 0.6 during the overall simulation.

RESULTS AND DISCUSSION

In the following, we expose some results from the achieved simulation. Results are accurate concerning the expected layers separation. The temperature field in the middle of each gallery 40 minutes after the ignition is presented on the Fig. 4. The hot and cold layers are well segregated, even after 40 minutes. The temperature in the hot layer reaches more than 500°C at the end of the combustion. It is in accordance with the simulated temperatures (Fig. 4). The thermocouple measures demonstrate that the cold layer is barely affected by the fire. The numerical results agree with this reality even if the numerical cold layer is less affected than in the experiment at the tree T1. It is not the case for all thermocouples trees.

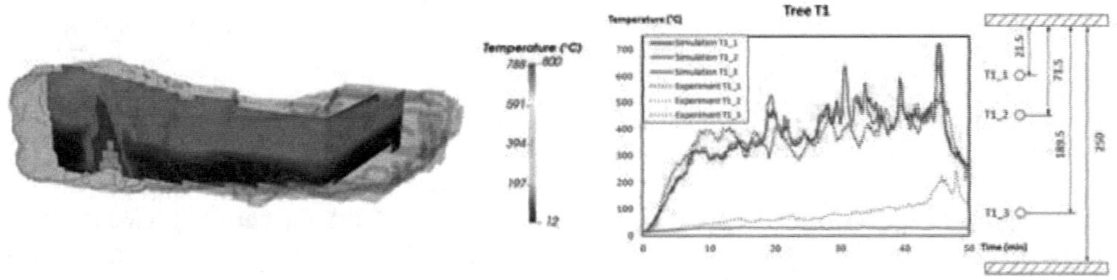

Fig. 4. On the left: numerical temperature field in the median sections of each gallery 40 minutes after the ignition of the combustion. On the right: Comparison between experimental and simulated temperatures at the tree T1.

As regards soot deposit, we have trouble simulating consistent results. The differences between experiment and simulation could be explained by the grid size (as studied in [16]) and by the convective heat transfer model itself. Better results should be given by wall-resolved LES. However, the prohibitive time consumption of this technique prevents its use. Besides, most similar simulations (default FDS simulations for instance), use empirical convective coefficients. Fortunately, because the interests are archaeological, only soot deposit locations are important in the study and the information about the amount of deposited soot is not critical.

IOP Conf. Series: Journal of Physics: Conf. Series **1107** (2018) 042017 doi:10.1088/1742-6596/1107/4/042017

CONCLUSION

Archaeological investigations about the Chauvet-Pont d'Arc cave have conducted to the achievement of an experimental localised wood fire in a natural gallery. The geometry of the experimental site makes the results applicable to general simulations in tunnels or mines. The quarry is instrumented in order to compare the experimental fire with a numerical simulation. The modelling is performed by FireFOAM [6]. This open source software can be modified thanks to its adequate framework. Accordingly, four improvements are added to the initial solver. To estimate the wall temperatures, an appropriate boundary condition based on energy balance is added. Then, in order to evaluate soot deposit, the Beresnev-Chernyak model is implemented. Because the experimental measurements correspond to the thermocouples temperatures, a gap exists with the surrounding gases. Therefore, a relevant thermocouple correction is performed after the calculation. Finally, the fractional effective dose is used to evaluate hazard during the experimental fire. The gases toxicity and the heat and radiation hazards are computed by means of existing correlations [12], [13].

It turns out that the simulation gives satisfactory results that are close to the experimental data except for soot deposit. The temperature field remains in accordance with the experiment and the thermocouple measurements validate the modelling of the fire. Other data regarding velocities, gas and particle concentrations demonstrate the same trend. Thereafter, this tool will be applied to archaeological matters. Several scenarii will be carried out in the Chauvet-Pont d'Arc cave, with various input parameters (amount of fuel, duration, fire numbers and supply).

REFERENCES

[1] A. Trouvé, Y. Wang, Large eddy simulation of compartment fires, International Journal of Computational Fluid Dynamics. 24 (2010) 449–466.

[2] S. Yuan, J. Zhang, Large eddy simulation of compartment fire with solid combustibles, Fire Safety Journal. 44 (2009) 349–362.

[3] C. Caliendo, P. Ciambelli, G. De, M.G. Meo, P. Russo, Simulation of fire scenarios due to different vehicle types with and without traffic in a bi-directional road tunnel, Tunnelling and Underground Space Technology. 37 (2013) 22–36.

[4] W.-S. Hsu, Y.-H. Huang, T.-S. Shen, C.-Y. Cheng, T.-Y. Chen, Analysis of the Hsuehshan Tunnel Fire in Taiwan, Tunnelling and Underground Space Technology. 69 (2017) 108–115.

[5] C. Ferrier, É. Debard, B. Kervazo, A. Brodard, P. Guibert, D. Baffier, V. Feruglio, B. Gély, J.-M. Geneste, F. Maksud, Les parois chauffées de la grotte Chauvet-Pont d'Arc (Ardèche, France) : caractérisation et chronologie, PALEO. Revue d'archéologie préhistorique. (2014) 59–78.

[6] FireFOAM: https://www.fmglobal.com/modeling

[7] OpenFOAM: https://openfoam.org

[8] K. McGrattan, S. Hostikka, R. McDermott, J. Floyd, C. Weinschenk, K. Overholt, Fire Dynamics Simulator User's Guide, Sixth edition 1019. NIST Special Publication, https://pages.nist.gov/fds-smv/manuals.html

[9] A. Yoshizawa, Statistical theory for compressible turbulent shear flows, with the application to subgrid modeling, PHYS. FLUIDS. 29 (1986) 2152–2164.

[10] A. Tewarson, Generation of Heat and Chemical Compounds in Fires. SFPE Handbook of Fire Protection Engineering, National Fire Protection Association (2002).

[11] B. Magnussen, B. Hjertager, On mathematical modeling of turbulent combustion with special emphasis on soot formation and combustion, Proc. Combust. Int. (1977) 719-729.

[12] D.A. Purser, J.L. McAllister, SFPE Handbook of Fire Protection Engineering, Chapter "Assessment of Hazards to Occupants from Smoke, Toxic Gases, and Heat" 5th ed., Springer-Verlag, New York, (2016)

[13] D.A. Purser, Physiological effects of combustion products and fire hazard assessment, Europacable Seminar "Safety during Fire" Brussels, (2009)

[14] X. Qu, Q. Meng, Z. Liu, Estimation of number of fatalities caused by toxic gases due to fire in road tunnels, Accident Analysis & Prevention, 50 (2013) 616-621.

[15] S. Beresnev, V. Chernyak, Thermophoresis of a spherical particle in a rarefied gas: Numerical analysis based on the model kinetic equations, Physics of Fluids. 7 (1995) 1743–1756.

[16] K.J. Overholt, J.E. Floyd, O.A. Ezekoye, Computational Modeling and Validation of Aerosol Deposition in Ventilation Ducts, Fire Technology, 52 (1), (2016), 149-166.

IOP Conf. Series: Journal of Physics: Conf. Series **1107** (2018) 042018 doi:10.1088/1742-6596/1107/4/042018

A Case Study of Far-Field Temperatures in Progressing Fires

Nils JOHANSSON

Division of Fire Safety Engineering, Lund University, Sweden

nils.johansson@brand.lth.se

ABSTRACT

The non-uniform conditions and potential progression of fires in larger spaces calls for modelling methods extending outside of the traditional compartment fire framework. Even though a handful of studies in the area exist there is still little guidance available on how progressing fires in large enclosures can be modelled. Three different methods to calculate far-field gas temperature in large enclosures are therefore reviewed in this paper with the help of a case study. The three methods used are: the analytical Alpert ceiling jet correlation, the Fire Dynamics Simulator (FDS) and a Multi-Layer Zone (MLZ) model.

The enclosure used in the case study is intended to represent a large open space, i.e. an office, warehouse or supermarket. Two different user defined fire scenarios, which represent two progressing fires, are analysed.

This is a comparative study and no experimental data is used to evaluate the models. However, it is considered reasonable to believe that FDS gives good predictions since it has shown to predict gas temperatures well in previous validation studies. The MLZ model has not been thoroughly evaluated, but it has been seen to overestimate experimental results slightly in a previous study.

The results from FDS and the MLZ model show that there is a temperature distribution (vertical and horizontal) in both fire scenarios, which means that there are non-uniform conditions in the gas layer. The results also indicate that there is a correspondence between the MLZ model and FDS; however, as the heat release rate increase the difference between the models increases. The analytical correlation results in much lower temperature predictions than both FDS and the MLZ model. The main reason for that is that the correlation assumes the flow to be unconfined, which is not the case in this study.

The main benefits of the MLZ model is the low computational time (about 1 minute on a laptop computer) compared to FDS (where the computational time was more than two days). Furthermore, the MLZ model provides temperature profiles both horizontally and vertically, which is not the case in conventional zone models.

INTRODUCTION

The turbulence caused by a fire will mix the hot gases in a small enclosure and the properties (like temperature and visibility) of the hot gas layer will be rather uniform both in the vertical and horizontal direction. This makes it possible to derive equations and find correlations for calculating e.g. gas temperatures. The first comprehensive work on this, so-called "compartment fire framework", was done by Kawagoe in the 1950s [1]. A lot of effort has been conducted within the area since then and the compartment fire has proven to be a robust and simple way of describing fire conditions under certain circumstances.

The time-temperature curves for post-flashover fires in Eurocode 1 [2] are based on the uniform temperature assumption, which is reasonable because the experimental tests that the method is based on were conducted in rather small enclosures. The situation becomes more complex in large enclosures where the hot gas layer cannot be regarded as uniform. For instance, it is not likely that flashover will occur in large enclosure in the same way as expected in the compartment fire framework. In the compartment fire framework the fire is fuel controlled and grows, at a certain time point the heat flux on the combustibles in the room is so intense that flashover occurs. Then all combustible material is involved in the fire and it becomes ventilation controlled, e.g. the heat release rate is controlled by the supply of oxygen. The terms regime I and regime II [3] are sometimes used to distinguish between ventilation controlled and the fuel controlled burning, respectively. It has been argued that fires in large spaces are likely to be within regime II [3][4], since the availability of air most likely will be high due to the presence of large openings and leakages in the boundaries. Traditional methods for calculating heat release rates in ventilation controlled fires are therefore inappropriate to use in larger enclosures where the fire progresses or travels, as it is termed by e.g. Stern-Gottfried and Rein [4].

That fires can progress in a large enclosure have been documented [4][5] and there are some examples when this have been studied in fire tests [5]. Kirby et al [6] did burns in a 23 m by 6 m enclosure and saw that flashover did not occur as according to the compartment fire framework, instead it progressed between fuel packages. Hidalgo et al [7] have recently presented an experimental study undertaken to characterise fires in large spaces. A dozen fire tests were performed Hidalgo et al in an 18 m by 6 m enclosure. A complete analysis of the data have not been published yet but the setup was well equipped with different types of sensors and the data from the tests will definitely add to the knowledge of enclosure fire dynamics in large spaces.

The non-uniform conditions and potential progression of fires in larger spaces calls for modelling methods extending outside of the compartment fire framework. Stern-Gottfried and Rein [4] presents the so-called traveling fires framework in which the thermal field induced by the fire is divided into two regions: the near field and far-field. The position and size of the regions are relative to the position of the fire, and moves within the enclosure as the fire progresses. The near field is the burning region of the fire, and the far-field is the region where no burning or flames are present, instead the smoke layer will provide a thermal exposure. The near field temperatures can be modelled with methods like the localised fire in Eurocode 1 [2] or with a worst case flame temperature of e.g. 1200°C as proposed by Stern-Gottfried and Rein. The far-field temperature is considered to be more challenging to model.

Rein et al [8] used the Computational Fluid Dynamics (CFD) model Fire Dynamics Simulator (FDS) to model the far-field temperatures but found it problematic due to the high computational cost. Therefore, later efforts to estimate far-field temperatures have focused on using the much simpler analytical methods like the ceiling jet correlation by Alpert [9]. Alpert assumed an axisymmetric fire plume beneath a flat, horizontal ceiling, unobstructed by walls and derived correlations for the ceiling jet temperature and velocity. The ceiling jet correlations are generally good for estimating gas temperatures in the early stages of fire. The problem with applying the Alpert correlation is that it is not applicable in an enclosed space when a hot gas layer, which will affect the ceiling jet temperature, forms. Furthermore, in the original work by Alpert [9] it was seen that at distances of 3 to 5 ceiling heights from the centre of the fire, the heat transfer to the ceiling were significant and this is not treated directly in the correlation. More recently promising efforts have been made by the research group in Edinburgh [5] to couple a model for localized fires with a simple zone model; however, the work is currently said to be on a conceptual stage.

Even though a handful of studies in the area exist there is still little guidance available on how progressing fires in large enclosures can be modelled. Three different methods to calculate far-field gas temperature in large enclosures will therefore be reviewed in this paper with the help of a case study.

APPLIED METHODS

The three methods used to calculate far-field temperatures are, the analytical Alpert ceiling jet correlation, a CFD model and a Multi-Layer Zone (MLZ) model.

Analytical model

Analytical models to estimate velocities and temperatures in ceiling jets have existed since at least the 1970s [9] but a continuous improvement and development have been done since then (see e.g. [10] and [11]). A ceiling jet is created when the buoyancy driven fire plume impinges on a ceiling and the hot gases spreads radially under the ceiling. The temperature in the ceiling jet will decrease as it moves away from the plume centreline due to entrainment of cold air and heat losses to the ceiling. If the ceiling jet is unconfined it will have a maximum thickness of about 5-13% of the total room height [12]. In a normal compartment fire this type of unconfined ceiling jet will only exist in the earliest stages of fire development before the hot gases accumulate in the compartment. Therefore, these types of models are normally used to estimate sprinkler and heat detector activation in pre-flashover fires. The Alpert ceiling jet correlation [9] (see Equation 1) is used in this paper since its been proposed to be used for modelling far-field temperatures [4]. In Equation 1, $\dot{Q}$ is the fire heat release rate, *r* is the radial distance from the plume centreline and H is the height to the ceiling.

$$T_g - T_a = 5.38\frac{(\dot{Q}/r)^{2/3}}{H} \tag{1}$$

CFD model

FDS developed by NIST [13], is often used in fire safety engineering. FDS is a CFD model where fire-driven fluid flows are simulated. The software solves the Navier–Stokes equations numerically with an emphasis on heat and smoke transport.

MLZ model

The MLZ concept has been described in previous publications [14][15]; therefore, no detailed description of the concept is therefore given in this paper. The conservation of mass and energy is used in zone models to calculate hot gas temperatures and flows between different zones are calculated with the help of pressure differences and the Bernoulli equation. The MLZ model uses these zone model principles. However, in contrast with the more established two-zone models (like CFAST [16]) where each room consists of two zones, each enclosure is divided into several regions (horizontal) and layers (vertical) in the MLZ model. This means that the enclosure is divided in to several smaller zones. The fire is specified as a heat release rate and the heat and hot gases rises upwards from the fire in a plume that enters the highest located layer in the fire region. Air and hot gases is also entrained in the plume from the layers that it passes through. Mass is transported horizontally to layers in adjacent regions due to hydrostatic pressure differences. There is also a flow of mass vertically between layers in each region, which is calculated based on the conservation of mass. Heat losses to boundaries and radiation between cells are also calculated. The MLZ model used in this paper extends in three dimensions and is based on the model used by Johansson [17]. Johansson made a minor evaluation study of the model and saw that it over predicted measured temperatures slightly.

ANALYSIS

The enclosure used in the case study is fictive and it is constructed to represents a large open space, i.e. an office, warehouse or supermarket. The enclosure is 20 m x 32 m x 4 m with one door (12 m x 3 m) that is placed in the middle of one of the short ends (see Figure 1 and Figure 2). The boundaries are made of 0.1 m thick lightweight concrete. Two measurement points (M1 and M2) are used in this analysis and both are placed in an area that is regarded as the far-field, e.g. no expected flames or combustion.

Design fires

Two simple well-defined fire scenarios are used in this study. The fuel load density determines the duration of fire in each fuel package. Two design fire scenarios that represent two progressing fires are modelled with the three methods. The first design fire scenario consists of four different fuel packages (each 2 m by 2 m large, see Figure 1) where each package burns according to a Fast (0.047 kW/s^2) alfa-t^2 during 235 seconds up to 2.6 MW. Fuel package 2 is ignited 120 seconds after fuel package 1, fuel package 3 is ignited 240 seconds after fuel package 1 and fuel package 4 is ignited 360 seconds after fuel package 1. Each fuel package has a fuel load density of 500 MJ/m^2; this results in that each fuel package burns out 920 seconds after ignited. The distance between each fuel package is 2 m. The second scenario includes nine fuel packages (see Figure 2) that are identical to the fuel packages in scenario 1. The nine fuel packages ignite every 120 seconds according to the sequence presented in Figure 2. The distance between the fuel packages is 2 m.

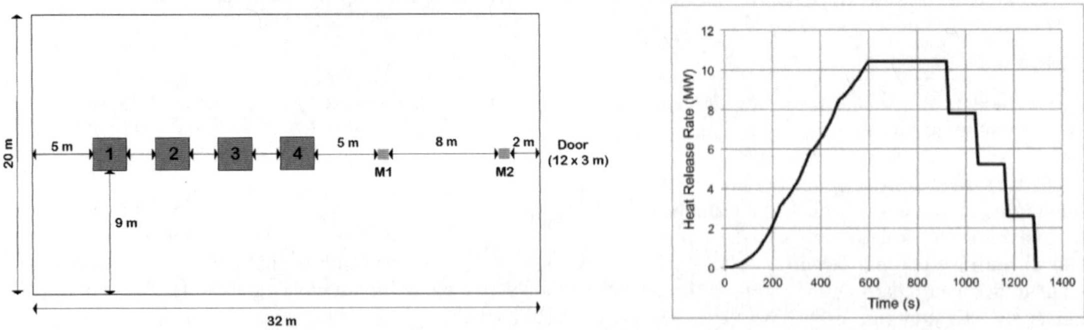

Figure 1: Layout of scenario 1 (left) and resulting total heat release rate form the four fuel packages (right).

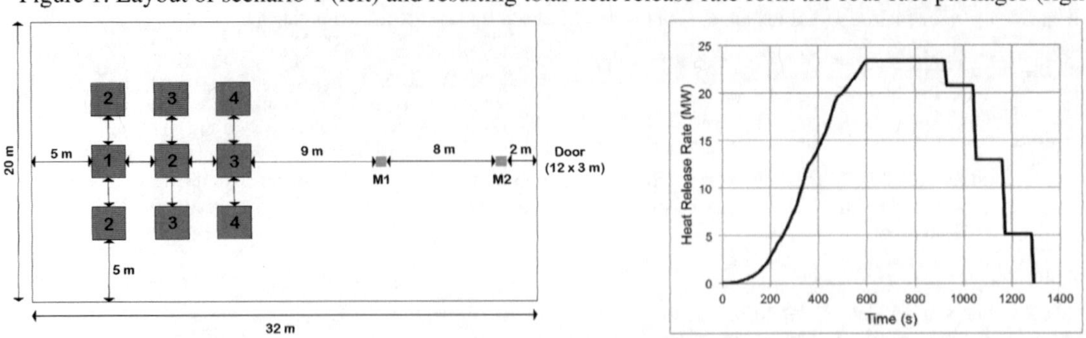

Figure 2: Layout of scenario 2 (left) and resulting total heat release rate form the nine fuel packages (right).

Setup of the models

The distance *r* in Alpert ceiling jet correlation (Equation 1) was taken as the distance to the location of the weighted average of the total heat release rate. Five different meshes were used in FDS, and a mesh size of 0.1 m was used around the fuel packages and 0.2 m in the rest of the domain. A total of 584,000 cells were used in scenario 1 and 2 while 1,200,000 cells were used in scenario 3. The enclosure was divided into 40 regions in the MLZ model and each region was divided into 8 layers, i.e. 320 cells in total.

Results

No effort is made here to compare the calculation times in detail. Even so, it is considered worth mentioning that the MLZ simulations were performed in less than 1 minute on a laptop computer while the FDS simulations took more than 60 hours each to perform on the LUNARC cluster [18]. The results are presented for different elevations at M1 in Figure 3, and as vertical temperature profiles in Figure 4-5. The results at measurement point M2 indicate similar trends as in Figure 3-5. Results from simulations with CFAST [16] are included in Figure 4-5 for reference.

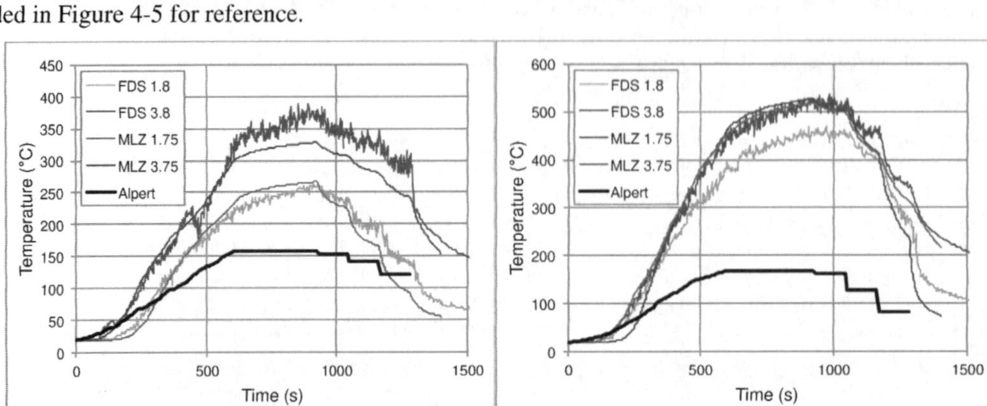

Figure 3: Temperatures in M1 in scenario 1 (left) and scenario 2 (right).

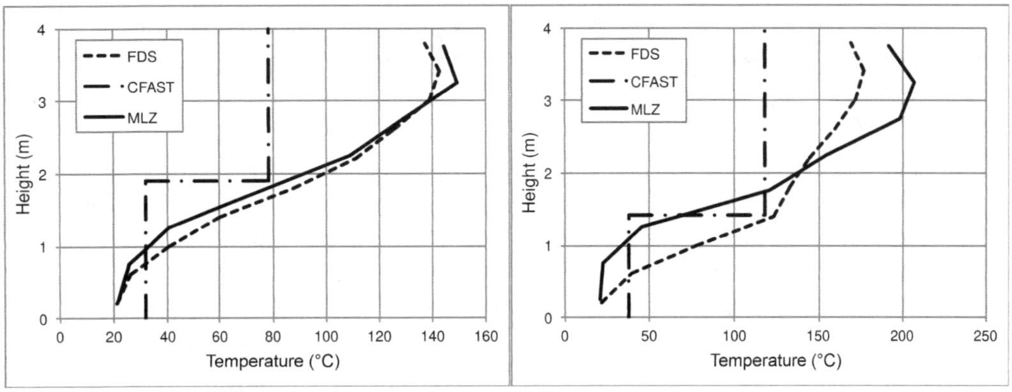

Figure 4: Vertical temperatures in M1 at 300 s in scenario 1 (left) and scenario 2 (right).

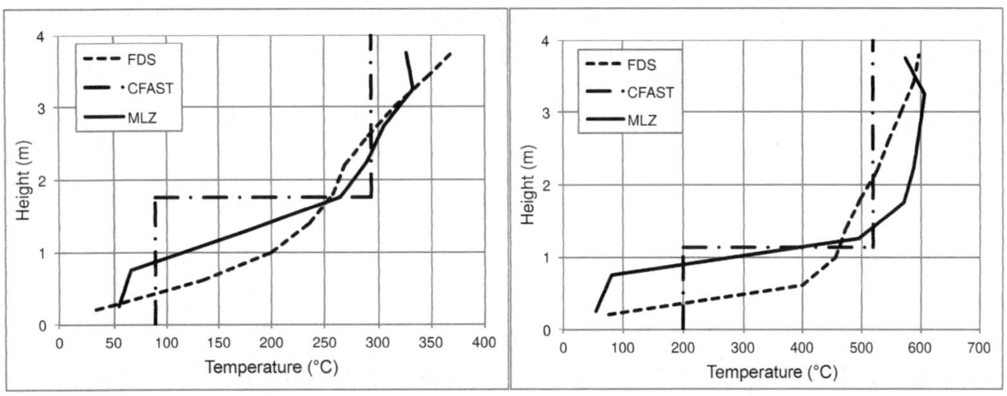

Figure 5: Vertical temperatures in M1 at 900 s in scenario 1 (left) and scenario 2 (right).

FDS and the MLZ model clearly shows that there is a vertical temperature distribution that is not possible to catch with the two-zone model like CFAST (see Figure 4 and Figure 5). Based on the results it can be seen that there is a correspondence between the MLZ model and FDS; however, as the heat release rate increase (as in scenario 2) the difference between the models increases. It is also clear from Figure 3 that the simple analytical formula by Alpert gives much lower temperatures than the numerical models.

DISCUSSION AND CONCLUSION

Three different methods to calculate far-field temperatures in progressing fires have been studied in this paper. No experimental data are available for the case analysed; consequently, it is not possible to state how accurate the methods are in predicting far-field temperatures. Even so, it is reasonable to believe that FDS gives resonable predictions since it has shown to predict gas temperatures well in previous validation studies [13]. However, when it comes to fully developed fires it has been seen that FDS can under predict temperatures [19][20]. The major drawback of FDS in this study is the long calculation time.

The results from both FDS and the MLZ models indicate that there is a vertical and horizontal temperature distribution, which is not possible to model with two-zone models like CFAST. The Alpert ceiling jet correlation gives a single temperature and it is much lower than the temperatures predicted by FDS and MLZ models. The main reason is that the Alpert correlation assumes the flow to be unconfined which is not the case in this study and probably seldom in any situations after the initial growth stage of a fire. If it cannot be guaranteed that the ceiling jet will be unconfined it will most likely underestimate the gas temperature.

The MLZ model that has been used in this study was verified as it was developed and written, however, the verification process has not been as thoroughly as that of FDS. Furthermore, in the previous evaluation study it has been seen that the model might overestimate gas temperatures slightly. Even so, the case study indicates that the MLZ model can estimate both vertical and horizontal far-field gas temperatures similar to FDS.

The main benefits of the MLZ model is the low computational time and that it provides temperature profiles both horizontally and vertically. It would be desirable to evaluate the MLZ model further against experimental data. The computer power available for fire safety designers are constantly improving, this means that the time to run a CFD models is constantly decreasing, which lessens the need for a quicker and less accurate tool like the MLZ model. However, the MLZ model is simple and easy to grasp which means that users are more likely to understand what they are doing and how the model works, which can result in less mistakes and more robust designs.

REFERENCES

[1] Kawagoe, K. (1958) Fire Behavior in Rooms. Report no. 27, Building Research Institute, Japan.

[2] EN 1991-1-2 (2002) *Eurocode 1: Actions on structures - Part 1-2: General actions - Actions on structures exposed to fire*. European Committee for Standardization.

[3] Torero, J.L., Majdalani A.H., Abecassis-Empis C. and Cowlard A. (2014) Revisiting the Compartment Fire. *Fire Safety Science* 11:28-45. http://dx.doi.org/10.3801/IAFSS.FSS.11-28

[4] Stern-Gottfried J, Rein G (2012) Travelling fires for structural design—part II: design methodology. *Fire Safety Journal* 54:96–112. http://dx.doi.org/10.1016/j.firesaf.2012.06.011

[5] Dai, X., Welch, S. and Usmani, A. (2017) A critical review of “travelling fire” scenarios for performance-based structural engineering, *Fire Safety Journal*, 91:68-578. https://doi.org/10.1016/j.firesaf.2017.04.001

[6] Kirby, B.R, Wainman, D.E., Tomlinson, L.N., Kay, T.R. and Peacock, B.N. (1999) Natural Fires in Large Scale Compartments. *Int. Journal on Engineering Performance-Based Fire Codes*, 1(2):43-58.

[7] Hidalgo, J.P. Cowlard, A. Abecassis-Empis, C. Maluk, C. Majdalani, A.H. Kahrmann, S., Hilditch, R., Krajcovic, M. and Torero J.L. (2017) An experimental study of full-scale open floor plan enclosure fires, *Fire Safety Journal* 89:22-40.

[8] Rein, G., Zhang, X., Williams, P., Hume, B., Heise, A., Jowsey, A., Lane, B. and Torero, J.L. (2007) Multi-storey fire analysis for high-rise buildings. In: *11th International Interflam*, London, UK.

[9] Alpert R. (1972) Calculation of response time of ceiling-mounted fire detectors. *Fire Technology* 8(3):181-195.

[10] Alpert R. (2011) The Fire-Induced Ceiling-Jet Revisited. In: *5th FireSeat Symposium*. Edingburgh, Scotland.

[11] Johansson, N., Wahlqvist, J. and van Hees, P. (2014) Numerical experiments in fire science: a study of ceiling jets. *Fire and Materials* 39(5):533-544. http://dx.doi.org/10.1002/fam.2253

[12] Karlsson B and Quintiere JG. (1999) *Enclosure Fire Dynamics*. Boca Raton, USA.

[13] Floyd, J., Forney, G., Hostikka, S., Korhonen, T., McDermott, R., McGrattan, K. and Weinschenk, C. (2013) *Fire Dynamics Simulator Technical Reference Guide - Volume 3: Validation*. NIST Special Publication 1018, National Institute of Standards and Technology, Gaithersburg, MD, USA.

[14] Suzuki, K., Harada, K. and Tanaka, T. (2002) A Multi-layer Zone Model For Predicting Fire Behavior In A Single Room. *Fire Safety Science* 7: 851-862. http://dx.doi.org/10.3801/IAFSS.FSS.7-851

[15] Suzuki, K., Harada, K. Tanaka, T. and Yoshida, H. (2004) An Application of a Multi-Layer Zone Model to a Tunnel Fire. In: *6th Asia-Oceania Symposium on Fire Science and Technology*.

[16] Peacock, R., Forney, G., and Reneke, P. (2017) *Consolidated Fire And Smoke Transport (Version 7) Volume 3: Verification and Validation Guide*. NIST Special Publication 1889v3, National Institute of Standards and Technology, Gaithersburg, MD, USA.

[17] Johansson, N. (2017) Estimating gas temperatures in large enclosures. In: *CONFAB 2017*, London, UK.

[18] Center for Scientific and Technical Computing, LUNARC, Lund University, Sweden http://www.lunarc.lu.se/resources/hardware/aurora/

[19] Pope, N.D. Bailey, C.G (2006) Quantitative comparison of FDS and parametric fire curves with post-flashover compartment fire test data, *Fire Safety Journal* 41(2):99-110. https://doi.org/10.1016/j.firesaf.2005.11.002

[20] Jahn, W., Rein, G., and Torero, J.L. (2011) A posteriori modelling of the growth phase of Dalmarnock Fire Test One, *Build. and Envir.* 46(5):1065–1073. https://doi.org/10.1016/j.buildenv.2010.11.001

IOP Conf. Series: Journal of Physics: Conf. Series **1107** (2018) 042019 doi:10.1088/1742-6596/1107/4/042019

Flow behaviour of the exchange flow through a ceiling vent in natural convection: A numerical approach using CALIF^{3}S-ISIS CFD software

Patricio BECERRA BARRIOS[1,3], Hugues PRETREL[2], Samuel VAUX[2] and Olivier VAUQUELIN[1]

[1] *Aix-Marseille Université (AMU), Laboratoire IUSTI, Laboratoire ETiC, France*
[2] *Institut de Radioprotection et de Sûreté Nucléaire (IRSN), Laboratoire ETiC, France*
[3] *School of Mechanical Engineering, University of Costa Rica (UCR), Costa Rica*

patricio.becerra@ucr.ac.cr

ABSTRACT

In the fire safety context, the exchange flow through a horizontal opening connecting two superposed zones receives a particular interest for developing predictive models and validation of computational codes. Its random behaviour over time could influence the fire source and the vertical spread of the fire smoke. This work is part of a research program developed at the French "Institut de Radioprotection et de Sûreté Nucléaire" (IRSN), and its main objective is to bring new knowledge on this subject by carrying out a numerical study with the computational open-source software named CALIF^{3}S-ISIS developed by IRSN. The configuration of the simulations is that of an experimental reduced-scale set-up, on which data have been obtained and could be used for validation. Large Eddy Simulation (LES) were performed (variation of wall boundary conditions, air injection temperatures, vent diameter and vent mesh refinement), and the bidirectional flow is properly estimated in comparison with the experiment. Particularly, it was determined that the increment of the geometric ratio L/D of the vent of diameter D and thickness L, also leads to an increment of a dimensionless exchange flow rate Froude number at the vent. However, a discrepancy was observed with the model proposed by Epstein based on his experiments in natural convection. The disagreement between the simulation results and the experiments is attributed principally to differences on the vertical temperature stratification of the lower room. Finally, it was investigated the influence of the vertical location within the vent over the mean flow behaviour through different levels of the vent, specifically an exchange of the downward and upward flow organization within the opening.

KEYWORDS: *bidirectional flow, horizontal vent, CALIF^{3}S-ISIS CFD code*

NOMENCLATURE LISTING

D	diameter (m)
Fr	Froude number (-)
g	gravitational acceleration (m/s^2)
L	heigth of the vent (m)
Q_v	volume flow rate (m^3/s)
Q_m	mass flow rate (kg/s)
T	temperature (K)
T	simulation time (s)
t	time (s)
z	vertical distance above floor level (m)
Greek	
ρ	density (kg/m^3)
$\bar{\rho}$	average density (kg/m^3)
subscripts	
$+$	upward flow rate
$-$	downward flow rate
∞	ambient condition

INTRODUCTION

Smoke flow through openings between different areas of a building represents a challenge to safety during a compartment fire. Its mass and energy transfer could have a significant impact on the control of contaminant particles and the spread of fire and smoke. In this study, we draw attention to the buoyancy-driven exchange flow through an horizontal opening in natural convection due to a confined fire source in a compartment, see Figure 1.

The earliest studies dealing with this problem were carried out by Brown [1] and Mercer & Thompson [2]. Brown [1] used air as working fluid and interpreted the exchange flow as a transfer phenomena and concluded that the exchange flow rates increase with the increase of the geometrical ratio L/D. Mercer & Thompson [2] using brine solutions and water in inclined and vertical pipes, reported the opposite behaviour observed by Brown and suggested a maximum exchange flow rate at an L/D between 0.66 and 3.5. Epstein [3] completed their works using water-brine as fluid and studied in detail the effect of the vent aspect ratio. Epstein proposed a correlation expressing the volume exchange flow Q_v in the form of a dimensionless Froude number at the vent of diameter D and thickness L, $Fr = Q_v/\sqrt{gD^5\Delta\rho/\bar{\rho}}$, where $\Delta\rho$ is the density difference between the top and the bottom of the opening, as a function of the geometric ratio L/D of the vent. He identified the exchange flow behaviour in four regimes from an oscillatory regime to a turbulent diffusion flow. Li [4] conducted full-scale experiments with fire using the laser Doppler velocimetry (LDV) technique to measure the flow at the vent. Li also performed CFD calculations using the standard $\kappa-\epsilon$ turbulence model and the Large Eddy Simulation (LES) technique. However, Li [4] concluded that only simulations using LES were in agreement with his experiments. Recently, Varrall [5] investigated the volume exchange flow using the stereoscopic particle image velocimetry (SPIV) technique as a non-intrusive measurement way. Varrall measured the velocity fields at the vent in a thermal steady state, after heating the air in the lower compartment. He determined an average flow organization with the upward flow occupying an egg-shaped central area and the downward flow occupying the parietal areas.

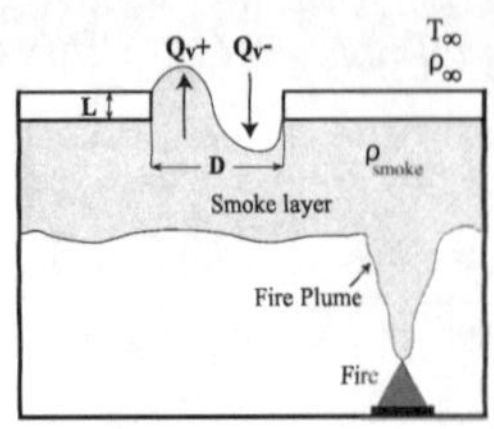

Fig. 1. Bidirectional flow exchange through the vent in an enclosure fire. Geometrical configuration.

Few numerical studies using the Computational Fluid Dynamics (CFD) focused on that problem. Spall & Anderson [6] investigated the exchange flow through a horizontal vent for a thin partition and indicated the presence of two dominating frequencies in the oscillatory regime. Harrison & Spall [7] studied the effects of the partition thickness to complete the work of Spall & Anderson [6] in the oscillatory flow regime. Sleiti [8] studied the effects of the geometrical ratio of the vent on the oscillatory regime using FLUENT software. His results indicated that the exchange flow and frequency increased when the aspect ratio decreased from 1.0 to 0.5, and as the vent ratio increased from 1.0 to 2.0, the exchange flow decreased and low frequency regime appeared.

The case examined herein is part of the study carried out at the French "Institut de Radioprotection et de Sûreté Nucléaire" (IRSN) in the context of a large research project program [9]. Specifically, this work focuses on the behaviour of the exchange flow at the horizontal vent observed in full-scale fire experiments. The aims of the present study is to evaluate the capability of the CALIF^{3}S-ISIS open-source CFD software developed by the IRSN to predict the bidirectional exchange flow through the vent and to investigate the flow inside the vent. The turbulence modelling used to simulate such flows relies on the LES technique. The simulated flows will be compared with the tendency of the function $Fr(L/D)$ of literature data, in which the exchange flow through the vent is a consequence of the density difference between the compartments. Finally, the experimental data obtained by Varrall [5] will be used for comparison with the numerical simulations based on the mean velocity field.

CASE STUDY

The case configuration (Figure 2a) is an experimental reduced scale set-up, which consists in two superposed compartments connected by a circular horizontal opening. The internal dimensions of the lower compartment are 1m x 1m x 1.5m, and the dimensions of the upper one are 1m x 1m x 1m. The length or height L of the partition between the compartments, which receive the vent, is 38mm thick. The steel framework allows different types of wall materials: glass, steel and insulating material. The fire source in the lower compartment was simulated experimentally by one electrical resistor of $2\,kW$ electrical power. A semi-elliptical fine wire mesh was located to reduce the ceiling jet effect on the flow through the horizontal opening produced by the thermal heat source.

The simulations presented hereafter were performed using the CALIF^{3}S-ISIS software (version 4.2) dedicated to the simulations of fire in confined and mechanically ventilated compartments [10]. The computational domain was based on the experimental set-up mentioned above. Four vent diameters were used: 127mm, 152mm, 191mm and 260mm. The buoyancy source was simulated by forcing a jet of hot air vertically upward into the lower compartment instead of hot plate as experimentally. The source was a 82 mm diameter nozzle located at 200 mm above the floor level. The hot air was injected at 0.43 m/s to maintain a Reynolds number close to 2000 at the injection. Two temperatures of hot air were tested: 363 K and 423 K to simulate the experimental vertical stratification temperature profile. The experimental vertical gradient profile was about $30°C/m$ at the steady state. Two wall boundary conditions were used: conduction and adiabatic walls. An exhaust on the opposite side of the nozzle at floor level was included to impose a mass flow conservation in the lower compartment, due to mass injection of the hot air by the nozzle. Also a screen to minimize the impact of the ceiling jet effect through the exchange flow was added to evaluate its influence over the exchange flow. Turbulence was modelled using the LES technique and WALE (Wall Adapting Local Eddy viscosity) subgrid-scale model. The Favre-filtered continuity, momentum and energy transport equations are solved. Time discretization is performed by a fractional step algorithm decoupling balance equations for the transport of energy and Navier-Stokes equations which are solved by a pressure correction technique. The horizontal opening was divided in three height levels: $z = 100.0\,cm$, $z = 101.9\,cm$ and $z = 103.8\,cm$, to investigate the flow behaviour inside the opening. The lowest slice coincides with the ceiling level of the lower compartment. Note that a convergence study has been performed in order to validate the refinement of the different meshes (M1: 1.2 million cells, M2: 1.9 million cells, M3: 2.7 million cells, M4: 3.5 million cells). Different vertical and horizontal grid spacings were tested (both in the rooms and within the vent). We considered convergence was reached as soon as the mean mass flow evolution at the vent was less than 5 % than the reference value obtained with the finest grid, M4. A structured Cartesian mesh was used with a mesh refinement applied at the vent, from 25 to 100 thousand cells. A schematic mesh of the configuration is shown in Figure 2b.

(a) Experimental set-up.

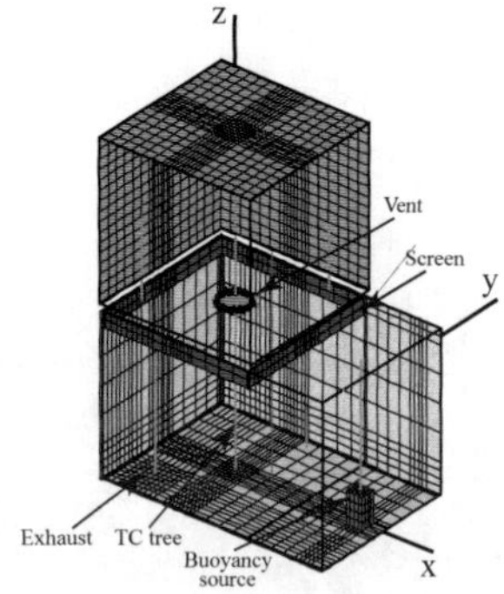

(b) Schematic mesh.

Fig. 2. Configuration of the experiment (a) and the corresponding mesh (b) for the simulation.

RESULTS

Features of the flow at the vent

The capability of the code to asses the bidirectional flow exchange at the vent is presented in Figure 3, where the net flow mass rate tends to zero after 300 *s* of simulation. Downward flow is plotted on negative values for visualisation purposes.

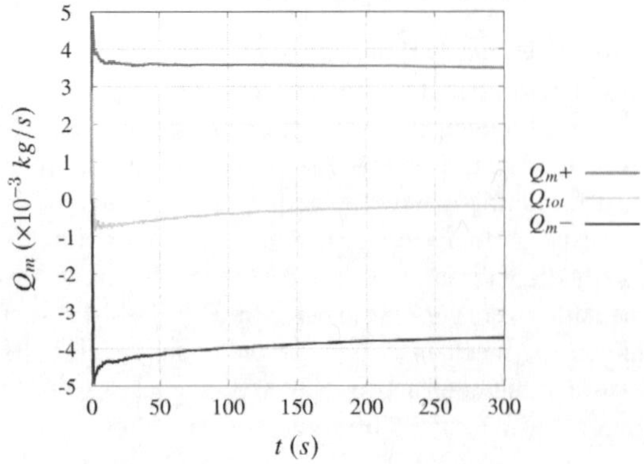

Fig. 3. Mean mass flow evolution for D = 260 mm at z = 101.9 cm.

Figure 4 presents instantaneous velocity fields at different simulation times. The positive values correspond to the upward flow and negative values to the downward flow. These figures show as observed experimentally by Varrall [5] the randomly distribution of the flow and there is no specific area for the upward or downward flow over time.

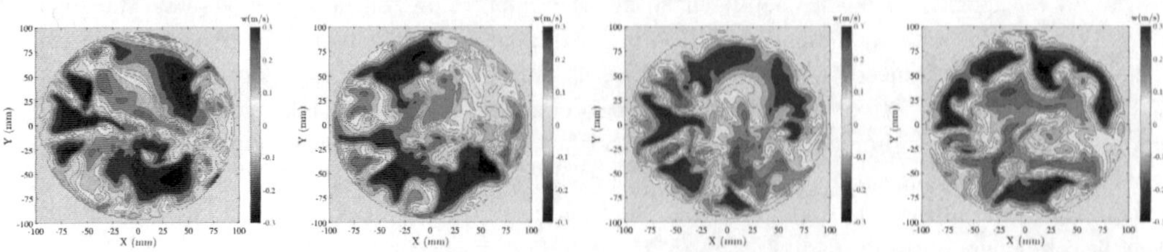

(a) t = 60 s (b) t = 120 s (c) t = 180 s (d) t = 300 s

Fig. 4. Instantaneous velocity field for D = 191 mm at z = 101.9 cm.

To understand the flow pattern on average over the simulations, Varrall's procedure [5] was used to determine the mean flow at the vent for the overall simulation time T. The simulated mean velocity fields for diameters D = 191mm (Figure 5a) and D = 260mm (Figure 5c) both at T = 300s were compared with their respective experimental results (Figure 5b and Figure 5d). The calculations of the average velocity field for both diameters showed an organization of the upward flow at the centre of the vent, while the downward flow was constrained to the periphery close to the edge of the vent with a shear zone between them, in which the mean velocity was null.

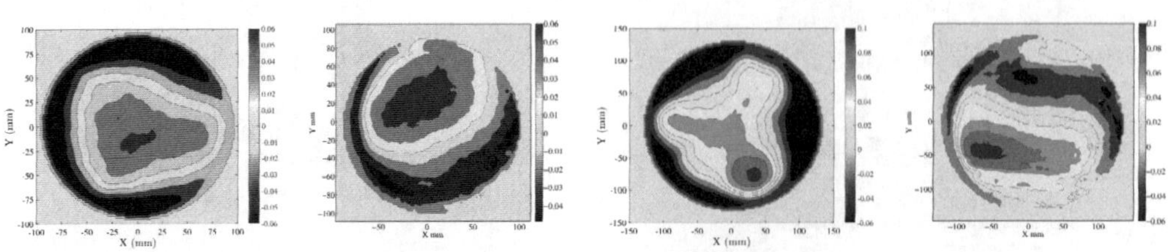

(a) Simulation (b) Varrall [5] (c) Simulation (d) Varrall [5]

Fig. 5. Mean velocity field comparative: (a) and (b) for D = 191mm, (c) and (d) for D = 260 mm at z = 103.8 cm.

Geometric ratio L/D effect and boundary conditions

It was observed that the Froude number Fr, expressed as Epstein [3], increases with the vent ratio for all the experiments simulated in the range of $0.1 \leq L/D \leq 0.5$. This is the case for the four geometric ratios simulated as shown in Figure 6a. This figure also presents a comparison between the simulations and some experimentally results available in the literature. The final cases simulated ($L/D = 0.15$ and $L/D = 0.20$) with higher refinement at the vent, use of the screen and conduction at walls, were closer to Epstein's correlation.

For the two wall boundary conditions used, the thermal stratification conditions observed experimentally in the lower compartment could not be maintained in steady condition, even by imposing a temperature profile at the start time in the lower room, as shown in Figure 6b. For the case using adiabatic walls, the vertical temperature profile became also quasi-constant for the entire height of the lower compartment by the end of the simulations.

The rise of the hot air injection temperature at the source between 363 K and 423 K increased about 5 K the temperatures of the lower room. Furthermore, the Froude number at the vent increased about 7% moving away from Epstein's correlation. For that reason, the lower injection temperature was retained.

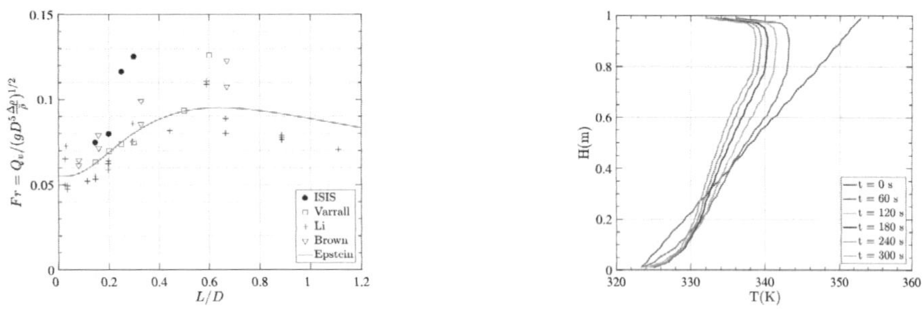

(a) (b)

Fig. 6. (a) Evolution of the experimental and numerical Froude number at the vent as a function of the ratio L/D. (b) Vertical temperature profile evolution for $D = 191\,mm$ ($L/D = 0.20$), conduction and $T_{inj} = 363\,K$.

Effect of vertical location within the vent

Figure 7 presents the influence of the horizontal slice height over the mean velocity field through the vent for two diameters and the same simulation time T. These velocity fields were not possible to obtain experimentally inside the vent because SPIV measurements were done at the top level of the vent. For both diameters, the lower level showed an opposite organization of the flow in comparison with the top level. At the lower level, the downward cold air occupied the central area while the upward hot air was observed close to the parietal zone. It has been observed thus an exchange of the flow pattern between the extreme levels of the vent. This behaviour of 'corkscrew' motion in vertical pipes have been observed in the core-annular flow of oil and water (see photos in Chap. VII [11]).

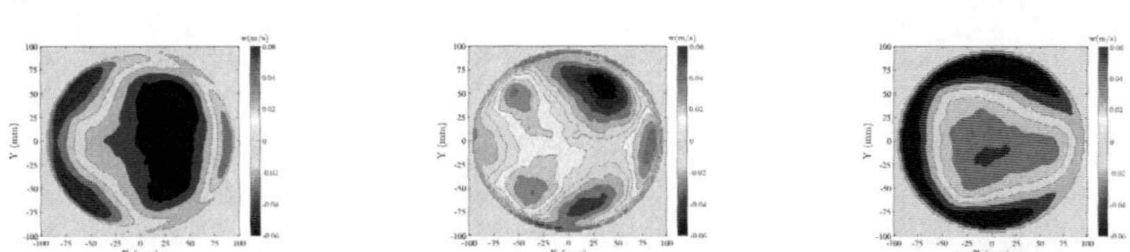

(a) $z = 100.0\,cm$ (b) $z = 101.9\,cm$ (c) $z = 103.8\,cm$

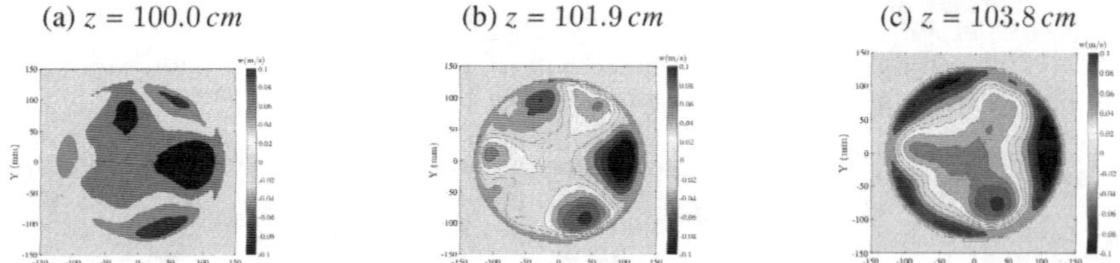

(d) $z = 100.0\,cm$ (e) $z = 101.9\,cm$ (f) $z = 103.8\,cm$

Fig. 7. Height influence for $D = 191\,mm$ (a,b and c) and for $D = 260\,mm$ (d,e and f), $T = 300\,s$.

IOP Conf. Series: Journal of Physics: Conf. Series **1107** (2018) 042019 doi:10.1088/1742-6596/1107/4/042019

CONCLUSION

In this study, numerical simulations (based on LES approach) of the flow at a vent connecting two superposed 'zones' of different density show the existence of a bidirectional flow, as observed experimentally. Four simulations were performed to examine the influence of the L/D vent ratio. It was observed that the increment of the aspect ratio L/D results in an increment of the Froude number. Two wall boundary conditions, conduction and adiabatic walls, and two hot air injection temperatures were employed in the simulations. A screen to limit the ceiling jet flow produced by the source was evaluated. Simulations with conduction at walls, a fine mesh at the vent and the use of the screen, showed results closer to Epstein's correlation. However, a discrepancy was observed with the model proposed by Epstein based on his experiments in natural convention. The disagreement between the simulation results and the experiments might be attributed principally to differences on the vertical temperature stratification of the lower compartment. Finally, an exchange of the flow organization between the upward and downward flow within the vent levels was observed, following a 'corkscrew' motion in the thickness of the horizontal opening.

REFERENCES

[1] Brown, W. G. (1962) Natural convection through rectangular openings in partitions – 2, *International Journal of Heat and Mass Transfer* 5: 869–878, DOI: 10.1016/0017-9310(62)90185-0.

[2] Mercer, A. and H. Thompson (1975) An experimental investigation of some further aspects of the buoyancy-driven exchange flow between carbon dioxide and air following a depressurization accident in a Magnox reactor, *Journal of the British Nuclear Energy Society* 14: 327–334.

[3] Epstein, M. (1988) Buoyancy-driven exchange flow through small openings in horizontal partitions, *Journal of Heat Transfer* 110: 885–893, DOI: 10.1115/1.3250589.

[4] Li, Z. (2007) "Characteristics of buoyancy driven natural ventilation through horizontal openings", PhD thesis, Aalborg University.

[5] Varrall, K. et al. (2016) Stereoscopic particle image velocimetry investigation of the bidirectional natural convection flow through a horizontal vent, *Fire Technology* 52: 2027–2041, DOI: 10.1007/s10694-016-0593-3.

[6] Spall, R.E. and E.A. Anderson (1999) A numerical study of buoyant, pulsating exchange flows through a vent in a thin horizontal partition, *Numerical Heat Transfer: Part A: Applications* 36: 263–272, DOI: 10.1080/104077899274750.

[7] Harrison, R.P. and R.E. Spall (2003) The effects of partition thickness on buoyant exchange flow through a horizontal opening, *Numerical Heat Transfer: Part A: Applications* 44: 451–462, DOI: 10.1080/713838255.

[8] Sleiti, A.K. (2008) Effect of vent aspect ratio on unsteady laminar buoyant flow through rectangular vents in large enclosures, *International Journal of Heat and Mass Transfer* 51: 4850–4861, DOI: 10.1016/j.ijheatmasstransfer.2008.02.027.

[9] Prétrel, H. et al. (2017) Experimental study based on large-scale smoke propagation fire tests through a horizontal opening connecting two mechanically ventilated compartments, *Fire Safety Journal* 90: 28–43, DOI: 10.1016/j.firesaf.2017.02.007.

[10] Suard, S. et al. (2011) Verification and validation of a CFD model for simulations of large-scale compartment fires, *Nuclear Engineering and Design* 241: 3645 –3657, DOI: 10.1016/j.nucengdes.2011.08.012.

[11] Joseph, D.D. and Y.Y. Renardy (1993) *Fundamentals of Two-Fluid Dynamics, Part II: Lubricated Transport, Drops and Miscible Liquids*, vol. 4, Springer Science & Business Media, DOI: 10.1007/978-1-4615-7061-5.

On the discrepancy of modelling the heat transfer for pure natural convection

Alexander BELT[1], Max BÖHLER[1], Leonie ROMMESWINKEL[2] and Lukas ARNOLD[1]

[1] *Civil Safety Research, Forschungszentrum Jülich, Germany*
[2] *Chemical Safety and Fire Defence, Bergische Universität Wuppertal, Germany*

a.belt@fz-juelich.de

ABSTRACT

In the present study, buoyancy driven flow fields in a bench scale apparatus are experimentally investigated by means of a non intrusive measurement technique, here the 2D-2C particle image velocimetry (PIV). An electrically heated copper block was used to induce a plume. As the temperature at the surface of the heating source is the main driver for the fluid dynamics, the surface temperature field was measured by thermography. The experimental results are compared to simulations with the Fire Dynamics Simulator (FDS). Simulations with a prescribed, homogeneous distributed surface heat flux and a 3D thermal heat conduction with a volumetric heat source were carried out. The impact of both approaches on the resulting flow fields showed no significant deviations, however, both do not agree with the experimental data. Whereas the surface temperatures of the heating source and its vicinity are in a good agreement for the 3D heat transfer approach. This leads to the conclusion that the modelled heat transfer mechanism does not provide valid predictions in this setup.

KEYWORDS

buoyancy driven flow, convective heat transfer, fire dynamics simulator (FDS), computational fluid dynamics (CFD), 3D thermal heat conduction, thermography, validation, particle image velocimetry (PIV), bench scale

IOP Conf. Series: Journal of Physics: Conf. Series **1107** (2018) 042020 doi:10.1088/1742-6596/1107/4/042020

INTRODUCTION

Convective heat transfer plays a major role in the cooling and heating of solids. In fires, the hot gases heat the ceiling and other structural parts of the building. If there is no forced convection, the dynamics of the flow is purely due to natural convection, which is driven by the heat flux in the near wall region, i.e. the boundary layer. As the boundary layer is in general not resolved in LES codes, like FDS [1], it needs to be modelled. The common approach is to use wall functions, which mimic the empirical correlations. In order to study the thermal convective flow, a basic experimental setup is chosen, which deliberately excludes phenomena like pyrolysis or combustion.

Meunders et al. performed a series of bench scale experimental investigations of buoyancy driven flows [2] in an enclosure. An electrically heated copper block was used as the heating source. The block induced turbulent or transitional plumes, depending on the heating power, and the velocity field was measured with the PIV technique. The comparison of the data from these validation experiments to the simulation results computed with FDS (version 6.2.0) showed deviations in the flow fields [3]. Initially, the deviations were suspected to be PIV measurement errors. The comparison of the PIV and additional laser Doppler velocimetry (LDV) results showed a good agreement and thus, an experimental error regarding the velocity measurements was excluded [4]. In the initial experimental setup, the block was just placed on the floor. Here, a significant influence of the geometry of the heating source on the turbulence induction, particularly at the upper edges, was observed [5]. To avoid the flow separation at the upper edges, the heating block was immersed in the floor.

For this new setup, preliminary FDS (version 6.5.1) simulations were run with a prescribed total heat flux. The results indicated the development of a plume with no indication of flow separation at the heating source and therefore the setup was physically implemented. Again, the focus was on the PIV measurements of the flow field and the thermal imagining of the surface of the heating source and its vicinity. However, the comparison of the experimental and the numerical velocity fields showed significant deviations. Furthermore, in contrast to the simulations the experimental plume was very unstable, which led to the assumption, that less heat is released at the surface than predicted by the model. This observation lead to the here presented experimental and numerical investigation of the surface temperatures.

In this paper we outline the comparison of the experimental results, i.e. velocity fields and surface temperatures, to FDS simulations. The error induced by the initially assumed boundary conditions in the preliminary simulations is discussed. Furthermore, the validation experiments and simulations for the 3D thermal heat conduction model in FDS are presented.

EXPERIMENTAL AND NUMERICAL SETUP

In figure 1 (left column) the technical drawing of the heating source and the surrounding insulation is presented. The copper block is heated by an electrical resistor, which is placed in a borehole in the symmetry axis. To avoid heat losses the block is insulated with a 20 mm thick calcium silicate plate, which has a low thermal conductivity. For mounting purposes the copper block and the insulation are pasted in a wooden shell. The material properties are listed in table 1. The heating source is mounted to a 760 mm × 10 mm × 760 mm stainless steel plate (X6CrNiMoTi17-12-2), which is the floor of the enclosure. For the PIV measurements the flow is to be seeded with particles. In this experimental setup DEHS particles were used, which is a synthetic oil, and are supposed to have good gas flow following properties [6]. The particles were generated by an atomizer, which is connected to an outlet at the floor plate of the enclosure. The floor plate has a second connection to the environment, which is used to remove the particles by a vacuum cleaner at the end of an experimental. During the run the connection is open, thus ambient pressure can be assumed in the enclosure. Both connections are located at the outer edge of the floor, so the influence on the flow field is assumed to be negligible. The experimental volume is enclosured by PMMA walls (width × height: 760 mm × 1440 mm). These walls are insulated with an extruded polystyrol (XPS) plate to reduce the impact of the environment on the boundary conditions.

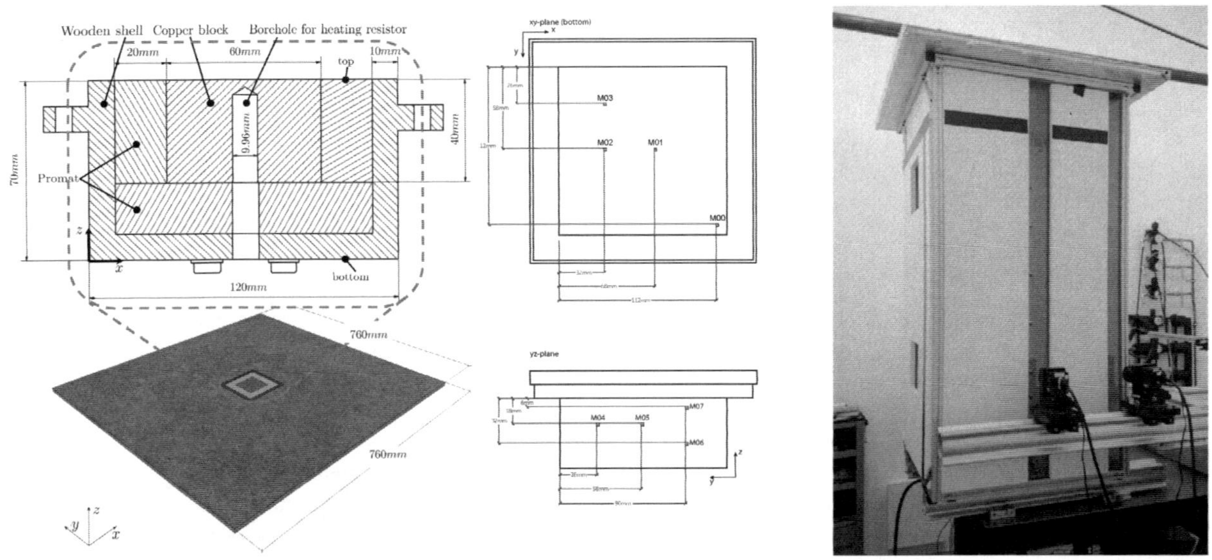

Figure 1: Technical drawing of the heating source (upper left), the floor plate of the experimental setup (bottom left), and the thermocouple instrumentation at the bottom and side of the wooden shell (middle column). A photo of the full experimental setup is on the right side.

One of the enclosure walls is equipped with optical openings for the IR-camera measurements. In the opposite side wall a small slot for the laser sheet inlet is added. Further openings in the insulation were created for the PIV cameras, which were mounted perpendicular to the light sheet. The top of the enclosure is designed as a plate heat exchanger, which is supplied by water at a constant temperature. So, the temperature at the ceiling is assumed to be constant. The full setup is depicted in figure 1 (right column).

Table 1: Material properties of the heating source and insulation.

	c_p / $\mathrm{kJ\,kg^{-1}\,K^{-1}}$	λ / $\mathrm{W\,m^{-1}\,K^{-1}}$	ρ / $\mathrm{kg\,m^{-3}}$
Copper	0.385	400	8920
Insulation (Promat)	0.95	0.083	450
Wood	1.7	0.3	1350
Stainless steel	0.5	15	8000

As the top of the enclosure is basically an aluminium plate, it could not be used as an optical access for the IR camera, the laser and the CCD cameras. Thus, the IR camera had to be mounted with an inclined view on the centred heating source. To keep the emissivity of the materials constant, chalk spray (= 0 9) was applied on the floor surfaces. The thermography's measurement uncertainty is stated with ±1 5 % or ±1 5 °C. The light sheet optics and the CCD camera of the PIV system were mounted on a traversing unit, so the velocity field could be observed in different heights. The measurement errors of the PIV measurements depend on various factors, that were studied by multiple authors [6, 7, 8]. Following these studies, we assume a measurement uncertainty of 6 % in the near field of the heating source, where the flow is assumed to be laminar.

Further temperature measurements, i.e. at one side of the wooden mounting device and the bottom of the wooden shell, have been instrumented with thermocouples (CO3-K). Four locations at the side and four at the bottom were selected. Figure 1 (middle column) shows the according positions (M00-M07). The thermocouples have a measurement uncertainty of ±0 75 % but at least ±2 2 °C. The sensor itself has a circumference of 0 25 mm, leading to a radius of 0 04 mm, which is the assumed positioning uncertainty.

Two kinds of FDS simulations were carried out: with a prescribed net heat flux at the surfaces and a 1D heat transfer in the solid (1DHT, full enclosure with the active cooling at the ceiling, 4 mm grid resolution), as well as with a prescribed volumetric heat source inside the copper block and a 3D heat transfer using FDS version 6.6.0 (3DHT, vicinity of the heat source, no enclosure, 2 mm grid resolution). The choice of these grid resolutions is based on sensitivity studies.

Regarding the 1DHT case, only a part of total heating power contributes to the heat flux through the floor into the enclosure. To quantify this fraction, detailed heat conduction simulations in ANSYS CFX (Version 18.1) have been computed. E.g. at an input power of $P_{el} = 75\ 4$ W the net heat flux is at least 60 % of the total electrical power.

In the 3DHT case, the modelling of the electrical resistor is based on a volumetric heat source. For this purpose, a centrally located region, extension of 4 mm × 4 mm × 38 mm, is defined inside the block. It delivers the previously defined heating power directly to the copper block. The dimensions of the thermal insulation as well as the wooden shell are identical to the experimental setup, as shown in figure 1 (left column). The material properties are set according to table 1. Although all relevant solid structures in the floor are modelled, the gas phase is modelled only up to a height of 70 mm.

COMPARISON OF EXPERIMENTAL AND SIMULATION DATA

The experiments have been performed within a heating power range of 1 4 W to 75 4 W, but in this contribution only a part of the experiments is presented.

The experimental and modelled velocity fields are compared along horizontal lines, cutting the plume, at three selected heights and for an experiment with an input power of $P_{el} = 75\ 4$ W. In figure 2a the profiles of the averaged velocity component in the z-direction (vertical) are shown, where the simulation data is based on computations with the 1D heat transfer model (solid line) as well as with the 3D (dashed line) one, but in a reduced computational domain. The data shows, that inside the simulated plume the peak velocity is at least twice as high as observed in the experiments (dotted line). As the convective heat transfer is the driver for the gas dynamics, this suggests, that the convective heat transfer in the simulation is overestimated. Therefore, the thermal boundary conditions, i.e. surface temperatures, are investigated more in detail.

Motivated by the discrepancy in the flow field, the thermocouple and thermographic measurements are compared to FDS simulations, where the 3D thermal heat conduction is taken into account. The block's top and its vicinity temperature are compared to the simulation results in figure 2b. Figure 3 compares the localised thermocouple measurements at the side and bottom of the block for three input power settings. Taking the measurement uncertainty into account, the modelling results are in good agreement with the experimental data. This observation leads to the conclusion, that the 3D thermal heat conduction model provides valid results in this setup. The small deviations are due to measurement uncertainties as well as the uncertainty of the exact material properties. However, small temperature differences at the solid surface may lead to significant changes in the gas phase temperature, as the ratio of the heat capacities of the solid and the gas is large. Which in turn results in a different gas dynamics.

CONCLUSION AND OUTLOOK

The temperature measurements at the surfaces of the copper block are in a good agreement with the simulations in FDS, taking the 3D thermal heat conduction model into account. This is also the case for the not presented power inputs. However, the comparison of the flow field shows significant deviations. As the convective heat transfer at the surface of the copper block is the main driver for the fluid dynamics, it may lead to the conclusion, that the modelling of the near wall phenomena by means of standard temperature and velocity wall functions may not be valid here. Hölling already showed, that the standard wall functions are not valid for natural convection at a vertical heated wall [9]. Experiments

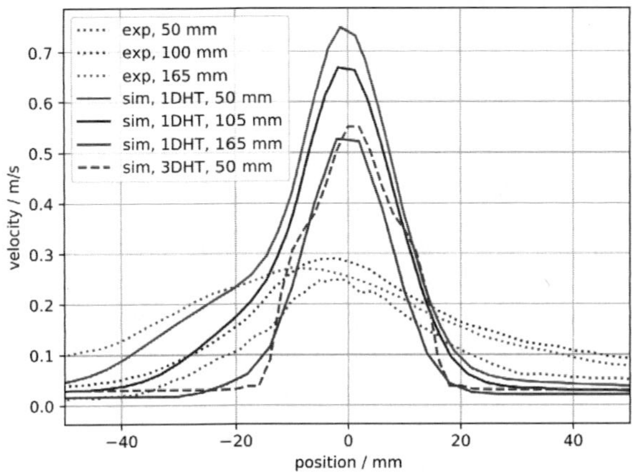

(a) Vertical velocity profiles through the plume at three heights: 50 mm, 105 mm, and 165 mm. Shown are experimental and simulation (1D and 3D heat transfer) results with an input power of $P_{el} = 75.4\,\mathrm{W}$.

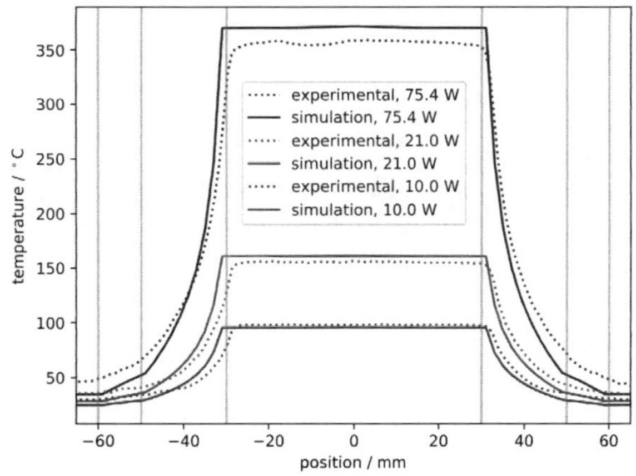

(b) Temperature profiles along the x-axis on the surface of the heating source for various heating powers. The vertical lines indicate the boundaries between the copper block ($|x| < 30\,\mathrm{mm}$), insulation ($30\,\mathrm{mm} < |x| < 50\,\mathrm{mm}$), wooden shell ($50\,\mathrm{mm} < |x| < 60\,\mathrm{mm}$) and the steel plate ($60\,\mathrm{mm} < |x|$).

Figure 2: Comparison of experimental and simulation results.

by Hundhausen et al. determined increasing deviations of the dimensionless velocity profile for mixed convection (buoyancy aided) conditions from the standard profile with increasing buoyancy effects, i.e. increasing Grashof number [10]. In the presented experimental setup, pure natural convection at a horizontal flat plate occurs. But the standard wall functions were developed for forced. Considering the transport equations, the main drivers for the convective heat transfer are the temperature and velocity gradients in the viscous sub-layer ($0 < y^+ < 5$). With increasing gradients the convective heat transfer rises. For natural convection the velocity gradient $\partial u/\partial y$ is lower than for forced conditions. Furthermore, the temperature profile is coupled to the velocity profile. Thus, the heat transfer for pure natural convection is overestimated, which might be an explanation for the deviation in the observed velocity fields. Considering the Nusselt correlation for a flat plate confirms the assumption, that FDS overestimates the convective heat transfer. The transfer coefficient h is computed in LES mode as a combination of natural and forced convection correlations [1]

$$h = \max\left(C|T_g - T_w|^{\frac{1}{3}}, \frac{k}{L}\mathrm{Nu}\right),$$

with the gas T_g and wall T_w temperature, a constant C [11], the gas thermal conductivity k, a characteristic length scale L and the Nusselt number Nu.

In the next steps, the validity of the above stated correlation in the presented setup will be investigated. Additionally, a direct numerical simulation (DNS) for a detailed analyses of the near wall phenomena at the heating source will be carried out and compared with the LES simulations. Furthermore, a sensitivity analyses regarding the influence of minor deviation of surface temperature on the quantities of the velocity field is planned.

OPEN ACCESS DATA

The experimental and simulation data presented in this contribution is publicly available [12].

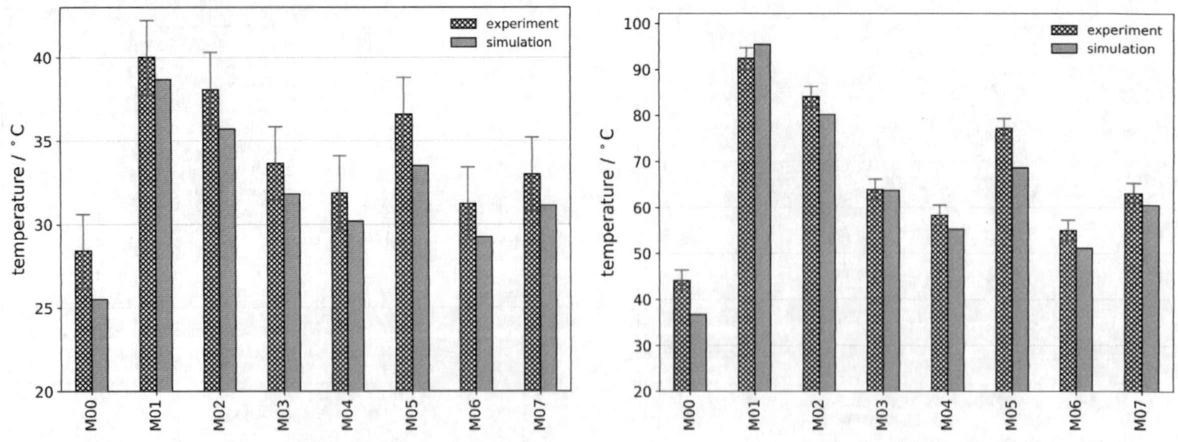

(a) $P_{el} = 10\,\mathrm{W}$ (b) $P_{el} = 75.4\,\mathrm{W}$

Figure 3: Thermocouple measurements at the surfaces of the wooden shell (bottom, side). The measurement uncertainties are indicated by the red error bars.

REFERENCES

[1] K. McGrattan et al., *Fire Dynamics Simulator - User's Guide*, Technical Report, National Institute of Standards and Technology, http://dx.doi.org/10.6028/NIST.SP.1019, 2017.

[2] A. Meunders et al., *Velocity measurement of a bench scale buoyant plume applying particle image velocimetry*, International Journal of Heat and Mass Transfer, 123C (2018) pp. 473-488

[3] A. Meunders, *A study on buoyancy-driven flows: Using particle image velocimetry for validating the Fire Dynamics Simulator*, PhD Thesis, Bergische Universität Wuppertal, Germany, 2016.

[4] A. Belt et al., *Temperature and laser Doppler velocity measurements of a spill plume in a small scale experiment with an electrical heat source*, IFireSS, Napels, Italy, 2017.

[5] A. Belt et al., *Temperature and velocity field measurements for a pure buoyancy driven plume in an enclosure*, Fire and Materials, San Francisco, USA, 2017.

[6] M. Raffel et al., *Particle Image Velocimetry - A Practical Guide - second edition*, Springer Verlag, 978-3-540-72307-3, Berlin, 2007.

[7] B. Wienecke, *PIV uncertainty quantification from correlation statistics*, Journal, Measurement Science and Technology, IOP Publishing Ltd, 2015.

[8] B. Simon, *Untersuchungen zur Wasserstoffrekombination unter Naturkonvektionsbedingungen*, PhD Thesis, Rheinisch-Westfälischen Technischen Hochschule Aachen, Germany, 2015.

[9] M. Hölling, H. Herwig *Asymptotic analysis of the near-wall region of turbulent natural convection flows*, Journal of Fluid Mechanics, 541, 383-397, doi:10.1017/S0022112005006300, 2005.

[10] A. Hundhausen et al., *CFD-grade measurements in a condensing boundary layer - results of the SETCOM facility*, NURETH, Xi'an, China, 2017.

[11] J.P. Holman, *Heat Transfer*, McGraw-Hill, New York, 7th edition, 1990

[12] A. Belt et al., *Dataset*, https://doi.org/10.5281/zenodo.1257565

Experimental parametric study on low-frequency oscillating behaviour of pool fires in a small-scale mechanically-ventilated compartment

Maxime MENSE[1], Yannick PIZZO[2], Hugues PRETREL[1], Jean-Claude LORAUD[2] and Bernard PORTERIE[2]

[1]*Institut de Radioprotection et de Sûreté Nucléaire (IRSN), PSN-RES, SA2I, Laboratoire commun ETiC, Cadarache, France*
[2]*Aix-Marseille Université, CNRS, IUSTI UMR 7343, Laboratoire commun ETiC, France*

maxime.mense@irsn.fr

ABSTRACT

The purpose of this work is to study the unstable oscillatory behaviour, with frequency in the order of few mHz, that has been occasionally observed in mechanically-ventilated compartment fires. To address this challenge, a series of experiments using a small-scale compartment was conducted using heptane and dodecane as fuels. Results show that, after the fire is fully-developed, unstable and stable combustion regimes can occur depending on the fuel type, the pool size and the air renewal rate of the compartment (ARR). A special regime of unstable oscillatory combustion with low frequencies (LF), accompanied by thermodynamic pressure and ventilation flow rate variations and displacement of the flame outside the pan, is observed. The occurrence and persistency of LF oscillations results from the competition between oxygen supply and fuel vapor supply due to the heat feedback from the flame and enclosure to the fuel tray. The LF oscillation period depends mainly on time required for storing enough fuel vapor for strong combustion. Whatever the fuel type, it is found that i) the range of ARR where LF oscillations appear and the oscillation amplitude increase with the pool size, and ii) the frequency increases, while amplitude decreases, with increasing ARR independently of the pool size. It is also found that the more flammable the fuel, i) the smaller pool size for which LF oscillations appear and the higher the frequency for the same ventilation conditions, and ii) the wider the range of ARR where LF oscillations appear for a given pool size.

KEYWORDS:

compartment fire; heat transfer; ventilation; low-frequency oscillations; unstable combustion; combustion regimes.

IOP Conf. Series: Journal of Physics: Conf. Series **1107** (2018) 042021 doi:10.1088/1742-6596/1107/4/042021

INTRODUCTION

The rate at which a compartment fire develops depends strongly on ventilation conditions (available oxygen). Most studies have focused on the effect of natural ventilation on compartment fire behaviour and fire extinction (see [1]-[6], to name but a few). For restricted air ventilation, unsteady under-ventilated fires can exhibit periodic oscillations and temporary flame quenching (e.g. [6]). In [4], Takeda conducted small-scale fire experiments, using poly-methyl-methacrylate as fuel. He showed that air vitiation results in a reduction of the fuel mass loss rate (MLR), with oscillations of ~1.33 Hz. Utiskul et al. [3] and He et al. [5] observed a similar flame oscillatory phenomenon, with different frequencies of around 1 Hz and 0.1-0.2 Hz, respectively. From small-scale experiments, Kim et al. [7] studied the effect of natural ventilation and fuel surface area on fire behaviour, using methanol as fuel. Oscillating combustion was observed before extinction, near the boundary between stable (beyond certain ventilation condition, combustion continues until the fuel was exhausted) and unstable (the balance between fuel vaporization and oxygen supply is broken, the fully-developed fire is not sustained, and extinction occurs) combustion regions, with frequency in the range of 0.02-0.2 Hz.
For fires in enclosures equipped with mechanical ventilation, such as in nuclear facilities, unstable low-frequency modes can be encountered, leading to pressure variations that can affect the confinement levels and hence the safety of the installation. Such a behaviour has been observed by Pretrel et al. [8] from experiments on heptane pool fires in the large-scale IRSN DIVA facility. Oscillations with frequency less than 10 mHz were recorded for a fire size (peak of heat release rate) of 0.5 MW fire and air renewal rate (ARR) from 12 to $17h^{-1}$. During this oscillatory phase, a drastic change in flame behaviour was observed: the flame emission intensity decreased, and a weakly blue flame left the pool surface and moved slowly through the compartment, before coming back above the pan. This phenomenon has been also observed in naturally-ventilated compartment fires (e.g. [2],[5], [8]-[12]). To improve the understanding of the unstable low-frequency behaviour of mechanically-ventilated fires, experiments using a 1:4 scale reproduction of DIVA were carried out, varying the type of fuel, the pool size and the ARR.

EXPERIMENTS

Fire experiments were conducted in a 1.5m×1.25m×1m mechanically-ventilated compartment, called NYX (Fig. 1). This compartment is a 1:4 scale reproduction of a room of the DIVA facility of IRSN [2]. Froude similarity criterion was used for scaling gas flow rates between DIVA and NYX. The walls E, S, N, B and F (see Fig. 1) are composed of two layers, a 0.002 m-thick steel layer and a 0.045 m-thick calcium silicate layer. The west wall is made from heat-tempered glass allowing visualization of the fire growth using a CCD camera positioned outside the compartment. Unlike the DIVA facility which is fully mechanically-ventilated, the inlet air flow is free whereas hot gases are extracted mechanically. The air renewal rate (hereafter labelled ARR in h^{-1}) is defined as the ratio between the inlet air flow rate before ignition and the compartment volume. A vertical rake of 5 thermocouples (K type, 0.5 mm) was placed in each corner of the enclosure to measure gas temperatures. Molar fractions of O_2, CO_2 and CO molar fractions were measured by probes inserted into the exhaust branch. McCaffrey probes [13] were used to evaluate gas velocity inside the ventilation ducts to determine the inlet air flow and gas extraction rates. The pressure difference between the gases in the compartment and the air outside the compartment was measured using a differential pressure transducer (Emerson ROSEMOUNT). All measurements were recorded at a frequency of 1 Hz.

Fig. 1: Experimental set-up NYX. The letters W, E, S, N, and B refer to west, east, south, north and back faces of the compartment, respectively.

A circular pan was filled with fuel up to a height of 0.043 m whatever the pool size and fuel type. Ignition was provided by a propane burner. An electronic balance was placed at the centre of the floor to measure the mass of the fuel over time. A series of 31 experiments was performed using two types of liquid fuels, namely heptane and dodecane, and varying the pool size and ARR (Table 1). Although heptane and dodecane have similar heats of vaporization at 25°C (365 vs. 293 kJ/kg), heats of combustion (45.0 vs. 44.6 kJ/kg), and flammability limits (1-7 vs. 0.6-4.7 % in vol.), they strongly differ in boiling temperature (98 vs. 216 °C) and flashpoint (-4 vs. 72°C) [14]. Therefore, dodecane is less volatile and less flammable than heptane, leading to discrepancies in fire behaviour.

RESULTS AND DISCUSSION

LF Oscillatory behaviour

Time evolution of the mass loss rate (MLR) over several cycles of LF oscillations for the 18-cm diameter heptane pool fire, with a renewal rate of 12.5 h^{-1}, is shown in Figure 2, along with instantaneous video captured frames. The analysis reveals a strong correlation between the burning rate oscillations and those in the fuel burning area. When the flame is positioned vertically above the pan, the heat feedback from the flame to the fuel surface increases. The combustion heat release induces a rapid pressure and temperature rise due to the increasing burning rate. This leads to a reduction in the air inlet flow rate. The oxygen supply is not large enough to sustain a strong combustion. The flame is lifted and leaves the pool surface or occupies only a part of it (the flame changes in appearance, intensity and colour, from yellow to bluish-yellow). The heat feedback, and thus the MLR, decrease, which is followed by a reduction in pressure and temperature in the compartment. The admission flow rate increases, providing more oxygen to support combustion. The flame recovers its position above the pan. For this case study, the frequency of oscillations is approximately 18 mHz. Due to the configuration of ventilation, it is worth noting that the flame migrates to the F-E corner when it leaves the pool surface. Although not shown here, an inversion of ventilation causes the flame to move to the B-W corner. As demonstrated by Thomas et al [15], thermal feedback from the enclosure can also significantly contribute to the fuel gasification rate and thus to the oscillating fire behaviour. From fire experiments under natural ventilation conditions, Kim et al. [7] showed that the gas temperature in the compartment became higher as the oscillation repeated, increasing the temperature of the fuel tray and thus the emission of fuel vapor. The air supply was not large enough to balance the fuel vapor supply and extinction occurred.

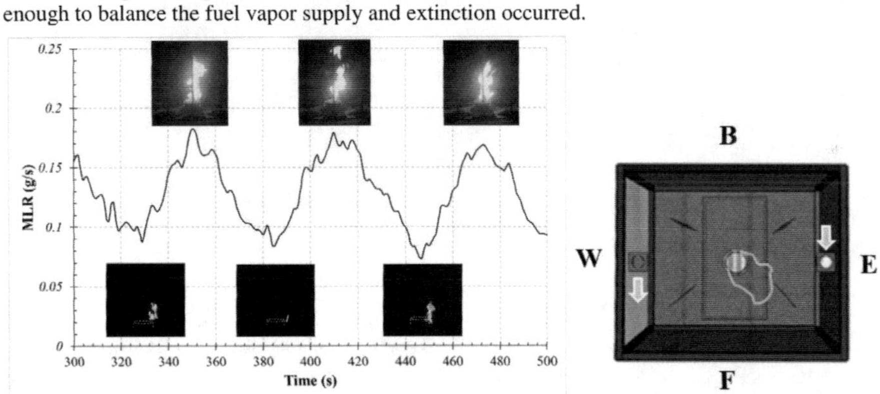

Figure 2: Left: MLR vs. time during the LF oscillatory phase for the 18-cm-diameter heptane pool fire with a renewal rate of 12.5 h^{-1}, and video captured frames after 329, 351, 385, 410, 447 and 473 s of fire; right: average position of the flame when it leaves the pool surface.

Regimes of combustion

Fig. 3 shows the time evolution of MLR for heptane and dodecane pool fires under forced ventilation conditions with different rates. In the first minutes, fire develops independently of the ARR (fuel-controlled fire). Subsequently, there are four distinct regimes of burning behaviour observed in the present study, depending on the fuel type, pool size and ARR. Three of them are unstable (extinction occurs before the fuel supply is completely exhausted) and one is stable (no extinction until burnout). Regime (1) corresponds to unstable combustion followed by rapid extinction due to filling. The compartment is filled nearly completely with smoke and the air renewal rate is not large enough to sustain a sufficient oxygen concentration (e.g. the red curve in Fig. 3b or the grey curves in Fig. 3e and Fig. 3f). In regime (2), where the fire is weakly ventilated, the flame stands upright, but it does not always occupy the entire fuel surface, which explains the decrease of MLR. A precarious balance is reached between air supply and fuel vaporization that can be broken at any time. The heat feedback from the flame and enclosure to the fuel tray is not sufficient to sustain combustion (not shown, gas temperatures remained relatively low and nearly constant throughout the test). Extinction occurs. This regime is observed for small pool diameters and ARRs (e.g. the orange and blue curves in Fig. 3a or the yellow curve in Fig. 3e). When the heat release rate of fire increases, the heat feedback is strong enough to sustain fuel evaporation. Unstable combustion, accompanied by low-frequency (LF) oscillations, takes place, which corresponds to regime (3). The oscillation period depends on time required for storing enough fuel vapor for strong combustion [7]. In this regime, extinction occurs due to an enhancement of the MLR resulting either from an increase of the heat feedback (e.g. the yellow curve in Fig. 3b) or from an increase of the heat transfer through the rim of the container, making easier the vaporization of the small quantity of liquid fuel remaining in the pan (e.g. the green, blue, red and black curves in Fig. 3c). Although not shown, a significant increase in gas temperature is observed over time. Regime (4) corresponds to stable combustion. This regime, marked by nearly steady burning, occurs for small pool sizes (e.g. the red, green and yellow curves in Fig. 3a or the black curve in Fig. 3e).

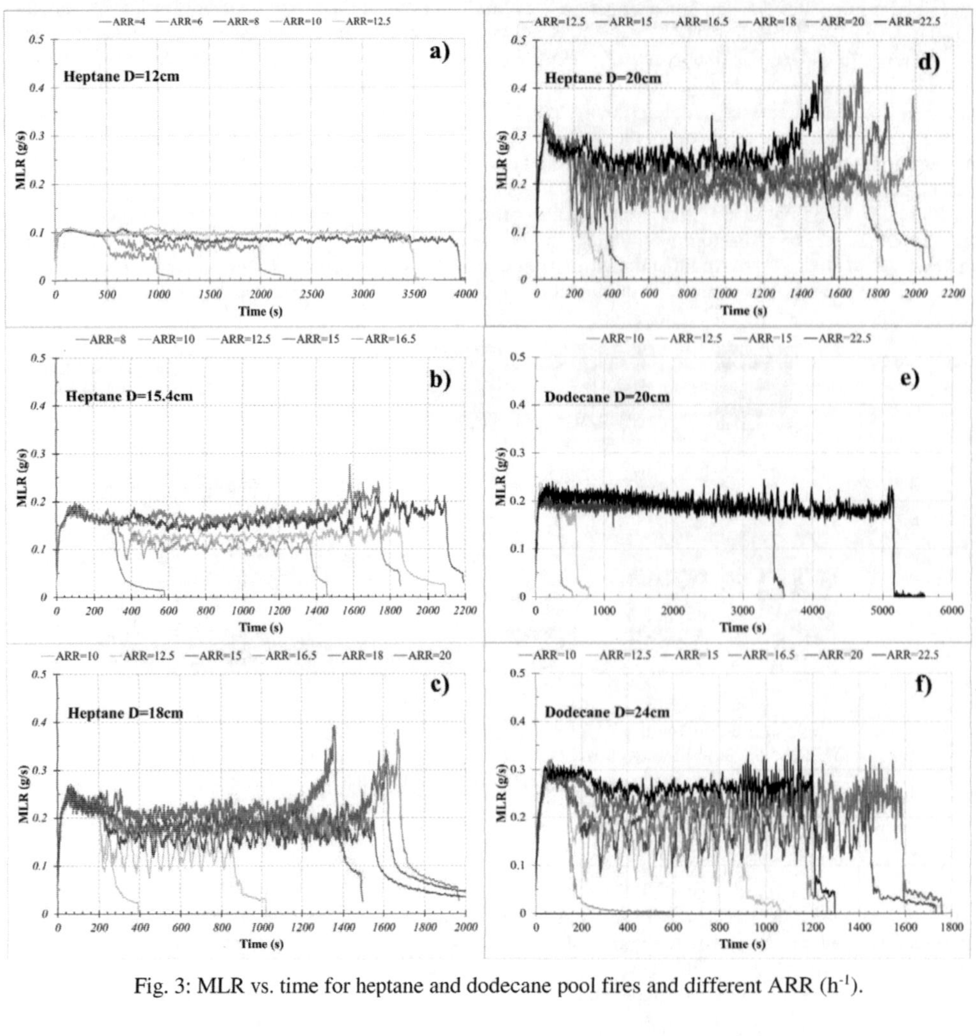

Fig. 3: MLR vs. time for heptane and dodecane pool fires and different ARR (h^{-1}).

Table 1 summarizes the results obtained from the experiments. The frequency of LF oscillations is calculated using a Fast Fourier transform (FFT).

For both fuels, some conclusions can be drawn. The frequency of LF oscillations increases, while amplitude decreases, with increasing ARR independently of the pool size (Table 1), as found by Pretrel et al. for large-scale experiments [8]. It is worth noting that the amplitude of LF oscillations increases with the pool size, as shown in Fig. 4 where the time evolution of MLR are plotted for heptane pool fires with ARR=12.5h^{-1}. Moreover, the range of ARR where LF oscillations appear and the oscillation amplitude increase with the pool size.

LF oscillation regime is strongly correlated to the flammability properties of the fuel used. For the same ventilation conditions, the more flammable the fuel, the smaller pool size LF oscillations appear and the higher the frequency. For example, for ARR=12.5h^{-1}, heptane pool fires exhibit LF oscillations (Regime (3)) from a diameter of 15.4 cm, whereas a larger diameter, i.e. 24 cm, is required for dodecane. Moreover, for the same pool size, the more flammable the fuel, the wider the range of ARR LF oscillations occurred.

All these results must still be confirmed by further experiments.

Table 1: Combustion regimes for heptane and dodecane compartment fires. D is the pool diameter. sf➔mf corresponds to a transition from single-frequency to multiple-frequencies oscillations during the test.

Fuel	D (cm)	ARR (h^{-1}) 4	6	8	10	12.5	15	16.5	18	20	22.5
Heptane	12	2	2	4	4	4	-	-	-	-	-
	15.4	-	-	1	3 sf=12.4mHz ➔mf	3 sf=16.5mHz ➔mf	3 mf	3 mf	-	-	-
	18	-	-	-	3 sf=16.1mHz	3 sf=18.1mHz	3 sf=22.4mHz ➔mf	3 sf=23.3mHz ➔mf	3 mf	3 mf	-
	20	-	-	-	-	3 sf=18.0mHz	3 sf=20.9mHz	3 sf=22.8mHz ➔mf	3 sf=26.0mHz ➔mf	3 sf=33.3mHz ➔mf	3 mf
Dodecane	20	-	-	-	1	2	3 mf	-	-	-	4
	24	-	-	-	1	3 sf=13.7mHz	3 sf=16.3mHz	3 sf=18.8mHz	-	3 mf	3 mf

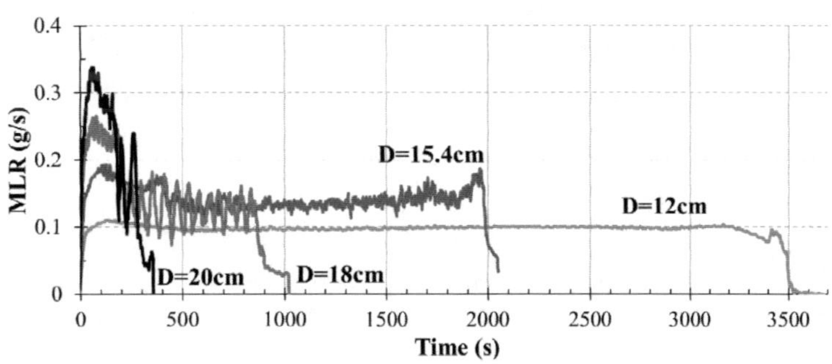

Fig. 4: MLR vs. time for different heptane pool sizes. Here, ARR=12.5 h^{-1}.

CONCLUSION

Small-scale experiments were conducted to investigate the low-frequency oscillating regime of hydrocarbon pool fires under forced ventilation conditions. This regime results from the competition between oxygen supply and fuel vapor supply due to the heat feedback from the flame and enclosure to the fuel tray. The influence of fuel type, pool size and air renewal rate on the occurrence and persistency of low-frequency oscillations is studied. However, further experiments are still required to complete the diagram of combustion regimes. It is found that, for both fuel types (heptane and dodecane): first, the range of ARR where the oscillatory regime appears and the oscillation amplitude increase with the pool size; and second, the frequency increases with the ARR, independently of the pool size, while the amplitude decreases. Results also show that the oscillatory regime is strongly dependent on the flammability properties of the fuel used, heptane fuel being more flammable than dodecane fuel. For a given ARR, the more flammable the fuel, the smaller pool size where LF oscillations appear and the higher the frequency. Moreover, for a given pool diameter, the range of ARR where LF oscillations appear is wider for heptane fires than for dodecane fires.

ACKNOWLEDGEMENTS

The authors are grateful to Provence-Alpes-Côte d'Azur for its financial support in the context of this thesis.

REFERENCES

[1] Peatross, M.J., Beyler, C.L., "Ventilation effects on compartment fire characterization," *Fire Safety Science -- Proceedings of the Fifth International Symposium*, International Association for Fire Safety Science, 1997, 403-414.

[2] Audouin, L., Rigollet, L., Prétrel, H., Le Saux, W., Röwekamp, M. (2013) OECD PRISME project, fires in confined and ventilated nuclear-type multi-compartments, *Fire Safety Journal* 62:80-101.

[3] Utiskul, Y., Quintiere, J.G., "Generalizations on compartment fires from small-scale experiments for low ventilations conditions," *Fire Safety Science -- Proceedings of the Eighth International Symposium*, International Association for Fire Safety Science, 2005, 1229-1240.

[4] Takeda, H. (1985) Oscillatory phenomenon and inverse temperature profile appearing in compartment fires, *Combustion and Flame* 61:103-105.

[5] He, Q., Li, C., Lu, S., Huang, S. (2014) Experimental study of pool fire burning behaviours in ceiling vented ship cabins, *Procedia Engineering* 71:462-469.

[6] Hu, Z., Utiskul, Y., Quintiere, J.G., Trouvé, A., "A comparison between observed and simulated flame structures in poorly ventilated compartment fires," *Fire Safety Science -- Proceedings of the Eighth International Symposium*, International Association for Fire Safety Science, 2005, 1193-1204.

[7] Kim, K., Ohtani, H., Uehara, Y. (1993) Experimental study on oscillating behaviour in a small-scale compartment, *Fire Safety Journal* 20:377-384.

[8] Prétrel, H., Suard, S., Audouin, L. (2016) Experimental and numerical study of low frequency oscillatory behaviour of large-scale hydrocarbon pool fire in a mechanically ventilated compartment, *Fire Safety Journal* 83:38-53.

[9] Sugawa, O., Kawagoe, K., Oka, Y., Ogahara, I. (1989) Burning behaviour in a poorly ventilated compartment fire – ghosting fire, *Fire Science and Technology* 9:5-15.

[10] Wakatsuki, K., Low ventilation small-scale compartment fire phenomena: ceiling vents, *MS Thesis*, Department of Fire Protection Engineering, University of Maryland, College Park, 2001.

[11] Bertin, G., Most, J.M., Coutin, M., Wall fire behaviour in an under-ventilated room, *Fire Safety Journal*, 2002.

[12] Matsuyama, K., Okinaga, S., Hattori, Y., Suto, H., Experimental study on fire behaviour in a compartment under mechanical ventilated conditions: the effects of air inlet position, *Fire Science and Technology*, 11-119, 2015.

[13] McCaffrey, B.J., Heskestad, G., A robust bidirectional low-velocity probe for flame and fire application, *Combustion and Flame*, Volume 26, 125-127, 1976.

[14] *SFPE Handbook of Fire Protection Engineering (3rd ed)*, DiNenno P.J (ed), National Fire Protection Association, Quincy, MA 02269, 2002.

[15] Thomas, P.H., Bullen, M.L., Quintiere, J.G., McCaffrey, B.J., Flashover and instabilities in fire behaviour, *Combustion and Flame*, Volume 38, 159-171, 1980.

Impact of the room geometry on the smoke filling time due to a fire plume

Samia HAOUARI HARRAK[1 2], El Mehdi KOUTAIBA[2], Rabah MEHADDI[1], Pascal BOULET[1] and Simon BECKER[1]

[1]*LEMTA, Université de Lorraine, CNRS, France*
[2]*Centre Scientifique et Technique du Bâtiment (CSTB), France*

samia.haouari-harrak@univ-lorraine.fr

ABSTRACT

In this paper, the influence of a room geometry on the smoke filling time with a fire plume is studied by means of full-scale simulations with the CFD model Fire Dynamics Simulator (FDS) and two-zone models. The results show that the room aspect ratio has little impact on the smoke filling times in the case of small surface rooms. Consequently, the zone models can be used to predict the evacuation time. In the case of large surface rooms, such as atria and hangars, the dynamic of the ceiling jet when it hits the side walls cannot be neglected. In that case, the zone models overestimate the smoke filling time which directly impacts the evacuation time.

KEYWORDS:

Compartment fires; smoke filling time; CFD model; zone models.

INTRODUCTION

Ensuring safe evacuation of people during fires is one of the main objectives of fire safety. To evaluate the Available Safe Evacuation Time (ASET), zone models are still used in fire safety engineering, mainly due to the short simulation time compared to CFD models. In compartment fire, the ASET is strongly affected by the smoke filling time. It is therefore crucial to correctly estimate it.

After the ignition of a fire, the hot gases produced by the combustion generate a rising plume that entrains the surrounding fresh air. When the plume hits the ceiling of the enclosure, it spreads radially. Then, a smoke layer forms and thickens with time due to the mass flow supplied by the plume. This filling process was first studied by Baines and Turner [1]. They developed a filling box model to predict the thickness of the stratified layer resulting from a point source of buoyancy in a confined environment. In their model, the dynamics of the ceiling jet and the outflow when it hits the side walls are neglected. They have assumed that when the plume hits the ceiling, at the initial time (t = 0 s), a layer of infinitesimal thickness forms. Then it gradually fills the enclosure due to the volume supplied by the plume.

By balancing the rate at which the buoyant fluid is supplied by the plume with the rate at which the upper buoyant layer deepens, Baines and Turner [1] obtained the following relation describing the position of the horizontal interface between the ambient and buoyant layers Z_{int}:

$$\frac{Z_{int}}{H} = \left(\frac{2}{3}\frac{t}{t_f} + 1\right)^{\frac{-3}{2}} \tag{1}$$

where $t_f = \frac{S}{C\, H^{\frac{2}{3}}\, B^{\frac{1}{3}}}$ is the filling box time scale, *S* is the surface area of the room, *H* is the height of the room, *C* is a constant which depends on the plume entrainment coefficient, and *B* is the buoyancy flux.

This approach is used in many zone models (see for instance [2]) to predict the smoke filling time. Actually, the fire plume is modeled using fire plume correlations such as Heskestad's model [3], Zukoski's model [4] and McCaffrey's model [5]. All these fire plume models are commonly based on the seminal work by Morton, Taylor and Turner [6]. To validate their model, Baines and Turner [1] performed experiments involving small scale salt plume. For rooms with aspect ratios $\Phi = R/H$ less than 1 (*R* and *H* being respectively half the width and height of the room), they observed that the dynamics of the ceiling jet resulting from the fire plume have a noticeable effect on the formation of the buoyant layer that cannot be neglected.

Following this work, several studies have been carried out to determine the filling process for room aspect ratios less than 1. Recently, Kaye and Hunt [7] [8] identified two sidewall flow patterns, depending on the room aspect ratio $\Phi = R/H$. They showed that for tall enclosures ($\Phi < 2/3$), when the ceiling flow hits the side-walls, it rolls over entraining an important amount of the ambient fluid into the buoyant layer (see Fig. 1-a). Indeed, this «Rolling mode» increases the smoke layer thickness and consequently reduces the filling time of the enclosure. However, for wide enclosures ($\Phi \geq 2/3$), they observed that when the ceiling flow hits the side-walls, it slumps back up into the buoyant layer entraining a small amount of the ambient fluid (see Fig. 1-b). Kaye and Hunt [8] proposed a correlation of the scaled time $\tau = t/t_f$, taken for the maximum buoyant layer depth *h* (see Fig. 1) to reach half height of the box, as a function of the room aspect ratio Φ which reads as:

$$\tau_{h=\frac{H}{2}} = \begin{cases} 0.8\,\Phi + 0.066, & \Phi < 0.6 \\ 0.55, & \Phi > 0.6 \end{cases} \tag{2}$$

The study by Kaye and Hunt [8] highlights the major influence of the room aspect ratio on the filling time of the room, especially for the small aspect ratios. Furthermore, this work has been conducted in the Boussinesq case, i.e. for marginal density (temperature) difference between the plume and its environment. The fire plume is a non-Boussinesq plume due to the large density contrast caused by the temperature difference between the fire plume and its environment. Thus, one can expect to observe differences with experiments involving real fires.

In the present paper, the aim is to study, by means of full-scale simulations, the effect of the room aspect ratio on the smoke filling time, and to evaluate the filling time predicted by the zone models based on the approach of Baines and Turner [1] for the prediction of the interface position during the smoke filling process.

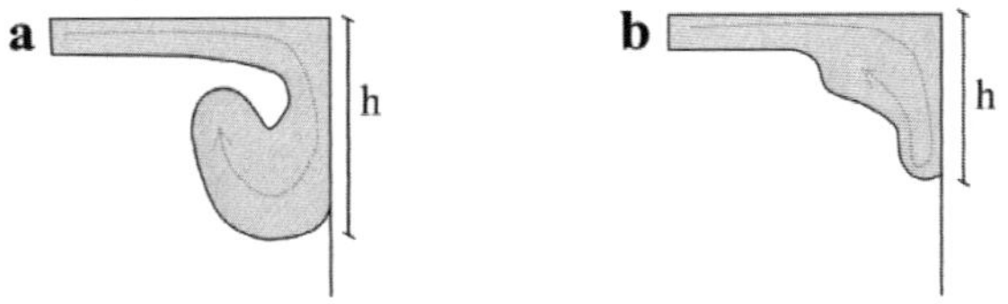

Fig. 1. A schematic of the two sidewall flow patterns: (a) «Rolling mode», (b) «Slumping mode».

TWO-ZONE MODEL

A two zone model is a simplified representation used to predict the smoke layer thickness that forms under the ceiling and its temperature. Fig. 2 presents the schematic of the studied configuration. We consider a square room of height H with a cross-sectional area S, connected to the outside by opening located at the bottom of the room. A fire, with heat release rate of $\dot{Q}_t$, is located at the floor level. We assume that, at the initial time (t = 0 s), a smoke layer of infinitesimal thickness forms and thickens with time due to the mass flow supplied by the plume $\dot{m}_p$. We also assume that the smoke layer is characterized by its average density ρ_{up}, its average temperature T_{up} and its thickness $h = H - z$. Note that T_0 and ρ_0 are, respectively, the temperature and the density of the ambient air.

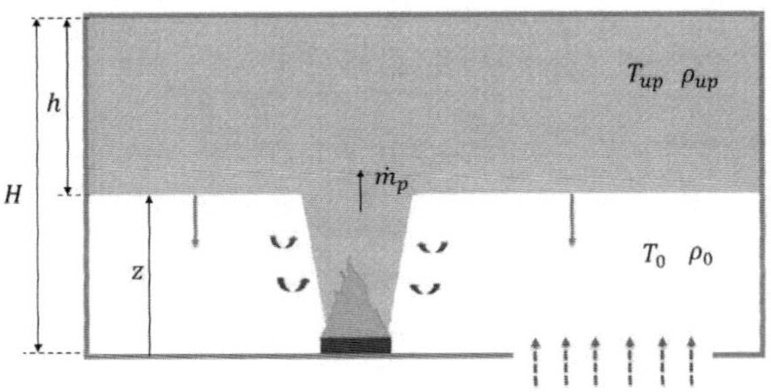

Fig. 2. Schematic of the studied configuration.

Assuming that the walls of the room, including the ceiling, are adiabatic, the conservation equations for mass and buoyancy flux can be written as:

$$\frac{d(\rho_{up} S\, h)}{dt} = \dot{m}_p \quad ; \qquad \frac{d(g\frac{\Delta\rho_{up}}{\rho_0} S\, h)}{dt} = B \tag{3}$$

where $\Delta\rho_{up} = \rho_0 - \rho_{up}$ and B is the buoyancy flux which is related to the convective heat release rate $\dot{Q}_c$ via the relation: $B = g\dot{Q}_c/(\rho_0 C_p T_0)$. C_p is the specific heat of air at constant pressure. Using the conservation equations (3), one obtains for the buoyant layer thickness, the following relation

$$\frac{dh}{dt} = \frac{\dot{m}_p}{\rho_0 S} + \frac{B}{gS} \tag{4}$$

To complete the model, a closure model is still required for the mass flow rate of the plume $\dot{m}_p$. It is generally evaluated using three different fire plume models, namely: McCaffrey [5], Zukoski [4] and Heskestad [3]. In what follows, equation (4) is solved numerically by a Runge-Kutta fourth order method to obtain the temporal evolution of the smoke layer thickness h. This evolution is subsequently compared to FDS numerical simulations.

FULL-SCALE SIMULATIONS

Fire Dynamics Simulator (FDS) version 6, is used here to study the effects of the room aspect ratio Φ on the smoke filling time. In what follows, a square room of height *H* with a cross-sectional area *S* is considered. The room is connected to the outside by openings located at the bottom of the room. A heptane pool fire, with heat release rate of 1 MW and surface area of 1 m², is located at the floor level in the middle of the room. All walls of the room, including the ceiling, are supposed to be adiabatic. A mesh size of 10 cm × 10 cm × 10 cm is adopted in all simulations based on mesh sensitivity study.

Two configurations are considered. In the first configuration, to vary the room aspect ratio, the surface of the room is kept constant (S= 25 m²) while its height is varied (see Table 1). In the second configuration, the surface area is varied while the height is fixed at 10 m (see Table 2). For each case, the time necessary for the maximum smoke layer depth *h* (see Fig. 3) to reach half height (H/2) of the room is evaluated and compared to the result of the two-zone model presented in the previous section.

Here, the three fire plume models are evaluated namely: McCaffrey [5], Zukoski [4] and Heskestad [3]. It should be noted that the height of the room is taken to be the distance between the virtual origin of the fire plume z_v and the ceiling, and the initial time (t= 0 s) is defined as the moment the plume hits the ceiling.

Table 1. Dimensions of the rooms simulated in series 1.

	S (m²)	*H* (m)	$\phi=L/(2(H-z_v))$
Case 1	25	2.5	1.07
Case 2	25	3.6	0.73
Case 3	25	8	0.32

Table 2. Dimensions of the rooms simulated in series 2.

	S (m²)	*H* (m)	$\phi=L/(2(H-z_v))$
Case 1	36	10	0.31
Case 2	64	10	0.41
Case 3	196	10	0.71
Case 4	400	10	1.02
Case 5	676	10	1.32
Case 6	1024	10	1.63

Configuration 1: variation of the height of the room (*S*= 25 m²)

To assess the influence of the aspect ratio Φ on the time taken for the smoke to reach half the height of the room, the time $t_{h=\frac{H}{2}}$ and the dimensionless time $\tau_{h=\frac{H}{2}} = t/t_f$ (t_f is the filling box time scale [1]) are plotted as a function of the aspect ratio Φ in Fig. 4. Fig. 4.a shows a clear dependency between the scaled time $\tau_{h=\frac{H}{2}}$ predicted by FDS and the room aspect ratio. It is also noticed that the results obtained by the two-zone models share similar qualitative behaviors with FDS results. In all cases, the two-zone models overestimate the filling time by a few seconds (Fig. 4.b).

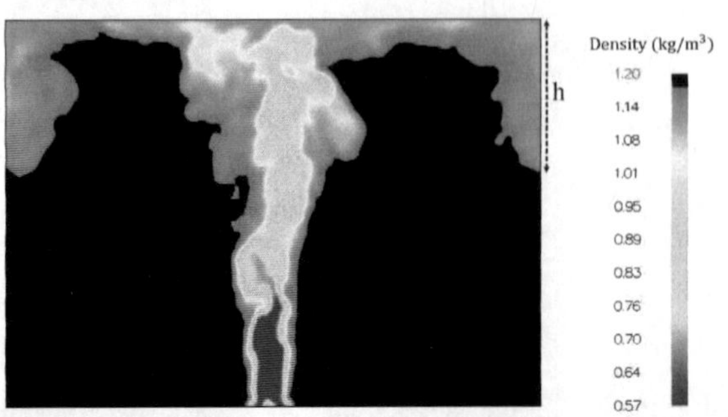

Fig. 3. Distribution of the density in the vertical median plane (Series 2 - case 3).

The behavior obtained by the two-zone models may be explained by the short smoke filling times. In fact, this fast filling time may hide the expected behaviors described by Kaye and Hunt [8]. Consequently, in this situation, the room aspect ratio has little impact on the smoke filling times.

Moreover, when comparing the FDS results with the experimental correlation by Kaye and Hunt [8], a complete mismatch is observed. These differences may be attributed to the fast filling time but also to non-Boussinesq effects (associated with a large temperature difference with the ambient), since the experiments of Kaye and Hunt are only related to Boussinesq plumes (salt water plume injected into fresh water). It can be concluded that the fresh-water / salt-water experiments may not be suitable for the description of the filling box with fire smoke.

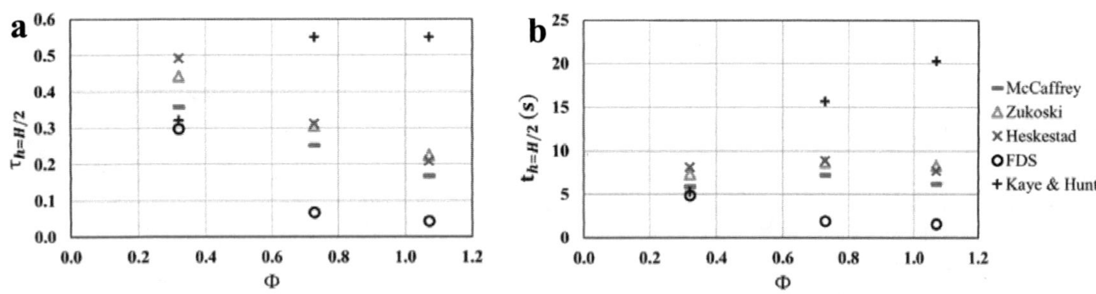

Fig. 4. Time taken for the smoke layer depth to reach half height of the room. (a) Variation of the scaled time with the aspect ratio, (b) Variation of the filling time with the aspect ratio.

Configuration 2: variation of the surface area of the room (H= 10 m)

The Fig. 5-a shows the variation of the scaled time $\tau_{h=\frac{H}{2}}$ with the room aspect ratio Φ. The results obtained by FDS show a strong dependency between the scaled time $\tau_{h=\frac{H}{2}}$ and the room aspect ratio, which is not the case for the two-zone models. Moreover, it is seen in Fig. 5-b that for a room aspect ratios $\Phi \geq 1$, the time required for the smoke to reach one half of the room height is very long compared to the time obtained by FDS, especially with the Heskestad and Zukoski plume models. These discrepancies increase with the room aspect ratio. These discrepancies can be attributed, in some cases, to the dynamics of the outflow at the side-walls where the maximum penetration depth h of the smoke can rapidly reach the half height of the room, before the smoke fills half of the whole room volume. In zone models, these dynamics of the outflow at the side-walls are not taken into account, which explains partially why the smoke filling times are overestimated in the case of an important surface area such as hangars and atria. Thus, the volume being very large, the time required to fill half of it will be important. Fig. 6 compares the smoke flow obtained by FDS (Fig. 6.a) and that obtained by the two-zone model (Fig 6.b) at t= 23 s after ignition and for the case: S=676 m² (Φ = 1.3). We can observe that the maximum penetration depth reaches 48 % of the room height while in the two-zone model the smoke layer reaches only 9 % of the room height. We can conclude that, in this case, zone models cannot be used to evaluate the ASET, especially in the first moments when the outflow dynamics at the sidewalls are important.

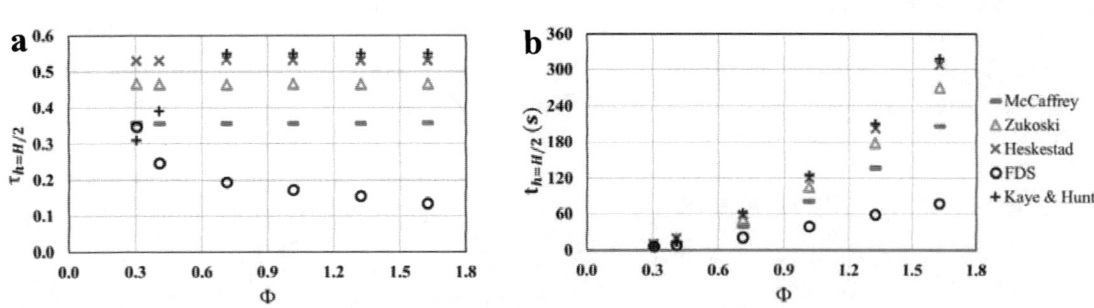

Fig. 5. Time taken for the smoke layer depth to reach half height of the room. (a) Variation of the scaled time with the aspect ratio, (b) Variation of the filling time with the aspect ratio.

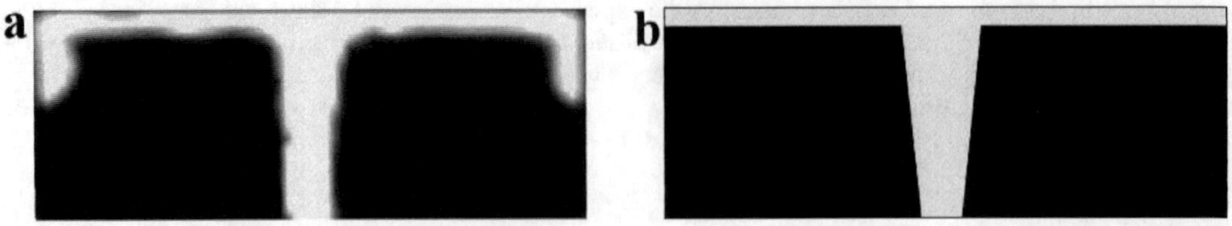

Fig. 6. Smoke configuration at t= 23 s: (a) obtained by FDS, (b) obtained by the two-zone model (using Heskestad's plume model).

CONCLUSION

In this study, full scale simulations with the CFD model FDS have been conducted to study the influence of the room aspect ratio on the smoke filling times, and to evaluate the smoke filling times predicted by the two-zone models. The results show that the room aspect ratio has little impact on the smoke filling times in the case of small room areas. Thus, the zone models can be used to predict the available evacuation time. However, in the case of a configuration with an important surface area, such as atria and hangars, the results found with the CFD model show that the maximum penetration depth of the smoke at the side walls can rapidly reach important values during its initial stage of development (after the outflow hits the side walls). This outflow dynamics are not considered in zone models, and consequently, the smoke filling times are overestimated. It can be concluded that zone models could be unfit to evaluate the available safe evacuation time when the flow dynamics at the side walls are important. This result does not call in question their usage in other situations, for example, where we are only concerned about the stationary phase of a filling emptying process.

REFERENCES

[1] Baines, W.D., and Turner, J.S. (1969) Turbulent buoyant convection from a source in a confined region, *Journal of Fluid mechanics 37*(1), 51-80, https://doi.org/10.1017/S0022112069000413

[2] Quintiere, J. G. (1989) Fundamentals of enclosure fire" zone" models, *Journal of Fire Protection Engineering*, 1(3), 99-119, https://doi.org/10.1177/104239158900100302

[3] Heskestad, G. (1984) Engineering relations for fire plumes, *Fire Safety Journal* 7(1), 25-32, https://doi.org/10.1016/0379-7112(84)90005-5

[4] Zukoski, E.E., Kubota, T., and Cetegen, B. (1981) Entrainment in fire plumes, *Fire safety journal* 3(3), 107-121, https://doi.org/10.1016/0379-7112(81)90037-0

[5] McCaffrey, B.J. (1983) Momentum implications for buoyant diffusion flames, *Combustion and Flame* 52, 149-167, https://doi.org/10.1016/0010-2180(83)90129-3

[6] Morton, B.R. and Turner, J.S. (1956) Turbulent gravitational convection from maintained and instantaneous sources. In Proc. R. Soc. Lond. A (Vol. 234, No. 1196, pp. 1-23). The Royal Society.

[7] Kaye, N.B., and Hunt, G.R. (2007) Overturning in a filling box, *Journal of Fluid Mechanics* 576, 297-323, https://doi.org/10.1017/S0022112006004435

[8] Kaye, N.B., and Hunt, G.R. (2007) Smoke filling time for a room due to a small fire: the effect of ceiling height to floor width aspect ratio, *Fire safety journal* 42(5), 329-339, https://doi.org/10.1016/j.firesaf.2006.12.003

IOP Conf. Series: Journal of Physics: Conf. Series **1107** (2018) 042023 doi:10.1088/1742-6596/1107/4/042023

Experimental investigation of externally venting flames geometric characteristics and impact on the façade of corridor-like enclosures

Konstantinos CHOTZOGLOU[1], Eleni ASIMAKOPOULOU[1], Jianping ZHANG[1], Michael DELICHATSIOS[1,2]

[1] *FireSERT, Belfast School of Architecture and the Built Environment, Ulster University, United Kingdom*

[2] *Currently: Northeastern University, Boston, USA and*

University of Science and Technology of China (USTC), Hefei, China

e.asimakopoulou@ulster.ac.uk

ABSTRACT

Understanding of the physics and mechanisms of fire development and externally venting flames (EVF) in corridor-like enclosures is fundamental to studying fire spread to adjacent floors in high-rise buildings. This work aims to investigate the burning behaviour of a liquid fuel pool fire in a corridor-like enclosure and to identify the key factors influencing EVF characteristics and its impact on façades. A series of experiments is conducted in a medium-scale corridor-facade configuration using ethanol pool fires. A new fuel supply system has been developed to keep the fuel level constant to minimize lip effects. The influence of fuel surface area and ventilation factor on the fire development is also investigated. Experimental measurements consist of mass loss, heat release rate, temperatures and heat fluxes inside the corridor and on the facade. Three distinct burning regions are observed and their characteristics depend on both the pan size and ventilation factor. A power dependence of EVF height in relation to excess external heat release rate has been found. The impact of EVF on the façade is investigated by measuring heat flux on the façade using thin steel plate probes. It is found that the characteristics of EVF strongly depend on opening dimensions and for large opening widths EVF tend to emerge from the opening as two separate fire plumes.

KEYWORDS:

liquid pool fire, corridor-like enclosure, façade fire, flame height, heat flux

IOP Conf. Series: Journal of Physics: Conf. Series **1107** (2018) 042023 doi:10.1088/1742-6596/1107/4/042023

INTRODUCTION

Understanding of the physics and mechanisms of fire development in enclosures and flames emerging through openings is fundamental to studying fire spread to adjacent floors or buildings. Recent high-rise building fires around the world highlight the importance of understanding the mechanisms of fire spread not only at the interior of buildings but also those due to Externally Venting Flames (EVF) [1,2]. Consequences of EVF induced fire events include loss of life and injuries, health impact through smoke exposure, property and infrastructure loss, business interruption, ecosystem degradation, soil erosion and huge firefighting costs [1]. Though substantial research has been conducted on fire characteristics in typical cubic-like compartments [3], there is still limited data on the development of enclosure fires and EVF in other geometries (e.g. long corridors, tunnels etc). For the few studies in corridor-like enclosures [4,5], gaseous fuels were used. As the mass flow rate (thus heat release rate) of gaseous fuels must be pre-defined, it does not consider the interaction between the flame, the hot gas layer and the pyrolysis rate of the fuel, which is of fundamental importance in growth and development of real fires. This work investigates the burning behaviour of a liquid fuel pool fire in a corridor-like enclosure and the heat impact of the subsequent externally venting flames (EVF) on the façade. The effects of the size and location of the pool fire and the dimensions of the opening on the burning characteristics of the EVF are examined. Experimental measurements consist of mass loss, heat release rate, temperatures and heat fluxes inside the corridor and on the façade. Correlations are developed for external flame height of EVF and heat flux on the façade and compared to the ones obtained using gaseous fuels [4,5].

EXPERIMENTAL DETAILS

Corridor-like enclosure and façade configuration

Experiments were conducted in a medium-scale corridor-facade configuration having internal dimensions 3 m x 0.5 m x 0.5 m and a 1.8 m x 1 m façade was attached to the front box. High tempered 40 mm resistant boards were used as inner corridor lining and exposed side of façade; the outer walls were constructed using 12 mm MDF boards. A schematic of the experimental facility, illustrating the locations of the employed measuring devices, is given in Fig. 1. A novel fuel supply system was designed to maintain constant fuel level to minimize lip effects; ethanol was used in a circular sandbox burner positioned at geometrical center of Box A or F. The fuel surface area and ventilation factor (by altering the dimensions of the door-like opening) were varied to examine their effects on the fire development. In total, more than 60 experiments were conducted to ensure repeatability. A summary of the main operational parameters, e.g. burner position, opening size (opening width W_o and opening height H_o), ventilation factor ($A_oH_o^{1/2}$, A_o corresponds to opening area), ventilation regime (Under or Over ventilated, indicated as U or O respectively), length scale ($l_1=(A_oH_o^{1/2})^{2/5}$) [4], total heat release rate (HRR) ($\dot{Q}_{exp}$) and excess HRR ($\dot{Q}_{ex}$) [4] for the experiments presented in this work is shown in Table 1.

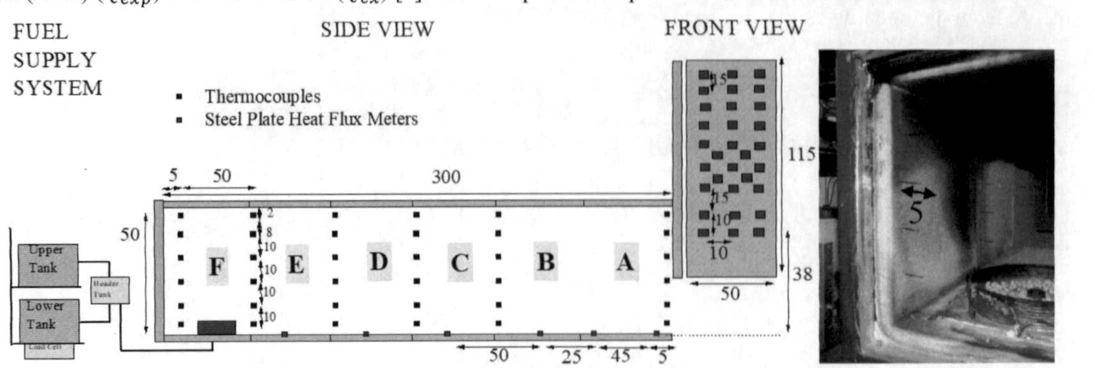

Fig. 1. Fire test configuration and experimental setup (left) and inner corridor lining (right).

Sensors and data acquisition system

Thirty-six K-type 1.5 mm diameter thermocouples are used to monitor gas temperatures inside the enclosure at every 6 s [4,5]. Six thermocouples were positioned 5 cm from the side wall, in each Box, to monitor gas temperature. Steel plates were used to monitor the heat flux on the façade, Fig.1. The experimental set-up was placed under a 3 x 3 m^2 hood to measure $\dot{Q}_{exp}$, production of CO, CO_2 and smoke. Videos were recorded by a CCD camera facing the façade, based on which an in-house developed image processing tool is used to evaluate the geometric characteristics of the façade fires by calculating the average flame probability (intermittency) [6]. The threshold limits for Red, Green and Blue color levels and luminosity were acquired through an extended statistical analysis of the flame ejected from the opening.

IOP Conf. Series: Journal of Physics: Conf. Series **1107** (2018) 042023 doi:10.1088/1742-6596/1107/4/042023

Table 1. Summary of main operational parameters for the examined test cases.

a/a	Test	Burner position	Pan diam.	Opening size	$A_oH_o^{1/2}$	Vent. Regime	l_1	$\dot{Q}_{exp}$	$\dot{Q}_{ex}$	Z_f
		Box	m	m x m	cm$^{5/2}$		-	kW	kW	m
1	FR20W20xH20	A	0.2	0.20x0.20	0.0179	O	0.20	12.1	-	0.11
2	FR20W25xH25	A	0.2	0.25x0.25	0.0313	O	0.25	21.5	-	0.21
3	FR30W10xH25	A	0.3	0.10x0.25	0.0125	U	0.17	22.2	-	0.22
4	FR30W20xH20	A	0.3	0.20x0.20	0.0179	O	0.20	17.0	-	0.13
5	FR30W25xH25	A	0.3	0.25x0.25	0.0313	O	0.25	31.0	-	0.34
6	FR30W30xH30	A	0.3	0.30x0.30	0.0493	U	0.30	56.1	-	0.49
7	FR30W50xH25	A	0.3	0.50x0.25	0.0625	U	0.33	75.1	-	0.48
8	FR30W50xH50	A	0.3	0.50x0.50	0.1768	O	0.50	69.5	-	0.33
9	BC20W25xH25	F	0.2	0.25x0.25	0.0313	U	0.25	39.0	2.5	0.22
10	BC20W30xH30	F	0.2	0.30x0.30	0.0493	U	0.30	68.5	11.1	0.51
11	BC20W50xH25	F	0.2	0.50x0.25	0.0625	U	0.33	74.0	9.0	0.37
12	BC30W25xH25	F	0.3	0.25x0.25	0.0313	U	0.25	39.5	2.9	0.19
13	BC30W30xH30	F	0.3	0.30x0.30	0.0493	U	0.30	70.0	14.0	0.58
14	BC30W50xH25	F	0.3	0.50x0.25	0.0625	U	0.33	77.0	9.0	0.64
15	BC30W50xH50	F	0.3	0.50x0.50	0.1768	O	0.50	105.0	-	0.30

RESULTS AND DISCUSSION

Heat release rate inside the enclosure

A parametric analysis is performed to identify the main parameters influencing the burning behaviour and subsequent EVF, namely pan size, pan location and ventilation factor. To demonstrate the main characteristic stages regarding the fire growth, the temporal evolution of measured *HRR*, $\dot{Q}_{exp}$, and theoretical HRR, $\dot{Q}_{th}$, are plotted in Fig. 2 for two characteristic underventilated test cases, BC20W30xH30 and BC30W30xH30. The theoretical HRR, $\dot{Q}_{th}$, is calculated by multiplying the measured fuel mass loss rate by the heat of combustion of ethanol, 26.78 MJ/kg. The maximum *HRR* in stoichiometric conditions inside an enclosure, $Q_{st,in}$, is calculated by multiplying the maximum air inflow rate by the heat released by complete combustion of 1 kg oxygen, which for most fuels is found to be approximately equal to 3000 kJ/kg [7]. To demonstrate the enclosure effect, free-burn ethanol pool fire experiments for the 30 cm (FB30) and 20 cm (FB20) pan are also depicted. Each experiment was terminated by stopping the fuel supply to the pan and closing the openings.

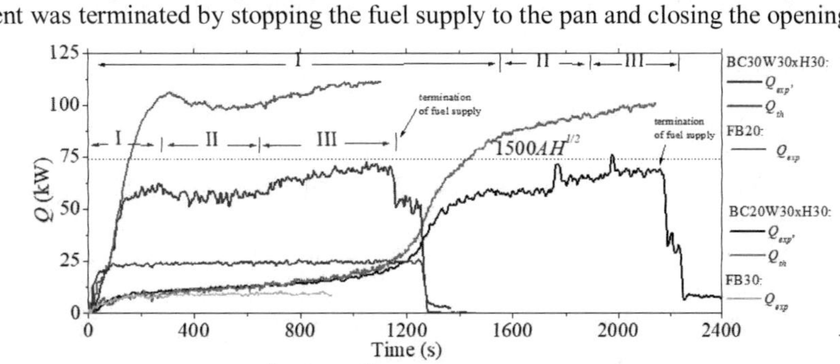

Fig. 2. Temporal evolution of $\dot{Q}_{exp}$ and $\dot{Q}_{th}$ for test cases BC30W30xH30 and BC20W30xH30.

For all under-ventilated fires with the pan located at Box F, as indicatively presented in Fig. 2, the fire behaviour is characterized by three distinct phases namely *Regions I, II* and *III* appearing in succession. *Region I* correspond to the fuel-controlled period (growth period), where the combustion efficiency is close to unity and thus $\dot{Q}_{exp}$, and $\dot{Q}_{th}$, are almost equal. After this phase, during *Region II*, fire gradually becomes ventilation-controlled, $\dot{Q}_{exp}$ reaches a plateau until flames ejects through the opening. Note that $\dot{Q}_{exp}$ does not reach the maximum heat released inside the enclosure approaching $1500A_oH_o^{1/2}$ value that has been found for rectangular compartments [3]. This value is decreased, calculated at approximately $1100A_oH_o^{1/2}$, indicating that the amount of air inflow in long corridors is less than in rectangular enclosures with the same opening geometry [8]. Flames ejection, as observed visually and through the image processing algorithm, associated with a sudden increase in the $\dot{Q}_{exp}$, indicates the beginning of *Region III*, where sustained external burning is observed until a plateau is formed near the end of the test indicating steady state conditions are established. As depicted in *Fig. 2*, not all the excess fuel burns at the exterior of the corridor as the combustion is incomplete and combustion efficiency in *Region III* is calculated at approximately 0.74 for both cases. Pan size has a significant effect on the time to reaching steady burning. In test cases using the smaller pan it takes longer for the walls and gases in the enclosure to be heated. For under-ventilated cases with the fuel pan located near the opening, at Box A, EVF appeared shortly after ignition and as a result the duration of *Regions I* and *II* is considerably reduced. For the under-ventilated cases, the external HRR, $\dot{Q}_{ex}$, is calculated as the difference between the average values of $\dot{Q}_{exp}$ during steady burning in *Region II* and *III* with the assumption that the burning inside the enclosure has reached the steady state before flame ejection.

Gas temperature inside the enclosure

Present analysis focuses on under-ventilated cases with EVF occurrence. Figure 3 shows the spatial mapping of the thermocouple gaseous temperatures inside the corridor at three characteristic time frames representing the three Regions discussed in previous sections for test cases BC30W30xH30 and FR30W30xH30. During *Region I*, low gas temperatures are observed in the lower layer as fresh air enters the enclosure through the opening, located at the right side of the corridor. In BC30W30xH30, during *Region II*, highest temperatures are observed at the vicinity of Boxes E and D indicating that combustion mainly takes place at these locations and flames gradually propagate towards the opening seeking for available oxygen [4, 5]. During *Region III*, the difference of gas temperatures between the upper and lower layers decreases towards the closed end of the corridor, but still, they cannot be assumed uniform inside the corridor. In *Region III*, flames fill the upper layer of the corridor extending towards the opening and eventually emerges from the opening when the *HRR* becomes sufficiently large. In test case FR30W30xH30, temperature stratification in the interior of the corridor is less evident as EVF emerge more quickly from the opening resulting in lower temperatures in Boxes C to F.

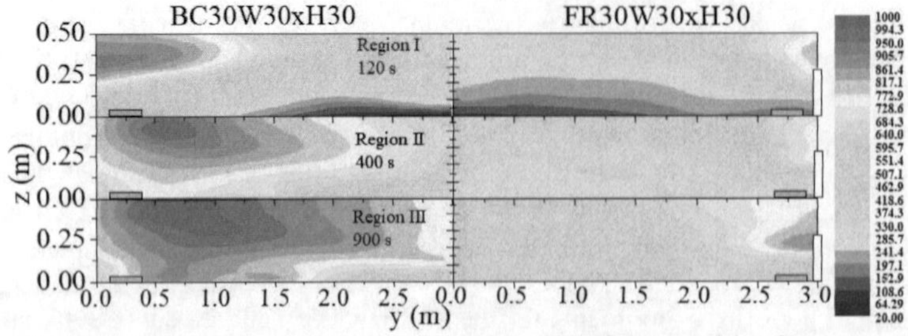

Fig. 3. Spatial visualization of gaseous temperature at the interior of the corridor for test cases F30W30xH30 (left) and B30W30xH30 (right) for 120, 400 and 900s.

EVF height

Figure 4 plots the dimensional flame height, Z_f/l_1, against the dimensionless external heat release rate $\dot{Q}^*_{ex}$ using experimental data from gas burner experiments in rectangular geometries [5], corridors [4] and liquid pool fires in corridors (current study). It is demonstrated that for small flame heights and larger openings the power dependence on the $\dot{Q}^*_{ex}$ calculated following [4] is 2/3, which is in accordance with the physical mechanism corresponding to "wall fire" in which flame is attached to the façade wall or a "half axisymmetric fire" in which flame is detached from the façade wall [5, 9, 10]. For $\dot{Q}^*_{ex}$>1, experimental results using gas burner from both rectangular and corridor geometries corroborate a 2/5 dependence on $\dot{Q}^*_{ex}$ [5,9,11].

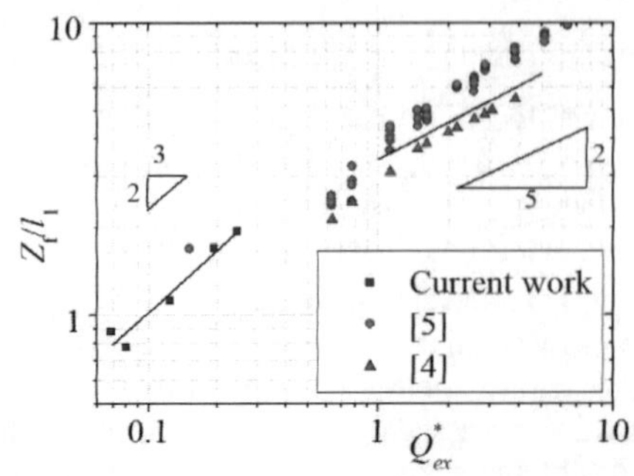

Fig. 4. Flame height correlation for gas burner experiments in rectangular geometries [5] corridors [4] and liquid pool fires in corridors (current study).

Heat fluxes

Figure 5 depicts the vertical distribution of the radiative heat fluxes measured using thin steel plate probes [5,12] at the centreline of the façade for all test cases. Measured heat fluxes decrease with increasing height as expected. The highest heat flux is found always along the centreline except for test cases when the front of the enclosure is completely open, i.e., BC30W50xH50 and FR30W50xH50. As it can be seen from their intermittency contours in Fig. 6, the resulted EVF for these two cases tend to emerge as two separate fire plumes and as a result the maximum heat flux values are off the centreline in contrast to other cases. This finding highlights the influence of the width of the opening, in addition to the ventilation factor, to the resulting EVF shape. For the same opening and pan size, higher heat flux values are observed when the fuel pan is positioned near opening at Box A.

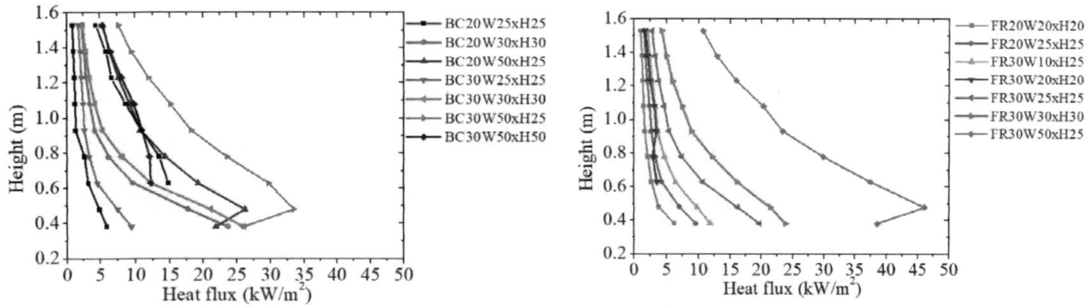

Fig. 5. Vertical distribution of heat flux at the centerline of the façade for the BC (left) and FR (right) cases.

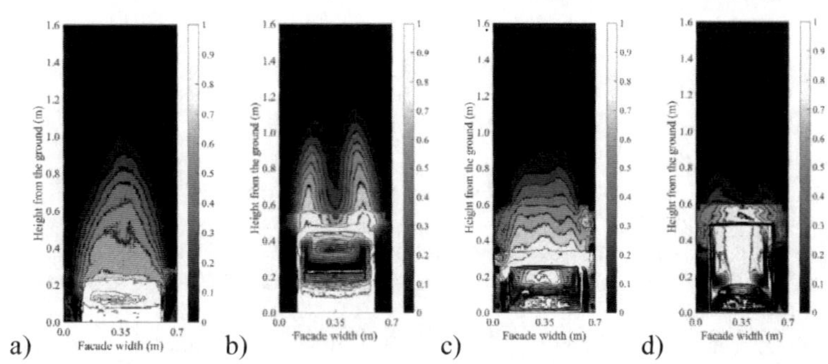

Fig. 6. Intermittency contours for a)BC30W50xH25, b)BC3W50x0H50, c)FR30W50xH25, d)FR30W50xH50.

Figure 7 shows the heat flux distributions for under-ventilated cases at the centerline above the opening on façade wall against the value of Z/Z_f, where Z is the location of each steel plate at the centreline as depicted in Fig. 1. Both are measured from the position of neutral plane corresponding to $0.5H_o$ above the bottom of the opening [7]. The residual effect of the aspect ratio of the opening is accounted for by using the exponential factor $\exp(0.60H_o/l_1)$ [4, 5]. It is shown that the measurements of heat flux in this work are mainly located in the intermittent flame and buoyant plume regimes where Z/Z_f>0.7. Despite the scattering of the current data, partially attributed to the different properties of the fuels used, qualitative analysis reveals that current results follow the same trend as Lee's [5] and Beji's [4] data.

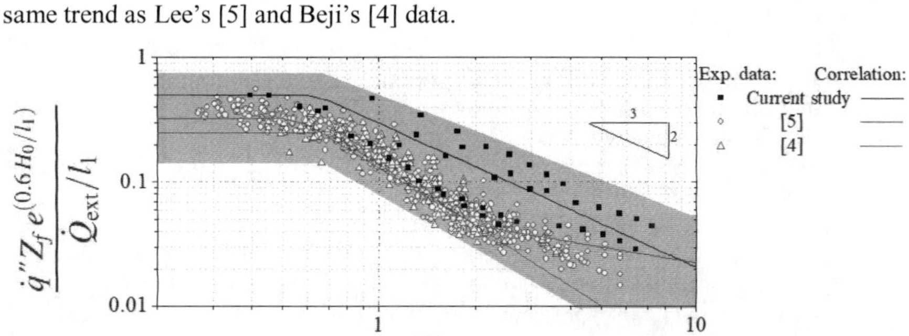

Fig. 7. Dimensionless heat flux against dimensionless height in rectangular geometries [5], corridors [4] and liquid pool fires in corridors (current study).

CONCLUSIONS

We presented an experimental work of liquid pool fires in a reduced-scale corridor-like enclosure, aiming with emphasis to examine the influence of the pool size, location and ventilation openings on the geometric characteristics of EVF and their impact on the façade. The main conclusions are:

1. Three distinct burning regions (*Region I*, *II* and *III*) were observed based on $\dot{Q}_{exp}$ corresponding respectively to fuel-controlled, ventilation-controlled and steady-state burning.
2. The location and size of the fuel pan has a strong impact on *HRR* and subsequent EVF characteristics.
3. For the present cases the dimensionless external heat release rate $\dot{Q}^*_{ex}$ is always less than one, for which a power dependence of the flame height of the EVF on $\dot{Q}^*_{ex}$ is found to be 2/3, compared to 2/5 for larger $\dot{Q}^*_{ex}$ values found for gaseous fuels in previous studies [4,5].
4. The centreline heat fluxes on the façade are nearly constant just above the opening where the consistent flame region is and subsequently decreases with height. The measurements in this work are mainly located in the intermittent flame and buoyant plume regimes when compared to previous studies using gaseous burner [4,5]. The dimensionless heat fluxes found in this work are generally higher than those reported in [4,5], owing mainly to the uncertainties in the determination of the excess heat release rate.
5. EVF of the two fully open cases exhibit distinctive characteristics and tend to emerge from the opening as two separate fire plumes resulting in lower heat flux values at the façade centreline.
6. The overall design of the medium-scale compartment-façade fire tests can be scaled-up provided certain non-dimensional parameters are preserved. Subsequently, correlations derived in the frame of the current research work can be used to enhance façade fire safety design methodologies and after detailed validation can be gradually incorporated in facades prescriptive fire safety regulations.

REFERENCES

[1] White, N., Delichatsios, M. (2014) Fire hazards of exterior wall assemblies containing combustible components, 1st ed., SpringerBriefs in Fire, Springer, New York, U.S.A.

[2] Asimakopoulou, E., Chotzoglou, K., Kolaitis, D., Founti, M. (2016) Characteristics of externally venting flames and their effects on the façade: a detailed experimental study, *Fire Technology* 52:2043-2069, https://doi.org/10.1007/s10694-016-0575-5

[3] Delichatsios, M.A., Silcock, G.W.H., Liu, X., Delichatsios, M., Lee, Y.P. (2004) Mass pyrolysis rates and excess pyrolysate in Fully Developed Enclosure Fires, *Fire Safety Journal* 39:1–21, https://doi.org/10.1016/j.firesaf.2003.07.006

[4] Beji, T., Ukleja, S., Zhang, J., Delichatsios, M.A. (2012) Fire behaviour and external flames in corridor and tunnel-like enclosures, *Fire and Materials* 36:636-647, https://doi.org/10.1002/fam.1124

[5] Lee, Y.P., Delichatsios, M.A., Silcock, G. (2007) Heat fluxes and flame heights in facades from fires in enclosures of varying geometry, *Proceedings of the Combustion Institute* 31:2521-2528, https://doi.org/10.1016/j.proci.2006.08.033

[6] Asimakopoulou, E., Kolaitis, D., Founti, M. (2016) Geometrical characteristics of externally venting flames: Assessment of fire engineering design correlations using medium-scale compartment-façade fire tests, Journal of Loss Prevention in the Process Industries 44: 780-790, http://dx.doi.org/10.1016/j.jlp.2016.09.006

[7] O'Connor, D., "The Building Envelope: Fire Spread, Construction Features and Loss Examples", SFPE Handbook of Fire Protection Engineering (5th ed.) Hurley M.J., National Fire Protection Association, Quincy, MA 02269, 2016, p 3242/3512.

[8] Yii, E., Fleischmann C., Buchanan, A. (2007) Vent flows in fire compartments with large openings, *Journal of Fire Protection Engineering* 17: 221-237, http://dx.doi.org/10.1177/1042391507069634

[9] Yokoi, S. (1960) Study on the prevention of fire spread caused by hot upward current, Japanese Ministry of Construction, Building Research Institute Report 34, 1960.

[10] Quintiere, J.G., Harkleroad, M., Hasemi, Y. (1986) Wall Flames and Implications for Upward Flame Spread, *Combustion Science and Technology* 48:191–222.

[11] Lu, K. H., Hu, L. H., Tang F., He L. H., Zhang, X. C., Qiu, Z. W. (2014) Heat flux profile upon building facade with side walls due to window ejected fire plume: An experimental investigation and global correlation, Fire Safety Journal 70:14-22, http://dx.doi.org/10.1016/j.firesaf.2014.08.001

[12] Zhang, J., Delichatsios, M. A. (2009) Determination of the convective heat transfer coefficient in three-dimensional inverse heat conduction problems, *Fire Safety Journal* 44: 681-690.

IOP Conf. Series: Journal of Physics: Conf. Series **1107** (2018) 042024 doi:10.1088/1742-6596/1107/4/042024

An experimental investigation of burning behaviour of liquid pool fire in corridor-like enclosures

Konstantinos CHOTZOGLOU[1], Eleni ASIMAKOPOULOU[1], Jianping ZHANG[1], Michael DELICHATSIOS[1,2]

[1] *FireSERT, Belfast School of Architecture and the Built Environment, Ulster University, United Kingdom*

[2] *Currently: Northeastern University, Boston, USA and*

University of Science and Technology of China (USTC), Hefei, China

e.asimakopoulou@ulster.ac.uk

ABSTRACT

This work aims to investigate the burning behavior of a liquid fuel pool fire in a corridor-like enclosure and to identify the key factors influencing fire development. A series of experiments is conducted in a medium-scale corridor-facade configuration using ethanol pool fires. A new fuel supply system has been developed to keep the fuel level constant to minimize lip effects. The influence of fuel surface area and ventilation factor on the fire development is also investigated. Experimental measurements consist of mass loss, heat release rate, temperatures and heat fluxes inside the corridor. Experimental results indicate that in corridor-like enclosures the fuel burning rate in ventilation-controlled conditions corresponds to about 2/3 of that observed in cubic-like enclosures. The fuel burning rate varies as the temperature distribution in the enclosure changes from uniform, in cubic-like enclosures, to layered, in corridors. The ventilation coefficient value used for the calculation of the inflow rate in corridor-like enclosures during post-flashover conditions is found to decrease with an increase of the ventilation factor.

KEYWORDS:

corridor, pool fire, ventilation factor, fuel burning rate, fuel pan size

INTRODUCTION

Substantial research has been conducted on fire characteristics in typical residential cubic-like compartments [1], [2], [3], but there is still limited data on the development of fires in other geometries, even though it has been demonstrated that the geometry of an enclosure affects significantly fire development [4]. Investigation of enclosure fires occurring in modern constructions that differ from typical cubic-like enclosures (e.g. long corridors, tunnels etc.) is thus essential [2], [3]. Studies [3], [5] have indicated the need to further progress knowledge related to the understanding of the physics of fire growth in corridor-like enclosures and the mechanisms that eventually may lead to fire spread to adjacent floors or buildings.
For under-ventilated compartment fires, the mass flow rate of the air entering the compartment is a key parameter in determining the maximum heat release rate inside the compartment. The air mass flow rate entering the compartment ($\dot{m}_a$) can be estimated using the ventilation factor, $A_oH_o^{1/2}$ [1, 2], where Ao and Ho are the area and height of the opening respectively. Kawagoe [1] applied the Bernoulli equation to calculate the air inflow through a single moderate opening when under-ventilated conditions prevail and found that $\dot{m}_a$ is proportional to $A_oH_o^{1/2}$, as shown in Equation (1), assuming uniform temperature distribution in the interior of a cubic enclosure [1]. The proportional constant C is referred to as the ventilation coefficient and takes values of 0.45 [1] or 0.5 [2] for cubic enclosures with moderate openings. Delichatsios et al. [3] proposed a correction for the expression of $\dot{m}_a$ for cubic enclosures, Equation (2), by subtracting Equation (1) by 0.5 $\dot{m}_T$, where $\dot{m}_T$ is the burning rate of fuel. It has been shown in [2], [4], [6] that the geometry of the enclosure has a significant effect on the burning rate of fuel in enclosure fires. Kawagoe [1] proposed that in rectangular configurations, $\dot{m}_T$, under ventilation-controlled conditions, can be attained by multiplying $A_oH_o^{1/2}$ by 0.1, Equation (3). However, limited information exists regarding corridor-like configurations.

$$\dot{m}_a = C \times A_o {H_o}^{1/2} \tag{1}$$

$$\dot{m}_a = 0.5A_o {H_o}^{1/2} - 0.5\dot{m}_T \tag{2}$$

$$\dot{m}_T = 0.1A_o {H_o}^{1/2} \tag{3}$$

The present research aims at investigating the differences in terms of fire development and burning behaviour between cubic- and corridor-like enclosures. Most of previous work in corridor-like enclosures [e.g., 6, 7, 8] has been performed using gaseous burners, with which the fuel supply rate must be pre-defined, therefore the interaction between the hot gases and the burning rate cannot be accounted for. In this work, a more realistic fire source is employed by using liquid pool fires. Two fuel pan sizes of constant fuel surface level and various opening sizes were used to investigate the influence of fuel surface area and ventilation factor on the burning rate and fire development.

EXPERIMENTAL SETUP

Corridor-like enclosure and façade configuration

The tests were conducted in a corridor-like enclosure having internal dimensions 3 m x 0.5 m x 0.5 m. The enclosure was constructed using six 0.5 m x 0.5 m cubic boxes and a 1.8 m x 1 m façade was attached to the front box, Box A, as indicated in Fig. 1, which shows a schematic drawing (side view) of the experimental configuration, depicting also the locations of the measurement devices. Opening dimensions, located in Box A, vary for each test case as summarized in Table 1.

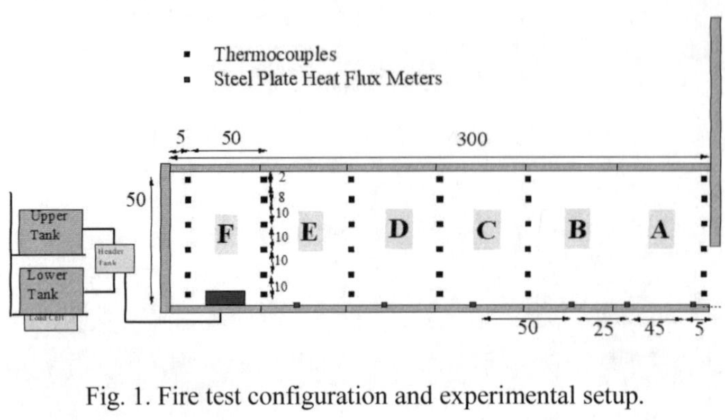

Fig. 1. Fire test configuration and experimental setup.

Fuel Delivery, Sensors and Data Acquisition System

In contrast to previous works using gaseous fuels in corridor-like enclosures [7], [8], ethanol pool fire is employed in this work to simulate more realistic fuel sources. A level maintenance system is designed to keep a constant fuel level (10 mm from the pan's rim) to minimize lip effects [9] and to establish steady-state conditions within the enclosure. Fuel is driven by gravity from the upper tank to the header tank as shown in Figure 1, with any excess fuel flushed to the lower tank. The fuel supply mechanism is placed on a balance with a maximum load of 36 kg and ±0.2 g accuracy. Two stainless steel circular pans, 0.06 m high and 0.2 or 0.3 m in diameter, representing the radiative heat transfer regime [10], are used. A water-cooling circuit wrapped around the pan helps retain constant pan temperature to reduce conductive heat losses and thus to retain constant burning rates [11], [12]. Small pebbles are placed inside the pan to suppress convective motion in the liquid and excess boiling of the fuel on the surface [13]. The burner is placed on the floor at the center of Box F. A total of thirty-six K-type thermocouples with a bead diameter of 1.5 mm are used to monitor gas temperatures inside the enclosure at every 6 s [6], [7], with six thermocouples positioned in each Box 5 cm from the side wall at 2, 10, 20, 30, 40 and 48 cm above the floor as shown in Figure 1. Six steel plate heat flux meters are located in the centerline at the floor level of the corridor-like enclosure [14]. The whole experimental set-up is placed under a 3 x 3 m^2 hood to measure heat release rate (*HRR*), production of CO, CO_2 and smoke. Footage from CCD cameras are used for visually determining flames emerging through the opening as discussed in [15].

Experiments

In total, sixteen different cases are investigated in the present work, excluding repeatability tests for each case. Table 1 summarizes all the experiment cases presented in this work. The effect of ventilation is investigated by altering the dimensions of the opening. Eight different door-like openings are used in the current study, with their dimensions shown in Table 1.

Table 1 Opening and pan dimensions of the test cases studied.

Test	Opening size (m x m)	Pan diam. (m)	Test	Opening size (m x m)	Pan diam. (m)
BC20W10H10	0.10 x 0.10	0.2	BC30W10H10	0.10 x 0.10	0.3
BC20W15H15	0.15 x 0.15		BC30W15H15	0.15 x 0.15	
BC20W10H25	0.10 x 0.25		BC30W10H25	0.10 x 0.25	
BC20W20H20	0.20 x 0.20		BC30W20H20	0.20 x 0.20	
BC20W25H25	0.25 x 0.25		BC30W25H25	0.25 x 0.25	
BC20W30H30	0.30 x 0.30		BC30W30H30	0.30 x 0.30	
BC20W50H25	0.50 x 0.25		BC30W50H25	0.50 x 0.25	
BC20W50H50	0.50 x 0.50		BC30W50H50	0.50 x 0.50	

RESULTS AND DISCUSSION

Effect of pan size

Significant differences in the burning behaviour are observed with different fuel pan sizes. To demonstrate the main characteristic stages regarding the fire growth, the temporal evolution of measured *HRR*, $\dot{Q}_{exp}$, and theoretical *HRR*, $\dot{Q}_{th}$, are plotted in Fig. 2 for two characteristic test cases, namely BC30W30H30 and BC20W30H30. The theoretical HRR, $\dot{Q}_{th}$, is calculated using Equation (4) based on the mass loss rate, as measured by the load cell and the heat of combustion, ΔH_c, of ethanol (26.78 MJ/kg). The maximum *HRR* in stoichiometric conditions inside an enclosure, $\dot{Q}_{st,in}$, is calculated by multiplying $\dot{m}_a$ by the heat released by complete combustion of 1kg oxygen, which for most fuels is found approximately equal to 3000 kJ/kg [2].

$$\dot{Q}_{th} = \dot{m}_T \, \Delta H_c \tag{4}$$

$$\dot{Q}_{st,in} = 3000 \times \dot{m}_a \tag{5}$$

As shown in Fig. 2, the fire behaviour is characterized by three district phases (*Regions I*, *II* and *III* as illustrated in the figures) appearing in succession. *Region I* corresponds to the fuel-controlled period (growth period), where the combustion efficiency is close to unity and thus $\dot{Q}_{exp}$ and $\dot{Q}_{th}$ are almost equal. In the case with the smaller pan, Region I period is substantially prolonged. This prolongation can be attributed to the radiation feedback to the fuel surface from the flame and the surroundings, which is less and results in smaller burning rates. During Region II, fire gradually becomes ventilation-controlled and $\dot{Q}_{exp}$ reaches a plateau until flames ejects through the opening. In test cases with large openings (e.g. W25xH25, W30xD30, W50xD25 and W50xD50), during this period, $\dot{Q}_{exp}$ is found to be less than $\dot{Q}_{st,in}$ indicating reduced air flow rate into the compartment. Flames ejection indicates the beginning of *Region III*, where sustained external burning is observed. In *Region III*, $\dot{Q}_{exp}$ continues to increase until a plateau is formed indicating that steady state conditions are established. Similar trends in *HRR* as depicted in Fig. 2 are generally observed in other test cases.

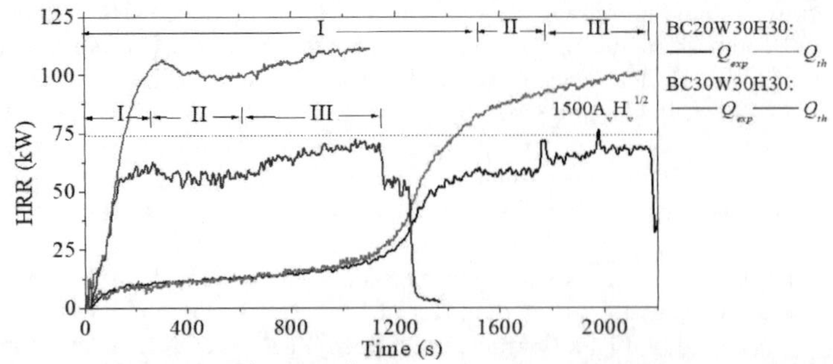

Fig. 2. Temporal evolution of $\dot{Q}_{exp}$ and $\dot{Q}_{th}$ for test cases BC30W30H30 and BC20W30H30.

To explain the difference in the *HRR*, gas temperature histories for test cases BC20W30H30 and BC30W30H30 are depicted in Fig. 3. During *Region I*, maximum temperatures are observed in Box F at the rear of the corridor as the fire is still fuel-controlled. For the case BC30W30H30, gas temperatures increase at much higher rates as a result of higher *HRR* and the duration of this region is much shorter compared to BC20W30H30. During *Region II*, maximum temperature is observed in Box E indicating that flames detach from the burner and gradually migrate towards the opening where oxygen availability is increased. At the same time, temperature in Box F decreases reaching a plateau, indicating steady-state conditions. In *Region III*, flames fill the upper layer of the corridor and appear to be anchored in Box E extending towards the opening, and eventually emerges from the opening when the *HRR* becomes sufficiently large.

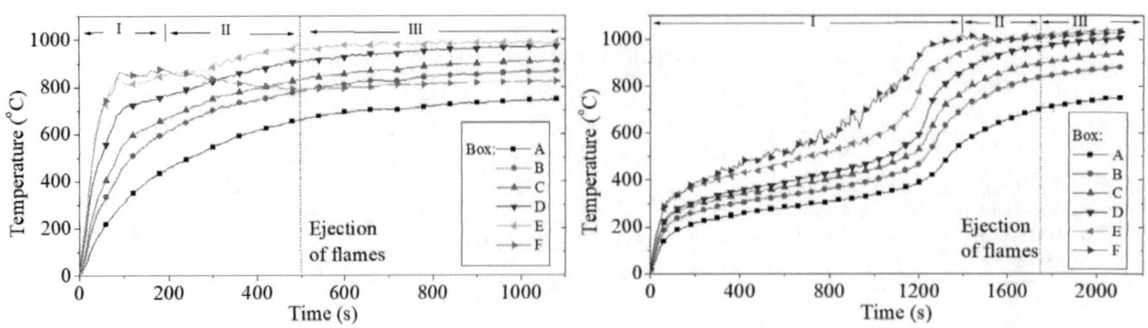

Fig. 3. Upper hot gas layer temperatures at each Box for BC30W30H30 (left) and BC20W30H30 (right).

Effect of ventilation factor

To examine the effect of the ventilation factor on the burning rate, Fig. 4 plots $\dot{m}_T$ against $A_oH_o^{1/2}$ both normalized by the fuel surface area for all test cases when the steady state condition is achieved. Additional data for cubic enclosures [16] are also included for comparison. Values of $\dot{m}_T$ are taken as the average of those at steady state conditions in Region III. As the ventilation factor increases, the normalized burning *rate*, $\dot{m}_T/A_f$, also increases until reaching a maximum value corresponding to the transition from ventilation- to fuel-controlled conditions, observed at about $A_oH_o^{1/2}/A_f$=2. A further increase in the opening factor results in a decrease in, $\dot{m}_T/A_f$, as the fire becomes fuel-controlled and finally, $\dot{m}_T/A_f$ approaches the free-burn burning rate. It is worth noting that the data for the cases with the smallest opening factor (W10xH10) is not include in Fig. 4, because the air flow rate is too limited in these cases to sustain burning and the fire was self-extinguished after a few minutes. The trend between $\dot{m}_T/A_f$ and $A_oH_o^{1/2}/A_f$ found in this work follows those obtained for cubic enclosures [1], [3]. Using experimental data from [1], [3] and [16] for different fuels, Delichatsios [3] found that the slope of the ventilation-controlled regime for cubic enclosures is 0.1, Equation (6). The present data indicate that in corridor-like enclosure a linear relation between $\dot{m}_T/A_f$ and $A_oH_o^{1/2}/A_f$ still exists, as shown in Fig. 4. However, the proportional constant is found to be 0.067, Equation (7), which is about 2/3 of the one observed in cubic-like enclosures.

$$\frac{\dot{m}_T}{A_f} = 0.1\frac{A_oH_o{}^{1/2}}{A_f} \tag{6}$$

$$\frac{\dot{m}_T}{A_f} = 0.067\frac{A_oH_o{}^{1/2}}{A_f} \tag{7}$$

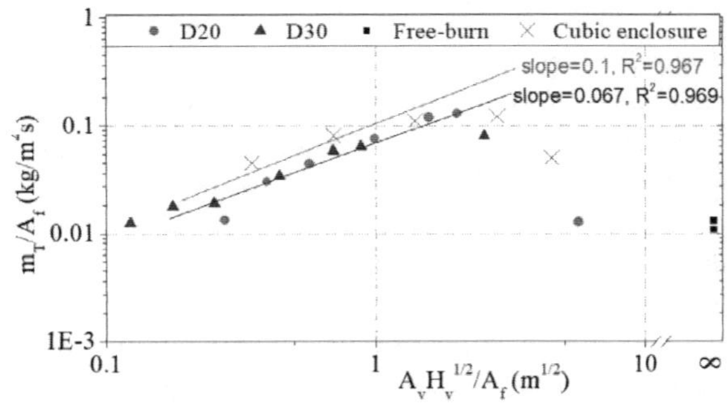

Fig. 4. Normalized values of $\dot{m}_T$ against $A_oH_o^{1/2}$ for cubic-[16] and corridor-like enclosures in *Regime III.*

The current results are in accordance with previous experimental studies in corridors [3] [4], [17], [18] demonstrating that the burning rate in corridor-like enclosures is less than that in cubic-like enclosures under ventilation-controlled conditions. This difference was attributed to a decrease of the air inflow rate or the ventilation coefficient, *C*, in corridor-like enclosures [4] and [19]. The typical values for *C* are 0.45 [1] or 0.5 [2], which have been deduced during post-flashover conditions in cubic enclosures with single moderate opening. Thomas et al. [4] and Yii et al. [19] examined the effects of opening dimensions on the air flow rate and concluded that assuming a constant *C* value overestimates the air inflow rate, especially when the opening width is the same as the full width of the enclosure. Yii et. al. [19] also showed that for large openings air entrainment dominates the vent flows. During *Region II*, all combustion takes place inside the enclosure. The ventilation coefficient can be calculated from the measured HRR in *Region II* as $C=\dot{Q}_{exp}/3000A_oH_o^{1/2}$ with the assumption that all oxygen is consumed in the enclosure in this region. This assumption is reasonable as (i) the *HRR* in this region is nearly constant and (ii) external burning occurs after the end of this region. Figure 5 shows the calculated *C* values for all test cases. It is found that *C* decreases with a decrease in the ventilation factor, which is in accordance to previous analysis of post-flashover fires [4], [19] indicating that $\dot{m}_a$ in long enclosures (e.g. corridors) is less than in rectangular enclosures with the same opening geometry.

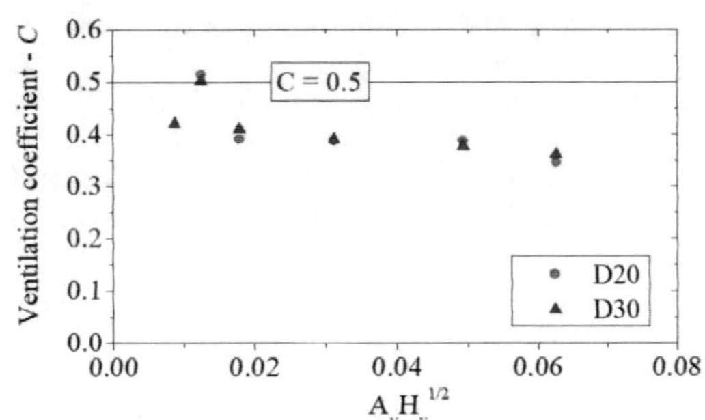

Fig. 5. Ventilation coefficient C for each test case against $A_oH_o^{1/2}$.

CONCLUSIONS

An series of medium-scale fire tests was performed using a liquid pool fire located at the rear of a corridor-like enclosure. Two pan sizes and eight opening sizes (thus ventilation factors) were used. The main conclusions of this work are:

1. For most cases, three distinct burning regions (*Region I, II* and *III*) have been observed, corresponding respectively to fuel-controlled, ventilation-controlled and steady-state burning. The duration of each regions depends on both the pan size and ventilation factor. In *Region II*, the heat release rate is nearly constant. The transition from *Region II* to *III* is indicated by the flame emerging from the opening. In *Region III*, the *HRR* continues to increase until a steady state is established.
2. For the cases when ventilation-controlled conditions are achieved, the normalized steady state mass burning rate, $\dot{m}_T/A_f$, is found to increases linearly with the normalized ventilation factor, $A_oH_o^{1/2}/A_f$, which is consistent with previous findings with cubic-like enclosures. However the proportional constant is found to be about 2/3 of that observed in cubic-like enclosures.

IOP Conf. Series: Journal of Physics: Conf. Series **1107** (2018) 042024 doi:10.1088/1742-6596/1107/4/042024

3. The effect of opening size on the air flow rate into the corridor was also examined, and the ventilation coefficient, *C*, for corridor-like enclosures during post-flashover conditions was found to decrease with an increase of the ventilation factor. These results are supported by the temperature measurements which show that the temperature in a corridor-like enclosure is not uniform even after ventilation-conditions are established. The temperature difference between the top and bottom locations increases with an increase in the ventilation factor.
4. The present work provides a framework towards the understanding the physics of the fire growth in corridors-shaped structures but future experiments should aim at further investigating the effect of corridor geometry (e.g. investigation of different aspect ratios and geometrical configurations), fuel type and positions within the enclosure. More detailed information regarding velocity distribution, gas and smoke concentration at the interior would further enhance the understanding of the phenomena involved.

REFERENCES

[1] Kawagoe, K., (1958) Fire Behaviour in Rooms, Report 27, Building Research Institute, Tokyo.
[2] Hurley, M.J., Performance-Based Design, SFPE Handbook of Fire Protection Engineering (5th ed.) Hurley M.J., National Fire Protection Association, Quincy, MA 02269, 2016, p 1233/1261.
[3] Delichatsios, M.A., Silcock, G.W.H., Liu, X., Delichatsios, M., Lee, Y.P. (2004) Mass pyrolysis rates and excess pyrolysate in Fully Developed Enclosure Fires, Fire Safety Journal 39:1– 21, https://doi.org/10.1016/j.firesaf.2003.07.006
[4] Thomas, I.R, Bennetts, I.D., (1999) Fires in Enclosures with Single Ventilation Openings-Comparison of Long and Wide Enclosures, Proceedings of the International Symposium of Fire Safety Science, 6:941-952, International Association for Fire Safety Science.
[5] Yii, E.H., (2002) Modelling the Effects of Fuel Types and Ventilation Openings on Post-Flashover Compartment Fires, PhD Thesis, University of Canterbury, Canterbury, New Zealand.
[6] Lee, Y.P., (2006) Heat Fluxes and Flame in External Facades from Enclosure Fires, PhD thesis, FireSERT, University of Ulster, Jordanstown, UK.
[7] Beji, T., (2009) Theoretical and Experimental Investigation on Soot and Radiation in Fires, PhD thesis, FireSERT, University of Ulster, Jordanstown, UK.
[8] S. Ukleja, S., (2012) Production of Smoke and Carbon Monoxide in Under-ventilated Enclosure Fires, PhD thesis, FireSERT, University of Ulster, Jordanstown, UK.
[9] Hall, A.R., (1972) Pool Burning: A Review, Report No. AD0781347, Rocket Propulsion Establishment: Westcott.
[10] Babrauskas, V., (1983) Estimating large pool fire burning rates, Fire Technology, 19:251–261.
[11] Blinov, V.I., Khudyakov, G. N. (1961) Diffusion burning of liquids, Report No. AERDL-T-1490-A, Army Engineer Research and Development Labs Fort Belvoir VA.
[12] Vali, A., Nobes, D.S., Kostiuk, L.W. (2013) Effects of Altering the Liquid Phase Boundary Conditions of Methanol Pool Fires, Experimental Thermal and Fluid Science, 44:786-791, https://doi.org/10.1016/j.expthermflusci.2012.09.023
[13] Luketa, A. (2010) Assessment of Simulation Predictions of Hydrocarbon Pool Fire Tests, Report No. SAND2010-2511, Sandia National Laboratories.
[14] Beji, T., Ukleja, S., Zhang, J., Delichatsios, M.A. (2012) Fire behaviour and external flames in corridor and tunnel-like enclosures, Fire and Materials, 36:636-647, https://doi.org/10.1002/fam.1124
[15] Asimakopoulou, E.K., Chotzoglou, K., Kolaitis, D.I., Founti, M.A. (2016) Characteristics of Externally Venting Flames and Their Effect on the Façade: A Detailed Experimental Study, Fire Technology 52(6):2043-2069, https://doi.org/10.1007/s10694-016-0575-5
[16] Bullen, M.L., Thomas, P.H. (1978) Compartment Fires with Non-Cellulosic Fuels, Proceedings of the Combustion Institute, 17:1139–48.
[17] Miyazaki, S., Watanabe, Y. (1998) An experimental study on fire phenomena of liquid fuel in a small-sized tunnel burning behaviour of n-heptane (Part 1), paper B-14 and an experimental study on fire phenomena of liquid fuel in a small-sized tunnel relation between inside temperature of the tunnel and burning rate of n-heptane (Part 2), paper A-15.
[18] Cooke, G.M.E. (1998) Tests to determine the behavior of fully-developed natural fires in a large compartment, Fire and Risk Science Fire Note 4, Building Research Establishment Ltd.
[19] Yii, E.H., Fleischmann, C.M., Buchanan, A.H. (2007) Vent Flows in Fire Compartments With Large Openings, Journal of Fire Protection Engineering 17 (3):211-237, https://doi.org/10.1177/1042391507069634

INDICATORS FOR THE QUALITY ASSESSMENT OF THE GRID RESOLUTION

Francesco Saverio CIANI[1] and Giordana GAI[2]

[1]*Guidi & Partners s.r.l., Italy*
[2]*Sapienza University of Rome, Italy*

francescosaverio.ciani@gmail.com

ABSTRACT

The choice of the grid resolution in thermo-fluid dynamics simulations is to be made as a balance between the quality of the results and the computational effort. When quick results are needed, coarse resolutions are often used, but in some cases this could be detrimental for the reliability of the numerical results. However, it is well known that reducing the size of the mesh affects dramatically the overall simulation time.

Several approaches can be used to ensure that the mesh size is appropriate to reproduce the fire and smoke dynamics for a given scenario with suitable computational effort and make reliable predictions.

The first and most used approach consists of limiting the ratio between the characteristic fire diameter D* and the grid size *dx* inside a range based on a series of experimental tests reported in literature. This approach is generally used as a first step in order to define a tentative grid resolution for the given model.

Other approaches consist of checking if some indicators are included in a literature-based range once the simulation has run. For example, the dimensionless parameter $y+$ is a measure of how well the near-wall field is solved and can be effectively useful when the roughness of the walls affects the overall flow field. The measure of turbulence resolution (MTR), instead, is a quantity related to the resolved i.e. calculated turbulent kinetic energy and gives an idea of how detailed the turbulence effects are accounted in the model.

These indicators are certainly a significant source of information regarding the quality of the model as well as the need of local refinement. However, they are not a guarantee for the achievement of convergent results of the quantity of interest (e.g. temperature, visibility, FED) which from the engineering point view are as valuable as the quality assessment of the model.

In the paper, the traditional indicators are reviewed from a theoretical point of view and the Pearson's coefficient is introduced as an attempt to add other statistical correlation principles. The purpose is to couple mesh quality metrics with a global convergence check of the quantity of interest of the fire safety analysis, driven also by aspects of conservativeness and computational cost. An application to a case study is also provided and shows pros and cons of the indicators.

KEYWORDS:

quality metrics, correlation, sensitivity analysis, grid optimization, turbulence, LES, fire compartments.

NOMENCLATURE LISTING

D^*	characteristic fire diameter (m)	T_∞	ambient temperature (K)
dx	grid size (m)	τ_w	viscous stress
K_{SGS}	modeled turbulence kinetic energy (subgrid)	δ_n	wall-normal cell dimension (m)
K_{LES}	resolved turbulence kinetic energy	δ_v	local viscous length scale
MTR	measure of turbulence resolution	$\tilde{u}_i$	component i of the velocity (m/s)
$y+$	viscous wall units	Q	total heat release rate
c_p	specific heat (J mol^{-1} K^{-1})	ρ_∞	Ambient density (kg/m^3)
u_τ	friction velocity (m/s)	μ	Viscosity (kg m^{-1}·s^{-1})
q_i	quantity of interest		

IOP Conf. Series: Journal of Physics: Conf. Series **1107** (2018) 042025 doi:10.1088/1742-6596/1107/4/042025

INTRODUCTION

The choice of the grid size in thermo-fluid dynamics simulations is to be made as a balance between the quality of the results and the computational effort. When quick results are needed, coarse resolutions are often used, but in some cases this could be detrimental for the reliability of the numerical results. However, it is well known that reducing the size of the mesh affects dramatically the overall simulation time. For instance, considering cubic cells, if the side of each cell is reduced by a factor of two, the simulation time will increase of approximately 16 times (23 for each axis and 2 for the time step, in order to satisfy the CFL condition). From the point of view of a fire practitioner, the main interest is to spend the least computational effort and provide reliable results in terms of output quantities e.g. temperature, radiative heat flux gas, visibility, FED, etc. This is not necessarily a guarantee for accurate results if comparisons are made with experimental tests, given the unavoidable limitations of the turbulence modeling [1]. However, considering that CFD allows for the most detailed description of fire and smoke phenomena, it might expect that a high accuracy from the simulation and reliable predictions for design purpose is achieved.

Today, the fire design of different types of buildings relies more and more on the fire safety engineering and the use of advanced modeling of fire is rocketing, also thanks to a renewed attention and openness of the fire authorities. FDS is currently the most used software for fire simulations for a wide variety of civil applications [2]. It works under the Low-Mach approximation, valid for the typical range of pressure showed during fire scenarios, and the turbulence formulation is a LES (Large Eddy Simulation) with implicit filtering. With this approach, a set of filtered equations is obtained from the complete Navier-Stokes equations, and as FDS adopts an implicit filter (filter ~ grid size) we know from the very beginning the results are affected by the grid resolution, as the governing equations are. Pope [3] identified in the adaptive LES the only way to make LES formulation complete e.g. free from flow-dependent specifications [4]. This can be theoretically obtained by adapting step by step the grid size (e.g. the filter) in order to maintain the level of turbulence resolution below the required tolerance and ensure in the whole domain the same accuracy and reliability of the results. However, this is not generally practiced in fire simulations and therefore a quality assessment must explicitly be carried out to check the correctness of the model and a balance among accuracy, conservative results and computing time.

METRICS FOR QUALITY ASSESSMENT

A review of some indicators helpful for the definition of the grid resolution is conducted. Some of them, e.g. rule of thumb, can be considered *a priori* methods, because they are applied to build the first tentative grid layout for the given model, with no information on the results. Others are instead applied a *posteriori,* e.g. MTR, once the results of the simulation are available and one can check the quality of the grid resolution.

Rule of thumb

The first and most used method consists of limiting the ratio between the characteristic fire diameter D^* and the grid size dx inside a range of 4-16 [2]. The characteristic fire diameter should be thought one of the many possible length scales that need to be resolved to capture accurately the fire and turbulence phenomena. It can be expressed via the relation (1) where Q is the total heat release of the fire, ρ_∞ is the ambient density, c_p is the ambient specific heat, $T\infty$ the ambient temperature and g is gravity.

$$D^* = \left(\frac{\dot{Q}}{\rho_\infty c_\infty T_\infty \sqrt{g}}\right)^{\frac{2}{5}} \tag{1}$$

In many instances, D* is comparable to the physical diameter of the fire, in which case, Q* is on the order of 1. The $\dot{Q}^*$ represent the square root of the Froude number and has been given by the following relation (2):

$$\dot{Q}^* = \frac{\dot{Q}}{\rho_\infty c_p T_\infty \sqrt{gD} D^2} \tag{2}$$

Since the formula (1) has derived from the formula (2) that defines $\dot{Q}^*$, it is important to define an optimum value for the engineering applications. Cox and Kumar [5] define that the dimensionless HRR ($\dot{Q}^*$) should be in the range of 0.3 to 2.5 for natural fires in buildings. The above mentioned range of the ratio D^*/ dx comes from the test series reported in NUREG 1824 [6] and therefore it is the range that worked well for those examined cases. Thus, this does not necessarily imply that the range is suitable for our model, considering the geometrical characteristics as well as the fire size and the ventilation conditions.

Near wall resolution *y+*

A method to check the quality of the grid resolution near the wall is to limit *y*+ in the range 30-1000 [2]. Similarly, Cox and Kumar [5] suggested that *y*+ must be greater than 400 to capture the main physical features of the flow The value of *y*+ is calculated as reported in (3), as a half of the ratio (since the velocity lives at the cell face center) between the wall-normal cell dimension δ_n and the local viscous length scale, δ_v. In addition, τ_w is the viscous stress evaluated at the wall and u_τ is the friction velocity.

$$y^+ = \frac{1}{2}\frac{\delta_n}{\delta_v} \; ; \; \delta v = \frac{\mu/\rho}{u_\tau} \; ; \; u_\tau = \sqrt{\tau_w/\rho} \tag{3}$$

This parameter *y*+ is a measure of how well the near-wall field is solved and can be effectively useful when the roughness of the walls affects the overall flow field. The *y*+ can be extrapolated from FDS as a boundary file or as a punctual device and the lowest is, the higher the resolution of the near wall field is.

Measure of turbulence resolution

The MTR is a quantity related to the resolved turbulent kinetic energy and gives a measure of the resolved turbulence in the domain. It is expressed as the ratio between the subgrid turbulent kinetic energy K_{SGS} and the total energy (sum of K_{SGS} and the solved energy K_{LES}) as in (4). By definition, MTR is always included between 0 (all solved) and 1 (all modeled) and satisfactory values should not exceed 0.2 for LES [3]. However, this threshold has been calculated in case of isotropic turbulence conditions, which are certainly not reached near the fire source or every time there is a strongly prevalent component of the velocity.

$$MTR = \frac{K_{sgs}}{K_{sgs}+K_{LES}} \tag{4}$$

Considering FDS, K_{SGS} can be obtained as a output quantity at a given point, while K_{LES} can be calculated by post-processing the three components of the velocity, considering mean values and fluctuations according to the equation (5) where $\tilde{u}_i$ is the component i of the velocity and $\langle\tilde{u}_i\rangle$ is its mean.

$$TKE = \frac{1}{2}(\tilde{u}_i - \langle\tilde{u}_i\rangle)(\tilde{u}_i - \langle\tilde{u}_i\rangle) \tag{5}$$

The MTR can be evaluated in the time range of interest e.g. when the flow field is rather stable.

THE NEED OF ADDITIONAL QUALITY INDICATORS

The methods described above are a valuable source of information for the quality of our grid resolution and the need of local refinement; nevertheless, in some cases they are not helpful to achieve a high quality configuration of the mesh. Two typical situations are identified:
- although the MTR is below the 0.2 threshold, by varying the grid resolution, the output quantities keep changing significantly (according to the tolerance required by the user);
- near the fire source and high gradient zones where the flow field is highly anisotropic it is not possible to achieve MTR lower than 0.2 i.e. MTR is not reduced with refinement. At the same time, however, there is no indication in literature on a different threshold that could be used to compare with and one could be forced to make predictions even if the results keeps changing and are affected by uncertainties.

Additional indicators taking into account directly the output quantities are reviewed below.

Pearson's Coefficient

Correlation coefficients are used in statistics to measure how strong a relationship is between two variables. There are several types of correlation coefficients: Pearson's coefficient is a correlation commonly used in linear regression [7]. It is possible to use the Pearson coefficient (r) to quantify the correlation (linear dependence) between two variables X and Y, where $\bar{X}$ and $\bar{Y}$ are the mean values of the samples.

$$r = \frac{\sum_{i=1}^{n}(X_i-\bar{X})(Y_i-\bar{Y})}{\sum_{i=1}^{n}\sqrt{(X_i-\bar{X})^2}\sqrt{(Y_i-\bar{Y})^2}} \tag{6}$$

As reported in the Fig. 1, the value of this coefficient are included in the range of -1 to +1. A value of 1 implies that a linear equation describes the relationship between X and Y perfectly, whereas a value of 0 implies that there is no linear correlation between the variables [7]. This Pearson's Coefficient (later called PC) could be used, during the sensitivity analysis, to assess the X-Y correlation where X and Y are the time-averaged values of the output quantities q_i (e.g. temperature, visibility, etc.) of the same fire scenario, but at different grid resolutions. The advantage of using such an indicator is that the correlation is checked directly on the q_i, therefore it could give an additional information compared to MTR and y+, which instead are ultimately the basis for local mesh refinement. By adjusting the grid resolution, the q_i is expected to converge to a stable trend and the PC to increase (0.75 is recommended as a good value [8]).

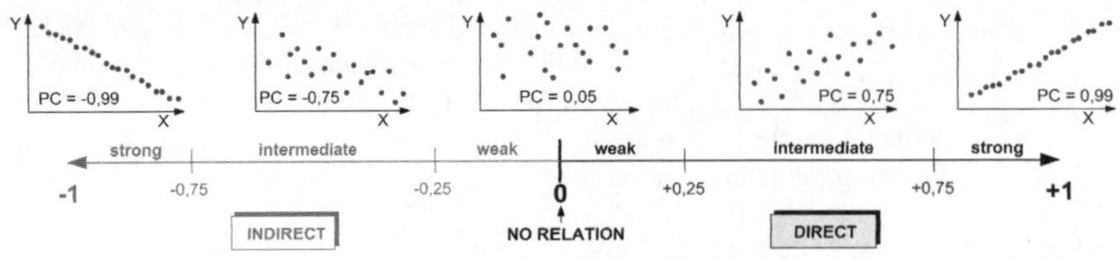

Fig. 1. Pearson coefficient values

Mean error

The mean error (ME) calculation can be coupled with the PC. For example, one may assure that although the PC is below the recommended value of 0.75, the mean error between the time-averaged trends is decreasing, implying that we are moving towards a stable prediction of the q_i. This calculation should be thought as an additional information for convergence of the model. The limitation of using ME is that it is not reliable when small values of the q_i are compared (large relative error, not representative of the overall model error).

TENTATIVE APPROACH

A tentative approach is reported in the Fig. 2 and commented on below. The drawback of the approach is that as a minimum three different configurations for the given simulation must be considered. The advantage is that it extends the zone where MTR is significantly higher than 0.2. A similar application can be done using *y+* instead of MTR (in case one is interested in near wall prediction).

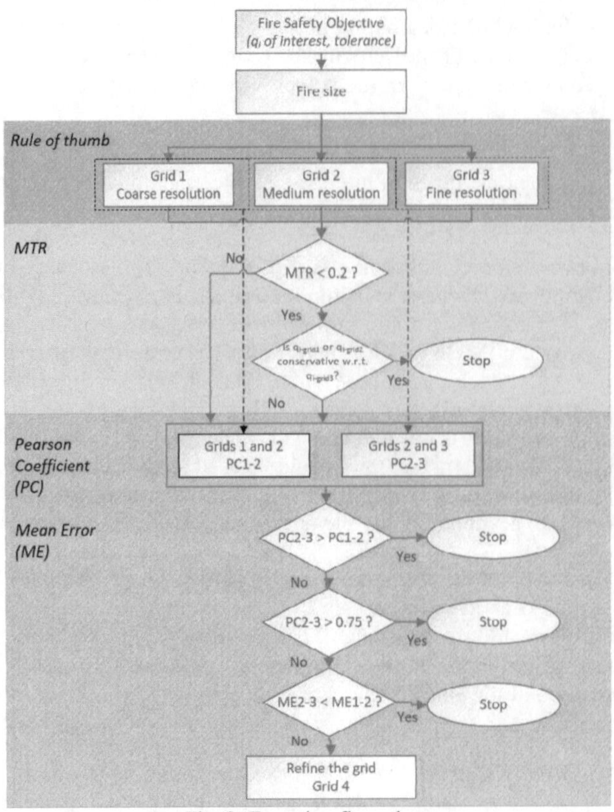

Fig. 2. Tentative flow chart

- Step 1. Define the objective of the analysis e.g. the fire size, the quantities of interest q_i and specifically the tolerance accepted when refining the mesh (e.g. 10 %, 20 % etc.);
- Step 2. Rule of thumb is applied for the choice of the grid resolution (coarse, medium and fine);
- Step 3. MTR (and/or *y+*) is calculated in the domain;
- Step 4. If MTR is lower than 0.2 and the q_i from the coarse or the medium resolutions are conservative with respect to the q_i of the fine resolution, one may consider the highest cell size for

making prediction (on the safe side with less computation cost). If the conservative q_i is that of finest grid, other indicators should be used;

- Step 5. Calculation of the PC for the q_i. If by refining the grid the correlation increases (higher PC), this implies that the time-averaged trends of the q_i are not diverging;
- Step 6. Although the PC is not increasing with the grid refinement, the finest grid can be considered as acceptable if the PC 2-3 is higher than 0.75 (recommended value for good correlation).
- Step 7. In case of weak correlation (< 0.75), it may be verified that there is no increment in the ME. Otherwise, an additional refinement is needed (grid 4).

CASE STUDY

The approach previously described is applied to a simple geometry (building size 20 m x 10 m x 3 m), lateral ventilation openings along the 10 m walls and central position of the fire (burner surface 2x1 m2). Table 1 shows that each one of the two scenario typologies have a different Heat Release Rate i.e. 2 MW and 4 MW, which is constant during the whole simulation. Three different grid resolutions are analyzed (50 cm, 25 cm, 10 cm). Furthermore, Table 1 shows the values of D^*/dx used for the initial calibration of the grid resolution. It is worth mentioning that both simulations consider value of $\dot{Q}^*$ that are inside the above mentioned range [6]. Specifically, the 4 MW case study has a $\dot{Q}^*$ of 1,1 whereas, in the 2 MW, it is 0.56.

Table 1. Case studies performed and values of D*/dx

Grid spacing (cm)	D*/dx	
	HRR: 2 MW	HRR: 4 MW
50	2.5	3.4
25	5.1	6.7
10	12.7	16.8

The MTR has been evaluated according to the equations (4)-(5) by post-processing the velocity components outputs in order to obtain TKE. The devices have been placed at two different heights from the ground i.e. 1 m and 2 m, and at different distances from the burner along the x and y axes (passing through the center of the compartment, where the burner is placed). Table 2 shows that, refining the grid, the MTR decreases below the threshold 0.2 recommended by Pope [3].
As previously commented, it is important to choose the mesh to capture the main physical features of the flow ensuring that the near-wall nodes satisfy the criterion y+ >400. In the 4 MW case the y+ reaches approximately values of 700, but it decreases by refining the grid up to 200 (grid spacing 10 cm), ensuring a satisfactory behavior in the near wall region (the results are only commented for lack of space).

Table 2. MTR along x and y direction for 2 and 4 MW cases

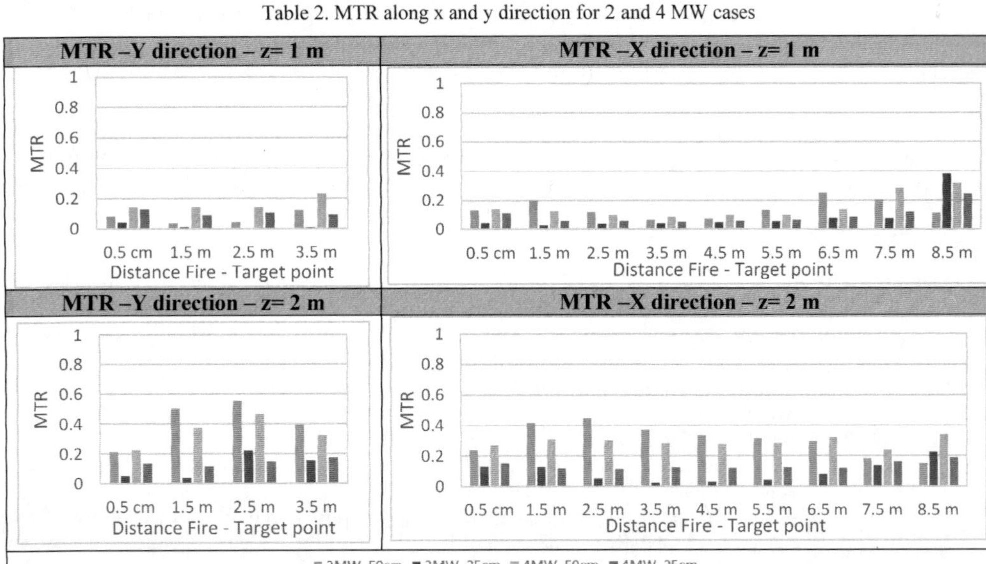

Table 3 shows instead the temperature values (time-averaged every 10 s) for different grid spacing and allows an overall assessment of conservativeness supporting the choice of the mesh, which normally accompanies the proper quality assessment of the model from practical engineering perspective. In this way, with a single graph and a unique value of PC we have an indication on which mesh is the best for our case without the need of plotting the trends of the output quantities at given points. In the 4 MW case, the correlation increases by refining the grid in agreement with the increment of the PC (from 0.88 to 0.9), so there is a convergence of the q_i. From this analysis, the 25 cm grid seems a reasonable choice for the simulation, because it provides higher (therefore, on the safe side) values compared to the 10 cm resolution but far less computational time. Nevertheless, depending on the aim of the fire safety analysis, it should be evaluated if the differences in the results are appreciable and hence if it is the case to refine the grid further.

Table 3. 4 MW fire at z = 2 m (average of significant points)

Temperature - Grid spacing 50 cm vs 25 cm	Temperature - Grid spacing 25 cm vs 10 cm
PEARSON COEFFICIENT = 0.88	**PEARSON COEFFICIENT = 0.90**
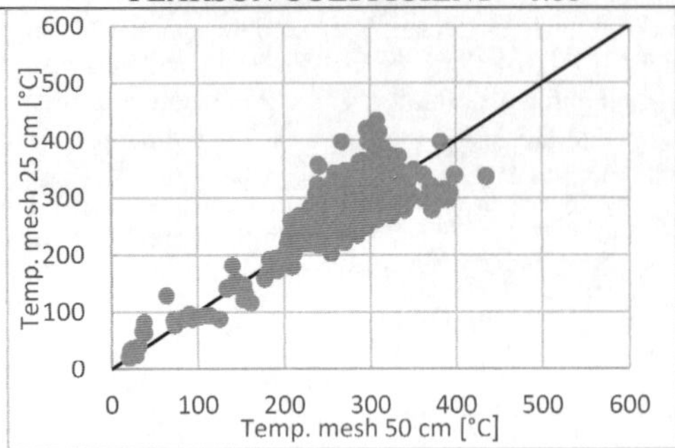	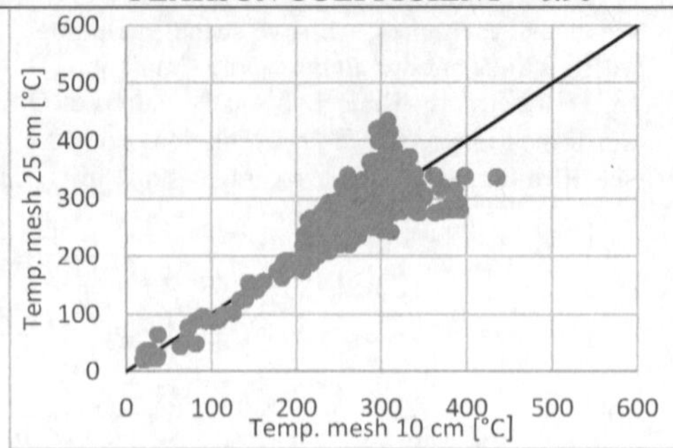

CONCLUSIONS

To wrap up, the following conclusions can be drawn.

- The traditional approaches used to assess the quality of the grid resolution are straightforward from the theoretical point of view and are generally a valid source of information in the assessment of our model. However, for fire conditions they can sometimes give poor information (e.g. when MTR is greater than 0.2, most of the times with high HRR and near the fire source).
- Additional possible indicators are identified e.g. the PC which can be used to analyze directly the q_i of different grid resolutions and check if, by refining the mesh, a global convergence has been reached for our model in the q_i, allowing us to make engineering predictions.
- The case study shows a possible use of the indicators reviewed and highlights that the PC is in agreement with the rule of thumb, *y+* and MTR (25 cm medium resolution is on the safe side). It is worth to notice that for more complex geometries and boundary conditions, higher HRR, etc. in which the traditional indicators might give no meaningful information in some region of the model, the PC may be even more helpful from a practical point of view as it is directly related to output quantities.

REFERENCES

[1] McDermott, R. "Quality assessment in the Fire Dynamics Simulator: a bridge to reliable simulations". *Proceedings of Fire and Evacuation Modeling Technical Conference,* Baltimore, Maryland, 2011.

[2] McGrattan, K., Hostikka, S., McDermott, R., Floyd, J., Weinschenk, C., Overholt, K. *Fire Dynamics Simulator User's Guide. FDS Version 6.6.0*, NIST, Gaithersburg, USA, 2017.

[3] Pope, S. B. *Ten questions concerning the large-eddy simulation of turbulent flows*. New Journal of Physics, 2004. doi: 10.1088/1367-2630/6/1/035.

[4] Pope, S.B. *Turbulent Flows*, Cambridge University Press, 2000.

[5] Cox, G., Sumar, K. "Modeling enclosure Fires Using CFD," *The SFPE Handbook of Fire Protection Engineering*, 3rd edition, Society of Fire Protection Engineers, 2002, pp. 3-205 till 3-207.

[6] U.S. Nuclear Regulatory Commission, Office of Nuclear Regulatory Research, Electric Power Research Institute (EPRI), Palo Alto, CA, NUREG- 1824 and EPRI 1011999, Verification and Validation of Selected Fire Models for Nuclear Power Plant Applications Volume 7 FDS, MD, 2007.

[7] Rodgers, J. L, Nicewander, W. A. Thirteen Ways to Look at the Correlation Coefficient, *The American Statistician*, Vol. 42, No. 1., February 1988, pp. 59-66.

[8] Evans, J. D. Straightforward statistics for the behavioral sciences. Pacific Grove, CA: Brooks/Cole Publ. 1996.

Preliminary Numerical Study of Fire-Induced Pressure Rise in a Passive House Compartment

Junyi LI[1], Tarek BEJI[1] and Bart MERCI[1]

[1]*Department of flow, heat and combustion mechanics, Ghent University, Belgium*

Junyi.Li@UGent.be

ABSTRACT

Passive houses are becoming more and more popular due to energy efficiency requirements. An important aspect is their airtightness, i.e., they have a very low leakage. The present paper describes a preliminary numerical study on fire-induced pressure variations in a structure that resembles a passive house compartment. The configuration without ventilation system is considered in this paper. The fire source used in the experiment consists of wood cribs that have been simulated in the Fire Dynamics Simulator (FDS), version 6.5.2, as a cube with a prescribed heat release rate (HRR) curve (based on experimental measurements). The grid sensitivity study is conducted by using different mesh cell size and comparing the results of pressure variation. A cell size of 0.1 m is considered sufficient in this preliminary study. The maximum over-pressure obtained without ventilation reached up to 870 Pa in the experiment. Such high over-pressure will definitely hinder the escape of occupants. The structural integrity may also be affected by the over-pressure. Using a constant leakage area, which is the default option, the pressure difference without ventilation rises up to 2027 Pa in the simulation. A good agreement with the experimental data can only be achieved if the simulated leakage area is not kept constant when the room pressure rises. Tuning the leakage pressure exponent in FDS significantly reduces the deviation between the simulation and experimental results. The default setting of reference pressure difference in FDS is used to examine the influence of this parameter on the pressure prediction. When applying the default reference pressure difference, a smaller leakage pressure exponent is needed to reproduce experimental results. Therefore, the reference pressure difference has a certain influence on leakage area and should be set according to the pressure used in determining the reference leakage area during experiments. The overall pressure evolutions are qualitatively well captured in the FDS simulations, as these strongly depend on the fire heat release rate evolution, which was prescribed to match the experimental profile. However, in order to improve the current level of accuracy and address the general validity of the observations made in the present study, more experimental and numerical work need to be carried out.

KEYWORDS:

fire-induced pressure; leakage; passive house fire; CFD modeling; FDS

1. INTRODUCTION

With the increase of energy-saving and environmental awareness, the development of low-energy buildings continues to grow [1]. A passive house is a relatively new type of low-energy house, which conforms to the highest energy standard for buildings. The passive house 'label' is a promise to spare up to 90% of heating (or cooling) energy compared to common buildings [2].

From a fire safety standpoint, the construction requirements of passive houses, in particular the very low leakage, may result in a substantial pressure increase in the event of a fire, due to the release of hot combustion products in a well-confined enclosure.

In experimental campaigns carried out in an air-tight apartment in Finland (study conducted by Aalto University and VTT), pressures of up to 1650 Pa have been reported [3,4]. Besides, in the framework of the OECD PRISME project [5,6], a series of experiments have been conducted in confined and mechanically ventilated compartments and over-pressures of more than 2500 Pa have been reported. These experimental results confirm that pressure variation is an important issue to be considered in a confined compartment fire [5,6]. Chow et al. [7] used the Fire Dynamics Simulator (FDS) to investigate the fire-induced pressure variation in a confined chamber and pointed out that simulation results agree reasonably well with experimental results. Wahlqvist et al [8] examined the influence of building leakages on the ventilation system during a fire. The fire-induced overpressure in a room of tight leakage class was found to be about 1900 Pa [8]. Bonte et al. [9] studied the capability of a zone model (CFAST) and a field model (ISIS) to predict the interaction between the mass loss rate and the total relative room pressure in the case of under-ventilated fire conditions. The over-pressure and under-pressure obtained in their study also reached thousands of Pascals.

Pressures between 100 Pa and 200 Pa would be enough to prevent door opening and evacuation of occupants. Thus, the reported pressure levels clearly indicate a threat to life safety. Moreover, a pressure rises up to 1000 Pa can also pose too strong an attack on the structural integrity. E.g., in one of the tests in [3] , a window was shattered. Obviously, this has a significant impact on the ventilation conditions and hence the fire dynamics.

The objective of the present paper is to study the fire-induced high-pressure level due to the airtightness, by comparing novel experimental and numerical results.

2. EXPERIMENTAL AND NUMERICAL SETUP

The CFD package FDS, version 6.5.2, is used with the default settings (unless specified otherwise). Lumped species and infinitely fast reaction model are used in FDS combustion simulation. The reader is referred to [10] for details on the FDS default settings and equations.

2.1 Geometry

The experimental facility considered in this research is in the region of Mons in Belgium and consists of two rooms connected by a door (see Fig. 1). The construction has the same inner dimensions as a 40 foot shipping container[11]. The fire room is 7.6 m long, 2.4 m wide and 2.4 m high and the adjacent room is 4.4 m long, 2.4 m wide and 2.4 m high. During the experiment, the partition door was closed but a small leakage area of 0.9 m × 0.01 m has been measured below the door. The configuration analyzed in this paper is the case without ventilation system (fresh air supply and gas extract) and the ducts were sealed.

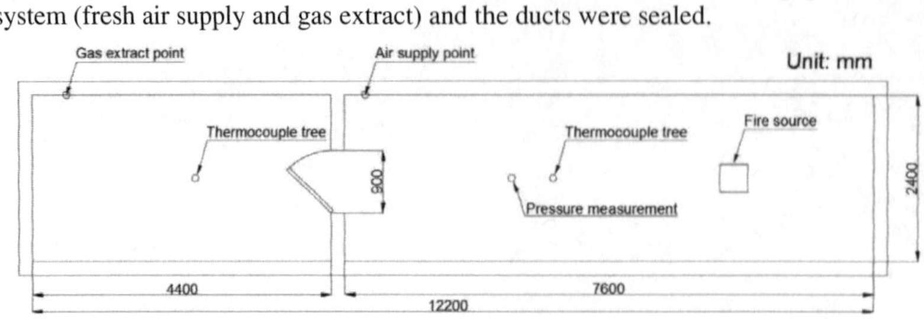

Fig. 1. Layout of experimental setup.

2.2 Fire load and Heat Release Rate

Experimental tests were carried out in the structure with overlapping wooden slats, see Fig. 2. The slats are 38 cm long and the section of a slat is 27 mm × 18 mm. There are 15 layers of slats. A stainless-steel cup 9.5 cm in diameter containing 100 ml of heptane is centered below the slats. In the FDS simulations, a 0.4 m × 0.4 m × 0.4 m cubic obstruction was set as a fire source.

The heat release rate (HRR) was calculated from the mass loss rate measured using a scale placed underneath the fire source, assuming complete combustion inside the compartment. The HRR obtained from the experiments was prescribed as HRR input in the FDS simulations. Fig. 3 confirms that the HRR as retrieved from the FDS output follows the input very well.

Fig. 2. Fire source in the experiments (overlapping wooden slats).

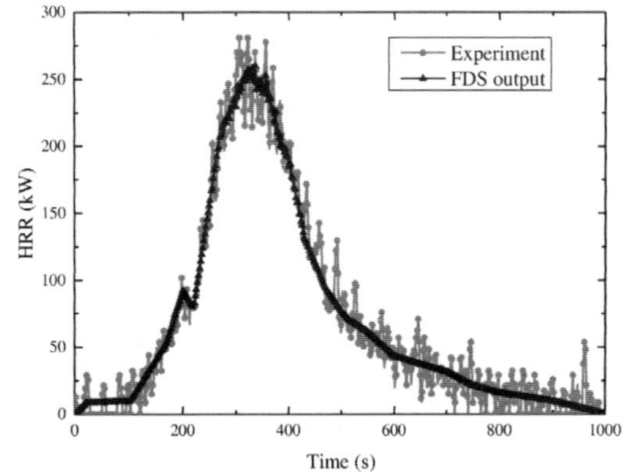

Fig. 3. Heat release rate evolution in time: experimental data and FDS output.

2.3 Mesh resolution

For the setting of mesh resolution, the non-dimensional expression $D^*/\delta x$ is used to calculate the suitable mesh size. D^* is the characteristic length scale associated with the fire HRR and δx is the mesh cell size. Equation 1 is used to calculate D^*.

$$D^* = \left(\frac{\dot{Q}}{\rho_\infty c_p T_\infty \sqrt{g}} \right)^{2/5} \tag{1}$$

where $\dot{Q}$ is the heat release rate (kW), ρ_∞ is the ambient air density (kg/m^3), c_p is the specific heat of fluid (kJ/(kg·K)), T_∞ (= 293 K) is the ambient air temperature and g is the gravitational acceleration (m/s^2).

The characteristic length scale is 0.5 m in this paper. This value is calculated based on the measured peak HRR value of 280 kW. The recommended ratio of $D^*/\delta x$ is between 4 and 16. According to the recommended ratio, the mesh cell size of 0.1 m, 0.075 m and 0.05 m were chosen to study the grid sensitivity.

2.4 Boundary conditions

The outer walls are made of 0.2 m concrete and finished with plaster on the inside to ensure the airtightness. Behind the plaster board an insulation layer is set with 0.05 m Rockwool. The partition wall is 0.2 m thick and has three layers: two gypsum layers (0.035 m) with Rockwool (0.13 m) in between. The partition door also has three layers: two steel layers (0.004 m) with foam (0.192 m) in between. The outer door was well sealed to meet passive house standards. The thermal properties of each material are listed in Table 1.

Table 1. Thermal properties of materials

Material name	Density (kg/m^3)	Conductivity (W/(m·K))	Specific heat (kJ/(kg·K))
Concrete	2200	0.7	0.75
Rockwool	45	0.035	0.84
Gypsum	1440	0.3	0.84
Foam	800	0.028	1.45
Steel	8050	50.2	0.49

2.5 Leakage settings

The leakage area of the experimental facility was determined through a leakage measurement (blower door test) with an imposed differential pressure of 50 Pa. During the test, a fan was used to pull air out of the compartment, creating and maintaining a negative pressure differential (50 Pa) between inside and outside. The air flow rate can be measured and the leakage area can be calculated by using Eq. 2 [12].

$$A_L = \sqrt{q_v^2 \rho / 2\Delta P} \tag{2}$$

where A_L is the leakage area (m^2), q_v is the air volume flow rate (m^3/h), ρ is the air density (kg/m^3), and ΔP is the pressure differential (Pa).

The leakage area obtained is 0.0026 m^2. The blower door test was conducted before each experiment and the leakage area was found to be constant [12]. This area was set as reference leakage area between the compartment and the ambient in FDS. There are two leakage simulation approaches in FDS: Pressure Zone Leakage (Bulk Leakage) and Localized Leakage [10]. The pressure zone leakage approach is intended to capture the bulk leakage that occurs through walls. The leak flow is uniformly imposed over all surfaces designated as leak path. The localized leakage approach is suitable for simulating small cracks with a known location. In the present study, the air leakage from the compartment to the ambient environment was simulated using the pressure zone leakage approach. The walls and ceiling were set as leak path. Since the partition door was closed during the experiment, two pressure zones (one for each room) were prescribed in FDS, in addition to the ambient pressure zone. The leakage area was evenly distributed over the walls and ceiling surface area and thus the leakage area of the fire room and adjacent room were 0.0016 m^2 and 0.001 m^2 respectively. Furthermore, the leakage between the fire room and the adjacent room under the partition door has been prescribed using the localized leakage approach.

3.RESULTS AND DISCUSSION

3.1 Grid sensitivity study

Finer cell size leads to more accurate simulation results but will also take more computational cost and storage space. The mesh grid sensitivity was studied based on the case in this paper, focusing on the pressure variation in the fire room of the compartment. The pressure curves with cell size of 0.1 m, 0.075 m and 0.05 m are shown in Fig. 4. The data shows a small deviation between the three cell sizes. The peak values for a cell size of 0.1 m, 0.075 m and 0.05 m are 2027 Pa, 2099 Pa and 2131 Pa, respectively. The relative difference between 0.1 m and 0.05 m is 5.1 %. During pressure drop and stationary phase, the three curves almost coincide. Thus, the cell size of 0.1 m is considered sufficient to ensure grid insensitivity of the results for this preliminary study.

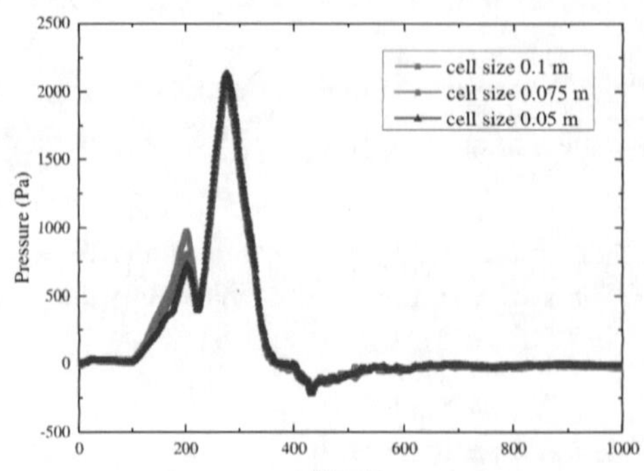

Fig. 4. Pressure variation with different cell size

3.2 Pressure variation

In the configuration without mechanical ventilation, the fire room pressure rises to a high level, see Fig. 5 (left), because the ducts were closed. The peak value of 2027 Pa, obtained in the first FDS simulation (performed with constant leakage area), was much higher than the experimental peak value of 870 Pa. A constant leakage area corresponds to the line $n = 0.5$ in Fig. 5, as explained below. Therefore, this over-prediction suggests that the leakage surface area may not be constant during the fire tests. The leakage area may grow as small gaps, cracks and other leakage paths open up when the room pressure increases. In the default settings of FDS, the leakage area is assumed to be constant. In order to simulate the non-constant leakage area, FDS provides a 'LEAK_PRESSURE_EXPONENT' parameter, n, which can be used to reproduce the variation of the leakage area:

$$A_L = A_{L,ref} \left(\frac{\Delta p}{\Delta p_{ref}} \right)^{n-0.5} \tag{3}$$

where A_L is the leakage area (m^2), $A_{L,ref}$ is the reference leakage area (m^2), Δp is the pressure difference (in Pa) between indoor and outdoor, Δp_{ref} is the reference pressure difference (in Pa) between indoor and outdoor, and n is the leakage pressure exponent. By default, $n = 0.5$, which means that the leakage area is independent of the pressure rise, and $\Delta p_{ref} = 4$ Pa. However, as the reference leakage area has been determined at a pressure difference of 50 Pa here, Δp_{ref} has been set to 50 Pa.

The pressure curves obtained with different exponents are shown in Fig. 5. The leakage pressure exponent does not affect the shape of the pressure evolution graph. This is not surprising, as it is primarily driven by the HRR evolution (Fig. 3). The absolute values reduce with increased values for n, due to the increased leakage area as the pressure exceeds Δp_{ref} , and the deviation between the numerical and experimental values of the pressure is reduced. This indicates that the leakage area measured under the pressure difference of 50 Pa in 'blower door tests' is insufficient to characterize the airtightness during fire conditions.

3.3 Influence of reference pressure difference on pressure variation

In order to illustrate the sensitivity of the results on the choice of reference pressure difference, the pressure variation under the reference pressure difference of 4 Pa (FDS default setting) is shown in Fig. 5. When n=0.5, the setting of reference pressure difference has no influence on the results, because the exponent (n-0.5) =0, see Eq. 2. When n increased to 0.6 and 0.7, the results become sensitive to the reference pressure difference. The simulation result match the experimental result when n=0.6. The reason is that with smaller reference pressure difference, $\Delta p / \Delta p_{ref}$ becomes larger under the same Δp, so lower value of n is required to be consistent with the experimental results. This means that the setting of reference pressure difference has a certain effect on the simulation results of pressure variation. The reference pressure difference should be set according to the pressure used in determining the reference leakage area.

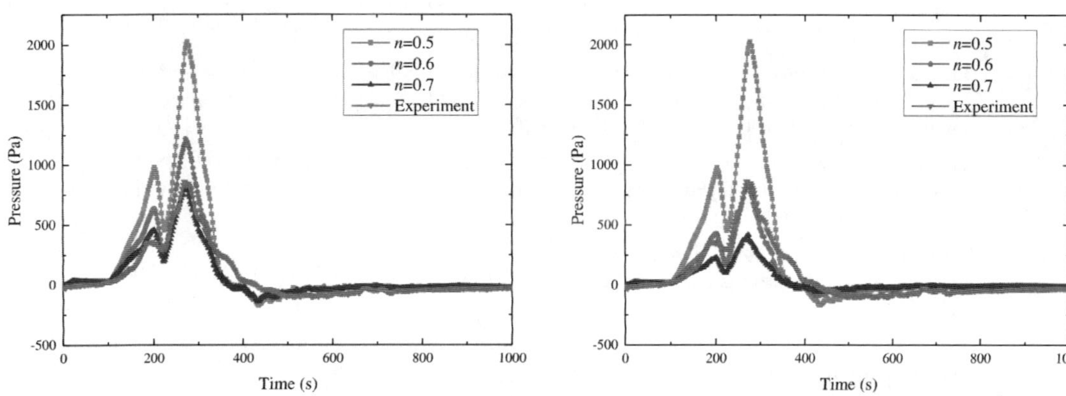

Fig. 5. Pressure variation as function of time with Δp_{ref} =50 Pa (left) and Δp_{ref} =4 Pa (right). The lines correspond to different values of pressure leakage exponent n, see Eq. 2.

4. CONCLUSIONS

A numerical study on the fire-induced pressure variation in a structure, resembling a passive house compartment, was conducted using FDS. Configurations without a mechanical ventilation system were considered during the research. The key results and findings of the present work are as follows.

The required mesh cell size is determined by conducting grid sensitivity study on the basis of fire-induced pressure variation. A cell size of 0.1 m is considered sufficient to ensure grid insensitivity of the results for this preliminary study.

Due to the airtightness, a fire occurring in a passive house could cause a high pressure rise, which can have a large impact on the occupant evacuation and fire rescue, of this type of buildings. The maximum over-pressure obtained without ventilation in the simulation is 2027 Pa. This pressure is high enough to impede evacuation and to affect the structural integrity of the compartment.

The leakage area during the course of a fire is not constant and will increase as pressure increases. Tuning the leakage pressure exponent in the FDS simulations strongly improves the level of agreement between simulation and experimental results.

The influence of reference pressure difference on pressure variation has been examined. The results show that with small reference pressure difference, which is 4 Pa in FDS default settings, smaller value for leakage pressure exponent is required to match experimental tests. Thus, the reference pressure difference does have an effect on the simulation results and should be set according to experimental conditions.

This study is a preliminary investigation on passive house fire without ventilation system. Although the simulations can capture the main aspects, the accuracy of results remains to be improved and it remains to be investigated how general the validity of the present conclusions are. More experimental and numerical studies with different conditions (HRR, leakage area, ventilation, et al) should be conducted and analyzed in depth.

ACKNOWLEDGEMENTS

The authors would like to thank the China Scholarship Council (CSC) for the financial support of the first author (Grant No. 201706420076). The experiment facility used during the research relates to a project initiated by Karel Lambert (Fire Fighter in Brussels) and supported by the KCC (Knowledge Centre), Ghent University and the University of Mons. Thanks to these institutions and to Sylvain Brohez (University of Mons).

REFERENCES

[1] Vinha, J., Manelius, E., Korpi, M., Salminen, K., Kurnitski, J., Kiviste, M. and Laukkarinen, A. (2015) Airtightness of residential buildings in Finland, *Building and Environment* 93:128-140, http://dx.doi.org/10.1016/j.buildenv.2015.06.011.

[2] Fourneau, C., Cornil, N., Delvosalle, C., Breulet, H., Desmet, S. and Brohez, S. (2012) Comparison of fire hazards in passive and conventional houses, *Chemical Engineering Transactions* 26:375-380, http://dx.doi.org/10.3303/CET1226063.

[3] Janardhan, R. K. and Hostikka, S. (2017) Experiments and numerical simulations of pressure effects in apartment fires, *Fire technology*, 53(3):1353-1377, http://dx.doi.org/10.1007/s10694-016-0641-z.

[4] Hostikka, S., Janardhan, R. K., Riaz, U. and Sikanen, T. (2017) Fire-induced pressure and smoke spreading in mechanically ventilated buildings with air-tight envelopes, *Fire Safety Journal*, 91:380-388, http://dx.doi.org/10.1016/j.firesaf.2017.04.006.

[5] Audouin, L., Rigollet, L., Prétrel, H., Le Saux, W. and Röwekamp, M. (2013) OECD PRISME project: Fires in confined and ventilated nuclear-type multi-compartments-Overview and main experimental results, *Fire Safety Journal*, 62:80-101, http://dx.doi.org/10.1016/j.firesaf.2013.07.008.

[6] Prétrel, H., Le Saux, W. and Audouin, L. (2012) Pressure variations induced by a pool fire in a well-confined and force-ventilated compartment, *Fire safety journal*, 52:11-24, http://dx.doi.org/10.1016/j.firesaf.2012.04.005.

[7] Chow, W. K. and Zou, G. W. (2009) Numerical simulation of pressure changes in closed chamber fires, *Building and Environment*, 44(6):1261-1275, http://dx.doi.org/10.1016/j.buildenv.2008.09.016.

[8] Wahlqvist, J. and Hees, P. (2017) Evaluating methods for preventing smoke spread through ventilation systems using fire dynamics simulator, *Fire and Materials*, 41(6):625-645, http://dx.doi.org/10.1002/fam.2404.

[9] Bonte, F., Noterman, N. and Merci, B. (2013) Computer simulations to study interaction between burning rates and pressure variations in confined enclosure fires, *Fire Safety Journal*, 62:125-143, http://dx.doi.org/10.1016/j.firesaf.2013.01.030.

[10] McGrattan, K., Hostikka, S., McDermott, R., Floyd, J., Weinschenk, C. and Overholt, K., *Fire dynamics simulator, user's guide*, NIST special publication 1019 (Sixth Edition), 2013, p. 113.

[11] Vanhaverbeke, D., *Fire development in passive houses: qualitative description and design of full-scale fire tests*, IMFSE Master thesis, 2015, p.41.

[12] Caravita, I., *Fire development in passive houses: experimental analysis and validation of theoretical models*, Università di Bologna Master thesis, 2017, p.58,74.

Full-scale smoke tests in a three-storey residential building

Gabriel GIOVANNELLI[1], El Mehdi KOUTAIBA[1], Elizabeth BLANCHARD[1], Samia HAOUARI HARRAK[1 2], Rabah MEHADDI[2], Pascal BOULET[2], Thomas ROGAUME[3] and Benjamin BATIOT[3]

[1]*Université Paris-Est, Centre Scientifique et Technique du Bâtiment (CSTB), France*
[2]*LEMTA, Université de Lorraine, CNRS, France*
[3]*Institut Pprime (UPR3346 CNRS), Université de Poitiers, ISAE ENSMA, France*

ElMehdi.KOUTAIBA@cstb.fr

ABSTRACT

This work presents summarized results based on three real-scale tests carried out on a three-story building. The main purpose of these tests is to get some information on how the smoke flows in a building with floors connected with a staircase in order to assess the ability of numerical codes to simulate the smoke flow in complex geometries with multiple rooms.

During these tests, a well-controlled source fire was used and measurements of temperatures and velocities were made all over the different stairs and rooms into the building. The heat release rate in the fire room was deduced from the measured mass loss rate and gas analysis. A comparison between numerical results using the two zones model CFAST [5] and the experimental results was also made.

The current work will provide a large database to evaluate software programs used in fire safety engineering.

KEYWORDS:

Full-scale tests; building fire; fire smoke; stratification; smoke propagation; zone model.

INTRODUCTION

An experimental campaign was carried out, in July 2016, on a 3 storey building in order to study the smoke flow in the case of a fire starting at the ground floor. This campaign was carried out at La Chataigneraie (Vendée, France) in collaboration with the P'PRIME Institute of Poitiers, LEMTA of Nancy and SDIS 85.

One main purpose of this work was to investigate how the smoke is traveling through the whole building, flowing between rooms and different storeys. Indeed, very few experimental campaigns that deal with such configuration exist in the literature, among them, we can cite the work by [6]. In this context, this article presents different measurements of temperatures, heat release rate and other physical quantities made during the smoke tests in the residential building in order to characterize the thermal conditions and stratification.

One of the main goals of this work was to build a database to evaluate the ability of software programs used in fire safety engineering to simulate smoke propagation in large and real domains. In the present study, the two zones model CFAST [5] was used and the numerical results were compared to experimental ones.

EXPERIMENTAL CONDITIONS

Presentation of the building

The experiments concern three apartments located on the ground, 2nd and 3rd floors of the building (the apartment on the 1st floor was kept closed and was not instrumented). These apartments are identical. Three rooms with opened doors are accessible: two bedrooms numbered 1 and 2 and a dining room/living room. The apartments are connected by a staircase. Access doors to the kitchen, sanitary facilities and a third bedroom are closed. The entrance to bedroom 1 of the 3rd floor apartment is also closed. The building configuration is shown in Fig 1.

The fire and air supply are located in bedroom 1 on the ground floor. The window of bedroom 2 on the third floor is kept open to ensure the natural smoke extraction.

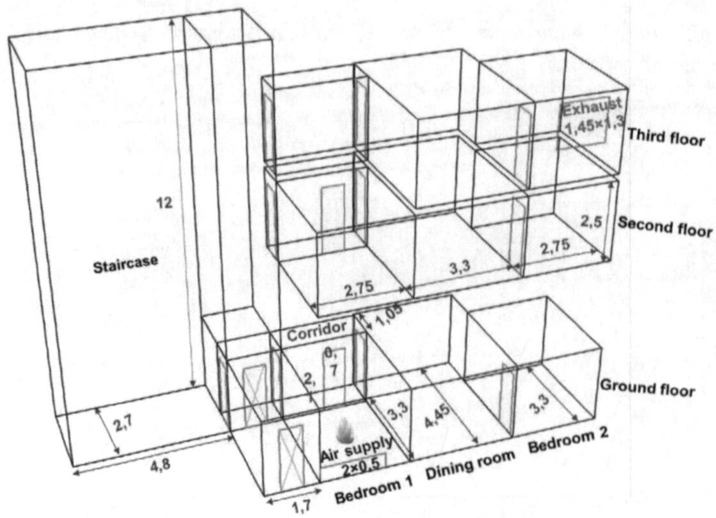

Fig 1 Test configuration (*dimensions in meters*)

Fire source

Inside the bedroom 1 on the ground level a calibrated wood crib is used as a fire source. It has been dimensioned using equations from the SPFE Handbook [1], in order to obtain a maximum value of heat release rate (HRR) of about 927 kW if the combustion is controlled by pyrolysis and 1.16 MW if the combustion is controlled by the ventilation. Note that the size of the doors of the rooms and the size of the vent area on the wall (0.5 x 2 m²) have been taken into account to calculate this later.

A preliminary test, performed at CSTB in the same conditions, suggested that the average HRR in the steady phase is 610-635 kW and that the collapse of the wood crib occurs 76 min 20 s after ignition. During the experiments the mass loss of the wood crib has been measured continuously.

IOP Conf. Series: Journal of Physics: Conf. Series **1107** (2018) 042027 doi:10.1088/1742-6596/1107/4/042027

EXPERIMENTAL AND NUMERICAL RESULTS

Three tests were made. The weather conditions were bad during the first with wind strength and direction preventing and stopping the smoke extraction. Moreover, during this test, the ceiling of the room fire collapsed on the fire source, so this test is not considered. The test results of the two other tests are similar and are presented in the following paragraphs.

Heat Release Rate

HRR is plotted in fig. 2. It is estimated from Thornton's rule and the measured mass loss rate. The averaged HRR during the steady phase is 790 to 935 kW. It is relatively close to the theoretical value of 927 kW. It is higher than the values of 610-635 kW measured during the preliminary test probably because of the room confinement. Thus, we obtain a burning time of 48 to 52-minutes which is less than the 76-minutes obtained in the pre-test.

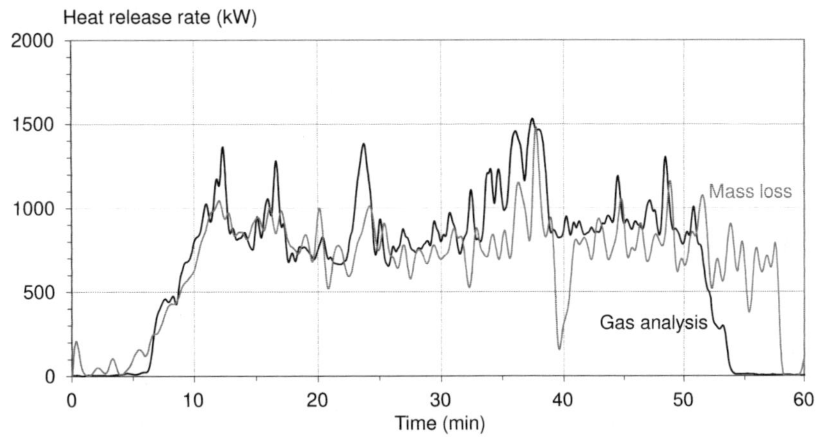

Fig. 2. Temporal evolution of the HRR

Temperature

We will mainly focus on the temperature measurements in the different rooms for every floor. The well-known Janssens and Tran [2] model was used to evaluate the temporal evolution of the smoke layer height and the lower and upper layers mean temperature in each room. The stratification criterion S proposed by Newman ([3]) is also calculated for every room. All the results are presented in fig 3, Fig 4 and Fig 5.

On the ground floor, the smoke interface stands at about 1 m from the floor in the fire room, 1.5 m in the dining room and 0.5 to 0.75 m in the bedroom 2. Newman's parameter S is above a value of 1.7 (considered by the author as a stratification threshold value) in bedroom 1 and dining room of the ground floor, suggesting that the flow is stable and stratified in these two rooms. The thermal stratification in bedroom 2 fulfills the threshold value criterion during the first 20 minutes. In the rooms on the 2nd and 3rd floors, the S parameter is less than 1.7, which shows an unstratified flow while the height of the smoke interface between the upper and lower layers varies between 0.4 and 0.75 m according to Janssens model.

Thresholds values usually used in fire safety engineering ([4]) are defined to evaluate the practicability of escape walkways in case of fire: air temperature, extinction coefficient, smoke layer height and radiative flux below respectively 40 °C, 0.4 m^{-1}, 1,8 m and 2 kW/m^2. The temperature is lower than the critical value of 40 °C near the floor (0.5 to 0.75 m) in bedroom 2 on the ground floor and in the hallway, and in the bedroom 1 and the dining room on the 2nd floor (see fig 4). The temperature levels in the other rooms do not allow evacuation of the building under acceptable conditions. It is also observed that the radiative fluxes measured on the ground floor, in the fire room and in the corridor quickly exceed the threshold value of 2 kW/m^2. Though, the radiative flux remains insufficient to spread the fire in the other premises.

Simulations were performed using the CFAST version 6.2 two-zone model. The HRR given in figure 2 (based on mass loss) is prescribed in the code. The radiative fraction is set at 0.33 (default value). The walls and ceilings are made of concrete for the fire room and gypsum in the other rooms. Fig. 4 shows that the temperature levels of the upper and lower layers are estimated by the model with acceptable accuracy. With the exception of the lower layer in the fire room, the difference between the temperatures, evaluated by the

CFAST code and the measurements, does not exceed 22%. We shall note that the temperature signals include the radiation on the thermocouples since they were not protected from radiations. The smoke layer height calculated with the zone model is significantly underestimated in the dining room on the ground floor. It is null in the bedroom 2 on the ground floor and the rooms of the 2nd and 3rd floors during the steady phase of the HRR. Note that in those rooms, Newman's parameter is lower than 1.7 (unstratified flow).

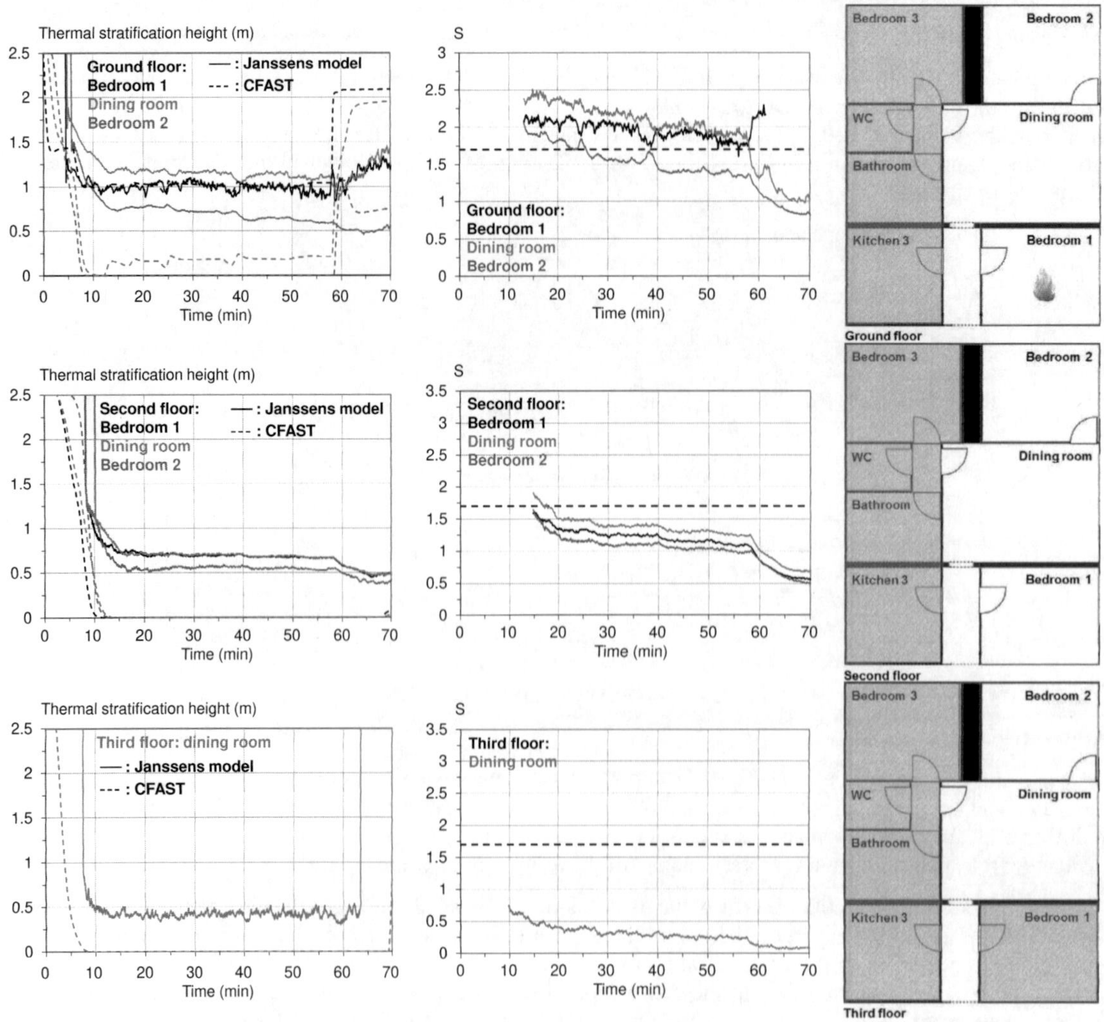

Fig. 3. Smoke layer height and stratification criterion compared to CFAST for the three floors.

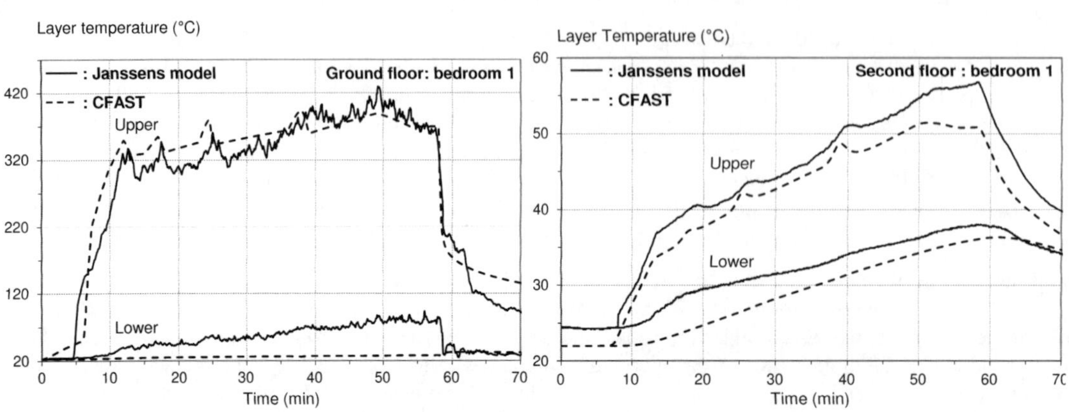

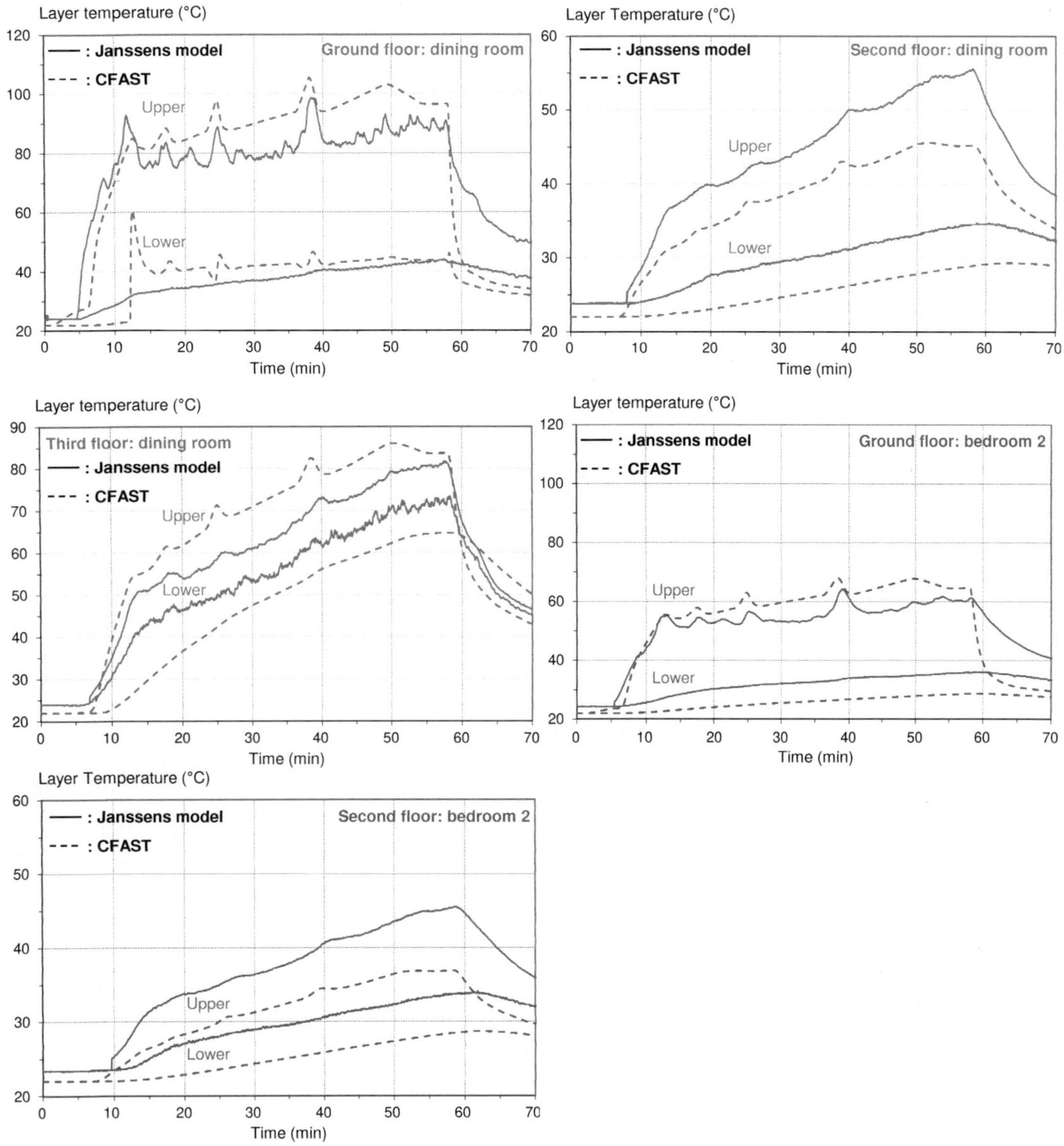

Fig. 4. Temperature measurements compared CFAST in different rooms.

The results are significantly identical between the two entrainment models proposed by the code (McCaffrey and Heskestad) and relatively insensitive to the soot and carbon monoxide production values generated by the combustion reaction. Using the version 7.2.4 of CFAST seems to improve the predictions of the smoke layer height when adding several assumptions: conductivity of the walls, considering leaks through the doors of the kitchen and sanitary area (using CFAST 7.2.4 with the real initial conditions gives an unreal physical behavior). However, the predictions of layer temperatures in 2nd floor rooms are less accurate than those obtained with version 6.2.

Smoke flow through the building

The velocities were measured at every door using propeller anemometers or McCaffrey probes linked differential pressure sensors. The mass flow rate flowing from the ground floor to the third floor is 2.6-2.7 kg/s during the steady phase of the HRR (see figure 5). The measurements suggest that the smoke fills all the open rooms of the building very fast. This requires less than 2 min 45 s.

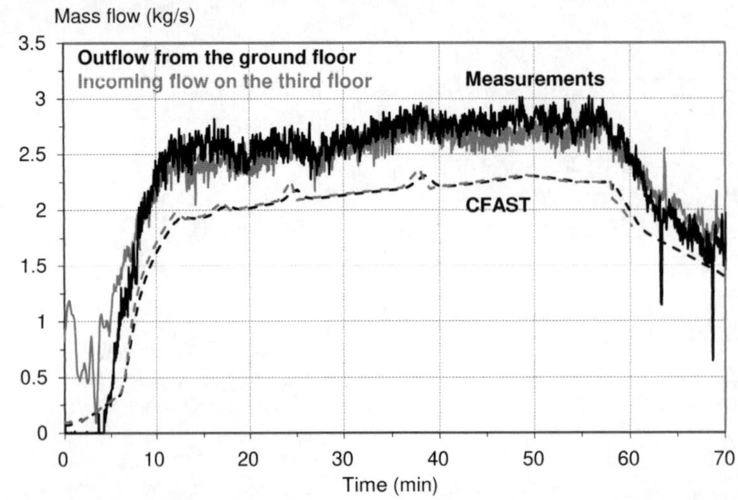

Fig. 5. Mass flow flowing from the ground floor to the 3rd floor

CONCLUSION

Smoke tests were carried out in a three-storey residential building. The tests involved three identical apartments located on the ground, 2nd and 3rd floors of the building, with a communication through a staircase. A wood crib was used as a fire source in the bedroom near the entrance gate to the ground-floor apartment. The measured HRR was about 790 to 935 kW.

The different measurements showed that the smoke filling time of all the premises was very fast and lasted less than 2 min 45 s. The mass flow rate from the ground floor to the top floor was evaluated at 2.6-2.7 kg/s during the steady phase of the HRR. The smoke layer height was found to be 1 m in the fire room, 1.5 m in the dining room on the ground floor and 0.4 to 0.75 m in the other rooms according to Janssens model [2]. According to Newman's criterion [3], the flow was stratified on the ground floor. The temperature was below the acceptability threshold value of 40 °C near the ground (0.5 to 0.75 m) in the bedrooms on the ground floor, the hallway and the dining room of the apartment located on the 2nd floor.

Upper and lower layer temperature levels were predicted by the CFAST version 6.2 [5] two-zone model with acceptable accuracy (a difference from experimental data less than 22%). The code underestimates the stratification smoke layer height in the dining room on the ground floor where it reaches its maximum according to the measurements.

ACKNOWLEDGMENTS

We thank our colleagues: Dominique Pardon, Jean François Moller, Silvio Renard Philippe Fromy and Karine Jacquemet who provided a great help to the experiments campaign and their expertise that greatly assisted the research.

REFERENCES

[1] Babrauskas, V., “Heat release rates” *The SFPE Handbook of Fire Protection Engineering (5th ed),* pp. 828-829, 2016.

[2] Janssens, M.L., and Tran, H.C., “Data reduction of room tests for zone model validation” *Journal of Fire Science*, 1992, pp. 528-555, vol. 218.

[3] Newman J.S., “Experimental evaluation of fire-induced stratification”, *Combustion and Flame*, 1984, pp. 33-39, vol. 57.

[4] “*Guide de bonnes pratiques pour les études d’ingénierie du désenfumage dans les établissements recevant du public*”, Laboratoire Central de la Préfecture de Police, 2017.

[5] Richard D. Peacock, Walter W. Jones, Paul A. Reneke, Glenn P. Forney, *CFAST – Consolidated Model of Fire Growth and Smoke Transport (Version 6)*, User’s guide, 2005

[6] Tarek Beji, Steven Verstockt, Rik Van de Walle, Bart Merci, “Global analysis of multi-compartment full-scale fire tests (‘Rabot2012’)”, *Fire Safety Journal*, 2015, pp. 9-18, vol. 76.

Review and comparative assessment of engineering correlations for the fire-induced air inflow rate through a compartment opening

Dimitrios PIKIOKOS[1] and **Dionysios KOLAITIS**[1,*]

[1] *School of Mechanical Engineering, National Technical University of Athens, Greece*

*dkol@central.ntua.gr

ABSTRACT

In a building fire, the initial fire compartment is rarely airtight, since it is quite common for some opening (e.g. door, window) to be present, either by chance (e.g. a window left open) or due to the failure of the glazing in a high temperature environment. The opening provides essentially a physical connection of the fire compartment to either the ambient environment or an adjacent room. The developing fire-induced buoyant flow is significantly affected by the presence of an opening, since an open vent provides a means for ambient air to enter the compartment (air inflow through the bottom section of the opening) and for gaseous combustion products, or even flames, to exit the compartment (plume outflow through the top section of the opening). In the context of building fire safety design, it is commonly required to estimate the characteristics of the fire-induced opening flow (or "vent flow"), since the ambient air inflow rate is the main parameter affecting fire ventilation conditions and can thus determine whether over- or under-ventilated fire conditions will prevail. As a result, the air inflow rate through an opening has a significant impact in several important fire development characteristics, such as fire spreading rate, compartment temperature, flashover or backdraft hazards etc.

There are several works available in the open literature that focus on developing appropriate fire engineering correlations to accurately estimate the fire-induced opening air inflow rate. These correlations are commonly used in simplified pre-flashover fire simulation tools (e.g. two-zone models). However, despite the importance of this parameter, a general consensus has yet to be achieved. The majority of the available correlations are semi-empirical in nature, where a general analytic opening flow model is combined with experimental results obtained in relevant fire tests. In addition, there are certain empirical parameters appearing in various correlations (e.g. flow discharge coefficient) or other parameters that require prior experimental observations, which may not always be available (e.g. thermal discontinuity height). Motivated by this observation, this work is aimed at evaluating a broad range of available fire engineering correlations for the estimation of the fire-induced air inflow rate through an opening; the correlations are validated and comparatively assessed by comparing the respective predictions with experimental data obtained in eight large-scale fire tests, all performed in fire compartments with a single door-type opening.

Nine different correlations available in the open literature are evaluated in this work; the original correlations are cast in a common format, using identical symbols to facilitate comparison. Aiming to evaluate the various correlations' performance, measurements of air mass flow rate through the opening, obtained in eight different large-scale fire tests, are used; the employed fire tests span a broad range of operational conditions, such as fire heat release rate (32 - 320 kW), burner position (back, centre, front), burner elevation (0 - 0.3 m) and ventilation factor (1.83 - 2.17 $m^{5/2}$).

In general, the investigated correlations are found to exhibit disparate levels of prediction accuracy. The majority of correlations over-predict the measured air inflow rate values, therefore yielding "conservative" results, appropriate for fire engineering design purposes. The correlation of Tewarson (1984) is found to provide the most accurate results (average error: 12.9%). Predictions are further analysed to determine the main physical parameters affecting the prediction accuracy; changing the burner location and elevation is found to significantly affect the quality of the obtained results.

KEYWORDS:

modelling; compartment fires; fluid dynamics; vent flow; fire-induced flow; fire engineering design; two-zone models

IOP Conf. Series: Journal of Physics: Conf. Series **1107** (2018) 042028 doi:10.1088/1742-6596/1107/4/042028

1. INTRODUCTION

In a building fire, the initial fire compartment is rarely airtight, since it is quite common for some opening (e.g. door, window) to be present, either by chance (e.g. a window left open) or due to the failure of the glazing in a high temperature environment. Openings provide a means of physical connection between the fire compartment and the ambient environment. A fire-induced buoyant flow is developed through the opening; in general, gaseous combustion products (or flames) exit the compartment (outflow) through the top section of the opening, whereas ambient air enters the compartment (inflow) through the bottom section of the opening. In the context of building fire safety design, it is commonly required to quantify the fire-induced opening flow (or "vent flow"); for instance, the ambient air inflow rate is the main parameter affecting fire ventilation conditions and can, therefore, determine whether over- or under-ventilated fire conditions will prevail. As a result, the air inflow rate through an opening has a significant impact in several important fire development characteristics, such as fire spreading rate, compartment temperature, flashover or backdraft hazards etc.

In a typical simplified pre-flashover fire safety engineering approach (e.g. two-zone model) there are two main opening flows taken into account, i.e. the air inflow rate and the gaseous combustion products outflow rate (Fig. 1). In order to estimate the air mass flow rate entering the fire compartment through an opening, a broad range of semi-empirical engineering correlations are available in the open literature. These correlations are commonly used in the context of fire safety engineering analysis, since the air mass flow rate determines the development of either over- or under-ventilated fire conditions inside the compartment, thus significantly affecting the fire growth rate. The semi-empirical nature of the available air mass inflow rate correlations means that the values of several parameters are either empirically determined (e.g. discharge coefficient) or require prior experimental measurements, which may not always be available (e.g. thermal discontinuity height).

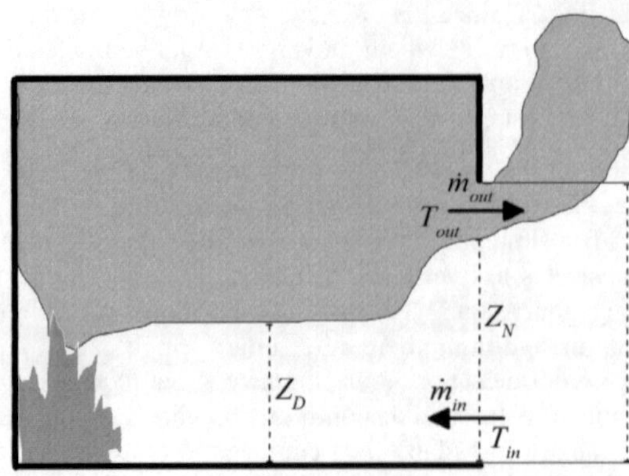

Fig. 1. Schematic drawing of the main characteristics of fire-induced opening flows.

In this context, this work aims to comparatively assess a broad range of semi-empirical correlations available in the literature, used for the estimation of the fire-induced air inflow rate through a compartment opening. Nine different correlations are evaluated by comparing their predictions to available measurements, obtained in eight large-scale fire compartment tests; all the considered cases corresponded to a single compartment with a single door-type opening.

2. FIRE-INDUCED OPENING FLOWS

A sketch of an idealised fire-induced opening flow in a pre-flashover compartment fire is depicted in Fig. 1. The initial transient stage is neglected, therefore it is assumed that steady-state conditions have been achieved. In this case, the compartment is essentially divided in two, vertically separated, "zones": the fresh ambient air, entering through the opening, is assumed to occupy the bottom part of the compartment, from the floor level up to the discontinuity height (Z_D), whereas the top part of the compartment, spanning from Z_D to the compartment ceiling height, is considered to be occupied by the combustion products. The interior of the compartment is assumed to be completely stratified, i.e. no mixing occurs between the two zones and each zone is considered to exhibit uniform temperature and density over its entire volume. Due to the developing thermal buoyant flow field, the warm combustion products are forced to exit through the top section of the opening, whereas the colder fresh ambient air enters the compartment through the bottom section of the opening. The height at the opening where the flow reverses its direction, corresponding to the interface between the incoming air flow and the outgoing combustion products flow, is commonly called the neutral plane height (Z_N). Experimental measurements have shown that the neutral plane height at the opening (Z_N) lies usually higher than the discontinuity height developing inside the compartment (Z_D).

By employing the standard incompressible flow equations (e.g. continuity equation, Bernoulli's law) and using a number of plausible assumptions, e.g. negligible fuel mass flow rate, physical properties of the gaseous products identical to the properties of air, a set of equations can be derived to estimate the steady-state mass flow rates of the two main flows entering and exiting the compartment [1, 2]. More specifically, estimations for the mass flow rate of the combustion products exiting the compartment (index: "out") and the ambient air entering the compartment (index: "in") are obtained using Eqs. (1) and (2), respectively [3, 4]. The discharge coefficient (C_d) is empirically determined; based on literature reports, a value of 0.68 is used in this work. The neutral plane height at the opening (Z_N) is estimated by using Eq. (3) [5], where H_o is the opening height. When values for the neutral plane height and the temperatures of the inflow and outflow streams are known, the discontinuity height inside the compartment (Z_D) is estimated by solving Eq. (4) [3]. Symbols "w" and "A" are used to represent the opening width and the total opening area, respectively.

$$\dot{m}_{out} = \frac{2}{3} C_d w \rho_{out} (H_o - Z_N)^{3/2} \sqrt{2g \frac{(\rho_{in} - \rho_{out})}{\rho_{out}}} \quad (1)$$

$$\dot{m}_{in} = \frac{2}{3} C_d w \rho_{in} Z_N^{\ 3/2} \sqrt{2g \frac{(\rho_{in} - \rho_{out})}{\rho_{in}}} \quad (2)$$

$$Z_N = \frac{H_o}{1 + \left(\frac{\rho_{in}}{\rho_{out}} \right)^{1/3}} \quad (3)$$

$$\frac{T_{in}}{T_{out}} \left(\frac{1}{Z_N} - 1 \right)^3 = \left(1 - \frac{Z_D}{Z_N} \right) \left(1 + \frac{Z_D}{2Z_N} \right)^2 \quad (4)$$

3. FIRE ENGINEERING CORRELATIONS FOR AIR INFLOW RATE THROUGH THE OPENING

A broad variety of engineering correlations for estimating the fire-induced opening air mass inflow rate, $\dot{m}_{in}$ (kg/s), has been developed and is available in the literature. Based on the assumptions used by the various research groups, each correlation presents either small or large discrepancies when compared against Eq. (2). The nine correlations evaluated in this work are presented in Table 1; they all refer to pre-flashover conditions, thus employing the two-zone model assumption. The only exception is correlation IN6, which is also able to account for post-flashover conditions, while a variation of correlation IN9 can also be used for post-flashover fires, when the combustion product gas temperature exceeds 573 K [11]. The parameter δ^* that appears in IN1 corresponds to an empirical layer depth correction term which, for door-type openings, is assumed to be equal to 0.023 [2]. Correlation IN5 is the only equation that utilises the mass flow rate of the fuel ($\dot{m}_{fuel}$); based on suggestions found in the literature [1, 12] this term can be considered negligible and is omitted here. Evidently, correlations IN8 and IN9 are essentially simplified forms of the main equation (Eq. (2)); they are formulated by directly substituting a set of typical values for the main physical quantities (e.g. g = 9.81 m^2/s, T_{in} = 20°C, T_{out} = 527°C, ρ_{in} = 1.2 kg/m^3, ρ_{out} = 0.435 kg/m^3). In certain correlations (e.g. IN1, IN2, IN7), the value of the thermal discontinuity height (Z_D) is required; in these cases, the respective value is estimated using Eq. (4).

Table 1. Engineering correlations for estimation of the fire-induced mass inflow rate through the opening.

Name	Correlation	Reference
IN1	$\dot{m}_{in} = \begin{cases} C_d\rho_{in}A\sqrt{2gH_o}\sqrt{\frac{\rho_{in}-\rho_{out}}{\rho_{in}}}\sqrt{\frac{Z_D}{H_o}-\delta^*}\left(\frac{Z_N}{H_o}-\frac{1}{3}\left(\frac{Z_D}{H_o}+2\delta^*\right)\right), Z_D \le Z_N \\ \frac{2}{3}C_d\rho_{in}A\sqrt{2gH_o}\sqrt{\frac{\rho_{in}-\rho_{out}}{\rho_{in}}\left(\frac{Z_N}{H_o}-\delta^*\right)^3}, Z_D > Z_N \end{cases}$	[2]
IN2	$\dot{m}_{in} = \frac{1}{3}C_d\rho_{in}w\sqrt{2g}\sqrt{\frac{T_{out}-T_{in}}{T_{out}}}\sqrt{Z_N-Z_D}\left(2Z_N+Z_D\right)$	[3]
IN3	$\dot{m}_{in} = \frac{2}{3}C_d\rho_{in}w\sqrt{2g}\sqrt{\frac{\rho_{in}-\rho_{out}}{\rho_{in}}Z_N^{\ 3}}$	[4]
IN4	$\dot{m}_{in} = 2.48w\sqrt{\frac{T_{in}\left(T_{out}-T_{in}\right)}{T_{out}^2}\left(H_o-Z_N\right)^3}$	[6]
IN5	$\dot{m}_{in} = \frac{2}{3}C_d\rho_{in}A\sqrt{2gH_o}\sqrt{\left(\frac{\left(T_{out}-T_{amb}\right)T_{amb}}{T_{out}^{\ 2}}\right)\left(\frac{H_o-Z_N}{H_o}\right)^3}-\dot{m}_{fuel}$	[7]
IN6	$\dot{m}_{in} = \frac{2}{3}C_d\rho_{in}A\sqrt{2gH_o}\sqrt{\frac{\rho_{out}\rho_{in}\left(\rho_{in}-\rho_{out}\right)}{\left(\rho_{out}^{1/3}+\rho_{in}^{1/3}\right)^3}}$	[1]
IN7	$\dot{m}_{in} = 0.66\rho_{in}A\sqrt{gH_o}\left(\frac{Z_N}{H_o}+\frac{Z_D}{2H_o}\right)\sqrt{\frac{T_{in}\left(T_{out}-T_{in}\right)}{T_{out}^{\ 2}}\left(\frac{Z_N-Z_D}{H_o}\right)}$	[8]
IN8	$\dot{m}_{in} = 0.13\rho_{in}A\sqrt{gH_o}$	[9]
IN9	$\dot{m}_{in} = 0.50A\sqrt{H_o}$	[10]

4. LARGE-SCALE TESTS OF FIRE-INDUCED FLOWS

Aiming to determine the accuracy of the correlation predictions, experimental results obtained in eight full-scale fire tests, available in the literature, are used for validation (Table 2); in all cases, a door-type opening was used.

Table 2. Main characteristics of the 8 large-scale compartment fire tests used for validation.

Test case	Fire power (kW)	Burner position	Burner elevation (m)	Ventilation factor ($m^{5/2}$)	References
1	32	Centre	0.3	2.17	[13], [14], [15]
2	160	Centre	0.3	2.17	
3	320	Centre	0.3	2.17	
4	62.9	Centre	0.0	1.83	[16]
5	105.3	Centre	0.0	1.83	
6	62.9	Back	0.0	1.83	
7	62.9	Front	0.3	1.83	
8	62.9	Centre	0.3	1.83	

The first three fire tests (test cases 1-3) were performed by Bryant [13-15] in an ISO 9705 compartment (3.60 m × 2.40 m × 2.40 m). The compartment had a single door-type opening, measuring 0.79 m × 1.96 m, located at the centre of one 2.40 m × 2.40 m wall. The 0.3 m jamb of the doorway, used to accommodate the PIV measurement system, extended beyond the exterior framework of the enclosure. A natural gas fired square burner (0.31 m × 0.31 m) was placed at the centre of the compartment's floor; the top surface of the burner was located 0.30 m above the floor. The three test cases considered here covered a broad range of fire heat release rate values (i.e. 32, 160 and 320 kW). The remaining five fire tests (test cases 4-8) were performed by Steckler et al. [16] in a 2.8 m × 2.8 m × 2.18 m compartment. A single door-type opening was also used in this case, measuring 0.74 m × 1.83 m. A methane burner was employed as a fire source; different burner positions and elevations have been investigated. In addition, the effect of the fire heat release rate was taken into account, by considering fire tests where different fire power values (62.9 and 105.3 kW) were used.

5. COMPARATIVE ASSESSMENT OF FIRE ENGINEERING CORRELATIONS

Predictions of the nine engineering correlations (c.f. Table 1) are evaluated by comparing the respective predictions against the experimentally determined air mass inflow rate values. A comparison of the obtained results against the available fire test data is presented in Fig. 2. It is evident that correlations IN4 and IN5 exhibit better levels of agreement with the experimental data, while predictions of the remaining correlations are quite spread out. However, it is important to note that the majority of correlations over-predict the air mass inflow rate, thus providing "conservative" results (higher values), appropriate for fire engineering design purposes. In Fig. 2 (right), prediction errors for all the investigated correlations are presented as a function of the fire heat release rate; in general, no statistically significant trend can be observed, apart from the fact that increasing the fire power results mostly in smaller prediction errors and reduced differences among the various correlations.

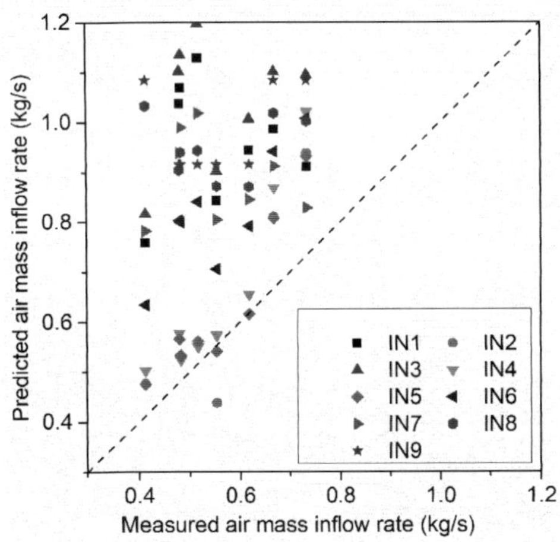

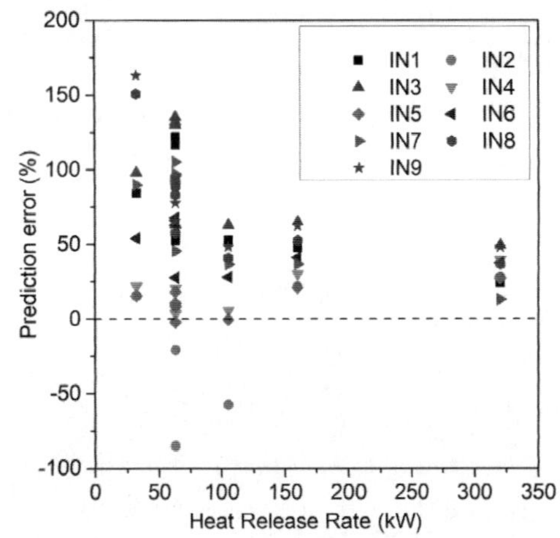

Fig. 2 Comparison of predicted and measured air mass inflow rates (left); effect of fire heat release rate on the prediction error (right).

Table 3. Absolute error values (%) of correlation predictions compared to air mass inflow rate measurements.

Test case	IN1	IN2	IN3	IN4	IN5	IN6	IN7	IN8	IN9
1	84.0	15.4	98.08	21.9	15.3	53.9	89.8	150.65	163.1
2	47.8	21.4	64.96	30.0	20.9	41.0	36.7	52.52	62.4
3	24.2	28.0	49.36	39.3	27.1	37.5	13.0	36.66	47.9
4	52.4	20.8	63.18	3.6	2.2	27.7	45.4	57.52	65.6
5	52.7	57.4	62.94	5.9	0.3	28.1	36.6	40.77	48.2
6	116.6	-	130.14	20.5	18.0	67.6	96.0	88.79	91.2
7	121.9	84.9	135.41	8.6	10.6	65.8	105.3	94.99	90.0
8	118.7	-	132.02	6.5	8.8	62.8	97.3	82.79	77.5
Average:	**77.29**	**37.98**	**92.01**	**17.04**	**12.90**	**48.05**	**65.01**	**75.59**	**80.74**

An overview of the absolute error for each correlation is given in Table 3; the last line corresponds to the arithmetic mean over all eight considered test cases. The presented percentage (%) error values are estimated as the fraction of the absolute difference between the predicted and experimental values over the corresponding experimental values. Missing values represent cases where the experimental measurements lied outside the "definition domain" of the respective correlation. It is evident that correlation IN5 exhibits the lower error values in the majority of the eight test cases and also the lower "average" error over all fire tests. A closer look at the obtained results reveals that certain fire tests (e.g. test cases 1, 6, 7 and 8) result in quite high prediction errors among all correlations; at a first glance, these test cases cannot be easily grouped in a single category. However, based on the test case characteristics (c.f. Table 2), it seems that the position of the burner may have a relatively large impact on the correlations' prediction accuracy. Test cases 1, 7 and 8 employ an elevated burner, while in test cases 6 and 7 the burner is positioned either at the "back" or the "front" side of the compartment. However, solid conclusions cannot yet be drawn; more test cases have to be considered.

6. CONCLUDING REMARKS

The aim of this study was to comparatively assess nine fire engineering correlations, available in the literature, that are commonly used (e.g. in two-zone compartment fire models) for the estimation of the fire-induced air inflow rate through a compartment opening. Experimental results, obtained in eight different large-scale compartment fire tests, covering a broad range of operational parameters (e.g. fire power, ventilation factor, burner position and elevation), were used to evaluate the obtained predictions. Correlation IN5 was found to provide the most accurate results, exhibiting an absolute average error of 12.9% over all the considered experimental test cases. The obtained prediction errors showed that the burner's position and elevation may affect the accuracy of the mass inflow rate calculations. Predictions of fire tests where the burner was located at the centre of the compartment flush to the floor, achieved, in most cases, higher levels of accuracy.

REFERENCES

[1] Karlsson, B. and Quintiere, J.G., *Enclosure Fire Dynamics*, CRC Press, 1999, Chapter 5.

[2] Prahl, J. and Emmons, H.W. (1975) Fire Induced Flow through an Opening, *Combustion and Flame* 25:369-385.

[3] Rockett, J.A. (1976) Fire Induced Gas Flow in an Enclosure, *Comb. Sci. Tech.* 12:165-175.

[4] Babrauskas, V. and Williamson, R.B. (1978) Post-flashover Compartment Fires: Basis of a Theoretical Model, *Fire and Materials* 2(2):39-53.

[5] ISO 16737 (2012) *Fire safety engineering - Requirements governing algebraic equations - Vent flows.*

[6] Quintiere, J.G., Rinkinen, W.J. and Jones, W.W. (1981) The Effect of Room Openings on Fire Plume Entrainment, *Combustion Science and Technology* 26:193-201.

[7] Tewarson, A., "Fully Developed Enclosure Fires of Wood Cribs", *Proceedings of the 20th Symposium (International) on Combustion*, 1984, pp. 1555-1566.

[8] Quintiere, J.G. and Wang, L. (2009) A General Formula for the Prediction of Vent Flows, *Fire Safety Journal* 44:789-792.

[9] Lee, Y.P., Delichatsios, M.A. and Ohmiya, Y. (2012) The Physics of the Outflow from the Opening of an Enclosure Fire and re-examination of Yokoi's Correlation, *Fire Safety Journal* 49:82-88.

[10] Babrauskas, V. (1980) Estimating Room Flashover Potential, *Fire Technology* 16(2):94-103.

[11] Johansson, N. and Van Hees, P., "A Simplified Relation Between Hot Layer Height and Opening Mass Flow", *Fire Safety Science - Proceedings of the Eleventh International Symposium*, International Association for Fire Safety Science, 2014, pp. 432-443.

[12] Tang, F., Hu, L.H., Wang, Q., Lu, K.H. and Yang, L.Z. (2012) An Experimental Investigation on Temperature Profile of Buoyant Spill Plume from Under-ventilated Compartment Fires in a Reduced Pressure Atmosphere at High Altitude, *International Journal of Heat and Mass Transfer* 55:5642–5649

[13] Bryant, R.A. (2009) A comparison of gas velocity measurements in a full-scale enclosure fire, *Fire Safety Journal* 44:793-800.

[14] Bryant, R.A. (2009) The application of stereoscopic particle image velocimetry to measure the flow of air into an enclosure containing a fire, *Experiments in Fluids* 47:295-308.

[15] Bryant, R.A. (2011) Evaluating practical measurements of fire-induced vent flows with stereoscopic PIV, *Proceedings of the Combustion Institute* 33:2481-2487.

[16] Steckler, K.D., Quintiere, J.G. and Rinkinen, W.J., "Flow Induced by Fire in a Compartment", *Proceedings of the 19th Symposium (International) on Combustion*, 1982, pp. 913-920.

IOP Conf. Series: Journal of Physics: Conf. Series **1107** (2018) 042029 doi:10.1088/1742-6596/1107/4/042029

Smoke spreading along corridor induced by outdoor wind: numerical simulation and parameter sensitivity analysis

Aijuan WANG[1], Brady MANESCAU[1], Khaled CHETEHOUNA[1] and Charles de IZARRA[1]

[1]*INSA Centre Val de Loire, University Orléans, PRISME, EA 4229, F-18020 Bourges, France*

aijuan.wang@insa-cvl.fr

ABSTRACT

During a fire in building of high or low height, 80% of deaths are caused by the inhalation of smoke. Considering the fact that outdoor wind is one of the important driving forces of the smoke movement of the fire, this work investigates its effects on the dynamics of smoke along corridor induced by compartment fire. This study is carried out using numerical simulation based on experimental data. The computational fluid dynamic code used is FDS. Through the analysis of mesh resolution, it is shown that the numerical results can agree well with the experimental data. Simulations are performed with different wind speeds. It can be seen that the outdoor wind has effects on the smoke stratification. Indeed, the more wind speed increase, the more the smoke stratification is disturbed. Under these conditions, smoke occupies entire corridor volume and thus presents a high risk of toxicity to people. It can be found that the temperature of the smoke increases suddenly and decreases to converge to a homogeneous value in the entire enclosure. Considering the importance of the input data used for the numerical simulation, a sensitivity analysis is made on parameters related to the material, fuel, combustion and extinction models. This analysis is carried out using a local sensitivity analysis method. It turns out that mass flux of fuel and activation energy as well as thermal conductivity of the walls have significant effects (more than 10%) on the smoke temperature near the celling.

KEYWORDS:

Smoke velocity, FDS, corridor, outdoor wind, sensitivity analysis

1 INTRODUCTION

In recent years, the number of high-rise building has been increasing rapidly in the world. While it brings convenience to people, it represents a major challenge for fire safety engineering due to the larger number of victims during fire. Thus, fire safety engineering of high-rise building has drawn many research interests. The researches [1][2] have shown that smoke is the main cause of death in fires due to the toxicity. Therefore, it is essential to study the behavior of smoke spreading inside the building. The behavior of smoke spreading can be affected by thermal buoyancy, expansion, wind speed, heat pressure and so on. For high-rise building, the wind velocity increases significantly with the height of the building and influences the smoke spreading along corridor in fires.

Therefore, researches have to be continued. Some experimental tests have already been done but they are very expensive and can not study many parameters. Several studies have been carried out on this issue [3][4]. In ref.[5], it is shown that the influence of the external velocity on the propagation of the smoke, and the influence of the fire power on the level of smoke stratification. In order to reduce the costs of implementing the experimental devices and to better understand the phenomena related to the smoke flow in an enclosure, numerical studies have been carried out[6][7][8].

Most of the numerical simulations focus on the propagation of smokes in a ventilated or not ventilated enclosure where the level of smoke stratification varies as a function of temperature profile or velocity. However, these studies do not take into account the effects of the outdoor wind on the smoke stratification in a corridor adjacent to a burning room with an opening. Using the data obtained by Li et al. [4], the aim of the present paper is to highlight the ability of Fire Dynamics Simulator (FDS 6) to reproduce the behavior of the smokes induced by outdoor wind.

Considering uncertainties in definition of input parameters in numerical modeling and uncertainties related to experimental measurements, a sensitivity analysis is applied in order to determine the effect of input parameters on the output [9][10].

In this work, a CFD approach is proposed to evaluate the effects of outdoor wind on the smoke spreading induced by an adjacent compartment fire. This paper is organized as follows: Section 2 presents the physical and numerical modeling, Section 3 deals with numerical results and sensitivity analysis on the results and Section 4 contains conclusions and perspectives.

2. PHYSICAL AND NUMERICAL MODELLING

2.1. Computational domain and boundary conditions

The computational domain used in this study represents 1/3 scale horizontal corridor with an internal dimension of 5.5 m (length) × 0.7 m (width) × 0.9 m (height) connected to a room having a dimension of 2.0 m (length) × 1.7 m (width) × 1.0 m (height) (cf. Fig.1). The dimension of the door connecting the room and the corridor is 0.7 m long by 0.3 m wide, and the dimension of the window opposite to the door is 0.5 m (width) × 0.5 m (height). The several walls are made in steel with 25 mm as thickness. The outdoor wind has modeled as a vent with a fixed flow at the level of window. In the current numerical study, outside has been modeled in two openings: the first in front of the corridor and the second beside the window as shown in the figure 1. This configuration represents the one used during experiment studies of Li et al. [4]. Several simulations are carried out with a fire located in the middle of the room with 0.6 m × 0.7 m as area. The fire source has modeled as a burner with a mass flow rate using liquefied petroleum gas as fuel. Three calculations were carried out with wind speeds varying from 0 to 3.46 m/s for Heat Release Rate of 92 kW.

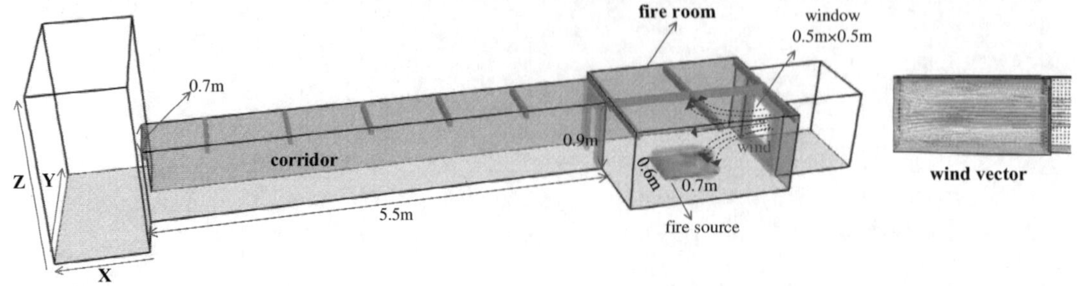

Fig. 1. Geometry of computational domain.

2.2. Numerical modelling approach

The simulations were carried out using the CFD code Fire Dynamics Simulator (FDS) 6. It solves the Navier-Stokes equations using an explicit finite difference scheme. Being a fire CFD code, it models the thermally driven flow with an emphasis on smoke and heat transport. It is a large eddy simulation (LES) model using a uniform mesh, and has parallel computing capability using Message-Passing Interface (MPI). In this work, the numerical modeling was made using the LES approach as turbulence model, the Eddy Dissipation Concept (EDC) as combustion model and thermal radiation model. The calculations are carried out from default models of turbulence and extinction but the combustion model used is based on the finite rate combustion. This combustion model uses Arrhenius parameters (A: pre-exponential factor and E: activation energy). The fuel used is the LPG and the mass flux given by the experimental data [4] has been used as input for all simulations.

2.3. Mesh resolution

The mesh resolution is carried out for four different mesh sizes: 20cm, 10cm, 5cm and 2.5cm. The choice of these mesh sizes is made by recommendations in the numerical studies[11][12]. Mesh resolution is made on temperature and smoke velocity at 70cm high. It is shown that for mesh size of 5cm, the prediction is similar with the experiment data (cf. Fig. 2). According to mesh size of 2.5 cm, the numerical results are better accurate. But in table 1, the time calculation of mesh size 2.5cm is 10 times longer than mesh size 5cm. In addition, the relative gap of the mesh size of 5cm is near to the mesh size of 2.5cm. So, reconciling precision and calculation time, the mesh size of 5cm will be used for the several calculations.

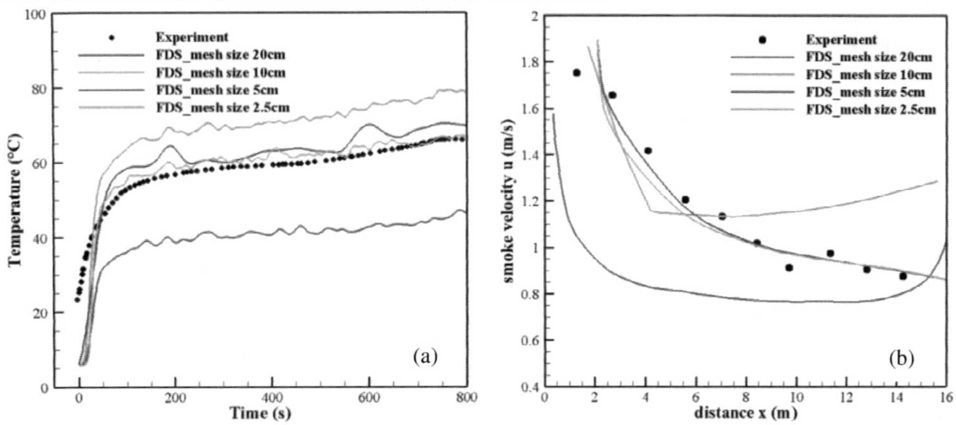

Fig. 2. Grid cells influence on: (a) temperature at 70cm high; (b) smoke velocity without wind at 70cm high.

Table 1. Results of different mesh size of numerical grid

Numerical grid	Number of cells	Relative Gap (%)		CPU time (h)
		Temperature (℃)	Smoke velocity (m/s)	
Mesh size 20cm	1685	31.06	33.17	1.2
Mesh size 10cm	11865	19.76	16.91	4.4
Mesh size 5cm	94920	6.06	6.23	9.6
Mesh size 2.5cm	759360	3.80	3.85	92.2

Relative Gap is obtained by:
$$RG = 100 \times \frac{\sqrt{\sum_{i=1}^{n}\left(y_{pre,i} - y_{\exp,i}\right)^2}}{\sqrt{\sum_{i=1}^{n}\left(y_{\exp,i}\right)^2}} \quad (1)$$

Where y_{pre} is a predicted value, y_{exp} is an experimental value and n is the number of experimental points.

Using the mesh size of 5cm, figure 3 presents the simulation of the smoke velocity field without wind velocity in the cross section y=0.5 m at 300 s. The cross section is the plan in the middle of the corridor. Taking this plan at the height of 70 cm, it is shown that the maximum value of the smoke velocity is near the door and

IOP Conf. Series: Journal of Physics: Conf. Series **1107** (2018) 042029 doi:10.1088/1742-6596/1107/4/042029

decreases with the distance as shown in the figure 2(b). In addition, taking account smoke stratification with a hot zone near the ceiling and a cold zone near the floor, it is observed that buoyancy effects give the reverse observation. Indeed, near the floor, the smoke velocity increases with the distance (distance from the doorway).

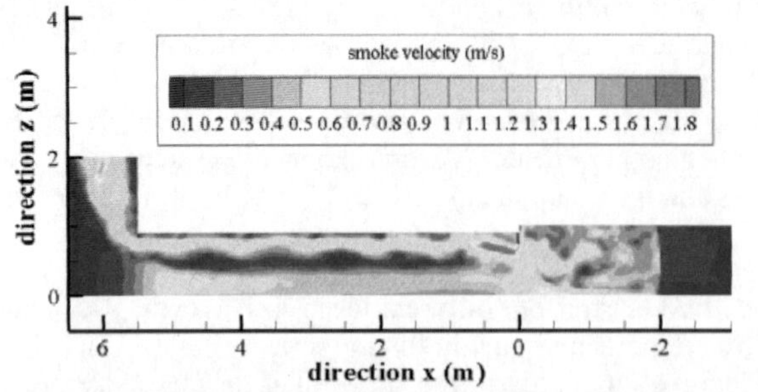

Fig. 3. Simulation of the smoke velocity field without wind velocity in the cross section y=0.5m at 300 s.

Considering tests with outdoor wind velocity of Vw=1.73 m/s and Vw=3.46 m/s, it is shown in figure 4 that prediction is similar with experiment. These comparisons highlight the ability of the solver to reproduce the smoke flow with the vortex induced by the recirculation zones.

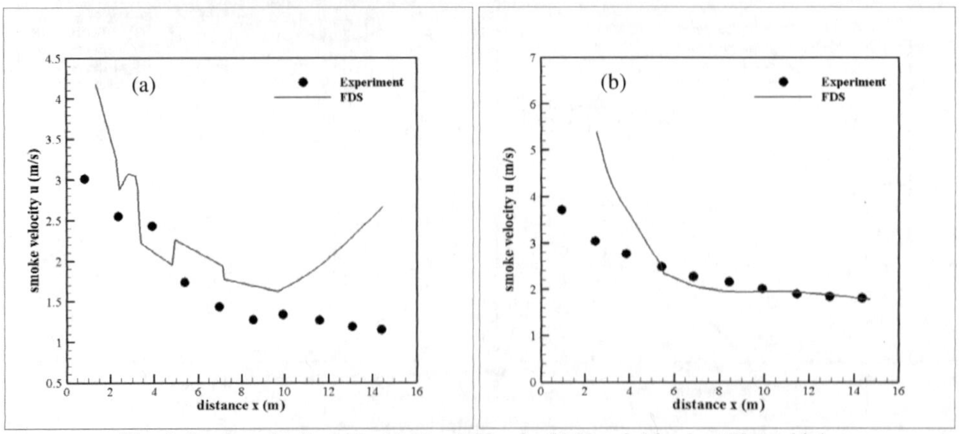

Fig. 4. Smoke velocity with: (a) Vw=1.73 m/s; (b) Vw=3.46 m/s at 70 cm high.

3. RESULTS AND DISCUSSION

3.1. Numerical results

In this section, the influence of wind velocity on smoke temperature is presented from numerical results. The calculations are carried out with three different wind velocities (Vw=0m/s,Vw=1.73 m/s and Vw=3.46 m/s). The choice of these three velocities is due to the fact that for velocities higher than Vw=3.46 m/s, the smoke has the same behavior.

From figure 4(a), it is observed that the smoke temperature increases suddenly with introducing of outdoor wind and converges to an average value little bigger than smoke temperature without wind. Moreover, it's presented that the more wind velocity increases, the more oxygen concentration increases. This contributes to the decrease of the smoke temperature near the ceiling.

Figure 4(b) presents the comparison between the smoke temperature near the celling and near the floor. It can be seen that the more wind speed increases, the more the smoke stratification is disturbed. When the wind velocity is 0m/s, smoke temperature near the floor is about of 5 °C and 60 °C near the ceiling. Regarding the wind velocity of 1.73 m/s, the smoke temperature near the floor is about 25 °C and 100 °C near the ceiling. From wind velocity of 3.46 m/s , the smoke temperature near the celling and floor are similar with an average of 76 °C.With these values, it is observed that the more wind velocity increases, the more smoke temperature near the floor increases. Then, it becomes similar with smoke temperature near the ceiling. In this context, it's possible to happen that smoke occupies entire corridor from a value of wind velocity. Under these conditions, it exist a high risk of toxicity for people. These different observations highlight the ability of FDS to reproduce the effects of wind on the smoke movement in an enclosure.

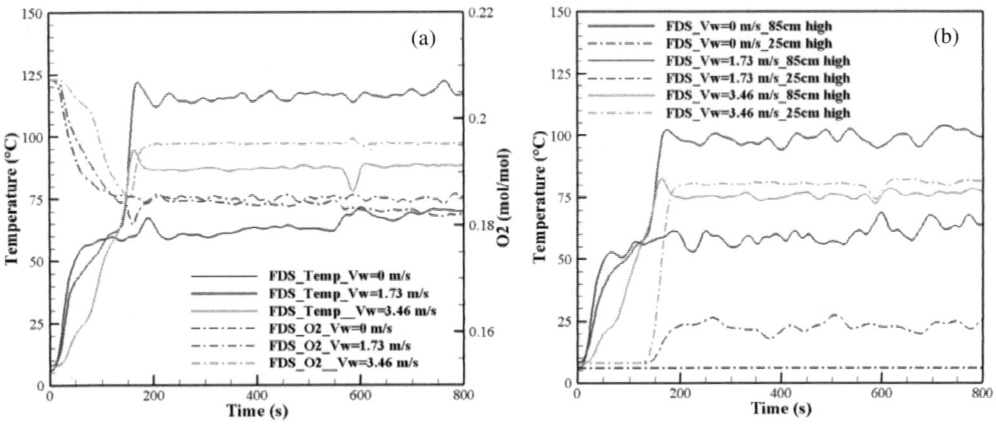

Fig. 4. Influence of wind velocity on: (a) temperature at 70cm high and (b) temperature stratification.

3.2. Sensitivity analysis on the results

A local sensitivity analysis (SA) proposed by Chaos [9] is carried out in this study in order to identify the input parameters which has an influence on the output data such as smoke temperature. This methodology is based on the determination of the local sensitivity coefficient S_j for different input parameters β. From the local sensitivity coefficient S_j, it is possible to calculate the area $S_{j,tot}$. For different parameters β, it is possible to calculate the area A_j representing the ratio between the area $S_{j,tot}$ and the sum of the different areas $S_{j,tot}$ of the different parameters β. These parameters are calculated from equations 2,3 and 4.

In the equation 2,3 and 4, f is the output data whose variation with effects of the input parameter are to be highlighted.

$$S_j = \frac{\beta^0{}_j}{f(t,\beta^0{}_j)} \frac{f(t,\beta_j+\Delta)-f(t,\beta_j-\Delta)}{2\Delta} \tag{2}$$

$$S_{j,tot} = \int \left| S_j(t) \right| dt \tag{3}$$

$$A_j = \frac{S_{j,tot}}{\sum S_{j,tot}} \times 100 \tag{4}$$

In this study, the input parameters β_j are fuel mass flux (MF), the material properties (conductivity λ, emissivity ε, density ρ and specific heat C_p), the Arrhenius parameters (A, E_a), the turbulence model parameter (Tur) and the extinction model parameter (Ext). Δ has been taken equal to 10% for calculation. The output data used to make the analysis sensitivity is the smoke temperature. From figure 5, it is shown that the values of Aj for nine input parameters are between 4% and 28%. Considering the fact that, an input parameter has an effect on the output data when its value of A_j is above of 10% (value of Δ). So, the input parameters which have the more effect on the smoke temperature are fuel mass flux, activation energy and conductivity. From this observation, it's important to define the value of MF, λ and E with a lot of precision in order to avoid big variation on the output data.

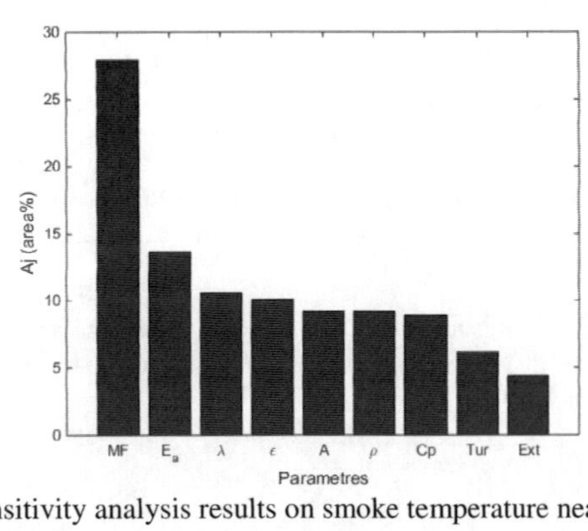

Fig. 5. Sensitivity analysis results on smoke temperature near the celling.

4. CONCLUSION

In this paper, a CFD approach is proposed to evaluate the effects of outdoor wind on the smoke spreading induced by an adjacent compartment fire. This approach is based on experiment data using FDS version 6. Simulations are performed for different wind velocities. It is found that the numerical results are similar to experimental data.

It is shown that the outside wind has noticeable impact on the smoke stratification. Indeed, with outdoor wind, the smoke temperature increases suddenly and decreases to converge to homogeneous value in the entire enclosure. The more wind velocity increases, the more the smoke stratification is disturbed. As a result, with a smoke present in the entire enclosure, a risk of toxic smoke can pose threaten to people's safety.

A local sensitivity analysis (SA) is carried out to highlight the influence of input parameters used in the modeling on output data such as the Heat Release Rate, temperature, velocity etc. It is found that mass flux of fuel and parameters related to combustion model can obviously affect output data such as smoke temperature near the ceiling.

From this work, it is proved that CFD FDS is able to provide information related to the smoke movement in a compartment and with the sensitivity analysis, it is shown that it's important to define correctly the input parameter such as fuel mas flux and the parameters of the combustion model.

REFERENCES

[1] K.T. Paul, T.R. Hull, K. Lebek, A.A. Stec, Fire smoke toxicity: The effect of nitrogen oxides, Fire Saf. J. 43 (2008) 243–251. doi:10.1016/j.firesaf.2007.10.003.

[2] T.R. Hull, D. Brein, A.A. Stec, Quantification of toxic hazard from fires in buildings, J. Build. Eng. 8 (2016) 313–318. doi:10.1016/j.jobe.2016.02.014.

[3] L.H. Hu, R. Huo, Y.Z. Li, H.B. Wang, W.K. Chow, Full-scale burning tests on studying smoke temperature and velocity along a corridor, Tunn. Undergr. Sp. Technol. 20 (2005) 223–229. doi:10.1016/j.tust.2004.08.007.

[4] S.C. Li, D.F. Huang, N. Meng, L.F. Chen, L.H. Hu, Smoke spread velocity along a corridor induced by an adjacent compartment fire with outdoor wind, Appl. Therm. Eng. 111 (2017) 420–430. doi:10.1016/j.applthermaleng.2016.09.086.

[5] M.H. Zhong, C.L. Shi, L. He, J.H. Shi, C. Liu, X.L. Tian, Smoke development in full-scale sloped long and large curved tunnel fires under natural ventilation, Appl. Therm. Eng. 108 (2016) 857–865. doi:10.1016/j.applthermaleng.2016.07.141.

[6] A. Li, Y. Zhang, J. Hu, R. Gao, Reduced-scale experimental study of the temperature field and smoke development of the bus bar corridor fire in the underground hydraulic machinery plant, Tunn. Undergr. Sp. Technol. 41 (2014) 95–103. doi:10.1016/j.tust.2013.11.009.

[7] G. Zhao, T. Beji, B. Merci, Study of FDS simulations of buoyant fire-induced smoke movement in a high-rise building stairwell, Fire Saf. J. 91 (2017) 276–283. doi:10.1016/j.firesaf.2017.04.005.

[8] Q. Wang, F. Tang, L.J. Li, X.C. Zhang, C.G. Fan, Large Eddy Simulation on the Effect of Smoke Exhaust Openings Arrangement on the Smoke Spread in Tunnel Fires, Procedia Eng. 135 (2016) 308–314. doi:10.1016/j.proeng.2016.01.135.

[9] M. Chaos, Application of sensitivity analyses to condensed-phase pyrolysis modeling, Fire Saf. J. 61 (2013) 254–264. doi:10.1016/j.firesaf.2013.09.016.

[10] Y. Jiang, K. Chetehouna, S. Zhang, J. Qin, W. Bao, N. Gascoin, Sensitivity analysis of fluid properties and operating conditions on flow distribution in non-uniformly heated parallel pipes, Appl. Therm. Eng. 130 (2018) 458–465. doi:10.1016/j.applthermaleng.2017.11.005.

[11] B. Magnognou, J.P. Garo, B. Coudour, H.Y. Wang, Risk analysis of unburnt gas ignition in an exhaust system connected to a confined and mechanically ventilated enclosure fire, Fire Saf. J. 91 (2017) 291–302. doi:10.1016/j.firesaf.2017.03.036.

[12] I. Sellami, B. Manescau, K. Chetehouna, C. de Izarra, R. Nait-Said, F. Zidani, BLEVE fireball modeling using Fire Dynamics Simulator (FDS) in an Algerian gas industry, J. Loss Prev. Process Ind. 54 (2018) 69–84. doi:10.1016/j.jlp.2018.02.010.

Experimental study of smoke propagation in a tilted tunnel

Romain HANOUZET[1,2], Olivier VAUQUELIN[2] and Samuel VAUX[1].

[1]*Institut de Radioprotection et de Sûreté Nucléaire (IRSN), Laboratoire ETiC, France*
[2]*Aix-Marseille Université (AMU), Laboratoire IUSTI, France*

romain.hanouzet@etu.univ-amu.fr

ABSTRACT

In the context of an underground repository, one of the main concerns lies within the management of the smoke produced by a fire. The propagation front of smoke is the first indicator received by the detectors and this front can be represented by a gravity current. In this study, air/helium experiments on the non-Boussinesq gravity currents generated from an instantaneous buoyancy source in horizontal and tilted tunnels have been carried out. A large range of density ratios $\gamma = \rho_c/\rho_0$ $(0.29 \leq \gamma \leq 0.88)$, between the density of the release (ρ_c) and the ambient (ρ_0) is investigated. First, similarly to the open ambient case, we note that the front of the gravity current propagates in three phases: the slumping, inertial and viscous phase. In each phase, a power-law exists between the front location of the gravity current (*x*) and the time (*t*). In the early inertial deceleration phase, $x \propto t^{2/3}$ is valid for both Boussinesq and non-Boussinesq gravity currents. Moreover, in the late deceleration phase, the viscous effects become more important and a power-law $x \propto t^{1/2}$ has been identified. Then, experiments on tilted tunnels with angles varying from 1% to 17 % have been performed to consider the influence of the slope. It is found that the transition between inertial and viscous propagation decreases when the slope increases. One possible explanation is that the mixing layer increases in the higher angle percentages which causes the buoyancy contained within the head of the density currents to decrease more rapidly. A high angle of slope can be seen potentially as a disadvantage for the propagation of the smoke.

KEYWORDS: *smoke spreading, tilted tunnel, non-Boussinesq gravity currents*

NOMENCLATURE LISTING

h_0	height of the initial box (m)	u	velocity (m/s)
w_0	width of the initial box (m)	**Greek**	
l_0	length of the initial box (m)	θ	slope angle of the tunnel (%)
H	total height of the tunnel (m)	σ	power law in viscous phase
W	total width of the tunnel (m)	ρ	density (kg/m^3)
L	total length of the tunnel (m)	$\Delta\rho$	density difference (kg/m^3)
x	front location (m)	γ	density ratio
t	time (s)	**subscripts**	
g	gravitational acceleration (m/s^2)	$_0$	ambient
k	constant	$_c$	current

INTRODUCTION

In case of an accident in a tunnel with a slope between 10 and 15% of an underground repository, the production of smoke can be instantaneous with a given volume of smoke released or continuously fed by a source. This work focuses only on the instantaneous release of a given volume of smoke. In this configuration, the propagation of the smoke front can be affected by the confinement, the density ratio $\gamma = \rho_c/\rho_0$ between the release and the ambient and by the slope angle of the tunnel. Therefore, there is a special need to quantify these effects on the kinetics of the smoke.

Gravity currents, or density currents, are flows driven by a horizontal density difference. This density difference may be produced by different factors such as, for example, the temperature difference, the presence of suspended sediments or dissolved materials. Several laboratory experiments [1], [3] and [4], and simulations [2], [11], have been performed to describe gravity current kinetics on low differences of density, i.e. for Boussinesq flows. Producing a gravity currents is relatively simple: a vertical gate is placed in a horizontal tunnel and each part is filled with two fluids with different densities. Then, removing the gate generates a natural flow. This experiment is called lock-exchange problem. The propagation of gravity currents has been largely studied using lock-exchange experiments in weak density difference, i.e. in the Boussinesq case (Simpson [10] and Shin & Linden [9]). The location of the front (*x*) of density currents as a function of time (*t*) is mainly studied. It has been observed that the gravity current propagates through three phases [1]: a slumping phase, an inertial-buoyancy phase and a viscous-buoyancy phase. Each phase is characterized by a different propagation rate. During the slumping phase, the front location of the gravity current is proportional to the time. Then, during the inertial phase, the current starts to decelerate and $x \propto t^{2/3}$ which can be explained by a balance between inertia and buoyancy. Finally, in the late deceleration phase, the inertial force tends to be less important than viscous forces and $x \propto t^{\sigma}$. Different values of σ have been found. Hoult [6] proposed $\sigma = 3/8$ by considering a balance between buoyancy and viscous forces from the interface, Huppert [7] revised his analysis and added viscous effects over a rigid horizontal surface and found $\sigma = 1/5$ and Dai [4] have determined $\sigma = 1/2$ by balancing gravity and viscous stress per unit volume.

In the case of fire hazard, smoke flows are mainly non-Boussinesq, i.e. exhibit large density differences. Only few experiments have been made on non-Boussinesq density currents [4], [8]. Recently, Lowe [8] performed lock-exchange experiments on non-Boussinesq heavy gravity currents within a large range of density ratios $\gamma = \rho_c/\rho_0$ ($0.61 \leq \gamma \leq 0.99$). He showed that for large density differences, heavy currents propagate faster than light currents, but in this study, propagation of density current is only studied in the slumping phase. Consequently, there is a need to understand the kinetics of a density current for larger density ratios and for a longer distance in a tunnel. Dai [4] extended the study on gravity currents for the case of propagation down a slope in deep ambient for angles varying in the range 0 to 9° and he found that the power-laws identified previously are robust.

This experimental study focuses on the influence of three parameters: the effect of the confinement, the large density difference between the release and the ambient and the slope of the tunnel, on the kinetics of gravity currents and especially on the three propagation phases described above.

EXPERIMENTAL SETUP

The experimental setup is a reduced-scale rectangular isothermal tunnel model, as in shown Fig. 1. Its internal cross-section is 0.25 m height (H) and 0.5 m width (W), the total length (L) is 10 m and the walls are transparent Perspex. The tunnel is supported by a steel frame which can be inclined, and the angle (θ) can be adjusted in the range ± 17 %. The initial volume of light fluid is fixed with 0.25 m height (h_0), 0.5 m width (w_0) and 0.5 m length (l_0). The light fluid is a helium-air mixture which is injected in the initial volume and the ambient fluid is air. The density of the light fluid is controlled by a helium concentration analyzer with an accuracy of 0.1 % volume concentration. In addition, seeding particles, obtained by a chemical reaction between ammonia and hydrochloride acid, are injected into the initial volume to provide flow visualization. The two fluids (light fluid and air) are separated by a gate and removing this gate set them into motion. Each experiment has been repeated three times to insure its repeatability. The initial volume remained fixed for all experiments presented hereafter.

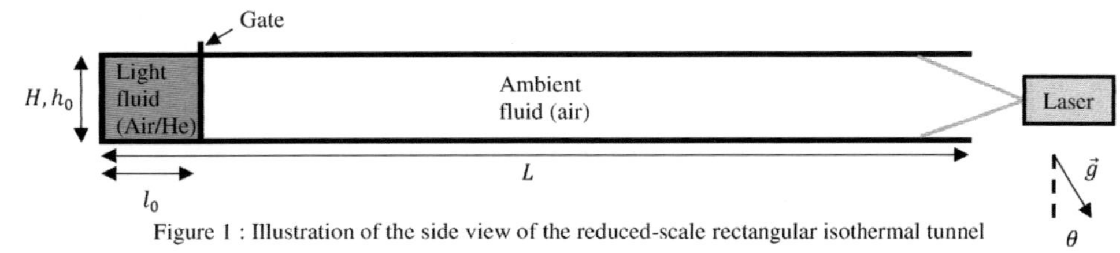

Figure 1 : Illustration of the side view of the reduced-scale rectangular isothermal tunnel

RESULTS

Effects of confinement on the kinetics of gravity currents

A first experiment is made to quantify the effects of confinement on the propagation regimes of the gravity currents for a Boussinesq case with a horizontal tunnel (θ =0 %). The initial density of the released fluid was set at 1.07 kg/m^3, corresponding to a density ratio $\gamma = 0.88$. The gravity current produced by our experiments propagates through the three phases described previously, as shown in Fig. 2. The confinement does not change the exponents of the power-laws found in the case of the propagation of a gravity current in an open ambient. In the late deceleration phase, we can note that we recover $x \propto t^{1/2}$ as suggested by Dai [3] in his experiments for Boussinesq gravity currents in an infinite ambient.

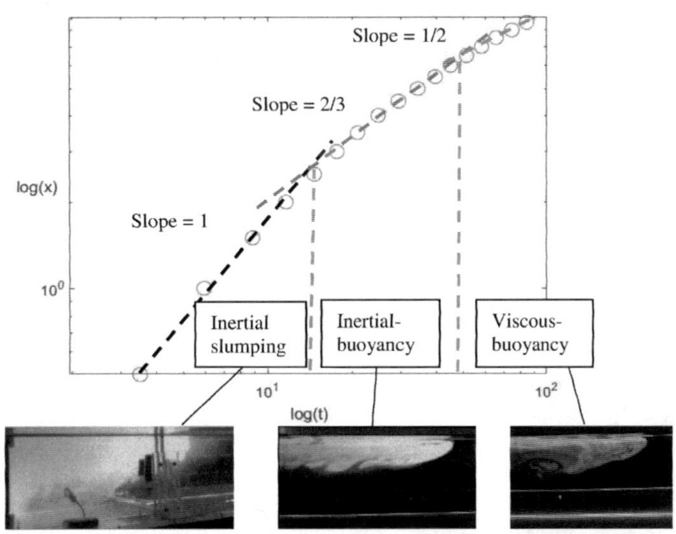

Fig. 2: The front location, x, versus time, t, on a log-log scale with an initial density of the light current $\rho_c = 1.07$ kg/m^3, i.e. $\gamma = 0.88$, on a 0 % slope, with 3 pictures of the front in each phase. The dashed lines represent the asymptotes for slumping, inertial and viscous propagation.

Effects of large density differences on the kinetics of gravity currents

Experiments on the kinetics of gravity currents on non-tilted tunnels are presented with a varying initial density in the range $0.35 \leq \rho_c \leq 1.07$ kg/m^3, corresponding to the density ratio range $0.29 \leq \gamma \leq$

0.88. As shown in Fig. 3, the front of the density current propagates faster when the density ratio is far from 1, i.e. when the density difference, $\Delta\rho$, is high. The mean velocity can be calculated experimentally and u_c varies in the range $0.13 \leq u_c \leq 0.63$ m/s.

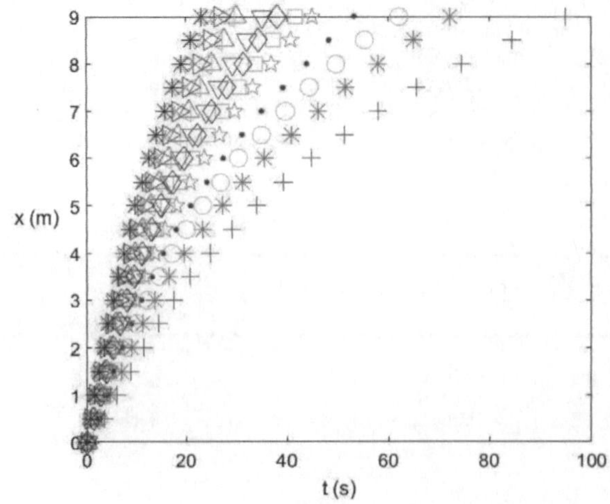

Fig. 3: The front location, x, versus time, t, on a 0 % slope for the gravity current produced from a released volume of initial density: (+) 1.07, (✳) 1, (○) 0.95, (•) 0.90, (☆) 0.85 (□) 0.77, (◇) 0.71, (▽) 0.65, (△) 0.60, (◁) 0.56, (▷) 0.53, (×) 0.50, (✡) 0.40, (✱) 0.35 kg/m^3.

The front location and the time were non-dimensionalized, noted respectively $\tilde{x}$ and $\tilde{t}$, as follow: $\tilde{x} = x/l_0$ and $\tilde{t} = t\sqrt{g\left(\frac{\Delta\rho}{\rho_c}\right)h_0}/l_0$. These dimensionless front location and time are convenient to describe the propagation in the slumping and inertial phases, as shown in Fig. 4 (a). In the late deceleration phase, the discrepancy of the data can be explained by the errors committed on both the experiment repeatability and the control of initial density. Nevertheless, in the inertial phase, for a large range of density ratio corresponding to non-Boussinesq gravity currents propagating in a horizontal tunnel, we can write, as shown in Fig. 4 (b):

$$\tilde{x} = k\tilde{t}^{2/3} \text{ with } k = 0.82^{+0.05}_{-0.07}.$$

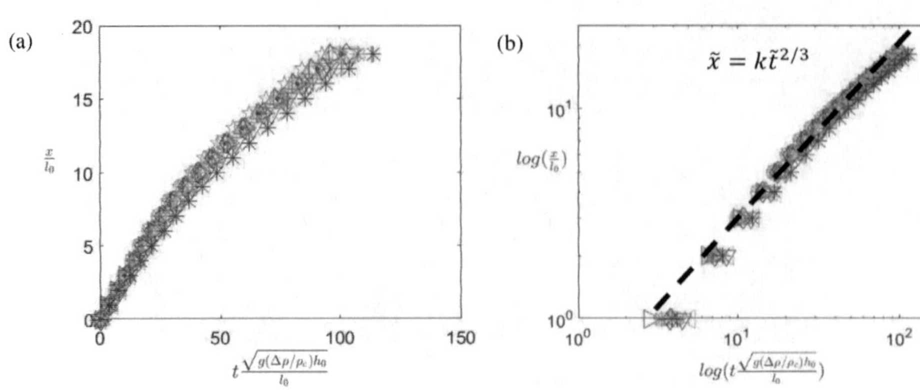

Fig. 4: The front location, $\tilde{x}$, versus time, $\tilde{t}$, on (a) a normal scale and (b) on a log-log scale. The front location, x, is non-dimensionalized by l_0 and time is non-dimensionalized by $l_0/\sqrt{g(\Delta\rho/\rho_c)h_0}$. In (b), the dashed line represents the inertial propagation: $\tilde{x} \propto \tilde{t}^{2/3}$.

It is important to note that the constant k "absorbs" the confinement and the geometry parameters of the tunnel. For example, if the cross section of the tunnel changes, k will be different, as suggested by Fannelöp [5].

Effects of slope on the kinetics of gravity currents

Results presented hereafter are made with a fixed initial density $\rho_c = 0.5$ kg/m³, i.e. $\gamma = 0.42$, on slopes varying within 1% and 17 %. Firstly, experiments carried out with 1, 10 and 17 % slope angles are presented in Fig. 5. Here we focus only on the slope influence on the propagation front in the inertial phase. In Fig. 5 is plotted the experimental front location $x^{3/2}$ versus time. When the experimental data can be fitted with a linear interpolation, the density current is in the inertial phase. As we can see in Fig. 5 (a), the flow propagates only in slumping and inertial phase for a 1% slope. In Fig. 5 (b) and (c), for respectively a 10% and 17% slope, the viscous phase appears sooner. On a 10 % slope, the transition time between the two phases is 21 s and on a 17 % slope, this time is 18 s.

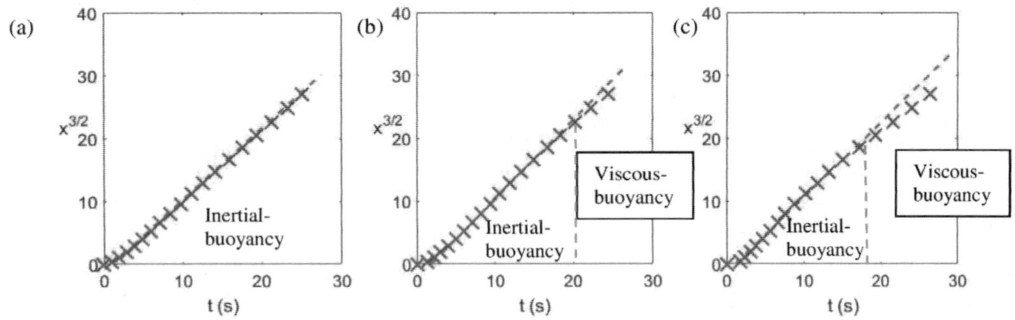

Fig. 5: Relationship between $x^{3/2}$ and time t on a (a) 1 % slope, (b) on a 10 % slope and (c) on a 17 % slope. The initial density of the released volume is $\rho_c = 0.5$ kg/m³. The dashed lines represent the linear interpolation existing between $x^{3/2}$ and t, corresponding to the inertial phase of the gravity current propagation.

Fig. 6 (a) shows variations of the front location, *x*, versus time, *t*, and Fig. 6 (b) shows the velocity of the gravity current, u_c, versus front location, *x*, for 1 and 17 % slopes. Surprisingly, for a quite long distance, a gravity current propagates faster at 1 % slope than at 17 %. Initially, on the acceleration phase, the front velocity is more important for a 17% slope, as shown in figure Fig. 6 (b). We note that the maximum speed for 17 % is around 0.9 m/s whereas it is 0.6 m/s for a 1 % slope. Then, in the late deceleration phase, the velocity decreases more rapidly for higher angles Qualitatively, the mixing region between fresh air and gravity current increases as the angle of the tunnel increases. Therefore, the buoyancy contained within the head of the gravity current may decrease more rapidly. Consequently, high slopes in tunnels can be a disadvantage for the kinetics of propagation of a fixed volume of gravity current released in a tunnel.

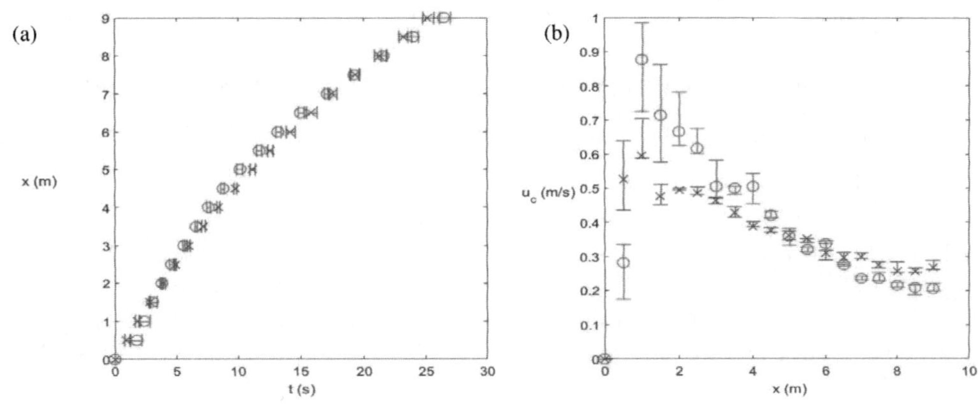

Fig. 6: (a) The front location, *x*, versus time, *t*, and (b) the speed of the current u_c versus the front location, *x*, on a (x) 1 % slope and on a (○) 17 % slope.

IOP Conf. Series: Journal of Physics: Conf. Series **1107** (2018) 042030 doi:10.1088/1742-6596/1107/4/042030

CONCLUSION

This study presents experimental investigation on the propagation of a gravity current produced by an instantaneous buoyancy source with a large density difference with the ambient in a tilted tunnel. The density current propagates through three flow regimes: slumping, inertial and viscous phases. In the inertial phase, the dimensionless front location ($\tilde{x}$) is proportional to the dimensionless time ($\tilde{t}^{2/3}$) for non-Boussinesq gravity currents in tunnels. In the early deceleration phase, this power-law relationship is robust for the entire range of the density ratio, $0.29 \leq \gamma \leq 0.88$, studied in this work. In the late deceleration phase, viscous effects became more important which resulted in another power-law: $x \propto t^{1/2}$. For gravity currents produced in a tilted tunnel, one noticeable result is that the viscous phase appears sooner when the slope increases. The transition time between the inertial and the viscous spreading decreases as the angle increases. Experimentally, we have observed that the mixing layer increases with the angle and that the buoyancy contained within the head of the gravity decreases more rapidly. Consequently, for high angles of tunnel, the slope can be disadvantageous for the propagation rate of the gravity current. Further works will be presented on the propagation of gravity currents produced by a continuous source.

REFERENCES

[1] Amy, L.A., Hogg, A.J., Peakall, J., & Talling, P.J., Abrupt transitions in gravity currents, *Journal of geophysical research,* 2005, vol. 110, F03001, doi:10.1029/2005JF000197

[2] Bonometti, T., Balachandar, S. & Magnaudet, J., Wall effects in non-Boussinesq density currents, *Journal of Fluid Mechanics*, 2008, vol. 99, pp. 449-475, doi: 10.1017/S002211200800414X

[3] Dai, A., Experiments on gravity currents propagating on different bottom slopes, *Journal of Fluid Mechanics*, 2013, vol. 731, pp. 117-141, doi: 10.1017/jfm.2013.372

[4] Dai, A., Non-Boussinesq gravity currents propagating on different bottom slopes, *Journal of Fluid Mechanics*, 2014, vol.741, pp. 658-682, doi: 10.1017/jfm.2014.5

[5] Fannelöp, T.K., *Fluid Mechanics for Industrial Safety and Environmental Protection,* 1994, Industrial Safety Series, volume 3, Elsevier edition

[6] Hoult, D.P., Oil spreading on the sea, *Annual Review of Fluid Mechanics*, 1985, vol. 4, pp. 341-368

[7] Huppert, H.E., The propagation of two-dimensional and axisymmetric viscous gravity currents over a rigid horizontal surface, *Journal of Fluid Mechanics*, 1982, vol. 121, pp. 43-58, doi: 10.1017/S0022112082001797

[8] Lowe, R.J., Rottman, J.W. & Linden, P.F., The non-Boussinesq lock-exchange problem. Part 1. Theory and experiments. *Journal of Fluid Mechanics*, 2005, vol. 537, pp. 101-124, doi: 10.1017/S0022112005005069

[9] Shin, J.O., Dalziel, S.B. & Linden, P.F., Gravity currents produced by lock-exchange, *Journal of Fluid Mechanics* 2004, vol.521, pp. 1-34, doi: 10.1017/S002211200400165X

[10] Simpson, J.E., Gravity currents in the laboratory, atmosphere and ocean, *Annual Review of Fluid Mechanics*, 1982, vol. 14, pp. 213-234, doi: 10.1146/annurev.fl.14.010182.001241

[11] Steenhauer, K., Tokyay, T. & Constantinescu, G., Dynamics and structure of planar gravity currents propagating down an inclined surface, *Physics of Fluid*, 2017, vol. 29, doi: 10.1063/1.4979063

IOP Conf. Series: Journal of Physics: Conf. Series **1107** (2018) 042031 doi:10.1088/1742-6596/1107/4/042031

Overpressure induced by fires in airtight buildings

Sylvain BROHEZ[1] and Irene CARAVITA[2]

[1]Chemical Engineering Department, Université de Mons, Belgium
[2]Chemical Engineering Department, Università di Bologna, Italia

sylvain.brohez@umons.ac.be

ABSTRACT

The need for sustainability and smaller ecological footprint leads to the construction of more airtight building envelopes with better thermal insulation in order to reduce the pollution and increase the energy efficiency, according to the Energy Performance of Buildings Directive (EPBD 2010/31/EU).
The specific fire hazards possibly associated with such structures raised questions amongst fire community in Belgium. Consequently, in 2010 the Belgian Ministry of Interior funded a study about the fire hazards possibly associated with Passive Houses in comparison to traditional houses. The University of Mons and ISSeP used zone modelling for investigating how much the characteristics of a passive house such as airtightness, ventilation and thermal insulation could affect the fire development. A significant difference in pressure rise (due to thermal expansion of fumes) was observed because of the airtightness in the passive house. However, due to a lack of validation from experimental data at large scale, the scientific monitoring committee of the project decided to not mention the potential problem that the forces of pressure could block the occupants during a certain period due to inwards-opening doors.

It was decided in 2015 to build a full-scale experimental facility in the region of Mons with the support of the KCC centrum and the Régie Provinciale du Hainaut. This facility was designed with the support of the Ghent University. The inner dimensions of the construction are the same as the inner dimensions of a 40-foot shipping container, but the outer shell consists of 20 cm concrete blocks and the inner shell is made of plaster board. The volume is divided into two rooms, which simulate two different rooms of a passive house. In one room air is extracted, while in the other one fresh air is supplied.

In this study, we present the tests carried out in this facility by burning pallets stack and wood cribs. Two different ventilation duct configurations were tested: one with mechanical ventilation on, the other with the ducts closed with an airtight metal cap (the mechanical ventilation being off). Measurements were made for gas pressure, mass loss rate, gas temperature, volumetric flow rate in the ducts and for some tests O_2, CO_2, CO and THC concentrations were also quantified. Over pressure peaks from 850 to 2035 Pa were measured without mechanical ventilation (ducts closed), while values from 420 to 750 Pa were observed with the mechanical ventilation on. Both the reverse flow of the fan supplying fresh air and the extra flow rate of fumes in the extraction duct due to the overpressure inside the apartment were not sufficient to prevent overpressure inside the rooms. These tests confirm there may be problems of overpressure in very airtight houses such as passive houses in case of fire. Especially, the occupants may not succeed in escaping during a period of several minutes due to the impossibility to open inward opening doors. This experimental study confirms the observations carried out during the reconstruction of the fire which occurred during the night of the 5 February 2013 in an passive apartment in Cologne, the occupant being blocked for about 2 minutes.

This experimental campaign was also used for validating the zone model CFAST predictive capability. The way to take into account the effective leakage which depends on the overpressure inside the building as well as the mechanical ventilation in CFAST for obtaining satisfactory simulation results are briefly presented in this paper. The use of validated software could be helpful to take into account the fire-induced overpressure in confined dwellings in fire safety design.

KEYWORDS

Passive house; compartment fires; fire growth; overpressure; modeling; CFAST

INTRODUCTION

The need for sustainability and smaller ecological footprint leads to the construction of more airtight building envelopes with better thermal insulation in order to reduce the pollution and increase the energy efficiency, according to the Energy Performance of Buildings Directive (EPBD 2010/31/EU) [1]. In such structures, one might wonder if there are specific risks in case of fire.

Some fatal accidents during the intervention of fire brigades were reported in the past due to a backdraft phenomenon in airtight and insulated renovated apartment [2]. On March 28, 1994, at 62 Watts Street in Manhattan, New York City, a fire took place in a three-story apartment that had been renovated by installing new windows and doors, minimizing air infiltration and setting up heavy thermal insulation [3]. Upon an emergency call, firefighters arrived and saw smoke coming from the chimney but no other signs of fire, so they decided to open the door of the apartment on the first floor where the fire took place leading to a backdraft and killing the team composed by three firefighters. A similar event occurred on September 14, 2002, in Neuilly-sur-Seine, near Paris, killing a team of five young firefighters [4]. Nowadays, the likelihood of backdraft is well recognised by fire brigades, new operational procedures were develop in order to take into account this particular hazard.

Thanks to the growing interest to build more airtight and insulated buildings, the specific fire hazards for the occupants raised questions amongst fire community in Belgium. In 2010 the Belgian Ministry of Interior funded a study about the fire hazards possibly associated with Passive Houses in comparison to traditional houses. The University of Mons and ISSeP used zone modelling for investigating how much the characteristics of a passive house such as airtightness, ventilation and thermal insulation could affect the fire development [5]. During the growing phase of the fire, similar results are obtained in terms of fumes temperatures and CO and HCN concentrations in the fire room for the same fire scenario and the same interior cladding. However, a significant difference in pressure rise (due to thermal expansion of fumes) was calculated due to the airtightness in the passive house, but the scientific monitoring committee of the project decided to not communicate about this potential problem because of a lack of validation from experimental data at large scale, even though inwards-opening doors could block the occupants during a certain period.

In the meantime, pool fire experiments were conducted in well-confined and force-ventilated enclosure for practical application in the nuclear industry [6]. Overpressure up to 3500Pa were measured leading to the flow inversions in the ventilation network. These results were also used to validate the capability of zone model and field model regarding heat release rate and pressure and showed that great care should be taken about the leakage area [7,8]. It can be noticed that the air change per hour fixed for the mechanical ventilation in these buildings are somewhat high for residential building and the area leakages are particularly low.

In 2015, full-scale experiments were carried out by van den Brink to study the effect of the building skin on the fire and the pressure behavior in well-insulated and airtight dwellings [9]. A 20.7 m^3 room was used for the test setup. An air change per hour of about 1 h^{-1} was measured at 50 Pa from a blower door test and bioethanol was used as fuel. From the experiments it appeared that pressures in the order of hundreds of Pa can occur within the first 23 seconds after ignition and may prevent the occupants from escaping.

More recently, a research about fire-induced pressures and smoke spreading in mechanically ventilated buildings with airtight envelopes has been carried out by Kallada Janardhan et al [10]. Heptane pool and polyurethane mattress fires were carried out inside a real 150 m^3 apartment. An air change per hour of about 2.9h^{-1} was measured at 50 Pa through a blower door test. The pressure ranges encountered were between 100 Pa to 1650 Pa for short periods and occurred in less than 50 s from the ignition. During this period of high over-pressure, it was not possible to open an inwards-turning door by pulling from inside. This experimental campaign was used for validating FDS predictive capability and gave good results even experimental uncertainties were encountered for the ventilation system configuration, the fan characteristics and the additional leakage due to high over pressures.

In 2015, a full-scale experimental facility was build up in the region of Mons with the support of the KCC centrum and the Régie Provinciale du Hainaut where the firefighters usually do their exercises and trainings. This facility was designed with the support of the Ghent University [11]. The objectives of this study were to quantify the overpressure induced by fires in airtight compartment with air change per hour close to the ones encountered in passive houses. It was decided to burn pallets stack and wood cribs instead of liquid fuel pan in order to have a more realistic fire scenario.

EXPERIMENTAL SETUP

Geometry - ventilation - measurements

The airtight building has the same inner dimensions of a 40-foot shipping container (12 m length, 2.38 m width and 2.44 m height). All the information about the test setup requirements are given in Vanhaverbeke [11]. The wall is composed by three layers of different materials: the inner shell consists of two plasterboards of 13 mm. Behind this layer, there is a 5 cm insulation shell made of a frame of steel studs with mineral wool. The outer shell is made of 20 cm concrete blocks finished with plaster on the inside to ensure the airtightness. The insulation does not need the thickness it should have in a passive house, since the heat front will not travel through this 5 cm insulation layer during the time of interest for the fire tests. The floor and the ceiling are made of 15 cm concrete slabs; the latter is insulated with 5 cm of mineral wool protected by 13 mm thick plasterboards to reduce the heat losses. The building has no windows but only one external door 2.15 m high, 0.9 m wide and 0.04 m thick composed of two layers of galvanized steel covering a stratum of polyurethane foam. The structure is divided into two compartments: the first room is 4 m long, the second one is 8 m. They are separated by a 0.12 m thick wall composed by a steel stud frame filled with mineral wool covered with plasterboards on both sides. There is also one door between the two rooms which is closed during fire tests. However, there is a space of about 1 cm height between the floor and the bottom of the door which is useful for air circulation between both rooms. The ventilation of the building is guaranteed by two fans branded Soler & Palau, type TD-250/100 SILENT, placed in the ducts outside the construction. In particular, the maximum static pressure was about 145 Pa and the volume flow rate corresponding to zero pressure was about 240 $m^3.h^{-1}$. For the experiments, it is assumed that the large room (which is the fire room) is dry such as a living room and air is supplied in it, whereas the small one is moist such as a bathroom and air is extracted from it. The Belgian government has established precise requirements for the minimal and the maximal flows for every different type of room [12]. An adjustable orifice plate (damper DIRU) was used to adjust the airflow by creating a pressure drop in the ducts. The installed value in this case is 75 m^3/h leading to an air change per hour of about 1 h^{-1}. The damper DIRU with flow meter offers measurements of pressure drop and temperature and thus the airflow rate can be obtained.

Measurements were made for gas pressure in fire room and between both rooms, gas temperatures in the center of each room by using 1 mm in diameter K-type thermocouples trees with 0.2 m vertical separation, fuel mass from load cell and airflow rate in the ducts (see Fig.1). For pallets fire tests, O_2, CO_2, CO and THC concentrations were also quantified. The mass loss rate was calculated from a smoothing technique and the HRR was then calculated. The heat of combustion was estimated through lab-scale experiments with a cone calorimeter.

Natural leakages

The airtightness of the set-up was measured using a blower door test (with the ducts closed with an airtight metal cap). A value of air exchange rate of $1h^{-1}$ was obtained at 50 Pa overpressure. This test at 50 Pa was made before each fire experiment in order to check the airtightness of the building. Since 2016, the airtightness is constant, so the building has not been damaged by the fires. Moreover, additional leakages due to high overpressure have been also well characterized in the range 0-600 Pa for validation model purpose.

All the elements of the setup such as ventilation system configuration, fan characteristics and additional leakages were well characterized in order to use the experimental data for zone and field models' validation.

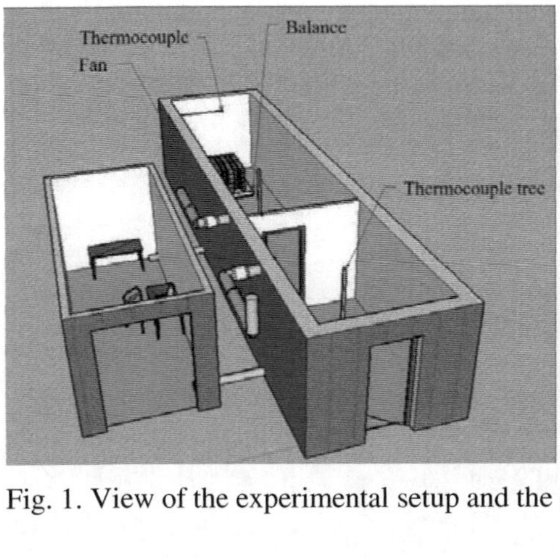

Fig. 1. View of the experimental setup and the (big) wood crib on the load cell.

IOP Conf. Series: Journal of Physics: Conf. Series **1107** (2018) 042031 doi:10.1088/1742-6596/1107/4/042031

EXPERIMENTAL RESULTS

In 2016, 4 preliminary tests were carried out with stack of 5 pallets but there were some problems of repeatability for the HRR (Fig.2). Consequently, in 2017 wood cribs of 15 layers of pine slats were used as fuel load. Two different cribs were used: a layer of 5 pine slats of 380x27x18 mm^3 (little crib) and a layer of 8 pine slats of 594x30x17 mm^3 (big crib). The ignition of the fuel load was carried out with heptane pans. Each fire load was tested with the two different ventilation duct configurations.

The measured HRR and overpressure are shown respectively in Fig.2 and 3. As can be seen, there is a good repeatability of the HRR curves for the wood cribs whatever the ventilation (on or off) . Values of about 250 kW (at 300s) and 500 kW (at 230s) are obtained for the HRR peak respectively for the little and the big cribs. As expected, the speed of the growing of the fire is a key parameter for the overpressure measured in an airtight compartment. Over pressure peaks from 850 to 2035 Pa were measured without mechanical ventilation (ducts closed), while values from 420 to 750 Pa were observed with the mechanical ventilation on. The order of magnitude of the pressure measured during the experiment without mechanical ventilation is higher than the one carried out with mechanical ventilation: the mechanical ventilation network succeed in lowering the pressure by helping the fumes going outside. At the beginning, a flow rate of 75 $m^3.h^{-1}$ is given at time zero by both fans. When the HRR increases, the pressure rises as well and pushes the fumes out of the compartment, the flow rate of the extraction duct grows, whereas the flow rate of the incoming air decreases and become negative when the overpressure is higher than 145 Pa, the maximum static pressure of the fan (Fig.4). Then due to a lack of oxygen in the fire room the HRR decreases and the pressure decreases consequently. Fresh air can thus be supplied via the fan when the overpressure has decreased below 145 Pa. For the little crib fire, a second peak of HRR is then observed, the wood crib collapsed and restarted to burn for a while leading to a second pressure peak (see Fig.4). For the big wood crib, a pulsating phenomenon can be observed 300 s after ignition of the fuel load for both ventilation configurations when the fire is under ventilated (when the oxygen depletion affect the heat release rate). These oscillations were not observed for the two others fuel loads, perhaps due to the HRR curves which are different. Since 150 Pa is the threshold value below which most people could open an inward door, it is important to notice that the period corresponding to a fire-induced overpressure higher than this threshold can last 3 minutes. These results confirm the observations carried out during the reconstruction of the fire which occurred in the night of the 5 February 2013 in a passive apartment in Cologne, the occupant being blocked during 2 minutes [13].

In Fig.5, one can see the gas temperature measured in the center of the fire room, as well as the CO_2 and O_2 concentration in the upper layer for the test carried out without mechanical ventilation with the pallets as fuel load. The temperature is quite uniform at a height greater than 1.4 meters and there is a strong stratification below this height. Concentration of about 10% and 11% are measured respectively for CO_2 and O_2 concentration 200 s after ignition. O_2 and CO_2 concentrations in the fumes measured when the ventilation was working were not relevant since the inlet flow rate from the fan affected the smoke environment near the gas-sampling probe.

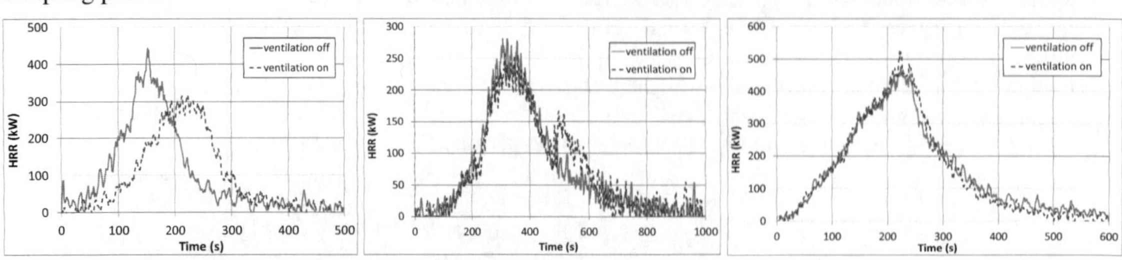

Fig. 2. Heat release rate (from left to right: pallets, little crib, large crib)

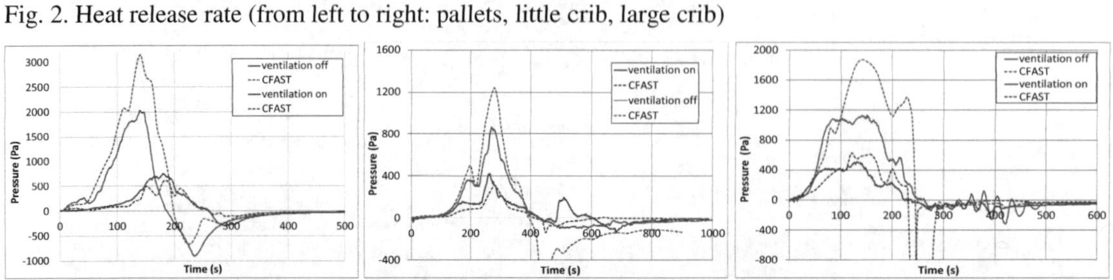

Fig. 3. Overpressure (from left to right: pallets, little crib, large crib)

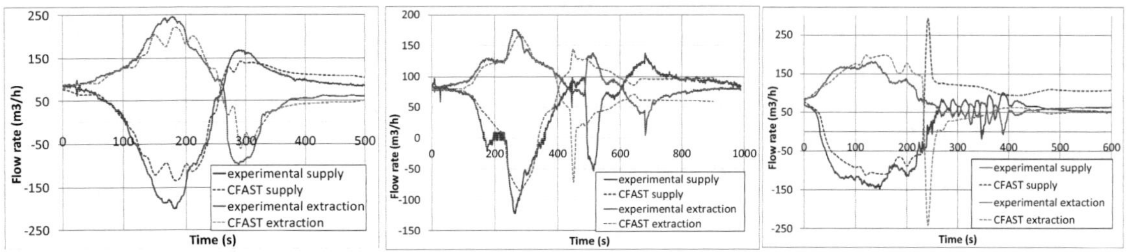

Fig. 4. Flow rate in the ducts when mechanical ventilation is working (from left to right: pallets, little crib, large crib)

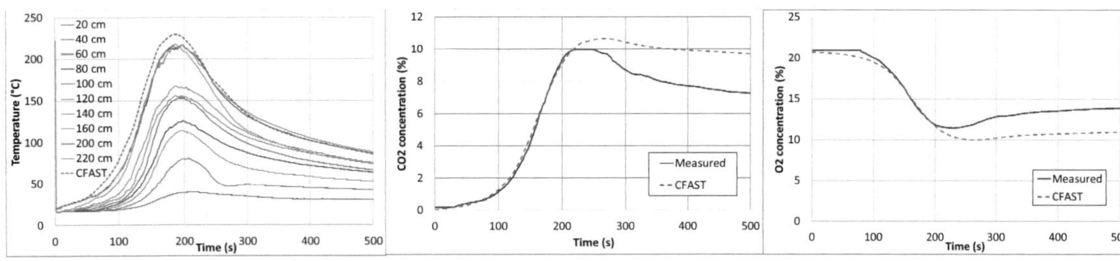

Fig.5. Temperature, CO_2 concentration and O_2 concentration in the fumes. Pallets fire with ventilation off

NUMERICAL SIMULATIONS WITH CFAST

The results obtained in the experimental setup were used to validate the capability of the zone model CFAST to predict major parameters such as pressure, temperature, volumetric flow rate in the ducts in an airtight house configuration. The HRR which is a key parameter was fixed as input data. The leakage modelling is another key parameter for the pressure calculation. At first, a constant leakage area of 26 cm^2 was fixed from the results obtained through the blower door test at 50 Pa. This approach was not satisfactory because the overpressure calculated by CFAST was over predicted by a factor of 4 when the mechanical ventilation was off. However, the overpressures measured during fire tests are an order of magnitude higher than the 50 Pa used during the blower door test. In a typical building, as the overpressure increases, the leakages area increases as small gaps, cracks and other leakage paths open up. The leakage areas were thus measured at a broader range pressure for improving the CFAST simulations. According to the ASTM E 779, overpressure range from 50 to 600 Pa was obtained from blower door test. A value of 0.7 was reached for the leak pressure exponent (n) and 26 cm^2 was obtained for the leak reference area at the leak reference pressure of 50 Pa according to the following correlation:

$$A_L = A_{L,ref}\left(\frac{\Delta p}{\Delta p_{ref}}\right)^{n-0.5} \tag{1}$$

This equation was used to create in CFAST a table of leakage area varying in function of time by taking into account the evolution of pressure during the fire [14].

For the tests carried out without mechanical ventilation, this time varying leakage improves the overpressure calculated by CFAST compared to the measured values and a difference of about 50% is observed (Fig.3). This error obtained under ventilation off condition is attributed to leakage model which locate the leakages in well-defined positions. Simulations carried with the Fire Dynamics Simulator (FDS) field model using the localized leakages model confirm this statement (in comparison to the pressure zone leakage approach) [14]. It can also be pointed out that the parameter values of equation 1 are validated in the range from 50 to 600 Pa while overpressure up to 2000 Pa are measured during the fire tests. CFAST leads also to fumes temperature, CO_2 and O_2 concentrations which are close to the measured ones (Fig.5).

For the tests carried out with mechanical ventilation, in case of high overpressure, CFAST does not take into account neither the reverse flow of the fan supplying fresh air nor the extra flow rate of the fumes extraction, the overpressure was either over predicted by a factor of 4. In order to overcome these limitations, an extra hole was added to a room connected to the mechanical ventilation when the pressure was higher than the maximum static pressure of the fan (145 Pa). The diameter of this hole was defined by the limiting element related to the pressure drop in the duct, that was the orifice plate. Thanks to this methodology, the sum of the fan flow rate and the volume flow rate through this hole obtained from CFAST is now similar to the measured flow rate in the duct (Fig.4). By considering all these improvements, CFAST leads to overpressures similar to the ones measured in the experimental setup for the tests carried out with mechanical ventilation.

The results of this experimental campaign were also used for validating the FSD model [14]. The FDS predictive capability to estimate overpressure induced by fire in airtight houses will be presented in a forthcoming paper.

CONCLUSIONS

Full-scale experiments were carried out in an airtight building in order to measure overpressure induced by fires. Over pressure peaks from 850 to 2035 Pa were measured without mechanical ventilation (ducts closed), while values from 420 to 750 Pa were observed with the mechanical ventilation on. Both the reverse flow of the fan supplying fresh air and the extra flow rate of fumes in the extraction duct due to the overpressure inside the apartment were not sufficient to prevent overpressure inside the rooms. All these experiments highlight that there may be problems of overpressure in very airtight houses in case of fire. Especially, the occupants may not succeed in escaping due to the inward door opening for period that can last 3 minutes. The experiments were also used for validating the zone model CFAST predictive capability.

ACKNOWLEDGEMENTS

The authors thank Philippe Staquet, Director of the Régie Provinciale Autonome (RPA) du Hainaut for the construction and the access to the experimental setup which was partially funded by the Centre Fédéral de Connaissances (KCCE). We thank the experimental team who participated to the project: Christophe Lheureux, Thierry Lestrade and Laure Piret-Gérard (Université de Mons), Pierre Bertelot (Ecole des Mines d'Alès) as well as Laurent Aprin (Ecole des Mines d'Alès) for the analyzers of O_2, CO_2 and CO.

REFERENCES

[1] Directive of the European Parliament and of the Council of 19 May 2010 on the energy performance of buildings. (Directive 2010/31/EU).

[2] Hume B. (2004). *Firefighting in under-ventilated compartments: Literature review*, Fire Research Technical Report 5/2005, Fire Statistics and Research Division, Office of the Deputy Prime Minister, London, UK.

[3] Bukowski, R. (1995). Modeling a backdraft: the fire at 62 Watts Street, *NFPA Journal, 89*: 85-89.

[4] Francetvinfo. (2017). Retrieved from www.francetvinfo.fr/faits-divers/mort-de-cinq-pompiers-a-neuilly-le-proces-s-ouvre-aujourd-hui_1669349.html, accessed November 2017

[5] Fourneau, C., Cornil, N., Delvosalle, C., Breulet, H., Desmet, S., Brohez, S. (2012). Comparison of fire hazards in passive and conventional houses. Chem Eng Trans 26:375–380.

[6] Pretrel, H., Le saux W., & Audouin L. (2012). Pressure variations induced by a pool fire in a well-confined and force-ventilated compartment. *Fire Safety Journal*, 52: 11-24

[7] Bonte F., Noterman N., & Merci B. (2013). Computer simulations to study interaction between burning rates and pressure variations in confined enclosure fires, *Fire safety Journal*, 62: 125-143

[8] Wahlqvist J., van Hees P. (2013). Validation of FDS for large-scale well-confined mechanically ventilated fires cenarios with emphasis on predicting ventilation system behavior. *Fire safety Journal*, 62: 102-114

[9] Van den Brink, V. (2015). *Fire safety and suppression in modern residential buildings.* Master thesis, University of Eindhoven.

[10] Kallada Janardhan, R., & Hostikka, S. (2017). Experiments and numerical simulations of pressure effects in apartment fires. *Fire Technology, 53*:1353-1377.

[11] Vanhaverbeke, D. (2015). *Fire development in passive houses: qualitative description and design of full-scale fire tests.* Master thesis, University of Edinburgh.

[12] NBN D 50-001 "Dispositifs de ventilation dans les bâtiments d'habitation" 1ère Ed. 1991.

[13] Brohez S., Duhamel P. (2018). Inwards doors blocked by fire induced overpressure in airtight apartment: a real case in Germany. *Chemical Engineering Transactions*. Accepted for publication.

[14] Caravita I., (2018). *Fire development in passive houses: experimental analysis and validation of theoretical models*. Master Thesis, outgoing Erasmus from Università di Bologna, University of Mons, B.

Experimental and Numerical Comparison of Small-scale Gaseous Fire Whirls

Monica Diab, Aaron Yip, Mahshid Hadavand, Jan Haelssig and Michael Pegg

Department of Process Engineering, Dalhousie University, Canada

Monica.Diab@dal.ca

ABSTRACT

Fire whirls can occur during urban fires, especially in intense fires in combustible building structures, and more often in forest or wildland fires. They are a special swirling diffusion flame characterized by significant enhancement in burning rates, flame heights and flame temperatures, along with a strong whirling motion of the flame. This whirling motion can pick up large firebrands and scatter them afar leading to spot fires. Many researchers have published experimental work on small- and medium-scale pool fire whirls and gaseous fuel fire whirls using split cylinders and various fixed-frame apparatus to investigate axial and tangential velocity profiles, axial and radial temperature distribution, burning rates, and flame heights. Likewise, several researchers have attempted to predict the experimental results of fire whirls using different modelling approaches and simulation software.

In this paper, experiments were undertaken to study the dynamics of propane gas fire whirls in a small-scale, square-based, fixed-frame apparatus. Measurements of flame height and temperature profiles (both axial centerline and radial) were made for a low initial momentum burner of 76.2 mm internal diameter. The burner was operated at a volumetric flow rate of 6 dm^3/min, which gave a heat release rate of 9.12 kW. Simulations using Fire Dynamics Simulator (FDS 6.6.0) and ANSYS Fluent 17.1 were performed to compare with the experimental measurements. Four separate mesh refinements were employed and four different sub-grid-scale (SGS) turbulence models were tested with FDS. The Deardorff, Wall-Adapting Local Eddy-viscosity (WALE), and dynamic Smagorinsky models, formed stable fire whirls for the two largest mesh refinements.

The temperature profiles were overpredicted at the core of the flame with FDS and underpredicted with Fluent. The FDS simulation prematurely predicts the peak temperature for the axial centreline profile, whereas with Fluent the axial temperature profile matches the general trend of the experimental measurements.

The visible flame height, determined through image processing, was approximately 0.88 ± 0.06 m, which corresponds to a measured temperature of ~500°C. The 500°C temperature contour was used as a rough approximation of the flame height in the numerical simulations. It was found that with Fluent the 500°C contour grew until the fire whirl stabilized and reached the top of the hood at 1.6 m, clearly overpredicting the flame height. The height estimates based on the predicted 500°C contours show a strong dependence on the mesh resolution. This is primarily due to increased instability resulting in more mixing and spreading of the temperature for the coarser mesh size. However, the simulated flame heights show less dependence on the SGS turbulence models.

KEYWORDS:

CFD; modelling; fire growth; firenado; fire tornado; fire whirl; swirling fire.

INTRODUCTION

Fire tornado, swirling fire, and firenado are all terms used to define the spiral formation of a fire whirl. Two recent examples have attracted media attention. The first is a video that recorded a fire whirl in April 2016 at St. Albert, AB, Canada [1]. A fire whirl appeared from a small grass fire, which spread so quickly fire fighters had to flee the site and one had to dive into the nearby lake to escape. The second is a video of a mass fire in August 2016 at Phelan, Southern California, USA in which a fire whirl forced 82 000 people to flee the area [2]. By the end of the day it had consumed 73 km^2 of vegetation and destroyed hundreds of homes. Disasters such as these have occurred throughout the world for many decades and because of the destructive effects of fire whirls have led to increased investigation of fire whirl formation and dynamics. Morton [3] has discussed the mechanisms that lead to both fire whirl formation and what sustains them. However, full-scale experiments on fire whirls are difficult to perform due to safety, costs and resources. A recent comprehensive review paper has summarized the research findings, primarily from laboratory-scale studies of liquid pool fire whirls and gaseous fuel fire whirls [4]. Many researchers have modelled fire whirls with varying degrees of success. Table 1 lists details of some of the modelling efforts to-date and references the experimental data they attempted to simulate with the type of fire whirl generator (FWG).

Table 1: CFD simulations of fire whirls

Authors [Ref] (Year)	Experimental [Ref], FWG	Transport model or software	Combustion model	Turbulence model
Satoh & Yang [5] (1997)	Satoh & Yang [6], 4 corner gaps	UNDSAFE-I	None, 20 kW non-uniformly distributed volumetric heat source, no radiation	-
Farouk *et al.* [7] (2000)	Analogous to Satoh & Yang [6], 4 corner gaps	Low Ma form of Navier-Stokes	None, Thermal Element Model (TEM) based on mixture-fraction theory	LES, Smagorinsky with $C_s = 0.14$
Battaglia *et al.* [8] (2000)	Emmons & Ying [9], Chigier *et al.* [10], Satoh & Yang [6], imposed circulation	NIST Fire Dynamics Simulator (FDS)	TEM	LES, Smagorinsky with $C_s = 0.14$
Meroney [10] (2003)	Bryam & Martin [11], 2 tangential slits; Satoh & Yang [6], 4 corner gaps	Fluent 6.0	None, 11.6 kW heat source at base for [11], 20 kW square-based vertical volume for [6]	LES, Smagorinsky-Lilly
Matsuyama *et al.* [12] (2004)	Full-scale Fire Lab at the Building Research Inst., Japan, 4 corner gaps	FDS 3.1	None, 10, 15 & 20 kW propane burner, mixture-fraction model	LES, Smagorinsky
Snegirev et al. [13] (2004)	Room enclosure at Training Centre of the Greater Manchester County Fire Service with 1 door and a ceiling vent	3D CFD model and code *Fire3D*, density averaged, low Ma, thermal radiation (Monte Carlo)	Single-step global reaction, reaction rate determined by eddy break-up model	Modified k-ε model
Forthofer *et al.* [14] (2009)	Soma & Saito [15], L-shaped fire in a cross wind (wind tunnel experiment)	FDS 5.0	None, specified heat-release-rate-per-unit-area	LES, SGS model not listed
Hartl *et al.* [16] (2014)	Di-methyl-ether burner, offset split cylinders	Spectral element–Fourier method [17]	None, heat input using flux boundary conditions	DNS
Satoh *et al.* [18] (2014)	Blinov & Kyudyakov [19], Hamins *et al.* [20], Satoh *et al.* [21–24], and Zhou *et al.* [25], 4 corner gaps	FDS 6.0	Mixture-fraction method	LES, SGS model not listed
Yuen *et al.* [26] (2016)	Chow & Han [27] 1 corner gap	FDS 6.1.2	Mixture-fraction method	LES, SGS model not listed

A successful model should capture both the qualitative and quantitative aspects of a fire whirl. Qualitative observations should include the transition from a buoyant diffusion gas flame to a stable fire whirl, and the stability of the fire whirl. If the qualitative properties of the simulation match experimental observations, then the quantitative properties (flame height, temperature and velocity profiles) should reasonably match the experimental data. The objectives of this study were to compare simulation results from FDS 6.6.0 and ANSYS Fluent 17.1 with data from gaseous fire whirl experiments, to identify limitations of and areas for improvement in the adopted modeling approaches, and to identify limitations in the experimental measurement techniques. The primary focus of this study was on fire whirl formation, stability, temperature profiles, and flame height.

EXPERIMENTAL

The dynamics of propane fire whirls was studied in a small-scale, square-based, fixed-frame apparatus, as shown in Fig. 1(a). The front and rear walls were tempered glass and the two sides were aluminum, all supported in tracks, which enabled them to slide to allow adjustment of the corner air gaps. For the experiments presented in this study, the air gap was set to 457 mm high × 64 mm wide. The base plate had an opening to accommodate gas burners of different diameters. Four low initial-momentum gas burners, each with a burner tube made of aluminum pipe (50.8, 76.2, 101.6, or 152.4 mm ID) packed with wire wool at the base and then filled with sand to just below the burner rim, were constructed. However, only the 76.2 mm ID burner was used in this study. The mass flow rate, and hence the heat release rate (HRR), was set by a mass flow meter at 6 dm^3/min, corresponding to a HRR of 9.12 kW. Temperatures were measured with type K thermocouples, centered in the frame. Eight thermocouples (T1–T8) were positioned at heights of 50, 100, 300, 500, 700, 900, 1000, and 1100 mm above the burner rim. Measurements were made at radial distances of 0 (centre of the burner), 20, 40, and 60 mm. A GoPro HERO4 Silver Camera was used to record the flame shape and height.

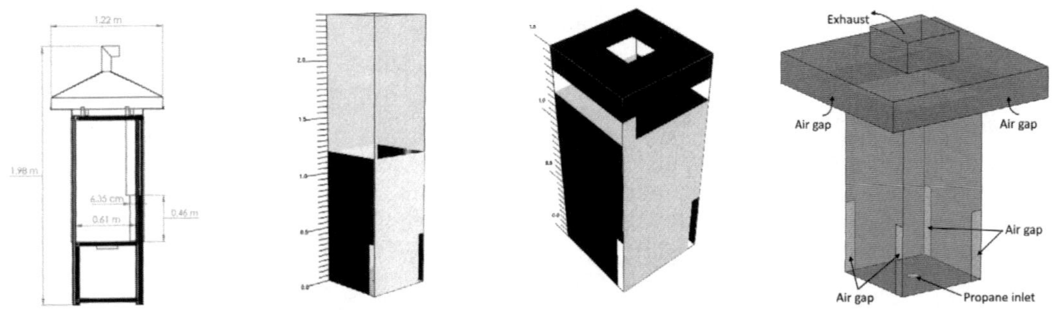

(a) Experimental (b) FDS geometry 1 (c) FDS geometry 2 (d) ANSYS Fluent

Fig. 1. Fixed-frame fire whirl apparatus and simulation geometries

NUMERICAL SIMULATIONS

The most common numerical approaches to fire modelling are RANS (Reynolds-Averaged Navier-Stokes), URANS (Unsteady RANS) and LES (Large-Eddy Simulation). In this study, a LES turbulence model was employed. In LES, the large-scale unsteady energy-carrying geometry-dependent anisotropic eddies are resolved, but the effect of the smallest scales in the flow are included through the sub-grid scale (SGS) model.

The two geometries shown in Fig. 1(b) and (c) were used for simulations with FDS 6.6.0. The geometry in Fig. 1(b) included an open region above the real outlet of the experimental apparatus to prevent boundary effects from impacting the simulation results. The geometry in Fig. 1(c) provided a simplified representation of the actual outlet, but is therefore also more sensitive to the boundary condition at the open face. The aluminum (black) and glass (blue) panels were 0.6 × 0.6 × 1.2 m, which corresponds to the size of the experimental apparatus [Fig. 1(a)]. An opening of 460 × 60 mm at the bottom right corner of each panel simulated the actual air gaps. A refinement study using four uniform computational meshes, containing 192 000 (M1), 648 000 (M2), 1 536 000 (M3), and 5 184 000 (M4) cells, was performed for geometry 1 to analyse the impact of mesh resolution on the numerical temperature predictions. Two meshes, which contained 1 280 000 and 4 320 000 cells, were tested for geometry 2. These meshes provide Plume Resolution Index values (as defined in the FDS User's Guide) between 8 and 25 for the simulations in this study. This is a similar range to many of the validation cases provided in the FDS Validation Guide. Thermocouple probes with the same properties as those in the experiments were used to record temperatures. Four different SGS turbulence models were tested using geometry 1 to study their impact on the simulation: Deardorff model ($C_V = 0.1$), Wall-Adapting Local Eddy-viscosity (WALE) model ($C_W = 0.6$), constant coefficient Smagorinsky model ($C_S = 0.2$), and dynamic Smagorinsky model. Van Driest damping was used for the Deardorff, and constant and dynamic Smagorinsky models to adjust the eddy viscosity in the cell layer immediately adjacent to walls. Only the Deardorff model was tested for geometry 2. The default combustion rate model in FDS 6.6.0 is a mixing-controlled, mixture-fraction model based on the Eddy Dissipation Concept (EDC). The default FDS combustion model settings were used in this study, and propane was specified as the fuel. The default FDS settings were retained for solving the Radiative Transport Equation (RTE). The burner was modelled as a square inlet having an equivalent cross-sectional area to the burner. All FDS simulations were run for at least 120 s. Data was averaged after the simulations has reached a pseudo-steady-state over at least 60 s.

The geometry used for ANSYS Fluent 17.1 simulations is shown in Fig. 1(d). The computational mesh was refined towards the center and contained a total of 351 581 grid cells. The hood dimensions on top of the main section are 1.22 × 1.22 × 0.2 m, and the connection to the exhaust is 0.2 × 0.2 × 0.2 m. As in the experimental apparatus, the bottom surface of the hood is not covered, rather there is an air gap opening as shown in Fig. 1(d). The Algebraic Wall-Modeled LES model (WMLES) was used as the turbulence model for this study. A non-premixed combustion model was applied using the Probability Density Function (PDF) approach to include the effect of turbulence-chemistry interaction. Radiation heat transfer was included using the P-1 model. The turbulent Schmidt and Prandtl numbers were set to 0.7. The fire whirl stabilized relatively fast in the Fluent simulations, and therefore data was averaged between 10 and 60 s.

RESULTS AND DISCUSSION

Fire Whirl Formation and Stability

In all experiments, a stable fire whirl was established within approximately 10 s following ignition of the propane. Once formed, the flame experienced small oscillations in its visible height and precessed somewhat irregularly around the burner. The Fluent simulation predicted fire whirl formation after approximately 3 s, followed by a small amount of precession around the burner. The FDS simulations using geometry 2 dod not yield stable fire whirls. For geometry 1, the FDS simulations using meshes M1 and M2 did not provide stable fire whirl predictions for any of the SGS models. The constant Smagorinsky model did not predict a stable fire whirl for mesh M3, and only a partially stable whirl was predicted for mesh M4. The Deardorff, WALE and dynamic Smagorinsky models with meshes M3 and M4 predicted stable fire whirl formation within 10 s. Following formation, these models predicted irregular precession around the burner, providing qualitative agreement with the experimental results. Therefore, only these three SGS models and two meshes were analyzed further.

Temperature Comparison

Figure 2 shows measured temperatures and simulation results from FDS and Fluent. The experimental data were averaged over the pseudo-steady-state period. The error bars in Fig. 2 indicate the root-mean-square error (RMSE) and provide a measure of the variability in the data due to flame oscillation and precession. The average axial centreline temperatures from the experiments, Fluent simulations, and FDS simulations using the Deardorff model with mesh M4 are compared in Fig. 2(a). Close to the burner, (below 10 cm) the FDS model overpredicted the temperature and as the height increased from 30 to 90 cm it underpredicted the temperature until the plume region. The simulation predicts the peak temperature (*i.e.* the temperature hump) prematurely. The likely reason for this is an overprediction of turbulence close to the burner outlet, resulting in unrealistically high combustion rate predictions. In reality, the first section of the flame should be relatively unmixed with low initial combustion rates, but properly resolving this rate would likely require a combination of higher mesh resolution, a better approximation of turbulence at the wall, and maybe also a better representation of the burner.

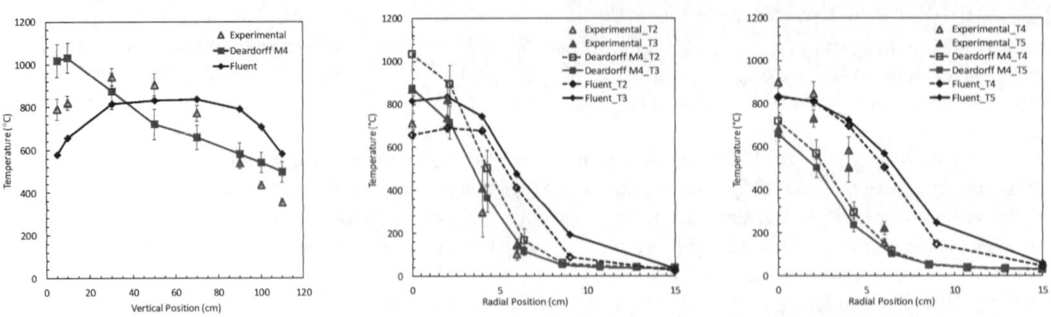

(a) Axial centreline (b) Radial for T2 and T3 (c) Radial for T4 and T5

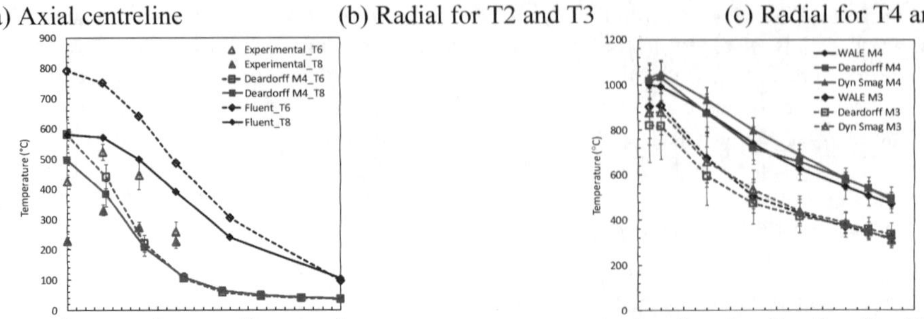

(d) Plume region for T6 and T8 (e) Axial centreline-comparison of SGS models for coarse and fine meshes

Fig. 2. Experimental and simulated temperature profiles

However, Fluent simulations underpredicted temperatures below 10 cm and overpredicted them above 30 cm. Nevertheless, the axial profile predicted by Fluent matches the general trend of the experimental profile more closely. The delayed onset of combustion is likely due to the use of the PDF, but it is apparent that the model would require further tuning to provide a closer prediction for the experiment. Figures 2(b–d) show the radial temperature profiles of the measured and numerical predictions. As expected from the axial profile, there are significant differences between the measured and predicted profiles. Although FDS and Fluent both over and underpredict temperatures in the flame at various positions, the shapes of the profiles generally follow the same trends as observed in the experiments. The combustion and/or turbulence modelling approaches would need further refinement to provide a better quantitative prediction of the data.

Height Comparison

Figure 3(a) shows the stable experimental fire whirl at 30 and 60 s. The visible flame height, as obtained by image processing through MATLAB, is approximately 0.88 ± 0.06 m. From the experimental data, the edge of the visible flame height corresponds to an average centreline temperature of approximately 500°C [Fig. 3(b)]. Since it is difficult to quantify the flame height in the numerical simulations, the 500°C contour was assumed to provide a representation of the flame height. Of course, this is only a crude approximation and should therefore not be interpreted as providing a definitive measure of the flame height. In the Fluent simulations, the 500°C contour grows until the fire whirl has stabilized and reaches the top of the hood at 1.6 m, which is an overprediction of the actual flame height. Again, this prediction could likely be improved through changes to the combustion model; however, more accurate radiation modelling may also improve predictions of the temperature. The predicted 500°C contours (height estimates) for the Deardorff and WALE SGS models for meshes M3 and M4 are provided in Fig. 3(c). Clearly there is strong link between the mesh resolution and the predicted height of the contour. As expected based on the temperature profiles presented in the last section, there is a much weaker dependence on the SGS model. The differences in the predictions between the two mesh resolutions is primarily due to increased instability, which leads to more mixing and spreading of the temperature for the coarser mesh. The surprising aspect is how much difference this makes in the time-averaged temperature contour. Considering that the finer mesh already contains over 5 million cells on the order of 5 mm in dimension, it will be necessary to explore model modifications (*e.g.* changes to the combustion model) that lead to better predictions without increasing computational requirements much further.

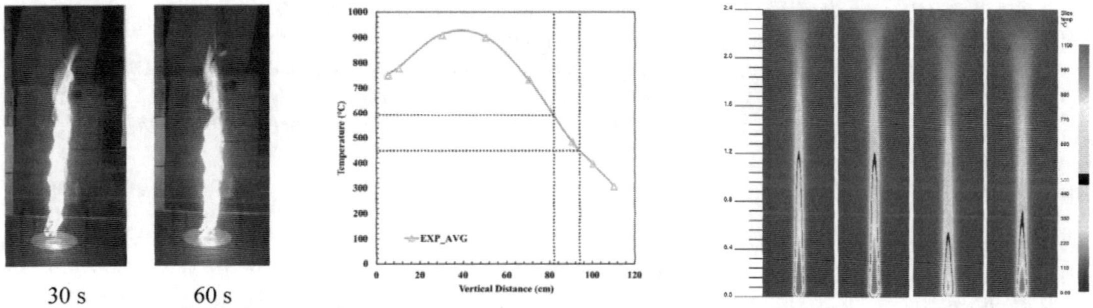

(a) Experimental observation of visible flame

(b) Temperature at visible flame height

(c) Time-averaged temperature contours for (left to right) Deardorff M4, WALE M4, Deardorff M3, and WALE M3

Fig. 3. Experimental fire whirl

CONCLUSIONS

Measurements of flame height, axial, and radial temperature profiles for a small-scale propane gas fire whirl were investigated experimentally and numerically. Simulations with FDS and Fluent have been compared with the experimental data for a 9.12 kW HRR. In the FDS simulation a stable fire was established after 10 s using the M3 and M4 mesh resolutions with the Deardorff, WALE, and dynamic Smagorinsky SGS turbulence models. The temperature profiles were overpredicted at the core of the flame with FDS and underpredicted with Fluent. The FDS simulation prematurely predicts the peak temperature for the axial centreline profile, whereas with Fluent the axial temperature profile matches the general trend of the experimental measurements. The 500°C temperature contour was used as a rough approximation of flame height in the numerical simulations. With Fluent the 500°C contour grew until the fire whirl stabilized at the top of the hood overpredicting flame height. With FDS, the height estimates based on the predicted 500°C contour show a strong dependence on the mesh resolution. However, the simulated flame heights show less dependence on the SGS turbulence models.

REFERENCES

[1] CBC (2016), available at: <http.cbc.ca/news/canada/edmonton/fire-tornado-st-albert-1.3539882>

[2] CBC (2016), available at: <http://www.cbc.ca/news/world/firenado-california-1.3724387>
[3] Morton, B.R. (1970) The Physics of Fire Whirls. *Fire Research Abstracts and Reviews* 12:1–19.
[4] Tohidi, A., Gollner, M.J. and Xiao, H. (2018) Fire Whirls. *Ann. Rev. Fluid Mech.* 50:1–28.
[5] Satoh, K, and Yang, K.T. "Simulations of Swirling Fires Controlled by Channeled Self-Generated Entrainment Flows," *Fire Safety Science -- Proceedings of the Fifth International Symposium*, International Association for Fire Safety Science, 1997, pp. 201–212. doi:10.3801/IAFSS.FSS.5-201
[6] K. Satoh and Yang, K.T. (1996) Experimental observations of swirling fires. *Proceedings of the ASME Heat Transfer Division,* 335:393–400.
[7] Farouk, B., Mcgrattan, K.B. and Rehm, R. "Large eddy simulation of Naturally Induced Fire Whirls in a Vertical Square Channel with Corner Gaps," *IMECE Proceedings.* Orlando, FL, USA, 2000.
[8] Battaglia, F., McGrattan, K.B., Rehm,R. and Baum,H.R. (2000) Simulating fire whirls. *Combustion Theory and Modelling*, 4:123–138. doi:10.1088/1364-7830/4/2/303
[9] Emmons, H.W. and Ying, S-J. (1967) The Fire Whirl. *Proceedings of the Combustion Institute*, The Combustion Institute, 11:475–488.
[10] Chigier, N.A., Beer, J.M. Grecov, D. and Bassindale, K. (1970) Jet Flames in Rotating Flow Fields, *Combust. Flame,* 14:171–180.
[11] Meroney, R. "Fire Whirls and Building Aerodynamics." *11th International Conference on Wind Engineering.* Lubbock, TX, USA, 2003.
[12] Byram, G.M. and Martin, R.E. (1970) The modeling of fire whirlwinds, *Forest Sci.*, 16:386–399.
[13] Snigerev, A. Yu, Marsden, J.A. Francis, J. and Makhviladze, GM. (2004) Numerical studies and experimental observations of whirling flames, *International Journal of Heat and Mass Transfer* 47: 2523–2539.
[14] Forthofer, J., Shannon, K. and Butler, B. "Investigating causes of large scale fire whirls using numerical simulation." *8th Symposium on Fire and Forest Meteorology.* 2009.
[15] Soma, S. and Saito, K. (1991) Reconstruction of fire whirls using scale modelling, *Combust. Flame,* 86:296–284.
[16] Hartl, K., Blackburn, H.M. and Smits, A.J. "Experimental and numerical Studies of a Laboratory Fire Whirl." *19th Australian Fluids Mechanics Conference.* Melbourne, AU, 2014.
[17] Blackburn, H.M. and Sherwin, S.J. (2004) Formulation of a Galerkin spectral element–Fourier method for three-dimensional incompressible flows in cylindrical geometries. *J. Comp. Phys.*, 197:759–778.
[18] Satoh, K., Liu, N., Xie, X. and Gao, W. "CFD Study of a Fire Whirl of Huge Oil tank – Burning rate, Flame Length, Distributions of *n*-Heptane and Oxygen in a Fire Whirl." *IMECE Proceedings,* Montreal, PQ, Canada, 2014.
[19] Blinov, V.T. and Kyudyakov, G.N. *Diffusion Burning of Liquids,* Akademii Nauk, SSSR, Moscow,1961 (Translated into English by Research Information Service for US Army Engineer Research and development Laboratories).
[20] Hamins, A., Yang, J.C. and Kashiwagi, T. *A Global Model for Predicting the Burning rates of Liquid Pool Fires*, NISTIR 6318, National Institute of Standards and Technology, September 1999.
[21] Satoh, K., Liu, N., Liu, Q. and Yang, K.T. "Numerical and Experimental Study of Fire Whirl generated in a 15 x 15 Square Array Fires Placed in cross Wind," *IMECE Proceedings.* Boston, MA, USA, 2008.
[22] Satoh, K., Liu, N., Liu, Q. and Yang, K.T. "Preliminary Study of Fire Spread in Cities and Forests, using PMMA Specimens as a fuel in CFD Simulations," *IMECE Proceedings.* Lake Buena Vista, FL, USA, 2009.
[23] Satoh, K., Liu, N., Zhou, K. and Xie, X. "CFD study of termination of fire whirls in urban fires," *9th Asia Oceania Symposium on Fire Science and Technology*, International Association for Fire Safety Science, 2012, pp. 1040–1049.
[24] Satoh, K., Liu, N., Xie, X., Zhou, K., Chen, H., Wu, J., Lei, J. and Lozano, J.S. CFD Study of Huge Oil Depot Fires – Generation of Fire Merging and Fire Whirl in Arrayed Oil Tanks," *Fire Safety Science -- Proceedings of the Tenth International Symposium*, International Association for Fire Safety Science, 2011, pp. 693–705.
[25] Zhou, K., Liu, N., Lozano, J.S., Shan, Y., Yao, B, and Satoh, K. (2013) Effect of flow circulation on combustion dynamics of fire whirl. *Proceedings of the Combustion Institute*, The Combustion Institute, 34:2617–2624.
[26] Yuen, A.C.Y., Yeoh, G.H., Yuen, R.K.K. and Lo, S.M. "Numerical Study on Small-Scale Fire Whirl using Large Eddy Simulation." *3rd International Conference of Fluid Flow, Heat and Mass Transfer.* Ottawa, ON, Canada, 2016.

IOP Conf. Series: Journal of Physics: Conf. Series **1107** (2018) 042033 doi:10.1088/1742-6596/1107/4/042033

Fire Safety Verification Method – The Australia Research Experience

Peter JOHNSON[1], and Nate LOBEL[1]

[1]Fire Safety Engineering, Arup, Australia

peter.johnson@arup.com

ABSTRACT

A methodology described as a Verification Method (VM) has been proposed in Australia as one way of evaluating the level of fire safety of a building design. It has been based on a similar VM developed in New Zealand, and has been issued by the Australian Building Codes Board (ABCB) for public comment with the possibility of adoption into the National Construction Code (NCC) in 2019.

As part of the VM development, a consortium of fire safety engineers and academics were invited to undertake some detailed research into calibration of eight different buildings using the proposed VM. In addition the researchers were asked to evaluate the buildings using quantitative risk assessment to determine the building risk levels against some risk tolerance criteria established by ABCB.

The research showed that none of the 8 buildings met the VM requirements, despite all complying with the NCC prescriptive building code requirements. All buildings as designed also had risk levels in excess of the ABCB risk tolerance criteria.

A Summary Report compiled by the research consortium was issued by ABCB as part of the Public Comment consultation process, and it shows that there would be some considerable uncertainties in using such a VM at this time for regulation of fire safety of buildings in Australia, not only for more complex buildings but even simpler retail, aged care and residential buildings.

The Summary Report suggests that more fire science and engineering research is required before such a VM can be fully adopted across Australia, and that in the interim, more concentration should be given to updating fire engineering guidance and strong attention be given to professional competence, education and accreditation of fire safety engineers.

KEYWORDS
Performance based design, fire safety engineering, risk assessment, verification method, fire modeling, fire statistics, competence, education, accreditation. regulatory system

INTRODUCTION

The cladding fires at the Lacrosse Building in Melbourne in 2015 [1], the Grenfell Building in 2017 [2], and many other external wall fires in high rise buildings around the world had created a degree of momentum in the examination of good practice of fire safety engineering and building design across Australia.
These events have led to many government inquiries, such as the Hackitt Inquiry [3] in the UK, the Victorian Cladding Taskforce [4], and the Shergold-Weir Inquiry for the Building Ministers' Forum (BMF) [5] in Australia. All are looking at the fire safety failures and questioning the whole building regulatory system as well as the competency, education and accreditation of fire safety engineers, and the adequacy of the science and engineering on which they base their designs. This situation is a major example of where research on regulatory policy, science and engineering need to come together to address some very significant community concerns.

Into the midst of this great state of flux, major issues are being raised about the ability of fire safety engineers to evaluate or quantify the level of fire safety afforded by a particular building design, and how such levels of safety can that be verified to ensure community safety objectives are met. This applies not only to buildings designed using the prescriptive Deemed-To-Satisfy (DTS) provisions, but also those based on performance based design, typically using a combination of DTS provisions and so called "Performance Solutions" or "Alternative Solutions" in the Australian National Construction Code (NCC).[6]

The Australian Building Codes Board (ABCB), which is responsible for the development of the NCC at the national or Commonwealth Government level, has a long term goal of using quantitative risk assessment and risk based acceptance criteria based on individual and societal risk using F-N curves to regulate all 17 hazards in the NCC, including for fire safety. It is part of their strategy to "measure" or "quantify" levels of safety in buildings and at the same time provide the more design freedom and encouragement of greater use of performance based building design in Australia, including the use of innovative materials and systems.

VERIFICATION METHOD FOR FIRE SAFETY

In the short term, ABCB decided that a key part of the way to start to "quantify" fire safety, and improve performance based design and building safety, was to adopt a Verification Method for fire safety (VM) [7] in the NCC that could be incorporated into building regulations in each State and Territory.

For the whole fire safety and regulatory community in Australia, the key questions raised by the proposed VM have been:
- Will its adoption in Australia lead to buildings of adequate safety, or create unacceptable risks?
- Is the VM rigorous and robust in terms of the science and engineering of fire safety?
- Will its adoption lead to much greater use of the NCC and performance based building solutions, rather than prescriptive DTS solutions, which is the stated ABCB national policy for the construction industry?

The ABCB developed VM was in part based on the VM from NZ [8], (also known as C/VM2), but has been changed through input from an ABCB working party.

The VM for fire safety in Australia is essentially a prescriptive process or methodology for a fire safety engineer to follow in evaluating whether the fire safety of a particular building design meets the Performance Requirements of the NCC with the following characteristics:
- There are 12 design fire scenarios which are prescribed in technical detail, along with stated data for many other inputs, which require fire engineering analysis in order to establish ASET/RSET outputs for each design fire.
- Fixed tenability and other acceptance criteria are prescribed
- Some very limited guidance is provided on fire and egress modelling as well as structural fire analysis.

- It is a process not unlike the International Fire Engineering Guidelines (IFEG) [9], but with many fixed input parameters or ranges of input data prescribed to be used in the VM.
- It does not attempt to generate or quantify safety levels or quantitative risk numbers for the overall building design, but sets out a process to "quantify" individual fire scenarios.

AUSTRALIAN VERIFICATION METHOD RESEARCH

VM Analysis

The authors of this paper were part of a team led by the Fire Protection Association of Australia (FPAA) which undertook a calibration study for ABCB into the robustness and scientific credibility of the VM that ABCB were proposing for NCC adoption in 2019 through a formal Public Comment process [10]. The research team involved a number of practicing fire safety engineers from major Australia consulting firms as well as academics from the University of Queensland led by Professor Jose Torero.

The VM was applied to 8 buildings representative of the major types of buildings constructed across Australia. These included retail, residential, aged care and warehouse buildings, as well as a major hospital and an 80 storey residential/hotel and mixed use occupancy.

Under instructions from ABCB, the buildings were designed to meet the current prescriptive DTS provisions, which would make them acceptable as meeting the NCC Performance Requirements and therefore be in compliance with the existing Australian building regulations. This immediately raised a significant issue as the DTS provisions were never based on fundamental fire safety engineering analysis or quantitative risk assessment.

Each of the 8 buildings were subject to detailed fire engineering and risk analysis, including:

- Smoke and fire modelling using all 12 VM scenarios, in many cases in multiple locations, to determine smoke and heat spread from the room of origin to adjacent rooms, escape routes and other floors of the building to determine time to untenable conditions and ASET for each scenario
- This modelling required use of prescribed smoke leakage rates, conditions for failure of smoke walls, and modelling of smoke spread from floors of fire origin into stairs and floors above, which is quite complex even for the most competent practitioners
- The VM only required the modelling only to be applied in spaces or escape routes serving more than 50 occupants, although this appears an arbitrary number without justification
- Egress analysis, using prescribed pre-movement times and travels speeds, including for a substantial percentage of the population with disabilities, to determination time to RSET
- Structural analysis, to determine time to burnout and potential for collapse, including an Unexpected Catastrophic Collapse scenario, and potential to collapse during evacuation and fire-fighting, or collapse on adjacent property
- These scenarios as prescribed in the VM included fires that blocked exits, fires in concealed spaces, high challenge/worst credible fires, and robustness checks with individual systems considered to have failed
- For each scenario and each design fire location involving life safety analysis, ASET was required to exceed RSET, but no safety factors were set or required by the VM.
- Comprehensive modelling was utilized using contemporary tools such as SFPE Handbook equations, and zone and FDS modelling, and sometimes all three were used for the same scenario to enable an inter-comparison of results, but data specified for wall failure and window breakage was not scientifically sound and no methodological support for analysis was provided.
- Each building was subject to quantitative risk analysis and assessment against ABCB risk tolerance criteria using traditional event tree and modelling inputs. In many cases, overseas probabilities of system performance and other risk factors had to be used, given Australian risk statistics are lacking.

The analysis and reports listed all the assumptions and limitation of the analysis, and a sensitivity analysis was undertaken for specific buildings and particular scenarios.

VM Summary Report Results

The results of these detailed investigations are captured in eight Fire Engineering Reports (FERs) and an overall Summary Report [10] which has been published and circulated by ABCB with the draft VM which has been issued for Public Comment.

The Summary Report drew some conclusions which are not only of concern to the fire safety engineering profession but raise important fire science and technology questions for the fire research community. They are as follows:

- For these relatively simple buildings, which comply with the DtS provisions of the current national building code or NCC, they did not meet the requirements or acceptance criteria of the VM.
- The buildings generally failed to meet the VM requirements for a significant number of the 12 design scenarios.
- The buildings generally had fire safety risk levels (individual and societal) which exceeded the risk tolerance criteria set by ABCB, even though they complied with the ABCB own NCC requirements.
- Based on the analysis undertaken, the buildings would appear to require additional fire safety provisions, including in some cases additional measures in relation to structural fire protection, compartmentation, external fire rated construction and active systems in order the meet the VM requirements. For a first order tool this seems appropriate and conservative, but does not meet ABCB's goal of "policy neutral".
- While the VM is apparently conservative in its analysis, the research showed that some choices of inputs, assumptions or modelling approaches under the VM could produce less conservative and buildings potentially lacking adequate safety
- Therefore, the VM as assessed by this research study, was not considered to policy neutral ie, yield safety levels equal to the current DTS or prescriptive requirements of the NCC at no additional cost, which was the hoped for research objective of our client ABCB.

A significant number of matters of interpretation and uncertainty associated with use of the VM arose in the application of the VM to analysis of these buildings. Study participants in this research, in attempting to carefully follow the proposed VM process as documented, found themselves in conflict about scenario application, analysis input, modelling approaches, tenability criteria, and other undefined process matters, despite having a high level of relevant expertise and experience. Another major issue was that depending on some of the inputs chosen or interpretations made, compliance with the VM requirements and the resulting risk levels could have been very different. This is because so many issues are left to the VM user and dependent on their competence, despite the VM trying to be a prescriptive process for "performance based design". In that sense it is no different to the IFEG [9], except ABCB are trying to mandate a prescribed VM process with no flexibility to use the latest science.

Public Comment Issues

Some of these concerns and modelling issues have flowed through into more detailed concerns with the VM if it was implemented in the NCC in its current form, with potentially unsafe designs, including:

- High rise (unlimited height) residential buildings without sprinklers and with a single stair, provided that smoke seals were provided to apartments and each storey had less than 50 occupants
- Potential to not have to assess facades on existing buildings undergoing refurbishment, unless there was a change in classification (The Grenfell Building issue)
- Arbitrary occupant limits for design scenarios raises the potential for unsafe buildings
- Buildings of unlimited height may be permitted to collapse due to fire, so long as they do so inwards

IOP Conf. Series: Journal of Physics: Conf. Series **1107** (2018) 042033 doi:10.1088/1742-6596/1107/4/042033

- Evacuation times which do not consider compartment layouts, travel distances, or actual time to prepare patients for evacuation in hospitals and aged care homes

CRITICAL ISSUES – FURTHER RESEARCH

Some of the critical scientific and engineering issues that emerged in the modelling and analysis of these 8 sample buildings and subsequent practitioner debate were:

- For any particular fire scenario, the VM approach to keep repeat modelling to determine the worst case location for maximum fire growth is impractical in a normal design/analysis engineering process.
- The fixed failure criteria of smoke walls set at 200oC and the values for leakage around doors appears to have no basis in the literature and there is no methodology to deal with window breakage at 300-500oC.
- There is no clear modelling guidance for smoke spread in long corridors, and between the fire floor, stairways and higher floors
- Insufficient VM guidance on key assumptions and modeling methods led to differences between engineers in calculating evacuation times, detection times, and overall Required Safe Egress Time (RSET) values
- Modelling of external fire spread up building facades incorporating combustible components
- What to do about use of average tenability criteria when using FDS or other CFD models which generate different values based on location?
- A lack of guidance to fire safety engineers on how to select material data and also critical inputs to fire and smoke models where not provided in the VM.
- With no robust statistics in Australia for rate of fire starts, spread between the room of origin, etc, and reliance overseas data meant large uncertainties in the quantitative risk assessment results
- No guidance on the limitations of the modelling or calculation methods and no requirements for sensitivity analysis.

This all suggests that much more fire research is required before a VM could be developed and promulgated into Australian regulations with the required level of certainty that, if a building met the VM requirements in full, it would have the required level of fire safety and be cost effective to construct and operate.

This level of uncertainty with the technology of verification methods is reflected in the significant changes to the application of the VM in New Zealand, where recently the VM has now been limited in its application to relatively simple buildings [11]. The idea that a high rise residential or commercial building could have only one stair and less than 15 minutes fire resistance, despite provision of sprinklers, has caused concern in Australia. The whole issue of robustness and resilience of design, and how those factors are evaluated in design and by approval authorities, is another area for rigorous fire research.

CONCLUSIONS

- A significant research study has been undertaken in Australia into a proposed Verification Method for fire safety evaluation of a building.
- The research showed some major issues with the VM which suggests it should not be adopted at this time.
- More research is required by the fire science and engineering communities to address the key issues holding back the applications of VMs into building codes and regulations
- In the meantime, updating fire engineering guidelines with the latest research results and lifting engineering competencies should be the priority.
- The research suggested that if ABCB want to move at this time with the VM, that a trial be conducted on a simple set of buildings such as warehouses, and a comprehensive research program to follow up on these buildings, their design and there compliance with the VM be followed.

IOP Conf. Series: Journal of Physics: Conf. Series **1107** (2018) 042033 doi:10.1088/1742-6596/1107/4/042033

- At the same time, there is a clear need in Australia and in other countries to lift the level of competence of fire safety engineers and other fire protection practitioners.
- This paper shows that policy reform of building and fire safety regulations cannot occur without the right engineering frameworks or without the backing of strong fire science and engineering research.

REFERENCES

[1] Genco, G. Lacrosse Building Fire 673 La Trobe Street, Docklands on 25 November 2014. Melbourne: City of Melbourne. Retrieved from https://www.melbourne.vic.gov.au/sitecollectiondocuments/mbs-report-lacrosse-fire.pdf

[2] Shirbon, E. (2017, November 16). Final death toll from London's Grenfell Tower fire is 71, say police. Retrieved from Sydney Morning Herald: https://www.smh.com.au/world/final-death-toll-from-londons-grenfell-tower-fire-is-71-say-police-20171116-gzn2v3.html

[3] Independent Review of Building Regulations and Fire Safety – Interim Report, Ministry of Housing, Communities and Local Government, Chair - Dame Judith Hackitt, UK December 2017

[4] 2017 https://www.planning.vic.gov.au/__data/assets/pdf_file/0016/90412/Victorian-Cladding-Taskforce-Interim-Report-November-2017.pdf

[5] Department of Industry, Innovation and Science. Assessment of the Effectiveness of Compliance and Enforcement Systems for the Building. Retrieved from BUILDING MINISTERS' FORUM: 2017/18 https://industry.gov.au/industry/IndustrySectors/buildingandconstruction/Documents/Terms-of-Reference.pdf

[6] National Construction Code, Volume One – Building Code of Australia, Australian Building Codes Board, Canberra 2016

[7] ABCB, Fire Safety Verification Method, Retrieved from Department of Industry, Innovation and Science - Australian Building Codes Board, Canberra, Feb 2018 https://www.abcb.gov.au/Consultation/Public-Comment-Draft

[8] C/VM2, Verification Method: Framework for Fire Safety Design. Ministry of Business, Innovation and Employment, New Zealand, 2013

[9] IFEG (2005), International Fire Engineering Guidelines; DBH, NZ; ABCB, Australia; NRC, Canada; ICE, USA.

[10] NCC 2019, Fire Safety Verification Method, Calibration Study Summary Report, Australian Buildings Codes Board, Canberra, Feb 2018

[11] BC Update 225: Revised Scope for C/VM2 Framework for Fire Safety Design, Building Performance – BC Update, MBIE, Wellington, New Zealand, 24 November 2017

On the Use of Risk Concepts in Fire Safety Engineering

Fredrik Nystedt

Division of Fire Safety Engineering, Lund University, Sweden

fredrik.nystedt@brand.lth.se

ABSTRACT

Design for fire safety may be carried out by two generic approaches – a set of prescriptive rules or by a performance-based approach where analytical tools are used to verify fire safety towards a set of functional requirements and performance criteria given by the building code. Normally, these two methods are mixed when design fire safety within a building. The option to apply fire safety engineering to the design of buildings has been available during the last 20 to 40 years, depending on which aspect of fire safety being considered. Still, the fire protection of a building too often relies on general recommendations rather than scientifically-based solutions, due to a lack of standardised verification methods, acceptance criteria and procedures to ensure high quality fire safety design. The concept of risk, i.e. the combination of the probability of a fire and a quantified measure of its consequence, has been thoroughly investigated in several fire safety engineering applications over the last decades. Although there are techniques available that allow designers to evaluate fire risks, risk acceptance criteria are missing in general. Structural fire safety design is the exemption having defined target reliabilities. Although these criteria only address the likelihood of collapse of structural element and not explicitly the characteristics of the failure.

Structural elements can be provided with fire resistance to control the spread of fire or to prevent structural collapse, or both and it is not uncommon to perform trade-offs between passive and active fire protection systems. But, very little effort has previously been made to understand the fundamental differences between these systems regarding their reliability and failure modes. Performance-based design of structural elements uses a heat exposure model to quantify the thermal load of the fire. The thermal load is characterised by the fire load (duration) and the intensity (air supply). Characteristic values of the fire load are found in various sources and commonly given in the building code, which ought to be used when designing for fire safety. A probabilistic approach was introduced in the 1980s where the probability of fire is expressed as a function of the probability of fire occurrence, the probability of a flashover and the probability of failure given a fully developed fire. Thus, the target probability of failure could be achieved by applying safety measures that alters the probability of any of these events. Currently, fire sprinklers do allow for a reduction in design fire load, but not other active safety system can be considered explicitly.

Passive as well as active system for fire safety could both be considered as appropriate provisions to achieve sufficient safety. Even though there are support of trade-offs between passive and active provisions, current regulations, guidance as well as practise do not treat the different aspect of risk related to these systems. By only considering the probability of collapse, the design could deviate from overall societal requirements on avoiding catastrophes or principles of robustness stating that consequences should not be disproportionate to their cause.

Traditionally, passive systems are assumed more robust. These findings are probably related to the concepts where target reliabilities are evaluated as the system is designed. Sprinklers are, on the other hand, assigned a probability of successful operation based on decades of statistics. This is an unfair comparison between the systems as a properly design sprinkler system always would prevent a fully developed fire, thus requiring no specific fire resistance on separating and structural elements. Naturally, this is not the path forward as the failure modes of both types of system must be treated and understood. Active systems could be argued to be more forgiven as the they do not care what mistakes are made to cause a fire, neither do they care if occupants act as planed or not. Passive systems are more sensitive to building use when e.g. doors are kept open.

Future performance criteria and risk acceptance criteria should not focus solely on probabilities. Emphasis must be put on establish criteria that measure the risk of the unwanted event considering type of initiating event, number of barriers, expected consequence, possibility of damage control, etc. Not until such criteria are available the full potential of performance-based fire safety design cannot be utilised.

KEYWORDS:

fire safety engineering, risk assessment; risk analysis, performance-based design; statistics; structural design; fire resistance, active systems, passive systems, building regulations.

IOP Conf. Series: Journal of Physics: Conf. Series **1107** (2018) 042034 doi:10.1088/1742-6596/1107/4/042034

INTRODUCTION

Design for fire safety may be carried out by two generic approaches. The designer could follow a set of simple prescriptive rules, which are defined as general recommendations in building codes. Or, the designer chooses to employ a performance-based approach. Normally, these two approaches are mixed within the building where code compliant deemed-to-satisfy solutions are combined with unique solutions verified towards the functional requirements and performance criteria given by the authorities having jurisdiction (AHJ). The option to use a performance-based approach were introduced to structural fire safety in the 1970s and for fire safety in general some twenty years later. However, the current performance-based fire safety design process has been under scrutiny over the last few years. Still, both designers and AHJs, put too much emphasis on the prescriptive design concepts rather than fundamentally managing and understanding fire risk. Alvarez et.al. [1] identifies several challenges with the design process, fundamentally that generic guidance is applied to specific projects and that levels of performance are compared between an engineering solution and one based on prescriptive requirements. The latter is also identified, by a recent Nordic initiative [2], as one of the major implications why performance-based fire safety design have not been overly successful. The fire protection of a building too often relies on general recommendations rather than scientifically-based solutions due to a lack of standardised verification methods, acceptance criteria and procedures to ensure high quality fire safety design. Frameworks for a future risk-informed fire protection design process has been proposed [3], [4]. Both concepts put emphasis on introducing well-defined performance criteria.

During the last 25 years, the concept of risk, i.e. the combination of the probability of a fire and a quantified measure of its consequence, has been thoroughly investigated in several fire safety engineering applications [5]. Although there are techniques available that allow designers to evaluate fire risks, risk acceptance criteria are still lacking. The designer is forced to compare risk levels with prescriptive design solutions even if such are not applicable to the specific case. Absolute fire risk criteria would allow for a more transparent design process that strives towards an acceptable safety level and that does not give any favours to either a prescriptive nor a performance-based solution. Currently such criteria are non-existent, exempt for structural safety where a set of target probabilities are established in the Eurocode [6]. However, these criteria mainly address the likelihood of collapse of structural element and not explicitly the characteristics of the failure in terms of pre-warning, consequence, partial failure, etc.

Structural elements and compartmentation can be provided with fire resistance to control the spread of fire or to prevent structural collapse, or both. This paper will address the concepts of risk in fire safety engineering with a focus on requirements on fire resistance. One of the top-level fire safety objectives in a building code is that the building should be provided with fire resistance for a certain time. This time is not precisely defined and could be the time required for occupants to escape and the fire service to intervene, or the full duration of a fire [7]. The requirements on fire resistance for most public buildings are related to withstand a complete burnout. However, it is not uncommon to perform trade-offs between passive and active fire protection systems. Active systems such as fire sprinklers will control the fire development and reduce the likelihood of a fully developed fire. This reduction will normally allow for lower requirements on fire resistance. The idea is to balance the two systems achieving the same safety level as if the active system wouldn't be present. Although, there are fundamental differences between passive and active provisions and their characteristic failure modes that need to be addressed. Without an understanding of these inherent features, it is hard to accomplish a properly balanced fire safety design solution. The objective of this paper is to increase the awareness risk regarding fire engineering applications related to fire resistance. The aim is to allow for a more balanced use of active and passive provisions to prevent fire spread and structural collapse. The paper has been prepared with financial support by Brandforsk, the Swedish Fire Research Board.

CURRENT DESIGN PRACTICE AND ITS IMPLICATIONS

Actions on structures related to fire

The guidance on designing structures to achieve proper fire resistance is extensive and there are three different methods that could used to verify that fire resistance exceeds fire severity [7]. Verification could be carried out the time domain, the temperature domain or the strength domain. The time domain refers to a prescriptive approach where an element is subjected to a standardised test and either pass it or fails. Elements are provided with ratings, e.g. R 60, EI 30, etc. depending on the outcome of the performance in the test. This prescriptive approach is not regarded as fire safety engineering and will not be discussed any further. In the 1970s a new set of methods were developed which allowed for an analytical approach to the design of structural fire safety [8]. Thermal exposure could now be quantified based on the conditions of a fully developed fire determined by the combustion characteristics of the fire load, the ventilation of the fire compartment and the thermal properties of

IOP Conf. Series: Journal of Physics: Conf. Series **1107** (2018) 042034 doi:10.1088/1742-6596/1107/4/042034

the enclosing structures. Later, so called parametric fire curves where derived based on these principles and included in the Eurocode [9]. These methods to express temperature within a fire compartment is referred to as the temperature domain. An element is considered to have a sufficient capacity if the temperature within the element or on the non-exposed side of a fire separating structure is less than the temperature that could cause failure, i.e. collapse or fire spread. The temperature domain is suitable to use for structures with an insulating or containing function and not for structural elements as it does evaluate structural performance in an adequate way [7]. Structural elements are preferably evaluated within the strength domain where the reduced load-bearing capacity of the heated element is compared with the applied load at the time of the fire. However, there is a strong link between the temperature domain and the strength domain as the outcome of the heat exposure estimated within the temperature domain is delivered as input to the assessment of sufficient load-bearing capacity in the strength domain. Heat exposure is primarily dominated by the available fire load (duration) and the air supply (intensity). Characteristic values on the fire load are given in the Eurocode [9] for diverse groups of occupancies such as dwellings, hospitals, hotels, offices and shopping centres and the design fire load is derived by multiplying a partial factor with the characteristic fire load. Normally, the partial factor is 1.0, i.e. the design fire load is equal to the characteristic value. A higher value of the partial factor would result in less likelihood that the actual fire load exceeds the design value. For tall buildings, a partial coefficient of 1.5 is used in Sweden to control risk as such buildings are considered to have a potential for a more catastrophic outcome in case of failure.

Probabilistic design approach

The CIB W14 undertook pioneering work on probabilistic design in the 1980s [10] and their definition of the probability that a fire-exposed structure fail is given in Eq. 1:

$$P_{failure} = P_{failure|flashover} \cdot P_{flashover|fire} \cdot P_{fire} \tag{1}$$

Thus, for a structure to fail due to fire, first a fire is required, secondly the fire must become fully developed and finally, failure must occur. An acceptable value on the probability of failure could be met by reducing the probability that a fire occurs, by reducing the probability that a flashover occurs once a fire has started or by reducing the probability of a structural failure in the case of a fully developed fire. Proposed safety levels i.e. target probabilities only implicit consider the consequences of a collapse [6]. Structural elements are assigned different target probabilities related to their importance regarding maintaining structural stability and the severity of a collapse, i.e. the potential consequence is managed by the magnitude of the target reliability. No target probabilities are given for other fire safety objectives than maintenance of structural stability.

Traditionally, structural fire safety design has been primarily concerned with the design of structures exposed to fully developed fires. Although codes allow for some possibilities to apply measures that prevent fires from becoming fully developed as the presence of fire sprinkler can lead to less thermal exposure to the structure and motivate a less severe design fire load. This idea of trade-offs between passive and active fire safety provisions were introduced in the 1970s [11]. The arguments behind such trade-offs are essential statistical where the active fire safety measure reduced the likelihood of a severe fire and thus can the structures could be design for less heat exposure. The Natural Fire Safety Concept [13] took this a step further and introduced a procedure of modifying the characteristic fire load based on compartment size, building type and the different active fire safety measures such as smoke alarm, sprinklers, rescue service, etc. The concept is included in Eurocode [9], but it is not allowed in all EU member countries due uncertainties of its implications. However, a reduction of the design fire load with a partial factor of app. 0.6 when sprinklers are present is well recognised.

Implications with current design practice

A comprehensive review of current design practice states that criticism has been raised regarding the heat exposure model being too unrepresentative of real fire conditions and the design practice being consequence-orientated rather than risk-informed [13]. Heat exposure is strongly dependent on the fire load. However, most fire load data are 40-50 years old and it is necessary to address the relevance of this data in relation to the fire loads today, given the increasing use of combustible materials in furnishing and construction. Fuel type and opening configuration also have a major influence on the temperature-time relationship in the fire compartment and large deviations between the parametric fire curve and full-scale tests have been observed [14], [15], [16].

Risk is mainly controlled by applying different partial factors to the characteristic fire load. However, current practice in Sweden offers two safety levels both related to heat exposure. Either the design fire load equals the characteristic value, or the characteristic value is increased by 50% for buildings taller than 4 stories. However, fire sprinklers are considered to sufficiently cover for the additional risk that an increased building height result in. Design fire load in a tall building with sprinklers has the same value as the design fire load in a lower, non-sprinklered building [9]. Risk is also controlled by assigning different target values for the probability of failure based on the importance of the structural element. Although, the target value is the same for most buildings. A reference value of the maximum individual risk is app. 10^{-6} per year [6]. Note that the target reliability is related to the system as designed, i.e. not built. Failure due to human error or ignorance and failures due to non-structural causes are not covered. Such failures are supposed to be covered by design review and other quality assurance procedures. Controlling risk only by changing the design fire load is strictly a probability-based approach that does not consider other aspects of risk than the likelihood of collapse. Neither the magnitude of the consequence nor when failure occurs is considered, which is insufficient in common practise of risk control [17].

RELIABILITY OF PROVISIONS FOR PREVENTING FIRE SPREAD AND COLLAPSE

The reliability of an active system, e.g. fire sprinklers is considered explicitly in the current design approach. Passive systems are, however, always assume to perform as intended. E.g., neither any ageing effects of sprayed passive protection nor the variability in performance of a board system are considered. Furthermore, doors are assumed to be closed and dampers to shut. The probability that a passive system operates as intended is assumed to be 100%, which of course is not true. Thus, there is an inherent imbalance within the system when performing trade-offs that favours passive provisions in front of active that needs to be addressed in future methodology development.

Passive provisions have other failure modes than active provisions in terms of how failure occurs. E.g. delayed or partial failure is more likely than complete failure of the barrier. Even if a building element cannot withstand the design thermal exposure and fail prior its design requirements are met, there is a likelihood that the capacity will be sufficient due to the variability in fire load. If a structural element is protected by other means than over-dimensioning, e.g. by spray-on systems, paint or by boards, there is increased uncertainty that the system will protect the structure with sufficient reliability. Intumescent paint is being exposed to ageing effects with a significant decrease in thermal resistance during the first service years [18]. Intumescent coating also performs differently depending on fire development [19]. The standardised test will produce results on the safe side for slower fires, but unsafe results for more rapid fires. Gypsum boards used to protect both structural and separating elements are also subject to uncertainty in terms of performance. E.g. loss of attachment due to heat exposure is hard to predict in general terms and need to be specified by the manufacturers [20].

Active provisions for fire safety, e.g. sprinklers show great record in minimising fire damage. Literature state a reliability of the system from a minimum of 70% to a maximum of 99.5% with most likely values in the range of 90% to 95% [21]. The U.S. experience with sprinklers states that flame damage was confined to the room of origin in 96% of fire when sprinklers were present, compared to 71% of fire without sprinklers [22]. Thus, sprinkler systems do account for a statistical proven reduction in the probability of a severe fire that could be used when balancing active and passive safety provisions. From a design point of view, a passive system would fail to operate if it does not withstand the pre-determined fire exposure. But, the performance might still be sufficient for escape and rescue service intervention. Hence, the failure might only be partial. This is not unique to passive systems as both fire sprinklers and smoke management system could fail partially having reduced performance compared to their design criteria. But, the performance still might be sufficient to control the fire and limit the upper layer temperature. Although, sprinkler systems are more likely to have complete failure as the dominant cause for sprinklers fail to operate is the water supply being shut off [22].

BALANCING FIRE RISK WITH PASSIVE AND ACTIVE PROVISIONS

Passive as well as active provisions for fire safety are both considered to be appropriate techniques to achieve sufficient safety, supported by the approach outlined in the Fire Safety Concepts Tree [24] where the two fire safety objectives on the top level are to prevent fire ignition and to manage fire impact. The fire impact could be managed by either managing the fire (e.g. with active system) or managing the exposed (e.g. with passive systems). The sprinkler system reduces the probability of a fully developed fire and this increase in safety opens for an optimisation of other measures, such as reduced capacity of structural and separating elements. By applying Eq. 1, one realises that the probability of failure given a fully developed fire could be increased with the same factor as the probability of flashover given fire is decreased. However, this approach does not consider the consequence of a failure any other way than that a failure occurs. From a societal point of view, it is of interest to extend Eq.1 with additional variables describing the effect of the failure. Such variables could be the probability that failure occur prior to successful escape or the probability that occupants and rescue service personnel will be notified on the upcoming failure in due time. The sensitivity of the design solution to common error could also be considered. Normally, active provisions are subject to more frequent inspections than passive provisions. Less effort is taken to ensure that self-closing doors are not blocked and that services passing through compartment walls or floors are adequately fire-stopped.

One other aspect that Eq. 1 not explicitly consider is the number of barriers in place to prevent an initiating event from progressing to an unwanted event, e.g. failure of a structural element. Fig. 1. illustrates two safety system. The first safety system has one barrier with a probability of failure of 0.001. The second safety system has three independent barriers each having a probability of failure of 0.1. The second safety system thus has the same total probability of failure as the first (i.e. $0.1^3 = 0.001$). But, could the performance of the two safety systems in Fig. 1 be considered equal? If measuring only the probability of failure, the answer is yes. A failure of 0.001 with one barrier is equal to the failure of three barriers having the same combined probability of failure. But if evaluating e.g. the time until failure, the result could differ a lot. If the barrier in the single barrier system is of on/off type, then it either operates with 100% effectiveness or it does not operate at all. Compare such a barrier to the multiple barrier safety system, where failure does not occur until all three barriers have failed. Such a system has a failure time that most likely is longer than what the single barrier system would have. From this point of view, these systems are not equal, despite having the same probability of failure.

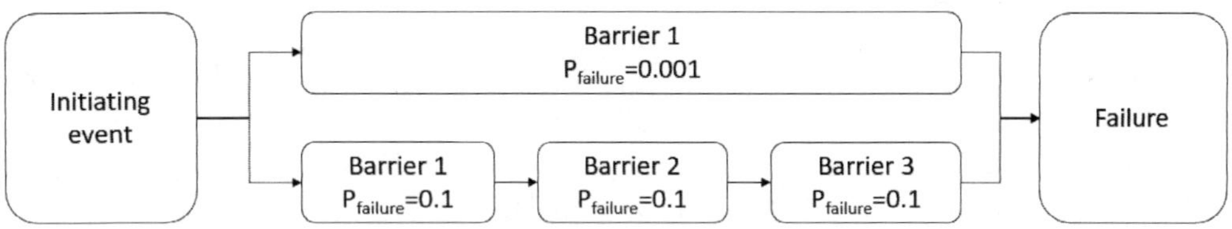

Fig. 1. Two safety systems providing the same risk of the unwanted event.

The possibility to reduce ratings on load-bearing structures in sprinklered buildings is a question like the one discussed above. In a building without fire sprinklers, the load-bearing structure will keep its capacity for the complete fire duration with a certain probability. When the capacity of the element is reduced in a sprinklered building there will be two different scenarios. When the sprinkler system operates effectively will have infinite load-bearing capacity as there is no heat exposure. But, when the sprinkler system fails the load-bearing capacity will be more sensitive to the actual fire load. Currently, there is no straightforward answer on how to decide on the required capacity in case of sprinkler failure.

CONCLUSIONS

Passive as well as active system for fire safety should both be considered as appropriate provisions to achieve sufficient safety. Even though there are support of trade-offs between passive and active provisions, current regulations and guidance as well as practise do not treat the different aspect of risk related to these systems. By only considering the probability of collapse, the design could deviate from overall societal requirements on avoiding catastrophes or principles of robustness stating that consequences should not be disproportionate to their cause.

Traditionally, passive systems are assumed more robust. These findings are probably related to the concepts where target reliabilities are evaluated as the system is designed. Sprinklers are, on the other hand, assigned a probability of successful operation based on decades of statistics. This is an unfair comparison between the systems as a properly design sprinkler system always would prevent a fully developed fire, thus requiring no specific fire resistance on separating and structural elements. Naturally, this is not the path forward as the failure modes of both types of system must be treated and understood. Active systems could be argued to be more forgiving as the they do not care what mistakes are made to cause a fire, neither do they care if occupants act as planed or not. Passive systems are more sensitive to building use when e.g. doors are kept open.

Future performance criteria and risk acceptance criteria should not focus solely on probabilities. Emphasis must be put on establishing criteria that measure the risk of the unwanted event considering type of initiating event, number of barriers, expected consequence, possibility of damage control, etc. The full potential of performance-based fire safety design cannot be utilised until such criteria are available.

REFERENCES

[1] Alvarez, A. Meacham, B.J., Dembsey, N.A., Thomas, J.R. (2013), Twenty years of performance-based fire protection design: challenges faced and a look ahead, *Journal of Fire Protection Engineering* 23:249–276, http://dx.doi.org/10.1177/1042391513484911.

[2] Mindykowski, P., Strömgren, M., *Fire Safety Engineering for Innovative and Sustainable Building Solutions*, Report 2017:42, RISE Research Institutes of Sweden, 2017.

[3] Alvarez, A., Meacham,B.J., Dembsey, N.A., Thomas, J.R. (2013) A Framework for Risk-Informed Performance-Based Fire Protection Design for the Built Environment, *Fire Technology* 50:161-181, http://dx.doi.org/10.1007/s10694-013-0366-1.

[4] Bjelland, H., Njå, O., Heskestad, A.W., Braut, G.S. (2014) The Concepts of Safety Level and Safety Margin: Framework for Fire Safety Design of Novel Buildings, *Fire Technology* 51:409–441, http://dx.doi.org/10.1007/s10694-014-0400-y.

[5] Yung, D., *Principles of Fire Risk Assessments in Buildings*, John Wiley & Sons, 2009, http://dx.doi.org/10.1002/9780470714065.

[6] European Standard, *Eurocode – Basis of structural design*, EN 1990, April 2002.

[7] Buchanan, A.H., Abu, A.K., *Structural Design for Fire Safety*. 2nd edition, John Wiley & Sons, 2017, http://dx.doi.org/10.1002/9781118700402.

[8] Magnusson, S.E. and Thelandersson, S. (1970) *Temperature-time curves of complete process of fire development; theoretical study of wood fuel fires in enclosed spaces*. Acta Polytechnica Scandinavica. Civil Engineering and Building Construction Series 65.

[9] European Standard, *Eurocode 1: Actions on structures - Part 1-2: General actions - Actions on structures exposed to fire*, EN 1991-1-2, November 2002.

[10] Kersken-Bradley, M., Pettersson, O., Schneider, U., Twilt, L., Vrouwenvelder, A., Witteveen, J. (1983), A conceptual approach towards a probability-based design guide on structural fire safety, *Fire Safety Journal* 6: 9-13, http://doi.org/10.1016/0379-7112(83)90038-3.

[11] Baldwin, R., Thomas, P.H. (1974), Passive and active fire protection – the optimum combination, *Fire Technology* 10:140-146, http://dx.doi.org/10.1007/bf02642517.

[12] European Commission (2003), *Natural Fire Safety Concept – Full-scale tests, implementation in the Eurocodes and development of a user-friendly design tool (final report)*, Report EUR 20580.

[13] Nystedt, F., Frantzich, H., "Risk and Uncertainty in Current Practice on Structural Design for Fire Safety", *Interflam 2013 -- Proceedings of the Thirteenth Fire Science & Engineering Conference*, Interscience Communication Ltd., 2013, pp. 1315-1326.

[14] Yii, Ee. H., Buchanan, A. H., Fleischmann, C. M. (2006), Simulating the effects of fuel type and geometry on post-flashover fire temperatures, *Fire Safety Journal* 41:62-75, http://dx.doi.org/10.1016/j.firesaf.2005.09.001.

[15] Feasey, R., Buchanan, A. (2002), Post-flashover fires for structural design, *Fire Safety Journal*, 37:83-105, http://dx.doi.org/10.1016/s0379-7112(01)00026-1.

[16] Barnett, C.R. (2007), Replacing international temperature–time curves with BFD curve, *Fire Safety Journal* 42:321-327, http://dx.doi.org/10.1016/j.firesaf.2006.11.001.

[17] Nordic Standard, *Fire Safety Engineering — Probabilistic Methods for Verifying Fire Safety Design in Buildings*, Technical Specification, prINSTA/TS 951:2018.

[18] Zhang, C., Li, G.-Q., Wang, Y.-C. (2014), Probabilistic analysis of steel columns protected by intumescent coatings subjected to natural fires, *Structural Safety* 50:16-26, http://dx.doi.org/10.1016/j.strusafe.2014.03.005.

[19] Cirpici, B.K., Wang, Y.C., Rogers, B. (2016), Assessment of the thermal conductivity of intumescent coatings in fire, *Fire Safety Journal* 81:74-84, http://dx.doi.org/10.1016/j.firesaf.2016.01.011.

[20] Kolarkar, P. & Mahendran, M. (2012). Experimental studies of gypsum plasterboards and composite panels under fire conditions. *Fire and Materials* 38:13-35, http://dx.doi.org/10.1002/fam.2155.

[21] Frank, K., Gravestock, N., Spearpoint, M., Fleischmann, C., (2013), A review of sprinkler system effectiveness studies. *Fire Science Reviews* 2:6, http://dx.doi.org/10.1186/2193-0414-2-6.

[22] Ahrens, M. (2017), *U.S. Experience with Sprinklers*, National Fire Protection Association, Quincy, MA.

[23] Rasbash, D. Ramachandran, G., Kandola, B., Watts, J., (2006), "Fire Safety Concepts Tree and Derivative Approaches". *Evaluation of Fire Safety*, Rasbash, D., et.al. (eds.), John Wiley & Sons, Ltd, http://dx.doi.org/10.1002/0470020083.ch16.

IOP Conf. Series: Journal of Physics: Conf. Series **1107** (2018) 042035 doi:10.1088/1742-6596/1107/4/042035

Flame properties of large kerosene fires

Giacomo EREZ[1,2], Gilles PARENT[1], Anthony COLLIN[1], Pascal BOULET[1], Mathieu SUZANNE[2], Eddie FAURE[2] and Aurélien THIRY-MULLER[2]

[1]*Laboratoire Énergies, Mécanique Théorique et Appliquée, Université de Lorraine/CNRS, France*
[2]*Laboratoire Central de la Préfecture de Police, Préfecture de Police de Paris, France*

giacomo.erez@univ-lorraine.fr

ABSTRACT

In order to provide experimental data to develop fire spread models, numerous tests were performed for a wide range of fire sizes and fuels. Here we focus on large scale kerosene tests (approx. 0.5 to 6 m²), with an emphasis on radiative flame properties and present a novel multispectral approach composed of an opacimetry setup and a FTIR spectrometer. This allowed to compute emittance values (based on transmittance measurements) and equivalent soot temperatures (by fitting the calculated intensities to the measured ones). These results, as well as mass loss rates and flame heights, show good repeatability and agree well with the literature, including with correlations predicting the burning rate, flame height and spectral dependence of the extinction coefficient. Generally it is shown that radiation emitted by the flame is not gray, and can only be considered black for very large flames (i.e. wider than 2.5 m for these tests). Additional data (smaller fires, other fuels, complementary parameters - e.g. flame shape) are still to be analyzed and used to model flame heat radiation.

KEYWORDS: *fire tests; scaling up; radiative heat transfer; flame height; opacimetry; spectrometry*

IOP Conf. Series: Journal of Physics: Conf. Series **1107** (2018) 042035 doi:10.1088/1742-6596/1107/4/042035

INTRODUCTION

Fire simulations require an accurate heat release rate (HRR), especially when used for fire investigation or reconstruction because results have to be reliable from the very beginning of the fire. One solution to obtain this input is to conduct full-scale tests, which provide readily usable data. However experimental conditions may sometimes be different from the desired scenario (e.g. free burning versus under ventilated). Moreover, these tests can be costly, and are more challenging regarding repeatability, reproducibility and safety. These limitations suggest that conducting small-scale tests and then using models to scale up the results would be a convenient alternative. However, it comes with many challenges, primarily understanding the changes that occur when going from small (milli- or centimeter) to large samples (meter). The study is focussed on radiative heat transfer as it is the driving force of flame spread [1,2], and its accurate modelling is thus crucial. The starting point of this work was to gather experimental data for various fire sizes as well as several fuels (solid and liquid). In addition to the classical measurements performed in fire tests (mass loss, room temperatures, heat fluxes), a special focus was put on characterising the flames from a radiative standpoint (innovative visible and infrared opacimetry, infrared imagery and spectrometry). In this paper we will describe the large scale tests (approx. 0.5 to 6 m²) and focus on some results for kerosene.

EXPERIMENTAL SETUP

Experiments

The large scale tests were performed in a former aircraft hangar (approx. 300 × 50 × 17 m^3, length × width × height), gracefully lent by Groupe ADP (Orly, France). It allowed performing well ventilated fires close to open conditions while being shielded from wind and rain. A total of nine experiments were performed with kerosene, as listed in Table 1.

Table 1. Kerosene test characteristics

Pool dimensions [m²]	Approx. fuel mass [kg]	Approx. water mass [kg]	Number of tests
0.70 × 0.70	25	50	3
1.00 × 1.00	50	100	3
1.75 × 1.75	120	200	2
2.50 × 2.50	250	300	1

The steel pans were square shaped with a height of 20 cm and a thickness of 4 mm. The lip size (height of the pan edge above the fuel surface) was approximately 5 cm at the beginning of the tests (the fuel height was not kept constant, and the tests were conducted until flame out). The fuel was laid on water to regulate combustion. The tests were started by igniting a small layer of ethanol poured beforehand on top of the kerosene.

Measurements

Most of the tests were performed using load cells to monitor mass loss. Only the largest pool fire (approx. 6 m²) could not be weighed. Numerous cameras were used to record the tests, mainly to study the visible flame shape and the smoke spread.

The radiative properties of the flames were characterized by infrared emission spectrometry and by opacimetry: a Fourier-transform infrared (FTIR) spectrometer (Matrix spectrometer by Bruker) allowed to record spectral radiation intensities during the tests (from 1000 cm^{-1} to 6000 cm^{-1}) and opacimetry measurements were made at different wavelengths. Regarding opacimetry, the light coming from laser diodes was modulated thanks to a chopper, and demodulated with a lock-in amplifier, to retrieve the transmitted intensity without the contribution from the flame emission. Four wavelengths were used simultaneously: 785, 1650, 2300 and 3800 nm for large scale tests, and 410, 520, 785 and 1650 nm for tests at smaller scale. We note that all these wavelengths are located outside the absorption bands of major combustion gases (CO_2 and H_2O) in order that only soot absorption was studied here. A 200 mm off-axis parabolic mirror was used to collect the four parallel light beams toward an integrating sphere. This allows to overcome the beam deviation due to the refractive index variations inside and around the flame. Bandpass optical filters were used to retrieve the signal corresponding to each wavelength. The measurements with the spectrometer and the opacimetry measurements were made along the same horizontal optical path (approx. 20 cm above the pan), to provide comparable data. Heat fluxes were also measured to study their vertical and horizontal distributions depending on the fire size and fuel. The results are presented elsewhere [2] and will be further discussed in later communications, along with comparisons with numerical model for example. Here we will focus on mass loss rate, flame height and soot radiative properties (emittance and equivalent temperature).

RESULTS AND DISCUSSION

Mass loss rates and flame heights

Mass loss rates (MLRs) for the 0.5 m² and 1.0 m² fires (repeated three times) are presented in Fig. 1.

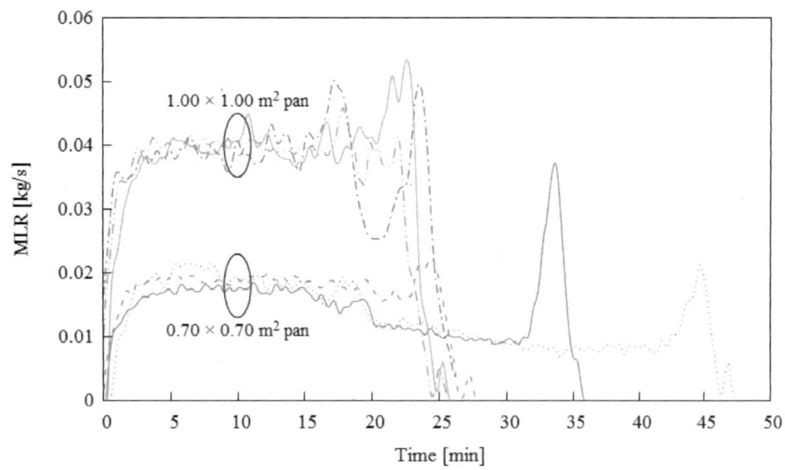

Fig. 1. Mass loss rates over time for two pan sizes.

These results show high repeatability for the growth and steady phases. The increase in MLR at the end of each test is due to water boiling and projecting fuel outside the pans. This behaviour is not an issue for the present study as we focus on the steady phase, which happens much earlier during the tests (approx. 5 min to 10 min after ignition) and is sufficiently long to gather representative time-averaged values.
Cameras were used to extract the visible flame shape and to evaluate its changes when scaling up. These visible flame shapes will later be used for radiation models. Images were processed to extract the flame contour and compute the flame height. The evolution of median mass flux and flame height with the pan diameter is shown in Fig. 2, and compared to the literature.

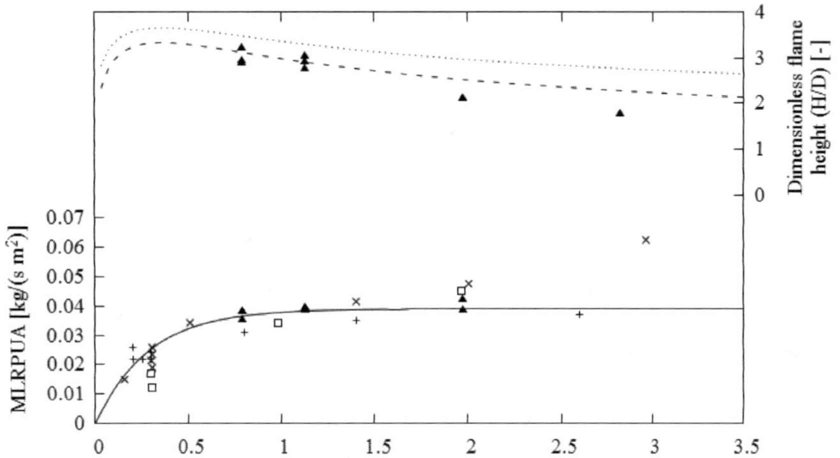

Fig. 2. Median mass loss rates per unit area (MLRPUA) and dimensionless flame heights (flame height divided by pan diameter), see ▲, compared to other kerosene tests (+ [4]; × [5]; □ [6]) and correlations from Babrauskas [4] (solid line), Heskestad [7] (dashed line) and Zukoski [8] (dotted line).

The agreement is good with experimental data from other sources, as well as common correlations to predict burning rate and flame height.

Flame radiative properties

In addition to flame shape, which is also a key factor of flame heat radiation, two other flame properties were determined, namely emittance and equivalent soot temperature.
The first quantity was computed from the opacimetry measurements, which allowed to determine transmittance τ across the flame at several wavelengths λ, as illustrated in Fig. 3.

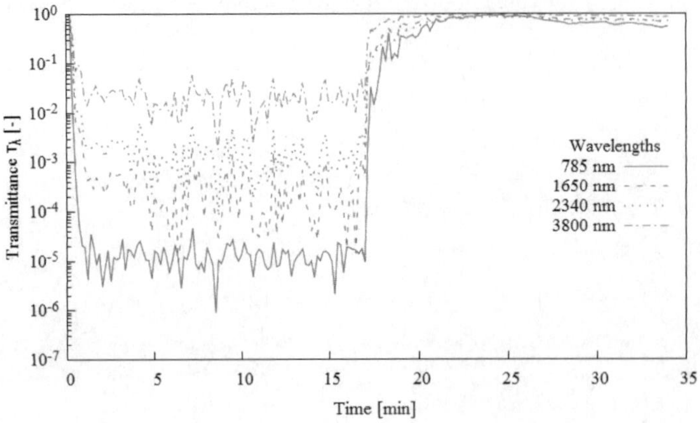

Fig. 3. Spectral flame transmittance for a 1.75 m wide pan at four wavelengths (1/100 points plotted).

Transmittance measurements can be translated to extinction coefficients β according to Beer-Lambert's law,

$$\beta_\lambda = \frac{-\ln(\tau_\lambda)}{e} , \quad (1)$$

where e is the length of the optical path ($\beta_\nu e$ is the optical thickness of the flame). The dependance of β on the wavelength is illustrated in Fig 4.

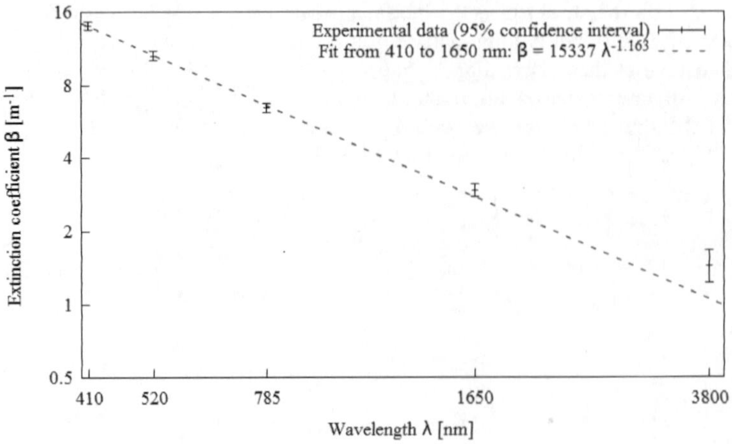

Fig. 4. Mean flame extinction coefficient as a function of wavelength for all fire sizes.
Base 2 log scale, fit from 410 to 1650 nm (plot includes data from additional tests, to be published).

The fit with the model [9]

$$\beta_\lambda = \frac{C1}{\lambda^{C2}} , \quad (2)$$

shows good agreement with the data. Here, constants $C1$ and $C2$ were estimated to be 15337 and 1.163 respectively, which is in the range of reported values ($0.7 < C2 < 2.0$ [9]). Equation 2 will later allow to easily take into account the spectral dependence of the flame radiative properties.
As the final objective of this work is to model flame heat radiation, predicting the flame emittance would be more appropriate. This can be done by considering that the particles in the flame are small in relation to the studied wavelengths (infrared), which implies that extinction is mainly due to absorption (i.e. negligible scattering), thus giving

$$\kappa_\lambda \approx \beta_\lambda \ , \tag{3}$$

with κ the absorption coefficient. Moreover, according to Kirchhoff's radiation law, for a given wavelength λ and direction Ω,

$$\epsilon_{\lambda,\Omega} = \alpha_{\lambda,\Omega} \ , \tag{4}$$

where ϵ is the emittance and α the absorptance. Consequently, with the hypothesis of a homogenous equivalent medium, we get the flame spectral emittance as a function of the extinction coefficient

$$\epsilon_\lambda = \alpha_\lambda = 1 - \exp(-\kappa_\lambda e) \approx 1 - \exp(-\beta_\lambda e).$$

Assuming an isothermal equivalent medium, this allows the calculation of the spectral intensity emitted by soot according to

$$I_\lambda = \epsilon_\lambda \, I_\lambda^0(T) \approx [1 - \exp(-\beta_\lambda e)] \, I_\lambda^0(T), \tag{5}$$

$I_\lambda^0(T)$ being the blackbody spectral intensity at temperature T. This model is valid for every wavelength if the emittance is known. Here measurements were done for wavelengths where soot is the primary emitter, i.e. outside the combustion gases wavelengths, consequently results are valid only for emission of soot. The spectral intensity being evaluated independently from the opacimetry measurement (i.e. with the infrared spectrometer), temperature is the only parameter to be determined. This is an advantage compared to classical emission measurements, for which two parameters (spectral emittance and temperature) are to be calculated from only one measurement, which is impossible (an infinite number of couples (ϵ_λ, T) can fit the measured intensity). The use of Eq. 5 to fit the intensities measured by the infrared spectrometer to obtain the soot equivalent temperature is illustrated in Fig. 5 with a test for each pan size.

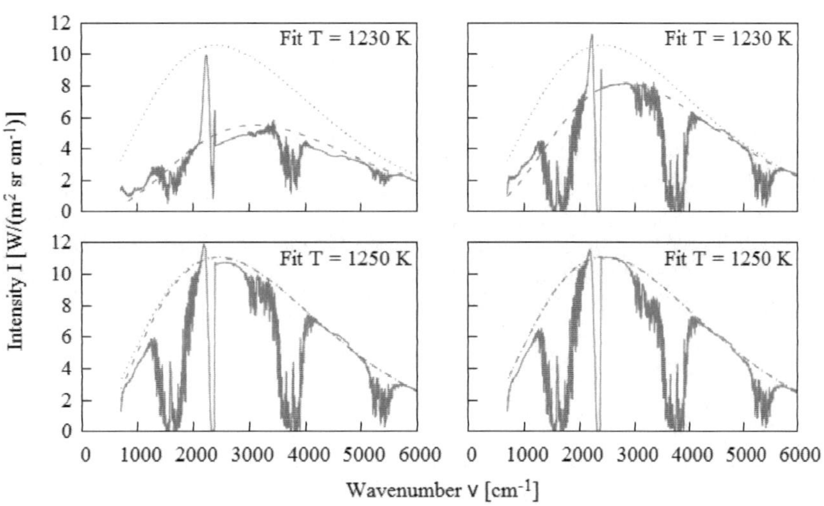

Fig. 5. Measured (solid) and modelled spectral intensities (dotted for blackbody, dashed for Eq. 5). From left to right, top to bottom: pan width of 0.70 m, 1.00 m, 1.75 m and 2.50 m (one test for each).

The large absorption bands seen in the experimental spectra (solid line) are due to cold atmospheric H_2O and CO_2, found between the pool and the spectrometer. In the tests performed during this campaign, the distance between the pool and the spectrometer was 23 m, which explains why the spectra are so much hollowed. Calculated spectra (dotted and dashed lines) do not take this absorption into account. Outside atmospheric absorption bands, there is a good agreement between measured and calculated spectra (solid vs. dashed line). Moreover, the fits give similar equivalent soot temperatures, namely 1230 K for the 0.70 m and 1.00 m pans and 1250 K for the 1.75 m and 2.50 m pans. We can also see that in the general case the radiation emitted by the flame is not gray, as stated by Eq. 2. It is also shown that the flame becomes black (i.e. dashed and dotted lines overlap) only for large flames, that is for a flame thickness e greater than 2.5 m. The authors emphasize that the present model gives an equivalent emittance and an equivalent temperature, not a mean emittance and a mean temperature, of an equivalent homogenous isothermal medium that emits a radiation similar to the mean radiation emitted by the flame in the direction of measurement. Our measurements conducted on a large range of flame thicknesses have shown that the extinction coefficient is weakly dependent on the flame thickness (results not shown here). Some additional measurements have to be performed at different heights in the flame. The goal is to see if a fully homogenous (i.e. same absorption coefficient and same temperature everywhere in the flame) volumetric flame model, can yield a good enough prediction of the radiative heat flux received either by the combustible or by any target around the flame.

CONCLUSION

The tests presented here are the first part of a larger experimental campaign aiming at investigating the changes in burning behaviour when scaling up. They are the biggest fires planned for this study and provide original data regarding some key flame properties, namely emittance and equivalent soot temperature. The results prove that spectrally resolved measurements are a valuable investment when studying flames and that further work is needed to accurately model them. Another key aspect that could not be addressed so far is flame inhomogeneity: complementary tests could be performed to produce "radiative maps" of flames, as done before [10–12] but with the multispectral approach presented here.
More generally, all results show high repeatability, and correlate well with common models (for MLR, flame height and extinction coefficient). In addition to these experiments, smaller fire sizes (down to 0.05 m^2) and other fuels (heptane, diesel, polyurethane foam) are being investigated. Further work is planned to study the evolution of other parameters (e.g. flame shape, flame spread over solids) when scaling up, and the differences from one fuel to another. These results will be used to assess various assumptions used when modelling radiative heat transfer (black or gray body, equivalent flame shape or temperature...) by comparing numerical results to the measured heat fluxes. This work will also provide valuable input data to develop fire spread models meant to predict full-scale burning behaviour based on small-scale tests.

REFERENCES

[1] De Ris J L (1979) Fire radiation—A review, *Symposium (International) on Combustion* 17 1003–16, https://doi.org/10.1016/S0082-0784(79)80097-1

[2] De Ris J L and Orloff L (1975) The role of buoyancy direction and radiation in turbulent diffusion flames on surfaces, *Symposium (International) on Combustion* 15 175–82, https://doi.org/10.1016/S0082-0784(75)80295-5

[3] Sautot C, Hanus F, Thauvoye C, Erez G and Thiry A (2018) Radiative Flux Affecting Vertical Steel Element Away from the Fire - Simplified Method LOCAFI, *10th International Conference on Structures in Fire* SiF 2018 (Belfast, UK), To be published

[4] Babrauskas V (1983) Estimating large pool fire burning rates, *Fire Technol* 19 251–61, https://doi.org/10.1007/BF02380810

[5] Souil J M, Vantelon J P, Joulain P and Grosshandler W L (1986) Experimental and Theoretical Study of Thermal Radiation from Freely Burning Kerosene Pool Fires, *Dynamics of Reactive Systems Part I: Flames and Configurations; Part II: Modeling and Heterogeneous Combustion* 105 388, https://doi.org/10.2514/5.9781600865794.0388.0401

[6] Koseki H (1989) Combustion properties of large liquid pool fires, *Fire Technol* 25 241–55, https://doi.org/10.1007/BF01039781

[7] Heskestad G (1984) Engineering relations for fire plumes, *Fire Safety Journal* 7 25–32, https://doi.org/10.1016/0379-7112(84)90005-5

[8] Zukoski E E (1986) Fluid Dynamic Aspects of Room Fires, *Fire Safety Science: Proceedings of the First International Symposium* (New York, USA: Hemisphere) pp 1–30, https://doi.org/10.3801/IAFSS.FSS.1-1

[9] Modest M F (2003) Radiative Properties of Particulate Media, *Radiative Heat Transfer* (Burlington: Academic Press) pp 361–412, https://doi.org/10.1016/B978-012503163-9/50012-6

[10] Markstein G H (1981) Scanning-radiometer measurements of the radiance distribution in PMMA pool fires, *Symposium (International) on Combustion* 18 537–47, https://doi.org/10.1016/S0082-0784(81)80059-8

[11] Raj V C and Prabhu S V (2018) Measurement of geometric and radiative properties of heptane pool fires, *Fire Safety Journal* 96 13–26, https://doi.org/10.1016/j.firesaf.2017.12.003

[12] Sudheer S and Prabhu S V (2012) Measurement of Flame Emissivity of Hydrocarbon Pool Fires, *Fire Technol* 48 183–217, https://doi.org/10.1007/s10694-010-0206-5

IOP Conf. Series: Journal of Physics: Conf. Series **1107** (2018) 042036 doi:10.1088/1742-6596/1107/4/042036

Fire Safety Protection Assessment of Industrial Technologies

Kamila Kempna[1], **Jan Smolka**[1], Milos Kvarcak[1], Ivan Kricfalusi[2], Zdenek Stejskal[3], Tomaz Hozjan[4] and Dionysis Kolaitis[5]

[1]*Faculty of Safety Engineering, VSB – Technical University of Ostrava, Czech Republic*
[2] *HSE, New Eltom, Czech Republic*
[3]*CEPRO, a.s., Czech Republic*
[4]*Faculty of Civil and Geodetic Engineering, University of Ljubljana, Slovenia*
[3]*Department of Chemical Engineering, National Technical University of Athens, Greece*

jan.smolka@vsb.cz

ABSTRACT

An accident in petrochemical industry can cause serious damages to human and animal health, property, environment and economy. That was proved by many accidents in the past. National laws, regulations, national, European or international standards or accepted protection principles define minimal requirements for safety and security protection of technologies. This restriction defines specific protections and periodical personal controls. Companies in petrochemical industry are often split in branches with specific types of technologies – pumping stations, storages, distribution lines, etc. Each of these technologies represents specific hazards for employees, environment, surrounding area or economy (national, organizational). Especially oil storages and farms represent strategical importance for national safety and continuity.

Increasing of protection is often organized individually in specific branches of the organizations. It can cause that the same organization has different protection of the same technology. That is the reason why many organizations design safety policy on higher level than is legally requested by insurance companies or by internal safety policy. The increase of policy is connected to increase of number of personal controls which can be implemented up to every hour per day, or to application technical protection. The technical protection can include leakage sensors, sprinkler systems, emergency reservoirs, process monitoring (pressure, temperature, mass flow, etc.) or CCTV systems. The uniform protection within the organization on the same technologies can increase employees working time efficiency, increase level of protection and decrease financial losses. For these purposes the evaluation of technical protection with impact on number of personal controls have been designed and described in this paper.

Presented system of assessment evaluates specific hazards and possible impact including possible damages to human health, property or environment with considered type of technology. Moreover, the building and technological protection is considered, as well as the fire protection systems, alarm systems, and monitoring cameras systems. Final number of personal controls of technology is based on type of organization, hazards and protection. A resulting recommended number is compared with required legal number of controls.

Main benefits of screening assessment are:

1) Allow planning of flat investments and development of technologies and its protection
2) Increasing efficiency of workforce
3) Defining critical areas for technical development – application of CCTV, modern detection systems, etc.
4) Increasing efficiency of working time of employees – e.g. replacement personal screening by CCTV systems
5) Decreasing costs of control and service actions

KEYWORDS:

explosion; risk assessment; detection; human behaviour

IOP Conf. Series: Journal of Physics: Conf. Series **1107** (2018) 042036 doi:10.1088/1742-6596/1107/4/042036

INTRODUCTION

The inherent part of oil and petroleum industry is storage and pipe transport. Manipulation with flammable liquids entails the risk of fire, explosion or leak and their consequences as injury, environmental pollution or material and financial losses. In order to prevent any incidents and ensure proper and safe operations, safety and security design and measures in petrochemical industry fulfil even higher standard level than required by laws, codes and insurance conditions in regular companies. Namely, consequences in case of accident can have big impacts on an organization, surrounding area, environment or in worst case entire nation [2][3]. A control screening assessment is a tool designed to estimate a number of personal controls based on type of technology, potential hazard and protection of technology. This tool respects all type of technologies, state of operation and their protection or law requirements.

NOMENCLATURE LISTING

R	Overall Impact	*Z*	Correction Index (%)
C	Technology Type Index	*S*	Protection Factor
E	Environmental Impact Index	*M*	Monitoring Index
F	Technical Failure Index	*P*	Technical Protection Index
I	Injury Index	*CCTV*	Closed Circuit Television

DESCRIPTION OF ASSESSED TECHNOLOGY

As a model organisation served CEPRO,a.s., is a joint-stock company with main activities in field of transport, storage and distribution of petroleum products. One of the main tasks is to maintain and protect national material reserves. The company fulfil request of European Union to ensure 90 days of average daily consumption of fuel by governing 17 fuel storages distributed throughout the country.[1] The site of storages includes underground and ground tanks, tanks for manipulation with fuel, filling platforms for railway and automobile tanks, workshops, fire brigade stations, administrative buildings, laboratories, fire-fighting tanks, electric substations, engine rooms, pumping stations, etc. The complexity of sites presents specific requirements for fire safety and fire protection. The assessment is prepared based on internal data and current fire risk assessment required and described by the Czech national law [5] with the most important information for following assessment.

GOALS OF ASSESSMENT

The main goal of this assessment is to get unified approach to assessment of the technology and defining number of personal controls for all branches within a company and therefore decrease man-hours, give a company tool how to improve fire safety and generally increase efficiency. Importantly also communication between specific branches and planning of horizontal development. Guidelines of Confederation of Fire Protection Associations - Europe (CFPA-E) and standards of NFPA [6] were used as source methodologies for assessment [3][4]. These guidelines are improved and implemented considering conditions of petrochemical industry. Source data are valuable for personal visits and fire safety documentation required by national laws [5].

Primary target of the assessment:

- Fire Risk Assessment and assessment of facility protection.
- United assessment approach of all branches in fire risk assessment and assessment of facility protection.
- Number of controls between branches for the same facilities unification.

Secondary impact of the assessment:

- Areas of technical development definition.
- Horizontal planning of organizational development.
- Estimating simple basic overview of technological protection.
- Efficiency increase of employees' working time.

STRUCTURE OF SCREENING ASSESSMENT

For efficient approach of the screening assessment, it is important to follow next specific steps for each technology during the process of assessment:

1) Basic description of technology
2) Risk Assessment
3) Protection Assessment
4) Correction
5) Result Evaluation
6) Comparing results with Law Requirements
7) Final recommended number of controls

IOP Conf. Series: Journal of Physics: Conf. Series **1107** (2018) 042036 doi:10.1088/1742-6596/1107/4/042036

Main Layout of Control Assessment Index

Procedure to conclude final number of controls depends on many factors and has to follow process as presented in Fig. 1. Overall Impact (R) includes influence of Technology type (C), possibility to harm Environment (E), Technical Failure (F), Injury (I), and Correction Factor (Z). Following, the Overall Impact is corrected by Safety correction (Z) and resulting value is adjusted by considering Actual level of protection (S) for both, Monitoring (M) and Technical Protection (P). For any requirements of national orders, laws or technical codes, resulting number of controls needs to be compared and then, final recommended number can be determined.

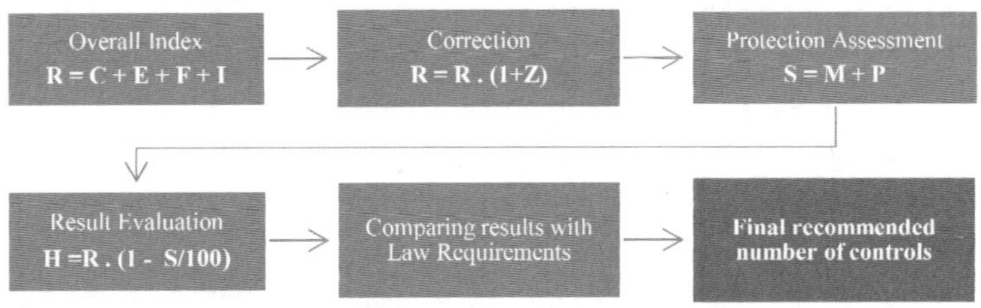

Fig. 1. Diagram of Assessment Procedure

Overall Impact Index R

As any safety measures primarily targets at people protection, it is necessary to perform initially assessment of impact of possible accident. Moreover, assessment includes operation of technology, its type, impact on the environment (E), employees and process continuity. Numbers in right columns in Table 1 are designed and evaluated in accordance with real company.

$$R = C + E + F + I \tag{1}$$

The final correction of Overall Impact is revised by responsible specialist who considers other conditions (e.g. accidents). This correction is adapted by Correction Index (Z) as given by Eq. (2).

$$R = R\,.\,(1 + Z) \tag{2}$$

Technology Index C

This index considers type of technology, its operational stated (e.g. non-stop operation or out of service), or presence of chemical substances. Values for specific sub-indexes can be found in Table 1.

$$C = C_1 + C_2 + C_3 + C_4 + C_5 \tag{3}$$

Index of Environmental Impact – E

In terms of environmental pollution, distinguishing whereas there is a risk of liquid leakage from evaluating technology needs to be done firstly. Next step is to consider possible amount of liquid leaked. Finally, index of environmental impact is calculated as with values given from Table 1 as stated by Eq. (4):

$$E = E_1\,.\,E_2 \tag{4}$$

Failure Index - F

This index specifies impact on continuity of organizational processes and connections of technology to other technologies or processes for a case of failure or accident as given by Eq. (3):

$$F = F_1 \tag{5}$$

Injury Index - I

There is a various presence of persons on different workplaces and technologies. Some technologies require permanent service and some are out of service and presence of persons is excluded, which is described by Eq. (6):

$$I = I_1\,.\,I_2\,.\,I_3 \tag{6}$$

All estimated values for specific indexes which are used for calculations and assessing the impact index (R) given by Eqs. (1) – (4) are presented in table 1.

Table 1. – Estimated values for indexes used for assessment of Impact Index - R

Index - C	State	Value
C_1 - *Category of Facility*	*1 - Technological*	10
	2 -Continuous	10
	3 - Other	1
C_2 - *Stored Liquid*	*Flammable Liquid (I. – III. class)*	10
	Flammable Liquid (IV. class)	5
	No Flammable Liquid	0
C_3 - *Technology*	*Open*	2
	Closed	4
	Not relevant	0
C_4 - *Type of Technology*	*Storage*	3
	Continual	5
	Not relevant	0
C_5 - *Operation*	*Active operation*	20
	Passive operation	10
	Out of Service	0
Σ		**Max. 50**
Index - E	**State**	**Value**
E_1 – *Hazard of liquid leakage out of technology*	*Yes*	1
	No	0
E_2 – *Quantity of possible leakage*	*Tens of litres*	3
	Leakage in technology and its accidental protection	10
	Out of accidental protection	20
Σ		**Max. 20**
Index - F	**State**	**Value**
F_1 – *Impact on process continuity*	*Without impact or with minimal limitation of processes*	0
	Limitation of processes for hours. Short-term affection	10
	Large impact - Affection of operation all company / areal / etc.	20
Σ		**Max. 20**
Index - I	**State**	**Value**
I_1 – *Facility service*	*Permanent service*	1
	Without permanent service	0
I_2 – *Possibility of persons injury in case of accident*	*No persons*	0
	One person	1
	Two or more persons	2
I_3 – *Which injury imminent*	*Minor injuries*	1
	Short-term hospitalization or convalescence	3
	Death, long-term hospitalization, invalidity	5
Σ		**Max. 10**

Correction Index - Z

The Correction Index (Z) is important for specific conditions which can be observed during practice e.g. maintaining of a technology or occurrence of specific hazards known by an attendant responsible for technology evaluation. The index allows increase in hazard index in range from 0% to 30%.

30 % - Accidental or different unaccepted state, working on technology by extraneous persons
20 % - Very frequent manipulation over the frame of unusual service in organization
10 % - Other risks
0 % - No correction.

Protection Assessment – S

The estimation of protection is based on assessment of monitoring by CCTV, sensors, detection which could substitute personal controls. This includes index *M* described by Eq. (7). The technical protection assesses possibility of protection technology in cased of non-standard situation. These factors are included in Index *P* (Eq. (8)).

Monitoring Index - M

$$M = (M_1 \,.\, (M_2 + M_3 + M_4)) + M_5 + M_6 \qquad (7)$$

Technical Protection Index -P

$$P = P_1 + P_2 + P_3 + P_4 + P_5 + P_6 + P_7 \quad (8)$$

All values for calculated indexes are summarized in Table 2., with description of specific values.

Table 2. – Table with estimated values for specific indexes for Protection Assessment - S

Index – M	State	Value
M_1 – CCTV	Rotate	2
	Fixed	1
	None	0
M_2– Does CCTV cover all areas?	Yes	3
	Partly	2
	No	1
M_3 – Is CCTV with permanent service?	Yes	5
	No	0
M_4 – Is CCTV equipped with night-vision / thermal vision?	Yes	5
	No	0
M_5 – Permanent service of operational values	Yes	8
	Partly	5
	No	0
M_6 – Indication of danger substances presence?	Yes	5
	No	0
Σ		Max. 29
Index – P	**State**	**Value**
P_1 – Critical places – Are they defined for the facility?	Yes	4
	No	0
	Not relevant	0
P_1 – Presence of electronical fire signalization	Yes	10
	No	0
P_2 – Equipped by Sprinkler / Dry Sprinkler System	Yes	10
	No	0
P_3 – Gas Detection	Yes	10
	No	0
P_4 - Drainage/ Separator/ Scumboart / Control Borehole	Yes	1
	No	0
P_5 – Bund / Collection Sump	Yes	5
	No	0
P_6 – Level Detection and Accident (high) Level Detection (gradient to Control System/ Out of Control System), Operational State of Technology Monitoring	Transmitting in Control System	3
	Local Signalization	1
	None	0
P_7 – Detection of Substance Presence in Sump (Interlayer), Transmitting in Control System / Local Signalization	Transmitting in Control System	3
	Local Signalization	1
	None	0
Σ		Max. 46

Calculation of Protection Assessment

Assessment of protection is based on evaluation of level of monitoring (M) and protection (P). The main reason is a possibility to substitute personal visiting of technology by employees and level of protection.

$$S = M + P \quad (9)$$

RESULT EVALUATION OF ASSESSMENT

The final calculation assesses estimated values. Number of personal controls (H) of technology for resulting values need to be estimated first with Eq. (10).

$$H = R\,.\,(1 - \frac{S}{100}) \quad (10)$$

For this evaluation it is important to account for the technologies in service state or out of service as well as technologies in non-stop operation with the highest level of protection. Number of controls and range of controls are estimated based on practice, safety requirements and impact assessment. Range moves from once a year (e.g. storages out of service with no chemical substances) to 12 per 24 hours (e.g. pumping stations in operation). These values are described in Table 3.

Table 3. – Resulting number of recommended screening controls

Final number of recommended number of controls					
Result Value of Assessment	**Number of Controls**	**Result Value of Assessment**	**Number of Controls per 24 h**	**Result Value of Assessment**	**Number of Controls per 24 h**
1 - 5	*1 per year*	16 - 28	*1*	67 - 75	*7*
6 - 15	*1 per month*	29 - 36	*2*	76 - 79	*8*
		37 - 44	*3*	80 - 84	*9*
		45 - 52	*4*	85 - 89	*10*
		53 - 59	*5*	90 - 94	*11*

Final Decision of Control Assessment

Specific number of personal controls in frame of fulfilling code or law requirements for specific technologies can be required. For this reason, the final decision needs to consider whereas determined number of personal controls includes required number of compulsory controls or not.

Calculated number of controls > Compulsory number of controls = **Calculated number of controls**
Calculated number of controls < Compulsory number of controls = **Compulsory number of controls**

CONCLUSIONS

While data from the assessment of the case study branch Sedlnice will be post-processed, the verification of efficiency will be evaluated. Additional work will identify possible initiating events and component failure modes with dynamic fault trees. These trees will be developed in Bayesian network. For evaluation of efficiency and development of assessment Bayesian network will be used as a technique considering investigation of possible scenarios with domino effects. A mapping of scenarios with analysis of probability and evaluation should be included [9][10]. The impact of possible accidental scenarios can be additionally investigated with numerical modelling software such as ALOHA. This software can provide data for defining impact of various scenarios e.g. jet-fire, boilover or leakage. Obtained data will be compared with case study real conditions and with designed assessment procedure described in this article.

REFERENCES

[1] COUNCIL DIRECTIVE 2009/119/EC - imposing an obligation on Member States to maintain minimum stocks of crude oil and/or petroleum products. In: . EU: Official Journal of the European Union, 2009, Year 9, 2009/119/EC. From: http://eurlex.europa.eu/LexUriServ/LexUriServ.do?uri=OJ:L:2009:265:0009:0023:EN:PDF

[2] ARGENTI, Francesca, Gabriele LANDUCCI, Genserik RENIERS a Valerio COZZANI. Vulnerability assessment of chemical facilities to intentional attacks based on Bayesian Network [online]. 2018, **169**(September 2016), 515–530. From: doi:10.1016/j.ress.2017.09.023

[3] Business Resilience: An Introduction to protecting your business. CFPA-E Guideline No 2:2013 N. Denmark, 2013. from: http://cfpa-e.eu/cfpa-e-guidelines/

[4] Introduction to qualitative fire risk assessment. CFPA-E Guideline No 4:2010 F. Denmark, 2010. From: http://cfpa-e.eu/cfpa-e-guidelines/

[5] Act No. 133/1985 Sb. About Fire Protection in Valid Wording. Czech Republic, 2018

[6] NFPA 30: Flammable and Combustible Liquids Code. 2018. USA: NFPA, 2018.

[7] EIDE, S. A., T. H. SMITH, R. G. PEATROSS a I. E. STEPAN. A Simplified Method for Quantitative Assessment of the Relative Health and Safety Risk of Environmental Management Activities. USA, 1996. From: http://www.iaea.org/inis/collection/NCLCollectionStore/_Public/28/043/28043271.pdf

[8] ARGENTI, Francesca, Gabriele LANDUCCI, Genserik RENIERS a Valerio COZZANI. Vulnerability assessment of chemical facilities to intentional attacks based on Bayesian Network [online]. 2018, **169**(September 2016), 515–530. From: doi:10.1016/j.ress.2017.09.023

[9] YANG, Xue, Stein HAUGEN a Nicola PALTRINIERI. Clarifying the concept of operational risk assessment in the oil and gas industry. *Safety Science* [online]. 2018, **108**(June 2017), 259–268. ISSN 0925-7535. From: doi:10.1016/j.ssci.2017.12.019

[10] XIN, Peiwei, Faisal KHAN a Salim AHMED. Dynamic hazard identification and scenario mapping using Bayesian network. Process Safety and Environmental Protection [online]. 2016, 105, 143–155. ISSN 0957-5820. From: doi:10.1016/j.psep.2016.11.003

Positive Pressurization and Ventilation for Fighting Fires in High-Rise Structures with Multiple Stairwells

Prabodh Panindre[1], N.S.Susan Mousavi[1], Sunil Kumar[1], and John Ceriello[2]

[1]*Department of Mechanical Eng., New York University, USA*
[2]*Fire Department of New York, New York, USA*

prabodh@nyu.edu

ABSTRACT

This study addresses the selection of the pressurization stairwell and the ventilation stairwell for fighting fires in a high-rise structure with multiple stairwells. The specific case of three stairwells, where one is regular and the other are paired scissor stairwells, has been considered in this study and various scenarios have been analyzed. Through simulations of a model that has been validated by on-site pressure tests, it is found that the most effective scenario which provides a safer environment for the rescue-operation is choosing the closest stairwell to the fire apartment as the pressurization stairwell, and the farthest stairwell to the fire apartment as the ventilation stairwell.

KEYWORDS:

Positive Pressure Attack (PPA), Positive Pressure Ventilation (PPV), Firefighter safety, Fire Dynamics Simulator (FDS)

IOP Conf. Series: Journal of Physics: Conf. Series **1107** (2018) 042037 doi:10.1088/1742-6596/1107/4/042037

INTRODUCTION

Flow of smoke, heat and other combustion products created by spread of fire can be efficiently driven away by pressurization of stairwells that creates safer environment for firefighters than without pressurization [1–4]. This is achieved by directing a significant amount of air into the building stairwell and public hallways by fan deployment that creates a positive pressure zone [5]. Optimization of positive pressurization for wind-driven high-rise fires in a typical high-rise building with one stairwell has been previously studied [6–8]. However, pressurization in a complex structure with multiple stairwells can be more complicated than that in a typical single stairwell scenario [9]. In a complex building with multiple stairwells, selection of attack (pressurization) stairwell where the fan is deployed and ventilation stairwell through which combustion products are vented can be critical [10]. Commonly found return stairs (also referred to "U return stairs") vertically returns to the same location of the building at each floor landing. In contrast, scissor stairwell contains two separate stairwells in one vertical shaft that are separated by a wall. Stairs exit on the opposite side of the building at each landing and the layout of the fire floor is the same as that of two floors below the fire [10]. Efficacy of the pressurization can vary greatly with respect to fire location, fan location, attack stairwell (stairwell that is pressurized) and ventilation stairwell (stairwell used to remove smoke and heat), as well as the sequence of venting tactics. Building on results of previous studies [8, 11], this paper analyzes the performance of positive pressurization tactics in high-rise fires in complex structures with return and scissor stairwells. The influence of concerned parameters as well as their relative impacts on the fire-rescue operation is investigated through simulations.

ON-SITE TESTS

In order to validate the simulation model, on-site pressurization tests, in the absence of fire, were conducted in a vacant modern seven-story residential high-rise structure with a scissor stairwell and a single typical return stairwell in Brooklyn, New York (average atmospheric condition: 76^0F/24.5^0C, average humidity: 80%, average pressure: 29.95 in/101.422 kPa). The apartments on each floor are connected via a long public hallway. A positive pressure zone was created utilizing pressurization fans (BD27-H-9.0 by Tempest Technology, http://www.tempest.us.com) placed 3 ft (0.9 m) away from the stairwell door such that the horizontal axis of the fan is parallel to the ground. Each fan blows approx. 24,000 CFM (11.23 m^3/s) air into the stairwell. To measure the increase in the static pressure, differential pressure sensors were located on floors 2, 5, and 7. The high port of a sensor was open to pressure inside the stairwell while the low port was open to atmospheric pressure via a flexible tube. The results of different fan arrangements were compared with those of the simulations. The bulkhead door (or the roof door) was closed to maintain desired pressure levels [8, 12]. Only parallel combinations (side-by-side) for the fan arrangements are considered subsequently since it was found that the series combination of the fans (back-to-back placement of the fans) is not efficient for increasing the volume of the air entering the stairwells [13]. Also, it was observed that two parallel fans at the first floor (i.e., ground floor) create better positive pressure zone inside the stairwell to effectively drive away the combustion products than placing them in the floors other than the first. Therefore, this strategy is used and tested with the fire scenes described below. As the simulations results are in good agreement (less than 10% difference) with those of the on-site tests, the simulation model is considered validated, and thereby it is used for further analysis.

MODELING

A seven-story building model (73m x 11m x 20.91m) consisting one scissor stairwell (Stairwells A and B in Figs. 1 and 2) and one single typical stairwell (Stairwell C), and a long public hallway is considered for the simulation. All stairwells have the dimensions of 5.6m x 2m x 20.91m, and each has eight landings with a door on each floor, the topmost door being the roof door. Each floor consists of doors of each stairwell leading to a public hallway that connects the apartments. It is assumed that the fire ignites on the sixth floor, noting that the ground level is considered as the first floor throughout this study. The apartment door opens into the living room which is connected to the kitchen, the bedroom, and the restroom/bathroom. There is a double window (2m x 1m) in the living room and a single window (1m x 1m) in the bedroom (see Figs. 1, and 2). All the doors have dimensions of (2.1m x 0.9m). There are apartments on both sides of the public hallway, but only one side is shown on Fig. 1. Also details of only three apartments layout are shown in the figure for clarity. The apartment has a typical fuel load that consistent with commonly used furniture items such as a bed, TV, sofa, wooden chairs, with thermophysical properties mentioned in the previous study [8]. The fuel load is same in all apartments for all simulations allowing a valid comparison of positive pressurization tactics [14]. As FDS allows only rectilinear meshing, pressurization fan is considered as a square surface [8, 11] of 0.5 m^2 area and is placed 3 ft (0.9 m) away from the stairwell door that provides air at the rate of approx. 24,000 CFM (11.23 m^3/s) into the stairwell. To perform the simulations, FDS 5.0 is used. Although the later versions of FDS are available, the accuracy of the results obtained by FDS 5.0 has been previously examined and proven [8].

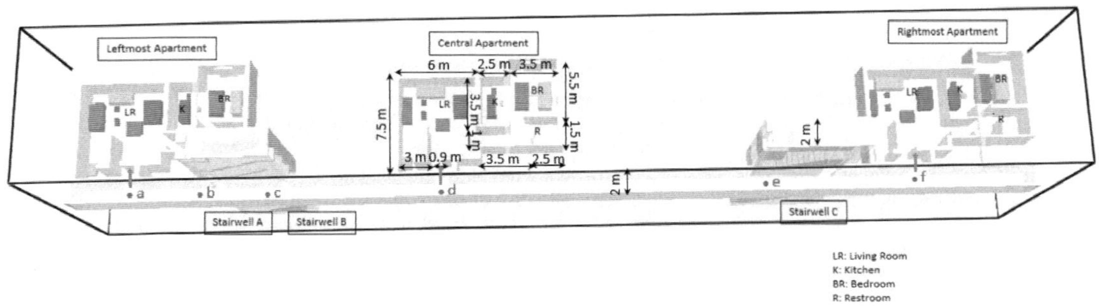

Fig. 1: 3-D model of the building structure, the layout of each floor consists of several apartments, of which three apartments are shown: leftmost, central, and rightmost one.

Meshing and boundary conditions

FDS is second order accurate in space and time. FDS can predict temperatures in similar fire scenarios to an accuracy of 5-20% in comparison to experimental results [15]. Grid convergence tests are performed by repeating the same scenario several times with increasing number of cells. The mesh is refined until the recorded values at all locations stabilized and the difference in final mean temperature (moving average of the fluctuations over 10 s) became relatively small (see Fig. 3). This procedure was also repeated with the soot density readings to ensure accuracy. Although multiple mesh schema are available in FDS, the model used for all simulations in this study has single uniform rectilinear mesh in all regions that includes 1,890,000 total number of cells of equal volume of 0.0088 m^3 (0.35m x 0.18m x 0.14m) and has the time-step of 0.143 second. The computational areas of interest include the stairwells, the public hallway, the apartment on fire, small regions around the fans, and a region outside the window through which wind conditions are introduced. The floors, ceilings, surfaces of furniture, walls, stairwell steps and all other solid surfaces of the structure are assigned as wall boundaries. The opening of windows and doors are simulated by creating holes in the surfaces at certain time instances. The computational domain has air at 25^{0}C and open boundary conditions in all directions except the ground.

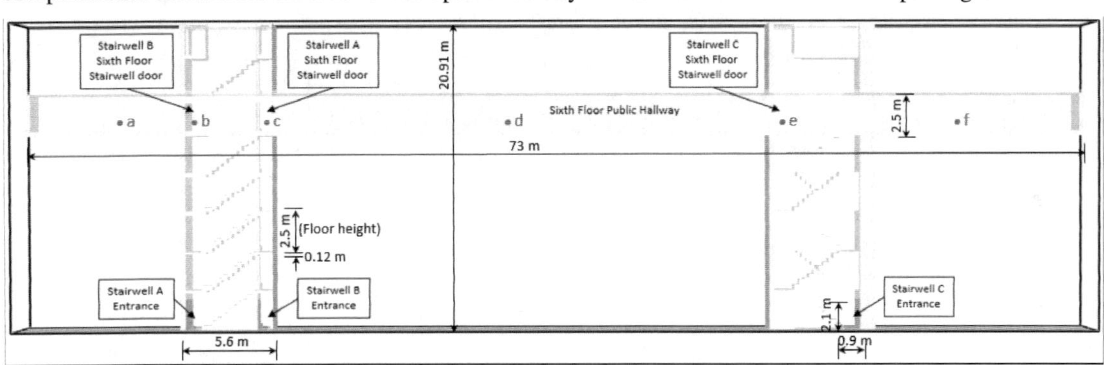

Fig. 2: Location of the temperature sensors in the 6th floor public hallway where fire is in one of the apartments.

Timeline of fire events

Although the unplanned sequence of random events in real-life fires can be much more complex than the simple timeline followed in the simulations, the following time line serves the purpose of this study. The attack stairwell (or pressurization stairwell) has the positive pressure zone allowing the firefighters to stretch the hose-line. The ventilation stairwell can be one of the stairwells other than the attack stairwell through which the heat and smoke is vented from the public hallway creating better thermal conditions as well as visibility for the firefighters. Depending on the fire location, attack stairwell, and the ventilation stairwell, several scenarios can be assumed as listed in Table 1. It is assumed that the fire ignites in the bedroom of an apartment on the sixth floor and the apartment door is left open by the occupants. Stairwell doors and roof doors are initially closed. As the fire continues to grow, windows in the bedroom fail at the 120th second allowing 10 mph (4.4 m/s) wind to enter the apartment which results in the rapid spread of the heat and smoke into the public hallway. Wind is introduced to consider more severe conditions and were simulated by inflow of air from a boundary surface [8, 11, 16]. However, similar trends are anticipated for non-wind conditions as the fresh air enters from the broken window. At the 180th second, a firefighters open the sixth floor door and start the fan at the attack (pressurization) stairwell. At the same moment, another firefighter opens the sixth floor stairwell door and the roof door of one of other two stairwells which is acting as ventilation stairwell. The simulation continues for another 70 seconds and is stopped at 250th second. This is considered enough time for firefighters to enter the fire floor with the hose-line through the pressurized attack stairwell and reach the door of the apartment on fire. The response time of fire departments vary to a great extent. Through another set of simulations, it has been verified that if firefighters open the door at later time, the temperature in hallway upon stairwell pressurization are relatively higher, but the overall trend of results obtained in this study remain valid.

The temperature variations inside the public hallway are recorded at six thermal sensor points (a-f) as shown in Figs. 1 and 2. Considering the crawling motion of firefighters during the search-rescue operation on fire floor, the measurement recording points, both temperature and soot density, are located in the middle of the public hallway outside the entrance doors of apartments and stairwells, at the mid-level of the apartment's height (i.e. 1.3 m above the floor level). As heat and smoke move upwards, the temperature increases from the floor towards the ceiling, this height is appropriate to indicate whether the temperatures are tenable for the firefighters or not.

RESULTS AND DISCUSSION

In a complex structure with multiple stairwells, there are several possibilities for attacking and venting fire as listed in Table 1. Ventilation of smoke inside the public hallway is performed through one stairwell as more than one venting stairwell can decrease the effectiveness of the pressurization at the fire apartment [8, 17]. Simulations are repeated for all corresponding cases assuming fire is ignited in the sixth floor in one of the following apartments: the central, the rightmost, or the leftmost one. The pressurization in all cases is provided by two parallel fans, based on on-site results, that are placed 0.9 m from the door on the first floor (i.e. ground floor). In all simulations, initial fire behavior is qualitatively similar to previous studies [8, 11]. When window fails it rapidly grows and spreads into the public hallway because of availability of more air from the window resulting in temperature increase at all locations. As the fuel in the apartment quickly becomes ventilation-limited, the temperature in the apartment begins to drop slightly. In all cases, the deployment of pressurization fan mitigates the spread of smoke and heat in the attack (pressurization) stairwell. The combustion products which are pushed back by pressurization in attack stairwell are vented through the ventilation stairwell or driven back into the fire apartment and escape through the broken window resulting in an increase in the temperature at those locations. In contrast, the temperatures at the pressurized stairwell drops creating a safer environment.

Table 1: Possible scenarios based on the choice of the apartment, attack stairwell, and ventilation stairwell.

Fire Apartment	Attack Stairwell	Ventilation Stairwell	Case No.
Leftmost Apartment	Stairwell A (scissor)	Stairwell B	1
		Stairwell C	2
	Stairwell B (scissor)	Stairwell A	3
		Stairwell C	4
	Stairwell C (return)	Stairwell A	5
		Stairwell B	6
Central Apartment	Stairwell A	Stairwell B	7
		Stairwell C	8
	Stairwell B	Stairwell A	9
Central Apartment	Stairwell B	Stairwell C	10
	Stairwell C	Stairwell A	11
		Stairwell B	12
Rightmost Apartment	Stairwell A	Stairwell B	13
		Stairwell C	14
	Stairwell B	Stairwell A	15
		Stairwell C	16
	Stairwell C	Stairwell A	17
		Stairwell B	18

For fire in each apartment, the effectiveness of pressurization varies depending on choice of attack stairwell and ventilation stairwell which has been shown in Figs. 4, 5, and 6. At the end of simulation (70 seconds after opening the door of attack stairwell), the recorded temperatures stabilizes to some extent as the constant pressurization flow continues to interact with the continuous generation of heat and smoke. Therefore, the final mean temperature (moving average of fluctuations over 10 seconds) in the public hallway (entry location for firefighters) are used to compare the efficacy of the possible pressurization/ventilation tactics listed in Table 1. As the main purpose of pressurization is to assist firefighters in making an effectively safer entry on the fire floor, the efficacy is primarily evaluated based on the heat and smoke conditions inside the public hallway. In all cases, it was found that pressurization is more effective if the attack stairwell is closer to the fire apartment and the ventilation stairwell is farther from the attack stairwell. For example, in case of rightmost location, firefighters would experience safer environment if they choose stairwell C (which is closer to the fire location) as the attack stairwell, and vent the heat and smoke in the public hallway through stairwell A or B (which are farther from attack stairwell) (see Cases 17 and 18 in Fig. 4). Whereas, for the leftmost location, firefighters should choose stairwell A or B as the attack stairwell, and vent the heat and smoke through stairwell C (see Cases 2 and 4 in Fig. 5). In case of fire in central location, firefighters can choose any of the stairwells as they are at almost the same distance from the fire location but the heat and smoke should be vented through the farthest stairwell from the attack stairwell (see Cases 8, 10, 11, and 12 in Fig. 6).

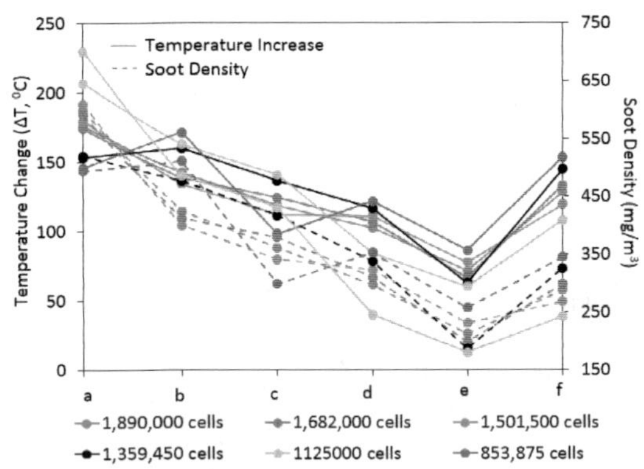

Fig. 3: Mesh convergence test using final mean temperature for Case 17 (see Table 1).

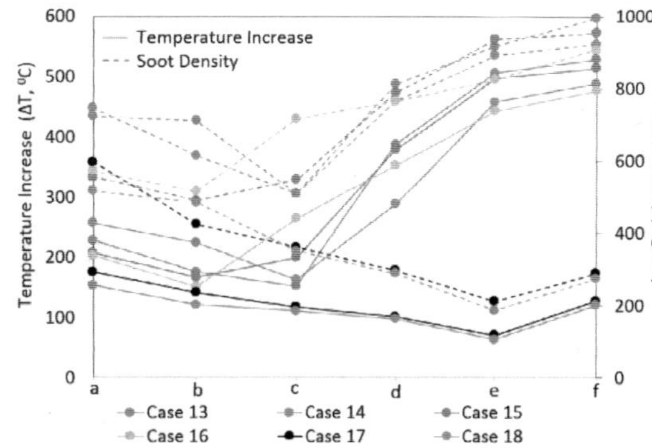

Fig. 4: Distribution of temperature increase and soot density in the public hallway at six temperature sensor locations a,b,c,d,e, and f when fire is in the rightmost apartment.

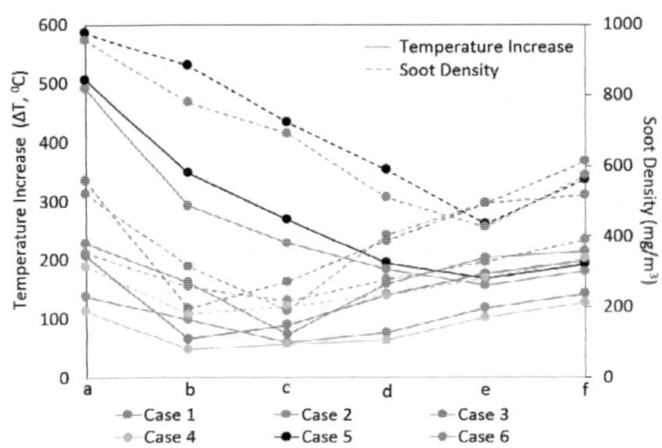

Fig. 5: Distribution of temperature increase and soot density in the public hallway at six temperature sensor locations a,b,c,d,e, and f when fire is in the leftmost apartment.

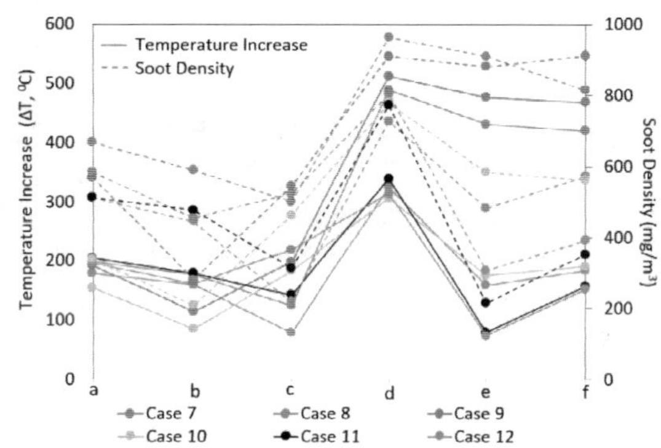

Fig. 6: Distribution of temperature increase and soot density in the public hallway at six temperature sensor locations a,b,c,d,e, and f when fire is in the central apartment.

Generally, in case of scissor stairwell, if other stairwells are available and accessible for use, firefighters should not use stairwell A as attack stairwell and B as ventilation stairwell or vice-versa. In this scenario, the pressure created in the attack stairwell does not aid in the venting of combustion products through adjacent stairwell, and the heat gets trapped in the public hallway. For example, in case of fire in the leftmost location, if attack is through stairwell A and venting from stairwell B (Case 1), the heat is pushed to the dead end on the other side as stairwell C is closed. As the ventilation stairwell is close to the attack stairwell, the positive pressure leaks through the ventilation stairwell and effectiveness of pressurization is reduced at the fire apartment. Similar analogies can be made for other relevant cases (Case 3, 7, 9, 13, and 15). When a stairwell that is farther from fire location is chosen as the attack stairwell, the pressure zone is not able to drive away combustion products as it has to pressurize long hallway (see Cases 5, 6, 13, 14, 15, and 16).

CONCLUSION

The present study examines fire behavior that only includes assumptions of essential firefighter actions such as initiating pressurization and opening stairwell doors on the fire floor. Other related operating procedures are not addressed, nor are possible evacuation plans, etc. It is anticipated that fire professionals will use the results of the study to understand fire behavior for various scenarios, and to guide fire ground procedures. Several possible fire scenarios with pressurization have been simulated for a large high-rise structure with return and scissor stairwells that provide multiple options for attacking and venting points. It was found that pressurization is more effective if the attack stairwell is chosen to be the one that is closer to the fire apartment while the ventilation stairwell is chosen as the one that is farthest from the fire apartment. The results are developed for buildings where three or more stairwells are available. Since only two stairwells are used in each scenario for pressurization and ventilation, the conclusion that the stairwell closest to the apartment on fire should be used for pressurization is also valid if only two stairwells are available.

ACKNOWLEDGMENT

This study is supported by the Assistance to Firefighters Grant Program from US Dept. of Homeland Security.

REFERENCES

[1] Garcia, K., Reinhard, K., Schelble, Raymond, and Schelble, Ray. (2006) Positive pressure attack for ventilation & firefighting. PennWell Books.

[2] Reick, M. (2009) Mobiler Rauchverschluss Erfahrungen aus dem Einsatzalltag, brandSchutz/Deutsche Feuerwehr-Zeitung, S. (7), pp. 563-569.

[3] Madrzykowski, D., Kerber, S., Kumar, S., and Panindre, P. (2010) Wind, fire & high-rises: firefighters and engineers conduct research to combat a lethal threat. Mechanical Eng.-CIME 132 (7) 22-28.

[4] Miller, R. S., and Beasley, D. (2009) On stairwell and elevator shaft pressurization for smoke control in tall buildings. Building and Environment 44 (6) 1306-1317.

[5] Zinn, B. T., Bankston,C. P., Cassanova,R. A., Powell, E. A., and Koplon, N. A. (1974) Fire spread and smoke control in high-rise fires. Fire Technology 10 (1) 35-53.

[6] Feng, X., and Li, Y. F. (2012) Evaluation of Positive Pressure Ventilation for High-Rise Buildings in Compartment Fire Situation, Advanced Materials Research, Vols. 446-449, pp. 2908-2913.

[7] Lambert, K., and Merci, B. (2014) Experimental study on the use of positive pressure ventilation for fire service interventions in buildings with staircases, Fire Technology 50 (6) 1517-1534.

[8] Panindre, P., Mousavi, N.S., and Kumar, S. (2017) Positive pressure ventilation for fighting wind-driven highrise fires: Simulation-based analysis and optimization. Fire Safety Journal 87, 57-64.

[9] Wang Y., and Gao F. (2004) Tests of stairwell pressurization systems for smoke control in a high-rise building.ASHRAE Trans 110, 185-93.

[10] McGrail, D. M. (2007) Firefighting operations in high-rise & standpipe-equipped buildings. PennWell.

[11] Panindre, P., Mousavi, N.S., and Kumar, S. (2017) Improvement of Positive Pressure Ventilation by optimizing stairwell door opening area. Fire Safety Journal 92, 195-198.

[12] Li, M., Gao, Z., Ji, J., and Li, K. (2018) Modeling of positive pressure ventilation to prevent smoke spreading in sprinklered high-rise buildings. Fire Safety Journal 95, 87-100.

[13] Kerber, S., Madrzykowski, D., and Stroup, D. (2007) Evaluating Positive Pressure Ventilation in Large Structures: High-Rise Pressure Experiments. National Institute of Standards and Technology, Building and Fire Research Laboratory, Gaithersburg, Maryland, NISTIR 7412.

[14] Shu, S.B., Chuah, Y.K., and Lin, C.J. (2012) A study on the spread of fire caused by the stack effects of patio-A computer modeling and a reconstruction of a fire scenario. In Building Simulation vol. 5, no. 2, pp. 169-178. Tsinghua Press.

[15] Smardz, P., and Novozhilov, V. (2006) Validation of Fire Dynamics Simulator (FDS) for forced and natural convection flows. University of Ulster pp. 16-20.

[16] Zhang, C. F., Chen, S. Y., and Chow,W. K. (2014) Wind effects on the smoke spread of high-rise buildings. The 2014 International Conference on Advances in Wind and Structures (AWAS14), Busan, Korea, 24-27 Aug 2014, paper no. W5B.3.WS116_407FR.

[17] Beal, C.M., Fakhreddine, M., and Ezekoye, O.A. (2009) Effects of leakage in simulations of positive pressure ventilation, Fire Technology 45 (3) 257-286

IOP Conf. Series: Journal of Physics: Conf. Series **1107** (2018) 042038 doi:10.1088/1742-6596/1107/4/042038

Experimental and numerical investigations in order to model the fire development and propagation for fire safety engineering studies

Jeremy COLOMBIANO[1, 2], Virginie DREAN[2], Thomas ROGAUME[1], Franck RICHARD[1], Benjamin BATIOT[1], Talal FATEH[2, 3], Ali NADJAI[3] and Eric GUILLAUME[2]

[1] *Institut Pprime (UPR3346 CNRS), Université de Poitiers, ISAE ENSMA, France*
[2]*Efectis France, France*
[3]*Fire SERT, Ulster University, Northern Ireland*

jeremy.colombiano@efectis.com

ABSTRACT

Fire safety engineering plays an important role in achieving safety objectives in a building and is in way of generalization in all the countries of the world, especially for reaction to fire. It is based on the use of numerical models. However, numerical simulations used for this particular field has got some limitations, for example due to the size of the domains to simulate, the very large number and the complexity of the materials involved, the multitude of the geometric configurations... that do not allow the representation of all the processes. In consequence, reaction to fire cannot be treated accurately in engineering studies and flame ignition or propagation cannot be represented due to the time of calculation. Nevertheless, their description is of first order. In this context, this work gives a first approach on technical bases in order to represent ignition, fire growing and development and flame propagation in engineering study. This description requires an understanding of the different processes involved in both solid and gaseous phases as well as their interface. A means to represent small-scale phenomena with engineering model limitations (time to compute, size meshing) should be performed.

For this, a multiscale study both experimental and numerical is in progress on two types of polymers: a pine wood and PVC. This approach is conducted in order to identify relevant phenomena that drive fire growing and flame propagation for different configurations: vertical, horizontal, co-current and counter-current. The study should lead to the definition of new sub-models, of numerical model recommendations (size meshing, wall laws, etc.) in order to describe those phenomena at large scales without too large time consumption.

KEYWORDS

Reaction to fire; Fire Safety Engineering; Fire growing; Flame propagation; Multiscale approach; Experimental study; Numerical simulation; FDS.

CONTEXT AND OBJECTIVES

Fire safety engineering (FSE) is an increasingly important part in achieving the definition of fire objective in a building. It enters in the so-called performance based approach as it matches objectives with conditions of a given local. The CFD code Fire Dynamics Simulator [1] (FDS) is widely used in FSE to model the main elements accounted in a study: building geometry, smoke extraction systems, detection systems, sprinkler system, main openings (doors, windows, outlets) and communications between volumes. Nowadays, FSE treats smoke removal and people evacuation studies, and is in way of development for reaction to fire. Reaction to fire is the fire behavior for a given material, including ignition, fire growing, propagation and extinction. It results in interactions between both the solid phase and the gaseous phase. It implies phenomena involved at larges scales as flow stream or thermal transport, but also at small scale, especially at the interface or in the reaction zone as mass and heat diffusion or chemical kinetic. Therefore, boundary layer flames features multi-physic phenomena, and these phenomena require specific near-wall model descriptions that so far have not been the focus of fundamental studies.

Gaseous phase [2][3] and solid phase [4][5] behavior are quite well understood separately, but the interaction between them isn't really understood. Generally, solid phase is seen as a porous burner slab, like the recent work of Ren et al. [6]. They present LES simulations of buoyancy-driven turbulent vertical wall flames and illustrate the potential of LES for wall fire simulations. However, the simulation is wall-resolved, with a near-wall mesh size of millimeters order, and the pyrolysis of the solid phase is not treated. LES is sensitive to the accuracy of the physical models that treat near-wall turbulence, combustion, thermal radiation and soot formation. In addition, the LES performance is adversely impacted when the computational grid resolution is not sufficiently fine. In [6], a near-wall grid resolution from 10 mm is not sufficiently fine to resolve wall gradients. A grid with that resolution is impossible to use for fire safety engineering because of time to compute and large domain sizes; therefore, it is necessary to adapt near-wall model and to develop technical bases in order to represent gradients and to perform engineering studies to account for reaction to fire.

Previous works, using similar multiscale approach have been carried out in order to study the thermal decomposition of a material. The first one was those of Bustamante Valencia [7]. He studied experimentally and numerically the thermal decomposition of a polyether polyurethane foam at three scales. He started at the matter scale using a TGA apparatus in order to determine the decomposition mechanism and integrated it at small scale and at the product scale. This integration leads to understand which parameters are needed in order to perform thermal decomposition studies. It was pointed out that the decomposition mechanism of the foam remains unchanged independently of the scale analysed. Another similar study was performed by Marquis [8]. He studied a composite sandwich using the same methodology as Bustamante Valencia. The findings were that this methodology is still applicable for composite material. Camillo [9] focused on the fire behaviour of a seat and a wall panel from railway transport system. This study starts at the matter scale, up to the real scale with a quite good agreement for the comparison of the experimental and numerical heat release rate (HRR). All these studies lead to the evaluation of the mass loss rate and of the HRR for a given material. They were focused on the solid thermal decomposition kinetics, therefore fire growing and flame propagation weren't pointed out by these studies.

In this context, the present work provides a new approach for technical bases in order to perform engineering studies including flame ignition development and propagation. It includes the study of the gaseous and the solid phases, with a specific attention at their interface. This study is focused on two materials in different configurations: vertical, horizontal co-current and counter-current. As numerous parameters and material properties are needed to model accurately the fire behavior at the material interface, the development of technical bases will include assessment of the relevant input parameters associated with the experimental tests needed. Numerical useful function and recommendation as wall laws or size meshing should be determined.

MATERIALS AND CONFIGURATIONS

The proposed methodology is applied on both a natural and a synthetic polymer that could be found in public buildings. The designated natural polymer is a cellulosic material, as pine wood. It will be considered as a partition panel, therefore in a vertical configuration. Wood exists through many varieties; its elementary composition is quite close from one to another. When heated, wood starts a pyrolysis process once the warm up and the dehydration is achieved. A carbonaceous layer appears at its surface, and constitutes the char. This layer plays an important role on wood decomposition as it becomes a heat shield for the virgin wood bellow its surface. In addition, the anisotropy of its properties makes it difficult to study experimentally and numerically. However, thermal decomposition of wood has been past studied in several sources [10][11]. Simple or detailed

decomposition models have already been proposed; also this current study is not focused on the development of thermal decomposition model.
The synthetic polymer is a polyvinyl chloride (PVC) coating that is usually found on floors in public buildings. PVC is the second most used plastic material and is composed of linear macromolecules with a general formula [CH2CHCl]n. The one studied is flexible and is used in floor application with a homogenous thickness with retardant added. We select a PVC floor coating with a Bfl-s2 reaction to fire ranking (Euroclasses). The support and the glue used must be taken into account to be as close as possible to real conditions. However, at small scale (matter scale and material scale) only the PVC is investigated. As the wood, its thermal decomposition has been past studied [12][13] and it is assumed to occur in two steps: the first one relates to the dehydrochlorination, with release of few hydrocarbons, and the second one corresponds to scission reactions, with release of toluene.
Because airflow plays an important role when a flame is spreading along a surface, co-current and counter-current flow phenomena will be investigated to be accounted in the numerical simulation. The heat flux emitted from the flame to the unburnt material surface depends on the configuration. Thus, in a co-current flow, flame can propagate easily as the heat flux received on the virgin surface is important. Generally, radiation is the dominant mode of heat transfer, although convection is present close to the ignition region. The flow is turbulent, and chemical time is of secondary importance compared to the flow time. In a counter-current situation, the flame cannot spread quickly; the virgin material is heated up mainly by diffusion close to the ignition region. The radiative part incoming from the flame can be neglected.
All these phenomena, involved at different scales, are characterized in order to understand the flame behaviour for each case. The more important achievement of this study will result in the investigation of a pathway between small, intermediate and full scales, leading to a better representation in numerical simulations in the context of fire engineering studies.

METHODOLOGY

Flame propagation depends on the local and assembly effects, the fluid dynamics, the flame geometry, the chemical reactions and the mass and heat transport mechanisms for the gaseous phase. For the condensed phase, it depends on the mass loss rate, mainly driven by the pyrolysis kinetics and the heat flux incoming to the surface. The thermal loads depend on the radiative part of the flame, and the understanding of these complex phenomena is a big challenge. Moreover, the chemistry of the condensed phase can largely influence the quantity of energy transmitted to the fuel by the apparition of a carbonaceous layer on its surface. Therefore, the state of each phase is strongly dependent to one another: the flame depends on the amount of fuel released and the release depends on the amount of heat transmitted to the condensed phase. For this reason, this study is mainly focused on processes involved at the interface.
Owing to the multiscale processes, a multiscale methodology is performed that provides insight into important features at multiple times and lengths of a physical phenomenon. The strategies proposed attempt to experimentally capture inherently important properties at various scales of a system and to numerically correlate them accurately to the system's macroscale properties by a multiscale validation. The numerical validation is admitted by comparison with experimental results: heat release rate, mass loss rate, spread rate, gas and solid temperature, etc. The proposed methodology is synthetized in the Fig. 1:

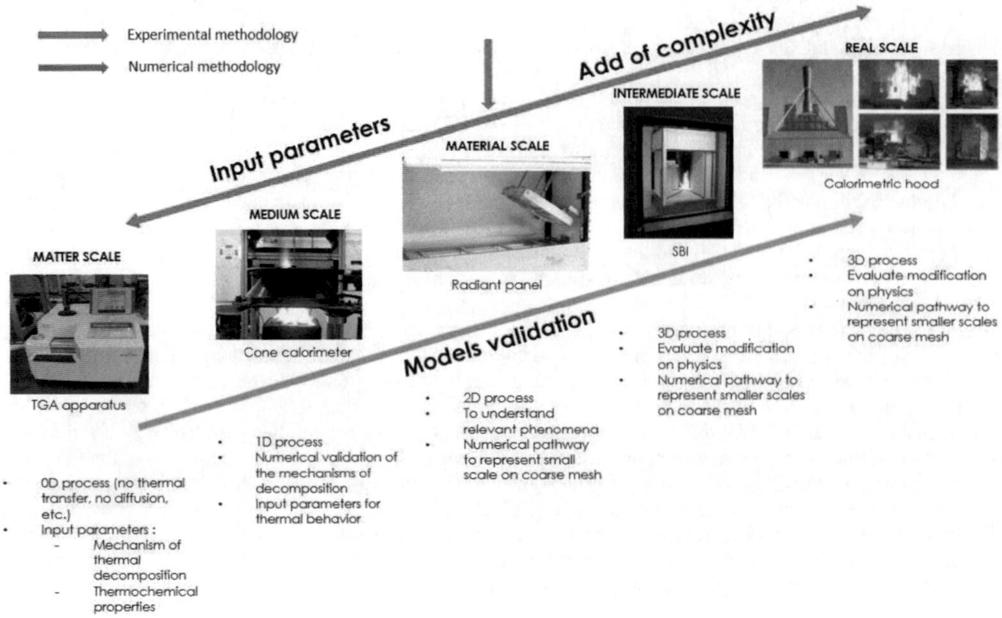

Fig. 1: Multiscale experimental and numerical methodology used.

Experimentally, the identification of relevant phenomena that must be modelled is performed at medium scale using radiant panel for the horizontal flame propagation of PVC, IMO-LIFT for lateral flame propagation, and a 1D vertical slab propagation (see Fig. 3) for wood. At this scale, all dependences on flame propagation described above are present; except the local and assembly effects.
The processes of interest are mainly located at the interface of the solid. The instrumentation of the 1D vertical slab facility is shown in Fig. 2.

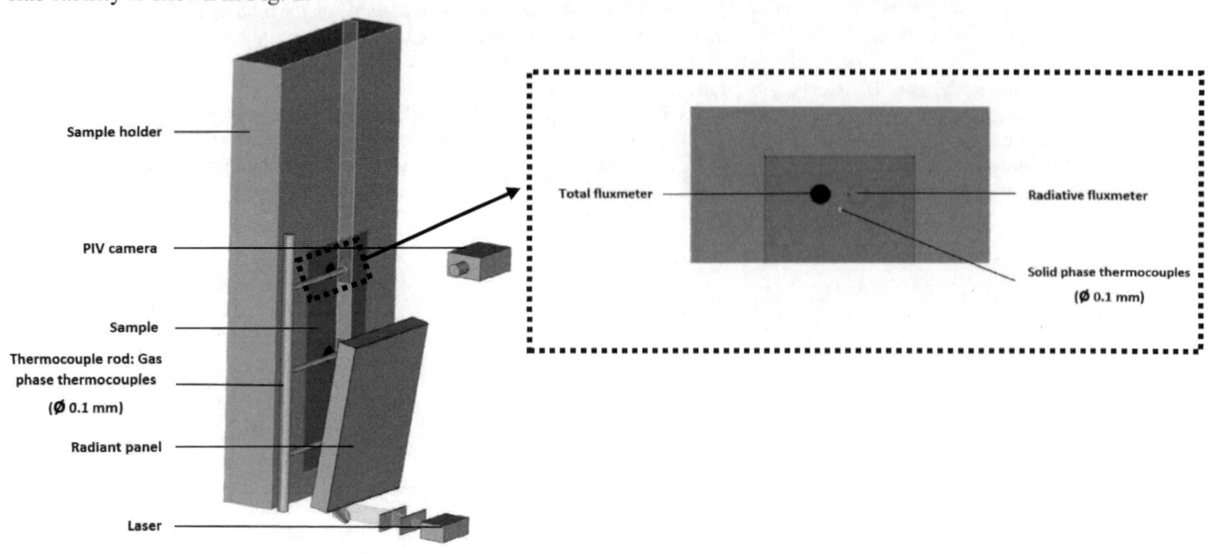

Fig. 2: Shematic representation of the 1D vertical propagation.

Similar instrumentation is used for the IMO-LIFT and radiant panel bench. Velocity and temperature measurements will be performed inside the reactive boundary layer. The average velocities (horizontal and vertical), as well as their fluctuations, will be measured thanks to the 'Particle Image Velocity' laser diagnostic and fine thermocouples (diameter: 0.1 mm) placed at different heights. Thus, two total and two radiative flux meters will be used at two different heights. This makes it possible to understand the phenomena controlling the mixing of reactants (convection, turbulent diffusion, mass diffusion), as well as the role of convection on heat transfer at the interface. The evaluation of the residence time, as well as the gas analysis using a Fourier Transform Infrared (FTIR), will lead to the determination of the chemistry regime. In addition, fine thermocouples placed along the surface and into the thickness of the solid phase will allow to accurately mapping the evolution of the temperature inside the slab. This step allows to observe and to evaluate key phenomena driving the fire kinetic that must be modelled.

Tests at smaller scale are performed mainly in order to get input data for simulations. Results achieved for TGA apparatus, coupled with a particle swarms analysis should give the mechanism and kinetic of decomposition with the Arrhenius parameters of the solid phase. First experimental results are shown in Fig. 3. The numerical validation of the model with an increase in scale and in complexity is in progress.

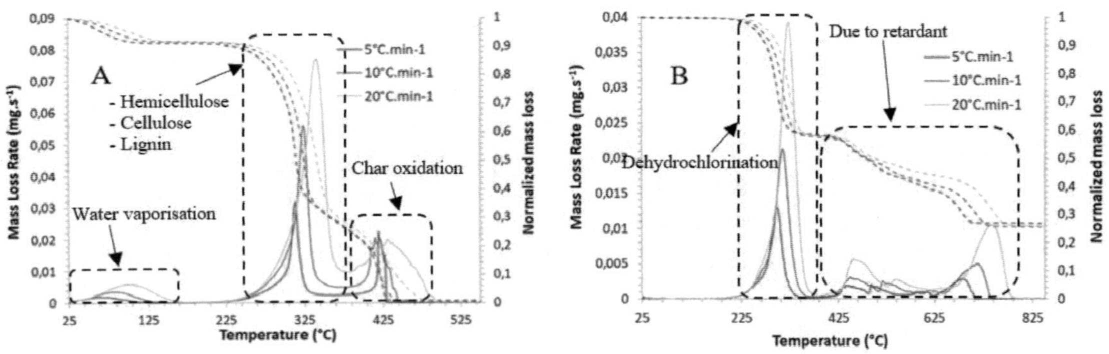

Fig. 3: TGA experimental results for pine wood (A), and PVC (B).

The model of pyrolysis is validated at material scale using the cone calorimeter. At this scale, a 1D thermal and mass diffusion is assumed in the solid phase. In the gaseous phase, the cone calorimeter permits to study flame development, without propagation. Three thermocouples (for pine wood only), placed at the surface, in the middle of the thickness and at the back of the solid will permits to validate the mechanism of decomposition taking into account the gradient of temperature into the sample. Also, an analysis of gas emitted before ignition should allow the understanding on gas concentration and on which gas is the precursor for ignition, this analysis is performed by a GCMS apparatus. This apparatus measures gas production closed to the solid. It differs from the FTIR analyser, which has a measurement point far from the solid. Once the numerical model validated at the material scale, with a good estimation of the HRR, the model will be extrapolated and validated at the medium scale by the implementation of a pathway representing small-scale phenomena on coarser mesh. The validation will lead to identify pertinent numerical models and code recommendations.

Experimental investigations using single burning item (SBI) and room corner benches will allow the identification on modifications of relevant physics due to material assembly modes, as well as angle effects. These experiments will lead to the adaptation and validation of numerical models that finally will be used to predict fire behaviour and HRR at real scale using large calorimetric hood, in construction type facilities as stairs or corridor. Experiments performed at this scale should permit to validate the models and to identify the main parameters used in FSE (Heat release rate, calorific charge, etc.).

All these steps could help to understand processes involved in fire propagation in different configurations. Unchanged processes involved from the medium scale to the real one should be identified and represented in numerical simulations taking into account code limitations. Therefore, a first step for technical bases can be proposed including relevant input parameters, experimental tests to perform at least to get them and model recommendation as mesh size.

CONCLUSION

To conclude, this beginning PhD work should give a first approach for reaction to fire in FSE. It will develops numerical models taking into account relevant phenomena driving the fire kinetic and relevant parameters with the way to assess them. This approach will allow FSE studies to estimate fire propagation of a given fire source and other secondary fire in order to define fire safety rules.

REFERENCES

[1] NIST, Fire Dynamics Simulator User's Guide Sixth Edition – Version 6.6.0

[2] Kent J.H., Bilger R.W. (1973) Turbulent diffusion flame, *Symposium (international) on combustion,* https://doi.org/10.1016/S0082-0784(73)80058-X

[3] Quintiere J.G. (2006) Fundamentals of fire phenomena, *John Wiley & Sons*

[4] Stoliarov S.I., Crowley S., Lyon R.E., Linteris T. (2009) Predicion of the burning rates of non-charring polymers, *Combustion and Flame*

[5] Kempel F., Schartel B., Linteris G., Stoliarov S.I. Lyon R.E., Walters R.N., Hofmann A. (2012) Prediction of the mass loss rate of polymer materials: Impact of residue formation, *Combustion and Flame*

[6] Ren N., Wang. Y., Vilfayeau S., Trouvé A. (2016) Large eddy simulation of turbulent vertical wall fires supplied with gaseous fuel through porous burners, *Combustion and Flame*, http://dx.doi.org/10.1016/j.combustflame.2015.12.008

[7] Bustamante Valencia L. (2009) Etude expérimentale et numérique de la decomposition thermique des matériaux à trois échelles : Application à une mousse polyéther polyuréthane utilisée dans les meubles rembourrés.

[8] Marquis D., Pavageau M., Guillaume E. (2012) Multi-scale simulations of fire growth on a sandwish composite structure, *Journal of fire sciences*

[9] Camillo A. (2014) Multi-scale investigation of fire behavior of a seat and a wall panel from European railway transport system.

[10] Batiot B. (2014) Etude expérimentale et numérique de la décomposition thermique du bois résineux.

[11] Fateh T., Richard F., Rogaume T. (2014) Modeling of the pyrolysis of plywood exposed to heat fluxes under cone calorimeter *Fire safety sciences-proceedings of the eleventh international symposium*

[12] Marcilla A., Beltran M. (1995) Thermogravimetric kinetic study of poly(vinylchloride) pyrolysis, *Elsevier* 219-229, http://dx.doi.org/10.1016/0141-3910(95)00050-v.

[13] Yu, J., Sun, L., Ma, C., Qiao, Y., and Yao, H. (2016) Thermal degradation of PVC: A review, *Elsevier* http://dx.doi.org/10.1016/j.wasman.2015.11.041.

IOP Conf. Series: Journal of Physics: Conf. Series **1107** (2018) 052001 doi:10.1088/1742-6596/1107/5/052001

Multi-scale modelling of wood degradation using thermally thin wood plates

Guillaume GERANDI, Virginie TIHAY-FELICELLI, Paul-Antoine SANTONI, Valérie LEROY-CANCELLIERI and Dominique CANCELLIERI

CNRS UMR 6134 SPE, University of Corsica, France

gerandi_g@univ-corse.fr

ABSTRACT

The mass loss rate is an indicator which defines the mass source term of combustible gases that supply the flames and influences the propagation of wildland fires. In this work, we investigated the thermal degradation of two woods (oak and eucalyptus) using a multi-scale approach. The originality of this work lies in the fact that we used thermally thin plates to carry out the experiments at different scales. At matter scale, experiments were conducted using a thermogravimetric analyser (TGA) under oxidizing atmosphere at several heating rates. We focused on temperatures ranging from 150°C to 650°C in order to study the thermal degradation of dry wood and thus, avoid the dehydration of free water. A kinetic mechanism was determined by using the thermogravimetric results. It consists of four steps, each one based on an Arrhenius law. This mechanism assumes that dry wood is composed of hemicellulose, cellulose and lignin. Each of these components degrades into char, which is then oxidized and releases ashes. The kinetic parameters of the four reactions were calculated by using the gradient descent algorithm method. The kinetic mechanism gives an accurate representation of the mass loss of both woods obtained with TGA. The kinetic mechanism was then tested at material scale using thermally thin wood plates. The experiments were carried out with a cone calorimeter at higher heating rates than used in TGA. Radiant heat fluxes ranging from 20 to 28 kW/m² were imposed at the top of the fuel sample in order to avoir igntion. The mass loss was recorded as well as the temperature at the back surface of the wood, which was supposed to be equal to the temperature of the whole plate. For a radiant heat flux of 20 kW/m², the predicted mass loss and mass loss rate are close to the experimental results for temperatures lower than 400°C. Beyond this value, the prediction underestimated the mass loss. For a radiant heat flux higher than 22 kW/m², the model did not correctly predict the thermal degradation of the plate. This difference is probably due to the assumption of an equilibrium temperature within the plate. Above this level of heat flux, the assumption of a thermally fine plate is no longer valid an it is necessary to take into account the temperature gradient within the plate.

KEYWORDS: *kinetic model, multi-scale, thermally thin plate, thermal degradation*

NOMENCLATURE LISTING

A	pre-exponential factor (s^{-1})
E_a	activation energy ($kJ.mol^{-1}$)
MLR	mass loss rate ($g.s^{-1}$)
m	mass (g)
m_0	initial mass (g)
m_f	final mass (g)
n	reaction order
p	proportion of constituent
R	gas constant ($J.K^{-1}.mol^{-1}$)
T	temperature (K)

Greek

α	Degree of conversion
$\dot{\omega}$	reaction rate (s^{-1})
ν	stoichiometric factor

Subscripts

0	initial
f	final
rad	radiative
c	convective

INTRODUCTION

The mass loss rate of materials is an important indicator that defines the mass source term of combustibles gases that supply the flames and leads the propagation of wildland fires. In detailed models of forest fires, more or less complex kinetic mechanisms are implemented to describe the thermal degradation of vegetative fuels. The kinetic parameters of these mechanisms are often calculated from thermogravimetric experiments at different heating rates [1,2]. During thermogravimetric analysis (TGA), the conditions of degradation are indeed well known and controlled. Unfortunately, the conditions of decomposition in the TGA apparatus are not representative of those encountered in actual fires. The materials are indeed generally in powder form and the heating rates are much lower than those of actual fires. Therefore, before using them in physics-based codes of forest fires, it is necessary to test them at a scale, for which the heating rates are closer to those encountered in fires and for which heat and mass transfers occur. The cone calorimeter is a good way to carry out experiments at the material scale [3–5]. It allows measuring the evolution of mass loss as well as the mass loss rate as a function of time by reaching realistic heating rates because of significant heat fluxes. At this scale, heat and mass transfer phenomena occur in studied sample, especially for thermally thick solid. The knowledge of the thermal properties of the sample is therefore necessary to model correctly the temperature gradient into the sample. This rises a problem since it is difficult to know all these quantities when the material is degrading. Fateh and al. [5], circumvented this difficulty in their multi-scale study on plywood by estimating numerically the thermal properties (conductivity, emissivity, density and specific capacity) using the genetic algorithm of Gpyro.

Our work deals with the development of a kinetic mechanism for modeling the thermal degradation of wood under oxidizing atmosphere at different scales, which can be used in physics-based codes of forest fires. Therefore, this kinetic mechanism must be representative of the degradation at all scales but it must remain as simple as possible to avoid a significant increase of the computational time. The originality of our approach consists of using thermally thin wood plates to perform the experiments at matter and material scales. First of all, the kinetic mechanism was determined from TGA experiments done with small samples cut from thermally thin plates. Then, the kinetic parameters were tested at material scale. To focus on the kinetic aspects, instead of using thermally thick material as usually [3–5], we decided to limit as much as possible the temperature gradients in the studied material by using thermally thin wood plates for the experiments with the cone calorimeter. The temperature of the back surface of the plate was recorded and it was considered equal to the temperature of the entire plate. This original approach has two benefits at material scale. First, the use of thermally thin plates reduces the superposition in time of the thermal reactions of degradation within the sample. Secondly, the problem complexity is reduced since it is not necessary to calculate the temperature inside the plate. The temperature being measured experimentally, there is no bias introduced by its calculation and the estimation of the thermal parameters. These experimental conditions, although not representative of actual fire conditions, allow us to focus on the study of the kinetics of wood degradation and to use heating rates that are much higher than those used with thermogravimetric analyzers.

1. MATERIAL AND METHOD

For the experiments, untreated wooden sheets of 0.6 mm thick employed for wood veneer were used as samples. Two kinds of wood were studied: oak (*Quercus alba*) and eucalyptus (*Eucalyptus globulus*). TGA experiments were performed using a PerkinElmer TGA Pyris 1. The samples were cut with a hollow punch in the thermal thin plate to obtain samples of $m_0 = 5$ mg and were placed in a 33 µl open platinum crucible. The furnace was continuously purged with a flowing atmosphere of air at 30 ml/min. The samples were first dried and then heated from 150 to 650°C in order to avoid the devolatilization of free water and to focus on the degradation of dry wood. Experiments were performed with the following heating rates: 2°C/min, 5°C/min, 10°C/min, 20°C/min and 30°C/min. The sample temperature, which is controlled by a thermocouple, did not exhibit any systematic deviation from preset linear temperature programs. All experiments were repeated three times.

The experiments at material scale were carried out with a cone calorimeter apparatus. The sample holder placed horizontally was an open square 10 cm × 10 cm basket made of stainless steel. Thermally thin plates of wood 100 × 100 × 0.6 mm^3 were used as studied material and were placed upon a ceramic wool in the sample holder. Radiant heat fluxes ranging between 20 and 28 kW/m^2 were imposed at the top of the fuel sample. These fluxes were chosen in order to obtain fast heating rates and to avoid flaming combustion. The level of radiant heat flux was check with a heat flux meter prior to each experiment. Before each set of experiments, the load cell was also calibrated and checked using standard weights. Its accuracy was 0.1 g and the sampling frequency was equal to 1 Hz. The smoke extraction was performed with an exhaust fan at a flow rate of 24 l/s. Two K-type thermocouples (sampling frequency of 25 Hz) were placed on the back surface of the plate: one on the center and one in the middle at 1 cm from the edge. As the plates were thermally thin, we considered a unique value of the temperature corresponding to the average of the values recorded by both thermocouples. Since the thermocouples induce pressure on the load cell, the measurements of mass loss and temperature were performed separately. At least three repetitions were carried out for each kind of measurement.

2. THERMAL DEGRADATION AT MATTER SCALE

2.1. Description

Figure 1 presents the evolution as function of temperature of the non-dimensional mass loss and mass loss rate (MLR) obtained by TGA for the two woods for a heating rate of 5°C/min, representative of the other conditions. For both woods, the thermal decomposition takes place in four steps. Table 1 presents the rate of change of the non-dimensional mass loss and temperature for these steps at different heating rates. The first step appears between 200 and 300°C and corresponds to the degradation of hemicellulose [6]. The second step takes place between 320 and 400°C. It leads to the most significant peak of mass loss rate and corresponds to the degradation of cellulose [6]. In the third step a shoulder, which corresponds to the degradation of lignin [6] is visible around 350 and 450°C. The last step leads to a peak beyond 450°C and it corresponds to the char oxidation. For both woods, the thermal degradation follows the same trend. However, for oak the thermal degradation of each component appears at lower temperatures than that for eucalyptus, excepted for the first step. With regard to the intensity of the peaks, the main difference occurs for step 1 (hemicellulose). The peak of mass loss of step 1 is indeed 1.7 times higher for oak than eucalyptus. According to literature [7,8], oak contains more hemicellulose than eucalyptus and this can explain the variation observed between the curves.

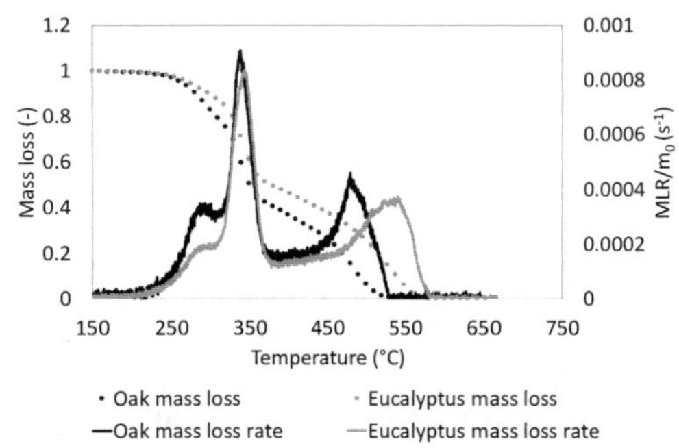

Fig.1. Evolution of non-dimensional mass loss and MLR with a heating rate of 5°C/min.

Table 1. Temperatures, MLR/m_0 (in parentheses) and residue obtained with thermogravimetric analysis.

Species	Heating rates (°C/min)	Temperature (°C) and MLR/m_0 (ms^{-1})				Residues (%)
		Step 1	Step 2	Step 3	Step 4	
Oak	2	282 (0.12)	321 (0.34)	367 (0.06)	465 (0.19)	0.62
	5	298 (0.32)	339 (0.85)	400 (0.13)	483 (0.42)	0.66
	10	310 (0.60)	358 (1.63)	406 (0.21)	518 (0.64)	0.80
	20	321 (1.27)	376 (3.51)	430 (0.38)	549 (0.95)	0.57
	30	331 (1.97)	383 (5.07)	450 (0.55)	551 (1.33)	0.72
Eucalyptus	2	282 (0.07)	335 (0.26)	386 (0.05)	515 (0.14)	1.65
	5	297 (0.18)	346 (0.76)	405 (0.12)	547 (0.31)	1.60
	10	305 (0.37)	360 (1.76)	431 (0.22)	573 (0.54)	1.95
	20	314 (0.76)	376 (3.48)	438 (0.41)	602 (0.89)	1.71
	30	325 (1.13)	385 (5.43)	450 (0.52)	660 (1.18)	1.72

2.2. Kinetic mechanism

Based on the previous results, a kinetic mechanism with four steps and based on an approach by constituents was chosen to model the thermal degradation of the woods. This approach has the advantage of being simple.

Moreover, the identification of the kinetic parameter on the mass loss curves of the wood allows considering the interactions between the constituents during the degradation. Therefore, the dry wood (DW) was assumed to consist of hemicellulose, cellulose and lignin:

$$DW = p_1\,Hemicellulose + p_2\,Cellulose + p_3\,Lignin \quad (1)$$

Where p_1 , p_2 and p_3 are the contribution of each constituent of the wood. The degradation of each component was then represented as follows:

$$Hemicellulose \xrightarrow{1} v_1 Char + (1-v_1)Gas \quad (2a)$$

$$Cellulose \xrightarrow{2} v_2 Char + (1-v_2)Gas \quad (2b)$$

$$Lignin \xrightarrow{3} v_3 Char + (1-v_3)Gas \quad (2c)$$

$$Char \xrightarrow{4} v_4 Ash + (1-v_4)Gas \quad (2d)$$

The degree of conversion α is defined as follows:

$$\alpha = \frac{m-m_0}{m_f-m_0} \quad (3)$$

Where m is the mass. The subscripts 0 and f correspond to the initial time and the final time respectively.

Each step of the mechanism corresponds to a global reaction described by the Arrhenius' law:

$$\dot{\omega}_i = (1-\alpha_i)^{n_i}.A_i.exp\left(-\frac{E_{ai}}{R.T}\right) \quad \text{for } 1 \leq i \leq 3 \quad (4a)$$

$$\dot{\omega}_4 = (p_1\alpha_1 + p_2\alpha_2 + p_3\alpha_3 - \alpha_4)^{n_4}.A_4.exp\left(-\frac{Ea_4}{R.T}\right) \quad (4b)$$

The total conversion rate was then obtained with the following equation:

$$\frac{d\alpha}{dt} = (1-v_1)\dot{\omega}_1 + (1-v_2)\,\dot{\omega}_2 + (1-v_3)\,\dot{\omega}_3 + (p_1v_1 + p_2v_2 + p_3v_3)\,\dot{\omega}_4 \quad (5)$$

The kinetic parameters (E_a, A, n and v) and the constituent proportions were determined by using the gradient descent algorithm method. Table 2 presents the values obtained for both woods. We can see that the kinetic parameters are in the same order of magnitude for these woods. The activation energies are around 160 kJ/mol, 190 kJ/mol and 169 kJ/mol for the degradation of hemicellulose, cellulose and lignin respectively. These values are in agreement with the literature [9,10]. Regarding the char oxidation, the activation energies are around 120 kJ/mol for both woods. These values are close to the results of Cancellieri and al. [11] but are much lower than those found by Conesa and al. [12] and Agrawal and al. [13], for which the activation energies for char oxidation are between 179 and 219 kJ/mol. Figure 2 presents the comparison of the degree of conversion and the conversion rate for oak and eucalyptus obtained experimentally at 5°C/min and calculated with the kinetic mechanism. The experimental data and the simulations are very close for both woods and it is also the case for the higher heating rates investigated here. Only a few variations appear at the maximum conversion rate and for the char oxidation, for which the kinetic model tends to slightly underestimate the conversion rate. This discrepancy is likely due to the use of a very simple mechanism to model this step. To improve the results at this scale, a more complex mechanism for char oxidation considering the oxygen level will be tested in the future. Despite these slight differences, the degree of conversion is well represented for the different stages of the thermal degradation. The best agreement was obtained for eucalyptus.

Table 2. Kinetic parameters for oak and eucalyptus using constituent approach.

Species	Reactions	n_i	$\ln(A_i)$ (s^{-1})	E_{ai} (kJ/mol)	v_i	p_i
Oak	1	1.425	29.87	160.99	0.407	0.262
	2	1.439	31.91	189.83	0.144	0.552
	3	1.461	21.82	168.39	0.157	0.186
	4	0.515	12.06	117.24	0.001	-
Eucalyptus	1	1.598	29.76	159.34	0.556	0.224
	2	1.466	31.93	191.06	0.179	0.546
	3	1.742	22.13	169.32	0.346	0.230
	4	0.498	11.63	121.45	0.001	-

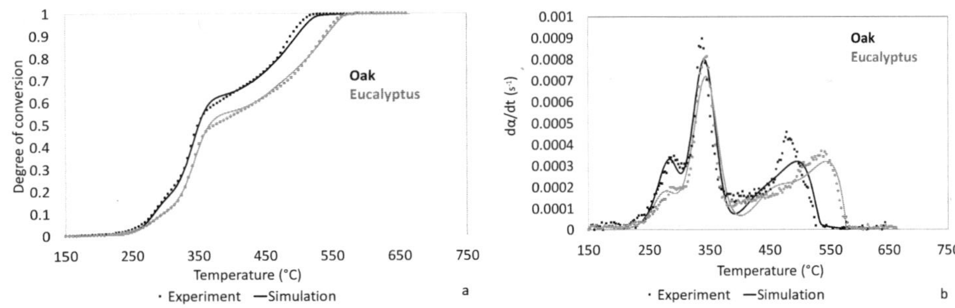

Fig. 2. a) Degree of conversion and b) Conversion rate – obtained by TGA for oak and eucalyptus at a heating rate of 5°C/min.

3. THERMAL DEGRADATION AT THE MATERIAL SCALE

3.1. Description

Figure 3 presents the time evolution of the temperature and mass loss for the plates of oak and eucalyptus for a radiant heat flux of 20 kW/m². The temperature of the plate increases rapidly to reach a maximum value of 540°C for eucalyptus and 580°C for oak. The heating rate is more significant at the beginning of the experiment with a maximum value of 600°C/min and it decreases after. The mass loss begins around 40 s which corresponds to a temperature of 260°C and it reaches a maximal value around 80 s at 360°C. These results are consistent with the TGA experiments. The mass loss of oak plates is more significant than that of eucalyptus plates. At this heat flux, eucalyptus plate only loses 73% of its initial mass. This behavior can be explained by the fact that the temperature of the plate does not exceed 540°C. According to TGA experiments, the maximal mass loss due to char oxidation occurs after 600°C for heating rates higher than 20°C/min (Table 1). Given the heating rates encountered during the cone calorimeter experiments (Fig. 3b), the char cannot therefore be completely oxidized.

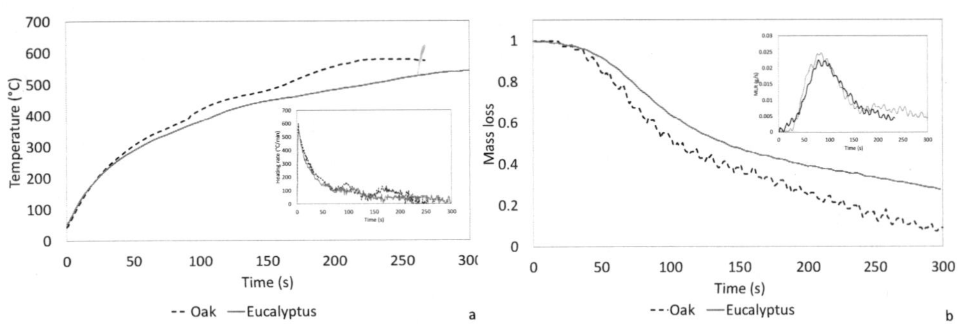

Fig. 3. Time evolution of a) the temperature and heating rate and b) the mass loss and mass loss rate – for a radiant heat flux of 20 kW/m².

3.2. Test of the set of kinetic parameters obtained with TGA experiments

The kinetic mechanism determined from TGA experiments was tested at the material scale. For this, the predicted mass losses were compared to the experimental values for the different heat fluxes and both woods. The same trends were observed for oak and eucalyptus. Figure 4.a and b present the results obtained for the plate of eucalyptus for respectively 20 and 28 kW/m². These heat fluxes were chosen since they are representative of two different behaviors. Until a value of 22 kW/m², the predicted mass loss is close to the experimental one for temperature lower than 400°C. Beyond this threshold, the model overestimates the mass loss. This discrepancy corresponds to the same phenomenon observed at matter scale and can be attributed to the model of char oxidation. For radiant heat fluxes higher than 25 kW/m², the predicted mass loss is delayed in comparison to the experiment. To explain this difference, the global Biot numbers [14] were calculated for each level of heat flux (Fig. 4c). Until 25 s (corresponding to Bi=0.11 and a temperature of 220°C), the evolution of the Biot as function of time is similar whatever the heat flux. After this threshold, two behaviors can be distinguished. For heat fluxes lower than 22kW/m², the Biot number increases quasi-linearly with time until it reaches an asymptotic value. For heat fluxes higher than 25 kW/m², a slope change appears leading to a significant increase of the Biot number. This change of behavior on the curves of Biot number coincides with the moment at which the predicted and observed mass losses diverge at 28 kW/m². Therefore, for a heat flux higher than 25 kW/m², the plates can no longer be considered as thermally thin and using the back face temperature is no more representative for modeling the plate mass loss. Thus, it is necessary to take into account the temperature gradient within the plate to model the degradation at material scale for such heat fluxes.

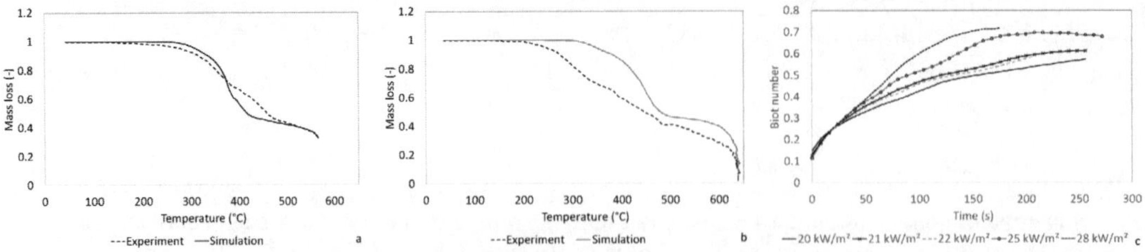

Fig.4. Evolution of a) the mass loss of eucalyptus for a heat flux of 20 kW/m² and b) the mass loss of eucalyptus for a heat flux of 28 kW/m² c) the Biot number for eucalyptus

CONCLUSION

The thermal degradation of two kinds of thermally thin wood plates was investigated by using a multi-scale approach. The following results were obtained:

- A kinetic mechanism with four steps was developed to model the thermal degradation of oak and eucalyptus. This model predicts efficiently the thermal decomposition for the levels of heating rates investigated, even if some differences occurred at the maximum conversion rate and for the char oxidation, which are underestimated.
- The use of thermally thin wood plates as samples is original and it has some benefits and drawbacks. It allowed studying the thermal degradation without the knowledge of the thermal properties of the wood and the solving of a thermal balance. However, the test of the set of kinetic parameters obtained from TGA experiments was only possible for levels of radiant heat flux lower than 22 kW/m².

REFERENCES

[1] V. Leroy, D. Cancellieri, E. Leoni, Thermal degradation of ligno-cellulosic fuels: DSC and TGA studies, Thermochim. Acta. 451 (2006) 131–138. doi:10.1016/j.tca.2006.09.017.

[2] T. Fateh, F. Richard, J. Zaida, T. Rogaume, P. Joseph, Multi-scale experimental investigations of the thermal degradation of pine needles, Fire Mater. 41 (2017) 654–674. doi:10.1002/fam.2407.

[3] C. Di Blasi, E.G. Hernandez, A. Santoro, Radiative Pyrolysis of Single Moist Wood Particles, Ind. Eng. Chem. Res. 39 (2000) 873–882. doi:10.1021/ie990720i.

[4] M.J. Spearpoint, J.G. Quintiere, Predicting the burning of wood using an integral model, Combust. Flame. 123 (2000) 308–325. doi:10.1016/S0010-2180(00)00162-0.

[5] T. Fateh, T. Rogaume, F. Richard, Multi-scale modeling of the thermal decomposition of fire retardant plywood, Fire Saf. J. 64 (2014) 36–47. doi:10.1016/j.firesaf.2014.01.007.

[6] M.G. Grønli, G. Várhegyi, C.D. Blasi, Thermogravimetric analysis and devolatilization kinetics of wood, Ind. Eng. Chem. Res. 41 (2002) 4201–4208.

[7] M.R. da Silva, G. de O. Machado, J. Deiner, C. Calil Junior, Permeability measuremens of brazilian Eucalyptus, Mater. Res. 13 (2010) 281–286. doi:10.1590/S1516-14392010000300002.

[8] R.C. Pettersen, The Chemical Composition of Wood, in: Chem. Solid Wood, American Chemical Society, 1984: pp. 57–126. doi:10.1021/ba-1984-0207.ch002.

[9] Z. Chen, M. Hu, X. Zhu, D. Guo, S. Liu, Z. Hu, B. Xiao, J. Wang, M. Laghari, Characteristics and kinetic study on pyrolysis of five lignocellulosic biomass via thermogravimetric analysis, Bioresour. Technol. 192 (2015) 441–450. doi:10.1016/j.biortech.2015.05.062.

[10] D.F. Arseneau, Competitive Reactions in the Thermal Decomposition of Cellulose, Can. J. Chem. 49 (1971) 632–638. doi:10.1139/v71-101.

[11] D. Cancellieri, E. Innocenti, V. Leroy-Cancellieri, WinGPYRO: A software platform for kinetic study of forest fuels, Fire Saf. J. 58 (2013) 103–111. doi:10.1016/j.firesaf.2013.01.005.

[12] J.A. Conesa, J. Caballero, A. Marcilla, R. Font, Analysis of different kinetic models in the dynamic pyrolysis of cellulose, Thermochim. Acta. 254 (1995) 175–192. doi:10.1016/0040-6031(94)02102-T.

[13] R.K. Agrawal, Kinetics of reactions involved in pyrolysis of cellulose I. The three reaction model, Can. J. Chem. Eng. 66 (1988) 403–412. doi:10.1002/cjce.5450660309.

[14] A. Lamorlette, F. Candelier, Thermal behavior of solid particles at ignition: Theoretical limit between thermally thick and thin solids, Int. J. Heat Mass Transf. 82 (2015) 117–122. doi:10.1016/j.ijheatmasstransfer.2014.11.037.

Experimental and numerical investigation of shrub combustion

Jean-Baptiste TRAMONI, Paul-Antoine SANTONI and Frédéric MORANDINI

CNRS UMR 6134 SPE, University of Corsica, France

tramoni_jb@univ-corse.fr; santoni_p@univ-corse.fr; morandini_f@univ-corse.fr

ABSTRACT

The use of physics-based models to simulate vegetation fires has been steadily increasing over the last decade. These models offer the possibility to study quickly and at a low cost the combustion of vegetation according to a wide set of parameters which are difficult to control experimentally. However, the validation and improvement of these models is still ongoing and coupled experimental / numerical approaches are needed.

In a previous work, a validation of the numerical code WFDS was done for the spreading of fires across pine needles fuel beds. It highlighted the ability of WFDS to reproduce the combustion dynamics and the related physical quantities (heat release rate, mass loss rate, rate of spread) during the growing and steady-state phases of the fire. The objective of the present work is to test WFDS for the prediction of the combustion of a natural shrub, which is a more complex fuel than a fuel bed. As a first step, 3 shrubs of rockroses (*Cistus monspeliensis*) were characterized. The distribution of the different classes of particles that compose the shrubs was determined and used as input for the shrub model. A series of 41 fire tests were conducted on the shrubs to investigate their flammability. Ignition was performed using radiant panel. The measurements of the main physical quantities were also used as a comparison basis for the numerical approach. A numerical model of shrub was developed from the characterization results. Then, the combustion of the shrub was simulated with WFDS. The grid mesh resolution needed for the simulation was estimated from the extinction length within the shrub crown and from the heat release rate (HRR) during the fire tests. Finally, experimental measurements and predictions were compared to determine the capability of WFDS to reproduce the combustion of the shrubs.

The experimental investigation of the shrub flammability showed four different regimes of combustion governed mainly by the location of ignition. Simulations were run for the flammability regime that exhibits the best reproducibility. The comparison between simulation and experimental results demonstrated a good ability of WFDS to assess the most relevant fire properties, namely the peak of HRR, flame duration and consumption rate. The main differences between the predictions and observations were identified during the ignition and extinction phases. A plateau of HRR is predicted shortly after the ignition, limiting the fire growth. In addition, the duration of the predicted smoldering phase is too short which causes a total extinction of the fire few seconds after the end of the flaming phase.

KEYWORDS: Flammability, Shrub combustion, Fire behavior

IOP Conf. Series: Journal of Physics: Conf. Series **1107** (2018) 052002 doi:10.1088/1742-6596/1107/5/052002

INTRODUCTION

Continuous efforts are being made towards the understanding of the spread of wildfires which increasingly threaten ecosystems, houses and people. The improvement of the computational resources and the ability of the physics-based models to predict the intricate processes involved in combustion, has led to an increase in the use of numerical approaches to study these fires. However, the development of these models is still pending and they need to be compared over a wide range of configurations in order to be improved. The validation tests are commonly performed across simple fuel structure, mainly fuel bed (needles) or grass consisting of a single type of particles.

In particular, the Wildland-urban interface Fire Dynamics Simulator (WFDS) was tested at field scale with large experimental grassland fires [1]. However, this first modelling approach considered a boundary fuel model. A first improvement of WFDS [2] has rendered the model capable of representing the spatial distribution of different fuel elements (leaves, twigs) within the vegetation. The char oxidation was nevertheless not considered at this stage. Then a modification of the thermal degradation model was done in order to include the char combustion processes and refine the gasification law [3]. In this work WFDS was validated for fire spreading across beds of pine needles.

In order to extend the testing of WFDS, the present study aims to investigate the combustion of a more complex living vegetation. Shrubs of rockroses were considered, which are highly porous media composed of different classes of particles with different thickness and moisture contents. The vegetation samples were characterized to obtain the mass proportion of the different classes of particle sizes as well as their spatial distribution. These results were used to develop a numerical model of a typical shrub of rockrose. Then, experiments were done in order to assess their flammability and to collect data on the most relevant physical quantities to test WFDS. Different burning behaviors depending on the ignition location, time to ignition and vegetation mass distribution, were put to the fore. They were classified into four distinct types of flammability. The simulations were done for one of the identified flammability group.

CHARATERIZATION OF THE SHRUB STRUCTURE

A good knowledge of the structure and mass distribution of the different classes of particles within the shrubs of rockrose is necessary to accurately predict their combustion with WFDS. Therefore, a characterization of 3 shrubs with similar size was performed. Their average height and crown diameter were 1.23 and 0.68 m, respectively. The shrubs were placed inside a 1.45 m high, 0.9 m long and 0.9 m wide cubic metallic mesh structure allowing the destructive sampling of the vegetation (252 cubes of 0.15 m side). The shrubs were cut off following the pattern of this grid. The particles in each sampling cube were then oven dried and separated following their thickness (leaves, dead twigs with 0-2 mm diameter, live twigs with 0-2 mm, 2-4 mm, 4-6 mm and 6-25 mm diameters, respectively). The dry mass of each particle class was measured. It should be pointed out that the bases of the shrubs (up to 0.25 m height) were not considered in this characterization since they were composed of roots and large particles (diameter greater than 25 mm) expected not to participate in the early combustion processes, for the fire intensities considered in the present study. The mass proportion of each particle class is denoted by:

$$\Gamma_k = \frac{m_{kd}}{m_d} \tag{1}$$

where Γ_k is the mass fraction of particle class k in the shrub, m_{kd} represents the dry mass of the particle class k and m_d is the total dry mass of the shrub. A similar mass distribution was measured for the 3 shrubs of rockrose. Low standard deviations were obtained for all particle classes. Experimental results exhibit that a shrub of rockrose is composed of 22.1 ± 0.8 % of leaves, 5.4 ± 0.8 % and 20.8 ± 1.3 % of dead and live twigs with a diameter between 0 and 2 mm respectively, 18.3 ± 0.9 % of live twigs with 2-4 mm diameter, 16.4 ± 1.6 % of live twigs with 4-6 mm diameter and 17.0 ± 4.0 % of live branches with 6-25 mm diameter. The shrub of rockrose is thus made of 48.3 ± 2.9 % of particle classes participating actively to the combustion [4].

The spatial distribution of the different classes of particles was determined following the height of the shrub, z. The mass distribution function is defined as:

$$\zeta_{kd}(z) = \frac{m_{kd}(z)}{m_d(z)} \tag{2}$$

IOP Conf. Series: Journal of Physics: Conf. Series **1107** (2018) 052002 doi:10.1088/1742-6596/1107/5/052002

where $m_{kd}(z)$ is the dry mass of the particle class k ranging between $z - \frac{\delta}{2}$ and $z + \frac{\delta}{2}$ ($\delta = 15$ cm) and $m_d(z) = \sum_k m_{kd}(z)$.

Three parts were identified in the shrub according to the height: the base (from 0.27 to 0.59 m high), the intermediate zone (from 0.59 to 0.91 m high) and the crown (from 0.91 to 1.23 m high). It was observed that 78% of the mass of the leaves is in the crown. The particle class with 0-2 mm diameter is present at 82% in the crown and in the upper half of the intermediate part. For the 2-4 mm diameter class, we observed that 58% of their mass is in the intermediate part of the shrub. The rest of the 2-4 mm family was distributed almost identically in the other two parts of the shrub. Particles of the 4-6 mm diameter class are mainly present in the lower half of the intermediate part (37% of the mass of these particles) and in the base (46% of the mass of these particles). Finally, 95% of the total mass of the class 6-25 mm is located in the base.

STUDY OF THE SHRUB COMBUSTION

The fire experiments were carried out under a Large-Scale Heat Release apparatus (LSHR) (Figure 1). This device allows the measurement of the HRR over time (sampling rate 1 Hz). The shrubs were placed on a load cell. The smoke was extracted through a 3 m × 3 m hood with a volume flow rate of 1 m^3/s. Two sets of two radiant panels, placed perpendicularly from each other to maximize the emitted heat flux, were used to preheat and ignite the shrub of rockroses. The radiant panels were heated at a temperature of 520°C which led to a radiant heat flux density of 20 kW/m^2. 41 fire tests were conducted and 29 were retained for the analysis. The shrubs had an average height of 1.25 m, a crown base height of 0.95 m and a crown diameter between 0.6 and 0.8 m. Their average mass was around 2 kg (including base). The shrubs were conditioned indoor at ambient temperature and humidity (25°C and 50%) for a period of 48 hours in order to obtain the expected range of moisture content (3-20 % for the leaves). The time to ignition was noted for each experiment. The flame duration and the rate of consumption of the fuel were also measured.

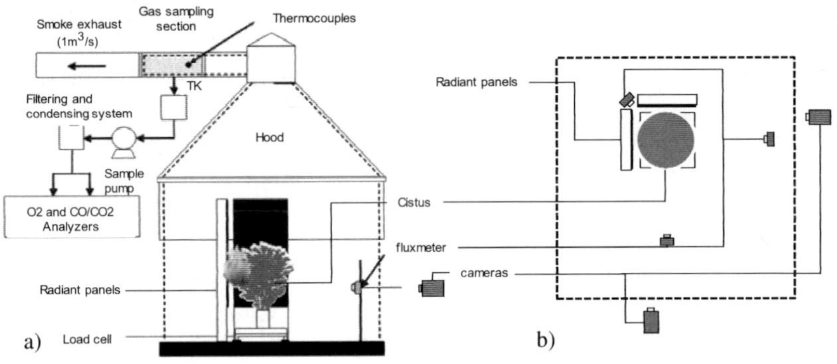

Figure 1 : Layout of the experimental set-up, a) side view b) top view

In order to study the flammability of the shrub four criteria were used namely ignitability, combustibility, sustainability and consummability. Then a Principal Component Analysis was done with these criteria and revealed the presence of four flammability regimes mainly dependant on the ignition position, mass distribution and time to ignition. An ignition zone in the higher part of the crown led to a weak or a moderate flammability. An ignition in the lower part of the shrub led to a high and fast flammability or to a high flammability with strong consumption of mass. As an example, the high and fast flammability regime presents a fast growing after the ignition with a HRR peak of 228 ± 15 kW and a flame duration of 44 ± 9 s. The rate of consumption of the shrubs was 25 ± 6 % which represents a crown consumption in the range of 57 - 97 %. The measurements of the HRR over time for the different tests exhibited a good repeatability during the phases of fire growth and flaming combustion but they showed a greater variability for the extinction phase. Indeed, the ignition of large diameter (4-6 mm) particles sometimes occurred increasing the duration of the extinction phase without impact on the peak HRR and growth time.

SIMULATION METHOD

Overview of WFDS

WFDS is a physics-based model of fire spread which purpose is to represent the fine mechanisms governing the fluid mechanics, the combustion and the heat transfers. [2]. The thermal degradation model was modified and coupled to the gas phase equations of heat and mass transfer. This degradation model is based on Arrhenius laws for the desiccation, pyrolysis and char oxidation [5]. Thus, the predicted HRR over time is the sum of the heat released by the flame (gas phase) and the HRR due to the char oxidation (solid phase). The modified degradation laws are:

$$R_{H_2O} = \rho_{b,H_2O} A_{H_2O} T^{-\frac{1}{2}} e^{-\frac{E_{H_2O}}{T_k}} \quad (3)$$

$$R_{pyr} = \rho_{b,dry} A_{pyr} e^{-\frac{E_{pyr}}{T_k}} \quad (4)$$

$$R_{char} = \frac{A_{char}}{\nu_{O_2,char}} \rho_g Y_{O_2} \sigma_k \beta_k e^{-\frac{E_{char}}{T_k}} (1 + \beta_{char}\sqrt{Re_k}) \quad (5)$$

where Re_k represents the Reynolds number, $Re_k = 2\rho_g |u| r_k/\mu$ with $r_k = \frac{2}{\sigma_k}$, σ_k and β_k are the surface to volume ratio and compactness of the class k of particles, respectively. The equation governing the temperature of the fine particles, T_k, was also modified as follow:

$$\rho_b c_{p,k} \frac{dT_k}{dt} = -\Delta h_{vap} R_{H_2O} - \Delta h_{pyr} R_{pyr} - \alpha_{char} \Delta h_{char} R_{char} - \langle \dot{q}'''_{c,b} \rangle_{Vb} - \langle \nabla . \dot{q}''_{r,b} \rangle_{Vb} \quad (6)$$

where the three first terms of the right-hand-side represent the desiccation (endothermic), the pyrolysis (endothermic) and the char oxidation (exothermic), respectively. The fuel element bulk contributions of convective and radiative heat transfer are represented by the fourth and fifth terms, respectively. The heats of vaporization, pyrolysis and char oxidation are $\Delta h_{vap} = 2.259 \times 10^3\ kJ\ kg^{-1}$, $\Delta h_{pyr} = 418\ kJ\ kg^{-1}$[5] and $\Delta h_{char} = -32 \times 10^3\ kJ\ kg^{-1}$, respectively [6]. The dimensionless parameter α_{char} is the fraction of the heat generated by the char oxidation which is absorbed by the solid fuel and $(1 - \alpha_{char})$ by the gaseous phase. In this case and in other studies, it is considered empirically that $\alpha_{char} = 0.5$ [5]. Further details on the WFDS and the modifications done are in Perez-Ramirez et al. [3].

Shrub model

The mass distribution obtained during the characterization provided reliable data for the modeling of the shrub of rockrose. However, the Cartesian metallic mesh structure used did not match perfectly with the shrubs envelope. As a result, an underestimation of the bulk density occurs on their edges. Thus, a more realistic model composed of 6 frustums of 15 cm height was used. These frustums were chosen to match at best as possible with the average height and diameter along height of the shrub. WFDS allows the use of several classes of particles. Thus, the numerical shrub was modeled as the superposition of six fuel layers, each composed of one class of particles identified in the previous section. Finally, the bulk density ρ_{bk} of each particle class (Table 1) was determined from the mass measurements of each family at the height z and the associated frustum volume $V(z)$:

$$\rho_{bk}(z) = \frac{m_{kd}(z)}{V(z)} \quad (7)$$

Concerning the vegetation, the surface-to-volume ratio for the leaves, the 0-2 mm, 2-4 mm, 4-6 mm and 6-25 mm classes were 2081, 1733 ,1000, 666 and 400 m^{-1}, respectively. The density of the particles was 478 and 961 $kg.m^{-3}$ for the leaves and the twigs, respectively. The mass fraction of consumed fuel turned into char and ashes was 30 % [7] and 13%. Finally, the radiative fraction (27 %) was estimated with the method detailed in[8]. The other model inputs ($A_{pyr} = 36300\ s^{-1}$, $E_{pyr} = 7250\ K$ ...) previously used [3] were kept in the present study.

Table 1 : Bulk density of each class of particles following the height

Height (m)	Leaves ($kg.m^{-3}$)	0-2 mm dead ($kg.m^{-3}$)	0-2 mm live twigs ($kg.m^{-3}$)	2-4 mm live twigs ($kg.m^{-3}$)	4-6 mm live twigs ($kg.m^{-3}$)	6-25 mm live twigs ($kg.m^{-3}$)
1.10 – 1.25	3.99	0	1.63	0.04	0	0
0.95 – 1.10	2.73	0.13	2.45	0.99	0.08	0
0.80 – 0.95	1.32	0.31	2.79	2.34	1.07	0
0.65 – 0.80	0.52	1.04	1.09	2.46	2.95	0.37
0.50 – 0.65	0.48	1.04	0.49	2.13	3.83	3.54
0.35 – 0.50	0.30	0.38	1.18	1.41	2.48	12.27

Mesh resolution within the shrub and the reacting flow

The grid resolution needed to perform the simulations was estimated from two characteristic length scales associated with two physical phenomena involved in the combustion. The first length scale corresponds to the extinction length, δ_R, which represents the absorption of radiation by vegetation. δ_R is given by:

$$\delta_R = \frac{4}{\beta_k \sigma_k} \tag{8}$$

The grid size used within the shrub, dx_b must be lower than δ_R. In this regard Mell et al. [3] proposed to use at most a grid cell size three times lower that δ_R ($dx_{b,\delta_R/3}$). A grid cell size five times lower that δ_R ($dx_{b,\delta_R/5}$) is also recommended. The second characteristic length scale allows to determine the grid size in the flow field for the buoyant plume and the flame regions. The grid resolution d_x is given by the dimensionless expression z_c/d_x proposed by McGrattan et al. [9] where z_c represents the characteristic fire diameter. z_c depends on the fire HRR, $\dot{q}$, the density ρ_∞, specific heat c_p, ambient temperature T_∞ and acceleration of gravity g:

$$z_c = \left(\frac{\dot{q}}{\rho_\infty c_p T_\infty \sqrt{g}}\right)^{\frac{2}{5}} \tag{9}$$

McGrattan et al. [4] suggested to choose a ratio z_c/d_x in the range of 4-16. The extinction length was calculated from the numerical shrub characteristics. A value of 0.144 m was obtained for δ_R which corresponds to a grid size lower than 0.048 m and 0.029 m for $dx_{b,\delta_R/3}$ and $dx_{b,\delta_R/5}$, respectively. Concerning the flow computation in flame and buoyant plume regions, the mesh size was calculated from the HRR peaks obtained for the four flammability regimes. The minimum value of z_c obtained for the low flammability, was 0.356 m for an average HRR peak of 100 kW. McGrattan et al. [9] suggested a mesh size in the range of 0.022 - 0.089 m. Based on these results, the mesh size was chosen as the minimum of both criteria (0.02 m) for the different domain regions.

COMPARISON BETWEEN THE PREDICTED AND MEASURED HRR

The test of the physic-based model WFDS presented in this study was conducted for the flammability regime with the best reproducibility, which corresponds to the "high and fast flammability". The LSHR domain, including the extraction hood and the radiant panels, were modelled to fully reproduce the experimental conditions. A preliminary calculation was carried out without the shrub in order to check the modelling of the radiant heat flux emitted by the radiant panels. Predicted and measured radiant heat flux density at a distance of 1 m from the panel were compared and a good agreement was found. Then, the numerical shrub model was configured in front of the radiant panels heated at 520°C. Under these conditions, the simulated amount of energy led to an auto-ignition of the shrub, but it was not enough to sustain the combustion and the spreading within the shrub. Thereby, a numerical piloted ignition, that consist of fuel element particles kept at a temperature of 1000°C, was added. The size of the ignitor was 10×20×20 cm^3. It was located in the lower part of the crown like in the experiments. The volume of the ignitor corresponds to the size of the flame observed in the experiment at the time of ignition and the duration was fitted in order to sustain the fire spread (15 s).

Figure 2 displays the predicted and measured HRR over time. The predicted and measured HRR peaks (mean) are in good agreement (230 kW and 228 ± 15 kW, respectively). In addition, the predicted flame duration (56 s) is slightly overestimated compared to the experimental one (44 ± 9 s). It should be noticed that the predicted total combustion duration (flaming and smouldering phases) is shorter than the observed one due to a too fast extinction of the char after the flameout. Concerning the fuel consumption, 32 % of the fuel was consumed in the simulation. This value is near the experimental one which was 25 ± 6 %. These overall results show the capability of WFDS to reproduce the main measured quantities related to the flammability. However, a plateau in the predicted HRR appeared shortly after the ignition and the extinction occurred too quickly after the flameout. This last phenomenon is probably due to a quicker cooling of the particles with fresh air after the flameout in the simulation compared to the experiments.

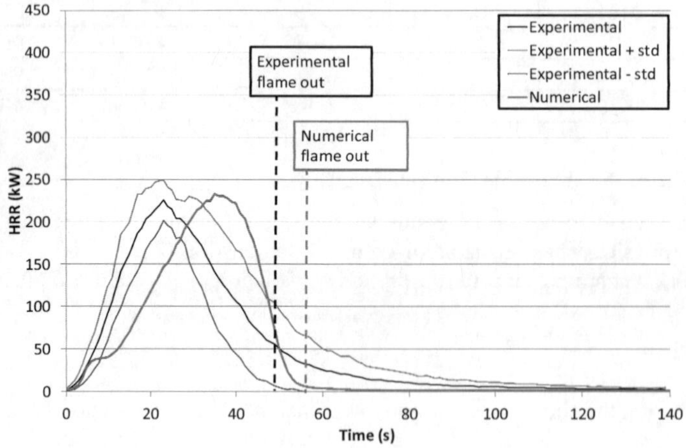

Figure 2 : Predicted and measured HRR of the "high and fast flammability" regime over time

CONCLUSION

In this paper the code WFDS was tested for the combustion of a living shrub. Shrubs of rockrose were characterized to obtain the mass proportion and distribution of different classes of particles in order to build a numerical model. Experiments carried out to study the flammability of these shrubs were used as a comparison basis. The numerical test was performed on the flammability that showed the best reproducibility. The mass distribution obtained during the characterization allowed us to develop a numerical shrub of rockrose. Finally, the simulations of the shrub combustion were performed. This study highlighted the capacity of WFDS to predict the peak HRR, flame duration and consumption rate. However, a plateau appeared after the ignition which alters the fire growth. The extinction of the smoldering phase was too fast which could be due to an overestimation of the convective cooling. An investigation on the causes of the plateau through more thorough investigation of the degradation laws (desiccation, pyrolysis and char oxidation) is currently pending.

REFERENCES

[1] W. Mell, M.A. Jenkins, J. Gould, P. Cheney, A physics-based approach to modelling grassland fires, Int. J. Wildland Fire. 16 (2007) 1. doi:10.1071/WF06002.
[2] W. Mell, A. Maranghides, R. McDermott, S.L. Manzello, Numerical simulation and experiments of burning douglas fir trees, Combust. Flame. 156 (2009) 2023–2041. doi:10.1016/j.combustflame.2009.06.015.
[3] Y. Perez-Ramirez, W.E. Mell, P.A. Santoni, J.B. Tramoni, F. Bosseur, Examination of WFDS in Modeling Spreading Fires in a Furniture Calorimeter, Fire Technol. 53 (2017) 1795–1832. doi:10.1007/s10694-017-0657-z.
[4] V. Tihay, P.A. Santoni, T. Barboni, L. Leonelli, Experimental and theoretical study of diameter effect on the ignition of cistus twigs, in: Adv. For. Fire Res., Imprensa da Universidade de Coimbra, 2014: pp. 179–189. doi:10.14195/978-989-26-0884-6_19.
[5] B. Porterie, J.L. Consalvi, A. Kaiss, J.C. Loraud, Predicting wildland fire behavior and emissions using a fine-scale physical model, Numer. Heat Transf. Part Appl. 47 (2005) 571–591. doi:10.1080/10407780590891362.
[6] A. Dahale, S. Ferguson, B. Shotorban, S. Mahalingam, Effects of distribution of bulk density and moisture content on shrub fires, Int. J. Wildland Fire. 22 (2013) 625. doi:10.1071/WF12040.
[7] V. Leroy, Contribution à la modélisation des feux de forêt: Cinétique de dégradation thermique et Cinétique de combustion des végétaux, PhD Thesis, Université Pascal Paoli, 2007.
[8] F. Morandini, Y. Perez-Ramirez, V. Tihay, P.-A. Santoni, T. Barboni, Radiant, convective and heat release characterization of vegetation fire, Int. J. Therm. Sci. 70 (2013) 83–91. doi:10.1016/j.ijthermalsci.2013.03.011.
[9] K. McGrattan, S. Hostikka, R. McDermott, J. Floyd, C. Weinschenk, K. Overholt, Fire dynamics simulator user's guide, NIST Spec. Publ. 1019 (2013) 6thEdition.

IOP Conf. Series: Journal of Physics: Conf. Series **1107** (2018) 052003 doi:10.1088/1742-6596/1107/5/052003

Laboratory Scale Experimental Study of Foam Suppression on Smouldering Combustion of a Tropical Peat

Nadhira Gilang Ratnasari[1], Afiri Dianti[1], Pither Palamba[1], Mohamad Lutfi Ramadhan[1], Gatot Prayogo[1], Agus Sunjarianto Pamitran[1], Yulianto Sulistyo Nugroho[1]

[1]*Department of Mechanical Engineering, Universitas Indonesia, Indonesia*

yulianto.nugroho@ui.ac.id

ABSTRACT

Solving peat fires problem continues to be a constant struggle in Indonesia as more hotspots are identified during the dry seasons. A number of research has been carried out to understand the most sufficient way to suppress peat fires using a range of different methods. This research was focused on the suppression of peat fire by using foaming-based suppression agent in the laboratory scale experiments. Experiments were carried out to explore the effect of foam suppression system on tropical peat fires. A solution of Class A Foam with a concentration of 3% was used to suppress Kalimantan peat fire with a density of ±0.3g/cm^3. Sample used in the experiments was taken from Tumbang Nusa Village, Pulang Pisau District, Central Kalimantan Province, with a coordinate of S: -3º47'34", E: 113º55'15". This experiment will explore how many foam layers of certain heights are needed in order to extinguish peat fire. Peat sample was put in a 10x10x10 cm^3 reactor, where a coil heater was turned on at 80 - 100W for 30 minutes to initiate a smouldering front. A set of type K thermocouples and infrared thermographs were used to explore the suppression mechanism that occurs. As the smouldering front moved away from the igniter to the other end of the reactor, foam with different thickness was poured on top of the peat to explore the effect of varying thickness on the suppression of peat fire. From the series of experiments, it was observed that there was a correlation between the thickness of the foam layer and the suppression of peat fires. The thickness of 10cm was found to be the most effective in terms of time at extinguishing peat fire. An average amount of 3.8L of foam per kilogram of Kalimantan peat was needed to extinguish peat fire. It was also found that the average amount of water needed to suppress peat fire using foam method was 3.4L/kg.

KEYWORDS:

smouldering, peat fire, fire suppression

INTRODUCTION

Wildland fires remain to be one of the most devastating environmental problems that occur across the globe. Not only it releases harmful substances into the air, such as carbon, but these chemicals also contribute highly to global warming. Peat is formed naturally as organic material goes through anaerobic decay in low pH conditions [1]. Indonesia is home to 20.6 million ha of peatlands. In 2016, the area of peatland in Indonesia was recorded to be as wide as 14.9 Million hectares which can be found across three main islands: Sumatera, Kalimantan and Papua [2].

Peat fire represents an example of smouldering fire that happens at low temperature, peaking at around 450 °C to 700 °C, and causes incomplete combustion [3]. This leads to a higher production rate of toxic gases and particulates compared to flaming combustion. Smouldering combustion is known to be flameless and moves slowly at a low and steady temperature [3]. The three main in situ methods to suppress peat fire have been identified by Hadden [4], which include cooling, smothering and burn-out. Water spray is an example of cooling method as it is able to bring down the temperature of burning fuels to below ignition temperature. However, the total amount of water needed to suppress peat fire can exceed 6 L/kg of peat [5]. Furthermore, this is caused by the change in peat characteristics when it is exposed to a very dry condition. Peat is known to have a hydrophilic characteristic which allows it to absorb water. However, in a very dry condition and the moisture content in peat lowers down to a certain point, peat is unable to absorb water and instead repels it [6]. With the lack of water in places where wildland fires occur, meeting the right amount of water needed to put out the fire may be problematic. On the other hand, Ohlemiller [7] finds that smothering method focuses on reducing the oxygen concentration below a critical level which ends the oxidation reactions.

Firefighting foam has long been a way of suppressing wildland fire. This research will explore the smothering mechanism to suppress peat fire using Class A Foam. Up until now there is still a lack of observation and research on how wildland fire, including peat fire, responds to foam application. There are many variables that may affect the performance of foam suppression system such as concentration of foam and the thickness of the foam layer. This research will observe the effect of the thickness of foam layer on the suppression of peat fires. Further research is being conducted to find out the effect of foam on peat fires suppression. Although the use of firefighting foam still sparks some controversy on the environmental impacts, it is still necessary to observe how smouldering peat reacts to foam.

EXPERIMENTAL METHODS

Peat Sample Preparation

The sample used in the experiment was taken from Tumbang Nusa Village, Pulang Pisau District, Central Kalimantan Province, with a coordinate of S: -3º47'34", E: 113º55'15". Peat sample was taken using a non-destructive method to provide a stable sample when it goes through testing. Peat sample from Kalimantan contains a lot of dead roots, twigs and branches as seen in Fig. 1(a) which shows that the sample used in the experiment is heterogonous. Before testing, peat sample was dried using electric oven at 104ºC for 24 hours. After drying, the moisture content of peat was checked using moisture balance. Peat sample used in the experiment had moisture content of less than 10%. The weight of the sample is maintained at ±300 grams, therefore the bulk density is approximately 0.3 gr/cm^3.

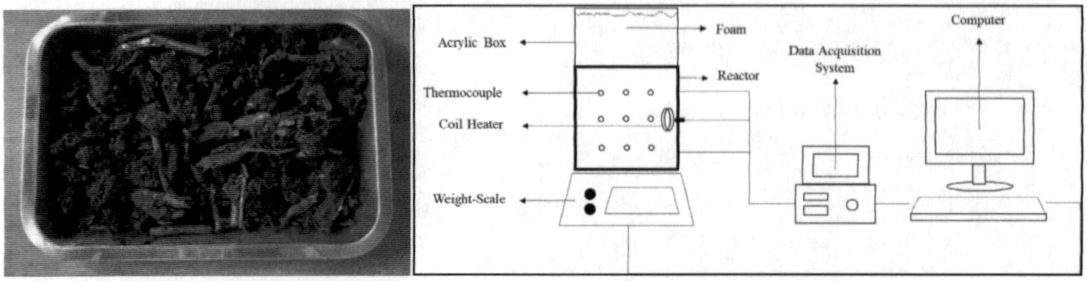

(a) (b)

Fig. 1. (a) Dried peat sample, (b) Experimental Setup.

Foam Suppression Experiment

A reactor, made from Stainless Steel, with a dimension of 10 x 10 x 10 cm was used in this experiment similar with the previous smouldering tests [4,5]. The reactor is also insulated to ensure that peat only comes in contact with oxygen at the top part. The sample was weighed beforehand to measure the mass loss after smouldering combustion occurs. To initiate smouldering front, coil heater was turned on at 100W for 30 minutes until combustion process was self-sustaining. Baseline experiment was carried out to observe smouldering combustion that occurs in the reactor and was left to self-extinguish. To examine the smouldering phenomenon, 9 type-K thermocouples are used at 3 different depths (25mm, 50mm, 75mm) at 3 distances (15mm, 50mm, 85mm from igniter) with a sampling frequency of 1Hz. The configuration of the experiment can be seen in Fig. 1(b). Apart from that, the mass of peat sample after smouldering combustion was recorded to obtain the mass loss during the test. Infrared camera, with an emissivity set at 95%, was also put in place to provide a qualitative visual representation of the smouldering combustion phenomenon. The photos were taken immediately after foam was poured onto smouldering peat from the side view.

After coil heater was turned off, peat sample was left to smoulder for 1 hour, which is the time required for the smouldering front to reach the other side of the reactor. At this point, suppression was carried out using Class A foam with a concentration of 1%. Foam is poured into an acrylic box, with varying height, was put on top the reactor to allow foam to spread evenly in top of the peat sample. The temperature at 9 different point at the reactor was observed to find out the effect of foam suppression method on the smouldering combustion of peat. The experiment was finished when all of the 9 thermocouples showed a temperature reading of below 80 ºC.[5] The mass of the peat sample was measured after the experiment to find the mass loss.

The main objective of this research is to quantify the effectiveness of foam in suppressing peat fire. The independent variable in this experiment was the thickness of the foam which was represented by the height of acrylic boxes with the variation as seen in Table 1. This experiment will explore how many foam layers of certain heights are needed in order to extinguish peat fire.

Table 1. Summary of Experimental Variation.

Thickness	7.5cm	10cm	5cm	2.5cm

RESULTS AND DICUSSION

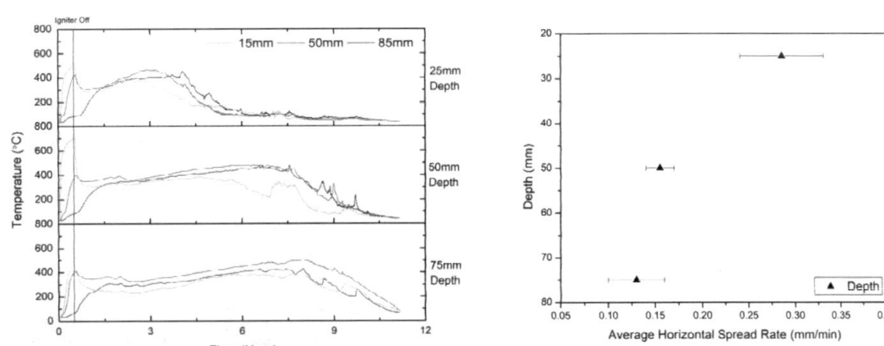

Fig. 2. Temperature measurement from baseline experiment.

Smouldering experiment was conducted beforehand to find out about the characteristics of smouldering combustion Kalimantan peat. As seen on Fig. 2(a), the first 30 minutes show a sharp increase in temperature due to the influence of the igniter. As the igniter was turned off, the temperature near the igniter plunges down to 400ºC from 700ºC. However, it slowly shows an increase measured by thermocouples at every depth and distance. Thermocouples at 25mm depth reach their maximum temperature at 466 ºC in 3.18 hours. On the other hand, thermocouples at 50mm depth require a slightly longer time, 6.4 hours, to reach their maximum temperature at 480.31ºC. Thermocouples at 75mm depth need the longest time, 8 hours, to reach their maximum temperature at 501.67 ºC. This phenomenon shows that smouldering peat at the upper part of the reactor smoulders at a higher rate than peat at the lower part of the reactor. To explore this phenomenon, Fig.2(b) shows that the horizontal spread rate at the depth of 25mm is higher, 0.33 mm/min, than the average horizontal spread at the depth of 50mm, 0.17mm/min, and 75mm, 0.16mm/min. A similar trend can be found at the experiment using Papuan peat where the spread rate decreased from 0.57 to 0.2 mm/min with increasing depth.[5]

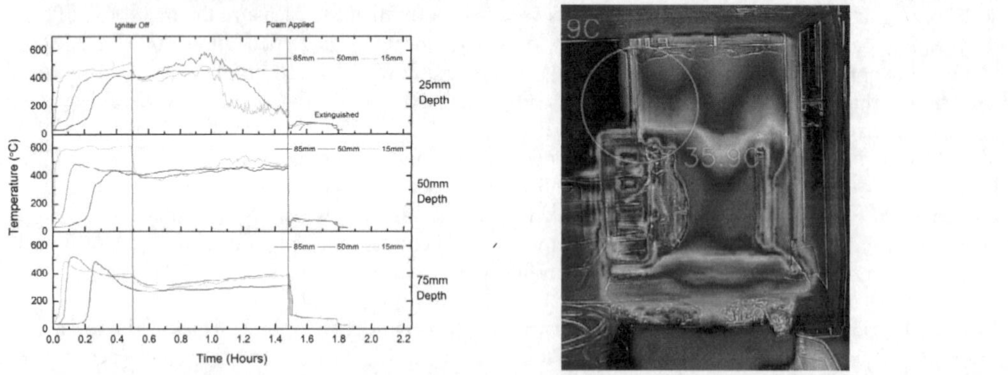

(a) (b)

Fig. 3. Temperature measurement of foam suppression experiment using (a) 10cm thickness and (b) qualitative infrared photography

To observe the effect of foam on peat fires, a layer of foam was applied directly onto the peat with varying thickness. Fig. 3(a) shows temperature measurement from foam suppression experiment using 10cm thick of Class A foam. From the graph it is known that the thickness of 10cm was able to extinguish smouldering peat inside the reactor in 12 minutes. As the foam layer was applied on peat fire, all thermocouples show a sharp decline in temperature. Thermocouples located in 25mm depth and 50mm depth can be seen reaching a temperature of below 50 ºC. On the other hand, thermocouples in 75mm depth show a more gradual decrease in temperature. This might have been caused by the foam still penetrating through the smouldering peat, therefore the time needed to extinguish the peat fire at the bottom of the reactor was slightly longer. Fig. 3(b) shows an infrared photograph of the experiment. As seen on the photograph, there was a distribution of temperature inside the foam layer which is shown by the gradient of colour with red indicating the hotter areas, and green showing a colder area. The lower part of the foam layer can be seen to be hotter than the upper part. This might be due to the heat released from the smouldering peat. As the foam layer also acted as a barrier to the atmosphere, the heat could not be released into the air causing an increase of temperature at the lower part of the foam layer. This may indicate that there was an increase in pressure inside the reactor as the foam was applied.

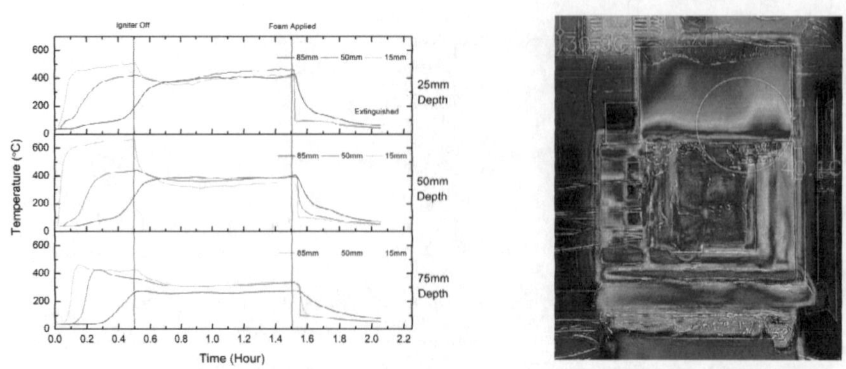

(a) (b)

Fig. 4. Temperature measurement of foam suppression experiment using (a) 7.5cm thickness and (b) qualitative infrared photography

Fig. 4(a) shows a temperature measurement from the experiment using a 7.5cm thick layer of foam. From the experiment, it was found that a 7.5cm thick layer of foam was able to extinguish smouldering peat inside the reactor. The time required for the peat fire to be put out was 30 minutes. Unlike the experiment using 10cm thick layer of foam, the experiment using 7.5cm thick layer of foam shows a more gradual decrease in temperature at every depth as the foam layer was poured on the peat fire. The thermocouples at 15mm and

50mm distance from the igniter at all depths show a quicker decline in temperature. However, the thermocouples at 85mm distance from the igniter need longer time to reach the temperature of extinction. This might have been caused by the peat at 15mm and 50mm distance from the igniter have been burning for longer amount of time than peat at 85mm distance from the igniter. Therefore, the peat sample left at 15mm and 50mm is not as much as the peat sample at 85mm distance from the igniter, which causes the temperature to decline faster at 15mm and 50mm distance from the igniter. Fig.4(b) shows an infrared thermograph of the foam layer during the experiment. Similar to the phenomenon shown in Fig.3(b), the foam layer can be seen to have different temperatures at the upper and lower part, shown by the red and green colour. This shows that the lower part of the foam layer is hotter than the upper part which may be due to the heat released from the smouldering peat inside the reactor.

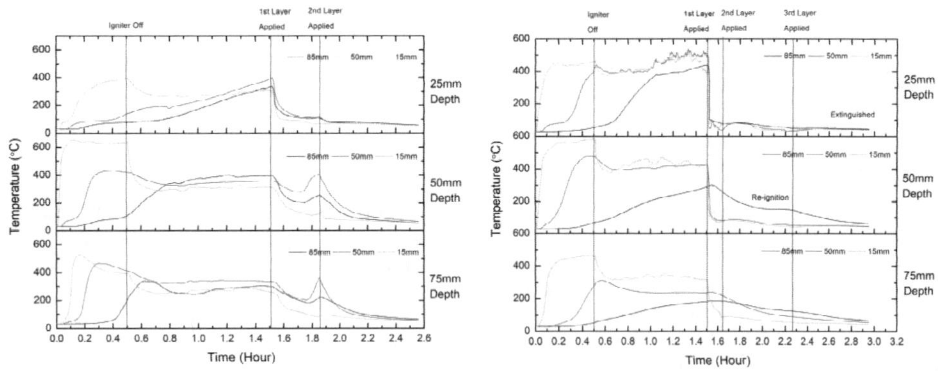

(a) (b)

Fig. 5. Temperature measurement of foam suppression experiment using a thickness of (a) 5cm and (b) 2.5cm.

Experiments were also carried out using foam layer with a thickness of 5cm. As seen from Fig.5(a), 2 layers of foam with a thickness of 5cm were needed to extinguish peat fire. The first layer of foam managed to bring down the temperature of smouldering peat up to 500 °C, which can be seen by the thermocouple at 15mm and 50mm from igniter and 25mm depth. After the first layer was applied, thermocouple at 15mm from igniter at the depth of 50mm, indicated an increase in temperature or re-ignition phenomenon. The second foam layer was applied after the first foam layer has completely disappeared. The second foam layer was able to extinguish the smouldering peat by bringing down the temperature to below 80 °C. The time required for the peat fire to be put out completely using 5cm thick foam layer was 47 minutes, which was longer than the time needed to extinguish peat fire using 10cm thick foam layer or 7.5cm thick foam layer.

Experiments were also conducted using foam layer with a thickness of 2.5cm. As seen from Fig.5(b), 3 layers of foam with a thickness of 2.5cm were needed to fully suppress smouldering peat inside the reactor. The first layer of foam did not last long, as it only took approximately 10 minutes for the foam layer to disappear. During that time, the thermocouples at all depths record a sharp decline in temperature. However, it was not enough to fully suppress the fire. The second layer of foam with the same thickness was applied immediately after the first layer disappears. The second layer of foam managed to stay for longer and bring the temperatures down to near 100 °C at 50mm depth and 75mm depth. However, as the second layer of foam disappeared, there was an indication of re-ignition as seen on Fig.5(a). The third layer of foam was applied and it managed to put out the peat fire. The time required for the burning peat to be completely suppressed was 71 minutes which was the longest time out of all the experiments.

Fig.6 shows the residual mass fraction of each experiment. As seen from the graph, the highest mass loss, 74.05%, can be observed during the baseline experiment which used 0cm thick of foam layer. As foam was applied on peat to explore how peat fire reacts to foam application, it can be seen that the total mass loss in the experiment where peat fire was suppressed (10cm, 7.5cm, 5cm, and 2.5cm) are significantly lower than that of the baseline experiment. The lowest mass loss can be seen at the experiment using 2.5cm thick of foam layer which is 31.4%. However, the difference in percentage of mass loss is not significant.

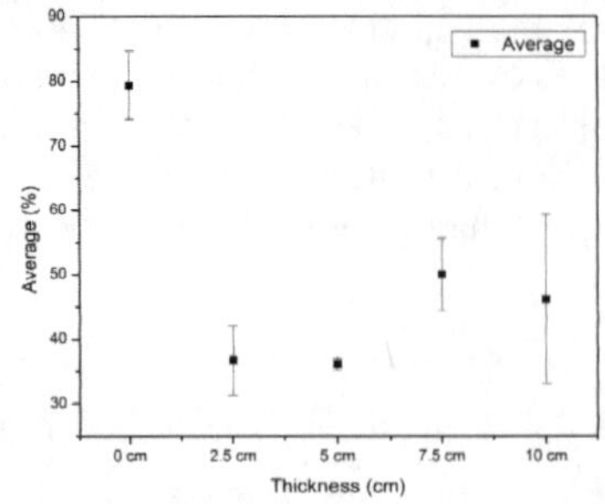

Fig. 6. Residual Mass Fraction of Foam Suppression Experiment.

CONCLUSION

Through the series of experiments, it can be found that there was a correlation between the thickness of foam layer on the suppression of peat fire. The thickness of 10cm was found to be the most effective in terms of time at extinguishing peat fire. An average amount of 3.8L of foam per kilogram of Kalimantan peat was needed to extinguish peat fire. It was also found that the average amount of water needed to suppress peat fire using foam method was 3.4L/kg. Foam layer acts as a barrier between smouldering peat and the air, therefore with the presence of foam layer, the combustion process inside the reactor is disturbed as there is no oxygen supply. Apart from that, Class A Foam also has the ability to wet the peat sample, causing a decrease in temperature. On the other hand, further studies on the interaction between foam and peat surface still need to be carried out to provide a better understanding of the suppression mechanism using foam.

ACKNOWLEDGEMENT

The authors would like to thank PTUPT grant programme of Ministry of Research, Technology and Higher Education Republic of Indonesia with the financial support.

REFERENCES

[1] W.H. Frandsen. (1991) Heat evolved from smouldering peat. International Journal of Wildland Fire, 1(3):197-204, https://doi.org/10.1071/wf9910197

[2] Wahyunto W, Nugroho K, Ritung S, Sulaeman Y., INDONESIAN PEATLAND MAP: METHOD, CERTAINTY, AND USES, 2014

[3] Rein G., "Smoldering Combustion," *SPFE Handbook of Fire Protection Engineering (5th ed),* Springer, Hurley, M.J. (ed), Aon Fire Protection Engineering Greenbelt, MD, USA, 2016, p. 581 – 582

[4] Hadden, R., Rein, G. (2011) Burning and Water Suppression of Smoldering Coal Fires in Small-Scale Laboratory Experiments. Coal and Peat fires: A Global Perspective, Volume 1: Coal-Geology and Combustion, pp. 317-326, http://dx.doi.org/10.1016/b978-0-444-52858-2.00018-9

[5] Ramadhan, M.L., Palamba, P., Imran, F.A., Kosasih, E.A., Nugroho, Y.S. (2017) Experimental study of the effect of water spray on the spread smoldering in Indonesian peat fires. Fire Safety Journal 91 671 - 679, http://dx.doi.org/10.1016/j.firesaf.2017.04.012

[6] Perdana, L.R., Ratnasari, N.G., Ramadhan, M.L., Palamba, P., Nasruddin, Nugroho, Y.S. (2017) Hydrophilic and hydrophobic characteristics of dry peat. IOP Conference Series: Earth and Environmental Science 105 012083, https://doi.org/10.1088/1755-1315/105/1/012083

[7] Ohlemiller, T.J., SFPE Fire Protection handbook, chapter 2-10: Smouldering Combustion, pp.2-201-2-210. National Fire Protection Association, Quincy, MA 02269, 3rd edition, 2002.

[8] United States Department of Agriculture. (1993). Foam vs Fire Class A Foam for Wildland Fires. National Wildfire Coordinating Group.

Assessment of the Fire Dynamics Simulator Modeling for the Heating and Evaporation of a Single Water Droplet at Moderate and High Temperatures

Tarek BEJI[1] and Bart MERCI[1]

[1]*Department of Flow, Heat and Combustion Mechanics, Ghent University, Belgium*

Tarek.Beji@UGent.be

ABSTRACT

The work described in this paper is undertaken with the purpose of providing a detailed assessment of the current modelling capabilities of the effects of fire suppression systems (e.g., sprinklers) in fire-driven flows. Such assessment will allow identifying key modelling issues and, ultimately, improving the reliability of the numerical tools in fire safety design studies. More specifically, we studied herein the heating and evaporation of a single water droplet. This rather 'simple' configuration represents the first step in a tedious and rigorous verification and validation process, as advocated in the MaCFP (Measurement and Computation of Fire Phenomena) working group (see https://iafss.org/macfp/). Such process starts ideally with single-physics 'unit tests' and then more elaborate benchmark cases and sub-systems, before addressing 'real-life' application tests. In this paper, we are considering the recently published comprehensive and well-documented experimental data of Volkov and Strizhak (Applied Thermal Engineering, 2017) where a single suspended water droplet of a diameter between 2.6 and 3.4 mm is heated up by a convective hot air flow with a velocity between 3 and 4.5 m/s and a temperature between 100 and 800°C. The high temperatures considered therein represent a strong element of novelty, since previous experimental studies on single droplets were limited to rather relatively 'moderate' temperatures, up to around 350°C. Furthermore, the monitoring of the time history of the droplet temperature field, in addition to the droplet lifetime, provides very useful information for model development and validation purposes. In this numerical study, 36 experimental tests have been simulated with the Fire Dynamics Simulator (FDS 6.6.0). The results show that the droplet lifetime is overpredicted with an overall accuracy of 31%. The accuracy in the range 300 to 800°C is even better, i.e., 7 %, whilst the cases of 200 and, more so 100°C, showed much stronger deviations. The measured droplet saturation temperatures did not exceed 70°C, even for high air temperatures of around 800°C, whereas the predicted values approached 100°C. Based on the current findings, further analysis is required on the modelling of the heat and mass transfer coefficients, and more specifically the sub-models for the Nusselt and Sherwood numbers.

KEYWORDS: *modeling; suppression; sprinklers; evaporation; Fire Dynamics Simulator.*

INTRODUCTION

An accurate water heating and evaporation model is essential for the assessment of the efficiency of active fire protection measures that are based on, for example, Early Fire Suppression Response (EFSR) sprinklers or water mist. In most of the Computational Fluid Dynamics (CFD) codes, droplet evaporation modelling is based on the so-called 'film theory' [1]. The film is a very thin layer at the interface between the liquid and the surrounding environment where heat and mass exchange as well as phase transformation occur. The thickness of the film is generally significantly (sometimes orders of magnitude) smaller than typical cell sizes used in fire dynamics (or combustion) simulations, except if Direct Numerical Simulations (DNS) are carried out, which could not be afforded for practical fire scenarios. Consequently, several correlations have been developed in the literature to estimate the convective heat and mass transfer film coefficients around droplets. These correlations are based on single suspended water droplet experiments where the droplet is subjected to a convective air flow (natural or forced) with a fixed velocity and temperature. The most widely used correlations date back to the early work of Ranz and Marshall for a forced convection flow [2] where the evaporation of water droplets in air is examined for a room temperature up to 220°C, a droplet diameter between 0.6 and 1.1 mm and a droplet Reynolds number, Re_d, between 2 to 200. Thanks to the significant advances in measuring technologies that could be used to characterize in detail the heating and evaporation process of a water droplet, more accurate data is available for model validation. For example, very recently, Volkov and Strizhak [3] published a very interesting set of data for the heating and evaporation of a water droplet with a diameter between 2.6 and 3.4 mm in a hot environment with free stream temperatures between 100 and 800°C and velocities between 3 and 4.5 m/s. The objective of this paper is to rely on this dataset to assess the capabilities of a CFD code, namely the Fire Dynamics Simulator (FDS 6.6.0), in the modelling of the heat up and evaporation process of single suspended water droplets [4-6].

NUMERICAL MODELLING

A detailed description of the mathematical modelling for droplet evaporation in FDS 6.6.0 is provided in [4-6]. Only the main equations for the case at hand are recalled herein for the sake of clarity. The mass and energy conservation equations for a single spherical liquid droplet (in the absence of interaction with solid boundaries) read [4-5]:

$$\frac{dm_d}{dt} = A_d \frac{Sh\, D_{lg}}{d_d} \rho \left(Y_v - Y_l\right) \tag{1}$$

$$m_d\, c_{p,l} \frac{dT_d}{dt} = A_d \frac{Nu\, k}{d_d}\left(T_g - T_d\right) + \frac{dm_d}{dt} L_v + q_r \tag{2}$$

where m_d, A_d, d_d and T_d are respectively the mass, area, diameter and temperature of the droplet, $c_{p,l}$ and L_v are the specific heat and latent heat of vaporization of water, ρ, k and T_g are respectively the density, thermal conductivity and temperature of the surrounding gas, q_r is a radiative source term, D_{lg} is the binary diffusion of water vapor in the surrounding gas (i.e., air), t is the time and Y_v and Y_l are the mass fraction of water vapor at the droplet surface and in the surrounding gas, respectively. The Sherwood, *Sh*, and the Nusselt, *Nu*, numbers are expressed as [2]:

$$Sh = 2.0 + 0.6\mathrm{Re}_d^{1/2}\, Sc^{1/3} \quad \text{and} \quad Nu = 2.0 + 0.6\mathrm{Re}_d^{1/2}\, \mathrm{Pr}^{1/3} \tag{3}$$

where Re_d is the droplet Reynolds number, Sc and Pr are respectively the Schmidt and the Prandtl numbers, taken as Sc = 0.6 and Pr = 0.7 (see page 87 in [6]). The droplet Reynolds number, Re_d, is calculated as:

$$\mathrm{Re}_d = \frac{d_d \left|\mathbf{u}_d - \mathbf{u}_g\right| \rho}{\mu\left(T_{film}\right)} \tag{4}$$

where $\mathbf{u}_d$ and $\mathbf{u}_g$ are respectively the droplet and gas velocity vectors and $\mu(T_{film})$ is dynamic viscosity of air at the film temperature, T_{film}, which is calculated using the one-third rule.

Equations (1) and (2), which are coupled to the equations for the gas temperature and the rate of change of vapor mass in the gas, are solved semi-implicitly over the course of a gas phase time step [5].

IOP Conf. Series: Journal of Physics: Conf. Series **1107** (2018) 062001 doi:10.1088/1742-6596/1107/6/062001

EXPERIMENTAL SETUP AND COMPUTATIONAL TESTS

Experimental setup and measurements

The experimental dataset relied upon herein for validation purposes has been obtained by Volkov and Strizhak [3]. The experimental configuration consists of a single water droplet (with a diameter between 2.67 and 3.37 mm) suspended in a hollow and transparent silica-glass cylinder of 0.1 m inner diameter. A hot air blower positioned below the cylinder blows hot air upwards with temperatures between 100 and 800°C and velocities between 3 and 4.5 m/s. The air temperatures are measured with a fast chromel-alumel (type K) thermocouple and the air velocity is controlled with the PIV technique.

Table 1. Experimental data [3].

T_a (°C)	U_a (m/s)	$d_{d,0}$ (mm)	$T_{d,0}$ (°C)[a]	t_d (s)	$T_{d,sat}$ (°C)[b]
100	3.0	2.67	30	87.2	40 ± 10
100	3.0	3.06	10	108.0	25 ± 10
100	3.0	3.37	(30)	145.0	-
200	3.0	2.67	(30)	61.8	-
200	3.0	3.06	35	74.9	40 ± 10
200	3.0	3.37	(30)	101.7	-
200	4.0	3.06	(30)	60.6	-
200	4.5	3.06	(30)	56.4	-
300	3.0	2.67	25	49.1	40 ± 10
300	3.0	3.06	25	54.8	40 ± 10
300	3.0	3.37	(30)	76.0	-
400	3.0	2.67	(30)	33.0	-
400	3.0	3.06	35	37.4	45 ± 10
400	3.0	3.37	(30)	49.7	-
400	4.0	3.06	(30)	32.1	-
400	4.5	3.06	(30)	30.2	-
500	3.0	2.67	30	23.7	50 ± 10
500	3.0	3.06	(30)	26.6	-
500	3.0	3.37	(30)	38.7	-
550	3.0	3.06	(30)	24.7	-
600	3.0	2.67	35	17.8	50 ± 10
600	3.0	3.06	35	22.8	50 ± 10
600	3.0	3.37	(30)	31.8	-
600	4.0	3.06	(30)	20.3	-
600	4.5	3.06	(30)	19.4	-
650	3.0	2.67	(30)	14.6	-
650	3.0	3.06	(30)	19.7	-
650	3.0	3.37	(30)	26.7	-
700	3.0	2.67	(30)	12.5	-
700	3.0	3.06	(30)	16.5	-
700	3.0	3.37	(30)	19.7	-
790	3.0	2.67	40	10.6	60 ± 10
790	3.0	3.06	40	14.4	50 ± 10
790	3.0	3.37	(30)	15.9	-
790	4.0	3.06	(30)	12.7	-
790	4.5	3.06	(30)	12.4	-

[a]These are estimates based on the profiles provides in [3]. A default value of 30°C (values between brackets) is assigned for cases where the information is not provided in [3].
[b]These are estimates based on the profiles provided in [3].

The time history of the droplet temperature field is obtained using the PLIF (Planar Laser-Induced Fluorescence) technique. At the moment of droplet placement on the symmetry axis of the glass cylinder, there exists already a temperature difference between the surface and the inside of the droplet, because the hot air flow is generated before the droplet placement. From a modelling perspective, since a uniform droplet temperature is assumed, an average of the initial temperature field is estimated based on the information provided in [3]. For the cases where such information is not available in [3], a default value of 30°C is taken. CCD images of the droplet are analysed during the heating process in order to monitor the time evolution of its radius until its complete evaporation, i.e., over the full droplet lifetime, t_d. It is stated in [3] that the material of the holder by which the water droplet is suspended does have an influence on the conditions of heating and thus t_d. In [3], the water droplet is held by a hollow metal rod. Nevertheless, two other types of material have been tested with a higher and a lower thermal conductivity. The hollow metal rod provided the medium droplet lifetime. The difference in the droplet lifetimes did not exceed 15%. Based on this, the experimental uncertainty considered herein is set to be 8 % (given that the metal rod has an intermediate thermal conductivity). The test conditions and the results in terms of t_d and droplet saturation temperature, $T_{d,sat}$, are displayed in Table 1.

Computational tests

The setup of the computational tests described in Table 1 is similar to the verification test case *water_evaporation_5*[1] described in [4] where stratification and noise (in the flow field) are turned off and the ambient pressure and temperature are fixed. For the case at hand, p_{amb} = 101325 Pa and T_a is fixed based on the values displayed in Table 1. Note also that the default radiation settings are turned on in this paper (as opposed to *water_evaporation_5*), given the relatively high gas temperatures that are involved. For example, at 790°C and without radiation the droplet lifetime increased by around 60% and the droplet saturation temperature decreased by about 10°C. An additional aspect that has been considered for the tests carried out herein and not for the test case *water_evaporation_5* is the setup of a uniform velocity field using specific options available in FDS and which allow 'forcing' the gas velocity field within the computational domain and 'freezing' that field throughout the calculation [6]. In other words, the flow field is not solved but imposed. Additional calculations that took into account turbulent fluctuations (by setting a velocity 'noise' of up to 20%) showed that the results varied by less than 1%.

RESULTS

Figure 1a shows an overall good agreement between the predicted and the measured droplet lifetimes. In fact, based on the methodology proposed in [7] for the quantification of the predictive uncertainty of complex models, the overall model uncertainty (over the full range [100 – 800°C] of air temperatures tested) is about 31% with a bias factor of about 1.71. Furthermore, one can clearly visualize in Fig.1a that the agreement is even better if the results for the low air temperatures are discarded. This is quantified in Table 2 which indicates that the model uncertainty reduces to 15% for the range [200 – 800°C] and 7% for the range [300 – 800°C].

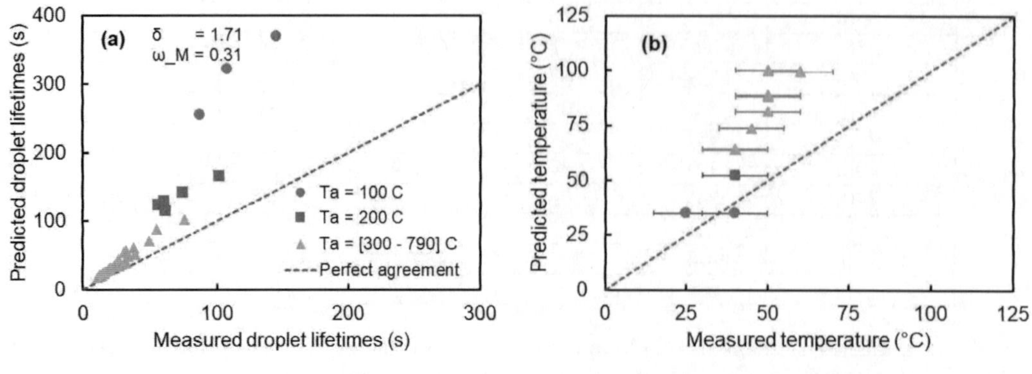

Fig. 1. Comparison between the predicted and measured values of the droplet (a) lifetime and (b) saturation temperature.

[1] https://github.com/firemodels/fds/blob/master/Verification/Sprinklers_and_Sprays/water_evaporation_5.fds

IOP Conf. Series: Journal of Physics: Conf. Series **1107** (2018) 062001 doi:10.1088/1742-6596/1107/6/062001

These results are quite surprising, because it is very often reported in the literature (e.g., in [8]) that a large temperature gradient between the droplet and the ambient air leads to a significant mass transfer reduction, which is generally accounted for by modelling a reduction factor for the Sherwood number. Based on the current results, this was not necessary for the case at hand although expressions (3) were used beyond their initial range of validity.

Table 2. Bias factor and model relative standard deviation depending on the air temperature range.

	Air temperature (°C)		
	[100 – 800]	[200 – 800]	[300 – 800]
Bias factor, δ	1.71	1.61	1.57
Relative standard deviation, ω_M	0.31	0.15	0.07

For the cases of T_a = 200°C, and even more so for T_a = 100°C, the droplet lifetimes are significantly overestimated, indicating an underestimation of the mass transfer process. This aspect has been discussed in [9] where an alternative to Eqs. (3) has been proposed, based on single droplet drying experiments over an air temperature range of 23-200°C and with Re_d from 30 to 100. The expression proposed in [9] for heat transfer reads:

$$Nu = 6.4 + 0.8\mathrm{Re}_d^{1/2}\,\mathrm{Pr}^{1/3} \tag{5}$$

The significantly higher value of 6.4 for the natural convection regime (and obtained after extrapolation), in comparison to the theoretical value of 2, is attributed in [9] to (1) a thermal gradient within the medium surrounding the droplet and which is induced by molecular diffusion, and (2) localized convection currents that increase heat and mass transfer. Equation (5) and other alternatives to Eqs.(3) need thus be implemented and tested against the experimental data used herein.

Regarding the droplet saturation temperature, there is, unfortunately, not enough data reported in [3] to undertake a statistical analysis similar to the droplet lifetimes, using the methodology developed in [7]. However, the results reported in Fig.1b and Fig.2 clearly show that the predicted saturation temperatures are generally higher than the measured values. This is particularly the case when the air temperature is high, as shown in Fig.2b where the measured droplet saturation temperatures did not exceed 60°C whereas the predicted values approached 100°C. This demonstrates the need to undertake a detailed analysis on the coupled processes of heat and mass transfer around the droplets in order to achieve a good agreement for the droplet lifetime and saturation temperature simultaneously. Furthermore, the present results may imply that the problems of potential numerical instability and super-saturation that have been addressed in [4] can be due (in some cases) to the physical model of heat transfer around the droplet, and which makes the droplet reach a too high saturation temperature.

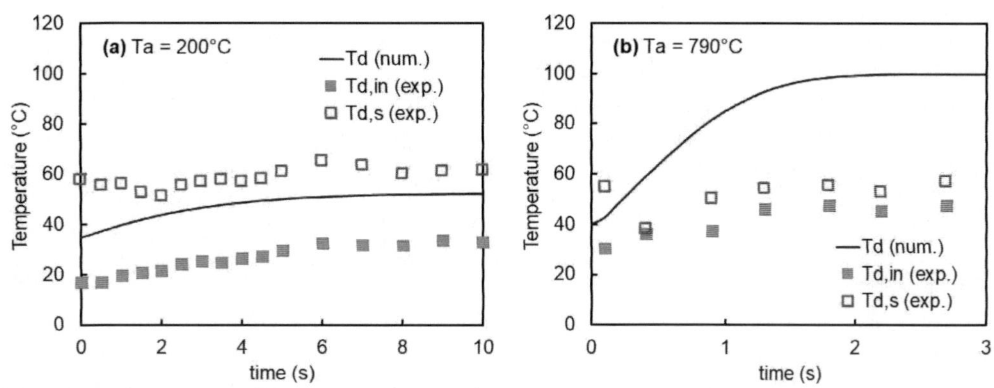

Fig. 2. Comparison between the time-evolution profiles of the measured temperature on the surface and inside the droplet and the predicted droplet temperature for $d_{d,0}$ = 3.06 mm, at U_a = 3 m/s and at different air temperatures. (a) T_a = 200°C. (b) T_a = 790°C.

IOP Conf. Series: Journal of Physics: Conf. Series **1107** (2018) 062001 doi:10.1088/1742-6596/1107/6/062001

CONCLUSIONS

In this paper we assessed the capabilities of the Fire Dynamics Simulator (FDS 6.6.0) in the modelling of the heat up and evaporation of a water droplets based on the experiments carried out by Volkov and Strizhak (Applied Thermal Engineering, 2017). In these experiments, a single suspended water droplet of a diameter between 2.6 and 3.4 mm is heated up by a convective hot air flow with a velocity between 3 and 4.5 m/s and a temperature between 100 and 800°C. The results, based on the simulation of 36 tests, show that the droplet lifetime is predicted with an overall accuracy of 31%. The accuracy in the range 300 to 800°C is even better, i.e., 7 %, whilst the cases of 200 and, more so 100°C, showed much stronger deviations that indicate an underestimation in the mass transfer rate for these moderate temperatures. Furthermore, the measured droplet saturation temperatures did not exceed 70°C, even for high air temperatures of around 800°C, whereas the predicted values approached 100°C. Thus, based on the current findings, further analysis is required on the modelling of the heat and mass transfer coefficients, and more specifically the sub-models for the Nusselt and the Sherwood numbers.

REFERENCES

[1] Sazhin, S.S. (2017) Modelling of fuel droplet heating and evaporation: Recent results and unsolved problems, *Fuel*, 196:69-101, https://doi.org/10.1016/j.fuel.2017.01.048

[2] Ranz, W.E. and Marshall, W.R. (1952) Evaporation from drops, *Chemical Engineering Progress*, 48 (Part 1) 141-146; (Part 2) 173-180.

[3] Volkov, R.S. and Strizhak, P.A. (2017) Planer laser-induced fluorescence diagnostics of water droplets heating and evaporation at high-temperature, *Applied Thermal Engineering*, 127:141-156, https://doi.org/10.1016/j.applthermaleng.2017.08.040

[4] Floyd, J. and McDermott, R. (2017) Development and evaluation of two new droplet evaporation schemes for fire dynamics simulations, *Fire Safety Journal*, 91:643-652, https://doi.org/10.1016/j.firesaf.2017.04.036

[5] McGrattan, K., McDermott, R., Hostikka, S., Floyd, J., Vanella, M., Weinschenk, C. and Overholt, K. (2017) Fire Dynamics Simulator Technical Reference Guide Volume 1: Mathematical Model, NIST Special Publication 1018-1 Sixth Edition, http://dx.doi.org/10.6028/NIST.SP.1018

[6] McGrattan, K., McDermott, R., Hostikka, S., Floyd, J., Vanella, M., Weinschenk, C. and Overholt, K. (2017) Fire Dynamics Simulator User's Guide, NIST Special Publication 1019 Sixth Edition, http://dx.doi.org/10.6028/NIST.SP.1019

[7] McGrattan, K. and Toman, B. (2011) Quantifying the predictive uncertainty of complex numerical models, *Metrologia*, 48:173-180, https://doi.org/10.1088/0026-1394/48/3/011

[8] Schwarz, J. and Smolik, J. (1994) Mass transfer from a drop – I. Experimental study and comparison with existing correlations, *International Journal of Heat and Mass Transfer*, 37:2139-2143, https://doi.org/10.1016/0017-9310(94)90315-8

[9] Walton, D.E. (2004) The Evaporation of Water Droplets. A Single Droplet Drying Experiment, Drying Technology, 22:431-456, https://doi.org/10.1081/drt-120029992

A Detailed Investigation on the Effect of the Sherwood and Nusselt Number Modelling for the Heating and Evaporation of a Single Suspended Water Droplet

Tarek BEJI[1] and Bart MERCI[1]

[1]*Department of Flow, Heat and Combustion Mechanics, Ghent University, Belgium*

Tarek.Beji@UGent.be

ABSTRACT

The work described in this paper presents a comprehensive analysis of the convective heat and mass transfer coefficient modelling around a single water droplet using an in-house code. The most widely used approach is to rely on sub-models for the non-dimensional heat and mass transfer numbers (called hereafter the Nusselt, Nu, and the Sherwood, Sh, numbers) and which are shown to take the theoretical and functional form of $\mathrm{Nu} = 2.0 + \mathrm{K}_2\,\mathrm{Re}_\mathrm{d}^{1/2}\,\mathrm{Pr}^{1/3}$ and $\mathrm{Sh} = 2.0 + \mathrm{K}_1\,\mathrm{Re}_\mathrm{d}^{1/2}\,\mathrm{Sc}^{1/3}$ where Re_d is the droplet Reynolds number, Pr and Sc are the Prandtl and Schmidt numbers of the surrounding gas and K_1 and K_2 are constants. This formulation, which is generally referred to in the literature as the Ranz-Marshall model (with $K_1 = K_2 = 0.6$), is the most used approach in Computational Fluid Dynamics (CFD) codes for fire safety engineering. In this paper, we first assessed this formulation based on 36 experimental tests carried out in [Volkov and Strizhak, Applied Thermal Engineering (2017)] and where a single suspended water droplet of a diameter between 2.6 and 3.4 mm is heated up by a convective hot air flow with a velocity between 3 and 4.5 m/s and a temperature between 100 and 800°C. The results showed that the overall model uncertainty in the droplet lifetime prediction is about 34% with particularly poor results when the air temperature is 100 or 200°C. The droplet saturation temperatures were overestimated by around 20 to 30°C. After this initial assessment, we performed a sensitivity analysis and selected a combination of values for K_1 and K_2 that provided an overall simultaneous good agreement for both the droplet lifetimes (model uncertainty of 5%) and the droplet saturation temperatures (around 10°C). This analysis showed that, for high air temperatures (i.e., $T_a \geq 300$°C), the value of $K_2 = 0.6$ remains suitable. However, for these cases, the value of K_1 needed to be increased (to 1.8 and up to 4) in order to promote the evaporation-induced cooling and improve the predictions in terms of droplet saturation temperatures. For $T_a = 100$°C, the 'best' combination was found to be $K_1 = K_2 = 1.8$. Such combination allowed to reduce the initially overestimated droplet lifetimes by promoting mass transfer (through an increased value of K_1) without slowing down the heating process that generates water vapor at the droplet surface (explaining the equally increased value of K_2). The case of $T_a = 200$°C appeared to be an intermediate case. The present results indicate that the 'classical' Ranz-Marshall approach with $K_1 = K_2 = 0.6$ is not optimal. A more thorough analysis, with eventually additional experimental data, is required.

KEYWORDS: *water droplet; heat transfer; mass transfer; evaporation.*

IOP Conf. Series: Journal of Physics: Conf. Series **1107** (2018) 062002 doi:10.1088/1742-6596/1107/6/062002

INTRODUCTION

An accurate water heating and evaporation model is essential for the assessment of the efficiency of active fire protection measures that are based on, for example, Early Fire Suppression Response (EFSR) sprinklers or water mist. Several correlations have been developed in the literature to estimate the convective heat and mass transfer film coefficients around droplets. The most widely used approach is to rely on sub-models for the non-dimensional heat and mass transfer numbers, namely the Nusselt, Nu, and the Sherwood, Sh, numbers, which are shown to take the theoretical and functional form of [1]:

$$\mathrm{Nu} = 2.0 + \mathrm{K}_2\,\mathrm{Re}_\mathrm{d}^{1/2}\,\mathrm{Pr}^{1/3} \quad \text{and} \quad \mathrm{Sh} = 2.0 + \mathrm{K}_1\,\mathrm{Re}_\mathrm{d}^{1/2}\,\mathrm{Sc}^{1/3} \tag{1}$$

where $\mathrm{Re_d}$ is the droplet Reynolds number, Pr and Sc are respectively the Prandtl and Schmidt numbers of the surrounding gas and K_1 and K_2 are constants. This formulation, which is generally referred to in the literature as the Ranz-Marshall model with $K_1 = K_2 = 0.6$ [1], is the most used approach in Computational Fluid Dynamics (CFD) codes for fire safety engineering. A myriad of similar correlations have been proposed in the literature [2-3] by modifying the coefficients K_1 and K_2, as well as the constant term 2, although the latter is suggested in [1] to remain unchanged in order to be consistent with the theoretical requirement of Nu = 2 and Sh = 2 at $\mathrm{Re_d} = 0$. The variety of the proposed coefficients stems essentially from the change in experimental conditions such as the nature of the experimental test (e.g., suspended or free-fall single droplet) or the velocity and temperature of the convective hot air flow. To the best of the authors' knowledge, most of the experimental studies on the evaporation of single water droplets in a hot environment are limited to 'moderate' free air stream temperatures, up to around 350°C (e.g., in [4]). However, very recently (in 2017), an experimental study have been reported on the heating and evaporation of a single suspended water droplet in a hot air environment with temperatures between 100 and 800°C [5]. The high temperatures considered therein represent a strong element of novelty and are particularly relevant for fire suppression applications. The objective of this paper is to develop an in-house code and rely on the dataset displayed in [5] in order to examine the accuracy of the numerical predictions depending on the choice for the values of K_1 and K_2 in Eq.(1).

NUMERICAL MODELLING

In this section the governing equations for mass and heat transfer are described along with the solution procedure that is used in the development in the in-house code.

Mass transfer

The evaporation rate of a spherical liquid droplet is expressed according to the Stefan-Fuchs model [2] as:

$$\frac{dm_d}{dt} = \pi d_d \frac{1}{\mathrm{Le}} \frac{k_m}{c_{p,v}} \mathrm{Sh} \ln\left(1 + B_M\right) \tag{2}$$

where d_d is the droplet diameter, Le is the Lewis number (taken as Le = 1), k_m is the mass weighted thermal conductivity of the mixture of water vapor and air at the air temperature and $c_{p,v}$ is the specific heat capacity of water vapor at constant pressure and droplet surface temperature. The variable B_M denotes the Spalding mass transfer number expressed as [2]:

$$B_M = \frac{Y_{v,s} - Y_{v,g}}{1 - Y_{v,s}} \tag{3}$$

where $Y_{v,s}$ is the mass fraction of the vapor at the liquid surface (calculated using the Clausius-Clapeyron equation) and $Y_{v,g}$ is its mass fraction in the surrounding gas.

Heat transfer

Based on the assumption of uniform temperature, T_d, across the droplet diameter, the energy conservation equation for the spherical liquid droplet reads:

$$\frac{dT_d}{dt} = \frac{1}{\tau}\left[\alpha_1\left(T_g^4 - T_d^4\right) + \left(T_g - T_d\right) - \alpha_2 \ln\left(1 + B_M\right)\right] \tag{4}$$

with

$$\tau = \frac{\rho_L d_d^2 c_{p,L}}{6k_m \,\mathrm{Nu}}, \quad \alpha_1 = \frac{d_d \varepsilon_L \sigma F_d}{k_m \,\mathrm{Nu}}, \quad \alpha_2 = \frac{\mathrm{Sh}}{\mathrm{Nu}} \frac{1}{\mathrm{Le}} \frac{L_v}{c_{p,v}} \tag{5}$$

where ρ_L, $c_{p,L}$, ε_L and L_v are respectively the density, specific heat, emissivity and sensible heat of vaporization of water and F_d is the view factor of the droplet (taken here as $F_d = 0.5$ [6]). The variables t and σ denote the time and the Stefan-Boltzmann constant.

Solution procedure

Similarly to [7], the heat transfer equation, i.e., Eq.(4), in combination with Eq.(2) for mass transfer, is discretized in time using a Crank-Nicolson scheme, along with a linearization of the radiation and natural logarithm terms using Taylor series.

EXPERIMENTAL SETUP AND MEASUREMENTS

The experimental dataset relied upon herein for validation purposes has been obtained by Volkov and Strizhak [5]. The experimental configuration consists of a single water droplet (with a diameter between 2.67 and 3.37 mm) suspended in a hollow and transparent silica-glass cylinder of 0.1 m inner diameter. A hot air blower positioned below the cylinder blows hot air upwards with temperatures between 100 and 800°C and velocities between 3 and 4.5 m/s. The air temperatures are measured with a fast chromel-alumel (type K) thermocouple and the air velocity is controlled with the PIV technique. The time history of the droplet temperature field is obtained using the PLIF (Planar Laser-Induced Fluorescence) technique. At the moment of droplet placement on the symmetry axis of the glass cylinder, there exists already a temperature difference between the surface and the inside of the droplet, because the hot air flow is generated before the droplet placement. From a modelling perspective, since a uniform droplet temperature is assumed, an average of the initial temperature field is estimated based on the information provided in [5]. For the cases where such information is not available in [5], a default value of 30°C is taken. CCD images of the droplet are analysed during the heating process in order to monitor the time evolution of its radius until its complete evaporation, i.e., over the full droplet lifetime, t_d. It is stated in [5] that the material of the holder by which the water droplet is suspended does have an influence on the conditions of heating and thus t_d. In [5], the water droplet is held by a hollow metal rod. Nevertheless, two other types of material have been tested with a higher and a lower thermal conductivity. The hollow metal rod provided the medium droplet lifetime. The difference in the droplet lifetimes did not exceed 15%. Based on this, the experimental uncertainty considered herein is set to be 8 % (given that the metal rod has an intermediate thermal conductivity). The test conditions and the results in terms of t_d and droplet saturation temperature, $T_{d,sat}$, are displayed in Table 1.

RESULTS

First, simulations are carried out using Eq.(1) and $K_1 = K_2 = 0.6$. This is referred to hereafter as the 'standard' Ranz-Marshall (R&M) model. A second series of simulations is carried out with modified values of K_1 and K_2, which is referred to as the modified R&M model.

Standard R&M model

Figure 1a shows an overall good agreement between the predicted and the measured droplet lifetimes. In fact, based on the methodology proposed in [8] for the quantification of the predictive uncertainty of complex models, the overall model uncertainty (over the full range [100 – 800°C] of air temperatures tested) is about 34% with a bias factor of about 1.13. Furthermore, one can clearly visualize in Fig.1a that the agreement is even better (with a model uncertainty of 5%) if the results for the low air temperatures, and more specifically 100 and 200 °C, are discarded. These results are quite surprising, because the initial range of applicability of the R&M model is up to 200°C and it is very often reported in the literature (e.g., [3]) that a large temperature gradient between the droplet and the ambient air leads to a significant mass transfer reduction, which is generally accounted for by modelling a reduction factor for the Sherwood number. Some of the proposed correlations have been reviewed in [2-3]. Based on the current results, this was not necessary for the case at hand although expressions (1) (with $K_1 = K_2 = 0.6$) were used beyond their initial range of validity in terms of free stream air temperature.

Table 1. Experimental data [5].

T_a (°C)	U_a (m/s)	$d_{d,0}$ (mm)	$T_{d,0}$ (°C)[a]	t_d (s)	$T_{d,sat}$ (°C)[b]
100	3.0	2.67	30	87.2	40 ± 10
100	3.0	3.06	10	108.0	25 ± 10
100	3.0	3.37	(30)	145.0	-
200	3.0	2.67	(30)	61.8	-
200	3.0	3.06	35	74.9	40 ± 10
200	3.0	3.37	(30)	101.7	-
200	4.0	3.06	(30)	60.6	-
200	4.5	3.06	(30)	56.4	-
300	3.0	2.67	25	49.1	40 ± 10
300	3.0	3.06	25	54.8	40 ± 10
300	3.0	3.37	(30)	76.0	-
400	3.0	2.67	(30)	33.0	-
400	3.0	3.06	35	37.4	45 ± 10
400	3.0	3.37	(30)	49.7	-
400	4.0	3.06	(30)	32.1	-
400	4.5	3.06	(30)	30.2	-
500	3.0	2.67	30	23.7	50 ± 10
500	3.0	3.06	(30)	26.6	-
500	3.0	3.37	(30)	38.7	-
550	3.0	3.06	(30)	24.7	-
600	3.0	2.67	35	17.8	50 ± 10
600	3.0	3.06	35	22.8	50 ± 10
600	3.0	3.37	(30)	31.8	-
600	4.0	3.06	(30)	20.3	-
600	4.5	3.06	(30)	19.4	-
650	3.0	2.67	(30)	14.6	-
650	3.0	3.06	(30)	19.7	-
650	3.0	3.37	(30)	26.7	-
700	3.0	2.67	(30)	12.5	-
700	3.0	3.06	(30)	16.5	-
700	3.0	3.37	(30)	19.7	-
790	3.0	2.67	40	10.6	60 ± 10
790	3.0	3.06	40	14.4	50 ± 10
790	3.0	3.37	(30)	15.9	-
790	4.0	3.06	(30)	12.7	-
790	4.5	3.06	(30)	12.4	-

[a]These are estimates based on the profiles provides in [5]. A default value of 30°C (values between brackets) is assigned for cases where the information is not provided in [5].
[b]These are estimates based on the profiles provided in [5].

Regarding the droplet saturation temperature, there is, unfortunately, not enough data reported in [5] to undertake a statistical analysis similar to the droplet lifetimes, using the methodology developed in [8]. Nevertheless, the results reported in Fig.1b clearly show that the predicted saturation temperatures are generally higher than the measured values. Furthermore, the deviations are more substantial for the highest air stream temperatures. For example, when the air temperature is around 800°C, the estimated droplet saturation temperature is around 60 ± 10°C, whereas the predicted value is 85°C.

This first series of simulations has shown that, when the standard R&M model is used, the droplet lifetimes are overpredicted for low air temperatures (100 and 200°C). This suggests an underestimation of the mass transfer coefficient (and thus, the Sherwood number) in these cases. This is qualitatively in line with the experimental findings reported in [9] where the heat and mass transfer processes have been found to be significantly more pronounced than what is suggested by the standard R&M model, although the range of conditions examined therein (i.e., air temperature range of 23-200°C, air velocities from 0.25 to 1.00 m/s and droplet Reynolds numbers from ~30 to 100 [9]) is within the range of the applicability of the R&M model proposed in [1]. In terms of droplet saturation temperature, deviations are more pronounced for high air temperatures, which might suggest an overestimation of the heat transfer coefficient and thus, the Nusselt number.

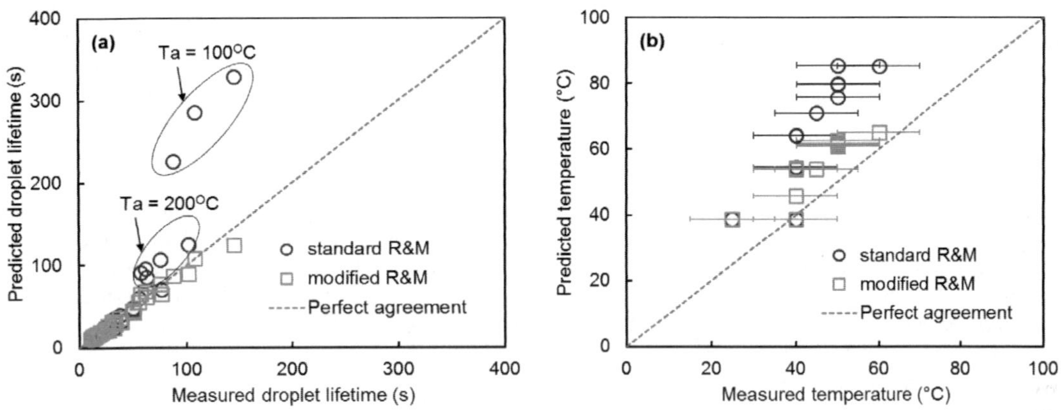

Fig. 1. Comparison between the predicted (with the standard and modified R&M model) and measured values of the droplet (a) lifetime and (b) saturation temperature.

Modified R&M model

In this paper, instead of examining the myriad of correlations proposed in the literature (see [2-3]), we have chosen to perform a sensitivity analysis on K_1 and K_2 (an alternative could have been an inverse modelling procedure) and remain consistent with the theoretical requirement of Nu = 2 and Sh = 2 at $Re_d = 0$. The purpose is not to propose a new correlation, but rather discuss the significance of the values in Table 2 that have been found to give a very good agreement for both the droplet lifetime and saturation temperature.

Table 2. Values of K_1 and K_2 in the modified Ranz-Marshall (R&M) model.

Air temperature (°C)	$\mathbf{K_1}$	$\mathbf{K_2}$
100	1.8	1.8
200	2.0	0.8
300	1.8	0.6
[400 – 650]	3.0	0.6
[700 – 790]	4.0	0.6

The improvement in the results is shown in Fig. 1, where a better agreement is obtained for the droplet lifetimes, including cases with air temperatures of 100 and 200°C (see Fig.1a). The overall model uncertainty, in this regard, has been reduced from 34% in the standard R&M model to 5% in the modified model. Furthermore, Fig.1b shows a clearly better agreement for the droplet saturation temperatures. Displaying these improved results is, *per se*, not the main goal of the paper. What is more interesting to discuss is the selection procedure of the coefficients K_1 and K_2. For T_a = 100°C, the high value of K_1 = 1.8 in comparison to the standard value of 0.6 can be easily explained by the need to increase mass transfer in order to reduce the overpredicted droplet lifetimes with the standard R&M model. However, for the same case, the need to substantially increase K_2 to 1.8 is not obvious. In fact, when only K_1 (and thus Sh) is increased, the coefficient α_2 (in Eq.(5)) increases and leads to a significant reduction in heat transfer due to evaporation-induced cooling (see Eq.(4)). The reduced droplet temperature results in a reduced concentration of water vapor at the droplet surface (by virtue of the Clausius-Clapeyron equation) and thus, a reduced driving force for evaporation. Consequently, the droplet lifetimes remain overestimated. Therefore, in the case of T_a = 100°C, to keep the influence of an increased K_1, the value of K_2 had to be increased simultaneously, maintaining the evaporation-induced cooling (see Eq.(4)) to the same level as for the case Sh = Nu. For the other end of air temperature range, i.e., T_a between 700 and 790°C, there is a different dynamics in the heat and mass transfer coupling. Based on the initial results with the standard R&M model, one would think that since the droplet lifetimes are well predicted and the droplet saturation temperatures are not, only the Nusselt number, and thus K_2, should be reduced and K_1 could be kept equal to 0.6. This 'optimization' strategy was not fruitful because a reduction in Nu yielded a strong reduction in the droplet temperature (see Eqs.(5) and (4)) and thus the droplet evaporation rate (the droplet lifetimes were then overestimated). However, by keeping $K_2 = 0.6$ and increasing the value of K_1 to 4, the evaporation-induced cooling was increased so that the droplet saturation temperatures are reduced to values that are closer to the experimental estimates. An increase of K_1 to 4 did not deteriorate the good predictions of the droplet lifetimes with the standard R&M model because at high temperatures the droplet Reynolds numbers are reduced (due to the reduced gas density). Thus, the influence of K_1 on the Sh is reduced. Consequently, the evaporation rates do not change substantially. The choice of the K_1 and K_2 values for the intermediate air temperatures between 100 and 790°C was based on a trade-off between the two selection strategies explained above (for the two bounds of the air temperature range).

CONCLUSIONS

In this paper we investigated the influence of the modelling for the Sherwood and Nusselt numbers on the heating and evaporation of water droplets. To this purpose, we developed an in-house code based on the 'film theory' concept that is typically used in Computational Fluid Dynamics (CFD). The assessment of the modelling and the analysis of the results is based on the experiments carried out in [5]. The results show that the standard values proposed in [1] give rather poor predictions for the droplet lifetimes when the air temperature is 100 or 200°C, but surprisingly good predictions for higher air temperatures. However, at high air temperatures, the droplet saturation temperatures are overestimated. A detailed analysis of the effect of the constants K_1 and K_2 on the accuracy of the predictions shows that they should be only equated for T_a = 100°C but with a higher value of 1.8. For higher air temperatures, a high value of K_1 promotes evaporation-induced cooling (and thus, reduced droplet saturation temperatures) without significantly affecting evaporation rates (and thus, the droplet lifetimes). In the light of the present results, the 'classical' Ranz-Marshall approach with $K_1 = K_2 = 0.6$ is clearly not optimal. A more thorough analysis, with eventually additional experimental data, is required.

REFERENCES

[1] Ranz, W.E. and Marshall, W.R. (1952) Evaporation from drops, *Chemical Engineering Progress*, 48 (Part 1) 141-146; (Part 2) 173-180.

[2] Sazhin, S.S. (2017) Modelling of fuel droplet heating and evaporation: Recent results and unsolved problems, *Fuel*, 196:69-101, https://doi.org/10.1016/j.fuel.2017.01.048

[3] Schwarz, J. and Smolik, J. (1994) Mass transfer from a drop – I. Experimental study and comparison with existing correlations, *International Journal of Heat and Mass Transfer*, 37:2139-2143, https://doi.org/10.1016/0017-9310(94)90315-8

[4] Downingm, C.G. (1966) The evaporation of drops of pure liquids at elevated temperatures: rates of evaporation and wet-bulb temperatures, *AIChE Journal*, 12:760-766, https://doi.org/10.1002/aic.690120424

[5] Volkov, R.S. and Strizhak, P.A. (2017) Planer laser-induced fluorescence diagnostics of water droplets heating and evaporation at high-temperature, *Applied Thermal Engineering*, 127:141-156, https://doi.org/10.1016/j.applthermaleng.2017.08.040

[6] Lin, S.X.Q. and Chen, X.D. (2002) Improving the glass-filament method for accurate measurement of drying kinetics of liquid droplets, *Chemical Engineering Research and Design*, 80:401-410, https://doi.org/10.1205/026387602317446443

[7] Floyd, J. and McDermott, R. (2017) Development and evaluation of two new droplet evaporation schemes for fire dynamics simulations, *Fire Safety Journal*, 91:643-652, https://doi.org/10.1016/j.firesaf.2017.04.036

[8] McGrattan, K. and Toman, B. (2011) Quantifying the predictive uncertainty of complex numerical models, *Metrologia*, 48:173-180, https://doi.org/10.1088/0026-1394/48/3/011

[9] Walton, D.E. (2004) The Evaporation of Water Droplets. A Single Droplet Drying Experiment, Drying Technology, 22:431-456, https://doi.org/10.1081/drt-120029992

IOP Conf. Series: Journal of Physics: Conf. Series **1107** (2018) 062003 doi:10.1088/1742-6596/1107/6/062003

Large eddy simulations of flame extinction with infinitely fast chemistry

Georgios MARAGKOS and Bart MERCI
Department of Flow, Heat and Combustion Mechanics, Ghent University, Belgium
georgios.maragkos@ugent.be

ABSTRACT

Large eddy simulations of flame extinction with N_2 as extinguish agent are performed focusing on combustion and radiation modelling with infinitely fast chemistry. The use of a EDC combustion model with dynamically determined coefficients, an enthalpy-based flame extinction model based on a locally variable critical flame temperature and the use of the WSGGM for radiation are employed in order to predict flame extinction in a turbulent CH_4 line burner. The numerical predictions of mean temperatures, combustion efficiencies and radiative fractions with different grid sizes are compared to the experiments by White et al. (2015).

Overall, the results from the numerical simulations agree well qualitatively and, to some extent, quantitatively with the experimental data when small grid sizes are employed. More specifically, the maximum values and the profile widths of the mean temperatures at two axial locations examined are reasonably well predicted. The decrease in the combustion efficiencies as the extinction limit is approached is reproduced by the numerical simulations. The decreasing trend in the radiative fractions as the oxidizer stream is diluted with N_2 is also captured by the simulations using a WSGGM model for radiation with a dynamically determined beam length which is calculated based on the local heat release rate. Nevertheless, the resulting radiative fractions are still over-predicted as the extinction limit is approached. Limitations of some of the typically used approaches regarding radiation modelling in flame extinction scenarios are outlined. Additionally, possible deficiencies on the use of a fixed critical flame temperature and/or re-ignition temperature in the flame extinction/re-ignition modelling are discussed and possible extensions of the modelling approaches are presented.

KEYWORDS:

flame extinction; LES; EDC; WSGGM; FireFOAM

IOP Conf. Series: Journal of Physics: Conf. Series **1107** (2018) 062003 doi:10.1088/1742-6596/1107/6/062003

INTRODUCTION

Flame extinction is a complex physical process which has received a lot of attention from the scientific community in the past and is inherently important when it comes to fire safety engineering since it deals with the protection of material property (i.e., industrial and/or residential) but also the life safety of the public. The use of Computational Fluid Dynamics (CFD) in this area of research is crucial and its advantages are multifold. The use of CFD can be used towards evaluating the performance of current fire detection systems, the design and testing of new ones but also the virtual simulation of a wide range of hypothetical fire scenarios that in reality would either be too expensive to conduct or would not be permitted for fire safety reasons.

Extinction can be broadly described as the reduction of the chemical reaction rate below a critical value for which self-sustained combustion cannot occur. In general, three different mechanisms of extinction can be defined: aerodynamic quenching (i.e., reduction of the flame residence times through flow perturbations), thermal quenching (i.e., radiative heat losses and/or water droplet evaporation) and quenching by dilution (dilution of the fuel and/or the oxidizer stream). Given the type of scenarios usually involved in fire applications, thermal quenching and quenching due to dilution are expected to be more the predominant mechanisms for flame extinction as opposed to aerodynamic quenching. CFD models should, in practice, be applicable and produce reliable and accurate predictions for a wide range of conditions/scenarios if they are to be used for practical applications of fire safety engineering. Currently, fire-related CFD codes mainly rely on the concept of a critical flame temperature or critical Damköhler number while other models have been recently presented as well such as, e.g., the reactive volume fraction model.

The work presented here aims to continue the authors' previous work [1] towards predictive fire modelling. The main objective of the paper is the application of advanced modelling approaches related to turbulence, combustion and radiation modelling along with the use of an enthalpy-based extinction model using a variable critical flame temperature and their evaluation when it comes to predicting flame extinction with infinitely fast chemistry. For this purpose, Large Eddy Simulations (LES) of a turbulent CH_4 line burner with N_2 as extinguish agent are presented and the results are compared to the experiments by White et al. [2]. The chosen experiments are very relevant for the fire research community since they are one of the target test cases of the newly established Measurement and Computation of Fire Phenomena (MaCFP) workshop (https://github.com/MaCFP), focusing on advancing predicting fire modelling, making them suitable for evaluation/validation purposes of CFD codes.

MODELLING

The standard version of FireFOAM 2.2.x (https://github.com/fireFoam-dev), developed by FM Global, is employed here. The code uses a fully compressible flow formulation and solves the Navier-Stokes equations, along with transport equations for species mass fractions and sensible enthalpy, using Favre-filtered quantities, assuming a unity Lewis number. The dynamic Smagorinsky model is used to model turbulence with a variable Prandtl number formulation [1]. Combustion is considered to be infinitely fast and it is based on the Eddy Dissipation Concept (EDC) [1] calculating the fuel reaction rate as:

$$\overline{\dot{\omega}_F'''} = \bar{\rho}\frac{\gamma^2\chi}{\tau(1-\gamma^3\chi)}min(\tilde{Y}_F, \tilde{Y}_{O_2}/s) \quad (1)$$

where $\gamma = C_\gamma(\nu\varepsilon_{sgs}/k_{sgs}^2)^{1/4}$ is the size of the fine structures, χ is the reacting part of the fine structures and $\tau = \min(\tau_{lam}, \tau_{turb})$ is a time scale which considers mixing under both laminar ($\tau_{lam} = \Delta^2/(C_{lam}\nu)$) and turbulent conditions ($\tau_{turb} = C_\tau(\nu/\varepsilon_{sgs})^{1/2}$). The constants C_γ and C_τ are determined dynamically based on the local Reynolds and Damköhler number similar to the work by Parente et al. [3]. The Damköhler number is defined as $Da = \tau/\tau_{chem}$ with the chemical time scale given by an Arrhenius equation and the constants for methane combustion taken from [4]. For radiation modelling, the radiative transfer equation is solved using the finite volume discrete ordinates model with the absorption/emission term modelled through the weighted-sum-of-gray-gases model (WSGGM). The path length, L, is calculated as $L = 3.6V/A$ with the volume, V, calculated by summing all the cell volumes where reaction takes place. Assuming a rectangular shape (due to the geometry of the line burner), then the corresponding surface area, A, can be calculated. This approach bypasses any modelling uncertainties related to prescribing a fixed path length (which will exist as the flame extinction limit is approached) and allows it to vary dynamically during the numerical simulations.

The modelling approach employed to account for local extinction considers an enthalpy balance based on the concept of a critical flame temperature. The criterion, examines whether a reactant mixture within a computational cell will have enough energy to raise its temperature above a critical flame temperature to sustain combustion and is expressed as [5]:

$$\hat{Y}_F\left(h_F(\tilde{T}) + \Delta H_{c,F}\right) + \hat{Y}_O h_O(\tilde{T}) + \hat{Y}_D h_D(\tilde{T}) > \hat{Y}_F h_F(T_{CFT}) + \hat{Y}_O h_O(T_{CFT}) + \hat{Y}_D h_D(T_{CFT}) \quad (2)$$

where $\hat{Y}$ is the mass fraction, $\hat{h}$ is the sensible enthalpy, the sub-scripts F, O and D denote the fuel, oxidizer (i.e., air) and diluents (i.e., inert gases or products of combustion) present in the reacting mixture, $\tilde{T}$ is the (filtered) initial temperature cell temperature, $\Delta H_{c,F}$ is the heat of combustion of the fuel and T_{CFT} is a critical flame temperature. The composition of the reactant mixture within a computational cell that can react is calculated based on the composition of the species present in that cell as $\hat{Y}_F = min(\tilde{Y}_F, \tilde{Y}_O/s)$, $\hat{Y}_O = s\hat{Y}_F$, $\hat{Y}_D = (\hat{Y}_O/\tilde{Y}_O)(\tilde{Y}_F - \hat{Y}_F + \tilde{Y}_D)$ where $\tilde{Y}_F, \tilde{Y}_O, \tilde{Y}_D$ are the (filtered) mass fractions of fuel, oxidizer and diluents in the computational cell and s is the oxidizer to fuel mass ratio. The enthalpy balance is based on the initial composition of the reactant mixture before any combustion is computed (i.e., after scalar transport but before any reaction). If the inequality is false then reaction within a computational cell is inhibited for the current time step. Based on the formulation of the model, any excess fuel within a cell acts as a diluent while any excess air does not. In general, the model is applicable to scenarios involving thermal quenching and quenching by dilution, since both mechanisms can be accounted for in the enthalpy balance.

For the evaluation of the flame extinction criterion, a critical flame temperature, T_{CFT}, is needed. A fixed value of 1700 K [6] is suggested by the CFD code Fire Dynamics Simulator (FDS). A sensitivity study with values between $T_{CFT} = 1500 - 1700$ K in [7] revealed some sensitivity in the numerical predictions when simulating the same test case. A different approach is employed here which allows the critical flame temperature, T_{CFT}, to vary depending on the local composition and residence time in every computational cell. The extinction temperatures as a function of the residence times, τ, are taken the from Perfectly Stirred Reactor simulations for different diluent agents (N_2, CO_2, H_2O) with varying concentrations reported in [8]. The local critical flame temperature on every computational cell is calculated as a mass-weighted average of the critical flame temperature based on the local inert (N_2, CO_2), given by a power law in the form of $T_{CFT,\ (N2,CO2)} = 1407.4\tau^{-0.073}$, and the critical flame temperature based on the non-inert diluents (H_2O) given by a power law in the form of $T_{CFT,\ (H2O)} = 1517.8\tau^{-0.034}$. The resulting critical flame temperatures are bounded between 1480 K and 1780 K, common minimum and maximum limits for most hydrocarbon fuels. In general, different approaches can be employed for determining T_{CFT} (e.g., from PSR simulations, opposed flow calculations with detailed chemistry, experiments) for different fuels when data are not available. Alternatively to what is employed in this study, T_{CFT} values as a function of strain rate can also be considered bringing this way some sub-grid scale effects in the flame extinction modelling. Nevertheless, these two approaches are not completely different from one another as small residence times in PSR calculations will result in T_{CFT} values similar to what you would get from e.g. opposed flow calculations with large strain rates and the other way around. Flame re-ignition is also an important aspect when it comes modelling flame extinction with infinitely fast chemistry. At any point in space and time, any unburned fuel in the computational domain can mix with the oxidizer and react regardless of the temperature of the mixture. This behavior is not realistic and can cause erroneous phenomena of re-ignition further downstream from the fire source where they should not be present. For this reason, an ignition temperature, T_{ign}, is specified in the numerical simulations which controls when any re-ignition phenomena can occur. This criterion essentially only allows any unburned fuel to react with the oxidizer if the temperature in the computational cell is above the specified T_{ign}value (and the enthalpy balance in Eq. (2) is satisfied). However, this approach cannot be globally applied within all the computational domain without affecting the primary ignition and combustion process. Following the work performed by White et al. [6], a small pilot flame area just above the fuel inlet, 25 mm in height, is specified in which combustion is allowed to take place without allowing for any flame extinction or re-ignition to occur (i.e., both phenomena essentially cannot occur there). Inside this region it is $T_{ign} = 0$ K. For the region outside the pilot flame area, $T_{ign} = 1200$ K is used, taken from extrapolated values of ignition temperatures of methane combustion at low strain rates, typically found in fires, based on [9]. This value is also close to the range of re-ignition temperatures suggested in [10] for the same test case. It is worth noting that this value is above the auto-ignition temperature of methane (i.e., 910 K). The use of this value is motivated by the fact that any re-ignition phenomena will occur for a lean methane mixture, diluted by hot air and combustion products, which will raise the ignition temperature of the mixture above the auto ignition temperature of methane. Nevertheless, a more advanced re-ignition modelling approach, that is not based on a user input parameter, is worth considering in future numerical simulations. A possible approach could be the use of a re-ignition temperature as a function of strain rate for different methane dilution ratios based on [9].

IOP Conf. Series: Journal of Physics: Conf. Series **1107** (2018) 062003 doi:10.1088/1742-6596/1107/6/062003

NUMERICAL SETUP

A computational domain of 1.6 m x 2 m x 1 m is used. The base mesh consists of 25 mm cells stretched towards the top boundary with an expansion ratio of 1.5. A grid refinement strategy is then employed up to the experimentally reported flame height. A box having dimension of 0.8 m x 0.8 m x 0.6 m refines the mesh to 12.5 mm while a second box of dimensions 0.6 m x 0.6 m x 0.4 m refines the mesh to 6.25 mm. The total number of cells is then approximately 886000. The methane flow rate is set to 1 g/s while the co-flowing oxidizer is provided at 85 g/s, according to the experiments. For the cases involving flame extinction, the oxygen mass fraction in the oxidizer port is initially maintained at $Y_{O2} = 0.233$ for 4 s and is subsequently decreased by 0.005/s (followed by equivalent increase in Y_{N2}). For the solution of the RTE, 48 solid angles are used for angular discretization. All simulations were set to run for 40 sec with a varying time step, limited by a maximum Courant number of 0.9. Time averaging over the last 35 s was used to produce time-averaged quantities. A second order backward scheme is used for time discretization. A second order filtered linear scheme is used for the convective terms which introduces small amounts of upwind if the solution is unbounded. For scalar transport, a second order TVD scheme using a Sweby limiter is applied.

RESULTS

The centerline mean and rms temperatures for different mesh sizes are first presented in Fig. 1(a) for a test case with $X_{O2} = 0.18$ in the oxidizer stream. The extinction model was not employed as the experimentally measured combustion efficiency was about 99%. Maximum flame temperatures of approximately 1240±40 K are predicted with all three mesh sizes, showing reasonable grid convergence, with an increase in the rms temperatures as the grid size decreases. The slightly lower maximum flame temperatures with the finer grid are attributed to a combined effect of turbulence, combustion and radiation. Even though the integrated reaction rates for all cases were the same (i.e., 1 g/s), the reaction rates right at the fuel source decrease and shift slightly downstream with decreasing grid size. Additionally, the predicted radiative fractions for the 25, 12.5 and 6.25 mm cases were 0.172, 0.196 and 0.212, respectively, while the experimental was about 0.195. The lack of influence of grid resolution on the mean and rms temperature profiles is attributed to the use of dynamic models that are able to self adjust their coefficients depending the local properties of the flow. Nevertheless, further investigation is needed in order to make more concrete statements regarding this aspect.

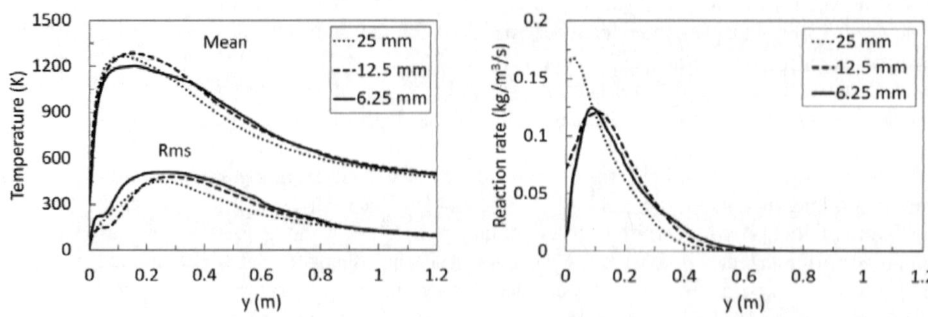

(a) (b)

Fig. 1. Centerline (a) mean and rms temperatures and (b) reaction rates for the $X_{O2} = 0.18$ case.

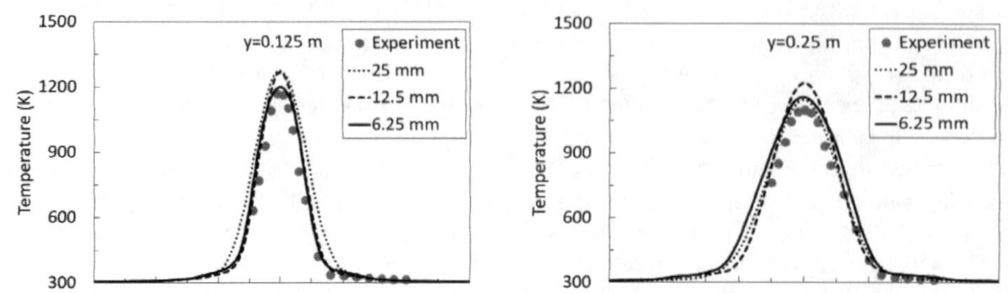

(a) (b)

Fig. 2. Mean temperatures at heights (a) $y = 0.125$ m and (b) $y = 0.25$ m for the $X_{O2} = 0.18$ case.

IOP Conf. Series: Journal of Physics: Conf. Series **1107** (2018) 062003 doi:10.1088/1742-6596/1107/6/062003

The radial profiles of the mean temperatures at two different heights for the $X_{O2} = 0.18$ case are presented in Fig. 2. Overall, the numerical predictions with the finest grid size employed (i.e. 6.25 mm) are able to reproduce quite reasonably both the flame width and the maximum flame temperatures at these two heights.

The predicted combustion efficiencies from the numerical simulations are presented in Fig. 3(a). It is interesting to note that the numerical predictions with the coarsest grid (i.e., 25 mm) are strongly affected by the extinction/re-ignition model and cannot accurately simulate the scenario. As the grid size is refined this influence diminishes with minimal influence on the 12.5 mm grid size (i.e., combustion efficiency slightly lower than 1 in the initial stage when $X_{O2} = 0.21$ in the oxidizer stream). Nevertheless, the predictions with both the 12.5 mm and 6.25 mm grid sizes are able to predict the oxygen concentration where extinction takes place (i.e., approximately $X_{O2} = 0.122$ based on the experimental data). The predictions with the 6.25 mm grid are in better agreement with the experiments, when compared to the 12.5 cm grid case, and are able to reproduce the experimental trends towards flame extinction even though the decay phase somewhat starts faster. Based on the results shown above, it is clear that with the current modelling approach a relatively fine grid is needed in the numerical simulations in order to accurately predict flame extinction. This can perhaps pose some limitations in cases where large-scale fire scenarios need to be simulated. It is also worth noting that the re-ignition temperature, T_{ign}, can influence the predictions and the resulting combustion efficiency (not shown here). Effectively an increase in the T_{ign} values results in flame extinction earlier than expected (i.e., for concentrations $X_{O2} > 0.12$). The start of the decay phase of the combustion efficiency is also partially affected by the choice of this value. Equally important can be the choice for the critical flame temperature, T_{CFT}, particularly in the case where a constant value is chosen and does not vary locally in every computational cell. The combination of different constant T_{CFT} and T_{ign} values can lead to equally accurate predictions of flame extinctions for perhaps the wrong reasons (i.e., compensating effects). It is obvious that accurate predictions of flame extinction with infinitely fast chemistry are directly linked to the combined modelling of combustion and radiation since it is the resulting (filtered) temperatures that will determine the degree of flame extinction (through comparison with T_{CFT} in the enthalpy balance and with T_{ign} in order to determine re-ignition). The influence of radiation/turbulence interaction should also be explored in the future.

Similar observations can be made by examining the predicted radiative fractions, χ_r, shown in Fig 3(b). Reasonably good predictions are obtained for the 12.5 mm and 6.25 mm cases with the simulations being able to reproduce the decrease in the heat released due to radiation as the flame extinction limit is approached. However, some over-predictions on the resulting radiative fractions on the finest grid case are still evident. It is worth noting that capturing this decrease in χ_r can be very important when trying to accurately predict flame extinction in any random fire scenario. For example, the use of a constant radiative approach cannot be directly employed, since it will lead to flame extinction faster than expected, nor can an a-priori tabulation of the radiative fractions with decreasing X_{O2} values be expected for any given scenario. Additionally, the use of the simpler gray-gas model is known to over-predict the resulting radiative fractions [11] unless a calibration constant is introduced which also puts some limitations in the application of the model to random scenarios. The accuracy of the WSGGM with a dynamically determined beam length in this scenario was satisfactory in terms of predicting the resulting radiative fractions. However, application of the model in more scenarios involving different fuels is needed before making concrete statements regarding its accuracy. Equally important, the use of a dynamically determined beam length will be more challenging when it comes to scenarios involving flame spread where a distinct flame shape cannot be so easily determined.

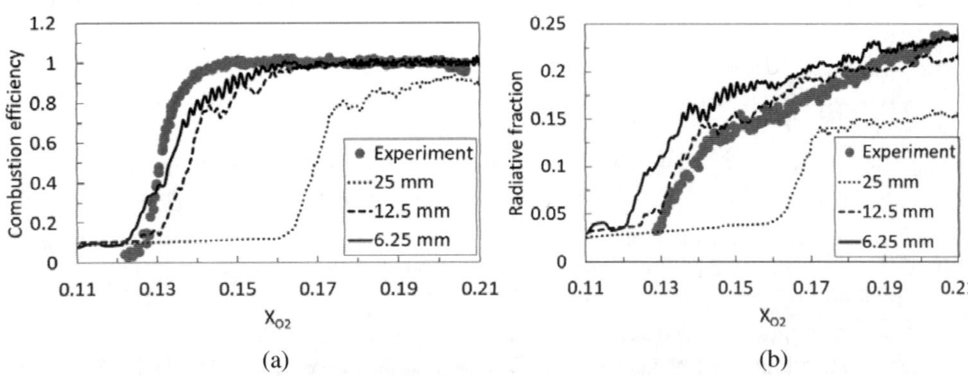

Fig. 3. Numerical predictions of the (a) combustion efficiencies and (b) radiative fractions.

CONCLUSIONS

Large eddy simulations of flame extinction with N_2 as extinguish agent were presented and the results were compared to the turbulent CH_4 line burner experiments by White et al. [2]. Use of advanced modelling approaches related to turbulence, combustion and radiation was made along with the application of an enthalpy-based extinction model with a locally varying critical flame temperature. The predictive capabilities of the models was analyzed by comparing the numerical predictions for different grid sizes to the experimental thermocouple temperatures, combustion efficiencies and radiative fractions.

The numerical predictions were, overall, in good agreement with the experimental data when fine grid sizes were used (i.e., in the order of 6.25 mm). More specifically, the resulting maximum flame temperatures and flame widths matched reasonably well the experimental profiles at two different heights examined. The numerical simulations were also able to reproduce the decreasing trend of the combustion efficiencies and radiative fractions with some discrepancies when the extinction limit was reached.

From the results of the current work, it is suggested that caution must be taken when modelling flame extinction with infinitely fast chemistry using an enthalpy-based extinction model based on the concept of a critical flame temperature. Accurate modelling of combustion (i.e., in terms of resulting mean and rms flame temperatures) and radiation (i.e., in terms of decreasing radiative fractions), along with relatively small grid sizes, is a pre-requisite for predicting flame extinction. Re-ignition modelling can also be important. The choice of a constant re-ignition temperature approach might be worth revisiting in the future by employing a more advanced re-ignition model based e.g. the local composition and strain.

ACKNOWLEDGEMENTS

This research has been funded by Ghent University (Belgium) through GOA project BOF16/GOA/004.

REFERENCES

[1] Maragkos, G., Beji, T., Merci, T. (2017) Advances in modelling in CFD simulations of turbulent gaseous pool fires, *Combust. Flame* 181:22-38, https://doi.org/10.1016/j.combustflame.2017.03.012

[2] White, J.P., Link, E.D., Trouvé, A.C., Sunderland, P.B., Marshall, A.W., Sheffel, J.A., Corn, M.L., Colket, M.B., Chaos, M., Yu., H.-Z. (2015) Radiative emissions measurements from a buoyant, turbulent line flame under oxdizer-dilution quenching conditions, *Fire Saf. J.* 76:74-84, https://doi.org/10.1016/j.firesaf.2015.05.003

[3] Parente, A., Malik, M.R., Contino, F., Dally, B.B. (2016) Extension of the Eddy Dissipation Concept for turbulence/chemistry interactions to MILD combustion, *Fuel* 163:98-111, http://dx.doi.org/10.1016/j.fuel.2015.09.020

[4] Bui-Pham, M.N. (1992) Ph.D. Dissertation, University of California, San Diego.

[5] Vaari J., Floyd, J., McDermott R. (2011) CFD Simulations on Extinction of Co-Flow Diffusion Flames, *Fire Safety Science* 10:781-793, 10.3801/IAFSS.FSS.10-781

[6] McGrattan, K., Hostikka, S., McDermott, R., Floyd, J., Vanella, M., Weinschenk, C., Overholt, K., Fire Dynamics Simulator User's Guide, NIST Special Publication 1019, Sixth Edition, http://dx.doi.org/10.6028/NIST.SP.1019

[7] White, J.P., Vilfayeau, S., Marshall, A.W., Truvé, A., McDermott, R.J. (2017) Modelling flame extinction and reignition in large eddy simulations with fast chemistry, *Fire Saf. J.* 90:72-85, https://doi.org/10.1016/j.firesaf.2017.04.023

[8] Sankaran, V., Jiang, H., Colket, M., Zhang, S., Corn, M. (2015) Flame Extinction Model for Fire Suppression Simulations, 9th U.S. National Combustion Meeting, Cincinnati, Ohio, May 17-20.

[9] Fotache, C.G., Kreutz, T.G., Law, C.K. (1997) Ignition of Counterflowing Methane versus Heat Air under reduced and Elevated Pressures, *Combust. Flame* 108:442-470, https://doi.org/10.1016/S0010-2180(97)81404-6

[10] Vilfayeau, S., White, J.P., Sutherland, P.B., Marshall, A.W., Truvé, A. (2016) Large eddy simulation of flame extinction in a turbulent line fire exposed to air-nitrogen co-flow, *Fire Saf. J.* 86:16-31, https://doi.org/10.1016/j.firesaf.2016.09.003

[11] Vilfayeau, S., PhD dissertation, University of Maryland, USA, 2015.

IOP Conf. Series: Journal of Physics: Conf. Series **1107** (2018) 062004 doi:10.1088/1742-6596/1107/6/062004

Control of Spontaneous Combustion of Sub-Bituminous Coal by Means of Heat Exchanger Submersion inside the Piles

Hanifa Khansa Zhafira[1], Anindia Setyo Widiani[1] and Yulianto Sulistyo Nugroho[1]

[1]*Department of Mechanical Engineering, Universitas Indonesia, Indonesia*

yulianto.nugroho@ui.ac.id

ABSTRACT

The low-rank coal, such as sub-bituminous and lignite has become a widely used alternative fuel for power generation. During the transportation process from mining location to power plant, coal on the barge might be stored for a long period of time. A control mechanism of spontaneous combustion of coal in barge was simulated in laboratory scale by means of submersing heat exchanger inside the coal bed. The low-rank coal is more prone to self-ignition (spontaneous combustion) which is quite difficult to detect, due to the complex event of spontaneous combustion involving many factors from both intrinsic and extrinsic factors, such as particle size, moisture content, ambient temperature, barometric pressure, *etc*. A laboratory scale experiment was set up to study the smouldering combustion phenomenon of coal samples and means of control by using heat exchanger submersion in a coal bed. The coal sample was placed inside a wire mesh cube with a coiled copper tube installed in the middle of the coal bed. Fresh water was flowed through the coiled copper tube with controlled intake temperature about 28 ºC. Approximately 65 grams of coal is heated in the oven at 210 ºC until the temperature in the middle of the pile reaches up to 320 ºC, then water is discharged through a copper tube to the middle of the pile and cooling it down. It is found that the temperature of coal sample can be brought down to 200 ºC depending on the water flow. Thermocouples are used to measure the coal's temperature in the middle of the pile. This method shows that the heat generated can be reduced to below the critical temperature for spontaneous combustion. The water flow rate has significant impact on temperature reduction of the coal inside the bed. This method has a potential benefit for controlling spontaneous combustion problem during barge transportation of coal.

KEYWORDS:
Coal; stockpiles; combustion; heat transfer; self-ignition; smouldering.

IOP Conf. Series: Journal of Physics: Conf. Series **1107** (2018) 062004 doi:10.1088/1742-6596/1107/6/062004

INTRODUCTION

Coal is one of the natural products that plays an important role in human life. Coal became one of the most widely used energy source in many countries. It is a natural resource that can be used in various fields, from fuel, gas producer, power source, even in industrial sector such as in making a component of-material and also support the state economy. Coal shipment often stored coal in barge for a long period of time, this event tends to lead to self heating of the coal and progress to combustion [1]. Several cases were reported regarding burning coal barges due to spontaneous combustion.

Shippers are required to provide a cargo declaration, especially if the cargo has a history of selfheating. One of the problems faced by coal producers is self-heating and spontaneous combustion [2,3]. Flaming is a chemical reaction that occurs between a fuel and an oxidizing agent, which results in the chemical energy stored in the outlet fuel being released as heat. Flaming is also called oxidation reaction because it involves Oxygen (O_2). During the transportation process from mining location to power plant, coal on the barge might be stored for a long period. The self-heating of stored coal can lead to spontaneous combustion that will reduce the coal quality, i.e. decreasing caloric content, volatile matter, and swelling capacities. Exothermic reactions that generate heat can reach the initiation temperature which ultimately forms hotspots in the hot spot of the coal if not lose [4].

The heat release occurs on the solid surface of the fuel and oxidation reaction and in the latter, these occur in the gas phase surrounding the fuel is the fundamental difference between smouldering and flaming combustion [5]. One of the most persistent types of combustion phenomena that leads to the largest fires on earth is a smouldering combustion. The slow, flameless, low-temperature burning of porous fuel [4]. Heterogeneous chemical reactions and the transport of heat, mass, and momentum in the gas and solid phases is fundamental combustion problem in smouldering [6]

The cooling unit affects the temperature gradient on the self-sustained smouldering combustion [7]. The energy transfer that occurs due to the different temperature between objects and material. A fluid passes over a solid surface of heat then energy is transferred to the fluid from the wall by heat conduction. These energies will be transported or conjugated, downstream by fluid, and diffused by fluid by conduction in the fluid them. This type of energy transfer process is called displacement heat convection. [8] The fluid motion raise heat transfer, then the higher velocity the higher heat transfer rate. This paper studies the application of heat exchanger submersion in the coal piles to control the spontaneous combustion occurrence at laboratory scale tests.

EXPERIMENTAL

This experiment uses sub-bituminous coal from Kalimantan, Indonesia. This type of coal is classified as medium calorie coal. The sub-bituminous coal contains more volatile matter and moisture than other bituminous coal types but has lower sulphur contents. The main properties of the coal are given in Table 1.

Table 1. Properties of the coal sample

	Particle size : ≤ 2000 mm		
Proximate Analysis	As Received	ADB	Dry
% Moisture	23.18	10.52	-
% Volatile matter	36.04	41.98	46.916
% Fixed Carbon	36.80	42.87	47.910
Ultimate Analysis			
% Ash	3.98	4.64	5.15
% C	52.58	61.25	68.45
% H	4.93	5.74	6.41
% N	0.85	0.99	1.11
% S	0.20	0.23	0.26
% O	14.29	16.64	18.60
Calorific Value (Cal/ gram)	5115	5958	6658

The samples were sieved using wire mesh 10 to obtain a sample with a particle size of ≤ 2000 mm. Figure 1 shows the coal sample used in this experiment. The apparatus consists of a temperature-controlled oven and wire mesh cube with an open top and 50 mm sides made from stainless steel 201 to hold the samples during experiment. The wire mesh cube was suspended in the re-circulating air oven. Two K-type thermocouples were used in this experiment. One thermocouple was placed in the middle of the wire mesh cube to measure the temperature of coal and one thermocouple was used to measure the temperature of the oven. The thermocouple was connected to a data acquisition device so that it can be read on a standard personal computer.

The coal sample was placed into the wire mesh cube that was then suspended in the oven. The oven was set to the temperature of 210 °C. The coiled copper tube with the diameter of 0.8 mm was installed inside the wire mesh cube. This tube was connected to a water supply from the disposable infusion set with bag. The configuration of a coiled copper tube and wire mesh cube is shown in Figure 2.

Fig. 1. Coal samples after sieving process

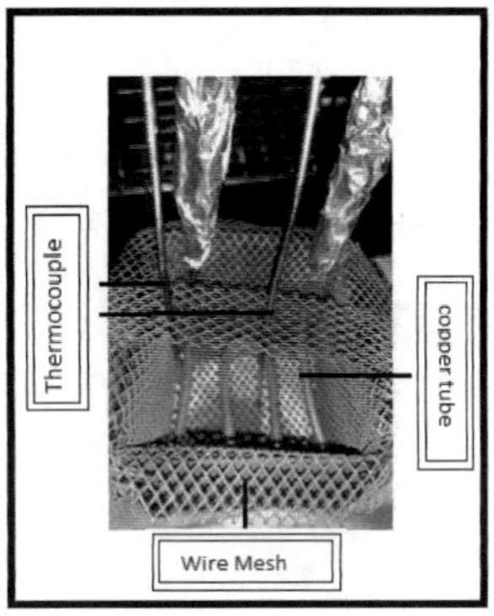

Fig. 2. Coiled copper tube installed inside the wire mesh basket

The experiment work was parted into two different procedures. The first procedure was developed in three stages. First, the coal inside wire mesh cube with installed coiled copper tube was suspended in the oven until the coal reach temperature of 320 ± 2 °C, then the water was distributed through the tube to the temperature differential of maximum 90 °C. When the coal's temperature was starting to increase, the water flow was stopped. Second, coal was suspended in the oven until temperature of 340 ± 2 °C. Later, water was distributed again like the first stage. The third stage was where the temperature of coal reach 360 ± 2 °C and water were distributed for ± 30 minutes until the coal's temperature was no longer reduced and has surpassed 360 °C. This procedure was repeated with a different flow rate. The variety of water flow rate that was used in this experiment are 40 mm/s; 55 mm/s; and 70 mm/s.

The second procedure was the coal inside wire mesh cube with installed coiled copper tube was suspended in the oven until coal reach temperature of 320 ± 2 °C, then the water was distributed through the tube continuously until the coal reach its lowest temperature.

For the third experiment the samples were sieved using wire mesh 10 to obtain a sample with a particle size of ≤ 2000 mm and wire mesh cube with an open top and 70 mm sides made from stainless steel 201 to hold the samples. Four K-type thermocouples were used in this experiment. One thermocouple was placed in the middle, one in the edge and one in the corner of the wire mesh cube to measure the coal's temperature and one thermocouple was used to measure the temperature of the oven. First, the coal inside wire mesh cube with installed coiled copper tube was suspended in the oven until the coal which in the middle of the wire mesh reach temperature of 250 ± 2 °C, then the water was distributed through the tube. In this experiment, the water flow which being used is 70 mm/s only.

IOP Conf. Series: Journal of Physics: Conf. Series **1107** (2018) 062004 doi:10.1088/1742-6596/1107/6/062004

RESULTS

The heating process that was applied to the coal cause self-sustained combustion on coal which makes its temperature took off to 320 °C. By distributed water in the tube, the heat inside coal was transferred to the water by conduction and convection. Figure 3 shows the decreasing of coal temperature when the water was distributed and when the water flow was stopped, the temperature was starting to rise rapidly.

For the water flow rate of 40 mm/s and the capacity of 0.5 mL/s, the temperature of coal decreased by 92 °C from 319 °C to 227 °C as the first stage. At the second stage, where the water started to distribute at coal temperature of 339 °C, the temperature difference of coal degraded to 86 °C. The temperature difference of the third stage is dropped for about 68 °C, from 359 °C to 291 °C and started to increase again, although the water was still distributed.

The same phenomenon was found in the experiment with water flow rate of 55 mm/s and the capacity of 0.7 mL/s. The temperature difference of the first stage, second stage, and third stage were 18 °C from 322 °C to 304 ° C; 9 °C from 342 °C to 333°C; and 7 °C from 362 °C to 355 °C.

For the water flow rate of 70 mm/s and the capacity of 0.9 mL/s, the temperature of coal decreased only by 6 °C at the first stage from 322 °C to 316 °C, and at the second stage was 4 °C from 342 °C to 338 °C. The interesting phenomenon happened at the third stage where the water started to distribute at coal temperature of 362 °C, the temperature of coal did not decrease. It can be seen as the heat absorbed by distributed water is lower than heat released by the coal. So whenever the water stops, the temperature will rise and it will not be lower than before.

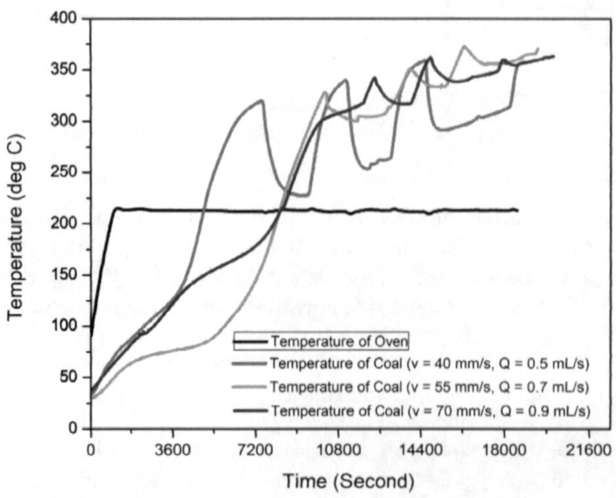

Fig. 3. The temperature of oven and coal during heating and water distributed stopped and go

The result of the second procedure experiment can be seen in Figure 4. For the water flow rate of 40 mm/s and the capacity of 0.5 mL/s experiment (Fig. 4A), the peak point is where the coal's temperature reached 318 °C, it is the coal's take-off temperature. Then the water was started to distribute through the copper tube. The water flows for about 6 hours continuously. By the end of the experiment, for the water flow rate of 40 mm/s and the capacity of 0.5 mL/s, the temperature difference of water is 16°C as well as the temperature difference of coal is 113 °C. For the water flow rate of 55 mm/s and the capacity of 0.7 mL/s (Fig. 4B), the temperature difference of the coal is 116 °C and the temperature difference of the water is 12°C. And for the water flow rate of 70 mm/s and the capacity of 0.9 mL/s, (Fig. 4C) the temperature difference of the coal is significantly high at 226 °C from 320 °C to 94 °C and the temperature difference of the water is 34°C. Whenever the higher flow rate of water flows continuously, the more heat is absorbed by the water, the heat absorbed by distributed water is higher than the heat released by coal.

IOP Conf. Series: Journal of Physics: Conf. Series **1107** (2018) 062004 doi:10.1088/1742-6596/1107/6/062004

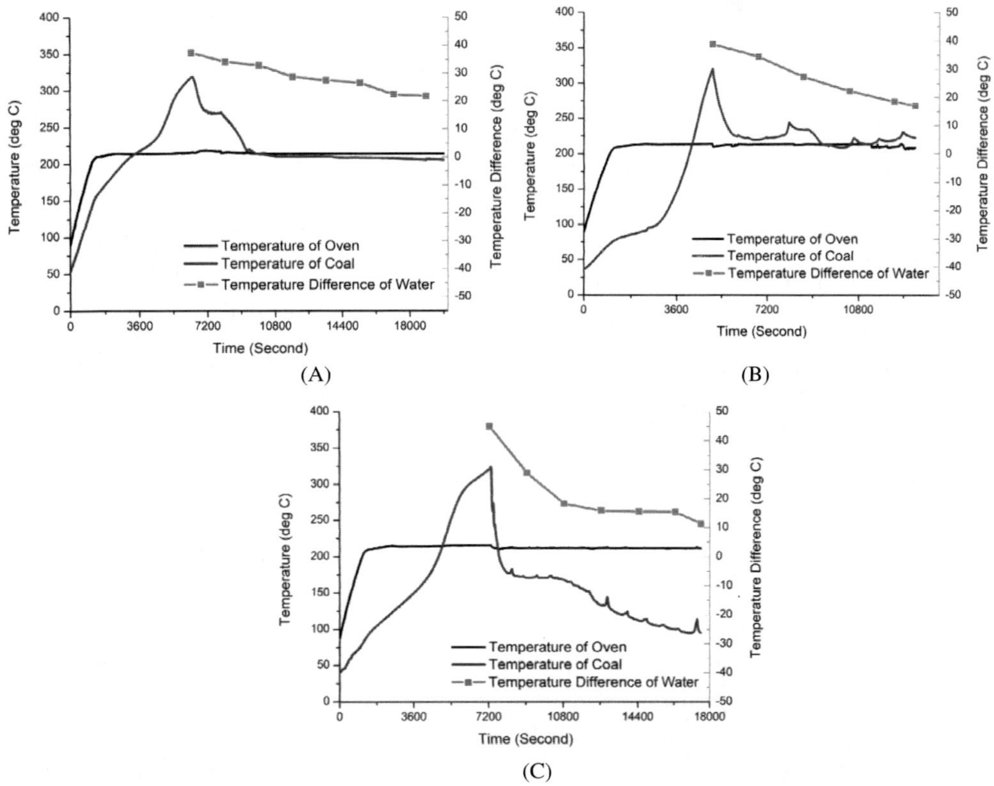

Fig. 4. The temperature of oven and coal during heating and continuous water distributed process; (A) water flow rate of 40 mm/s and the capacity of 0.5 mL/s; (B) water flow rate of 55 mm/s and the capacity of 0.7 mL/s; (C) water flow rate of 70 mm/s and the capacity of 0.9 mL/s

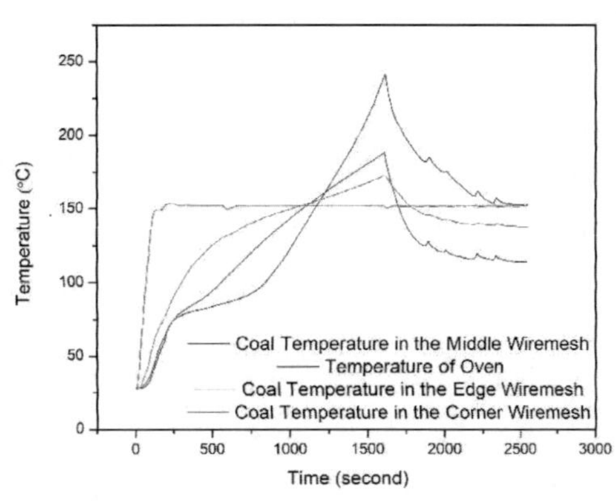

Fig. 5. The temperature of oven, coal temperature in the middle, edge and corner during heating and continuous water distributed process

The result of the third procedure can be seen in Figure 5. For the water flow rate of 70 mm/s and the capacity of 0.9 mL/s (Fig.5), the peak point is where the coal's temperature reached 240 °C. It is the coal's take-off temperature. The water was started to distribute through the copper tube. The water flows for about 6 hours continuously. By the end of the experiment, for the water flow rate of 70 mm/s and the capacity of 0.9 mL/s, the temperature difference of water is 57°C.

CONCLUSION

The result of the experiment shows that as water flowed through a pipe inside the coal bed temperature decreased significantly. When the water in the tube was stopped and released in sequence, a trend of temperature rise was observed. Flow rate of the cooling water inside the heat exchanger played a great role in controlling the development of the hotspots within the coal wire mesh basket. The effectiveness of cooling mechanism depended upon the flow rate and the initial temperature of the coal when water flow was started. By using a larger wire mesh basket with a dimension of 7 cm sides, it was observed that the critical temperature for spontaneous combustion of coal decreased from 240 °C (5cm sides basket) to 159 °C. Application of the heat exchanger for larger basket shows that temperature at the corner of the wire mesh that did not pass by the pipe can decrease significantly. As the effect of self-heating phenomenon that leads to spontaneous combustion in the coal pile is affected by the heat balance and the availability of oxygen from the surrounding atmosphere, it is critical to define the best position of the heat exchanger submersion inside the coal bed for real application of this approach in real stockpile or coal barge design.

REFERENCE

[1] Ran Vijay Kumar Singh (20130 Spontaneous heating and fire in coal mines. Procedia Engineering 62 (2013) 78 – 90
[2] Sujanti, Wiwik., Zhang, Dong-ke. (1999) A Laboratory Study of Spontaneous Combustion of Coal: The Influence of Inorganic Matter and Reactor Size. *Fuel* 78:549–556
[3] Y.S. Nugroho1, A.C. McIntosh, B.M. Gibbs (2000) Low-temperature oxidation of single and blended coals. *Fuel* 79:1951–1961
[4] Falcon, Rosemary. (1986) Spontaneous Combustion of The Organic Matter in Discards From The Witbank Coalfield.*Journal of the South African Institute of Mining and Metallurgy* 86:243-250
[5] Rory M. Hadden, Guillermo Rein, Claire M. Belcher. Study of the Competing Chemical Reactions in the Initiation and Spread of Smouldering Combustion in Peat. *Proceedings of the Combustion Institute* 34:2547–2553
[6] Guillermo Rein. (2009) Smouldering Combustion Phenomena in Science and Technology. International Review of Chemical Engineering. 1:3-18
[7] Ragni Fjellgaard Mikalsen, Smoldering combustion-from pulsations to extinguishment. Poster.
[8] Stoecker, Wilbert F., Jones. 1982. *Refrigeration and Air Conditioning*, 2nd Edition. McGraw-Hill Inc. Urbana-Champaign.

IOP Conf. Series: Journal of Physics: Conf. Series **1107** (2018) 062005 doi:10.1088/1742-6596/1107/6/062005

Evaluation of Oxygen Reduction System (ORS) in Large-scale Fire Tests

Xiangyang Zhou, Yibing Xin and Sergey Dorofeev

FM Global, Research Division, USA

xiangyang.zhou@fmglobal.com

ABSTRACT

An oxygen reduction system (ORS) is a fire prevention system that uses a low-oxygen environment to reduce, if not eliminate, the potential for ignition and fire propagation in a protected space. The key parameter for ORS design is the limiting oxygen concentration (LOC), defined as the lowest O_2 concentration that can support combustion for a given fuel. The present work examines LOCs of solid fuels at large scale in a configuration representative of a rack-storage and discusses the results in relation to the design concentration for ORS.

To simulate ORS applications in engineering practice, a two-tier fuel array of standard commodities in rack storage configuration was set up in an enclosure. A constant N_2/Air mixture flow was supplied to the enclosure at a desired oxygen concentration. The oxygen concentration was varied from 9% up to 17%. A premixed flame with a constant heat release rate was used as the ignition source to maintain repeatable test conditions. This premixed flame ignitor represents potential heat sources such as electric arc and hot work that are not sensitive to oxygen level. The tested materials included five standard commodities: Class 3, Cartoned Unexpanded Plastic (CUP), Cartoned Expanded Plastic (CEP), Uncartoned Unexpanded Plastic (UUP) and Uncartoned Expanded Plastic (UEP). This paper reports detailed test results at different oxygen levels only for Class 3. The impact of the test conditions on fire propagation was examined and the results showed that the oxygen concentration was the major parameter to control fire propagation. When successful flame spread is initiated by the igniter, the fire size tends to be larger as the igniter is sustained for a longer time. The results of fire propagation success were obtained for the five standard commodities under different oxygen concentrations with a sustained igniter (hard limits) and without a sustained igniter (soft limits). The LOC was defined as an oxygen concentration at 0.05 probability of flame spread by using statistical analysis. The resulting LOC values measured for different commodities in a two-tier rack storage are:

- Cartoned (Class 3, CUP and CEP) – hard limit 11.1%,
- Uncartoned (UUP and UEP) – hard limit 13.0%,
- Cartoned (Class 3, CUP and CEP) – soft limit 13.8%,
- Uncartoned (UUP and UEP) – soft limit 14.7%.

These LOCs are generally lower than the oxygen design concentrations recommended by existing standards including VdS 3527 and EN 16750 due to different test conditions. The hard limits resemble fundamental LOC for gases and vapors and do not depend significantly on the ignition duration and array size, while the soft limits vary significantly with the size and configuration of the fuel array and ignition duration. It is concluded that the hard limits are more suitable for ORS design purposes.

KEYWORDS: *oxygen reduction system; limiting oxygen concentration, rack-storage fire*

INTRODUCTION

An oxygen reduction system (ORS) is a fire prevention system that uses a low-oxygen environment to reduce the likelihood of ignition and minimize fire propagation in a protected space. A typical ORS consists of an on-site nitrogen generator located outside the protected space, piping and pump network to provide an N_2/Air mixture of the desired oxygen concentration, and multiple sensors measuring oxygen concentrations within the protected space. A control unit located outside the protected space monitors the signals from the oxygen sensors to adjust the nitrogen production and supply.

The ORS uses the reduction of the oxidizer concentration as the protection strategy. If ignition and fire propagation can be greatly reduced, if not entirely prevented, the damage from heat, water and smoke becomes minimal, leading to favorable fire protection for high-value occupancies and other occupancies that are sensitive to water and smoke damage. However, when the O_2 level is too low, life safety becomes a concern even for a primarily unoccupied space. As a result, proper ORS design requires to know not only the key parameter, the limiting oxygen concentration (LOC), but also the fire behavior at oxygen concentrations close to the LOC.

The LOC is the lowest O_2 concentration that can support combustion for a given fuel. There have been extensive studies of LOC levels for premixed and non-premixed flames of gaseous fuel and vapors in various configurations including microgravity conditions [1-3]. It has been shown that the LOC depends on the competition between heat release and heat losses at the scale of the laminar flame thickness. The LOC decreases as the strain rate induced by turbulent or laminar flows decreases. It reaches a minimum at a low strain rate condition and then increases with even lower strain due to radiation losses becoming dominant. This minimum LOC for a given fuel, temperature, and diluent is a fundamental property that results in flame quenching at any flow condition [3, 4].

For solid fuels such as plastics and packaging, the determination of LOC is more complicated. One way to tackle the problem is to refer to the LOC of the products of thermal decomposition of solid fuels, i.e., the pyrolysis gases. The composition of the pyrolysis gases depends on the nature of the fuel and on the thermal conditions experienced by the fuel, such as the heating rate, maximum temperature, etc. The fundamental LOCs of these fuels are expected to be in the range of 6-13% by volume [2].

Another way to consider the LOC problem for solid fuels is to assess the competition between heat release and heat losses at a macro level for the problems of heterogeneous ignition and flame spread. The rate of heat release in heterogeneous reaction depends on the heat flux to the fuel surface from an ignition source or spreading flame, and on the oxygen concentration. The rate of heat losses depends on the convective flow over the fuel surface and the radiative loss. The ignition source creates an ignition kernel, which generates energy roughly in proportion to its volume and loses heat roughly in proportion to its surface. The larger the size of an ignition source, the lower is the portion of heat lost by the ignition kernel. With a positive heat balance the ignition kernel expands due to flame spread and it may become sufficiently large that the removal of the ignition source doesn't change the positivity of that balance. Therefore, with a longer duration of the ignition source, the size of the ignition kernel becomes larger, making it less dependent on the continuation of the ignition source. Moreover, any configuration with fuel surfaces facing each other over a narrow gap traps radiation and reduces the radiation loss. The physical effects described above suggest that the macro-level LOC depends significantly on the overall scale, fuel configuration, the size and duration of the ignition source.

The LOCs of several solid fuels were studied using laboratory-scale experiments in the Fire Propagation Apparatus (FPA), where additional heating was provided to model large-scale fire test conditions [5]. It was found that, for common fuels such as corrugated cardboard, wood, polystyrene (PS) and polyethylene (PE), the LOCs are in the range of 10-13%. This level of oxygen concentration is lower than the 15-16% range given by the testing prescribed by standards such as VdS 3527 [6] and EN 16750:2017 [7]. This difference is expected considering the effects described above and the test conditions such as ignition and fuel configuration [8].

The present work examines LOCs of solid fuels at large scale in a configuration representative of rack-storage and discusses the results in relation to the design concentration for ORS.

EXPERIMENTAL SETUP

The large-scale fire test was designed to simulate ORS applications with sufficient low oxygen air supply. Figure 1 shows the schematic of a simulated ORS environment and a photo of the test enclosure under a 5 MW calorimeter. The test enclosure included two parts: a lower plenum space for the supply of the nitrogen/air mixture, and an upper controlled-atmosphere (CA) volume as test space. A steel perforated plate with 13% opening area was installed at the separation between the lower plenum and the upper CA room. The upper CA room was enclosed using fire resistant gypsum board. The tested commodities including the wood pallets were placed on the perforated plate at the center of the enclosure in a 1×2 and 2-tier high rack-storage configuration. The flue space between the two commodities was 0.15 m. Five standard commodities of Class 3, Cartoned Unexpanded Plastic (CUP), Cartoned Expanded Plastic (CEP), Uncartoned Unexpanded Plastic (UUP) and Uncartoned Expanded Plastic (UEP) [9] were tested and only the results for the Class 3 commodity are reported in this paper. Below the commodity platform was the perforated floor that was designed to achieve a uniform co-flow boundary condition. The nitrogen-diluted air supply was provided through a 0.3-m duct discharging downward to the floor of the lower plenum to generate a uniform air/N_2 upward flow. A liquid nitrogen tank with vaporizer was used in this work to provide gaseous N_2.

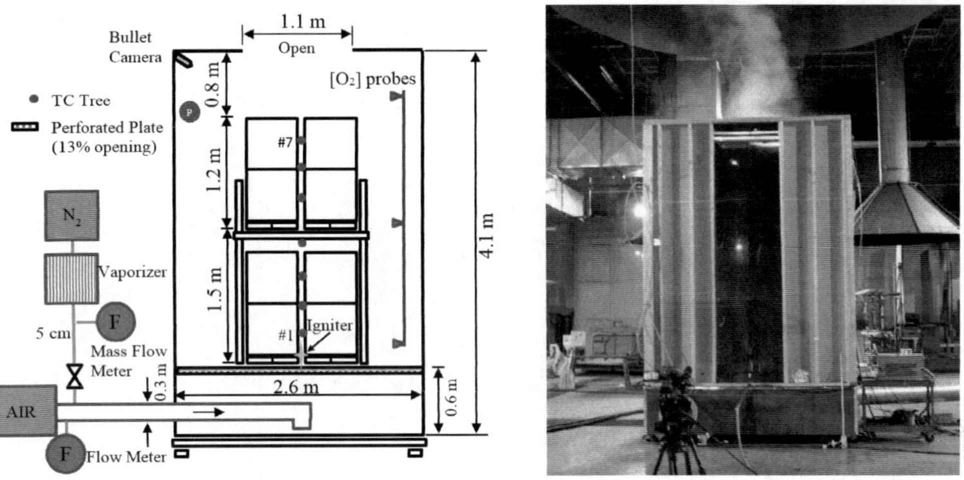

Fig. 1. Elevation view of the simulated ORS test design and a photo of the test enclosure.

The ignition location of all tests was selected at the center of the fuel array. To maintain repeatable initial conditions, a premixed flame (propane/air) with a constant heat release rate (HRR) of 33 kW was used, which is consistent with that of two half igniters used routinely in sprinkler fire tests under normal air conditions. This premixed-flame ignitor also represents potential heat sources such as electrical short or arc and hot work that are not sensitive to oxygen level. The O_2 concentration in the enclosure was measured by three gas sampling probes installed at different elevations. High-Definition cameras were positioned in front of the observation window to record the test. Seven thermocouples [TCs, Type K, gage 28 (0.33 mm), bare-bead] were installed vertically along the centerline of the fuel array to monitor flame propagation, as shown in Fig. 1. The enclosure air was purged by supplying the N_2/air mixture to achieve the specified O_2 design concentration before ignition for each test. The O_2 concentrations used in this work were between 9% and 17% by volume.

RESULTS AND DISCUSSIONS

To illustrate the fire development, Fig. 2 shows a series of flame images recorded at different times for Class 3 commodity with [O_2] at 11.4%. Prior to ignition, the O_2 concentration in the enclosure was maintained around the target value for at least one minute. The average gas supply rates were 0.32±0.01 m^3/s for N_2, and 0.430±0.001 m^3/s for air, which were adjusted toward the target O_2 concentration. Before ignition, the temperature of the air/N_2 mixture entering the enclosure was 17.3±0.1 °C monitored by TC #1 near the perforated floor. The average velocity of air/N_2 mixture in the enclosure was estimated to be 0.11 m/s based on the total gas flow rate and the enclosure cross-section area. As shown in Fig. 2, the time of 0 s denotes the ignition event when the premixed flame was pushed to the flue center of the fuel array. At 20 s after ignition, the flame spread upward along the external cardboard surface with exfoliation of several large pieces of paper. At 40 s, the flame propagated to the 2nd tier. At 70 s, the flame height exceeded the top of the fuel array and started to exit through the opening at the top of the enclosure. In the present work, successful fire propagation was indicated by the flame height exceeding the top of the fuel array.

Once the flame height exceeded the top of the fuel array, the igniter was shut off at 79 s by stopping the supply of propane/air. Figure 2 shows that the flame disappeared at 82 s, indicating that the fire was extinguished three seconds after the igniter was shut off. For this case, the test result is considered unsuccessful propagation after igniter shut-off. Following these flame images, Fig. 3(a) shows the gas temperatures measured in the flue center

IOP Conf. Series: Journal of Physics: Conf. Series **1107** (2018) 062005 doi:10.1088/1742-6596/1107/6/062005

by using seven TCs. Except for TC #1 near the igniter, the other TCs showed that gas temperature increased with time to 800°C or a higher value. Using the temperature of 400°C to denote flame arrival, the time required for fire propagation was 23 s from TC #2 to #4, and 9 s from TC #4 to #7. The flame propagation speed was initially slow in the 1st tier and then increased in the 2nd tier. After the igniter was shut off at 79 s, Fig. 3(a) shows that all the gas temperatures dropped rapidly with time, indicating a decaying fire.

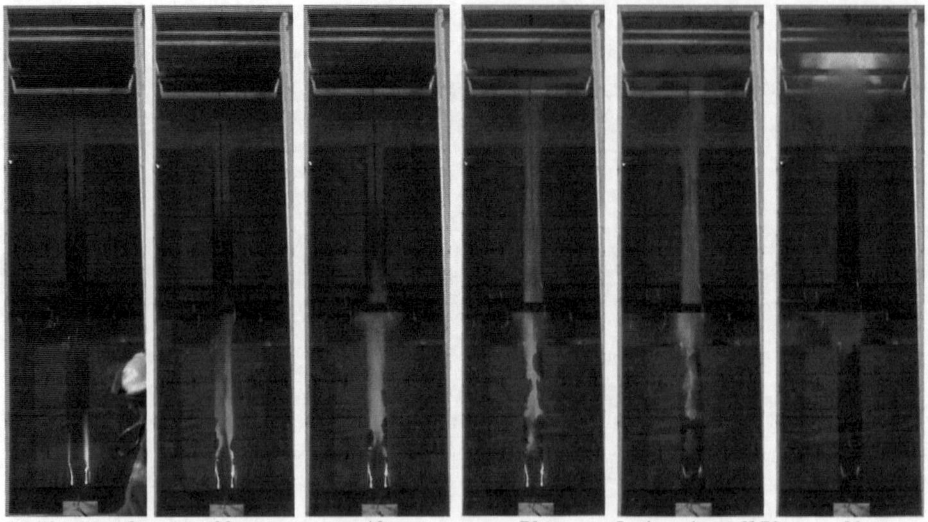

Fig. 2. Images of fire development for Class 3 commodity with 11.4% O_2.

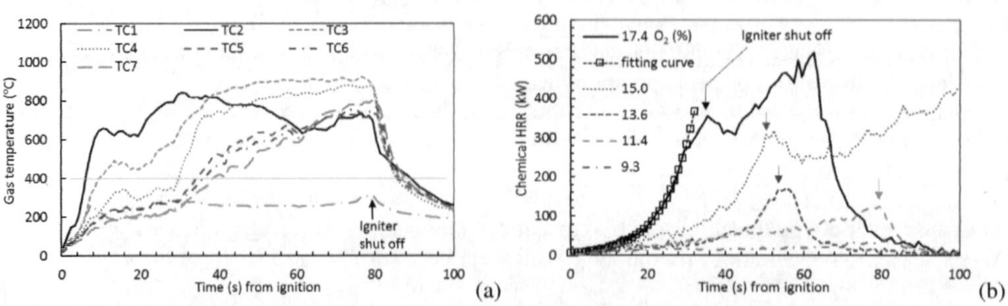

Fig. 3. (a) Gas temperatures measured in the flue center along the height for 11.4% O_2, and (b) chemical heat release rates (HRR) measured for Class 3 commodity under different O_2 concentrations.

Figure 3(b) shows the chemical HRRs measured for Class 3 under different O_2 concentrations, where the arrows mark the time when the flame propagated to the top of the fuel array and the igniter was shut off. For 17.4% O_2, Fig. 3(b) shows a fast HRR growth (10 kW/s) from ignition to the first peak of 360 kW when the igniter was shut off at 35 s. This is considered a propagating fire. After the igniter was shut off at 35 s, the HRR slightly reduced and then increased again to the second peak of 520 kW at 63 s. The fire was manually extinguished at 63 s. This fire at 17.4% O_2 was sustained without the igniter. For a lower oxygen level of 15.0%, Fig. 3(b) shows a similar HRR growth trend. The test results are that the flame propagated both with/without sustained igniter. It is noted that the peak value of HRR with 15.0% O_2 is slightly lower than that of 17.4% O_2. For the other two tests with 13.6% O_2 and 11.4% O_2, the HRRs increased with time and then decreased after the igniter was shut off. Both tests are marked as propagated for flame spread with igniter and non-propagated when the igniter was shut off. For 9.3% O_2, the Class 3 commodity could not be ignited, and Fig. 3(b) shows a flat HRR with time. The test results with 9.3% O_2 are non-propagated for both cases of flame spread with/without igniter.

The arrows in Fig. 3(b) also denote the times required for the flame to propagate to the top of the fuel array. Following these arrows, Fig. 3(b) shows that the fire size (HRR) and the fire propagation speed decrease with the oxygen level. The combustion theory discussed in previous work [5] has shown that the chemical reaction rate and the flame temperature will reduce with the oxygen level. Therefore, as shown in Fig. 3(b), the fire propagation was delayed due to lower flame temperature and thus reduced flame heat fluxes to the solid fuel at lower oxygen levels. To quantify the fire growth rate, as shown in Fig. 3(b) for 17.4% O_2, the exponential function of $HRR(t) = a_0e^{bt}$ was used to fit the initial HRR growth during 0 – 30 s, in which b (s^{-1}) denotes the growth rate parameter and a_0 is a fitting coefficient. For the oxygen level of 17.4% shown in Fig. 3(b), the obtained growth rate parameter is b = 0.13 (s^{-1}). Using the same fitting function for the other oxygen levels, the growth rate parameters are 0.09 s^{-1} for 15.0% O_2, 0.056 s^{-1} for 13.6% O_2, 0.03 s^{-1} for 11.4% O_2, and near zero for 9.3% O_2.

When the oxygen level is reduced to a limit, such as 9.3% O_2 in Fig. 3(b), the chemical heat release of both the ignition source and reactions between pyrolysis gas and oxygen cannot overcome the heat losses from the combustion zone to sustain continuous ignition along fuel surfaces. When the igniter is shut off, the heat release rate from the reaction between the pyrolysis gases and oxygen needs to be sufficiently high to exceed the heat loss and sustain burning. As shown in Fig. 3(b), the fire can only sustain at higher oxygen levels (15.0% and 17.4%), but not at low oxygen levels (13.6% and 11.4%), after the igniter is shut off. This observation is specific to the size of the fuel array and the duration of the ignition (of the order of 1 min) as noted in the introduction.

Since the duration of the ignition source may vary significantly in real fire events, its impact on fire growth deserves further investigation. For a target oxygen concentration of 13%, Fig. 4 shows the chemical HRRs measured in five tests with different igniter shut-off times. The arrows in Fig. 4 indicate the time of igniter shut-off. Clearly, the fire continued to grow while the igniter was on. All fires started to decrease in intensity after the igniter was shut off. For the earliest shut-off at 55 s, Fig. 4 shows that the HRR dropped in only 7 s to 1/10 of its maximum. When the igniter was shut off at 210 s, the HRR remained above 600 kW for more than two minutes and the HRR decay rate became slower. Finally, all the fires extinguished with time. This behavior, however, may be due to the limited array size impacting the outcome of the tests, denoted by the red and black curves in Fig. 4. It is seen in Fig. 4 that, when successful flame spread is initiated by the igniter, the fire size tends to be larger as the igniter is sustained for a longer time.

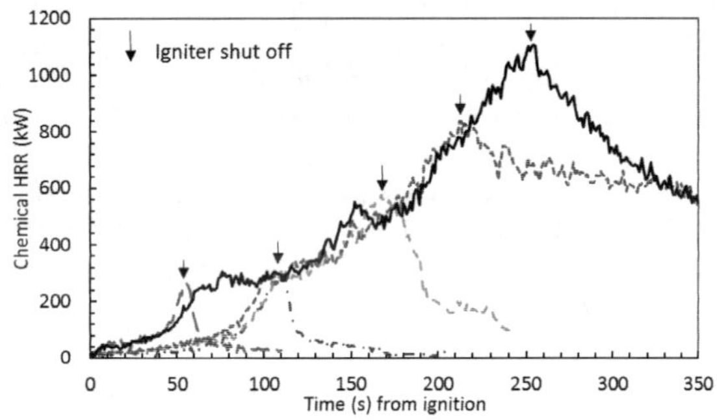

Fig. 4. Chemical HRRs measured for Class 3 commodity at 13% O_2 with different igniter shut-off times.

The test results of fire propagation with/without the sustained igniter were also obtained for the other commodities at different oxygen levels [10]. Since the oxygen level was not adjusted continuously to precisely identify the flame extinction condition, and given the natural variability in the results, statistical analysis was used to estimate the LOC value from the data. Exact logistic regression is used to model binary outcome variables. Figure 5 shows the fire propagation probability curve with respect to the oxygen volume fraction with continuous igniter and after igniter shut-off, where the black circles are the test results denoted as zeros (non-propagation) and ones (propagation). The cartoned commodities (Class 3, CUP and CEP) are placed in the same group because the initial fire propagation takes place on the external corrugated cardboard. The statistical analysis showed that the oxygen level was the dominant variable to control fire propagation.

Figure 5 also shows the cutoff points of the probability curve at the 0.05, 0.5 and 0.95 probabilities of fire propagation. To apply a small margin of safety, the LOC is estimated here to the cutoff points as the oxygen volume fraction corresponding to the 0.05 probability. This value hence provides a 95% confidence level in the LOC results. The resulting LOC values measured for different commodities in a two-tier rack storage are:

- Cartoned (Class 3, CUP and CEP) with a sustained ignitor 11.1%,
- Uncartoned (UUP and UEP) with a sustained ignitor 13.0%,

- Cartoned (Class 3, CUP and CEP) with ignitor shut-off after ignition 13.8%,
- Uncartoned (UUP and UEP) with ignitor shut-off after ignition 14.7%.

It should be noted that the limits with a sustained ignitor approach the fundamental LOC values for gaseous fuels such as 11.1% - 12.0% for methane [2] as described in the introduction. These limits are thus the hard limits that do not depend significantly on the fuel configuration, size and ignitor duration. The limits with ignitor shut-off after ignition are the soft limits. They are obtained for relatively short (~1 min) ignition duration and are significantly dependent on the ignition duration and the size of the fuel array. It would, therefore, appear that the hard limits should be used for ORS design purposes.

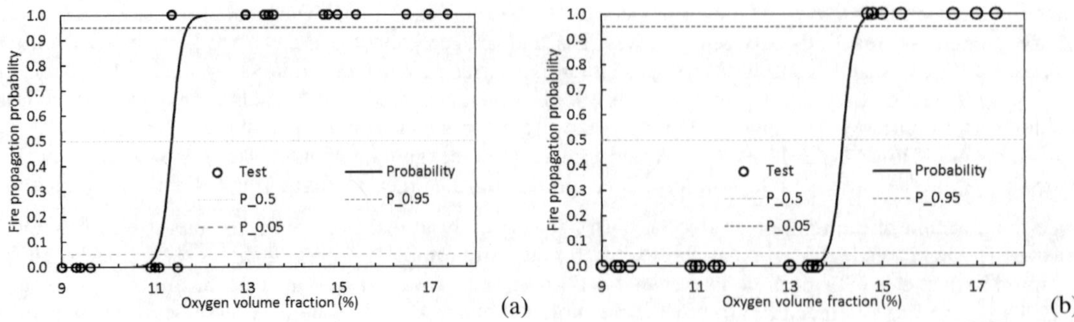

Fig. 5. Fire propagation probability by oxygen level estimated for cartoned commodities (Class 3, CUP and CEP): (a) with continuous igniter, and (b) after igniter shut-off.

SUMMARY

This work has evaluated the LOC limits for oxygen reduction system (ORS) in large-scale fire tests. A two-tier fuel array of standard commodities was set up in a rack-storage configuration within an enclosure. A constant nitrogen/air mixture flow was supplied to the enclosure at a desired oxygen concentration. The oxygen concentration was varied from 9% to 17%. A premixed propane ignitor was used as ignition source. The tested materials included five standard commodities of Class 3, CUP, CEP, UUP and UEP. Results for fire propagation success were obtained for the tested commodities under different oxygen concentrations with/without a sustained igniter to determine the limiting oxygen concentration (LOC) to support fire. These LOCs have been found to be generally lower than the oxygen design concentrations recommended by existing standards including VdS 3527 and EN 16750:2017 due to different test conditions. The difference between the hard and soft limits have been discussed, with the conclusion that the lower hard limit should be used for ORS design.

REFERENCES

[1] Maček, A. (1979) Flammability Limits: A Re-examination, *Combustion Science and Technology* 21:43-52.
[2] Zlochower, I. A. and Green, G. M. (2009) The Limiting Oxygen Concentration and Flammability Limits of Gases and Gas Mixtures, *J. Loss Prev. Process Ind.* 22 (4): 499–505.
[3] Snegirev, A. Yu. (2015) Perfectly stirred reactor model to evaluate extinction of diffusion flame, *Combustion and Flame* 162:3622–3631.
[4] Dorofeev, S. B. (2017) Thermal quenching of mixed eddies in non-premixed flames, *Proc. Combust. Inst.* 36:2947-2954.
[5] Xin, Y. and Khan, M. M. (2007) Flammability of Combustible Materials in Reduced Oxygen Environment, *Fire Safety Journal* 42:536-547.
[6] Vds 3527en: Inerting and Oxygen Reduction Systems, Planning and Installation, 2007-01 (01).
[7] EN 16750:2017 Fixed Firefighting Systems - Oxygen Reduction Systems - Design, Installation, Planning and Maintenance, 2017-09-13.
[8] Nilsson, M. and Hees, P. (2014) Advantages and challenges with using hypoxic air venting as fire protection. *Fire and Materials* 38(5):559-575.
[9] FM Approvals LLC., "Approval Standard for Automatic Control Mode Sprinklers for Fire Protection, Class Number 2000," Norwood, MA 02062, USA, March 2006.
[10] Zhou, X. and Xin, Y., "Evaluation of Oxygen Reduction System (ORS) in Large-Scale Fire Tests," FM Global Research Report, January 2018. http://www.fmglobal.com/research-and-resources/research-and-testing/research-technical-reports

IOP Conf. Series: Journal of Physics: Conf. Series **1107** (2018) 062006 doi:10.1088/1742-6596/1107/6/062006

Comparison of sprinkler activation times under flat and corrugated metal deck ceiling

Marcus RUNEFORS[1], Per BOSTRÖM[2] and Erik ALMGREN[2]

[1]*Division of Fire Safety Engineering, Lund University, Sweden*
[2]*Bengt Dahlgren Fire & Risk AB, Sweden*

marcus.runefors@brand.lth.se

ABSTRACT

The distance of sprinklers below ceiling is of fundamental importance since it has a great influence on activation time. Accepted distances are specified in different sprinkler standards such as NFPA 13, FM 2-0 and EN 12845 for different configurations. This distance is unambiguously defined for flat smooth ceilings, but for corrugated ceilings, such as corrugated metal deck ceilings found in many warehouses, the distance can be measured from either the top or the bottom of the deck. NFPA 13 states that it should be measured from the bottom if the thickness of the deck is below 75 mm and from the top if it is higher. The requirement in NFPA 13 has been adopted from the FM datasheets, but the authors have been unable to identify any background to this requirement.

In this paper, the activation time of sprinkler heads at different distances below a corrugated ceiling with thickness of 125 mm was simulated using FDS 6.6.0 and compared to distances below a flat smooth ceiling. The case was based on a warehouse with 6-tier storage of Uncartoned Expanded Plastics (UEP) with a ceiling height of 11.2 m and clearance to storage of 2.6 meters. Only the activation of the first sprinkler was simulated since it was expected that the drag from droplets of previously activated sprinkler heads would thicken the ceiling jet, making the activation time of additional sprinkler less sensitive to the distance from ceiling.

The results show that the corrugated ceiling has a large influence on the ceiling jet allowing sprinklers further from the ceiling to activate.

In the case studied, the activation time of sprinklers below the corrugated ceiling more closely resembled those under a flat smooth ceiling when the distance below the corrugated ceiling was measured from the bottom of the deck. More simulations and experiments are however needed before any guidance on changes in sprinkler standards can be given.

KEYWORDS:

suppression; CFD; sprinkler activation; corrugated ceiling

INTRODUCTION

The distance of sprinklers below ceiling is of fundamental importance for effective operation since it has a great influence on activation time. If the sprinkler is located too far from the ceiling, it will not be affected by the ceiling jet of the fire and will therefore fail to activate according to the fire tests on which the design is based. When the sprinkler eventually activates, the fire might be too large for the specified sprinkler flow and the system will fail to achieve its intended goal.

For flat smooth ceilings, the distance below ceilings are unambiguously defined, but for corrugated ceilings such as corrugated metal deck ceiling the distance can be defined as either the top or bottom of the deck. NFPA 13-2017 [1] specifies in point 8.5.4.1.2 that this depends on the depth of the corrugated ceiling, if this is below 75 mm the distance should be measured from the bottom and if it is above 75 mm it should be measured from the top. This was adopted to the 2007 edition of NFPA 13 based on a proposal to the committee [2]. The requirement was adopted from a similar requirement in FM datasheet 2-0 [3]. The authors have been unable to identify any empirical background to this requirement.

The depth of the corrugated metal deck is chosen to achieve the needed stiffness of the design which is, among other things, dependent on snow load. Therefore, at least in many northern countries, the depth of the corrugated ceiling is almost always above 75 mm and therefore the distance from ceiling should be measured from the top of the metal deck ceiling. This often poses a practical challenge for sprinkler design since the designer must put the pipe below beams and other similar obstacles.

The flow patterns in non-flat ceilings are rather complex and continuous obstructions close to the ceiling is known to push the ceiling jet downwards towards the sprinkler head and therefore sprinklers are generally allowed to be placed further from the ceiling in such situations (defined as obstructed ceilings in NFPA 13). This effect can be expected to occur to some extent also in corrugated metal deck ceilings, but the extent is unknown.

In this paper numerical experiments [4] are performed using FDS to investigate the influence of the corrugation of the ceiling on the sprinkler activation time.

METHODOLOGY

Two different cases were compared where one had a corrugated ceiling and the other had a flat smooth ceiling. The comparison was performed using simulation in FDS version 6.6.0 [5]. The quantity of interest was the activation time of the first sprinkler since after the first activation, the turbulence of the ceiling jet is increased by the drag of the sprinkler droplets. This will lead to a thickening of the ceiling jet and therefore the activation time of the following sprinkler can be assumed to be less dependent on the distance from ceiling compared to the first sprinkler.

Since only the time of activation of first sprinkler is sought, the cooling of the sprinkler by other sprinklers are not needed. Further, the heat conducted from the bulb to the to the sprinkler and pipe are neglected. Therefore the sprinkler bulb temperature can be calculated using the equation below [6].

$$\frac{dT_l}{dt} = \frac{\sqrt{|\mathbf{u}|}}{RTI}(T_g - T_l) \quad (1)$$

Where T_l is the temperature of the sprinkler bulb and T_g and $\mathbf{u}$ are the gas temperature and velocity vector respectively. The quantity RTI is a measure of the thermal inertia of the sprinkler bulb.

Geometry and boundary conditions

The geometry used represents a part of a warehouse with a ceiling height of 11.2 meters with a six-tier rack storage (8.6 m). The distance between the boxes was 0.4 meters in both the horizontal and vertical direction. The corrugated ceiling had dimensions according to below.

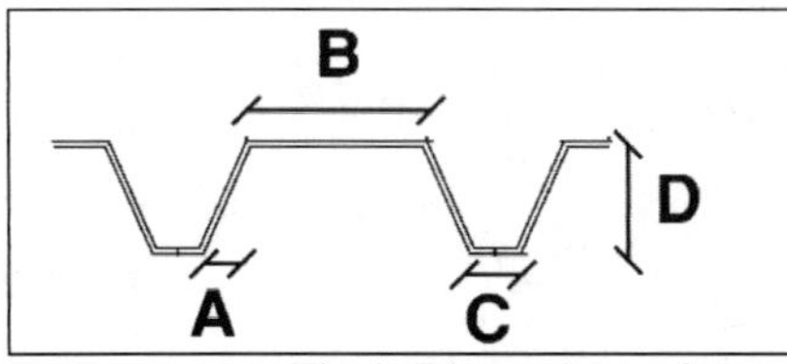

Measurement	Distance
A	25 mm
B	125 mm
C	75 mm
D	125 mm

Fig 1. Dimensions of corrugated ceiling

The simulated domain was 7.0 x 5.5 meters with open vents in all vertical boundaries. The thermal boundary condition for the ceiling and floor was set to a fixed value (ambient, 20°C).

Since mesh boundaries in regions with high velocity gradients is not recommended due to limitations in the pressure solver [7] the cell size was identical from the floor to the ceiling in region containing a 4 x 4-meter floor area. The chosen cell size was based on the need to resolve the geometry of the corrugation and was set to 25 mm. A grid independence study was performed for a limited case using a 12.5 mm grid. Outside this well resolved area, a crude mesh with 50 mm cell size was simulated spanning 1.5 meters in three directions outside the well resolved mesh to locate the open boundaries away from the region of interest. The forth direction (away from the sprinkler head) was judged not to affect the results and therefore the open boundary was placed directly on the well resolved mesh.

The sprinkler had an activation temperature of 68°C and an RTI of 36. The location is furthest away possible from the point of ignition given a 3 x 3 m spacing. The values are chosen based on that they are representative for the type of occupancy considered. The sprinkler was located under the higher part of the ceiling which was found to have a very limited effect on activation time, but slightly conservative for the comparison.

Fire

A study made for rack storage fires shows that the fire growth for a 5-tier (7,3 m) storage of Uncartoned Exposed Plastics (UEP) is 860 kW/s [8]. The current storage configuration consists of a 6-tier storage rack, which could result in a slightly higher fire growth rate. However, the expected fire growth rate is difficult to anticipate since no tests has been made for a 6-tier rack storage fire. Furthermore, a slightly higher fire growth rate is only assumed to affect the sprinkler activation time in absolute numbers and not the comparison between corrugated and flat roof. Therefore, a fire growth rate correspondent to a 5-tier rack storage fire is selected. The heat release rate per unit area was set to 350 kW/m^2.

The heat release rate of the different surfaces of the boxes was prescribed based on an analysis of videos from actual experiments and was found to be approximately according to below. The heat release rate per unit area was assumed to be the same for all surfaces once they ignite.

- 0-4 seconds Inside of tier 1
- 4-8 seconds Inside of tier 2
- 8-16 seconds Inside of tier 3 and outside of tier 1
- 16-24 seconds Inside of tier 4 and outside of tier 2
- 24-32 seconds Inside of tier 5 and outside of tier 3

After 32 seconds the heat release rate is 27.5 MW and sprinkler activation is assumed to have occurred and the simulation is terminated.

RESULTS

In this chapter, results from both the grid validation study and the comparison between flat and corrugated ceiling are presented.

Grid validation study

Since it was not feasible to run the entire case with a finer grid, a limited case was simulated which had the same geometry apart from that the ceiling height was reduced to 2 meters and the fire was simulated as horizontal, radially growing, fire at floor level following an ultra-fast curve (0.188 kW/s^2). This was performed to verify that the simulations adequately captured the flow patterns in the corrugated ceiling. The grid size is very well resolved compared to the fire size and therefore no grid sensitivity analysis relating to the fire was needed.

The quantities that are compared are gas temperature and velocity since those are the quantities, together with sprinkler temperature rating and RTI, that determine the sprinkler activation time (see Eq. 1). The location of measurements is indicated in the figure below.

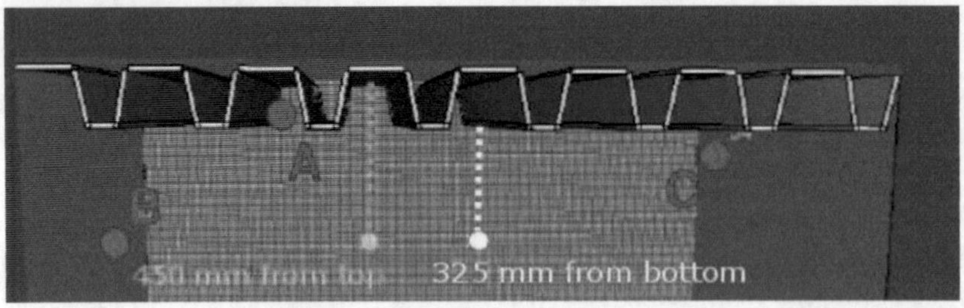

Fig 2. Roof configuration and location of measurement probes

The points are located 100 mm (A), 450 mm (B) and 200 mm (C) from the top of the deck ceiling. The results are shown in the figures below. Note the different scales in the different diagrams.

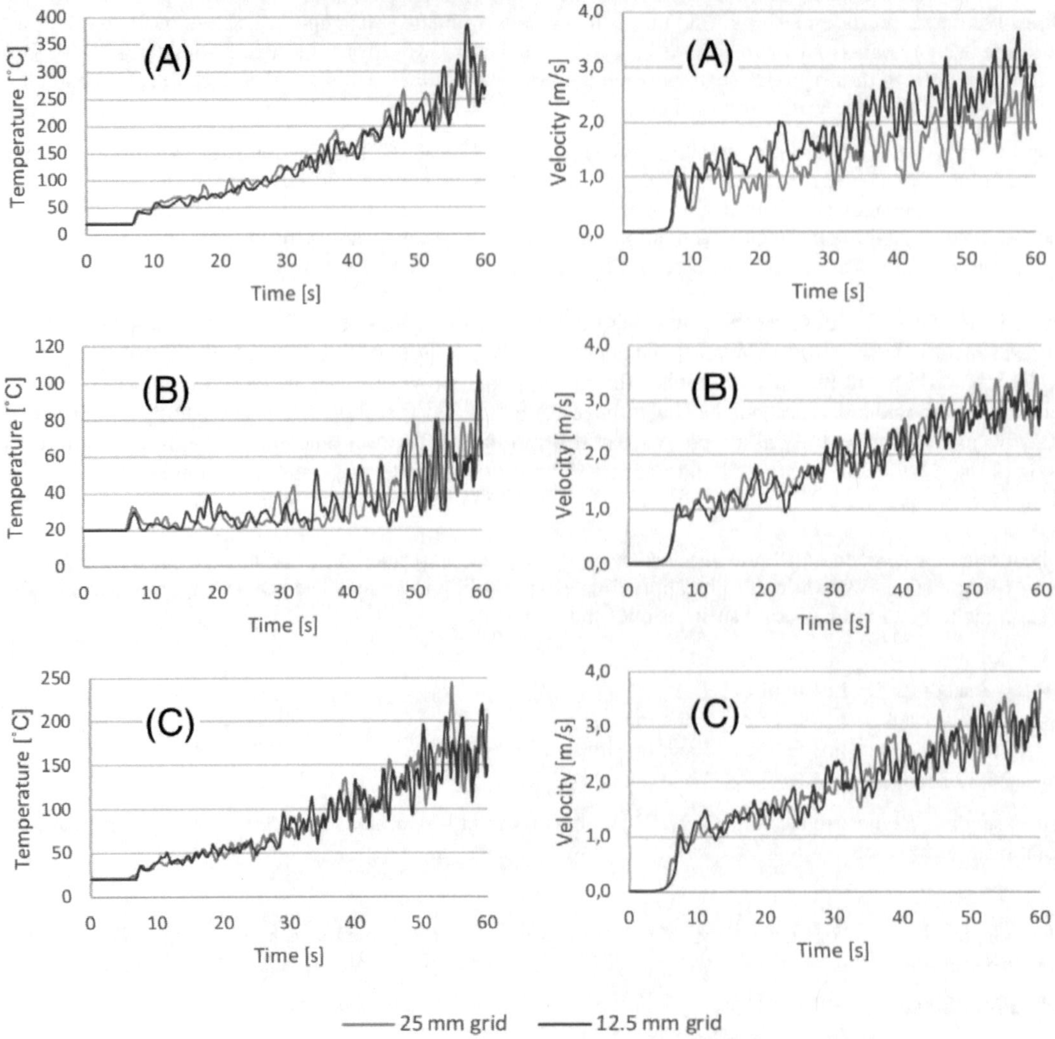

Fig 3. Impact of grid resolution on temperature and velocity on different locations A-C in fig. 2

Comparison between flat and corrugated ceiling

Below a representative figure of the sprinkler link temperature under a flat and corrugated ceiling can be found.

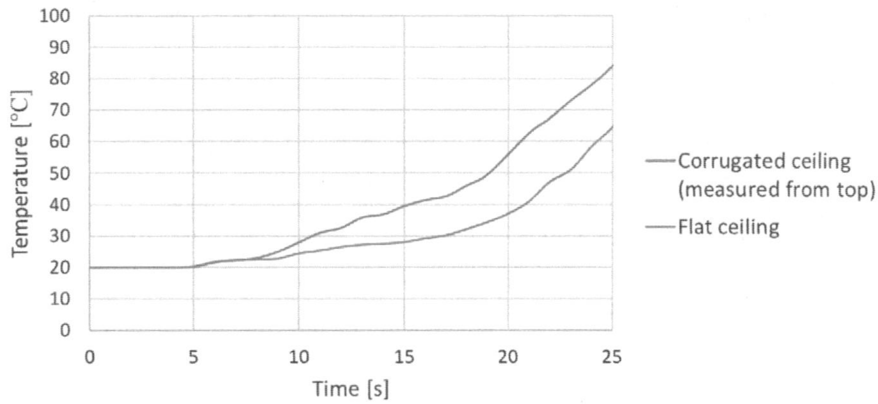

Fig 4. Typical diagram of the sprinkler link temperature at a certain distance from ceiling (in this case 450 mm) depending on type of ceiling.

The time to activation of first sprinkler can be found below.

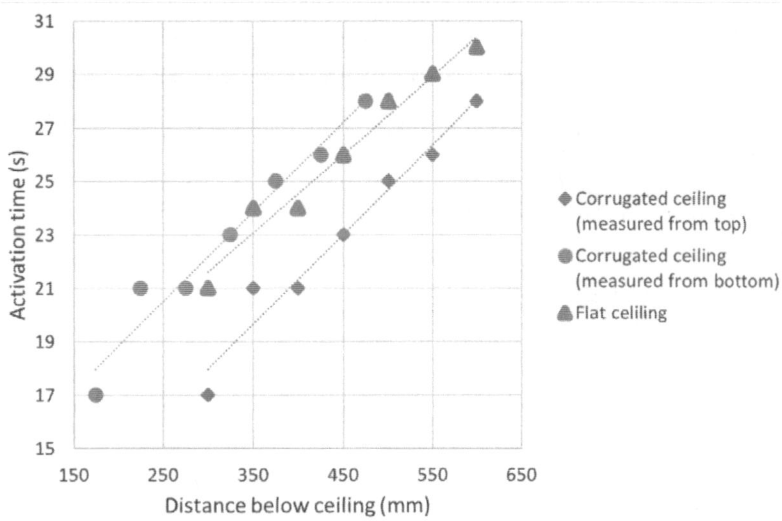

Fig 5. Activation time for sprinkler under flat and corrugated ceiling depending on distance from ceiling

DISCUSSION

The chosen grid resolution (25 mm) was found to be sufficiently grid independent (fig 3) apart from velocity inside the rills of the corrugated ceiling. Given the small dimensions of this volume, this was expected and can be accepted since the area of interest are further away from the ceiling as given by location "B" and "C" in fig 2.

The expected phenomena were the ceiling jet are pushed down by the corrugated ceiling was identified and did have a significant effect on the sprinkler link temperature (fig 4) and thus the sprinkler activation time (fig 5).

The results in fig. 5 indicates that the activation time is significantly lower at the same distance from flat and corrugated ceiling when measuring the distance in according to NFPA 13 and FM datasheet 8-9 (top of deck). The results are much closer when an alternative method of measuring the distance to the bottom of the deck are employed. The tested configurations are, however, too limited to give any conclusive results to the generalizability of the results to other similar configurations.

CONCLUSION

The results indicate that measuring the distance to sprinklers from the top of corrugated ceilings provide very conservative results and, at least in the configuration tested, the distance from the bottom of the corrugated ceiling gave results that more closely resembles those of flat ceilings. The tested configurations are, however, too limited in scope to give any guidance to possible changes in sprinkler standards. More numerical and experimental work should be conducted in the future to assess the generalizability of the observation. These studies should assess both the effect of ceiling geometry and sensitivity to, for example, ceiling height and fire growth rate.

REFERENCES

[1] NFPA, *NFPA 13 – Standard for the Installation of Sprinkler Systems*, 2016 edition, National Fire Protection Association, 2016

[2] NFPA, *Report on Proposals A2006*, NFPA 13, National Fire Protection Association, 2006, p.13-72

[3] FM Global, *Installation Guidelines for Automatic Sprinklers*, FM Global Property Loss Prevention Data Sheet 2-0, January 2018

[4] Johansson, N., Wahlqvist, J. & van Hees, P. (2014) *Numerical experiments in fire science: a study of ceiling jets*, Fire and Materials, 39(5), pp.533-544, DOI:10.1002/fam.2253

[5] NIST, *Fire Dynamics Simulator User's Guide*, Sixth Edition, NIST Special Publication 1019, 2017

[6] NIST, *Fire Dynamics Simulator Technical Reference Guide – Volume 1: Mathematical Model*, Sixth edition, NIST Special Publication 1018-1, 2017

[7] McDermot, R., *FDS Road Map*, Accessed March 15th 2018, https://github.com/firemodels/fds/wiki/FDS-Road-Map

[8] Xin, Y. (2017). *Effects of storage height on critical delivered flux of representative fuels*. Fire Safety Journal 91, pp 624-632. DOI: 10.1016/j.firesaf.2017.03.044

Discharge Characteristics of a Portable Compressed Air Foam System

Oluwadamilola OKUNROUNMU[1], Paul LHOTSKY[2] and George HADJISOPHOCLEOUS[1]

[1]*Department of Civil and Environmental Engineering, Carleton University, Canada*
[2]*CIVELEC Consultants, Canada*

oluwadamilola.okunrounmu@carleton.ca

ABSTRACT

Existing portable foam extinguishers generate fire-fighting foam at high pressures with the aid of an air aspirating nozzle. This system could encounter several limitations at the point of application such as poor foam quality due to the use of fire contaminated air for foam generation and insufficient momentum to reach the seat of fire. Research has shown that by incorporating compressed air into a portable foam system, the integrated foam system could generate superior quality foam with high momentum when properly installed with the right components. Several studies had been conducted on the extinguishing performance of compressed air foam systems on multiple fire types, both for small and large fires. Compressed air foam systems mitigate exposure of the operator to heat and provides faster knockdown of the fire plume as compared to air-aspirated foam because of its stronger stability and rheology. Since the expansion ratio of the foam can be regulated to combat specific fire types and sizes, compressed air foam systems can be utilized in protecting a variety of equipment of varied sizes.

The aim of this study is to investigate the discharge characteristics of a portable compressed air foam at low pressure. For this study, the requirements of NFPA 10 and CAN/ULC-S508 for a new system were used to determine the feasibility of the system.

The effect of air pressure on the expansion ratio of the foam was investigated with foam concentrate ranging from 2% to 4% for three different hoses with lengths of 1-m, 2-m and 3-m. Pressure used ranged from 1.72 bar to 5.52 bar. The 3% and 4% solution for the 2-m hose and 3-m hose exhibited similar trend of a rise and fall with pressure by generating fluid foam of medium expansion ratio in the range of 19 to 28. However, the expansion ratio of 3% solution and 4% solution for the 1-m hose increased monotonically with increasing pressure and generated wet foam of low expansion ratio in the range of 8 to 15. While low expansion foams are effective in extinguishing liquid pool fires, medium expansion foams are used for structural protection due to its slow drainage time and its ability to adhere to sloped, vertical, horizontal and slippery surfaces.

Discharge range tests were conducted to investigate the horizontal projection of the foam from the nozzle at a height of 0.9m above the ground. The test was conducted in an open space with little interference of wind. Visual record of the maximum discharge range was taken at intervals. The foam from the 1-m hose projected from 1 m at 2.42 bar to 2.4 m at 5.52 bar while the foam from the 2-m hose projected from 1.8 m at 2.42 bar to 4.5 m at 5.17 bar. Likewise, the foam from the 3-m hose with an initial discharge of 1.85 m at 2.41 bar increased progressively to above 4.5 m at 4.83 bar. The tests demonstrated the relationship between pressure and the momentum of the foam, showing that an increase in pressure leads to an increase in the range covered. Furthermore, flow rates at different pressures were investigated using 3% foam solution with a 2 m hose. The flow rate of the foam ranged from 8 g/s to 20 g/s at 1.93 bar and 5.24 bar respectively, indicating linear progression with pressure. The flow rates correspond to application times of 244 and 102 seconds respectively for the 2-liter solution. Overall, all foams tested met the requirements of the CAN/ULC-S508 standard.

KEYWORDS: *low pressure; compressed air foam; portable extinguisher; discharge range test*

IOP Conf. Series: Journal of Physics: Conf. Series **1107** (2018) 062007 doi:10.1088/1742-6596/1107/6/062007

INTRODUCTION

The use of portable fire extinguishers is an important part of the fire protection system in a building. Portable fire extinguishers are fire safety equipment used to extinguish or suppress small fires before they become a threat to life and property. The available combustible materials in a building govern the type of fire that could occur and hence the selection of portable fire extinguisher is important to ensure that it can suppress the expected class of fires. The use of an extinguisher not suitable for a class of fire may contribute to the fire hazard. For instance, the combustible materials in the dining room (such as sets of furniture, interior finishes, and light partitions) would typically generate Class A fires while in the kitchen a fire involving cooking oil is a class B fire. Therefore, it is ideal to have a portable fire extinguisher that is applicable on multiple fire types.

While some of the existing fire extinguishers can be used for two or more fire classes, their suppression effectiveness and adverse consequences after application is a concern. For example, water extinguishers can be used for Class A fires materials but could create a hazard when used on electrical equipment, or in laboratories containing flammable liquids and combustible metals [1]. Likewise, it contaminates chemical materials by reacting with the substance and water damages to water sensitive equipment. Therefore, it is advisable to be utilized on only unreactive materials, ordinary combustible materials and in locations that don't require major clean-up. On the other hand, dry powder extinguisher is suitable for electrical equipment combustible liquids but not applicable to cooking oil fire due to its poor cooling property [2]. In addition, application of dry powder type fire extinguisher in large quantity might reduce visibility in the environment and might require thorough clean-up process.

Several studies had been conducted on the extinguishing performance of compressed air foam system on multiple fire types, both for small and large fires. NRC Canada investigated the feasibility of using a fixed pipe CAF system for the protection of aircraft hangar in place of foam-water sprinkler system [3] while Rie et al [4] investigated the suppression performance of a CAF system at different air-to-aqueous foam solution mixing ratios on class B fires. In other studies, Weinschenk et al [5] investigated the suppression capacity of CAF for interior firefighting. Since the expansion ratio of the foam can be regulated to combat specific fire types and sizes, compressed air foam systems can be utilized in protecting a variety of equipment of varied sizes.

Existing portable foam extinguishers generate fire-fighting foam at high pressures with the aid of an air-aspirating nozzle [6]. However, the system could encounter several limitations at the point of application such as poor foam quality due to the use of fire contaminated air for foam generation and insufficient momentum to reach the seat of fire. Kim and Crampton [7] compared the extinguishing performance of a manually operated CAF system with hose stream application of water only and water-foam solution on a full-scale compartment fire tests. The results showed that the mobile CAF system was more effective in suppressing the fire compared to the other two systems. The mobile CAF system generated superior quality foam with high momentum when properly installed with the right components. In addition, it reduced exposure of the operator to heat and provided faster knockdown of the fire plume as compared to air-aspirated foam because of its stronger stability and rheology. However, the mobile CAF generating system operated at high pressure with little consideration to the quantity of foam solution used.

While the existing foam systems operate at high pressures, the study of operating portable CAF system at low pressure has not been studied. The aim of this study is to investigate the discharge characteristics of a portable compressed air foam at low pressure. For this study, the requirements of NFPA 10 and CAN/ULC-S508 for a new system was used to determine the feasibility of the system.

DESCRIPTION OF TESTS

The tests were conducted outdoors in a clear weather with little interference of wind. The temperature was fairly consistent, averaging about 18 degrees Celsius at the time of the experiment. The compressed air foam generating system for this study is shown in Figure 1. The set-up consists of an air compressor, 3-L container, 8-mm diameter plastic hose, nozzle and a tarp. A 2-L premixed foam solution in the designated proportion was prepared and poured into the 3-L container. The 3-L container was then pressurized by air at different pressures to create compressed air foam. The compressed air foam passes through an 8-mm diameter plastic hose, which was expected to be fully developed inside the hose and finally discharged through a nozzle. The foam concentrate used for the tests was Alcohol-Resistant Aqueous Film Forming Foam (AR-AFFF) and the concentrates was varied from 2 to 4%.

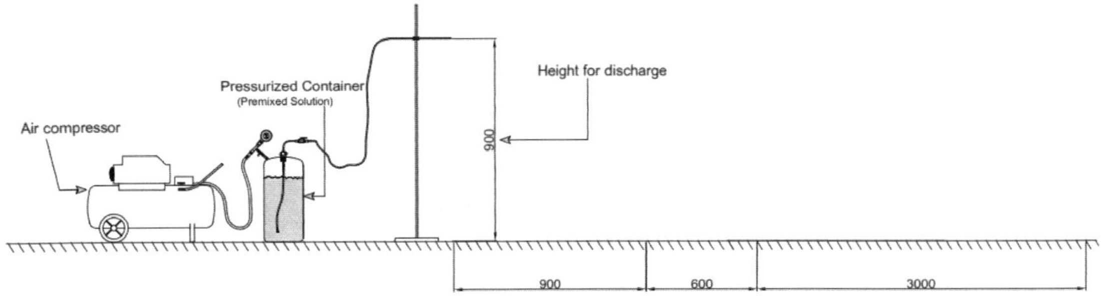

Figure 1: Schematic diagram of the discharge range test

To achieve a 3% foam solution, 60 mL of foam concentrate was combined with 1940 mL of water to obtain 2000 mL foam solution. The 2-L foam solution was poured into the 3-L container and pressurized at the designated pressure to create compressed air foam. The generated foam was formed inside the plastic hose and ejected through the nozzle. This procedure was repeated for the 2% solution test and 4% solution tests.

A weighing scale was used to determine the foam expansion ratio the foam expansion ratio, which is the ratio of the volume of the expanded foam to the volume of the foam solution. The expanded foam was collected in a 2000 mL calibrated container and filled to the brim while excess foam was removed by sliding a smooth plank over the top of the container. The weight of the expanded foam was determined on a calibrated scale to a precision of 0.5 g. Subsequent data was collected to ascertain the result. The obtained data was used to classify the foam into low, medium or high expansion ratio while visual observation was used to describe the physical properties of the foam.

The discharge range test was conducted to investigate the horizontal projection of the foam. The discharge nozzle was placed horizontally at a height of 0.9 m above the ground. Specific locations were marked on a tarp to measure the horizontal projection of foam at different pressures as illustrated in Figure 1. Two digital video cameras were positioned around the experiment to obtain visual records, while high resolution digital cameras were used to capture images of the foam and the effect of the wind. The dispersion of the foam was slightly affected by the wind which altered the horizontal discharge of the foam with minor variation. The discharge test was conducted in accordance to CAN/ULC-S508, *"Rating and Fire Testing of Fire Extinguishers"*, [8]. It is required that an extinguisher with a capacity under 2.3 kg of agent solution should have an initial discharge of not be less than 1.5 m from the nozzle and a minimum of 90 percent of the discharged foam agent solution shall be effectively discharged beyond a point of 0.9 m from the nozzle. This enables the system to generate foams with sufficient momentum to penetrate a fire plume and reach the fuel surface as well as provide heat exposure protection to the operator. The discharged foam was evenly dispersed from the nozzle and the foam flow rate was measured at different pressures.

RESULTS AND DISCUSSION

This section describes a series of test conducted to study the discharge characteristics of a portable CAF system. The experiment investigated the foam expansion ratio of different hose length, discharge range of the foam and foam flow rate as a function of pressure.

Effect of air pressure on expansion ratio

A series of tests were conducted to determine the effect of air pressure on the expansion ratio of the foam. Foam solutions ranging from 2% to 4% with three different hoses with lengths of 1-m, 2-m and 3-m were subjected to pressures in the range of 1.72 bar and 5.52 bar. The results of these tests are illustrated in Figures 2 and 3.

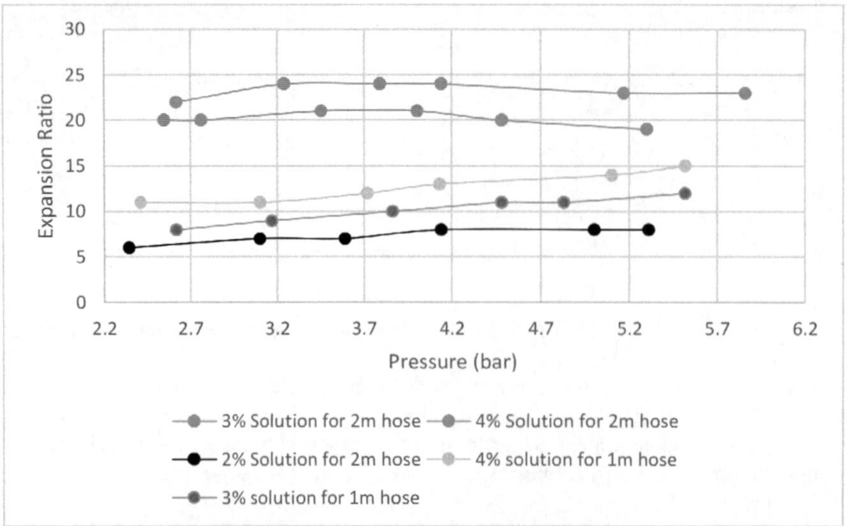

Figure 2: Expansion ratio for 1-m and 2-m hose

Figure 1 shows the foam expansion ratio of 1-m and 2-m hose as a function of pressure. As shown above, the expansion ratio of 3% solution and 4% solution for the 2-m hose were higher than the other combinations for all pressures with the expansion ratios in the range of 19 to 24. However, the 2% solution generated a wet foam with very low expansion ratio ranging from 6 to 8. This illustrates the influence of foam concentration on the quality of the generated foam. While the 3% and 4% solution for the 2-m hose generated fluid foam with medium expansion ratio, the 2% solution generated a low expansion foam. The 3% and 4% solutions for the 2-m hose generated foam with higher expansion ratios than the 1-m hose of the same solution. The expansion ratio of both 3% solution and 4% solution for the 1-m hose increased monotonically with increasing pressure while the same solution for the 2-m hose experienced an initial increase and then decreased with pressure. Low expansion foams are effective in extinguishing liquid pool fires.

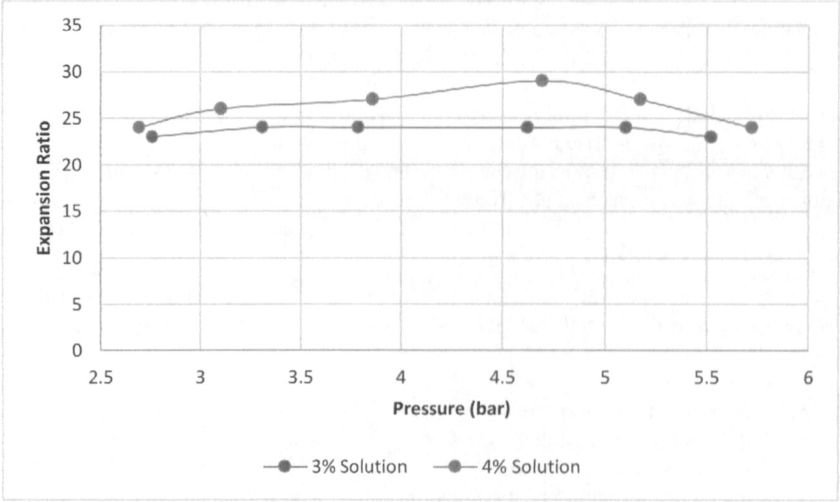

Figure 3: Expansion ratio for 3-m hose

Figure 3 shows the results of the tests with the 3-m hose. The expansion ratios of both solutions show similar trends as the results of the tests with the 2-m hose but with higher expansion ratios in the range of 19 to 28. The generated foam can be categorized as a fluid foam with medium expansion ratio. This demonstrates the influence of hose length on the expansion ratio as the hose length increases, the more the fluidity of the foam resulting in higher expansion ratios. Medium expansion foam can be used for structural protection due to its slow drainage time and its ability to adhere to sloped, vertical, horizontal and slippery surfaces. The 3% and 4%

solution of the 2-m hose and 3-m hose generate steady and uniform compressed air foam at low pressures. Hence, they are both suitable for a portable fire extinguisher.

Discharge range tests

Discharge range tests were conducted to investigate the horizontal projection of the foam from the nozzle at a height of 0.9 m above ground. The test was conducted in an open space with little interference of wind. Figure 4 shows a photo captured during a live test with the nozzle located at a height of 0.9m above ground. The yellow lines represent distances of 0.9 m and 1.5 m from the nozzle. Visual record of the maximum discharge range was taken at intervals. For the 1-m hose, the initial discharge at 2.42 bar was in the range of 1.0 m from the horizontal nozzle and increased progressively to 1.9 m at 3.45 bar. The discharge steadily increased to 2.25 m at 4.48 bar and finally to 2.4 m at 5.52 bar. However, the initial discharge exhibited a longer projection at 2.42 bar for the 2-m hose by starting off at 1.8 m and increased to 2.7 m at 3.45 bar and to 4.5 m at 5.17 bar. A similar pattern was observed for the 3-m hose with the initial discharge of 1.85 m at 2.41 bar, 2.6 m at 3.31 bar and above 4.5 m at 4.83 bar. Overall, the discharge range test for the three hoses met the requirements of the CAN/ULC-S508 standard.

Figure 4: Photo of a discharge range test

Flow Rate Tests

A series of tests was performed to investigate the flow rate of the compressed air foam at different pressures. These tests were conducted using a 3% foam solution with a 2 m hose. As expected, the flow rate of the foam increased with increase in pressure ranging from 8 g/s to 20 g/s as shown in Figure 5. These flow rates correspond to application times for the 2-liter foam solution of 244 and 102 seconds respectively.

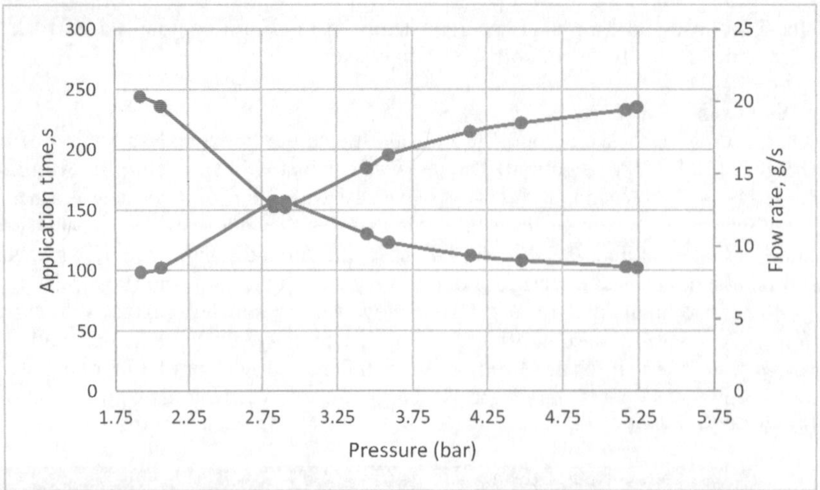

Figure 5. Effect of pressure on application time and flow rate

CONCLUSION

This study investigates the discharge characteristics of a portable CAF system by considering foam expansion ratio of varied hose length, discharge range of the foam and foam flow rate as a function of pressure. Compressed air foam was generated by injecting air under pressure into a foam solution stream.

The result showed that with the right combination of foam concentrate, hose length and nozzle, the portable CAF system can discharge over a long distance at low pressure. The CAF system generated a uniform and steady foam with expansion ratio of over 20 by using a combination of 3% foam solution and a 2-m hose at a pressure of 2.41 bar. With this pressure, the flow rate of the foam was at 12 g/s resulting in an application time of 150 seconds. Hence, it is suitable for a portable system.

From the results of the discharge tests performed, it can be concluded that the portable fire extinguisher meets the discharge requirements of the CAN/ULC-S508 standard for new extinguisher.

REFERENCES

[1] NFPA 10. (2002). Standard for portable fire extinguishers. *NFPA Codes and Standards*, Quincy, MA.: National Fire Protection Association.

[2] Liu Z, Kim AK, Carpenter D (2006). A study of portable water mist fire extinguishers used for extinguishment of multiple fire types. *Fire Safety Journal* 42:25–42. doi: 10.1016/j.firesaf.2006.06.008

[3] Kim, A. K., & Crampton, G. P. (2009). Evaluation of the Fire Suppression Effectiveness of Manually Applied Compressed-Air-Foam (CAF) System. *Fire Technology*,48(3), 549-564. doi:10.1007/s10694-009-0119-3

[4] Rie, D., Lee, J., & Kim, S. (2016). Class B Fire-Extinguishing Performance Evaluation of a Compressed Air Foam System at Different Air-to-Aqueous Foam Solution Mixing Ratios. *Applied Sciences*,6(7), 191. doi:10.3390/app6070191

[5] Weinschenk, C. G., Madrzykowski, D. M., Stakes, K., & Willi, J. M. (2017). Examination of Compressed Air Foam (CAF) for interior firefighting. *National Institute of Standards and Technology*. doi:10.6028/nist.tn.1927

[6] Neumeir, A., & Effenberger, R. (2000). *U.S. Patent No. US6543547B2*. Washington, DC: U.S. Patent and Trademark Office

[7] Kim, A. K., & Crampton, G. P. (2009). Evaluation of the Fire Suppression Effectiveness of Manually Applied Compressed-Air-Foam (CAF) System. *Fire Technology*,48(3), 549-564. doi:10.1007/s10694-009-0119-3

[8] UL 8. *Standard for Safety for Water Based Agent Fire Extinguisher.* UL 8, Seventh Edition, 2016

IOP Conf. Series: Journal of Physics: Conf. Series **1107** (2018) 062008 doi:10.1088/1742-6596/1107/6/062008

Evaluation of the justifiable investment in residential sprinkler system installations using the J-value methodology

Michael SPEARPOINT[1] and Danny HOPKIN[1]

[1]*Olsson Fire & Risk UK Ltd*

michael.spearpoint@olssonfire.com

ABSTRACT

It is common for agencies around the world to conduct some form of cost-benefit analysis (CBA) when proposing an investment in a safety measure. Since resources are limited, expenditure on one particular safety measure effectively means that there are not the resources available for other measures. Due to finite financial resources, an investment in a safety scheme will require a trade-off elsewhere. As such, the objective of a CBA is to assess whether the proposed measure provides a net benefit to society or whether resources would be better directed elsewhere.

Fire safety measures, such as sprinkler systems, are one means to mitigate the risk posed by fire in buildings. Sprinkler systems have a long record of property and life protection but there is a cost associated with their installation and ongoing maintenance. For many years various agencies have examined whether sprinkler systems should be mandated in some or all buildings within their jurisdiction, and as a result have conducted different forms of CBA. However, there are a number of challenges when carrying out such assessments particularly when it comes to placing an economic value on human life.

This paper revisits three previous CBA available in the literature for the installation of residential (domestic) sprinkler systems in single family dwellings. The paper applies recent work to extend the J-value methodology which employs the life quality index (LQI) concept as a mean of valuing the benefit of increasing life expectancy through the mitigation or reduction in the occurrence rate of a hazard. Similar to the previous studies investigated, the paper finds that the unilateral installation of sprinkler systems in single family homes does not provide a net benefit to society. This conclusion is reached through the adoption of a novel and objective judgement scalar that is easy to implement and interpret.

KEWORDS

cost-benefit analysis, residential sprinkler systems, J-value, life quality index

INTRODUCTION

In the light of events, such as the Grenfell Tower incident, there has been considerable discussion in the UK regarding the installation of sprinkler systems into high-rise residential buildings. Some parties, such as the London Assembly Planning Committee [1], suggest that national building regulations be changed to mandate that sprinkler systems are fitted in such buildings if they meet specific criteria related to their height, etc. Although sprinkler systems are not fully reliable [2], there is little doubt that sprinklers can have a beneficial impact on mitigating the fire hazard to people and property [3], [4]. However, the installation of sprinkler systems has various costs that need to be accounted for, such as installation and maintenance. It is therefore reasonable to suggest that any proposal to mandate widespread installation of fire protection measures should assess whether that investment provides a net benefit to society, or whether finite financial resources would be better invested elsewhere.

Recent work by Hopkin, *et al.* [5] has extended the J-value methodology [6] to fire safety investments. The work examined how the costs and benefits associated with the reduction in fatalities, injuries and fire damage can be included as part of a life quality index (LQI) assessment [7]. The study also shows how discounting of the investment cost can be incorporated into the analysis. An exemplar evaluation of a fire safety scheme was included in the work but it was not specific to a particular type of system or jurisdiction. This paper revisits three cost-benefit analysis (CBA) studies from the literature and applies the J-value methodology to the values previously employed in those studies to examine whether the installation of sprinkler systems provides a net benefit to society. In the process, the limitations of some of these studies, specifically the value of a statistical life, are addressed through the LQI and in cognisance of society's capacity to commit resources.

J-VALUE

In simple terms, the J-value provides a scalar measure of the efficacy of a safety investment through normalising the life-time cost of a scheme against the benefits realised in terms of improving life expectancy for those affected, reducing injury and loss of assets / damage. The total (implying an interest beyond solely improving life expectancy) J-value is.

$$J_T = \frac{C}{\Delta D_f + \Delta D_i + M_R \cdot \Delta D_d} \cong \frac{C}{\Delta D_f + \Delta D_i + \Delta D_d}$$

where C is the cost of the safety scheme, M_R is a risk scalar [-], ΔD_i is the change in losses due to injury, ΔD_d is the change in losses due to damage, and ΔD_f is the safety investment that is warranted to protect people's lives. The parameter, M_R, will begin at approximately unity for an organisation with assets large enough to dwarf a financial loss. However, it will rise to a higher value if the loss that may occur is comparable with its total assets. For the former case, $J_{fi,T}$ simplifies to the right-hand formulation.

At the core of the J-value is a need to quantify the benefit of improving life expectancy through investments in safety provisions. In the past many CBA studies have adopted the value of a preventable fatality (VPF) concept. However, VPF is based upon so-called 'revealed preferences' where Sunstein [8] emphasises that there are large variations in 'observed' VPF with respect to gender, race and social status. In the J-value, this is achieved via a proportionality metric derived from the LQI, described as the societal willingness to pay (SWTP) value, given by

$$SWTP \approx \frac{GC_x}{q}$$

where G is the societal wealth, as measured by GDP per capita, q represents work-life balance [-], and, C_x a proportionality constant [-] specific to a given demographic profile. Simply, SWTP is the investment that should be made to avert a single fatality in cognisance of the given society's capacity to commit resources. Such is the LQI's traction over dated and subjective concepts like VPF, SWTP values can now be found in international standards such as ISO 2394 [9]. It follows that the SWTP can be multiplied according to the number of fatalities averted to arrive at a valuation ΔD_f of the benefits of improving the life expectancy of a given group (of size N) affected by a potential hazard. The change in annual expected number of fatalities Δ_f arising from the fire safety measure is

$$\Delta_f = \frac{N_{fatal}\chi_f}{N_{units}} \quad \text{and} \quad \Delta D_f = SWTP \cdot \Delta_f$$

where N_{fatal} is the number of fatalities per fire before the safety investment, χ_f is the reduction in fatalities due to the fire safety measure and N_{units} is the number of properties (e.g. buildings or apartments etc.)

affected. Correspondingly, the annual injury ΔD_i and damage reduction benefits ΔD_d are similarly derived as

$$\Delta D_i = \frac{N_{injury}\chi_i}{N_{units}}\zeta_i \quad \text{and} \quad \Delta D_d = \frac{N_{fires}\chi_d}{N_{units}}\zeta_d$$

where ζ_i and ζ_d are the costs of the prevented injury and property damage, respectively. χ_i and χ_d are the corresponding reductions in injuries and property damage as a result of the safety investment.

As the safety investment relates to the reduction of future risk, the J-value analysis requires that future costs and benefits incurred at different times are annualised using a continuous discount rate (γ) over the service life of the safety scheme (L). Where the safety investment relates to a single upfront investment, then the future benefit terms need to be discounted such that

$$\Delta D_\gamma = \frac{\Delta D_f + \Delta D_i + \Delta D_d}{\gamma} \cdot (1 - e^{-\gamma L})$$

The total discounted cost (C) of the associated safety investment typically includes an upfront sum (c_0) and a maintenance sum (m). For sprinkler systems, this work has accounted for the costs for its installation (c_{ins}) and the connection to the water supply (c_{ws}) as separate items, leading to a combined value in c_0

$$C = c_0 + \sum_{t=1}^{L} \frac{m}{(1+\gamma)^t} \quad \text{and} \quad c_0 = c_{ins} + c_{ws}$$

The corresponding total fire safety J-value is then given by

$$J_{fi,T} = \frac{C}{\Delta D_\gamma}$$

If the value for $J_{fi,T}$ exceeds unity then the safety measure is deemed to result in a net dis-benefit to society as the realised benefits would be less than a society's capacity to commit resources. Correspondingly, a $J_{fi,T}$ less than unity implies that the safety investment is justified, leading to a net benefit to society. Where a safety scheme is solely motivated by improved mortality, $\Delta D_i + \Delta D_d$ can be disregarded. However, this generally leads to a lesser need to invest as other societal benefits that are not linked to mortality are disregarded.

PREVIOUS SPRINKLER COST-BENEFIT ANALYSIS STUDIES

Welsh case

In 2012 the Welsh Government engaged the Building Research Establishment (BRE) to undertake a domestic sprinkler regulatory impact assessment. This was in response to changes in Welsh Law in 2011 requiring that domestic sprinklers be provided in all new and converted residential premises. The assessment took the form of a CBA, with consideration of reductions in fatalities, injury and property damage. In valuing a statistical life, road traffic accident figures were adopted by BRE due to noted limited alternatives at the time.

To derive the impact of sprinkler installations on mortality, injuries and damage, BRE adopted statistics from fire and rescue service incident report forms. An indirect computation of impact on mortality, damage and injury was adopted through correlating fire burn area, and the noted metrics. It was subsequently presumed that sprinklers, where successfully operational, capped the burn area, leading to a change in the observed fatalities, injuries and damage extent. A full explanation can be found in Fraser-Mitchell and Williams [10].

The BRE study considered a range of possible residential tenancies, such as single occupancy houses, flats, hostels, care-homes, etc. However, this study is limited to an analysis of single occupancy houses. As such, mean values for reduction in fatalities, injuries and damage of 90, 64 and 93%, respectively, are adopted. Valuations for the costs of injury prevented, and costs of fire damage are adopted per the BRE report, and represent mean values across all dwelling fire incidents.

At the time, the BRE study concluded that: (a) fitting sprinklers in all new Welsh residential premises was not cost beneficial, (b) mandating sprinklers in care-homes is cost effective, and (c) sprinklers may be marginally cost effective in new blocks of flats.

New Zealand case

A comparative analysis of the J-value is conducted using the study by Duncan, *et al.* [11]. Their study examined the cost-benefit of installing domestic sprinkler systems in New Zealand to the then current standard NZS 4515:1995 [12] and the implications of proposing a more cost-effective system design. Details of their analysis and the proposed system designs are given in Duncan, *et al.* and not repeated here but they found that domestic sprinkler systems, constructed to NZS 4515:1995, were not cost-effective at a cost per life saved.

Duncan, *et al.* give costs for the installation, water connection and annual maintenance of systems. However, they do not express the total fatalities and injuries explicitly but quote values of 6 and 40 per 1,000 fires, respectively. Using a value of 5,967 fires from New Zealand Fire Service statistics over the five-year period from 1993 to 1997, results in 36 fatalities and 239 injuries per year. Similarly, the reduction in fatalities, injuries and property damage are not stated directly by Duncan, *et al.* but values for cases with and without sprinkler installation are given. Hence, values are derived for the analysis shown here.

Australian case

Beever and Britton [13] conducted research into the cost-effectiveness of various fire safety measures in buildings in Australia including the installation of residential sprinkler systems. The research made no recommendation to extend the building codes in Australia to require the installation of sprinkler systems in domestic dwellings. The Beever and Britton study was followed-up by Hasofer and Thomas [14] who carried out an analysis of the benefit-cost of installing sprinklers using the LQI method, similar to the work presented here. Hasofer and Thomas found that the installation of sprinklers in single family dwellings was not an economical proposition.

Although the work presented here is similar to that by Hasofer and Thomas, the exact formulation of the J-value and the necessary input parameters differ in their details. Thus, the original study from Beever and Britton has been examined in order to extract input parameters relevant to this study. Using data from the 1991 Australian census, Beever and Britton stated that 14,610,000 people lived in single family detached and semi-detached homes. Using the quoted average number of people per home of 2.99 suggests there were 4,886,288 single family homes in Australia at that time. Beever and Britton estimated that around 36,600 dwelling fires occurred over a four-year period from 1989/90 to 1993/4 which gives 9,150 fires per year. Using the number of single family homes, this works out to be 1.87 fires per 1,000 homes per year, consistent with Beever and Britton.

In dwellings without sprinklers there were 7 fatalities per 1,000 fires, i.e. 64 fatalities per year and 70 injuries per 1,000 fires, i.e. 641 persons per year. Beever and Britton assessed that the installation of sprinklers led to there being a reduction ranging from 3.89 fatalities to 1.46 fatalities per 1,000 fires. In order to give the highest benefit to the installation of sprinklers, this paper uses the lower 1.46 value giving a reduction in fatalities due to sprinklers of 79 %. With sprinklers installed Beever and Britton quote injury rates of around 15 to 30 per 1,000 sprinklered fires and again using the lower statistic gives a reduction of 79 %. Property damage with sprinkler and without sprinklers installed were taken to be $AU 3,900 and $AU 24,000 respectively, giving a percentage reduction of 84 %. Finally, Beever and Britton assumed a 20-year life for the sprinkler system, and also that the mains water pressure was sufficient and did not give a separate water connection cost.

ANALYSIS

Table 1 summarises the J-value input values that have either been directly sourced from the original reports or derived, as necessary. Currency values are given in the local unit for which each original study was conducted (where conventionally the scarab symbol ¤ is used to designate a general currency unit), namely pounds sterling (£) for the Welsh study, New Zealand dollars ($NZ) and Australian dollars ($AU), respectively. SWTP values for Wales (UK) and Australia are taken from ISO 2394 [9], with the change in mortality due to life risk reduction uniformly distributed across the age distribution. A 3 % discount rate is applied, with SWTP expressed within ISO 2394 in terms of common purchasing power parity (international dollars). For New Zealand, the UK value has been used, adjusted for the currency exchange rate, since the two countries have a similar GDP and demographic profiles. Care should be taken comparing costs since exchange rates and inflationary adjustments have not been made herein. The New Zealand case includes inputs from Duncan, *et al.* [11] for sprinkler system installations to NZS 4515:1995 and for the proposed revisions.

Table 2 shows the derived quantities using the previously presented equations and the appropriate values from Table 1 where it is noted that the reduction in fatalities for all three cases are of the same order of magnitude. Hasofer and Thomas expressed their result as benefit to cost ratio of 0.122 which is equivalent to 8.20 as a cost to benefit ratio similar to the 9.04 result found in this study for the Australian case. For the New Zealand case, a J-value of 6.16 is obtained for a system designed to NZS 4515:1995 compared with 3.56 for the proposed draft revisions. For the Welsh case, the J-value is 5.93, which is similar to the NZS 4515 result. As all values are in exceedance of unity, this implies a net dis-benefit to society in all evaluated cases.

Table 1. Input values to J-value analysis.

Input	Symbol	Metric		
		Wales	New Zealand[1]	Australia
Number of single occupancy units (dwellings)	N_{units}	≈1,200,000	1,318,800	4,886,288
Number of fires in single occupancy dwellings (fires / yr)	N_{fires}	1,421	5,967	9,150
Fatalities due to single occupancy dwelling fires (persons / yr)	N_{fatal}	14	36	64
Injuries due to single occupancy dwelling fires (persons / yr)	N_{injury}	345	239	641
Cost of injury prevented (¤)	ζ_i	19,960	30,000	21,100
Cost of damage per unsprinklered fire (¤)	ζ_d	8,800	17,200	24,000
Reduction in fatalities due to sprinkler introduction in single occupancy dwellings (%)	χ_f	90	80	79
Reduction in injuries due to sprinkler introduction in single occupancy dwellings (%)	χ_i	64	63	79
Reduction in damage due to sprinkler introduction in single occupancy dwellings (%)	χ_d	93	83	84
System installation cost (¤)	c_{ins}	1,950	6,500 (4,070)	2,550
Water supply costs (¤) [2]	c_{ws}	1,125	200	n/a
Annual maintenance cost (¤)	m	96	635 (280)	500
Discount rate (%)	γ	3	8	5
Average sprinkler system life (yrs)	L	45	50 (30)	20
Societal willingness to pay (¤)	SWTP	2,260,272	5,321,904	5,807,698

[1] Values in brackets are for changes to NZS 4515:1995 proposed by Duncan *et al.* [11].
[2] Beever and Britton [13] did not explicitly provide a cost for the water supply.

A sensitivity analysis on the input values shows that, similar to Hasofer and Thomas, varying the injury rate from 15 to 30 per 1,000 fires made almost no difference to the Australian J-value analysis, only changing the result by 0.05. However, using the upper value of 3.89 fatalities per 1,000 fires where sprinklers are installed changes the reduction in fatalities to 44 % and consequently the J-value increases to 12.29.

Table 2. Derived quantities.

Derived quantity (unit)	Symbol	Metric		
		Wales	New Zealand[1]	Australia
Reduction in fatalities through safety measure (per / dwelling / yr)	Δ_f	1.1×10^{-5}	2.2×10^{-5}	1.0×10^{-5}
Life preservation utility (¤ / yr)	ΔD_f	23.73	115.58	60.25
Injury reduction utility (¤ / yr)	ΔD_i	3.67	3.39	2.17
Damage reduction utility (¤ / yr)	ΔD_d	9.69	64.25	37.64
Discounted benefit over life, L (¤)	ΔD_γ	916	2,248 (2,083)	1,265
Upfront investment in safety measure (¤)	c_0	3,075	6,700 (4,270)	2,550
Discounted cost of safety measure (¤)	C	5,429	13,849 (7,422)	8,887
J-value for sprinkler system installation (-)	$J_{fi,T}$	5.93	6.16 (3.56)	9.04

[1] Values in brackets are for changes to NZS 4515:1995 proposed by Duncan *et al.* [11].

DISCUSSION AND CONCLUSIONS

Conducting any CBA can be a challenging exercise since the assessment needs to identify who is incurring the costs and/or accruing the benefit, and have all of the costs and benefits been considered. For example, considering sprinklers as a provision in isolation inherently presumes that other precautions are effective in the manner intended. Beever and Britton [13] discuss the potential reduction in domestic insurance claims if sprinkler systems were installed and conclude any savings would be insignificant compared to the other costs, such as the annual maintenance. Hasofer and Thomas [14] note that insurance is the transfer payments from one group in society to another group which does not affect a benefit to cost ratio societal indicator.

Beever and Britton include in their study a consequent reduction in the fire brigade response if sprinkler systems were installed. Here they recognised that only a proportion of fires would be in single family homes, that the brigade would still likely be called to sprinklered buildings and that less than half of brigade callouts are to fires. Clearly any change in regulations, whether applied to new buildings or to a retrofitting programme, can take many years to have an effect. Even if it was accepted there was a case to reduce the fire brigade's capability with the installation of sprinklers, this would not be possible until some future date.

In this work there has been no assessment of water damage from accidental sprinkler activations as well as the difference in potential water damage from fire brigade hose streams in the event of a fire. In a wider context, there is the issue of the potential for trade-offs that come with over investment in a safety scheme. That is, given a society's finite resources, where are cost-cuts made to subsidise widespread sprinkler adoption? Whilst not the subject of this paper, the acknowledgement of trade-offs is an important CBA consideration.

This paper shows that applying the J-value methodology to the installation of sprinkler systems into single family dwellings does not provide a societal benefit for the three cases evaluated. This result is consistent with Fraser-Mitchell and Williams [10], Duncan, *et al.*[11], Beever and Britton [13], Hasofer and Thomas [14]. Work is ongoing to apply the J-value assessment methodology to the proposal made by the London Assembly Planning Committee [1] for high-rise residential buildings to understand its economic implications.

REFERENCES

[1] Greater London Authority, (2018) Never again: Sprinklers as the next step towards safer homes, London Assembly Planning Committee, 50 pp.

[2] Frank, K., Gravestock, N., Spearpoint, M., and Fleischmann, C. (2013) A review of sprinkler system effectiveness studies. Fire Science Reviews, 2:6. 10.1186/2193-0414-2-6

[3] Melinek, S. J. (1993) Effectiveness of sprinklers in reducing fire severity, Fire Safety Journal 21(4), 299-311. 10.1016/0379-7112(93)90018-L

[4] Hall, J. R. (2010) U.S. experience with sprinklers and other automatic fire extinguishing equipment, National Fire Protection Association, Quincy, MA.

[5] Hopkin D., Spearpoint, M., and Van Coile, R. (2018) The J-value and its role in evaluating investments in fire safety schemes, Fire Technology, 10.1007/s10694-018-0752-9.

[6] Thomas, P. J. and Jones, R.D, (2010), Extending the J-value framework for safety analysis to include the environmental costs of a large accident, Process Safety and Environmental Protection, 88(5), 297-317. 10.1016/j.psep.2010.03.007

[7] Pandey, M.D., Nathwani, J.S., and Lind, N.C. (2006). The derivation and calibration of the life-quality index (LQI) from economic principles, Structural Safety, 28(4), 341-360. 10.1016/j.strusafe.2005.10.001

[8] Sunstein, C.R. (2005). Laws of fear: Beyond the precautionary principle. Cambridge University Press. Cambridge, UK.

[9] ISO 2394:2015, General principles on reliability of structures. International Organization for Standardization.

[10] Fraser-Mitchell, J. and Williams, C. (2012) Cost benefit analysis of residential sprinklers for Wales, BRE Report 276803, BRE Global, Garston, Watford, UK.

[11] Duncan, C. R., Wade, C. A, and Saunders, N. M. (2000) Cost effective domestic fire sprinkler systems. New Zealand Fire Service Commission Research Report Number 1, 2000.

[12] Standards New Zealand (1995) NZS 4515:1995 Fire sprinkler systems for residential occupancies (including private dwellings), Standards New Zealand, Wellington, New Zealand.

[13] Beever, P., and Britton, M. (1999) Research into cost-effective fire safety measures for residential buildings. Building Control Commission, Melbourne, Victoria, Australia.

[14] Hasofer, A. M., and Thomas, I. R. (2008). Cost benefit analysis of a fire safety system based on the life quality index. Fire Safety Science, 9, 969-980. 10.3801/IAFSS.FSS.9-969

IOP Conf. Series: Journal of Physics: Conf. Series **1107** (2018) 062009 doi:10.1088/1742-6596/1107/6/062009

Experimental Study of a Downward Directed Water Mist suppressing a Diffusion Flame

Karthikeyan Rajasekar[1], Amit Kumar[1] and Satyanarayan Chakravarthy[1]

[1]*Department of Aerospace Engineering, NCCRD Indian Institute of Technology Madras, India*

karthikeyan@smail.iitm.ac.in

ABSTRACT

Interaction of fires with water have been a focus of study for human kind from the times immemorial. With the recent advances in water-based suppression, and restrictions put on halon usage, sprinklers and water-mist based fire suppression systems have become popular for industrial usage. Of this water mist is an interesting candidate, for it has advantages and to name a few would be less water requirement, non-wetting nature, gas phase action on fire. Water mist, as an effective fire suppressant, works on the principle of oxygen dilution and flame cooling.

In this work, we are studying the action of water mist on a diffusion flame. The water mist was generated from a Delavan nozzle, fed with high pressure water, which is then directed downward into the raising flame. A circular burner fed with Methane in a controlled manner generated the diffusion flame. The reason to prefer a gas burner is to have a control over heat release rate which will help in simulating various fuel sources of different burning rate. A glass cubicle was used to isolate the experiments from ambient disturbances. A Digital SLR camera was used to record the flame and subsequently the images were processed to obtain flame height. Particle Image Velocimetry was used to study the effect of the flame on the water mist., by imaging the Mie scattered laser light from the droplets. Parameters that were varied were the injection pressure and flow rate of the Methane into the burner.

Visual imaging from the camera indicated that there were occasional flame tip flattening during the suppression mechanism. The flame luminosity is diminished and the diffusion flame height is reduced due to the action of the water mist. Moreover, this effect is more pronounced for smaller flame than larger flame, indicating the role of the buoyancy in fire suppression effectiveness. The PIV results show that the droplets are slowed down very close to the nozzle due to the flame. Also, a recirculation zone is observed at the periphery of the spray cone, very close to the nozzle. It is a toroidal vortex that is generated due to the interaction, and the droplets thrown away from the spray cone gets trapped into this flow. The flow structure is present in a highly turbulent region and is identified in the mean field. Another finding is that close to the burner, fine droplets that could make it to the base, gets entrained into the flame. It is expected that the droplets tend to be entrained into the larger flame, as the entrainment is stronger for large flame, compared to the smaller flame. But the PIV results indicate otherwise, which is attributed to the larger size of droplets making its way close to the burner.

We believe that these results will enlarge our understanding of the suppression mechanism of water mist and could form a basis for validating various simulation models.

KEYWORDS:

Water mist; fire suppression; PIV; Flame height; Image Processing

IOP Conf. Series: Journal of Physics: Conf. Series **1107** (2018) 062009 doi:10.1088/1742-6596/1107/6/062009

INTRODUCTION

Water mist is the name given to a water spray in which all droplets are of the diameter less than 1000μm [1]. Water has been a firefighting material since time immemorial. But water in the form of mist has been studied for fires fighting only in the last few decades. There have been tests carried out to assess the effectiveness of water mist in the late 50s and 60s [2]. But the momentum for scaling the technology died out due to various reasons, mainly due to the prevalence of Halon. A renewed interest in the water mist has been observed lately. There has been a lot of factors that fuelled this. The Montreal protocol banned Halons, which called for a swift identification of a suitable substitute, and water mist has been a top contender.

The mechanism of fire suppression by water droplets has been well studied. The droplets in the water mist enter the fire and evaporate which cools the fire as well as dilutes the ambient of Oxygen due to rapid phase change from liquid to gas. There have been speculations on the degree of the role played by water mist in the kinetics of the flame suppression but this is yet to be fully studied [3]. In the past, few studies on fire suppression have been carried out by loading the co-flow with water mist that enters the flame [4][5]. This case is not very representative of real fires as entrainment occurs from the stagnant ambient air. Also, the water mist may not enter perfectly from the entrainment. Studies have been carried out where the water mist was generated from high-pressure nozzle systems. For practical purposes, the convenient mounting and orientation of mist injection would be from the ceiling, in continuance with sprinkler system. This motivated few studies where the mist was directed downward into the flame [6]. Few pool fire suppression studies using the same orientation has also been undertaken [7][8].

Some experiments that have been carried in the past shows the use water mist consisting droplets of diameter ranging from 5μm [9] **to** few hundreds of microns. The parameters that were studied in large extent were loading fraction, temperature and species. Many of them were intrusive and there is a paucity of non-intrusive diagnosis. Piezoelectric discs have been employed to generate ultra-fine mists, which has been commercialized too. But there is quite an uncertainty as to what is the exact loading, diameter and velocity that will quench or suppress the fire [10]. Droplets with large momentum might penetrate the flame front without cooling the flame, while droplets with less momentum might evaporate before reaching the flame front. A lot of work has gone into the numerical simulation too [11], that have attempted to validate the results.

In this study, a set of experiments were carried out to understand more about the interaction of downward directed fine water droplets of the mist with flame. In line with real fires, a diffusion flame is set up with no co-flow. A circular burner fed with Methane was used for this. Also, being gas fed gives the advantage to control the heat release rate and the scale of the fire with which the water mist is interacting to suppress it. Water mist was generated from a simplex nozzle and they spray was directed downward into the flame. Particle Image Velocimetry was used to deduce the velocity field within the spray cone. A Digital SLR camera was used to study the effect of water mist on the flame.

The height between the water mist injection point and the flame was fixed at a height of 900mm. The injection pressure of water into the nozzle was varied between **7bar** and **10bar**. The flow rate of gas into the burner was adjusted to provide an equivalent heat release rate of 5kW (10 slpm of CH4), and 10kW (20 slpm of CH4). The hear release rate is estimated by making use of CH_4's heat of combustion.

We believe that the results from these experiments will facilitate as a baseline for validating and comparing simulation results.

METHODOLOGY

Experimental setup

The schematics of the experimental setup is shown in Fig.1. A Delavan nozzle of the model CT-1.5–30 B was used to generate the water mist. The nozzle generated a solid cone spray with a spray angle of 30°. The flow rate at 7bar(100psi) gauge pressure was around 0.084 LPM(1.5GPH). The nozzle was mounted on a 600mm X 600 mm Aluminium plate, which houses a machined nozzle holder. A stainless-steel water pressure tank was pressurised in the range of 7-10bar which was connected to the nozzle. Compressed N2 was used for pressurising the water tank.

A 50mm internal diameter burner of height 200mm forms a diffusion flame. CH4 was fed through Alicat MFC into the burner to set a controlled flame. CH4 flows through a set of four meshes with progressively reducing wire and mesh sizes, which sets up a uniform profile at the outlet at the burner mouth. A gas fed diffusion flame burner is the closest analogy to a real-life fire encountered in fire accidents.

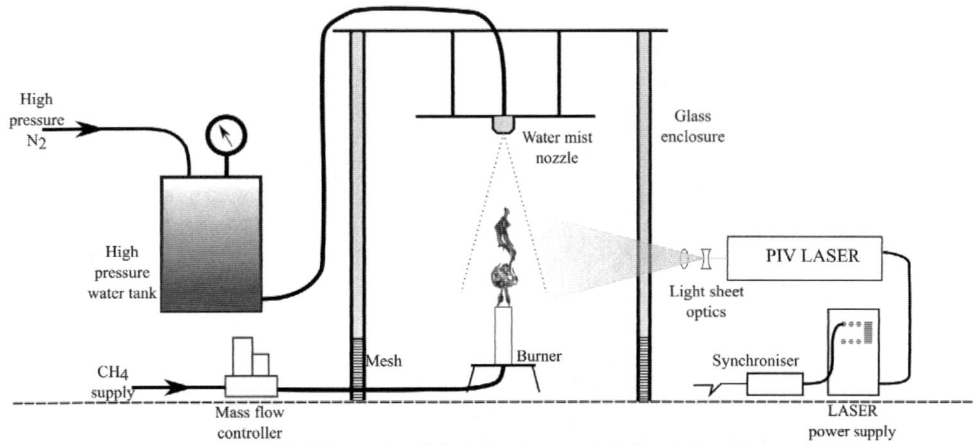

Fig.1.Schematic of the experiment; the sketch is not to scale.

An enclosure made of 6mm tempered glass contained the nozzle and the burner at its centre and all experiments are conducted within it. The motive was to avoid any disturbances from entering the experimental region and hence provide good control over the ambient flow conditions. This enclosure is of size 2m X 2m X 2m. A 400mm opening is provided at the base, all around the enclosure for entrainment of ambient air required for combustion of the flame set up by the burner. A fine mesh is installed all around it to help a draught proof entrainment, hence the effect of the ambient flow disturbances is reduced inside the enclosure. An active hood of 150 cubic feet per minute flow rate, is present at the top of the enclosure.

For visual imaging of the flame, Nikon D5200 DSLR was used, fitted with an 18-55mm Nikon lens. Videos were recorded using the DSLR at 60 frames per second with a frame size resolution of 1920px X 1080px. A Litron twin cavity Nd: YAG laser is used for Particle Image Velocimetry studies. The 532nm laser beams were pulsed at a repetition rate of 12Hz and the beams passed through a set of sheet optics to create a thin sheet. Mie scattering from the droplets was imaged directly to measure the velocity field. LaVision's HighSpeedStar camera was used to image the scattered light, which is fitted with a Sigma 70-300mm Lens and a 532nm(±3nm) narrow bandpass filter to cut out on ambient light. To Synchronize the camera and laser pulse, a simple Arduino was used, to ensure the two pulses straddles between the frames. All the above systems were set out of the enclosure to not affect the experiments as well as to protect these systems from the dispersing fine mist.

Data processing techniques

The DSLR videos were read using Matlab and each frame was thresholded to obtain a binary image which isolated the flame from the dark background region. The binary images were ensemble averaged and the flame height was estimated as the vertically the farthest point on a contour line corresponding to the value of 0.5. This method of obtaining flame height has been reported in works of Zukoski and Cetegenzu [12]. These binary images were also used in further processing to obtain.

The high-speed camera captures the Mie scattered light from the spray and while the spray interacted with flame. These images were later processed for obtaining data on the velocity field. A window close to the nozzle and another close to the burner were chosen for PIV. The camera was moved to each of these regions of interest by means of a traverse. The window size was large in those regions where the droplet concentrations are less, say close to the burner, and the window size was small in regions where concentrations are high. Also, the delay between the two-laser pulse was adjusted according to the window size and average velocity expected in that region [13].

The images acquired were 1024px X 1024px with bit-depth of 12-bit per pixel. These images were scaled to 16-bit images and were processed using LaVision's Davis software for obtaining PIV results. A multi-pass scheme was used from the cross-correlation with an initial interrogation window size of 128px X 128px to a final 32px X 32px. The last pass had 50% overlap of windows. A 3X3 smoothening operation was performed before the PIV processing, as this will help in increasing the efficiency of correlation. For this work, the parameter of interest in the mean velocity field, hence the results are averaged and is presented here.

IOP Conf. Series: Journal of Physics: Conf. Series **1107** (2018) 062009 doi:10.1088/1742-6596/1107/6/062009

RESULTS AND DISCUSSIONS

Flame Height variation

The flame height was calculated for the flame with and without mist suppression. The experiments were repeated three times on two occasions, recreating the same conditions. Instantaneous image from the video revealed occasional flame tip flattening, during suppression, which is shown in Fig.2a. Also, the flame luminosity was significantly diminished during the process of suppression. This is shown in Fig.2b obtained from ensembled average of images extracted from 60fps video recorded over 30 seconds. Furthermore, flame heights were calculated as described in the above section. The trend of the flame height is depicted in Fig.2c. The weaker flame tends to get suppressed a lot faster than the stronger flame. This could be attributed to the hot and large momentum of the buoyant plume present in the 20 slpm flame, which carries away and evaporates a lot of the mist droplets, while the droplets could reach well into the smaller 10slpm flame and suppress it.

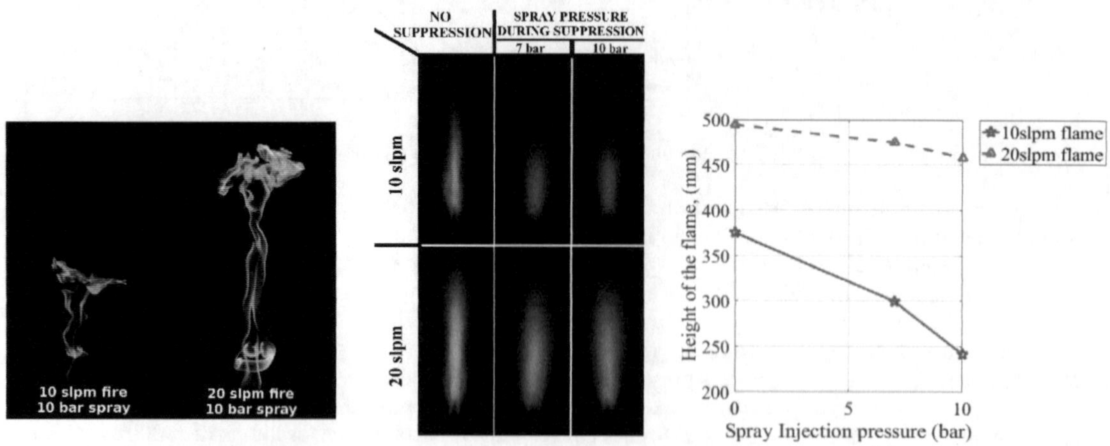

(a) (b) (c)

Fig. 2a Instantaneous Flame tip flattening, Fig.2b Changes to the flame luminosity, Fig.2c Flame height trend

Heskestad's work [14] on flame height indicate a direct correlation between heat release rate and flame height. In future, calorific studies will be carried out to quantify the relation between the flame height under suppression and the associated reduction in heat release rate.

Velocity field

The origin of the coordinate system coincided with the centre of the nozzle exit. The mean velocity was estimated by averaging the results spanning 50-60s. The presence of the flame affected the spray significantly close to the nozzle. The flow rapidly slows down along the vertical direction and increases radial velocity. This is clearly seen from Fig 3. Also, the light bands that pad about the central solid lines are the standard deviation values. Clearly, the flame increases the fluctuations and hence the turbulence in the spray field.

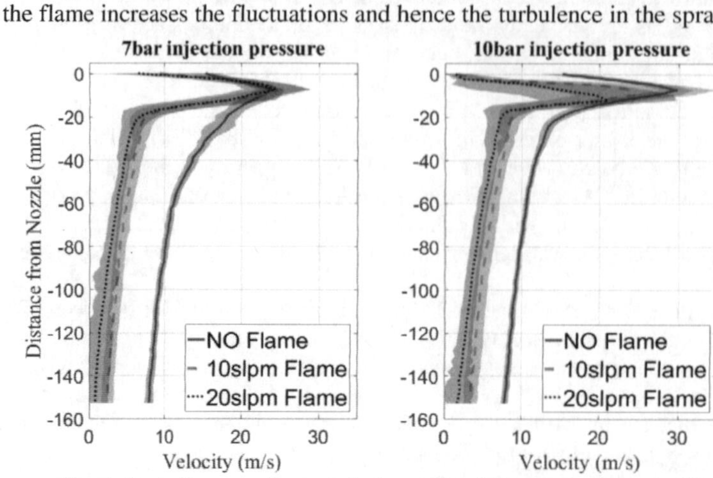

Fig. 3. Centreline velocity magnitude profile of the spray, with and without flame.

IOP Conf. Series: Journal of Physics: Conf. Series **1107** (2018) 062009 doi:10.1088/1742-6596/1107/6/062009

Another effect of the flame on the spray is that the droplets are thrown away from the spray cone. Due to the strong entrainment of ambient into the spray cone, the dispersed droplets partly re-enter the spray region, close to the nozzle. This causes a re-circulation zone in a shape of a toroidal vortex to be formed very close to the nozzle and on the periphery of the spray cone. Again, for representative purposes, the flow close to the nozzle is depicted in Fig.4(a), with streamlines false coloured according to the velocity magnitude. A clear vortex is noticeable and the shape is supposedly toroidal, due to circular symmetry. The size and location of this re-circulation zone are related to the injection pressure and the flame strength. The variation of the location and size (diameter of the toroid's axis) is depicted in Fig 4(b). The bars indicate the root-mean-squared values of the position, which was obtained by performing a moving average over successive 10 seconds of PIV results. The spread indicates that the structures are not stationary with time and are turbulent in nature. This vortical structure is not observable in any instantaneous velocity realisation but is present in the mean field.

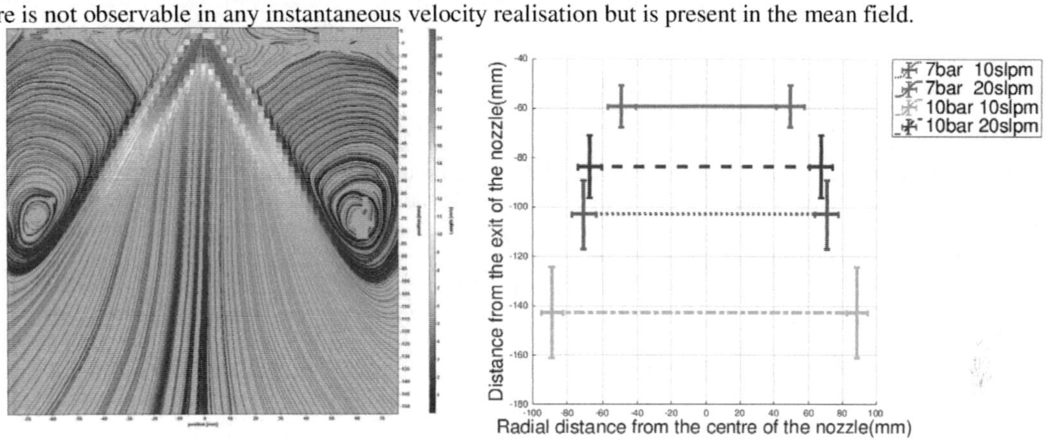

(a) (b)

Fig.4. Representation of the re-circulation zone for 7bar spray suppressing a 20slpm flame and the location-size statistics of the vortex.

We also would like to predict that the presence of this vortex, however minimal, should to some degree, enhance fire suppression as the re-entering combustion products tend to be inert and lacking in Oxygen. Detailed studies will be carried out in future to decipher the degree of this spray vortical structure's effect on flame suppression.

Similarly, studying the radial velocity component of the droplet field, close to the burner as shown in Fig.5., one can observe that there is a strong tendency for the droplets to move towards the burner. This is caused by the entraining ambient fresh air into the flame. The fine water droplets are easily affected by a weak entrainment flow of air into diffusion flame. It is expected that with an increase in strength of the flame, the stronger the entrainment and hence larger the tendency of the droplets to move towards the flame. These entraining droplets too will suppress the fire. But the observation is contrary to the prediction, with droplets moving slowly towards the flame when the flame is stronger. This might be because small droplets could have been carried away by the strong plume and only large drops make it to the base of the burner mouth. Droplet size measurement will be carried out in future to validate this. Also, one can observe, the radial velocities are larger for higher injection pressure indicating that when droplets are with larger momentum, they can reach the burner mouth. The large peaks present at the centre of the profile are due to the soot scattering from the flame.

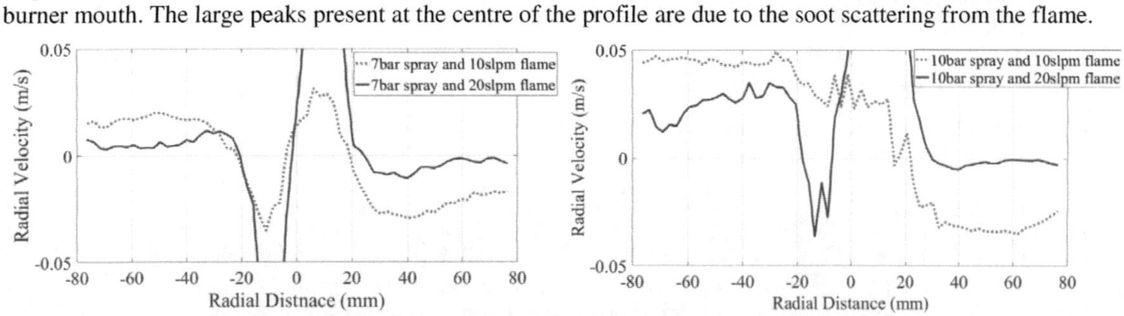

Fig.5.Radial profile of the Radial velocity component, 80mm above the burner.

IOP Conf. Series: Journal of Physics: Conf. Series **1107** (2018) 062009 doi:10.1088/1742-6596/1107/6/062009

SUMMARY AND CONCLUSION

To understand the action of the water mist on diffusion fires, plain imaging and PIV experiments were carried out to understand the flame behaviour and to measure the velocity field of the water droplets, respectively. Commercially available Delavan nozzle was used at two different pressures (7bar and 10bar) to generate fine, atomized water mist. A circular burner of 50mm diameter generated a buoyant diffusion flame which was fed using Methane. The intensity of the flame was controlled by altering the flow rate of Methane. Two flow rates, 10 and 20 slpm of Methane were used in this study. The mist from the nozzle was directed downward into the rising flame. Visual imaging of the flame by means of DSLR aided in estimating the flame heights. By means of PIV, selected regions of the spray cone were studied. The Mie scattered light from the droplets was captured and was processed using LaVision's Davis software.

Flame tip flattening and flame height reduction were observed when the videos were analysed. The flame height of the weaker flame(10slpm) is significantly reduced in comparison to the stronger flame(20slpm), which could be because of the strong buoyant plume carrying away the droplets, rendering the suppression to be ineffective. Results from the velocity field indicate the development of a toroidal vortex at the other edge of the spray cone, close to the nozzle, in the presence of the flame. The strength and the injection pressure dictates the size and location of the vortex. Owing to large turbulence in both the flow field, the vortex's location and size keep on changing. Also, fine droplets which made it close to the burner gets caught up by the slow-moving entrainment flow and these droplets enter the flame at the burner mouth. Despite a stronger entrainment for the larger flame, the droplet's radial velocities were small compared to a smaller flame, possibly indicating that only larger droplets make it to the base of the flame

To take this work further, studies could include the statistics of the droplet sizes which will indicate the droplet evaporation, multiphase PIV, high-speed PIV to investigate the turbulence, temperature measurement of the flame field, etc. Subsequently, the results from these future work, will add to the already existing body of knowledge of fire suppression and further refine our understanding of the action of water mist on fire.

REFERENCES

[1] SFPE, *SFPE Handbook of Fire Protection Eng*. 2015.

[2] D. J. Rasbash, Z. W. Rogowski, and G. W. V. Stark, "Mechanisms of extinction of liquid fires with water sprays," *Combust. Flame*, vol. 4, pp. 223–234, 1960.

[3] A. K. Liu, Z., Kim, "A Review of water mist fire suppression systems - Fundamental studies," *Fire Prot. Eng.*, vol. 10, no. 3, pp. 32–50, 2000.

[4] C. C. Ndubizu, R. Ananth, P. a. Tatem, and V. Motevalli, "On water mist fire suppression mechanisms in a gaseous diffusion flame," *Fire Saf. J.*, vol. 31, no. 3, pp. 253–276, 1998.

[5] J. Liu and G. Liao, "Procedia Engineering Experimental study of the effect of water mist on CH 4 / air non-premixed flames," *Procedia Eng.*, vol. 26, pp. 1279–1286, 2011.

[6] B. Downie, C. Polymeropoulos, and G. Gogos, "Interaction of a water mist with a buoyant methane diffusion flame," *Fire Saf. J.*, vol. 24, no. 4, pp. 359–381, 1995.

[7] G. Heskestad, "Extinction of gas and liquid pool fires with water sprays," *Fire Saf. J.*, vol. 38, no. 4, pp. 301–317, 2003.

[8] X. Huang, X. S. Wang, and G. X. Liao, "Characterization of an effervescent atomization water mist nozzle and its fire suppression tests," *Proc. Combust. Inst.*, vol. 33, no. 2, pp. 2573–2579, 2011.

[9] K. C. Adiga, R. F. Hatcher, R. S. Sheinson, F. W. Williams, and S. Ayers, "A computational and experimental study of ultra fine water mist as a total flooding agent," *Fire Saf. J.*, vol. 42, no. 2, pp. 150–160, 2007.

[10] R. Ananth and R. C. Mowrey, "Ultra-Fine Water Mist Extinction Dynamics of a Co-Flow Diffusion Flame," *Combust. Sci. Technol.*, vol. 180, no. 9, pp. 1659–1692, 2008.

[11] K. Prasad, C. Li, and K. Kailasanath, "Optimizing water-mist injection characteristics for suppression of coflow diffusion flames," *Symp. Combust.*, vol. 27, no. 2, pp. 2847–2855, 1998.

[12] E. E. Zukoski, B. M. Cetegen, and T. Kubota, "Visible structure of buoyant diffusion flames," *Symp. Combust.*, vol. 20, no. 1, pp. 361–366, 1985.

[13] R. D. Keane and R. J. Adrian, "Optimization of particle image velocimeters. II. Multiple pulsed systems," *Meas. Sci. Technol.*, vol. 2, no. 10, pp. 963–974, 1991.

[14] G. Heskestad, "Luminous heights of turbulent diffusion flames," Fire Saf. J., vol. 5, no. 2, pp. 103–108, 1983.

A survey on the factors affecting horizontal assisted evacuation in hospitals

Lamija Catovic[1], Chatarina Alniemi[1], Enrico Ronchi[1]

[1]Department of Fire Safety Engineering, Lund University, Sweden

enrico.ronchi@brand.lth.se

ABSTRACT

This paper presents the results of a survey on assisted evacuation conducted with 120 hospital staff members in Sweden. Staff members included people with different roles in the hospital (e.g. doctors, nurses, department heads, etc.). The survey consisted of a questionnaire aiming at investigating the factors which affect assisted evacuation in hospitals and the current status of assisted evacuation training in Sweden. The survey included a list of questions concerning factors which were previously identified through a literature review and that they may affect the effectiveness of horizontal assisted evacuation in hospitals. Hospital staff had to evaluate the importance of these factors through a Likert-scale questionnaire. Three factors were identified in the survey by hospital staff as the most important in horizontal assisted evacuation, namely 1) coordination/collaboration between staff, 2) the patient need for assistance and 3) Training on evacuation routines. According to the questionnaire results, occupants who are in need for life-sustaining equipment can be the most difficult to evacuate. Results also show that only 52% of respondents had ever taken part into an evacuation drill and 50% had practiced how to handle evacuation aid tools needed for assisted evacuation, with rescue mattresses being the most used evacuation aid. 78% of respondents were aware of the person responsible for arranging tasks in case of an emergency situation.

KEYWORDS:

Assisted evacuation, hospital, human behaviour, training, fire safety, survey.

INTRODUCTION

Evacuation plans and procedures in hospitals can be particularly critical given the need for assistance that some occupants may have. Occupants may in fact not be able to evacuate independently (i.e. to perform self-rescue activities), thus needing assistance. Patients may vary in their self-rescue abilities, ranging from ambulant patients with reduced mobility to non-ambulant patients who need assistance with a wheelchair or other carrying device (e.g. stretcher, blankets, mattresses, etc.) [1], [2]. The rescuing abilities of hospital personnel can play a key role in the effectiveness of a hospital evacuation procedure. These individuals are generally responsible for assisting patients in their relocation to a safe place. In particular, hospital personnel may be responsible for the relocation of a number of people with a repeated procedure, which may generate physical fatigue [3]. This is also linked to the number of staff available for these tasks. This number may depend on several factors, such as the type of care provided by the hospital, time of the day in which an emergency take place, etc.

The investigation of the factors affecting the effectiveness of assisted evacuation procedures can be performed in several ways. One methodology would be to perform evacuation experiments and full scale drills. Nevertheless, experiments in hospitals may be difficult to arrange given the need to not interrupt the regular hospital operations as well as logistic and financial issues. Therefore, a limited number of experimental studies are available in the literature [3], [4] and they mostly focus on vertical evacuation, i.e., the usage of evacuation aid tools for assisted egress in multi-storey buildings.

Total evacuation strategies may not be practical in hospitals. For this reason, it is often recommended to relocate people horizontally to the nearest safe fire compartment [5]. The present paper focusses then on the issues associated with horizontal assisted evacuation, i.e. patient relocation procedures that take place within one floor [6]. Horizontal evacuation procedures in hospitals are generally an established process, and they include information concerning the prioritization of the patients [1], [7]. Staff must ensure that all patients needing assistance are relocated to a safe place and that this process does not hinder other staff intervention [7]. In this context, national [8] and International regulations and guidelines [9] give provisions and recommendations to account for the issues associated with evacuation planning, in some instances including information on the personnel actions that are expected in case of a fire emergency. Nevertheless, an evaluation of the evacuation procedures in case of the presence of people in need for assistance need to be done in light of the actual status of current knowledge and training of fire evacuation issues of the hospital personnel.

To address this issue, a questionnaire study has been arranged to investigate the status of current staff training experience in Sweden as well as to assess which factors are considered by hospital personnel as the most important ones during horizontal assisted evacuation procedures. It should be noted that the focus of the questionnaire is on the staff rather than on the patients as their role has been considered of primary importance in assisted evacuation scenarios. The scope of the work was to provide a context to current provisions and fire safety evacuation requirements concerning hospitals and understand the hospital personnel's perspective on the factors which may (either positively or negatively) affect assisted evacuation.

THE SURVEY

As a first step of the work, a literature review has been conducted in order to identify a list of factors which may impact horizontal assisted evacuation in hospitals. The scope was to identify a set of commonly identified assisted evacuation issues which will subsequently be ranked by the survey respondents with a Likert-Scale questionnaire. Different sources were used to collect information on those factors, such as guidelines focusing on general evacuation strategies in Sweden and internationally [8]–[10], evacuation plans adopted for Swedish hospitals [11]–[13] and research studies investigating human behaviour during hospital evacuation [2], [3], [6], [7], [14]–[16]. This analysis also included the review of different routines adopted in relation to the type of patients and the department in the hospital that are involved in an evacuation. In addition, a set of most commonly used evacuation aid devices for assisted evacuation have been identified and their features have been reviewed (e.g. carry chairs, wheelchairs, evacuation chairs, rescue sheets and mattresses, stretchers, etc.)

Based on the literature review, the following 10 factors have been identified and included in the survey questions:

1. The patients need for assistance (e.g., depending on whether the patient is older / pregnant / has a disability / in need of life-support equipment)
2. The distance to which the patient has to be moved
3. The number of staff per patient (ratio between staff/patients) available in evacuation scenarios

4. The physical abilities of the staff
5. Coordination/collaboration between staff members
6. The consistency between written instructions on evacuation procedure and the actual implementation (e.g. who does what, prioritization of patients, etc.)
7. Training on evacuation routines (e.g. rescue, alert, alarm, extinguishment, assembly)
8. The amount of rescue equipment available
9. Training in the usage of evacuation aid devices
10. Movable and fixed physical objects that are present on the evacuation route depending on the type of room (e.g. beds, equipment attached to the patient, furniture, locked doors)

The survey respondents were asked to grade the importance of each factor with a 5-point Likert scale ranging from (1= not important at all to 5=very important). Additionally respondents were asked to list what they thought they are the most important factors affecting assisted evacuation (giving them the possibility to explain their answer). Participants were also given the possibility to mention additional factors which may affect assisted evacuation procedures as well as reviewing 1) experience of hospital evacuation drills, 2) experience of actual fire evacuations, 3) knowledge of people responsible for the evacuation procedure in the hospital, and 4) experience with training with individual evacuation aid devices. Background questions to the survey participants included information on their role in the hospital (e.g. nurses, doctors, department heads, secretaries, etc.), time spent working in the current hospital, age and gender.

Survey dissemination

The survey was prepared in a spreadsheet form and distributed both through online mailing (approximately 300 emails were sent with a Google form link) as well as distributing the questionnaires in person in hospitals (approximately 100 paper copies were distributed). The target distribution list included a variety of roles within hospitals and a variety of hospitals all over Sweden (varying in size and location, i.e. from smaller to larger Swedish cities/towns).

Participants

As the refusal rate increases, the potential volunteer bias increase [17]. For this reason, the questionnaire was made as short and simple as possible to decrease rejection rate and to avoid volunteer bias. A total of 120 filled questionnaire forms were collected (30% response rate). This included 76% of women and 24% of men. This gender ratio is in line with current hospital staff gender ratio in Sweden, where the total number of female staff in the three largest categories of personnel (nurses, doctors and physiotherapists) are 77% of the total [18]. Approximately 50% of the respondents were aged between 40-60 years old, 30% of the respondents were aged between 20-40 years old and 20% over 60 years old. The majority of the respondents had worked in the hospitals for more than 5 years. The geographical distribution of the respondents is presented in Table 1.

Table 1. Geographical distribution of survey respondents.

Geographical area of respondents	Respondents [n]	Respondents [%]
Norrland (Northern part of Sweden)	7	6
Svealand (Central part of Sweden)	31	26
Götaland (Southern part of Sweden)	78	65
Unknown	4	3

The roles of the hospital staff who responded to the survey is presented in Figure 1. This included nurses, heads of hospital departments (intended as the chiefs of different hospital units), assistant nurses, doctors and other (this can include administrative staff, or other personnel working at the hospital).

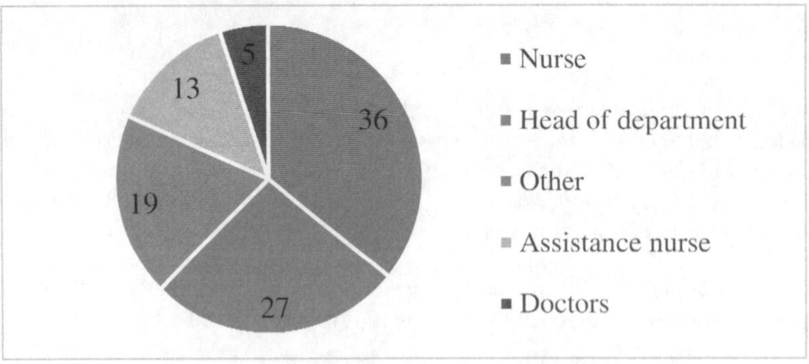

Fig. 1. Distribution of hospital staff roles of the respondents

RESULTS

Results are here presented both in terms of the factors affecting assisted evacuation as well as the four questions concerning the previous experience/knowledge of hospital staff concerning evacuation procedures. Results show that 52% of the respondents had performed an evacuation drill in a hospital. Among the respondents, 16% had experienced an actual evacuation in a hospital. 22% of the respondents were not aware of the person responsible to hand out information in case of an emergency evacuation. A 50/50 split was obtained concerning the number of respondents who had practiced the use of an evacuation aid device. Among the respondents who have practiced the use of an evacuation aid device, the majority of them had experience with rescue sheets/mattresses, followed by stretchers (see Figure 2).

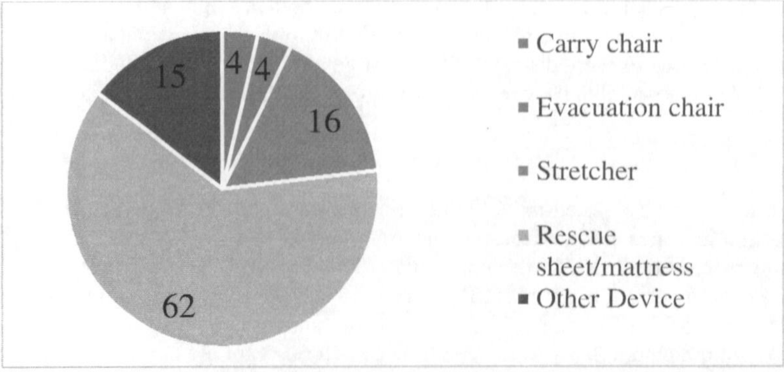

Fig. 2. Type of evacuation aid device which respondents had practiced with

The descriptive statistics of the Likert-scale responses to the questions concerning the most important factors in case of assisted evacuation in hospitals are presented in Table 2 (presented as percentages). Results show that most of the factors received quite high scores. This is not surprising as the factors have been defined through a literature review aiming at identifying the most important factors affecting assisted evacuation.

In order to investigate which differences between the different factors were significant, a Wilcoxon-Signed Rank Test was used (corrected critical value applying the Bonferroni correction was 0.0025). This non-parametric test was used comparing all factors against each other, with the aim to provide a ranking of factors. The resulting ranking of factors has permitted the identification of the three most important factors, namely coordination/collaboration between staff, the patient need for assistance and Training on evacuation routines.

It should be noted that the responses are here presented in an aggregated form (i.e. responses from all survey respondents are presented together). This is mostly due to the limited sample size of the survey. Nevertheless, a qualitative analysis of the responses given by people belonging to different groups has also been performed. Some of the key findings of this analysis show that while 100% of the head of departments are aware of who is responsible for the evacuation procedures, this percentage is reduced to 71% for the nurses, 33% for doctors and 78% for the category "other". In contrast, the number of experienced emergencies is overall consistent among different groups while the number of people that have practiced the use of evacuation aid devices corresponds to approximately 60% for both the head of departments and the category "other", 33% for doctors and 42% for nurses.

Table 2. Responses (expressed in percentages) to Likert-scale questions on the factors affecting assisted evacuation in hospitals.

Factor number	Factor	Likert-scale response				
		5 (very important)	4	3	2	1 (not important at all)
1	The patients need for assistance	85	8	6	0	0
2	The distance to which the patient has to be moved	53	28	18	0	0
3	The number of staff per patient	63	28	7	2	0
4	The physical abilities of the staff	40	27	28	6	0
5	Coordination/ collaboration between staff members	81	18	1	0	0
6	The consistency between written instructions on evacuation procedure and the actual implementation	68	24	8	0	0
7	Training on evacuation routines	69	23	8	0	0
8	The amount of rescue equipment available	48	29	13	8	2
9	Training in the usage of evacuation aid devices	49	29	14	6	2
10	Movable and fixed physical objects that are present on the evacuation route	66	24	8	2	0

Additional open questions conducted during the actual survey showed that the greatest proportion of respondents claimed that one key issue is the limited number of staff per patients in case of emergency. An additional factor that was mentioned by respondents included the fact that often it is unknown the number of relatives that are actually inside hospitals and this may have dual impact on the evacuation effectiveness (they might be of help during evacuation procedures or they may hinder the evacuation).

DISCUSSION

The findings of the survey give information concerning the experience and training that hospital personnel has on assisted evacuation and their responses might be useful to provide guidance both on future safety improvements for assisted evacuation scenarios in hospitals as well as future research studies.

The collaboration and coordination between hospital staff members has been identified as one of the most important factors in assisted evacuation procedures, especially given the fact that a procedure relies on the correct handling of the roles that each staff member has in case of emergency. Clearly communicated instructions on the roles of each staff member should be provided and they should improve coordination of rescuing activities. This is somehow in contrast with the fact that 22% of the survey respondents were not able to identify who is the person responsible for giving instructions in case of a fire evacuation scenario. Evacuation drills and training on the use of evacuation aid devices might be of help to increase the effectiveness of an assisted evacuation procedure. Nevertheless, the survey showed that only half of the personnel have taken part in evacuation drills and approximately half has ever practiced the use of an evacuation aid device. The survey shows that this seems to be associated also with the fact that not all staff is on site when such type of training is performed, thus part of the hospital personnel do not receive any evacuation training.

The assessment of the resources for patients needing assistance relies on a correct evaluation of patient's health conditions and self-rescue ability (or inability). Future work should focus on increasing both patients and staff awareness of their emergency evacuation procedures in order to facilitate the staff performance of assisted evacuation.

The survey findings also indicate that the most common device in use for training in Swedish hospitals is rescue sheets/mattresses. It should be noted that an experiment conducted by Hunt et al [14] showed that rescue sheets resulted in the slowest average horizontal transportation speeds (0.9 m/s) if compared with stretchers, evacuation chairs and carry chairs. This finding though should also take into consideration the preparation time associated with each patient and device, i.e. how long it takes to place patients in an evacuation aid device in relation to the type of patient and the device itself.

Despite the present survey has as main limitation the limited sample size of the respondents (120 people), it allows to identify some potential key issues affecting the effectiveness of assisted evacuation in hospitals. In particular, some of the results demonstrated that a significant part of the hospital personnel (50%) had experienced neither an evacuation drill nor practiced the use of an evacuation aid device. This finding should lead to a larger discussion on the need to take actions to improve the awareness on the benefits that evacuation drills and training may have. On this matter, future studies should assess quantitatively the relationship between the hospital staff training and the time that could be gained in case of an emergency given a more trained personnel.

REFERENCES

[1] V. Alonso-Gutierrez and E. Ronchi, "The simulation of assisted evacuation in hospitals," presented at the Fire and Evacuation Modelling Technical Conference, Malaga, Spain, 2016.

[2] J. Poppenborg and S. Knust, "Modeling and optimizing the evacuation of hospitals based on the MRCPSP with resource transfers," *EURO J. Comput. Optim.*, vol. 4, no. 3–4, pp. 349–380, 2016.

[3] A. L. E. Hunt, "Simulating hospital evacuation," PhD Thesis, University of Greenwich, 2016.

[4] A. Hunt, E. R. Galea, and P. J. Lawrence, "An analysis of the performance of trained staff using movement assist devices to evacuate the non-ambulant," in *5th International Symposium. Human Behaviour in Fire*, 2012, pp. 328–339.

[5] V. Alonso, *Egress modelling in health care occupancies*. The Fire Protection Research Foundation, USA, Retrieved July, 2014.

[6] N. Akıncıtürk, "An Evaluation of Hospital Evacuation Strategies with an Example," *Int. J. Appl.*, vol. 5, no. 3, 2015.

[7] C. Johnson, "Using computer simulations to support a risk-based approach for hospital evacuation," *Dep. Comput. Sci. Brief.*, 2006.

[8] Boverket, "Boverkets byggregler (föreskrifter och allmänna råd) [Building rules (regulations and general guidelines]." Boverket, 2017.

[9] National Fire Protection Association, *NFPA 101, life safety code*. 2018.

[10] E. Ronchi and D. Nilsson, "Fire evacuation in high-rise buildings: a review of human behaviour and modelling research," *Fire Sci. Rev.*, vol. 2, no. 1, p. 7, 2013.

[11] S. Vårdgivare, "Brandskydd [Fire Safety]." Vårdgivare Skåne [Scanian Health care provider], 2017.

[12] MSB, "Statens räddningsverks allmänna råd och kommentarer om systematiskt brandskyddsarbete [General state rescue work. Recommendations and comments on systematic fire safety work]." Swedish Contingency Agency (MSB), 2017.

[13] U. Sahlgrenska, "Systematiskt brandskyddsarbete [Systematich fire safety work]." Sahlgrenska University Hospital, 2017.

[14] A. Hunt, E. R. Galea, and P. J. Lawrence, "An analysis and numerical simulation of the performance of trained hospital staff using movement assist devices to evacuate people with reduced mobility," *Fire Mater.*, vol. 39, no. 4, pp. 407–429, 2015.

[15] A. Rahouti, S. Datoussaïd, and R. Lovreglio, "A sensitivity analysis of a hospital evacuation in case of fire," presented at the Fire and Evacuation Modelling Technical Conference, Malaga, Spain, 2016.

[16] D. A. Purser and M. Bensilum, "Quantification of behaviour for engineering design standards and escape time calculations," *Saf. Sci.*, vol. 38, no. 2, pp. 157–182, Jul. 2001.

[17] "Volunteer Bias," in *Encyclopedia of Research Design*, 2455 Teller Road, Thousand Oaks California 91320 United States: SAGE Publications, Inc., 2010.

[18] S. Socialstyrelsen, "Statistik om legitimerad hälso- och sjukvårdspersonals arbetsmarknadsstatus [Statistics on certified health care personnel, work market status]." Sveriges officiella statistik, 2016.

IOP Conf. Series: Journal of Physics: Conf. Series **1107** (2018) 072002 doi:10.1088/1742-6596/1107/7/072002

Study of Sound Direction Evacuation

Wenhui Dong, Chunyu Yu, Mei Zhibin

Shenyang Fire Research Institute, China

ycyu01@aliyun.com

ABSTRACT

Nowadays, many buildings have exit signs which are used to guide occupancies to the exit when an emergency happens. However the assistance given by signage based on visual means may be reduced or totally ineffective in a smoke filled environment. And occupants may not be familiar with the building and its associated emergency egress routes and may have to rely on exit maps, markings, or signage to guide them to an exit or other means of evacuation. There are other limitations of signage-based visual means. Persons with visual disabilities will have difficulty with emergency information that relies on visual cues. Other weaknesses are that the illuminated exit signs may often compete with nearby bright lights or other visual elements. Exit signs may be inadvertently blocked from view due to renovations or remodeling of building.

Directional sound is a new kind of signage which can help occupants to pinpoint instantly where the sound is coming from and find the exit quickly. Used as a directional sound pattern, broadband sound can enable localization cures due to all binaural transfer function and precedence effect mechanisms. And broadband sound is distinctive from the audible sounds of bells and fire emergency voice speakers.

The fire alarm sounders and the fire emergency voice speakers are two important sounds in a fire emergency evacuation. So the compatibility of the directional sounders with fire alarm sounders and fire emergency voice speakers is very important in application. The three sound used in a fire emergency evacuation should work timesharingly and alternately.

Ambient sound levels may greatly affect the performance of the directional sounders, for the ambient sound may possibly have wide spectral range. A directional sound test room is built according to Annex B Directional test for sounders of the British standard BS PAS 41:2003[1] Directional Sounders – Requirements and tests. And tests were made to see the performance of different frequency distribution sounds used for evacuation direction under ambient sound levels.

KEYWORDS:

evacuation ; directional sound technology ; broadband sound.

IOP Conf. Series: Journal of Physics: Conf. Series **1107** (2018) 072002 doi:10.1088/1742-6596/1107/7/072002

INTRODUCTION

In most buildings, exit routes are identified by visual means, exit signage. But the signage is easily obscured with even low level of smoke. Besides, exit can also be hard to find in visually cluttered areas, such as supermarkets and shopping malls. Many behavioral studies [2,3,4] have repeatedly shown that one of the most natural instincts in the event of a fire is people evacuate a building by the route through which they entered. This is rarely the quickest or most appropriate way. Many people fail to spot nearby exits and in some cases walk straight past visible fire exits.

Directional sound is a new technology which offers a solution to improve the exit signage for building occupants. This technology can help occupants to pinpoint instantly where the sound is coming from and relocate the exit. The technology was first proposed by professor Deborah Withington from University of Leeds. She performed numerous evacuation tests[5,6,7,8] involved scenarios such as with and without directional sound, evacuations with low visibility conditions created by use of theatrical smoke, and variety of room and complex corridor arrangements. In smoke-filled scenarios with little or no visibility of exit signs, directional sound demonstrated its benefit to provide cues that significantly reduced the time for subjects to find the nearest exit. Boer L C[9,10] designed a new sound signal on the basis of professor Deborah Withington's work and proposed some evacuation tests in tunnel and ship environment.

The directional sound technology is widely used in many countries. NFPA72 National Fire Alarm Code of America[11] has provided guidance on installing and maintaining directional sounders from the 2007 Edition. And the BSI published the standards BS 8465-2005 Code of practice for design and installation of directional sounder evacuation systems[12] and BS PAS 41:2003 Directional Sounders – Requirements and tests[1].

THE PRINCIPLE OF SOUND DIRECTION EVACUATION

The ability to pinpoint the location of a sound source is based on the physics of sound and the physiology of the human hearing mechanism. The human brain processes a large amount of signals, some of which provide cues to the sound source's location.

The horns, bells and loudspeakers installed in buildings to produce the alarm signal and voice messaging typically work at the same time to alert people to an imminent danger as wide an area as possible but give no information concerning the direction to, or location of, the nearest exits.

People are able to hear sound ranging from about 20Hz to 20000Hz. The primary localization cures are provided by interaural time differences, interaural intensity differences and head-related transfer function(HRTF). At lower frequencies the time delay between arriving sound signals is detectable. Interaural time difference is most evident in frequencies below about 1000Hz. At higher frequencies the loudness/intensity differences between the ears is more noticeable because of partial shielding of the more distant ear by the head. Interaural intensity differences are most evident for frequencies above 3000Hz. HRTF describes the transforming effect of the head, torso and external ear on sound as it travels from the sound source to the ear canals. The shape of the ear attenuates some frequencies and amplifies others like a filter. The HRTF changes depending on sound source location, providing an additional localization cue. And HRTF is most effective in the 5000Hz to 10000Hz range.

All of the above cues can be utilized simultaneously when the source signal is broadband sound containing a range of low and high frequencies. The combination of different cues provides reinforcement and redundancy of information to enhance the ability to locate the sound source. So the broadband sound is used as directional sound signal to help occupancy locate exit points and for guidance in complex evacuation routes in a large building. Directional sounders contain frequency content extending over almost all of the entire range discernible by the human ear. This wide band type of frequency content is often referred to as "white noise".

Since the development of directional sound technology, numerous studies have been conducted that compare occupancy egress time assisted by directional sounders to egress time based on conventional visual signals and signage. These studies have yielded a large body of data that indicates that the use of directional sounders consistently results in a significant reduction of egress time.

TESTS OF LOUDSPEAKERS

As far as concerned, broadband sound is used in directional sound technology[11]. Theoretically, this sound should have a bandwidth ranging from about 20Hz to 20000Hz based on human hearing range. But in fact, few loudspeakers in the market are capable of presenting such a broadband sound and with acceptable

IOP Conf. Series: Journal of Physics: Conf. Series **1107** (2018) 072002 doi:10.1088/1742-6596/1107/7/072002

price. Also, the performance of these loudspeakers is uneven in currant market. Used as a unit for directional sounder, the size of a loudspeaker is generally not bigger than 4 inches since a directional sounder with bigger size usually consumes more power and has aesthetic problems. So the effective range of the frequency bandwidth of the sounder is needed to be identified. The ability of producing broadband sound of the loudspeakers available in the market with the size below 4 inches is tested in an acoustical sound enclosure with an audio analyzer.

Fig. 1. Acoustical sound enclosure

The results are showed in Table 1. The effective frequency range of most loudspeakers tested reaches 500-10000Hz, but not the broadband frequency range 20-20000Hz. Whether this effective frequency range good enough for localization needs to be tested.

Table 1. The effective frequency range of chosen loudspeakers

Number	Effective frequency range(Hz)	Size(inch)	Number	Effective frequency range(Hz)	Size(inch)
1	500～10000	4	10	1600～20000	2
2	400～12620	3	11	2000～16000	2
3	2000～10000	2	12	400～10000	2
4	500～16000	2.5	13	1600～16000	2
5	1600～20000	2.5	14	1600～5000	2
6	400～10000	1.5	15	1600～20000	1
7	2000～16000	1.5	16	1600～20000	1
8	500～12620	2	17	1800～14000	1.5
9	2000～20000	1.5	18	500～20000	4

DIRECTIONAL SOUND TEST ROOM

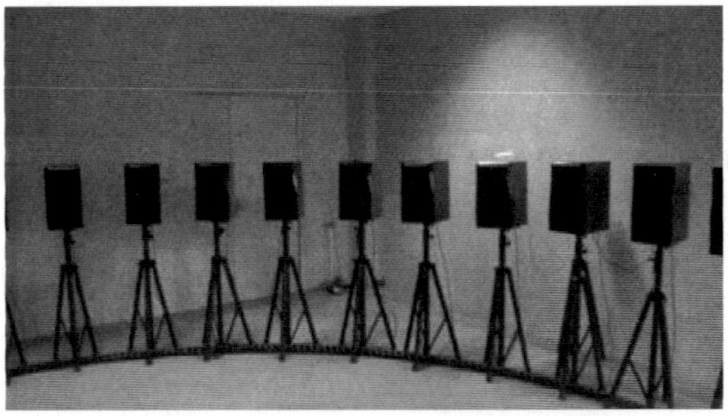

Fig. 2. Directional sound test room

A directional sound test room is built according to Annex B Directional test for sounders of the British standard BS PAS 41:2003 Directional Sounders – Requirements and tests[1]. The performance of the directional sound signals will be tested in this room. The dimensions of the test room are shown in Fig.2. The room is geometric symmetry in order to provide acoustically symmetric condition around the subject.

IOP Conf. Series: Journal of Physics: Conf. Series **1107** (2018) 072002 doi:10.1088/1742-6596/1107/7/072002

Ambient sound levels may greatly affect the performance of the directional sounders, for the ambient sound may possibly have wide spectral range.

To test the localizing of a sound signal, test subjects with specified age distribution[1] will determine the location of bursts of sound from numerous possible positions. The test subjects should have normal hearing for their age.

36 full frequency sound boxes are placed in a circle with 10 degrees interval and 3 meters diameter around the subject. A digital audio signal is produced from a computer with sound card and transformed to a sound signal, which is amplified by a power amplifier before played by one of the sound boxes.

DIRECTIONAL SOUND TEST

Ten test subjects participated in the directional sound test. The age distribution is shown in Table 2.

Table 2. Age bandings of subjects

Age	Number
18-27	2
28-37	4
38-47	2
48-57	1
58-67	1

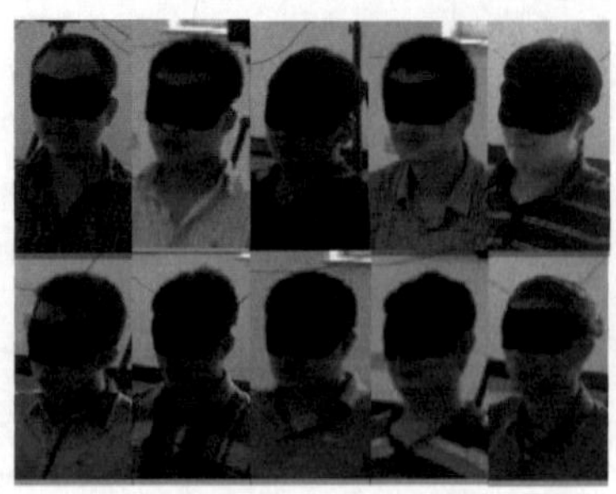

Fig. 3. Test subjects

Broadband sound with different frequency bandwidth is tested. Besides, pure tone signals are also tested for reference. The tested sound level is set 75dB. Appendix A to NFPA's Fire Alarm Code[11] shows examples of what average ambient sound levels may be for different applications. But in China, the ambient sound levels of most possible applications may be higher. The fire alarm sound and fire emergency voice speaker sound should be at least 15dB higher than ambient sound levels. Besides, directional sound should be at least 15dB lower than the fire alarm sound considering the compatibility between the two sounds. So two ambient sound levels are considered in the tests. One is below 60dB, and the other is 75dB.

As in the test, 36 sound boxes are placed in a circle with 3 meters diameter around the subject. The circle is marked from 0°to 360°. Each sound box is placed in a position with specified degree.

Prior to commencing the tests, the subject needs to know:

1) Do not move head during the tests.
2) There are a number of different sounds to be played, and he or she will be require to identify the location of the sound.
3) Each sound will be played 10 times with each time from one random sound box. A gap of 5s is given between each sound and he or she has to make a decision on location in the 5s.
4) In the noise effect tests, a noise sound will be played continuously, while he or she will distinguish and tell the location of the directional sound signal.

For each sound, there are 100 test results. In each test, the direction pointed by the subject has a deviation with the sound box which gives the audible signal. And if the deviation is bigger than ±5°, the percent correct number will diminish by one. The results are listed in Fig.4.

Background noise has some impact on the locating of the sound source. When the background noise level is closer to that of the test sound, pure tone sound patterns tend to be harder to locate, while broadband sound with bigger frequency range is still easier to pinpoint.

Besides, it's easier to pinpoint the location of broadband sound, but the frequency bandwidth should be as wide as possible. According to the British standard BS PAS 41:2003, test subjects shall correctly identify the location of the sounder at least 75% of the time. In Fig.4, when ambient sound level is below 60dB, three sounds with bandwidth 0.02-20kHz, 0.3-16kHz, and 0.5-10kHz meet the above requirement. But when ambient sound level is 75dB, only two sounds with bandwidth 0.02-20kHz and 0.3-16kHz meet the requirement. More work should be done to choose the proper loudspeaker which can give more broadband effective frequency range.

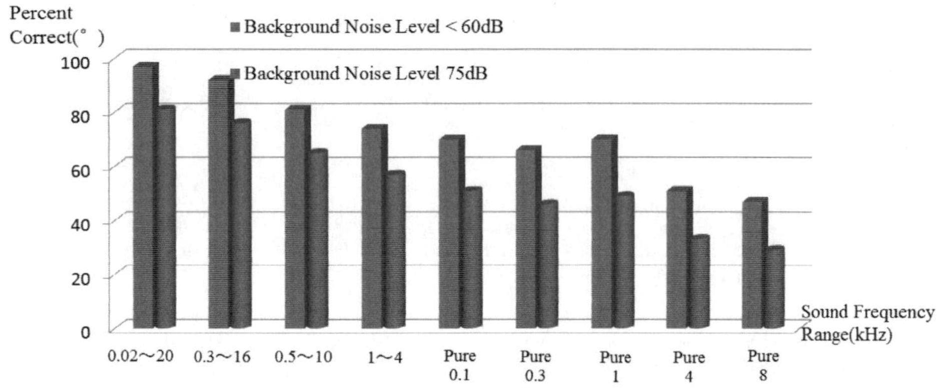

Fig. 4. Results of the localization tests

COMPATIBILITY WITH FIRE ALARM SOUNDERS AND FIRE EMERGENCY VOICE SPEAKERS

The fire alarm sounders and the fire emergency voice speakers are two important sounds in a fire emergency evacuation. So the compatibilities of the directional sounders with fire alarm sounders or fire emergency voice speakers are very important in application.

The fire alarm sound can comfortably exceed most common background noises. And from the above test results, if the bandwidth of directional sound is broad enough, the percent correct can meet the requirement 75% of the standard BS PAS 41:2003 even if the background noise is louder. Fire alarm sound has a strong tonal content as opposed to the more even, broadband content of a directional sounder. DANIEL J [13]compares the spectrum of a typical fire alarm signal at a high setting with the spectrum of the directional sounder at a lower power setting. Although the fire alarm may have an overall A-weighted sound level 20 dBA or more higher than the overall A-weighted sound level of the directional signal, both sounds are audible.

But where directional sounders and voice evacuation systems employing speakers are utilized within the same occupancy the directional sounders can reduce the speech intelligibility of fire emergency voice speakers. So it is recommended [14] that when both systems are utilized in the same installation that the two systems are not operated simultaneously.

The fire alarm sound and fire emergency voice speaker sound should work timesharingly and alternately. And the time of each fire alarm may be 10s to 20s, and 10s to 30s for that of each fire emergency voice speaker. Considering the compatibilities of the three systems, the directional sounders should work the same time with fire alarm sounders with a lower overall A-weighted sound level, but timesharingly and alternately with fire emergency voice speakers.

DISCUSSION

The engineering application of directional sound technology will face some problems which will be discussed here.

Compared with the widely recognized fire alarm sound signal, the broadband sound may cause confusion to Chinese residents without training when an emergency happens. Occupants in buildings may do not know how to react when they hear the sound. So an enhancement with the addition of voice messaging is necessary. The voice message after the directional sound will tell occupants of what action to take when they approaching a directional sounder[9]. This message will allow them to react quickly and confidently without training or prior knowledge of this new evacuation signage.

Traditional fire alarm and voice evacuation system both give out sound signals when an emergency happens in buildings. The directional sound is an additional evacuation aid to be used, but not a replacement for the above two traditional systems.

The directional sounders should be installed at the exit. But sounders only at the exit may not provide directional sound loud enough to reach the majority of the occupants. Besides, in large, open areas, there may be multiple exits, but multiple directional sounders can cause interference with each other. So the installation of the directional sounders should be further investigated.

ACKNOWLEDGMENT

The study was conducted under the project 2016YFC0800602 supported by the state key development program of China.

REFERENCES

[1] PAS 41:2003, Directional Sounders-Requirements and tests. BSI, 17 February 2003.

[2] Sime, Jonathan. Movement Toward the Familiar Person and Place Affiliation in a Fire Entrapment Setting. Environment and Behavior 17:6(1985)697-724.

[3] MeClintock, T., et al. A Behavioral Solution to the Learned Irrelevance of Emergency Exit Signage. 2d International Symposium on Human Behavior in Fire, March 26-28, 2001, Boston, MIT InterScience Communications, 23-33.

[4] SFPE Engineering Guide, Human Behavior in Fire, Society of Fire Protection Engineers, Bethesda, Maryland, 2003.

[5] D. J. Withington, M. Lunch. Directional Sound Evacuation- An Improved System for Way Guidance From Open Spaces and Exit Routes on ships. RINA Passenger Ship Safety Conference, London, UK 2003.

[6] D. J. Withington. Faster evacuation from ferries with sound beacons[J]. Fire, 2000: p. 30.

[7] Withington D J. Life saving applications of directional sound[J]. Pedestrian and Evacuation Dynamics (Berlin/New York: Springer), 2002.

[8] Withington D J. Localisable alarms[J]. Human factors in auditory warnings, 1999: 33-40.

[9] Van Wijngaarden S J, Bronkhorst A W, Boer L C. Auditory evacuation beacons[J]. Journal of the Audio Engineering Society, 2005, 53(1/2): 44-53.

[10] Boer L C, van Wijngaarden S J. Directional sound evacuation from smoke-filled tunnels[J]. Safe & Reliable Tunnels. Innovative European Achievements, Prague, Czech Republic, 2004.

[11] NFPA72-2013, National Fire Alarm and Signaling Code. NATIONAL FIRE PROTECTION ASSOCIATION OF America. August 29, 2012.

[12] BS8456-2005, Code of practice for design and installation of directional sounder evacuation systems. BSI, 18 August 2005.

[13] DANIEL J. O'CONNOR, P.E., FSFPE. The psychoacoustic basis and implementation of directional sounders. 2006, National Fire Protection Association.

[14] Directional Sounder with Voice Messaging, Application Guide by System Sensor, 2007.

A Time-continuous Compartment Model for Building Evacuation

Bertrand MAURY[1,2], Sylvain FAURE[1,2], Jacques ANGELÉ[2] and Rémi BACHIMONT[2]

[1] *Laboratoire de Mathématiques d'Orsay, Univ. Paris-Sud, CNRS, Université Paris-Saclay, 91405 Orsay, France*
[2] *Signactif, 260 avenue Aristide Briand - 92220 Bagneux - France.*

Bertrand.Maury@math.u-psud.fr

ABSTRACT

We propose here a general framework to estimate global evacuation times of complex buildings, and to dynamically investigate the dependence of this evacuation time upon various factors. This model relies on a *network*, which is in some way the skeleton of the building, the nodes of which are the bottlenecks or exit doors. Those nodes are connected by edges which correspond to portions of egress paths located within a given room. Such models have been proposed in a discrete setting. The model we propose takes the form of a continuous evolution equation of the differential type. It relies on a limited number of variables, namely the number of people gathered upstream each node, together with the number of people on their way from a node to the next one. The basic parameters of the model are the capacities of doors, and the time needed to walk from one node to the next one. In spite of its macroscopic character (the motions of pedestrians are not described individually), this approach allows to account for complex and nonlinear effects such as capacity drop at bottlenecks, congestion induced speed reduction, and possibly some dispersion in evacuees behaviors. We present here the basic version of the model, together with the numerical methodology which is used to solve the equations, and we illustrate the behavior of the algorithm by a comparison with experimental data.

KEYWORDS :

modeling; human behavior; risk assessment; evacuation times; exit capacities;

IOP Conf. Series: Journal of Physics: Conf. Series **1107** (2018) 072003 doi:10.1088/1742-6596/1107/7/072003

1 Introduction

Providing accurate and robust estimates for evacuation times in complex buildings is a long-term challenge in public safety. A common dilemma resides in the opposition between microscopic and macroscopic approaches. Microscopic descriptions (see e.g. [5, 10, 11] allow for a precise description of evacuees interactions, possibly accounting for non uniformity of individual behavior (social tendencies, speed, ...), but they lead to higher, possibly prohibitive, computational times. Besides, they call for an accurate knowledge (at least statistically) of people characteristics, which is most of the time out of reach. On the other hand, macroscopic models ([6, 8]) handle the crowd as a continuum, represented by a local density. The evolution of this density typically follows a conservative transport equation which expresses the "people conservation", and the core of the model lies in the manner the effective velocity is determined, based on individual tendencies and local density. This approach makes it possible to account for very large number of people at a reasonable computational cost. Yet, most models of this type are not able to reproduce some observable effects, like the Faster-is-Slower effect ([4]), the Capacity Drop phenomenon ([1]), or the fluidizing role of an obstacle ([12]). Both approaches rely on a fine description of the behavior of individuals and their interactions with neighbors (finite number of finite-size individuals for the microscopic setting, infinitely many point particles for the macroscopic setting).

If one aims at predicting the evacuation time of complex building, it may be of interest to use a coarse grain description of the crowd, based on quantities that are directly observable and measurable. This approach, which we may call *systemic*, is based on a decomposition of the building into various *compartments*, corresponding to distincts areas (like rooms, halls, or corridors). Those compartments are connected by doors, and the balance of global headcounts in compartments is driven by fluxes between them. The model we propose here is based on a network defined as follows: the nodes of the network are the exits of the various compartments. Each compartment has a certain number of entrances, each of which is the exit of an compartment upstream. Each entrance of a given room is connected by an oriented edge to the exit. We shall follow the convention that the exit points to the entrances, in such a way that the arrows (oriented edges) express a dependence relation. More precisely, a node i points to a node j if i is *influenced* by the situation of j (possibly in the past). The model which is presented here is not new in its constitutive principles. In particular the Capacity Constrained Routing Approach presented in [7] is based on the same type of network. The same type of model is also proposed in [3] to represent car traffic networks. The main novelty of the approach lies in the nature of the model, which is *continuous in time*, whereas previous approaches were essentially discrete. This continuous character gives a sound theoretical character to the approach, which can be used in particular to design rigorous methodologies for parameter identification. It also makes it possible to tune up the time step, which is in our setting a discretization parameter, depending on the situation which is considered.

While allowing for faster-than-real-time computations, the large granularity of the description level a priori rule out the possibility to properly describe small scale interactions between individuals, and this model may not be used to investigate in any way the cause of the aforementionned phenomena. Yet, the most relevant parameters (in particular capacities and node-to-node travel time) can be made dynamic, and allow for an account of phenomena like the Capacity Drop, or the reduction of the walking speed in case of congestion.

IOP Conf. Series: Journal of Physics: Conf. Series **1107** (2018) 072003 doi:10.1088/1742-6596/1107/7/072003

2 Model description, mathematical formulation

Let us start with a toy problem: A certain quantity of people is accumulated upstream a single exit door. Since we aim at setting a continuous model, applicable to large numbers of entities, we represent this quantity by a *real number* $N \in \mathbb{R}_+$. The capacity of the exit, that is the maximal number of individuals which can go through it per unit time is denoted by C. We denote by $f = f(t) \geq 0$ the incoming flux, that is the upstream flux of pedestrians, and by $\Phi = \Phi(t) \geq 0$ the instantaneous flux through the door. The incoming flux is assumed to be known. The balance at the door writes $dN/dt = f - \Phi$. The core of the model relies in the expression of Φ as a function of the dynamic variables N, f, and the static parameter C. By definition of the capacity it holds that $\Phi \in [0, C]$, and $\Phi = C$ whenever $N > 0$. When $N = 0$, Φ lies between 0 and C. Its value is f when $f < C$, but it may happen that Φ saturates to C if $f > C$. To sum up, the evolution problem can be written

$$\left\{ \begin{array}{rcl} \dfrac{dN}{dt} & = & f - \Phi \\ \Phi & = & \left| \begin{array}{lll} C & \text{if} & N > 0 \\ \arg \min\limits_{\tilde{\Phi} \in [0,C]} \left| \tilde{\Phi} - f \right| & \text{if} & N = 0 \end{array} \right. \end{array} \right. \tag{2.1}$$

The extension to a many-room building is built in a similar manner, by accounting for the fact that people flowing through some passage node reach the next one after some transit time which is a parameter of the model. We denote by R the number of room / compartments, m_i the number of inlet accesses to room i. For $n = 1, \dots, m_i$, we denote by α_i^n the index of the room upstream the access n to room i. The time spent by an individual to walk (in room i) from entrance n to the exit is T_i^n. Finally, Φ_i is the instantaneous flow rate of people through the exit of i. The problem writes

$$\left| \begin{array}{rcl} \dfrac{dN_i}{dt} & = & -\Phi_i + \sum\limits_{n=1}^{m_i} \Phi_{\alpha_i^n}(t - T_i^n)\ ,\ \ i = 1, \dots, R, \\ \Phi_i & = & \left| \begin{array}{lll} C_i & \text{if} & N_i > 0 \\ \arg \min\limits_{\tilde{\Phi} \in [0,C_i]} \left| \tilde{\Phi} - \sum\limits_{n=1}^{m_i} \Phi_{\alpha_i^n}(t - T_i^n) \right| & & \text{for } i = 1, \dots, R. \end{array} \right. \end{array} \right. \tag{2.2}$$

3 Numerical solution

In spite of non-trivial mathematical issues (see [9]), designing numerical algorithms to solve this sort of problems is straightforward. Let $\tau > 0$ denote a time step. To simplify the presentation, we shall assume that the transfer times are whole multiples of this time step, and denote by $\tilde{T}_i = T_i/\tau \in \mathbb{N}$ the corresponding dimensionless times. We shall furthermore assume that people are initially gathered in the neighborhood of exits. The approximation of N_i at time $t^k = k\tau$ is denoted by N_i^k.

The scheme is actually simpler than the continuous model, since it simply expresses that, at some time t^k, the number of individual walking through exit i between t^k and t^{k+1} is $C_i\tau$ whenever there is enough people available to achieve this full capacity, and N_i otherwise. In other words, the average flux in this time interval is either C_i or N_i/τ.

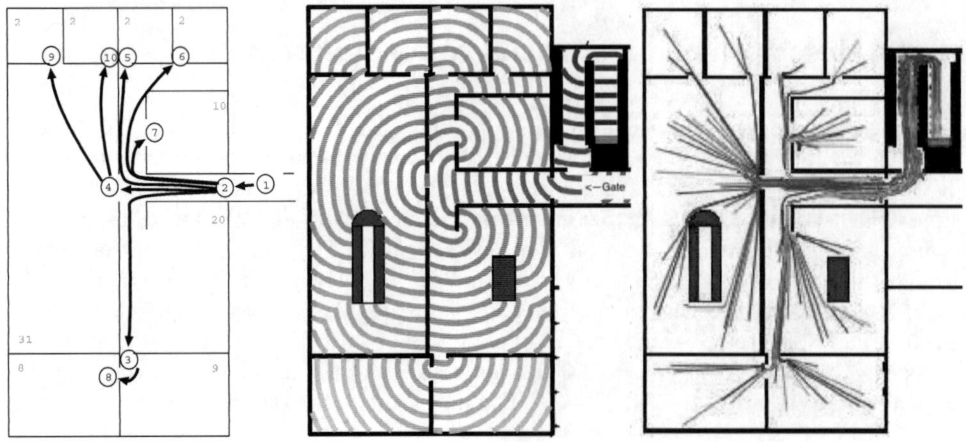

Figure 3.1: Topography: network, geodesic distance to the gate, and typical paths

$$\left| \begin{array}{rcl} \Phi_i^{k+1} & = & \min(C_i, N_i^k/\tau)\ ,\ \ i = 1, \ldots, R, \\ N_i^{k+1} & = & N_i^k - \tau\, \Phi_i^{k+1} + \sum_{n=1}^{m_i} \tau\, \Phi_{\alpha_i^n}^{k+1-\tilde{T}_i^n}\ ,\ \ i = 1, \ldots, R, \end{array} \right. \tag{3.1}$$

where Φ_i^k is set to 0 for all $k < 0$ (no evacuation before the initial time), and R is the number of rooms. Assuming, as we did in the continuous setting, that all people are initially gathered upstream doors, we supplement this system with initial conditions $N_1^0, \ldots, N_R^0$. We may add an extra equation to keep an account of people who have evacuated the building at time k:

$$N_0^{k+1} = N_0^k + \tau\, \Phi_1^{k+1}.$$

It is also straightforward, for each room i, to keep an account of the number $\overline{N}_{i,n}$ of individuals who are on their way to the exit of room i, coming from the exit of room α_i^n:

$$\overline{N}_{i,n}^{k+1} = \overline{N}_{i,n}^k + \tau \Phi_{\alpha_i^n}^{k+1} - \tau \Phi_{\alpha_i^n}^{k+1-\tilde{T}_i^n}.$$

Global people balance is straightforwardly obtained by summing up all discrete equations:

$$N_0^k + \sum_{i=1}^{R} N_i^k + \sum_{i=1}^{R} \sum_{n=1}^{m_i} \overline{N}_{i,n}^k = \text{Constant}.$$

4 Illustration: comparison to experimental data

The model is tested in the configuration presented in [2], were a full set of experimental evacuations is presented. The topography is represented in Fig. 3.1 (left): it is made of 10 interconnected compartments. The underlying oriented network is represented on the left-hand side of the figure. Transit times are automatically computed as follows: the common target is defined as the gate upstream thestairs, denoted by 2 on the figure. The geodesic distance to the target, which corresponds to the length of the shortest path (accounting for walls and obstacles) to the target, is quasi-instantaneously computed by mean of a fast marching algorithm. More precisely, distances are computed at the center of cells of a cartesian grid which covers the whole domain, in a frontal way, starting from the target where is it set at 0, and then propagating backward upstream the domain. We refer to [9] (Chapter 8) for methodological and implementational details concerning those computations. The travel times are computed from those distances by setting the speed of pedestrian at 1 m/s. Isolines of this geodesic distance is represented in Fig. 3.1 (middle). To illustrate the evacuation, we also represent some egress paths computed with a microscopic model ([10]), based on an initial random distribution of individuals (Fig. 3.1, right).

The time needed to get from one node to the other can then be computed as the difference between the geodesic distances divided by the speed of pedestrians. The initial number of individual is 86, distributed over the rooms 2 to 10 (circled black indexes in Fig. 3.1, left). The headcount in each room is indicated by the numbers in red. The effect of various gate widths have been experimentally studied, and we reproduce with the present model the two scenarios, referred to as scenario 4 and scenario 5, respectively, like in [2]. In scenario 4, the exit gate is wide open (1.24 m), and the corresponding capacity is estimated at 2.78 Ps^{-1}. In scenario 5, the door is reduce to half its width (0.62 m), with a capacity of 1.72 Ps^{-1}. We run the numerical model is both situations, and we illustrate the results by representing the computed number of people gathered upstream the exit gate versus the time. Fig. 4.1 represents the plots in the two scenarios. The model makes it possible to recover the "state" of this sensitive point when the evacuation time goes on. As expected (see the figure at the top), this number first decreases. Then, around time 8 s, people coming from upstream rooms start to reach this node. The gate continues to work at full capacity, but the incoming fluxes are higher, which explains the increase. The little irregularities on the curve between times 8 and 17 corresponds to instants at which the first people coming from some given upstream room reach the gate node. Then, in a final phase (after time 25), all people still in the building are gathered at the gate, and the crowd flows out at full door capacity. The second plot corresponds to a smaller capacity: the initial decreasing phase is less efficient, then, in the second phase, the slope is higher (the incoming flux is the same as previously, but the outflow smaller). We recover, like in the experiments, a larger evacuation time, around 50 s, to compare to the 32 s of Scenario 4.

References

[1] E. M. Cepolina, Phased evacuation: An optimisation model which takes into account the capacity drop phenomenon in pedestrian flows, *Fire Safety Journal* 44, 2009, pp. 532–544,

[2] T. Chen, L. Pan, H. Zhang, S. Narayanan, N.Soldner, Experimental Study of Evacuation from a 4-storey Building, *Procedia Engineering* Volume 62, 2013, pp. 538–547.

[3] C. Daganzo, The cell transmission model: network traffic, *Transp. Res.* Part B 29, 1995, pp. 79–93.

[4] A. Garcimartín, I. Zuriguel, J.M. Pastor, C. Martín-Gómez, D.R. Parisic, (2014), Experimental evidence of the"Faster Is Slower" effect, *Transportation Research Procedia* 6, October 2014, pp. 760–767.

[5] D. Helbing, A. Johansson, Pedestrian, Crowd and Evacuation Dynamics, *Encyclopedia of Complexity and Systems Science*, 2010, pp. 6476–6495, Springer New York.

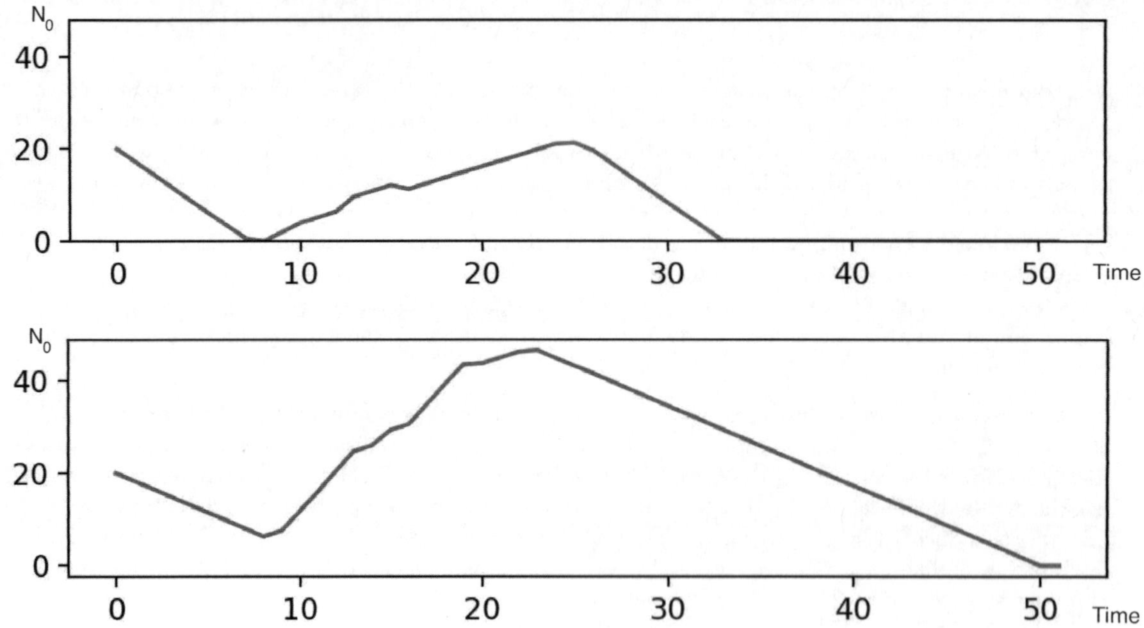

Figure 4.1: Computed number of people gathered upstream the exit gate vs. time, for scenarios 4 (top) and 5 (bottom)

[6] R.L. Hughes, A continuum theory for the flow of pedestrians *Transportation Research* Part B 36, 2002, pp. 507–535.

[7] Q. Lu , Y. Huang, S. Shekhar, (2003) Evacuation Planning: A Capacity Constrained Routing Approach. In: Chen H., Miranda R., Zeng D.D., Demchak C., Schroeder J., Madhusudan T. (eds) Intelligence and Security Informatics. ISI 2003. *Lecture Notes in Computer Science*, vol 2665. Springer, Berlin, Heidelberg.

[8] B. Maury, A. Roudneff-Chupin, F. Santambrogio, A macroscopic crowd motion model of gradient flow type, *Mathematical Models and Methods in Applied Sciences*, **20**, 2010, pp. 1787–1821.

[9] B. Maury, S. Faure, *Crowds in Equations*, London: World Scientific Publishing Europe Ltd, 2018.

[10] Maury, B., Venel, J., A discrete contact model for crowd motion, *ESAIM: M2AN*, Volume 45, Number 1, 2011.

[11] A. Schadschneider, A. Seyfried, Empirical results for pedestrian dynamics and their implications for cellular automata models, in *Pedestrian Behavior*, Ed.: H. Timmermans, Emerald, p. 27, 2009.

[12] I. Zuriguel, J. Olivares, J.M. Pastor, C. Martín-Gómez, L.M. Ferrer, J.J. Ramos, A. Garcimartín, Effect of obstacle position in the flow of sheep through a narrow door, *Phys. Rev. E*, 94, 2016.

IOP Conf. Series: Journal of Physics: Conf. Series **1107** (2018) 072004 doi:10.1088/1742-6596/1107/7/072004

People evacuation in tunnel fires: a cross evaluation of two methodologies

Benjamin TRUCHOT[1], Christophe WILLMANN[2], Joel GUIVARCH[2]

[1]*INERIS, Verneuil en Halatte, France*
[2]*CETU, Bron, France*

Benjamin.truchor@ineris.fr

ABSTRACT

In France, safety assessment of tunnels is mainly based on a risk analysis called the specific hazard investigation. One of the key issues of these studies consists in modeling people evacuation. The commonly used method for computing the people evacuation consists in evaluating their displacement velocity based on smoke opacity. However, it is obvious that other fire consequences parameters, thermal effects and toxicity, will affect people displacement.

To evaluate whether considering those effects is relevant and could modify conclusions of specific hazard investigations in tunnels or not, two approaches were used in a comparative manner. The first approach is based on currently used method and considers only the smoke opacity. The second innovative approach introduces thermal and toxic consequences on the displacement velocity. It enables considering the toxic dose impact on human being during the evacuation process and the thermal effect together with the visibility.
This new approach also introduces a new method for the evaluation of lethality and user evacuation which are managed with thresholds in the current method. This difference was also considered in the comparison.

Those two approaches were applied on many fire scenarios to compare the conclusions in terms of people trapped or killed by the fire. For each of those scenarios, consequences were evaluated based on a CFD model, for different tunnel geometries, design fires and ventilation strategies. The two most relevant of those scenarios are presented in the present paper.

This comparison shows only minor differences for the specific case of tunnel fire safety. One way to consider the effect of the new methodology should consist in keeping the existing method but modifying the corresponding thresholds for the different quantities relative to fire consequences.

KEYWORDS: EVACUATION, HUMAN BEHAVIOR, FIRE IN TUNNEL, FIRE CONSEQUENCES

IOP Conf. Series: Journal of Physics: Conf. Series **1107** (2018) 072004 doi:10.1088/1742-6596/1107/7/072004

INTRODUCTION

In France, every owner of a tunnel longer than 300 m should have a safety documentation in which the tunnel is described, the safety level is assessed and the operational aspects are explained. The safety assessment is mainly based on a risk analysis, so called specific hazard investigation.
Evaluation of fire consequences on people during the evacuation process is a key issue of specific hazard investigation. Regarding fire impact, several aspects are considered including temperature rise, radiation, smoke toxicity and smoke opacity. All those effects influence the people evacuation process, first through their influence on the displacement velocity and next, for evacuation and lethality. The current method [1] used for Specific hazard investigation only includes threshold to deal with these aspects.

Under a joint project research initiated by the CETU (Centre d'Etude des Tunnels, a national center for tunnel safety) and involving INERIS (Institut National de l'Environnement Industriel et des Risques, a national center for risk industrial risk protection), a new approach was elaborated for determining these impacts. This new approach was based on the last available input of the state of the art, especially the dose concept.

This new approach was compared to the existing one within the study of several scenarios to determine it added value. This article presents this work and it conclusion.

CURRENTLY USED APPROACH

The currently used approach is based on recommendation [1] and best practices of the engineering company.

In the current method, the walking velocity is computed by considering only the local visibility. The impact of CO and temperature is then considered through the possible death of people.

Walking velocity during evacuation process

As mentioned above, in the currently used recommendation, the user walking velocity is calculated based on the local visibility using the following assumption[1]:

- Walking velocity is 1 m/s if local visibility is higher than 20 meters;
- Walking velocity is 0,5 m/s if local visibility is between 5 and 20 meters;
- Walking velocity is 0,3 m/s if local visibility is lower than 5 meters.

While having determine the walking velocity, the evacuation direction is the required to compute the people displacement. People are assumed to go directly to the closest safe evacuation door while the local visibility stays higher than 5 m. When the local visibility becomes lower than this threshold, 50% of users is assumed choosing the wrong way, for instance run toward the fire instead of toward the nearest emergency exit.

User lethality and evacuation

Temperature and toxic fire consequences are used to estimate the available duration for the evacuation process [1]. Fire toxicity is considered through the CO local concentration assuming that users couldn't walk more than 15 min if the CO concentration is higher than 3000 ppm. Regarding thermal effect, following the same approach, it is assumed that user cannot walk more than 15 minutes if local temperature is higher than 80°C.

Concerning the lethality, not considered in [1], additional hypothesis is commonly used and consists in assuming that user die if the CO concentration becomes higher than 5000 ppm or the temperature becomes higher than 100°C. This death is assumed to be instantaneous.

NEW APPROACH

This new approach consists in considering a walking speed as a function of local toxic, thermal and visibility conditions. It mainly also considers the Fractional Effective Dose approach [2].

User walking speed and direction

The proposed approach is based on the international work on the fire consequences on people regarding their evacuation capability [2]. This international standard gives information about the impact of fire on evacuation capability. It clearly shows that both toxic and thermal fire consequences will affect people during the evacuation process. This standard however does not link the fire consequences parameters with displacement velocity. This link was proposed by some authors [3] for different application based on the following formulae:

$$v = v_0.f_1(K_s).f_2(\rho).f_3(T)$$

Such a formula enables to consider all the three parameters of fire consequences on people displacement velocity. In this formula, $f_1(K_s)$ is a velocity correction function depending on the extinction coefficient, $f_2(\rho)$ is a similar function that depends on toxicity and, finally $f_3(T)$ is the function that gives the modification as a function of temperature.

The visibility effects on people walking velocity is based on existing data [5] and considering that smoke from fire in tunnel are irritant ones. $f_1(K_s)$ can then be expresses as follow:

$$f_1(K_s) = \max(f_{1_{\min}}, -67.154.K_s^3 + 55.033.K_s^2 - 14.876.K_s + 2.3133)$$

The extinction coefficient is then based on CO_2 concentration and local temperature [1]: $K_s = 83000.[CO_2]/T_g$. One of the key issue regarding the impact of visibility is the minimum value when opacity is the only effect on people. Regarding available data [5], the minimum appears to be 0.3 m/s. A minimum value of 0.3 m/s appears as the most relevant one regarding the present objective and the tunnel context.

The objective of the toxic influence on people moving velocity is to consider not only CO impact but all other compounds that could be in smoke as hydrogen chloride, sulfur dioxide and others. Following ISO 13571 approach [2] it is then possible to build an equivalent CO toxicity to be considered in the toxic impact on people evacuation. Considering that velocity becomes 0 when the FED [2] reach 1, that corresponds, for this specific case of application to a toxic dose of 35 000 ppm.min. In the current approach, the considered dose is 3000 ppm during 15 min, this means 45 000 ppm.min. the evolution of velocity correction as a function of CO concentration can be written as follow:

$$f_2(d_{CO}) = \min(1; -0.289.\ln(d_{CO}) + 3.0604)$$

Following [3][4], the temperature correction function could be written as follow:

$$f_3(T) = \begin{cases} v_0 & T_0 < T < T_{\sigma1} \\ v_0 + \dfrac{(v_{\max} - v_0)\left(\dfrac{T - T_{\sigma1}}{T_{\sigma2} - T_{\sigma1}}\right)^2}{v_{\max}} & T_{\sigma1} < T < T_{\sigma2} \\ v_{\max}.\left[1 - \left(\dfrac{T - T_{\sigma2}}{T_d - T_{\sigma2}}\right)^2\right] & T_{\sigma2} < T < T_d \end{cases}$$

Then, thresholds should be defined. The first one, $T_{\sigma1}$ corresponds to the minimum velocity that could affect people, it was fixed to 30°C. Then, in this approach, the velocity is supposed to increase because peoples become aware about the risks while the temperature is under $T_{\sigma2}$, that was fixed to 60°C. Then, between $T_{\sigma2}$ and the lethal temperature, T_d, the velocity is supposed to decrease because of the temperature effect on human beings. T_d was fixed to 120°C. One main difference with the model described in [3] is the maximum velocity, originally fixed at 5 m/s, it was reduced, for the present approach to 1.6 m/s which appears more coherent with feedback experience concerning road tunnel evacuation. It is also assumed that, if the visibility is lower than 5 meters, 50% of people head in the wrong direction.

User lethality and evacuation regarding heat and radiance

Considering lethality implies computing both thermal and radiative effects. However, in the context of road tunnel evacuation process, it can be demonstrated that radiative effects should influence people only in the surrounding of the fire. Radiation from the smoke could, in some specific case, influence people but it is currently impossible to take such an effect into account since the impossibility to model their effect with precision enough with existing models in the context of road tunnels. Consequently, only the convection impact is used, its effect is evaluated through the following formula [2]:

$$X_{FED}^{th} = \int_{t_1}^{t_2} \left(\frac{1}{t_{conv}}\right).dt \text{ with } t_{conv} = (4.1.10^8).T(t)^{-3,61}$$

In this equation X^{th}_{FED} represents the thermic dose that impacts user between times t_1 and t_2. In the 13571 ISO standard, the thermic dose is only related to the direct effect on the body for instance burns. When building the new method for evacuation velocity computing, it is then assumed that the evacuation is not possible anymore when $X^{th}_{FED} = 1$.

Based on a similar approach than incapacitating approach, lethal effects should consider both convective and radiative effect. The corresponding thresholds are based on a temperature of 120°C for convective effects and a radiance of 5 kW/m² for radiative effect [9].

User lethality and evacuation regarding toxicity

Evaluating the fire toxic effect on people requires first defining the nature of the impact, incapacitation, non-reversible effects or lethality. Several methods were then developed for those different applications as the ISO 13571 [2] one for incapacitation or the large bonfire non-reversible and lethal effects in the context of land use planning [7]. When dealing with fire in tunnels, which means in confined space, the distinction between the different types of gas, asphyxiant and irritant, should be considered as proposed in [2]. This approach consists in computing through the Fractional Effective Dose, FED, for asphyxiant gases, and Fractional Effective Concentration, FEC, for irritant gases. This evaluation is based on specific thresholds to determine the incapacitation and should be adapted for lethality.

Then, to consider lethality, the lethal dose of CO, $5{,}09.10^9$ $ppm^{2,27}$.min, shall be considered. However, while CO is not the only gas to be produced by the fire, other gases as to be considered by modifying the CO source term as a function of real fire emissions following, for example the methodology described in [8]. This method consists in increasing artificially the CO emissions by a factor representative of the relative toxicity of other gases. To illustrate, if a gas with a lethal threshold twice the one of CO, the CO equivalent will be twice this gas quantity. The main limit of such an approach is that it considers a constant ratio between toxic gases along the whole fire. It is also important to note that, in this dose evaluation, the breathing rate increase due to CO_2 is not considered.

COMPARAISON BETWEEN THE TWO APPROACHES

Several scenarios were chosen and analyzed with the two methods and the results have been evaluated (mainly number of deaths).

Choose of scenarios

In this project among twelve fire scenarios studied and 6 were chosen for their relevance for the comparison. A scenario is relevant when consequences for users are realistic and sever enough to involved the assumptions of the two methods and evaluate their impact on the results (mainly number of death). The scenarios modelled were based on two geometries. The first corresponds to the French Talant tunnel, nearby Dijon and is assumed to be longitudinally ventilated. Different fire standard curves were modeled for this tunnel, from 30 to 200 MW with different initial conditions and ventilation strategy. The second is a fictitious square section tunnel with a transvers ventilation system. Only 100 MW fire was modeled for this tunnel with considering 2 fire positions. While the objective of the present paper is not to evaluate the method but to provide some examples of application, only two scenarios are presented in paragraphs hereafter, Talant-2 and Talant-4.

Complementary common hypothesis for the two approaches

While the existing and new approach use different hypothesis for people walking velocity and people lethality, additional required hypothesis should be identical for a comparison purpose. For each approach, the required time for people to evacuate their car, after having decided to do so, is fixed to 90 s [1], this time is supposed reaching 300 s for a bus evacuation.

Then, the alarm process should be defined. In the present article, people are supposed to decide to evacuate their car when the visibility decreases near the ground 150 m in front of us.

Modeling and graphic representation

Consequences modeling were achieved using the FDS, version 6, fire code [6]. The tunnel geometry was fitted to be representative of the Talant existing tunnel, nearby Dijon in France. This circular shape tunnel is 630 m long, 10 m width and the maximum height is 8 m. One of the specificities of this tunnel is the 6% slope, considering that the traffic is going down. For numerical modeling cells were $0.5 \times 0.25 \times 0.25$ m^3. For the analysis, all quantities were considered 1,5 m above ground.

The Talant-2 scenario analysis

IOP Conf. Series: Journal of Physics: Conf. Series **1107** (2018) 072004 doi:10.1088/1742-6596/1107/7/072004

This case corresponds to a 100 MW fire with an initial counter pressure about 20 Pa that induces a 1 m/s air velocity in the opposite direction of the traffic. Considering a congestion case, the fire ventilation was chosen to maintain the smoke stratification, which means a fire ventilation of 1 m/s in the direction of the traffic. The people evacuation process is represented on Figure 1. On these pictures, continuous line represents people who evacuate in the correct direction, i.e. in the opposite of the fire; dotted line represents persons who evacuate in the wrong direction because of the low visibility; the vertical line marks the death for the person considering lethal criteria are reached.

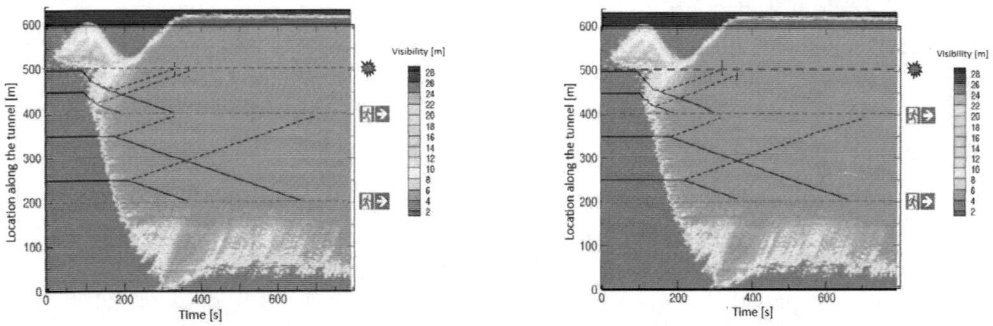

Existing approach | New approach

Figure 1 : Visibility along time and space together with people evacuation for both evacuation model.

As highlight by the table below, this first comparison shows that, for this case, a modification of the evacuation hypothesis does not affect the conclusion relative to the capability of people to evacuate. The deaths are only users who have headed in the wrong direction because of visibility conditions.

		deaths number (time)	
		current approach	New approach
Talant-2	incapacity	6 (≈300 s)	6 (≈300 s)
	lethality	6 (≈300 s)	6 (≈300 s)

Table 1 Talant 2 - deaths number with the current and the new approach

The Talant-4 scenario analysis

This case corresponds to a 200 MW fire with an initial counter pressure about 20 Pa that induces a 1 m/s air velocity in the opposite direction of the traffic. It is considered that no fire ventilation is activated for that case. The people evacuation process is represented on Figure 2. As for figure 1, on these pictures, continuous line represents people who evacuate in the correct direction, i.e. in the opposite of the fire; dotted line represents persons who evacuate in the wrong direction because of the low visibility; the vertical line marks the death for the person considering lethal criteria are reached.

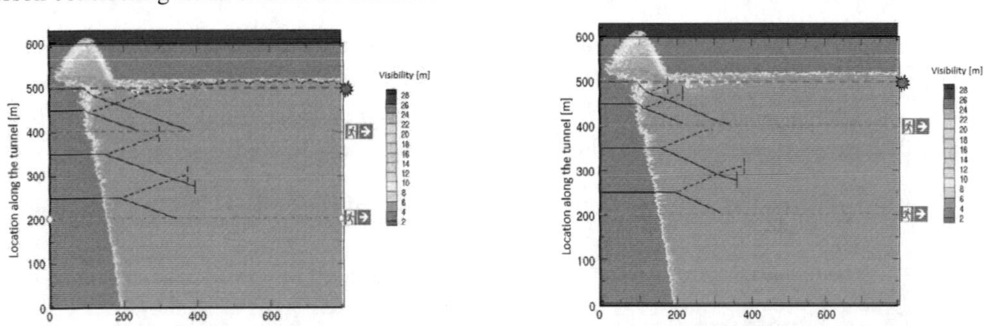

Existing approach | New approach

Figure 2 : Visibility along time and space together with people evacuation for both evacuation model.

For this second configuration, the deaths number is also the same for the current approach and the new approach. It must however be pointed out that the deaths are only users who have headed in the wrong direction because of visibility conditions.

		deaths number (time)	
		current approach	New approach
Talant-4	incapacity	34 (≈400 s)	34 (≈200 to 400 s)
	lethality	34 (≈400 s)	34 (≈200 to 400 s)

Table 2 - Talant 4 - deaths number with the current and the new approach

CONCLUSIONS

Regarding the state of the art, the new approach enables an improvement in terms of considering main fire consequences (visibility, temperature and toxicity) on people evacuation and lethality.

This new approach however is more complicated than the current one. Especially, it cannot be easily implemented in specific hazard investigations, the risk analysis on which safety tunnel evaluation is based. Indeed, to implement it, this approach should be considered together with the fire dynamic evolution.

In addition, the study of several scenarios chosen to highlight the impact of fire consequences parameters on users have shown only minor differences in terms of lethality between the current approach and the new one. This could be explained by the specificities of road tunnel evacuation and the general assumptions made in the specific hazard investigations, having in mind that these general assumptions are identical for both approaches. Indeed, with these specificities and assumptions, people do not stay long enough in average conditions of temperature or toxicity to have their velocity or survival conditions affected regarding the concepts of the new approach (dose concept for instance).

It is also crucial, for such modelling, to be aware about human behavior and movement capability in smoke.

Furthermore, these minor differences could be reduced by modifying the thresholds used in the current approach. CETU currently works on these modifications by involving the stakeholders of safety hazard investigations.

REFERENCES

[1] CETu, Guide to road tunnel safety documentation - booklet 4: specific hazard investigations, 2003.

[2] ISO standard 13571: Life-threatening components of fire - Guidelines for the estimation of time available for escape using fire data, 2010.

[3] Milke, J., "Evaluating the early development of smoke hazard from fires in large spaces", ASHRAE, vol 106, 2000

[4] Wei, Z., Hui, D., Tong, W., "The application of fire spread and evacuation simulation technology in large stadium", Stoch Environ Res Risk Assess 23:433–439, 2009.

[5] Yamada, T. and Akizuki, Y., Visibility and Human Behavior in fire smoke, The SFPE Handbook of Fire Protection Engineering – 5th Edition, NFPA, 2016.

[6] Mc Grattan, K, et al. Fire Dynamics Simulator User's guide, NIST Special Publication 1019, 6th edition, 2013.

[7] INERIS Ω16 document, Toxicité et dispersion des fumées d'incendie, Phénoménologie et modélisation des effets, in French, 2005.

[8] Truchot, B., Fouillen, F. and Collet, S., An experimental evaluation of toxic gas emissions from vehicle fires, Fire Safety Journal, 2018.

[9] French regulation, « Arrêté du 29 septembre 2005 relatif à l'évaluation et à la prise en compte de la probabilité d'occurrence, de la cinétique, de l'intensité des effets et de la gravité des conséquences des accidents potentiels dans les études de dangers des installations classées soumises à autorisation ».

IOP Conf. Series: Journal of Physics: Conf. Series **1107** (2018) 072005 doi:10.1088/1742-6596/1107/7/072005

A Review of Current and Historical Occupant Load Factors for Mercantile Occupancies

Michael SPEARPOINT[1] and Charlie HOPKIN[1, 2]

[1] *Olsson Fire & Risk UK Ltd*
[2] *The University of Manchester*

michael.spearpoint@olssonfire.com

ABSTRACT

The determination of the occupant load of a building is often necessary as part of a fire engineering design assessment. Designers often use occupant load factors (or other equivalent terminology) to carry out this calculation typically using values taken from various published guidance or deemed-to-satisfy documents. This paper examines past and current values for mercantile occupancies that are specified in documents that are used in various jurisdictions around the world. In general three approaches are used; the occupant load factors are defined by the floor level of the building, by the type of mercantile activity within the spaces or through the application of uniform values throughout the occupancy. A simple six-storey exemplar building is used to illustrate the differences in the total occupant load and the relative number of occupants per floor when the different occupant load factor approaches are employed. Even for this simple building the occupant load differs by a factor of 4.1 and the distribution of the occupants varies noticeably between upper and lower levels depending on the particular document selected for the calculation.

KEYWORDS

occupant density; mercantile; occupant load.

INTRODUCTION

Background

In doing an egress calculation for a fire engineering design assessment it is necessary to determine the occupant load of a building and the distribution of the people throughout the building's spaces. The definition of occupant load is not universal. In NFPA 1 [1] it is described as "the total number of persons that might occupy a building or portion thereof at any one time". The New Zealand Acceptable Solutions [2] defines it as "The greatest number of people likely to occupy a particular space within a building". The Scottish Technical Handbook [3] states "...the appropriate number of occupants in each space for normal circumstances".

Typically occupant loads are calculated from occupant load factors (or other equivalent terminology) specified in various 'prescriptive documents'. In some cases these documents are deemed-to-satisfy methods to meet a regulatory requirement and other cases they are provided as a form of guidance. However, Authorities Having Jurisdiction "...might specify a higher occupant load than calculated using occupant load factors... to provide an additional safety margin" [4] or occupant loads can be derived from measured values.

This paper examines the adopted occupant load factors for mercantile spaces given in a number of internationally available documents, although it is beyond the scope of this work to attempt to review every jurisdiction around the world and trace their development.

Measuring Occupant Density

It would be expected that occupant load factors used in design are based on historical data from measured occupant densities. Although measuring occupant density appears to be a straight-forward process it does come with temporal, dimensional and cultural complexities. The traditional method is to count the number of occupants at a given instant in time and then relate that to the footprint of the space in question. In these instances, the density depends heavily on the time and placement of the measuring area. To reduce the impact of these factors, more complex methods can be adopted, such as by using Voronoi diagrams [5]. Videographic methods can also be employed although these are not always precise. By taking such measurements in buildings of the same type, at different times of the day and year, gives an indication of the variation in occupant density and what might be expected as the maximum density.

Occupant density and occupant load factors are both quoted either in terms of the number of people per unit area or the area taken up per person (pp). One dimensional aspect is that these areas may depend on the body size of the individuals across the sampled population, the type of clothing worn, whether people are carrying items etc. [6]. The issues of what footprint should be used, whether a gross or net area be adopted and if a net area, then what elements are discounted from the calculation, are subject to interpretation.

In terms of temporal factors, the Post-War Building Studies in the UK [7] notes the difficulty of counting people in shops and department stores as well as changes that occur during seasonal sales. There could be changes in society over longer time scales in the way that mercantile occupancies are configured and used by people. Finally, there may be variations in social and cultural practices that mean the way in which mercantile spaces are used vary in different societies. The application of occupant load factors from one country or region may not be applicable to other parts of the world.

BASIS OF PUBLISHED OCCUPANT LOAD FACTORS

Notwithstanding the previous definitions of occupant density, a rational expectation is that published occupant load factors represent reasonable worst-case scenarios to allow for an acceptable level of conservatism, such that they correspond to some upper percentile related to a distribution of actual values that occur over the life of the space. However, the original underlying data used to determine the basis of the factors is difficult if not impossible to find in many cases. Clearly the maximum number of people a space can hold is related to an ultimate packing density and is not practical in real-world occupancy situations. Aside from record breaking attempts to try to fit as many people in a space that is physically possible, occupant density values up to 9 pers./m^2 [8] (0.11 m^2 pp) have been reported although ref. [4] notes that at an occupant density of 0.28 m^2 pp movement generally reaches a virtual stop.

From a regulatory and practical perspective, typically buildings are divided into different occupant uses with their own occupant load factors although this is not universally true and the approach to determine occupant load in the Republic of Korea does not follow a building use classification system [9], for example. Occupant uses are generally divided into assembly, mercantile etc. with sub-classifications. Modern retail developments are often complex multi-use buildings containing shops plus food halls; circulation areas, ancillary spaces;

IOP Conf. Series: Journal of Physics: Conf. Series **1107** (2018) 072005 doi:10.1088/1742-6596/1107/7/072005

storage; back-of-house administration etc. This paper only considers the specific mercantile activities generally referred to as 'shops' that are accessible to the public. However, shops themselves can be classified by various types including department stores selling clothes and homewares; showrooms for bulky items such as furniture; food supermarkets; specialised traders and/or mixed sales areas. The application of such definitions can be subject to interpretation.

Of the aforementioned general space use classifications, those for offices have been documented in some detail regarding the level of apparent conservatism. Schulte [10] quotes the proceedings of the 26th Annual [NFPA] meeting in 1922 discusses the use of applying 100 sq. ft pp (9.3 m^2 pp) for offices from surveys conducted by the Building Exits Code Committee. It is noted that there is the statement "...the Committee felt that this figure of 100 square feet was conservative" although there is no further comment on to what extent it is considered conservative. However, when it comes to mercantile space there are less comments on the level of conservatism noted in the historical literature. Nevertheless, in the recent work by De Sanctis et al. [11] the probability of exceedance of design occupant load densities in NFPA 101 [12], Approved Document B (ADB) [13] and New Zealand Acceptable Solutions, C/AS [2] compared to their measurements ranged between 5.9 % and 0.01 %. In the work of Angerd and Frantzich [14] the maximum occupant density from their analysis of mercantile spaces in Sweden did not exceed 0.28 pers./m^2 net area (3.6 m^2 pp) and their best-fit log-normal distribution has an average value of 12.5 m^2 pp such that the ADB value of 2 m^2 pp would be at the 99.9 % upper limit. The occupant load factors used in the various documents are discussed in the next section.

PUBLISHED VALUES

Uniformly Applied Values

The simplest method to specify an occupant load factor is to use a single value throughout an occupancy type and this approach is used in documents for several occupancy classifications. Specifically for mercantile occupancies it appears that the occupant load factor used in Japan and Switzerland is 2 m^2 pp according to Youn et al. [9] and De Sanctis et al. [11] respectively whereas in the International Building Code (IBC) [15] 2015 edition, a gross mercantile occupant load factor of 60 sq. ft pp (5.6 m^2 pp) is given. The IBC also provides an 'occupant formula' to determine the occupant load factor that increases between 2.8 and 4.6 m^2 pp for malls with a specified range of gross leasable areas. BS 5588 [16] Part 11 suggests a floor space factor of 4 m^2 pp for retail premises where the occupancy is not known but notes that this could be 2 to 10 m^2 pp where the occupancy is known and the value can be ascertained by reference to similar premises.

Based on Floor Level

As an alternative to using a single occupant load factor for mercantile spaces some documents vary their factors as a function of floor level. Historically for department stores in the US Schulte [10] quotes the 22nd Annual (NFPA) meeting in 1918 in which "... an average of one person for each thirty-two square feet of floor space (irrespective of counters, etc.) on the upper floors, and one person for each twenty-five square feet on the ground floor and basement" (i.e. 3.0 m^2 pp and 2.3 m^2 pp respectively). In 1935 NBS M151 [17] gives 30 sq. ft pp (2.8 m^2 pp) for street floors and basements and 60 sq. ft pp (5.6 m^2 pp) for the other floors for the gross area of store (retail) spaces. The 2012 edition of the IBC followed NBS M151 whereas NFPA 101 [12] provides gross mercantile sales area values of 2.8 and 5.6 m^2 pp depending on the relationship of the floor to street level corresponding to NBS M151 but with an additional factor of 3.7 m^2 pp for sales areas on two or more street floors. NFPA 101 also includes a method for calculating the occupant load factor that increases between 2.8 and 5.1 m^2 pp over a range of gross leasable areas. The New York Building Code [18] in 1938 gave net values that slightly differ from NBS M151 / NFPA 101 with 25 sq. ft pp (2.3 m^2 pp) for first floors and basements and 60 sq. ft pp (5.6 m^2 pp) for the other floors. However, the 1968 edition gave 50 sq. ft pp (4.6 m^2 pp) for other floors and by 2008 gross areas values of 30 sq. ft pp (2.8 m^2 pp) for first floors and basements and 60 sq. ft pp (5.6 m^2 pp) for the other floors are specified, matching NBS M151.

In New Zealand, NZS 1900 [19] gave values for the ground floor of shops as 20 sq. ft pp net or 30 sq. ft pp gross (1.9 m^2 pp and 2.8 m^2 pp respectively) and for other floors 40 sq. ft pp and 60 sq. ft pp (3.7 m^2 pp and 5.6 m^2 pp respectively) where the gross values correspond to those given in NBS M151. As discussed in the next section, New Zealand no longer uses this approach.

Qatar [20] has three classes of occupancy that depend on the number of storeys and building area but mercantile occupant load factors are the same in each case and follow NFPA 101. Similarly, the 2007 edition Saudi Building Code [21] gives gross occupant load factors of 2.8 m^2 pp for basement and grade floor areas and 5.6 m^2 pp for areas on other floors but no occupant load factor for areas on two or more street floors.

Canada has had a National Building Code [22] since 1941 but it has not been possible to access early editions. Some provinces and municipalities implement the code exactly as is, while others use a slightly modified version. Occupant load factors for mercantile uses in Ontario are 3.7 m^2 pp for basements, first storeys and second storeys having a principal entrance from a pedestrian thoroughfare or parking area. For other storeys, a value of 5.6 m^2 pp is used. These values have been used from at least 1997. The 1998 edition of the British Columbia Fire Code adopts the same values as with Ontario. In either case it is not clear whether values are net or gross. Comparing with M151 it would appear that the 'other storey' value is a gross value. However, the basement, first floor etc. does not correspond to the M151 gross value of 2.8 m^2 pp but does match the net value given in NZS 1900 [19] for other floors. Whether this match is coincidental is unknown to the authors.

Occupant load factors in Hong Kong, Australia and India all follow a similar procedure. In the 2011 edition (and the 1996 edition) of the Hong Kong Code of Practice for Fire Safety in Buildings [23], basement, ground, first and second floors of retail shops and department stores are specified as 3 m^2 pp. For the third floor and higher the value is 4.5 m^2 pp. The Australian National Construction Code [24] specifies 3 m^2 pp levels entered direct from the open air or any lower level. A value of 5 m^2 pp is then given for all other levels. The Indian National Building Code, along with the Delhi building bye-laws [25], specifies 3 m^2 pp for the street floor and sales basement and a value of 6 m^2 pp is then given for upper sales floors.

In the UK, occupant load factors are discussed in ref. [6] in which an occupant load factor of 75 sq. ft pp for retail shops and upper floors of department stores and 10 sq. ft pp for department store "sales basements, ground floor and any upper floor used for bazaars or special sales displays" are recommended. These equate to 6.7 m^2 pp and 0.9 m^2 pp respectively and are at the upper and lower extremes of similar environments found elsewhere. While ref. [6] distinguished occupant load factors between floor levels, no existing UK document referred to in this paper has adopted this approach.

Based on Occupancy Type

Rather than defining mercantile occupant load factors with regard to building storey an alternative (or supplementary) method is to give occupant load factors for specific mercantile activities.

In the UK, the 2013 edition (consistent back to 2000) of ADB [13] in England and Wales, the 2017 edition (consistent back to 2005) of the Scottish Technical Handbook [3] and the 2005 edition of Technical Handbook E for Northern Ireland recommend a floor space factor of 0.7 m^2 pp for a queuing area or shopping mall, 2 m^2 pp for shop sales areas including supermarkets and department stores and 7 m^2 pp for shop sales areas trading in furniture and other 'bulky goods'. For shopping malls, ADB also makes reference to BS 5588 [16] Part 10, which recommends an occupant load calculation based on mall widths, with 0.75 m^2 pp applied for mall widths up to 8 m, and 2 m^2 pp for any additional mall width beyond 8 m. This approach is also described in annexes in both the BS 9999:2008 and BS 9999:2017 editions.

The now superseded BS 9999:2008 [26] listed 63 different occupancy types in which 7 m^2 pp for showrooms, 2 m^2 pp for shops and bazaars were given. In comparison BS 9999:2017 has streamlined floor space factors and gives 2, 4 and 7 m^2 pp for normal (clothing store), medium (supermarket) and low (furniture showroom) density conditions respectively. The generic term 'shops' in BS 9999:2008 no longer exists in BS 9999:2017 but could be represented by the 'normal' or 'medium' categories and also that the value for supermarkets here does not correspond to other UK guidance discussed above.

In 1991 the Republic of Ireland's Building Regulations [27] gives an approach equivalent to BS 9999:2008 however in the 1997 and 2006 editions BS 5588 [16] parts 10 and 11 are adopted and no specific factors are listed for mercantile occupancies.

The 2006 edition of the New Zealand Acceptable Solution C/AS1 [2] specified gross area mercantile occupant load factors as 0.3 pers./m^2 for shop spaces, malls and circulation areas, 0.1 pers./m^2 for furniture and large appliance stores or similar and 0.2 pers./m^2 for showrooms (equivalent to 3.3, 10 and 5 m^2 pp respectively). In the equivalent 2012, C/AS4 document these values were given as 3, 10 and 5 m^2 pp respectively where the value for shop spaces etc. has been modified to 3.5 m^2 pp in the 2017 edition. The New Zealand Verification Method C/VM2 [28] adopts the same values as C/AS4, 2017 although C/VM2 does not explicitly state whether these are for gross or net area.

The 2000 edition of the Russian Construction Standards and Regulations [29] is the only reference reviewed in which the occupant load factor varies between urban and rural locations. A value of 1.35 m^2 pp is given for sales areas of shops in towns and urban settlements; 2.0 m^2 pp for rural shops and 1.6 m^2 pp for sales areas.

Finally, some published documents give occupant load factors for specific mercantile activities in addition to those as a function of building storey. In Hong Kong [23] markets, supermarkets, showrooms, jewellery and goldsmith shops, pawn shops and money changers are specifically identified with a value of 2 m^2 pp. In Australia, ref. [24] specifies 5 m^2 pp for showrooms which are described as display areas, covered malls or arcades. In Ontario, dining, alcoholic beverages and cafeteria areas are specified within the mercantile category as 1.1 m^2 pp however British Columbia does not list these within its mercantile use category.

APPLYING THE PUBLISHED VALUES TO DESIGN

To illustrate the variation between recommendations, occupant load factors have been applied to a simplified hypothetical retail design to determine an overall building occupant load. This is not intended to encapsulate all possible variations in published values, but only to provide a single example. The building is assumed to have six storeys including a ground level, two basement levels and three upper levels each with a gross area of 1000 m^2. The ground and first storey (Level 1) are assumed to be accessible from a street. In cases where it is not clearly specified whether a gross or net floor area factor should be assumed then the gross has been applied. Where there is a 'choice' of occupant load factors, the value which results in the greatest occupant load has been selected for conservatism – an approach often adopted by designers.

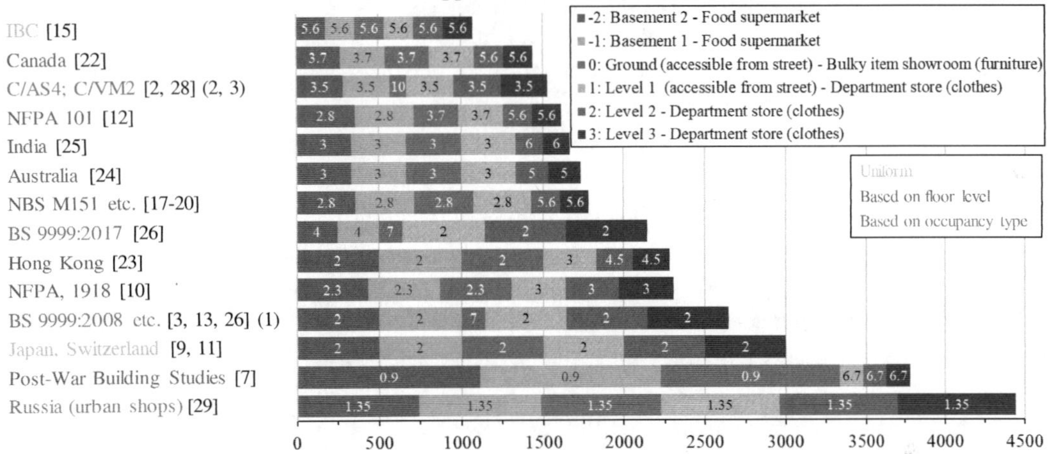

(1) BS 9999:2008 only has shops at 2 m^2 pp whereas 2017 edition distinguishes between supermarkets and clothing stores.
(2) C/AS4 and C/VM2 separate 'showrooms' from areas with furniture using factors of 5 m^2 pp and 10 m^2 pp respectively.
(3) Occupant load changes from 1629 to 1867 if the 3.5 m^2 pp or 3 m^2 pp factors using 2017 or 2012 editions of C/AS4 respectively.

Fig. 1. Total calculated occupant loads using published occupant load factors (numbers in m^2 pp).

Fig. 1 shows the number of occupants per level using the indicated occupant load factors. Values for BS 9999:2008, ADB and Scottish Technical Handbook have been consolidated as have NBS M151, NYC (2008), Qatar and gross values from NZS 1900. The variation in the proportion of the occupant load in the basement, for example comparing the Post-War Building Studies with BS 9999 is illustrated. For the exemplar building the occupant load varies from 1074 to 4444 but clearly larger and more complex retail buildings are being designed, where for these designs variations between occupant loads could be even greater.

CONCLUSION

In general there are three approaches for determining the occupant load in mercantile spaces; by floor level, by occupancy type or as uniform values. In all of the documents reviewed only BS 5588 allows for a calculation that considers the circulation areas rather than the shop areas. In addition, occupant load factors for mercantile spaces not only differ by jurisdiction but also over time within some jurisdictions.

The authors have not been able to source the original data used to select the occupant load factors, and therefore have been unable to identify the basis for these factors and to what distribution percentile they correspond. The difficulty in obtaining this information can make it hard for designers to determine the relevance of occupant load factors when applied to modern design and why there can be large disparities between jurisdictions. Whether these variations are solely down to cultural differences is not clear but it would seem the basis of current occupant load factors have origins that go back more than 60 years. By performing this review, a need to revisit mercantile floor space factors for new data has been highlighted. Future work should take into consideration the different cultural practices between jurisdictions.

IOP Conf. Series: Journal of Physics: Conf. Series **1107** (2018) 072005 doi:10.1088/1742-6596/1107/7/072005

REFERENCES

[1] NFPA 1, Fire Code, National Fire Protection Association, Quincy, MA, USA, 2012.

[2] MBIE, Acceptable Solutions, Wellington, New Zealand, 2006 through to 2017.

[3] Technical Handbook – Non-domestic. Scottish Government, 2005 and 2017.

[4] Cote, R. and Harrington, G., (eds.), *NFPA 101 Life Safety Code Handbook*, 12th ed., National Fire Protection Association, Quincy, MA, USA, 2006.

[5] Steffen, B., Seyfried, A. (2010) Methods for measuring pedestrian density, flow, speed and direction with minimal scatter, *Physica A*, 389(9): 1902-1910, 10.1016/j.physa.2009.12.015

[6] Predtechenskii, V.M. and Milinskii, A.I., *Planning for Foot Traffic Flow in Buildings*. Amerind Publishing Company, Inc, New Delhi, 1978.

[7] Fire grading of buildings, Post-War Building Studies, No. 29, HMSO, London, 1952.

[8] Xiaoping, Z., Jiahui, S. and Yuan, C. (2010) Analysis of crowd jam in public buildings based on cusp-catastrophe theory, *Building and Environment*, 45(8): 1755-1761, 10.1016/j.buildenv.2010.01.027

[9] Youn, H., Hwang ,Y. and Kwon, Y. (2011) A study of density of the person in a classroom for building evacuation safety regulations in Korea. In: Peacock R., Kuligowski E., Averill J. (eds) Pedestrian and Evacuation Dynamics. Springer, Boston, MA

[10] Schulte, R.C., Fire protection history – Part 250: 1918, (Design occupant factors - Department stores), Building Code Resource Library, 2014.

[11] De Sanctis, G., Kohler, J. and Fontana, M. (2014) Probabilistic assessment of the occupant load density in retail buildings. *Fire Safety Journal*, 69:1-11, 10.1016/j.firesaf.2014.07.002

[12] NFPA 101, Life Safety Code, National Fire Protection Association, Quincy, MA, USA, 2015.

[13] Fire Safety Approved Document B, Buildings Other than Dwellinghouses, 2000 through to 2013, UK.

[14] Angerd. M. and Frantzich, H. The use of uncertainty analysis in performance based design. *4th Int. Conference on Performance-based Codes and Fire Safety Design Methods*, 2002, pp.179-189.

[15] International Building Code (IBC), International Code Council (ICC), 2015.

[16] BS 5588. Fire Precautions in the Design, Construction and Use of Buildings – Part 10 and Part 11. British Standards Institution, 1991 and 1997 respectively.

[17] Courtney, J., Houghton, H. and Thompson G., Design and construction of building exits. NBS Publication M151, National Bureau of Standards, Washington, 1935.

[18] New York Building (Construction) Code, New York City, 1938, 1968 and 2008 editions.

[19] NZS 1900. Model Building Bylaw, Chapter 5, Fire Resisting Construction and Means of Egress, New Zealand Standards Institute, 1963.

[20] Civil Defence Regulation, Ministry of Interior, State of Qatar, 2012.

[21] Saudi Building Code, Kingdom of Saudi Arabia, 2007.

[22] The National Building Code of Canada (NBC), National Research Council Canada, 2015.

[23] Code of Practice for Fire Safety in Buildings, Buildings Dept., Hong Kong, 1996 and 2011.

[24] The Building Code of Australia, Class 2 to Class 9 Buildings, Volume one, Australian Building Codes Board, Canberra, ACT, Australia, 2007.

[25] Puri, V., *Compendium of Delhi building bye-laws and development regulations.* JBA Pub., 2007.

[26] BS 9999, Fire safety in the design, management and use of buildings – Code of practice, The British Standards Institution, 2008 and 2017 editions.

[27] Building Regulations, Technical Guidance Document B – Fire Safety, Dublin, Republic of Ireland, 1991, 1997 and 2006.

[28] C/VM2 Ministry of Business, Innovation and Employment, Verification method: Framework for fire safety design, Wellington, New Zealand, 2014.

[29] Construction Standards & Regulations, SNiP 2.08.02-89, Moscow, Russian Federation, 2000.

Small-scale Studies on Evacuation Characteristics of Pedestrians with physical, mental or age-related Disabilities

Paul Geoerg[1], Robert Malte Polzin[2], Jette Schumann[3], Stefan Holl[3] and Anja Hofmann[1]

[1]Bundesanstalt für Materialforschung und -prüfung, Germany
[2] MathConsult GmbH, Linz, Austria
[3]Forschungszentrum Jülich GmbH, Germany

Paul.Geoerg@bam.de

ABSTRACT

Evaluation of evacuation process is common separated by two phases: pre-movement and movement-phase. Individual capabilities of pedestrians have an impact on both and therefore on the time required for safe escape. In view of evaluating the evacuation process, characteristics of realistic movements of pedestrians involved are needed, but so far such data are mainly available for individuals with unrestricted evacuation capabilities only.

Due to the lack of available data, parameter studies in an assisted living for people with disabilities were performed to explore the movement during evacuation. Preparation times for assistance (n=10), unimpeded walking speeds (n=34) depending on the assistance devices in the plane as well as walking speeds on stairs (n=11) were focussed.

Time to prepare an assistance device for utilisation depends significantly on the kind of device type. Further on, practical experience in handling the devices for assistance has an impact on the preparation time. In contrast to the results of the preparation times, any relation on assistance devices were investigated for the unimpeded walking speed in the plane. A remarkable decrease of walking speed on stairs was observed which is independent from the utilised assistance device.

The presented data improve and extend the insufficient engineering data base and raised up questions regarding the complexity of considering pedestrians movement with impairments in evacuation planning.

KEYWORDS

evacuation; human behaviour; egress; disabilities; pre-movement phase; movement phase; preparation time; unimpeded walking speed; walking on stairs; need for assistance

NOMENCLATURE LISTING

10MWT	10 metre walking test	r	Radius semicircle /m
ASET	Available safe egress time	RSET	Required safe egress time
$dist_{landing}$	Resulting walking distance on landing /m	SiME	Safety for people with disabilities
		$stairtread_{width}$	Width of a stair tread /m
gap_{width}	Gap between two stairs /m	$v_{landing}$	Velocity on landing /ms^{-1}

IOP Conf. Series: Journal of Physics: Conf. Series **1107** (2018) 072006 doi:10.1088/1742-6596/1107/7/072006

INTRODUCTION

Reliable data of pedestrians' movement in infrastructures are required for a proper evaluation of egress in case of emergency. Due to the complexity of infrastructures, performance-based building regulations become relevant to predict the available safe egress time for providing an appropriate level of safety [1, 2]. Significant research has been carried out into investigating pedestrians' movement and human behaviour of evacuating occupants (see [3] for a comprehensive overview). Due to the challenges in organisation and regarding to financial limitations, most of the studies analyse the movement of young and unimpaired participants. Thus, the studies do not represent the real society with a wide variety of individual disabilities. This fact will further increase as the World Health Organisation estimates that 15 % of the world's population is on ageing [4] and live with some form of disability. This amount will increase in the future due to ageing and global increase in chronic health conditions. Currently there is a lack of comprehensive data for pedestrians with physical, mental or age-related disabilities [5]. For this purpose, comprehensive parameter studies on the individual behaviour in an appropriate home for people with disabilities were conducted. The objective of the study on small-scale is to observe the individual evacuation capabilities involving the needs of assistance and to analyse the individual movement behaviours of people with physical, mental or age-related disabilities through controlled video studies. The preparation time for assistance, individual unimpeded walking speeds and walking speed on stairs were performed.

METHODS

In the following we provide an overview of the conducted experimental research on evacuation behaviour involving individuals with disabilities, including the participant recruitment for the studies.

Participants recruitment

Due to the limited number of varying multiple disabled participants, a configuration of participants based on suitable single disabilities was improper. Gear the challenging individual characteristics towards the real life, we focused on voluntary participants and did not take any classification into account. All participants are characterised by multiple disabilities regarding mentally, mobility and aged disabilities with essential need for assistance in daily life (see Tab 1).

Tab 1: Description of the individual characteristics of the participants. Letters help to distinguish individuals.

Age	Frequency	Characterisation (Identifier)
$0 < a \leq 40$	2	Walking and cognitive disability, including spasm, evacuation incapable (Impaired B); unrestricted participant without disabilities (Unimpaired B)
$40 < a \leq 60$	4	Two participants with restricted physique and cognition, limited evacuation capabilities (Impaired D, Impaired F, Impaired G); participant without impairments (Unimpaired A)
$60 < a \leq 70$	2	Participant with restricted physique in a wheelchair and mental disability (Impaired A); participant with limited walking abilities with walking frame (Impaired H)
$a > 70$	2	Participants with age-related disabilities and walking frames (Impaired C, Impaired E)

In total, ten participants aged between 21 and 71 years including two unrestricted study participants without any disabilities have taken part in these studies investigating the egress time (Fig 1). The cooperating association Lebenshilfe Bergisches Land supports participation of people with disabilities in working life including accommodation and has chosen suitable participants with disabilities for the studies on evacuation behaviour. Participation was voluntary for everybody and a cancellation without any negative consequences at any time was possible. Three participants were limited in contractual capability. For those cases, participants and their legal representatives agreed the participation.
For participation all attendees have been paid 25 € per half day. Only anonymous data were used for the experiments. The ethics committee of the University of Wuppertal (Germany) has approved the conduction without any ethical concerns.

IOP Conf. Series: Journal of Physics: Conf. Series **1107** (2018) 072006 doi:10.1088/1742-6596/1107/7/072006

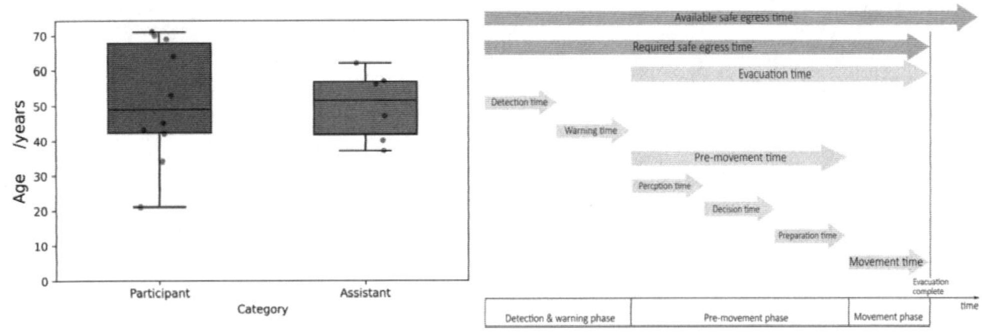

Fig 1. Ageing characteristics of participants and assistants (left) and engineering timeline (right).

Setting and data collection

We focussed the preparation time for assistance, the unimpeded walking speed and the (average) walking speed on a stair depending on assistance for investigation. Studies were captured by high definition cameras. The collected video data was synchronised by time and used to extract the preparation and movement times of the participants.

Firstly, the preparation time for repositioning from lying in bed to sitting in wheelchair was determined. In addition, changeovers from one wheelchair to wheelchair and evacuation chair were observed. Considering preparation as most important part of the pre-movement time (Fig 1, right) for people with disabilities, the objective was to measure time requirements for prepare an assisted or an autonomous evacuation. Start and end of preparation runs were defined by a reference point on the floor near the door (Fig 3 (b)). Two assistants were involved and changed after two conducted runs. Performed actions were spoken aloud by the assistants, recorded and documented.

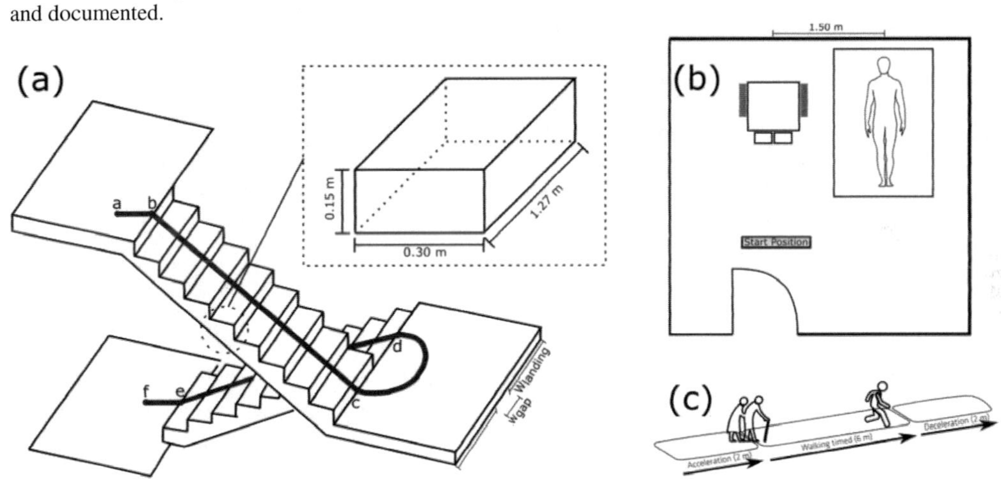

Fig 2. Sketches of the studies on movement on stairs (a), preparation time (b) and unimpeded walking speed (c). Note that the length of the measurement area in the 10MWT has been modified to 5.6 m.

Secondly, the unimpeded walking speeds (v_0) depending on the kind of assistance device (evacuation chair, wheelchair and escape mattress) were measured. A well-documented, clinical measurement method [6] for walking speed over a short time – called 10 metre walking test (10MWT) – was used. Participants were instructed to walk a distance of six meters while the time was measured. For acceleration and deceleration additional space was given, which was not included in the measurement (Fig 3 (c)). The distance covered is divided by the time it took the individual to walk that distance. Due to a technical camera shift, the length of the measurement area in the 10MWT has been modified to 5.6 m for analysis.

Thirdly, movement on stairs (Fig 3 (a)) was observed for using the evacuation chair and escape mattress. Both stair sections have nine steps respectively with a tread width of 0.30 m and a riser height of 0.15 m. The result in the diagonal yields 3.02 m width gradient angle of 26.6°, which is decisive for the sliding speed on steps. The width of the stair treads was 1.37 m, as measured between the handrails on either side. The path of the emergency staircase turns anticlockwise in descent. Four video cameras were used to record the movement on stairs. One camera was installed in front of the stairs on the upper floor, another above the landing area and two cameras on the lower floor from frontal and lateral perspective. To distinguish movement on stairs and movement on landings, both parts were considered separately.

For calculation of the velocity on stairs the walking distance between the points *b* - *c* and *d* - *e* was considered. Movement on the landing between the points *c* - *d* was estimated by the circumference of a semicircle. The resulting walking distance on landing ($dist_{landing}$) is calculated by Eq 1.

$$\mathrm{dist}_{\mathrm{landing}} = \pi r \qquad \text{(Eq 1)}$$

Two assistants were involved in every run and switched the positions after one run.

RESULTS AND DISCUSSUION

Resulting preparation times, (individual) unimpeded walking speeds in the plane and walking speeds on stairs are presented and discussed in the following.

Studies on preparation of assistance

The required preparation times are presented in Fig 4. Overall, we measured mean preparation time of 68.8 ± 66.0 s in a wide range from 15.1 s up to 253.4 s. Starting from lying needs less preparation time and has a less variance for preparation (33.3 ± 15.6 s) than starting from seating (83.4 ± 78.5 s). Repositioning of participants to the evacuation chair is more difficult. The temporal range for the participant Impaired A is 18.0 s. An improvement over time caused by training in the assistance can be detected. The participant is raised, then moved into a stable position by hugging one of the assistants, afterwards rotated and finally positioned on the evacuation chair. For this run, the movements are executed one after the other, which leads to shorter preparation times.

Examining the kind of assistance, long preparation times using the escape mattress are striking (between 138.0 s and 254.0 s). While preparation using an evacuation chair or a wheelchair takes less time than approx. 60.0 s, utilisation of an escape mattress took more than 120.0 s for preparation (Fig 3, right), which applies for disabled and non-disabled participants. Analysing the captured videos it is emphasised that handling of this kind of assistance device is not intuitive for the staff and the usage was not known. It took a lot of time to prepare the escape mattress for usage. Assistants had difficulties fixing the participant on the mattress in the process. Furthermore, the proper method for turn the participants lying in an escape mattress was unknown by all associates. It has shown that it is useful to rotate the participant sideways on bed, so that the participant can be placed on the escape mattress in shorter time.

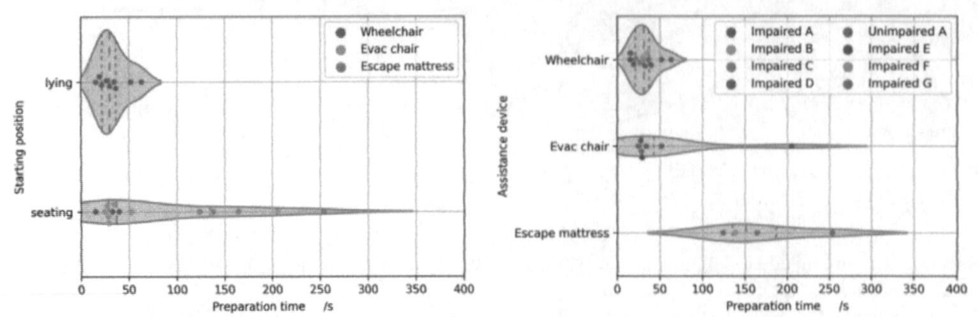

Fig 3. Preparation times depending on starting position (left) and type of assistance device (right).

Studies on movement in the plane

Unimpeded walking speeds of participants are presented in Fig 5. Overall a mean walking speed of 2.01 ± 0.49 m/s (0.76 up to 2.69 m/s) was measured. Unimpeded walking speed is high influenced by individual abilities and boundary conditions (like age, gender, health status, motivation, facility, familiarity, day time) [7, 5]. Recommendations in the order of 1.20 m/s are embedded in international design codes [8, 9]. Wide variations (0.3 m/s up to 2.5 m/s and higher) in the walking speed of non-disabled pedestrians are well documented [3, 5, 10]. Take pedestrians with disabilities into account leads to comparable ranges and variations of unimpeded walking speeds (0.23 m/s up to 1.95 m/s) [11].

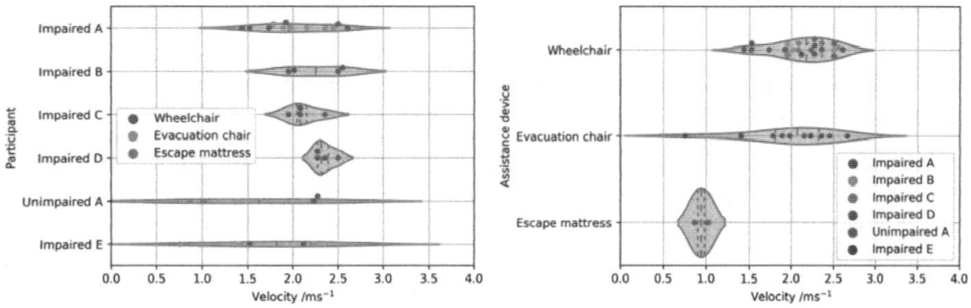

Fig 4. Unimpeded walking speed depending on participant (left) and assistance device (right).

As shown for the preparation times, a slower movement is striking out, if the escape mattress is utilised (Fig 4, right). Pull the mattress over the ground is the principle of movement using this assistance device. In consequence the speed depends on friction and there is a high demand on physical strength to move on. The unimpeded walking speed depends on the person moving the wheelchair or evacuation chair. The type of impairments (participant) does not affect the unimpeded walking speeds.

Studies on movement in on stairs

Walking on stairs was investigated by a lot of researchers during the last decades with focus on quantitative (e.g. walking speed) and qualitative (e.g. merging, overtaking) aspects (see [3] for a comprehensive overview). Walking without any disabilities and without assistance devices downstairs leads to decreased walking speeds in a scale of 0.30 to 0.70 m/s (e.g. [12 - 15]) because of the increased complexity of movement, the impact of individual strength and stamina. Unimpeded walking speeds on stairs depending on the assistance device are presented in Fig 6. Overall a mean walking speed on stairs of 0.38 ± 0.07 m/s (0.24 up to 0.46 m/s) and a mean walking speed on the landing stairs of 0.17 ± 0.06 m/s (0.09 up to 0.29 m/s) was measured. For comparison, Hunt et al. [16] measured walking speeds in a (significantly longer) stairwell for an evacuation chair (0.83 ± 0.04 m/s) and rescue sheet (0.67 ± 0.15 m/s). Walking on the landing between the two stairs is more widely distributed. This is caused by manoeuvrings because of the untrained usage of the assistance devices.

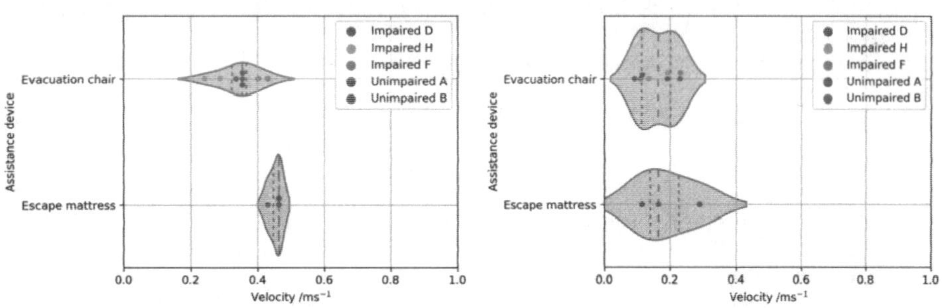

Fig 5. Walking speed on stair (left) and walking speed on landing depending on assistance device (right).

The evacuation chair is used for people who are not able to walk down the stairs. The sliding system of the evacuation chair provides a braking effect on the steps. This braking effect is achieved by the interaction of the weight of the evacuated person with the pressure, which is exerted by the assistance in the direction of steps. Downstairs one assistant is sufficient and overtaking is possible (but rarely observed [17]). Carry handles should be mounted behind the armrests for a better balance.

SUMMARY

Parameter studies under laboratory conditions on the pre-movement and movement phase in an assisted living for people with disabilities were performed. Preparation times, unimpeded walking speed in the plane and moving on stairs were focussed. The time to prepare an assistance device depends significantly on the kind of device. Small difference between starting position lying or seating was observed. Practical experiences in handling the devices for assistance have a positive impact on the preparation time. Our studies show the distribution of unimpeded walking speeds and the dependence of the assistants moving the wheelchair or evacuation chair. The movement does not depend on the assistant device but on the capability of the pusher. As expected, a significantly decreased walking speed on stairs was observed, which is independent from the utilised assistance device. Manoeuvrings caused by the large and unwieldy escape mattress leads to slow walking speeds on the landing. The large variety of impairments (of which we have investigated a small sample only) and the diversity of assistance devices make it difficult to obtain general movement characteristics. Even though the small sample size of participants, the studies improve the insufficient engineering data base leading to the conclusion that the training of the assisting person in proper usage of the assistance devices has a significant impact on pre-movement and movement-phase.

ACKNOWLEDGEMENTS

Part of this work has been performed within the research program Safety for people with physical, mental or age-related disabilities (SiME) supported by the German Federal Ministry of Education and Research - BMBF (FKZ starting on 13N13946). PG thanks the SFPE Foundation for financial support with the Dr. Guylène Proulx, OC Scholarship.

REFERENCES

[1] Gwynne, Steven M. V., E. Kuligowski, M. Spearpoint, E. Ronchi, Bounding defaults in egress models, Fire Mater. (2013), pp. 335-352.

[2] M.J. Hurley, E.R. Rosenbaum, Performance-Based Design, in: M.J. Hurley (Ed.), SFPE handbook of fire protection engineering, Springer, New York, 2016, pp. 1233–1260.

[3] S. Gwynne, K.E. Boyce, Engineering Data, in: M.J. Hurley (Ed.), SFPE handbook of fire protection engineering, Springer, New York, 2016, pp. 2429–2551.

[4] United Nations, World Population Prospects: The 2015 Revision. Comprehensive Tables, 2015.

[5] K. Boyce, Safe evacuation for all - Fact or Fantasy?: Past experiences, current understanding and future challenges, Fire Safety Journal (2017).

[6] R.W. Bohannon, Comfortable and maximum walking speed of adults aged 20-79 years: reference values and determinants, Age and ageing 26 (1) (1997) 15–19.

[7] U. Weidmann, Transporttechnik der Fussgänger, Transporttechnische Eigenschaften des Fussgängerverkehrs Literaturauswertung, Zürich, IVT, Institut für Verkehrsplanung, Transporttechnik, Strassen- und Eisenbahnbau, 1992.

[8] British Standards Institution, Application of fire safety engineering principles to the design of buildings, London.

[9] RiMEA e.V., Guideline for Microscopic Evacuation Analysis (RiMEA): Version: 3.0.0, 2016.

[10] S. Holl, Methoden für die Bemessung der Leistungsfähigkeit multidirektional genutzter Fußverkehrsanlagen. Dissertation, Wuppertal, 2016.

[11] P. Geoerg, S. Gwynne, K. Boyce, F. Berchtold, S. Holl, A. Hofmann, Engineering egress data considering pedestrians with impairments: (unpubished, in preparation), N.N. (2018).

[12] R.D. Peacock, B.L. Hoskins, E.D. Kuligowski, Overall and local movement speeds during fire drill evacuations in buildings up to 31 stories, Safety Science 50 (8) (2012) 1655–1664.

[13] V.V. Kholshchevnikov, D.A. Samoshin, A.P. Parfyonenko, I.P. Belosokhov, Study of children evacuation from pre-school education institutions, Fire Mater. 36 (5-6) (2012) 349–366.

[14] A.R. Lárusdóttir, A. Dederichs, Behavioural aspects of movement down stairs during elementary school fire drills, in: T.J. Shields (Ed.), Human behaviour in fire: 5th international symposium September 2012, Cambridge, UK ; [proceedings of the 5th International Symposium on Human Behaviour in Fire 2012 ; held at Downing, Cambridge, UK, 19 - 21 September 2012], Interscience Communications, London, 2012, pp. 120–127.

[15] S. Burghardt, A. Seyfried, W. Klingsch, Performance of stairs – Fundamental diagram and topographical measurements, Transportation Research Part C: Emerging Technologies 37 (2013) 268–278.

[16] A. Hunt, E.R. Galea, P.J. Lawrence, An analysis and numerical simulation of the performance of trained hospital staff using movement assist devices to evacuate people with reduced mobility, Fire Mater. (2013), pp. 407-429.

[17] T.J. Shields, K.E. Boyce, N. McConnell, The behaviour and evacuation experiences of WTC 9/11 evacuees with self-designated mobility impairments, Fire Safety Journal 44 (6) (2009) 881–893.